UNIVERSE

THE SOLAR SYSTEM

Third Edition

UNIVERSE

THE SOLAR SYSTEM

Third Edition

Roger A. Freedman
University of California, Santa Barbara

William J. Kaufmann III
San Diego State University

W.H. Freeman and Company • New York

Publisher: Susan Finnemore Brennan
Acquisitions Editor: Valerie Raymond
Development Editor: David Chelton
Project Editor: Kerry O'Shaughnessy/Matrix Publishing Services
Media and Supplements Editor: Amy Thorne
Marketing Manager: Anthony Palmiotto
Design Manager: Diana Blume
Production Coordinator: Paul W. Rohloff
Senior Illustration Coordinator: Bill Page
Illustrations: Imagineering Media Services
Photo Editor: Bianca Moscatelli
Photo Researcher: Elyse Rieder
Editorial Assistant: Whitney Clench
Composition: Matrix Publishing Services
Printing and Binding: RR Donnelley

Guest Essay Photo Credits

p. 20, Don Fukuda/National Optical Astronomy Observatories;
p. 44, courtesy of James Randi; p. 63, courtesy of
Mark Hollabaugh; p. 208, courtesy of Geoff Marcy; p. 299,
courtesy of John Grotzinger; p. 401, courtesy of
Scott S. Sheppard; p. 28-17, courtesy of Kevin Plaxco.

Library of Congress Control Number: 2006938227
ISBN-13: 978-0-7167-9563-6
ISBN-10: 0-7167-9563-9

Printed in the United States of America

First printing, 2007

W. H. Freeman and Company
41 Madison Avenue
New York, NY 10010

www.whfreeman.com

Chapters 17–27 can be found in *Universe,* 8th edition, of which this book is an abbreviated version.

To Lee Johnson Kaufmann and
Caroline Robillard-Freedman,
strong survivors

and to the memory of
S/Sgt. Ann Kazmierczak Freedman, WAC

About the Authors

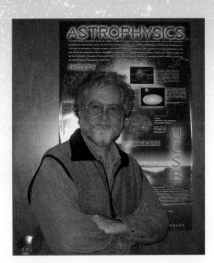

Roger A. Freedman is on the faculty of the Department of Physics at the University of California, Santa Barbara. He grew up in San Diego, California and was an undergraduate at the University of California campuses in San Diego and Los Angeles. He did his doctoral research in nuclear theory and its astrophysical applications at Stanford University under the direction of Professor J. Dirk Walecka. Dr. Freedman came to UCSB in 1981 after three years of teaching and doing research at the University of Washington.

 Dr. Freedman holds a commercial pilot's license, and when not teaching or writing he can frequently be found flying with his wife, Caroline. He has flown across the United States and Canada.

William J. Kaufmann III was the author of the first four editions of *Universe*. Born in New York City on December 27, 1942, he often visited the magnificent Hayden Planetarium as he was growing up. Dr. Kaufmann earned his bachelor's degree *magna cum laude* in physics from Adelphi University in 1963, a master's degree in physics from Rutgers in 1965, and a Ph.D. in astrophysics from Indiana University in 1968. At 27 he became the youngest director of any major planetarium in the United States when he took the helm of the Griffith Observatory in Los Angeles. During his career he also held positions at San Diego State University, UCLA, Caltech, and the University of Illinois. Throughout his professional life as a scientist and educator, Dr. Kaufmann worked to bridge the gap between the scientific community and the general public to help the public share in the advances of astronomy. A prolific author, his many books include *Black Holes and Warped Spacetime, Relativity and Cosmology, The Cosmic Frontiers of General Relativity, Exploration of the Solar System, Planets and Moons, Stars and Nebulas, Galaxies and Quasars,* and *Supercomputing and the Transformation of Science.* Dr. Kaufmann died in 1994.

Contents Overview

To Lee Johnson Kaufmann and
Caroline Robillard-Freedman,
strong survivors

and to the memory of
S/Sgt. Ann Kazmierczak Freedman, WAC

About the Authors

Roger A. Freedman is on the faculty of the Department of Physics at the University of California, Santa Barbara. He grew up in San Diego, California and was an undergraduate at the University of California campuses in San Diego and Los Angeles. He did his doctoral research in nuclear theory and its astrophysical applications at Stanford University under the direction of Professor J. Dirk Walecka. Dr. Freedman came to UCSB in 1981 after three years of teaching and doing research at the University of Washington.

Dr. Freedman holds a commercial pilot's license, and when not teaching or writing he can frequently be found flying with his wife, Caroline. He has flown across the United States and Canada.

William J. Kaufmann III was the author of the first four editions of *Universe*. Born in New York City on December 27, 1942, he often visited the magnificent Hayden Planetarium as he was growing up. Dr. Kaufmann earned his bachelor's degree *magna cum laude* in physics from Adelphi University in 1963, a master's degree in physics from Rutgers in 1965, and a Ph.D. in astrophysics from Indiana University in 1968. At 27 he became the youngest director of any major planetarium in the United States when he took the helm of the Griffith Observatory in Los Angeles. During his career he also held positions at San Diego State University, UCLA, Caltech, and the University of Illinois. Throughout his professional life as a scientist and educator, Dr. Kaufmann worked to bridge the gap between the scientific community and the general public to help the public share in the advances of astronomy. A prolific author, his many books include *Black Holes and Warped Spacetime, Relativity and Cosmology, The Cosmic Frontiers of General Relativity, Exploration of the Solar System, Planets and Moons, Stars and Nebulas, Galaxies and Quasars,* and *Supercomputing and the Transformation of Science.* Dr. Kaufmann died in 1994.

Contents Overview

Chapters 17–27 can be found in *Universe*, 8th edition, of which
this book is an abbreviated version.

Contents

I Introducing Astronomy

II Planets and Moons

7-5 Small chunks of rock and ice also orbit the Sun

In addition to the eight planets, many smaller objects orbit the Sun. These fall into three broad categories: asteroids, which are rocky objects found in the inner solar system; trans-Neptunian objects, which are found beyond Neptune in the outer solar system and contain both rock and ice; and comets, which are mixtures of rock and ice that originate in the outer solar system but can venture close to the Sun.

> The smaller bodies of the solar system contain important clues about its origin and evolution

New "Motivation Statements" connect detailed discussion to the broader picture

As students read a science text, they often wonder "What is the key idea? How does it relate to the discussion on previous pages? What motivated scientists to make these observations and deductions?" To address these questions, I've added to each section a free-standing **Motivation Statement**—a brief sentence that shows students how that section's material fits in with the larger picture of astronomy presented in the chapter. Some statements deal with the scientific method. Others compare topics in one section with those in another section or chapter; still other statements give a quick summary of a section's conclusion.

Streamlined coverage of planets focuses the student on comparative planetology and highlights the interplay of planetary surfaces and atmospheres

Two chapters on comparative planetology were introduced in the Seventh Edition: Chapter 7, *Our Solar System,* and Chapter 8, *The Origin of the Solar System.* I've revised and updated these chapters with the latest information, including the new official distinctions between planets and dwarf planets. In addition, I've combined the chapters on Mercury, Venus, and Mars into a single Chapter 11 on the terrestrial planets, and I've added comparative planetology figures to several other chapters. As in previous editions, each planet's surface and atmosphere are discussed in tandem, where other texts address them separately. This underscores their interrelationships.

Preface

Astronomy is one of the most dynamic and exciting areas of modern science. Recent discoveries have shown us previously unseen worlds in the frigid outer depths of our solar system, given us insight into the dramatic ways in which the most massive stars can collapse into black holes, and allowed us to see the universe as it was more than thirteen billion years ago. But studying astronomy is much more than just learning a collection of amazing facts: It is also discovering the nature of the scientific way of knowing and understanding how scientists work and think.

Bill Kaufmann wrote the first four editions of *Universe* with the goal of giving students real insight into the nature of science. Since taking over stewardship of *Universe* in 1996, I've continued Bill's program and added an additional goal of helping students develop problem-solving skills. With these goals in mind, I've introduced some new features to this Eighth Edition while strengthening well-received features from earlier editions.

New features of the Eighth Edition help students focus on key topics

Students need to find specific topics when studying for exams or doing homework. They also benefit from reminders of what the key topics are within a section of the book. New features of the Eighth Edition are designed to help students with both of these.

New COSMIC CONNECTIONS figures summarize key ideas visually

Many students learn more through visual presentations than from reading long passages of text. To help these students I've added large figures, called *Cosmic Connections*, to most chapters in the Eighth Edition. These give an overview or summation of a particular important topic in a chapter. Subjects range from the variety of modern telescopes to the formation of the solar system; from the scale of distances in the Universe to the life cycles of stars; and from the influence of gravitational tidal forces to the evolution of the Universe after the Big Bang. The *Cosmic Connections* figures also convey some of the excitement and adventure that lead students to study astronomy in the first place.

Contents

II Planets and Moons

Preface

Astronomy is one of the most dynamic and exciting areas of modern science. Recent discoveries have shown us previously unseen worlds in the frigid outer depths of our solar system, given us insight into the dramatic ways in which the most massive stars can collapse into black holes, and allowed us to see the universe as it was more than thirteen billion years ago. But studying astronomy is much more than just learning a collection of amazing facts: It is also discovering the nature of the scientific way of knowing and understanding how scientists work and think.

Bill Kaufmann wrote the first four editions of *Universe* with the goal of giving students real insight into the nature of science. Since taking over stewardship of *Universe* in 1996, I've continued Bill's program and added an additional goal of helping students develop problem-solving skills. With these goals in mind, I've introduced some new features to this Eighth Edition while strengthening well-received features from earlier editions.

New features of the Eighth Edition help students focus on key topics

Students need to find specific topics when studying for exams or doing homework. They also benefit from reminders of what the key topics are within a section of the book. New features of the Eighth Edition are designed to help students with both of these.

New COSMIC CONNECTIONS figures summarize key ideas visually

Many students learn more through visual presentations than from reading long passages of text. To help these students I've added large figures, called *Cosmic Connections*, to most chapters in the Eighth Edition. These give an overview or summation of a particular important topic in a chapter. Subjects range from the variety of modern telescopes to the formation of the solar system; from the scale of distances in the Universe to the life cycles of stars; and from the influence of gravitational tidal forces to the evolution of the Universe after the Big Bang. The *Cosmic Connections* figures also convey some of the excitement and adventure that lead students to study astronomy in the first place.

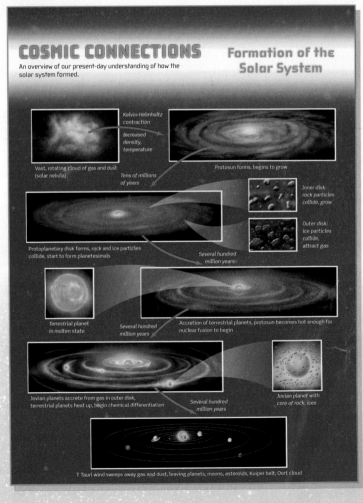

COSMIC CONNECTIONS — **Formation of the Solar System**
An overview of our present-day understanding of how the solar system formed.

7-5 Small chunks of rock and ice also orbit the Sun

In addition to the eight planets, many smaller objects orbit the Sun. These fall into three broad categories: asteroids, which are rocky objects found in the inner solar system; trans-Neptunian objects, which are found beyond Neptune in the outer solar system and contain both rock and ice; and comets, which are mixtures of rock and ice that originate in the outer solar system but can venture close to the Sun.

The smaller bodies of the solar system contain important clues about its origin and evolution

New "Motivation Statements" connect detailed discussion to the broader picture

As students read a science text, they often wonder "What is the key idea? How does it relate to the discussion on previous pages? What motivated scientists to make these observations and deductions?" To address these questions, I've added to each section a free-standing **Motivation Statement**—a brief sentence that shows students how that section's material fits in with the larger picture of astronomy presented in the chapter. Some statements deal with the scientific method. Others compare topics in one section with those in another section or chapter; still other statements give a quick summary of a section's conclusion.

Streamlined coverage of planets focuses the student on comparative planetology and highlights the interplay of planetary surfaces and atmospheres

Two chapters on comparative planetology were introduced in the Seventh Edition: Chapter 7, *Our Solar System,* and Chapter 8, *The Origin of the Solar System.* I've revised and updated these chapters with the latest information, including the new official distinctions between planets and dwarf planets. In addition, I've combined the chapters on Mercury, Venus, and Mars into a single Chapter 11 on the terrestrial planets, and I've added comparative planetology figures to several other chapters. As in previous editions, each planet's surface and atmosphere are discussed in tandem, where other texts address them separately. This underscores their interrelationships.

7
Comparative Planetology I: Our Solar System

Fifty years ago, astronomers knew precious little about the worlds that orbit the Sun. Even the best telescopes provided images of the planets that were frustratingly hazy and indistinct. Of asteroids, comets, and the satellites of the planets, we knew even less.

Today, our knowledge of the solar system has grown exponentially, due almost entirely to robotic spacecraft. (The illustration shows an artist's impression of one such robotic explorer, the *Mars Reconnaissance Orbiter* spacecraft, as it approached Mars in 2006.) Spacecraft have been sent to fly past all the planets at close range, revealing details unimagined by astronomers of an earlier generation. We have landed spacecraft on the Moon, Venus, and Mars and dropped a probe into the immense atmosphere of Jupiter. This is truly the golden age of solar system exploration.

In this chapter we paint in broad outline our present understanding of the solar system. We will see that the planets come in a variety of sizes and chemical compositions. There is also rich variety among the moons of the planets and among smaller bodies we call asteroids, comets, and trans-Neptunian objects. We will investigate the nature of craters on the Moon and other worlds of the solar system. And by exploring the magnetic fields

Learning Goals
By reading the sections of this chapter, you will learn

7-1 The important differences between the two broad categories of planets: terrestrial and Jovian

7-2 The similarities and differences among the large planetary satellites, including Earth's Moon

7-3 How the spectrum of sunlight reflected from a planet reveals the composition of its atmosphere and surface

7-4 Why some planets have atmospheres and others do not

11
Mercury, Venus, and Mars: Earthlike yet Unique

Mercury	Venus	Mars
R I V U X G	R I V U X G	R I V U X G

(Mercury: NASA; Venus: NASA/JPL, MIT, and USGS; Mars: J. Bell/Cornell University, and M. Wolff/SSI)

These images show the three planets that share the inner solar system with our Earth. All three have solid surfaces, and, in principle, a properly protected astronaut could stand on any of them. The differences between these three worlds, however, are pronounced.

Mercury is small, airless, and extensively cratered. It is also mysterious: Half of its surface has never been viewed at close range, and it has a perplexing and unexpected magnetic field. It also rotates in a manner unique in the solar system, spinning three times on its axis for every two orbits around the Sun.

Venus is nearly the same size as Earth, but it is shrouded by a perpetual cloud cover that hides its surface from view. To penetrate the clouds and learn what lies beneath, two advanced technologies were needed: radar, which allowed astronomers to "see"

through the clouds, and unmanned spacecraft, which orbited Venus at close range and even landed on its surface. The false-color topographic map shown here reveals that Venus has no true continents but merely highlands (shown in red) that rise gently above the planet's lower-lying areas (shown in blue). We have also learned that Venus's atmosphere is thick and searing hot, and that its clouds contain droplets of corrosive sulfuric acid.

Mars has captivated the popular imagination like no other planet. But rather than being the abode of warlike aliens, Mars proves to be an enigmatic world. Some parts of its surface are drier than any desert on Earth, while other locations show evidence of having been underwater for extended periods. Our challenge is to understand how these three nominally Earthlike worlds evolved to be so unique and so different than our own Earth.

Learning Goals
By reading the sections of this chapter, you will learn

11-1 What astronomers have learned by observing the terrestrial planets from Earth

11-2 The radically different ways in which Mercury, Venus, and Mars rotate on their axes

11-3 The outstanding features of Mercury, and why its magnetic field came as a surprise

11-4 How the advent of the space age transformed our understanding of Venus and Mars

11-5 How geologic activity took a very different form on Venus than on Earth, and why it essentially stopped on Mars

11-6 The key differences among the atmospheres of Earth, Venus, and Mars

11-7 How the atmospheres of Earth, Venus, and Mars evolved to their present states

11-8 The evidence that there was once liquid water on Mars

11-9 What we know about the two small satellites of Mars

255

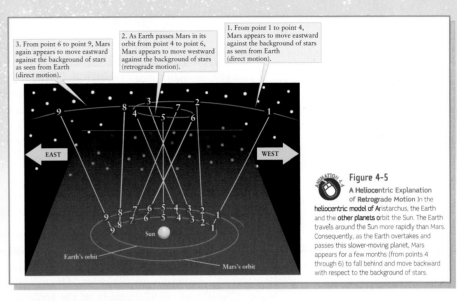

ANIMATION 4.5 **Figure 4-5**
A Heliocentric Explanation of Retrograde Motion In the heliocentric model of Aristarchus, the Earth and the other planets orbit the Sun. The Earth travels around the Sun more rapidly than Mars. Consequently, as the Earth overtakes and passes this slower-moving planet, Mars appears for a few months (from points 4 through 6) to fall behind and move backward with respect to the background of stars.

An expanded program of explanatory art

Many new figures in the Eighth Edition continue the practice of adding balloon captions to help students interpret complex figures. Many photos also contain brief labels to point out the most important relevant features.

Up-to-date information shows the cutting edge of astronomy

I've brought *Universe* up to date with the latest astronomical discoveries, ideas, and images. These include:

- New material on using transits to study extrasolar planets (Chapter 8)

Figure 8-18
A Transiting Extrasolar Planet If the orbit of an extrasolar planet is nearly edge-on to our line of sight, like the planet that orbits the star HD 209458, we can learn about the planet's (a) diameter, (b) atmospheric composition, and (c) surface temperature. (S. Seager and C. Reed, Sky and Telescope; H. Knutson, D. Charbonneau, R. W. Noyes (Harvard-Smithsonian CfA), T. M. Brown (HAO/NCAR), and R. L. Gilliland (STScI); A. Feild (STScI); NASA/ JPL-Caltech/D. Charbonneau, Harvard-Smithsonian CfA)

- Updated discussion of global climate change, including historical data on global temperatures and CO_2 concentrations (Chapter 9)
- New information on the history of Martian water from the *Mars Exploration Rovers* (Chapter 11)
- *Cassini* observations of the moons of Saturn (Chapter 13)
- New discoveries about trans-Neptunian objects (Chapter 15)
- Updated information about hypernovae and the origin of gamma-ray bursts (Chapter 20)
- Chandra X-Ray Observatory images revealing the presence of supermassive black holes in galaxies (Chapter 24)
- Observations of the infrared background due to the first stars (Chapter 27)

Figure 9-30
Atmospheric CO_2 and Changes in Global Temperature This figure shows how the carbon dioxide concentration in our atmosphere (upper curve) and the Earth's average surface temperature (lower curve) have changed since 1000 A.D. The increase in CO_2 since 1800 due to burning fossil fuels has strengthened the greenhouse effect and caused a dramatic temperature increase. (Intergovernmental Panel on Climate Change and Hadley Centre for Climate Prediction and Research, U.K. Meteorological Office)

45. Use the *Starry Night Enthusiast*™ program to examine the Jovian planets Jupiter, Saturn, Uranus, and Neptune. Select each of these planets from the **Solar System** submenu in the **Favourites** menu. If you desire, you can remove the image of the astronaut's feet by selecting **Feet** in the **View** menu. Position the mouse cursor over the planet and click and drag the image to examine the planet from different views. Describe each planet's **app**earance. Which has the greatest color contrast in its cloud tops? Which has the least color contrast? What can **you** say about the thickness of Saturn's rings compared to their diameter?

A revised selection of Observing Projects

The CD-ROM that comes free (upon request) with this text contains the *Starry Night Enthusiast*™ planetarium software from Imaginova Inc. Every chapter of the Eighth Edition of *Universe* includes one or more Observing Projects that use this software package; many of these projects are new or revised. As in the Seventh Edition, many projects are accompanied by an *Observing Tips and Tools* box to provide help and guidance.

Additional features have been retained from previous editions

Three versions of the text meet the needs of different instructors

In addition to the complete 28-chapter version of *Universe*, two shorter versions are also available. *Universe: The Solar System*, Third Edition, includes Chapters 1-16 and 28; it omits the chapters on stars, stellar evolution, galaxies, and cosmology. *Universe: Stars and Galaxies*, Third Edition, includes Chapters 1-8 and 16-28; it omits the detailed chapters on the solar system but includes the overview of the solar system in Chapters 7 and 8.

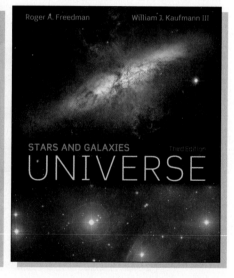

Cautions—Confronting misconceptions

CAUTION! The Jovian planets are sometimes called "gas giants." It is true that their primary constituents, including hydrogen, helium, ammonia, and methane, are gases under normal conditions on Earth. But in the interiors of these planets, pressures are so high that these substances are *liquids*, not gases. The Jovian planets might be better described as "liquid giants"!

Many people think that the Earth is closer to the Sun in summer than in winter and that the phases of the Moon are caused by Earth's shadow falling on the Moon. However, these "common sense" ideas are incorrect (as explained in Chapters 2 and 3). Throughout *Universe*, paragraphs marked by the **Caution** icon alert the reader to conceptual pitfalls such as these.

Analogies—Bringing astronomy down to Earth

ANALOGY To appreciate just how tiny the nucleus is, imagine expanding an atom by a factor of 10^{12} to a diameter of 100 meters, about the length of a football field. To this scale, the nucleus would be just a centimeter across—no larger than your thumbnail.

When learning new astronomical ideas, it can be helpful to relate them to more familiar experiences on Earth. Throughout *Universe*, **Analogy** paragraphs make these connections. For example, the motions of the planets can be related to children on a merry-go-round, and the bending of light through a telescope lens is similar to the path of a car driving from firm ground onto sand.

Tools of the Astronomer's Trade and S.T.A.R.— A problem-solving rubric

All the worked examples in *Universe* (found in boxes called **Tools of the Astronomer's Trade**) follow a logical and consistent sequence of steps called **S.T.A.R.**: assess the Situation, select the Tools, find the Answer, and Review the answer and explore its significance. Feedback from instructors shows when students follow these steps in their own work, they more rapidly become skillful problem solvers.

BOX 5-6 | Tools of the Astronomer's Trade

Applications of the Doppler Effect

Doppler's formula relates the radial velocity of an astronomical object to the wavelength shift of its spectral lines. Here are two examples that show how to use this remarkably powerful formula.

EXAMPLE: As measured in the laboratory, the prominent H_α spectral line of hydrogen has a wavelength $\lambda_0 = 656.285$ nm. But in the spectrum of the star Vega (Figure 5-21), this line has a wavelength $\lambda = 656.255$ nm. What can we conclude about the motion of Vega?

Situation: Our goal is to use the ideas of the Doppler effect to find the velocity of Vega toward or away from Earth.

Tools: We use the Doppler shift formula, $\Delta\lambda/\lambda_0 = v/c$, to determine Vega's velocity v.

Answer: The wavelength shift is

$$\Delta\lambda = \lambda - \lambda_0 = 656.255 \text{ nm} - 656.285 \text{ nm} = -0.030 \text{ nm}$$

The negative value means that we see the light from Vega shifted to shorter wavelengths—that is, there is a blueshift. (Note that the shift is very tiny and can be measured only using specialized equipment.) From the Doppler shift formula, the star's radial velocity is

$$v = c\frac{\Delta\lambda}{\lambda_0} = (3.00 \times 10^5 \text{ km/s})\left(\frac{-0.030 \text{ nm}}{656.285 \text{ nm}}\right) = -14 \text{ km/s}$$

Review: The minus sign indicates that Vega is coming toward us at 14 km/s. The star may also be moving perpendicular to the line from Earth to Vega, but such motion produces no Doppler shift.

By plotting the motions of stars such as Vega toward and away from us, astronomers have been able to learn how the Milky Way Galaxy (of which our Sun is a part) is rotating. From this knowledge, and aided by Newton's

universal law of gravitation (see Section 4-7), they have made the surprising discovery that the Milky Way contains roughly 10 times more matter than had once been thought! The nature of this unseen *dark matter* is still a subject of debate.

EXAMPLE: In the radio region of the electromagnetic spectrum, hydrogen atoms emit and absorb photons with a wavelength of 21.12 cm, giving rise to a spectral feature commonly called the *21-centimeter line*. The galaxy NGC 3840 in the constellation Leo (the Lion) is receding from us at a speed of 7370 km/s, or about 2.5% of the speed of light. At what wavelength do we expect to detect the 21-cm line from this galaxy?

Situation: Given the velocity of NGC 3840 away from us, our goal is to find the wavelength as measured on Earth of the 21-centimeter line from this galaxy.

Tools: We use the Doppler shift formula to calculate the wavelength shift $\Delta\lambda$, then use this to find the wavelength λ measured on Earth.

Answer: The wavelength shift is

$$\Delta\lambda = \lambda_0\left(\frac{v}{c}\right) = (21.12 \text{ cm})\left(\frac{7370 \text{ km/s}}{3.00 \times 10^5 \text{ km/s}}\right) = 0.52 \text{ cm}$$

Therefore, we will detect the 21-cm line of hydrogen from this galaxy at a wavelength of

$$\lambda = \lambda_0 + \Delta\lambda = 21.12 \text{ cm} + 0.52 \text{ cm} = 21.64 \text{ cm}$$

Review: The 21-cm line has been redshifted to a longer wavelength because the galaxy is receding from us. In fact, most galaxies are receding from us. This observation is one of the key pieces of evidence that the universe is expanding, and has been doing so since the Big Bang that took place almost 14 billion years ago.

BOX 4-3 | The Heavens on the Earth

Newton's Laws in Everyday Life

In our study of astronomy, we use Newton's three laws of motion to help us understand the motions of objects in the heavens. But you can see applications of Newton's laws every day in the world around you. By considering these everyday applications, we can gain insight into how Newton's laws apply to celestial events that are far removed from ordinary human experience.

Newton's *first* law, or principle of inertia, says that an object at rest naturally tends to remain at rest and that an object in motion naturally tends to remain in motion. This explains the sensations that you feel when riding in an automobile. When you are waiting at a red light, your car and your body are both at rest. When the light turns green and you press on the gas pedal, the car accelerates forward but your body attempts to stay where it was. Hence, the seat of the accelerating car pushes forward into your body, and it feels as though you are being pushed back in your seat.

Once the car is up to cruising speed, your body wants to keep moving in a straight line at this cruising speed. If the car makes a sharp turn to the left, the right side of the car will move toward you. Thus, you will feel as though you are being thrown to the car's right side (the side on the outside of the turn). If you bring the car to a sudden stop by pressing on the brakes, your body will continue moving forward until the seat belt stops you. In this case, it feels as though you are being thrown toward the front of the car.

Newton's *second* law states that the net outside force on an object equals the product of the object's mass and its acceleration. You can accelerate a crumpled-up piece of paper to a pretty good speed by throwing it with a moderate force. But if you try to throw a heavy rock by using the same force, the acceleration will be much less because the rock has much more

mass than the crumpled paper. Because of the smaller acceleration, the rock will leave your hand moving at only a slow speed.

Automobile airbags are based on the relationship between force and acceleration. It takes a large force to bring a fast-moving object suddenly to rest because this requires a large acceleration. In a collision, the driver of a car not equipped with airbags is jerked to a sudden stop and the large forces that act can cause major injuries. But if the car has airbags that deploy in an accident, the driver's body will slow down more gradually as it contacts the airbag, and the driver's acceleration will be less. (Remember that *acceleration* can refer to slowing down as well as to speeding up.) Hence, the force on the driver and the chance of injury will both be greatly reduced.

Newton's *third* law, the principle of action and reaction, explains how a car can accelerate at all. It is not correct to say that the engine pushes the car forward, because Newton's second law tells us that it takes a force acting from outside the car to make the car accelerate. Rather, the engine makes the wheels and tires turn, and the tires push backward on the ground. (You can see this backward force in action when a car drives through wet ground and sprays mud backward from the tires.) From Newton's third law, the ground must exert an equally large forward force on the car, and this is the force that pushes the car forward.

You use the same principles when you walk: You push backward on the ground with your foot, and the ground pushes forward on you. Icy pavement or a freshly waxed floor have greatly reduced friction. In these situations, your feet and the surface under you can exert only weak forces on each other, and it is much harder to walk.

The Heavens on the Earth—Revealing the universal applicability of the laws of nature

The **Heavens on the Earth** boxes illustrate how the same principles astronomers use to explain celestial phenomena can also explain everyday behavior here on Earth, from the color of the sky to why diet soft drink cans float in water.

Wavelength tabs—Astronomers observe the sky using many forms of light

Astronomers rely on special telescopes that are sensitive to nonvisible forms of light. To help students appreciate these different kinds of observations, all the images in *Universe* appear with wavelength tabs. The highlighted letter on each tab indiates whether the image was made with **R**adio waves, **I**nfrared radiation, **V**isible light, **U**ltraviolet light, **X** rays, or **G**amma rays.

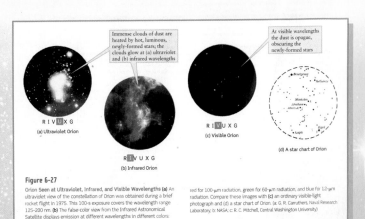

Immense clouds of dust are heated by hot, luminous, newly-formed stars; the clouds glow at (a) ultraviolet and (b) infrared wavelengths

At visible wavelengths the dust is opaque, obscuring the newly-formed stars

R I **U** X G
(a) Ultraviolet Orion

R I V U X G
(c) Visible Orion

R **I** V U X G
(b) Infrared Orion

(d) A star chart of Orion

Figure 6-27

Orion Seen at Ultraviolet, Infrared, and Visible Wavelengths **(a)** An ultraviolet view of the constellation of Orion was obtained during a brief rocket flight in 1975. This 100-s exposure covers the wavelength range 125–200 nm. **(b)** The false-color view from the Infrared Astronomical Satellite displays emission at different wavelengths in different colors: red for 100-μm radiation, green for 60-μm radiation, and blue for 12-μm radiation. Compare these images with **(c)** an ordinary visible-light photograph and **(d)** a star chart of Orion. (a: G. R. Carruthers, Naval Research Laboratory; b: NASA; c: R. C. Mitchell, Central Washington University)

End-of-Chapter material has been revised and consolidated

Key Words

A list of Key words appears at the end of each chapter, along with the number of the page where each term is introduced.

Key Ideas

Students can get the most benefit from these brief chapter summaries by using them in conjunction with the notes they take while reading.

Questions

Items from these sections are designed to be assigned as homework or used as jumping-off points for class discussion. Some questions ask students to analyze images in the text or evaluate how the mass media portrays concepts in astronomy. Advanced questions are accompanied by a *Problem-Solving Tips and Tools* box to provide guidance. Web/eBook questions challenge students to work with animations and interactive modules on the Universe Web site or eBook (described below) or to research topics on the World Wide Web. Near the end of the book is a section of **Answers to Selected Questions.**

Key Words

active optics, p. 140	Newtonian reflector, p. 136
adaptive optics, p. 140	objective lens, p. 131
angular resolution, p. 139	objective mirror (primary
baseline, p. 140	mirror), p. 135
Cassegrain focus, p. 137	optical telescope, p. 129
charge-coupled device (CCD),	optical window (in the Earth's
p. 142	atmosphere), p. 147
chromatic aberration, p. 134	photometry, p. 142
coma, p. 138	pixel, p. 142
coudé focus, p. 137	prime focus, p. 136
diffraction, p. 139	radio telescope, p. 144
diffraction grating, p. 143	radio window (in the Earth's
eyepiece lens, p. 131	atmosphere), p. 147
false color, p. 140	reflecting telescope (reflector),
focal length, p. 130	p. 135
focal plane, p. 131	reflection, p. 134
focal point, p. 130	refracting telescope
focus (of a lens or mirror),	(refractor), p. 131
p. 130	refraction, p. 130
imaging, p. 142	seeing disk, p. 139
interferometry, p. 140	spectrograph, p. 143
lens, p. 129	spectroscopy, p. 143
light-gathering power, p. 132	spherical aberration, p. 137
light pollution, p. 141	very-long-baseline
magnification (magnifying	interferometry (VLBI),
power), p. 132	p. 146
medium (*plural* media),	
p. 130	

Key Ideas

Refracting Telescopes: Refracting telescopes, or refractors, produce images by bending light rays as they pass through glass lenses.

• Chromatic aberration is an optical defect whereby light of different wavelengths is bent in different amounts by a lens.

• Glass impurities, chromatic aberration, opacity to certain wavelengths, and structural difficulties make it inadvisable to build extremely large refractors.

Reflecting Telescopes: Reflecting telescopes, or reflectors, produce images by reflecting light rays to a focus point from curved mirrors.

Charge-Coupled Devices: Sensitive light detectors called charge-coupled devices (CCDs) are often used at a telescope's focus to record faint images.

Spectrographs: A spectrograph uses a diffraction grating to form the spectrum of an astronomical object.

Radio Telescopes: Radio telescopes use large reflecting dishes to focus radio waves onto a detector.

• Very large dishes provide reasonably sharp radio images. Higher resolution is achieved with interferometry techniques that link smaller dishes together.

Transparency of the Earth's Atmosphere: The Earth's atmosphere absorbs much of the radiation that arrives from space.

• The atmosphere is transparent chiefly in two wavelength ranges known as the optical window and the radio window. A few wavelengths in the near-infrared also reach the ground.

Telescopes in Space: For observations at wavelengths to which the Earth's atmosphere is opaque, astronomers depend on telescopes carried above the atmosphere by rockets or spacecraft.

• Satellite-based observatories provide new information about the universe and permit coordinated observation of the sky at all wavelengths.

Questions

Review Questions

1. Describe refraction and reflection. Explain how these processes enable astronomers to build telescopes.
2. Explain why a flat piece of glass does not bring light to a focus while a curved piece of glass can.
3. Explain why the light rays that enter a telescope from an astronomical object are essentially parallel.
4. With the aid of a diagram, describe a refracting telescope. Which dimensions of the telescope determine its light-gathering power? Which dimensions determine the magnification?
5. What is the purpose of a telescope eyepiece? What aspect of the eyepiece determines the magnification of the image? In what circumstances would the eyepiece not be used?
6. Do most professional astronomers actually look through their telescopes? Why or why not?
7. Quite often advertisements appear for telescopes that extol their magnifying power. Is this a good criterion for evaluating telescopes? Explain your answer.

Robotic Geologists and the Search for Habitable Environments on Mars

by John Grotzinger

Scientists know that which exists; engineers create that which never was.

Theodore von Kármán, founder of NASA's Jet Propulsion Lab

Humans are explorers by nature. The great quests of humanity have presented grand challenges that by necessity have inspired creative design of ships of exploration in the search to understand the history of water on Earth and its potential role in enabling the evolution of life here or on a far lifelike terrestrial planet, human imagination has no limit to be fulfilled. Since humans cannot yet travel to other planets we have developed robots that can survive interplanetary travel and the harsh surface conditions on other planets. These robotic explorers are designed to act as geologists, such as was the case for the 2004 Mars Exploration Rover.

In early 2003, two golf cart–sized rovers destined to land on the martian surface were launched from Cape Canaveral, and began their 300-million-kilometer journey to the red planet. This mission succeeded beyond anyone's expectations, and in 2004, 2005, and 2006 three of the greatest years in the history of space exploration. The Mars Exploration rovers, Spirit and Opportunity—were designed to survive the harsh martian surface conditions and drive no farther than 300 m. Remarkably, at the time this article was written, each rover has operated for more than two martian years, and driven a combined distance of more than 10 km. The rovers have had to survive nighttime temperatures below –90 °C, dust devils that could have tipped them over, winter seasons that diminished their solar power, and driving rocky slopes of almost 30° and through piles of

treacherous windblown sand and dust. The rovers have also discovered a treasure trove of geologic wonders. These discoveries include compelling evidence for water on the ancient martian surface—a necessary condition for life, at least as it exists on Earth, and can be contemplated for Mars.

In particular, the rover Opportunity has discovered evidence from the Meridiani Planum area that water may have pooled in shallow lakes, known as playas, and infiltrated pores in sediments and soils to cause aqueous alteration. So we now know that water was present in the shallow subsurface environment and was at least intermittently present at the surface as shallow lakes and streams. But the evidence for water alone is not enough to demonstrate that life may have existed. One must also ask if the water would have had physical attributes that would have allowed life to develop.

Mars today is a forbidding place. Temperature and atmospheric pressure lie near the triple point of water—indeed, liquid water is not stable on the present-day martian surface. The surface is also chemically harsh and subject to strong radiation. It is doubtful that organisms thrive today at the martian surface. Opportunity's discoveries indicate that surface environments may have, unfortunately, been biologically challenging for most of Mars's history. The salty and transient streams and lakes that covered Meridiani 3 to 4 billion years ago indicate that while chemical weathering and erosion provided many of the elements required for life, ambient environments were arid, acidic, and oxidizing. Terrestrial ecology suggests that microorganisms could survive many aspects of the inferred Meridiani environment, but habitability would depend critically on the time scale of sufficiently high water activity to support cell biology—a time scale that is currently unknown.

Whether Meridiani is broadly representative of the martian surface 3 to 4 billion years ago is unknown, but remote sensing from orbiting spacecraft suggests that it could be. The sulfate minerals that are so widespread at Meridiani seem to be widespread in others areas of the planet, particularly the equatorial lowlands. However, there appear to be hints that other suites of minerals—some dominated by clays—may be associated with the oldest rocks on Mars. From an astrobiological perspective this is encouraging because it allows for less acidic conditions. All considered, the formerly water-drenched

John Grotzinger is a field geologist interested in the evolution of Earth's surficial environments and biosphere. He received a B.S. in geoscience from Hobart College, an M.S. in geology from University of Montana in 1981, and a Ph.D. in geology from Virginia Tech in 1985. He served as a geologist on the Mars Exploration Rover team. He was the first to conduct ground-based exploration of the geology of another planet, resulting in the discovery of sedimentary rocks formed in aqueous depositional settings.

(continued on the next page)

New Guest Essays

Several chapters in the Eighth Edition end with essays written by scientists involved with some of the recent discoveries described in the text. These include essays by John Grotzinger on the exploration of Mars; Neil Gehrels and Peter Leonard on gamma-ray bursters; and Kevin Plaxco on astrobiology. For a full list of essays, see the Contents Overview on pages viii–ix.

Media and Supplements Package

FOR STUDENTS

The eBook: Affordable, Innovative, Customizable

The *Universe*, Eighth Edition, eBook is a complete online version of the textbook. It provides a rich learning experience by taking full advantage of the electronic medium. The online eBook integrates all the existing media resources and adds such unique features as:

- Easy access from any Internet-connected computer via a standard Web browser

- Quick, intuitive navigation to any section of subsection, as well as any printed book page number

- Integration of all student Web site animated tutorials and activities

- In-text self-quiz questions and links to all glossary entries

- Text highlighting, bookmarking, and a powerful Notes feature that allows students to add notes to any page

- A full glossary and index

- Full-text search, including an option to also search the glossary and index

The eBook is available FREE with printed copies of the text (see the back cover for the ISBN) or can be purchased online at www.whfreeman.com/universe8e.

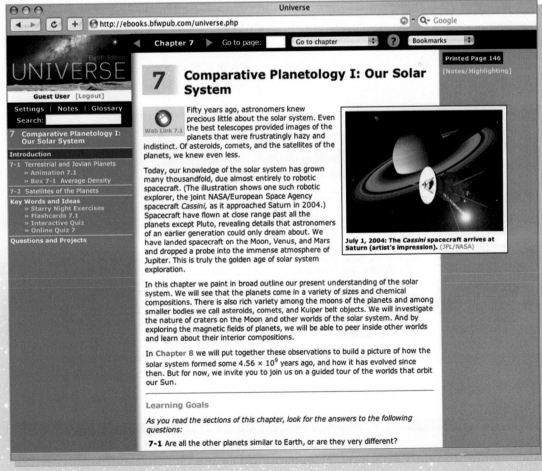

Starry Night Enthusiast™

Starry Night Enthusiast™ 5.8 is a brilliantly realistic planetarium software package. It is designed for easy use by anyone with an interest in the night sky. See the sky from anywhere on Earth or lift off and visit any solar system body or any location up to 20,000 light years away. View 2,500,000 stars along with more than 170 deep-space objects like galaxies, star clusters, and nebulae. You can travel 15,000 years in time, check out the view from the International Space Station, and see planets up close from any one of their moons. Included are stunning OpenGL graphics. You can also print handy star charts to explore outside. The *Starry Night Enthusiast* CD is available at no extra charge with the text upon request; see the back cover for the ISBN.

The Free Student Companion Web site

www.whfreeman.com/universe8e serves as a study guide and includes these features:

- **Learning Goals** for each chapter help students formulate study strategies

- **Online Quizzing** offer randomized questions and answers with instant feedback referring to specific sections in the text, to help students study, review, and prepare for exams. Instructors can access results through an online database or they can have them e-mailed directly to their accounts.

- **Animations and Videos,** both original and NASA-created, are keyed to specific chapters.

- **Web links** provide a wealth of online resources for the student.

- **Active Integrated Media Modules** take students deeper into key topics from the text. Topics include:

  - Determining Distances to the Stars and Beyond
  - Doppler Effect
  - Nearest Stars
  - Relativistic Redshift

- **Interactive Drag and Drop Exercises** based on text illustrations help students grasp the vocabulary in context.

- **Flashcard exercises** offer help with vocabulary and definitions

- **Looking Deeper** articles in PDF format extend topics discussed in the text. These include the search for gravitational waves, how to interpret the shapes of spectral lines, and discoveries about the inner core of the Earth.

Astronomy Portal

A powerful teaching and learning environment that combines three learning tools: the **Interactive eBook, Conceptual Learning Resources,** and the **Assignment Center** for homework and tutorials. These tools are unified within a proven and widely used course management system, creating a convenient and customizable system for instructors and students.

Observing Projects Using *Starry Night Enthusiast™*

ISBN 1-4292-0074-X, T. Alan Clark and William J. F. Wilson, University of Calgary, and Marcel Bergman.

Available for packaging with the text, and compatible with both PC and Mac, this book contains a variety of comprehensive lab activities for *Starry Night Enthusiast™* 5.8.

FOR INSTRUCTORS

Instructor eBook

The *Universe,* Eighth Edition eBook offers instructors flexibility and customization options not previously possible with any printed textbook. Instructors have access to:

- **eBook Customization:** Instructors can choose the chapters that correspond with their syllabus, and students will get a custom version of the eBook with the selected chapters only.

- **Instructor Notes:** Instructors can create an annotated version of the eBook by adding notes to any page. When students in the instructor's course log in, they will see their instructor's notes.

- **Custom Content:** Instructor content can include text, Web links, and images, allowing instructors to place any content they choose exactly where they want it.

Instructor's Resource CD-ROM

ISBN 1-4292-0076-6

To help instructors create lecture presentations, course Web sites, and other resources, this CD-ROM contains the following resources:

- All images from the text

- Instructor's Manual

- Test Bank

- Online resources from the text's companion Web site

Instructor's Manual

ISBN 1-4292-0075-8, by Mark Hollabaugh, Normandale Community College.

This manual includes solutions to all end-of-chapter problems, detailed chapter outlines, and classroom-tested teaching hints and strategies. The extensive resource guide covers student and instructor reading materials, audiovisual material, and discussion/paper topics.

Test Bank CD-ROM

Windows and Mac versions on one disc, ISBN: 1-4292-0077-4; printed version ISBN 1-4292-0072-3, by T. Alan Clark and William J. F. Wilson, University of Calgary, and Thomas Krause, Towson University.

More than 3,500 multiple-choice questions are section-referenced. The easy-to-use CD-ROM version includes Windows and Mac versions on a single disc, in a format that lets you add, edit, resequence, and print questions to suit your needs.

Online Course Materials (WebCT, Blackboard)

As a service for adopters, we will provide content files in the appropriate online course format, including the instructor and student resources for this text. The files can be used as is or can be customized to fir specific needs. Course outlines, prebuilt quizzes, links, activities, and a whole array of materials are included.

PowerPoint Lecture Presentations

A set of online lecture presentations created in Powerpoint allows instructors to tailor their lectures to suit their own needs using images and notes from the textbook. These presentations are available on the instructor portion of the companion Web site.

Overhead Transparencies

ISBN 1-4292-0073-1

100 full-color transparencies of key illustrations, photos and tables from the text.

Acknowledgements

I would like to thank my colleagues who have carefully scrutinized the manuscript of this and earlier editions. *Universe* is a stronger and better textbook because of their conscientious efforts:

Richard Bowman, *Bridgewater College*

Richard A. Christie, *Okanagan College*

Eric Collins, California State University, *Northridge*

George J. Corso, *DePaul University*

Steven Desch, *Arizona State University*

James Dickinson, *Clackamas Community College*

Juhan Frank, *Louisiana State University*

Joshua Gundersen, *University of Miami*

Melinda Hutson, *Portland Community College*

Douglas R. Ingram, *Texas Christian University*

Adam G. Jensen, *University of Colorado, Boulder*

Darell Johnson, *Missouri Western State University*

Michael Joner, *Brigham Young University*

Lauren Jones, *Denison University*

Steve Kawaler, *Iowa State University*

William Keel, *University of Alabama*

Charles Kerton, *Iowa State University*

H. S. Krawczynski, *Washington University at St. Louis*

Lauren Likkel, *University of Wisconsin, Eau Claire*

Franck Marchis, *University of California, Berkeley*

Edward M. Murphy, *University of Virginia*

Donna Naples, *University of Pittsburgh*

Gerald H. Newsom, *The Ohio State University*

Brian Oetiker, *Sam Houston State University*

Ronald P. Olowin, *Saint Mary's College*

Michael J. O'Shea, *Kansas State University*

J. Douglas Patterson, *Johnson County Community College*

Charles Peterson, *University of Missouri*

Richard Rand, *University of New Mexico*

Frederick Ringwald, *California State University, Fresno*

Caroline Simpson, *Florida International University*

Larry K. Smith, *Snow College*

Charles M. Telesco, *University of Florida*

I'm particularly grateful to Neil Gehrels, John Grotzinger, Peter Leonard, Kevin Plaxco, Lisa Randall, and Scott Sheppard, whose new essays for this edition greatly expand its coverage, and to Bob Kirshner and Geoff Marcy, who were kind enough to update their essays from the Seventh Edition in light of recent developments.

Many others have participated in the preparation of this book and I thank them for their efforts. I am particularly grateful for the guidance I have received from Valerie Raymond, my gracious acquisitions editor; her talented assistant, Whitney Clench; and David Chelton, my insightful development editor. Amy Thorne, my media and supplements editor, supervised the creation of the superb Web site and eBook. Special thanks go to Kerry O'Shaughnessy, my project editor, who skillfully kept the book on track and on target; and to Paul Rohloff for coordinating the vast array of resources needed to produce a book such as this. Diana Blume deserves full credit for the stunningly beautiful cover design as well as the handsome look and feel of the book. I also thank Bill Page and Imagineering Media Services for coordinating and producing the excellent artwork, and extend my gratitude for Bianca Moscatelli for finding just the right photographs. Louise Ketz deserves a medal for wading through my prose and copyediting it into proper English.

On a personal note, I would like to thank my father, Richard Freedman, for first cultivating my interest in space many years ago, and for his personal contributions to the exploration of the universe as an engineer for the Atlas and Centaur launch vehicle programs. Most of all, I thank my charming wife, Caroline, for putting up with my long nights slaving over the computer!

Although I have made a concerted effort to make this edition error-free, some mistakes may have crept in unbidden. I would appreciate hearing from anyone who finds an error or wishes to comment on the text.

Roger A. Freedman
Department of Physics
University of California, Santa Barbara
Santa Barbara CA 93106
airboy@physics.ucsb.edu

To the Student

How To Get the Most From *Universe*

If you're like most students just opening this textbook, you're enrolled in one of the few science courses you'll take in college. As you study astronomy, you'll probably do relatively little reading compared to a literature or history course—at least in terms of the number of pages. But your readings will be packed with information, much of it new to you and (I hope) exciting. You can't read this textbook like a novel and expect to learn much from it. Don't worry, though. I wrote this book with you in mind. In this section I'll suggest how *Universe* can help you succeed in your astronomy course, and take you on a guided tour of the book and media.

Apply these techniques to studying astronomy

- **Read before each lecture** You'll get the most out of your astronomy course if you read each chapter *before* hearing a lecture about its subject matter. That way, many of the topics will already be clear in your mind, and you'll understand the lecture better. You'll be able to spend more of your listening and note-taking time on the more challenging ideas presented in the lecture.

- **Take notes as you read and make use of office hours** Keep a notebook handy as you read, and write down the key points of each section so that you can review them later. If any parts of the section don't seem clear on first reading, make a note of them, too, including the page numbers. Once you've gone through the chapter, re-read it with special emphasis on the ideas that gave you trouble the first time. If you're still unsure after the lecture, consult your instructor, either during office hours or after class. Bring your notes with you so your instructor can see which concepts are giving you trouble. Once your instructor has helped clarify things for you, revise your notes so you'll remember your new-found insights. You'll end up with a chapter summary in your own words. This will be a tremendous help when studying for exams!

- **Make use of your fellow students** Many students find it useful to form study groups for astronomy. You can hash out challenging topics with each other and have a good time while you're doing it. But make sure that you write up your homework by yourself, because the penalties for copying or plagiarizing other students' work can be severe in the extreme. Some students find individual assistance useful too. If you think a tutor will be helpful, link up with one early. Getting a tutor late in the course, in the belief that you'll be able to catch up with what you missed earlier on, is almost always a lost cause.

- **Take advantage of the Web site and eBook** Take some time to explore the *Universe* Web site (**www.whfreeman.com/universe8e**). There you'll find review materials, animations, videos, interactive exercises, flashcards, and many other features keyed to chapters in *Universe*. All of these features are designed to help you learn and enjoy astronomy, so make sure to take full advantage of them. Your instructor may also have asked you to purchase the eBook version of this textbook (www.ebooks.bfwpub.com), which combines the complete content of the book and the Web site in a convenient online format.

- **Try astronomy for yourself with your star charts** At the back of this book you'll find a set star charts for the each month of the year in the Northern Hemisphere. (For a set of Southern Hemisphere star charts, see the *Universe* Web site.) Star charts can get you started with your own observations of the universe. Hold the chart overhead in the same orientation as the compass points, with southern horizon toward the south and western horizon toward the west. (To save strain on your arms, you may want to cut these pages out of the book.) Depending on the version of this textbook that your instructor requested, this book may also include a CD-ROM with the easy-to-use *Starry Night Enthusiast*™ planetarium program, which you can use to view the sky on any date and time as seen from anywhere on Earth.

Before you study *Universe,* take this quiz

Universe has many features designed to help you succeed in your study of astronomy. To get the most from it, understanding these features and knowing how to use them are essential. To make sure that you do, first read through the Preface on the preceding pages. Then take this brief quiz. If you can answer all the questions, you're ready to begin studying astronomy! (You can check your answers on page Q-I.)

1. Which specially labeled paragraphs alert you to common misconceptions and conceptual pitfalls?

2. Which specially labeled paragraphs draw analogies between ideas in astronomy and aspects of everyday life?

3. In many chapters, in addition to the numbered sections, you will also find material set off in Boxes. Which type of Box provides extra help with solving mathematical problems? Which relates astronomical principles to phenomena here on Earth?

4. Throughout the book you will encounter icons labeled "Web Link," "Animation," "Video," "AIMM," or "Looking Deeper." Where should you look to find the information to which these icons refer?

5. Many of the figures in this book are accompanied by the letters R I V U X G, with one of the letters highlighted. For instance, Figure 6-32*a* on page 152 has the letters R I V U X G, while Figure 6-32*d* has R I V U X G. What is the significance of the highlighted letter?

6. Where can you find self-tests and review material for each chapter of *Universe*?

7. Refer to the Appendices at the back of this book. On which page(s) of *Universe* would you look to find the following? (a) the value of the Stefan-Boltzmann constant; (b) the average orbital speed of Mars; (c) the average distance from the center of the planet Saturn to its moon Titan; (d) the distance in light-years to the star Proxima Centauri.

8. Refer to the Index at the back of this book. On which page(s) of *Universe* would you find the following terms described? (a) spicule; (b) refraction; (c) tidal force; (d) aphelion.

9. Refer to the Answers to Selected Questions at the back of this book. What is the answer to Question 45 of Chapter 4? (**NOTE:** Your instructor may assign as

homework some of the questions whose answers can be found in the Answers to Selected Questions. If so, your instructor will expect you to write out and explain your calculations to show how this answer is obtained.)

10. Where in this book can you find star charts for each month of the year?

Here's the most important advice of all

I haven't mentioned the most important thing you should do when studying astronomy: Have fun! Of all the different kinds of scientists, astronomers are among the most excited about what they do and what they study. Let some of that excitement about the universe rub off on you, and you'll have a great time with this course and with this textbook.

In preparing this edition of *Universe,* I've tried very hard to make it the kind of textbook that a student like you will find useful. I'm very interested in your comments and opinions! Please feel free to send me e-mail or write me, and I will respond personally.

Best wishes for success in your studies!

Roger A. Freedman
Department of Physics
University of California, Santa Barbara
Santa Barbara CA 93106
airboy@physics.ucsb.edu

UNIVERSE
THE SOLAR SYSTEM
Third Edition

1

Astronomy and the Universe

Imagine yourself in the desert on a clear, dark, moonless night, far from the glare of city lights. As you gaze upward, you see a panorama that no poet's words can truly describe and that no artist's brush could truly capture. Literally thousands of stars are scattered from horizon to horizon, many of them grouped into a luminous band called the Milky Way (which extends up and down across the middle of this photograph). As you watch, the entire spectacle swings slowly overhead from east to west as the night progresses.

For thousands of years people have looked up at the heavens and contemplated the universe. Like our ancestors, we find our thoughts turning to profound questions as we gaze at the stars. How was the universe created? Where did the Earth, Moon, and Sun come from? What are the planets and stars made of? And how do we fit in? What is our place in the cosmic scope of space and time?

Wondering about the universe is a key part of what makes us human. Our curiosity, our desire to explore and discover, and, most important, our ability to reason about what we have discovered are qualities that distinguish us from other animals. The study of the stars transcends all boundaries of culture, geography, and politics. In a literal sense, astronomy is a universal subject—its subject is the entire universe.

1-1 To understand the universe, astronomers use the laws of physics to construct testable theories and models

Astronomy has a rich heritage that dates back to the myths and legends of antiquity. Centuries ago, the heavens were thought to be populated with demons and heroes, gods and goddesses. Astronomical phenomena were explained as the result of supernatural forces and divine intervention.

The course of civilization was greatly affected by a profound realization: *The universe is comprehensible.* This awareness is one of the great gifts to come to us from ancient Greece. Greek astronomers discovered that by observing the heavens and carefully reasoning about what they saw, they could learn something about how the universe operates. For example, as we shall see in Chapter 3, they measured the size of the Earth and were able to understand and predict eclipses without appealing to supernatural

Learning Goals

By reading the sections of this chapter, you will learn

1-1 What distinguishes the methods of science from other human activities

1-2 How exploring other planets provides insight into the origins of the solar system and the nature of our Earth

1-3 Stars have a life cycle—they form, evolve over millions or billions of years, and die

1-4 Stars are grouped into galaxies, which are found throughout the universe

1-5 How astronomers measure the positions and sizes of celestial objects

1-6 How to express very large or very small numbers in convenient notation

1-7 Why astronomers use different units to measure distances in space

1-8 What astronomy can tell us about our place in the universe

1

forces. Modern science is a direct descendant of the intellectual endeavors of these ancient Greek pioneers.

The Scientific Method

Like art, music, or any other human creative activity, science makes use of intuition and experience. But the approach used by scientists to explore physical reality differs from other forms of intellectual endeavor in that it is based fundamentally on *observation*, *logic*, and *skepticism*. This approach, called the **scientific method**, requires that our ideas about the world around us be consistent with what we actually observe.

The scientific method goes something like this: A scientist trying to understand some observed phenomenon proposes a **hypothesis**, which is a collection of ideas that seems to explain what is observed. It is in developing hypotheses that scientists are at their most creative, imaginative, and intuitive. But their hypotheses must always agree with existing observations and experiments, because a discrepancy with what is observed implies that the hypothesis is wrong. (The exception is if the scientist thinks that the existing results are wrong and can give compelling evidence to show that they are wrong.) The scientist then uses logic to work out the implications of the hypothesis and to make predictions that can be tested. A hypothesis is on firm ground only after it has accurately forecast the results of new experiments or observations. (In practice, scientists typically go through these steps in a less linear fashion than we have described.)

> Hypotheses, models, theories, and laws are essential parts of the scientific way of knowing

Scientists describe reality in terms of **models**, which are hypotheses that have withstood observational or experimental tests. A model tells us about the properties and behavior of some object or phenomenon. A familiar example is a model of the atom, which scientists picture as electrons orbiting a central nucleus. Another example, which we will encounter in Chapter 18, is a model that tells us about physical conditions (for example, temperature, pressure, and density) in the interior of the Sun (Figure 1-1). A well-developed model uses mathematics—one of the most powerful tools for logical thinking—to make detailed predictions. For example, a successful model of the Sun's interior should describe what the values of temperature, pressure, and density are at each depth within the Sun, as well as the relations between these quantities. For this reason, mathematics is one of the most important tools used by scientists.

A body of related hypotheses can be pieced together into a self-consistent description of nature called a **theory**. An example from Chapter 4 is the theory that the planets are held in their orbits around the Sun by the Sun's gravitational force (Figure 1-2). Without models and theories there is no understanding and no science, only collections of facts.

CAUTION! In everyday language the word "theory" is often used to mean an idea that looks good on paper, but has little to do with reality. In science, however, a good theory is one that explains reality very well and that can be applied to explain new observations. An excellent example is the theory of gravitation (Chapter 4), which was devised by the English scientist Isaac

Figure 1-1   R I V U X G

Our Star, the Sun The Sun is a typical star. Its diameter is about 1.39 million kilometers (roughly a million miles), and its surface temperature is about 5500°C (10,000°F). A detailed scientific model of the Sun tells us that it draws its energy from thermonuclear reactions occurring at its center, where the temperature is about 15 million degrees Celsius. (NSO/AURA/NSF)

Newton in the late 1600s to explain the orbits of the six planets known at that time. When astronomers of later centuries discovered the planets Uranus and Neptune and the dwarf planet Pluto, they found that these planets also moved in accordance with Newton's theory. The same theory describes the motions of satellites around the Earth as well as the orbits of planets around other stars (see Chapter 8).

An important part of a scientific theory is its ability to make predictions that can be tested by other scientists. If the predictions are verified by observation, that lends support to the theory and suggests that it might be correct. If the predictions are *not* verified, the theory needs to be modified or completely replaced. For example, an old theory held that the Sun and planets orbit around a stationary Earth. This theory led to certain predictions that could be checked by observation, as we will see in Chapter 4. In the early 1600s the Italian scientist Galileo Galilei used one of the first telescopes to show that these predictions were incorrect. As a result, the theory of a stationary Earth was rejected, eventually to be replaced by the modern picture shown in Figure 1-2 in which the Earth and other planets orbit the Sun.

An idea that *cannot* be tested by observation or experiment does not qualify as a scientific theory. An example is the idea that there is a little man living in your refrigerator who turns the inside light on or off when you open and close the door. The little

Figure 1-2

Planets Orbiting the Sun An example of a scientific theory is the idea that the Earth and planets orbit the Sun due to the Sun's gravitational attraction. This theory is universally accepted because it makes predictions that have been tested and confirmed by observation. (The Sun and planets are actually much smaller than this illustration would suggest.) (Detlev Van Ravenswaay/Science Photo Library)

man is invisible, weightless, and makes no sound, so you cannot detect his presence. While this is an amusing idea, it cannot be tested and so cannot be considered science.

Skepticism is an essential part of the scientific method. New hypotheses must be able to withstand the close scrutiny of other scientists. The more radical the hypothesis, the more skepticism and critical evaluation it will receive from the scientific community, because the general rule in science is that extraordinary claims require extraordinary evidence. That is why scientists as a rule do not accept claims that people have been abducted by aliens and taken aboard UFOs. The evidence presented for these claims is flimsy, secondhand, and unverifiable.

At the same time, scientists must be open-minded. They must be willing to discard long-held ideas if these ideas fail to agree with new observations and experiments, provided the new data have survived evaluation. (If an alien spacecraft really did land on Earth, scientists would be the first to accept that aliens existed—provided they could take a careful look at the spacecraft and its occupants.) That is why scientific knowledge is always provisional. As you go through this book, you will encounter many instances where new observations have transformed our understanding of Earth, the planets, the Sun and stars, and indeed the very structure of the universe.

Theories that accurately describe the workings of physical reality have a significant effect on civilization. For example, basing his conclusions in part on observations of how the planets orbit the Sun, Isaac Newton deduced a set of fundamental principles that describe how *all* objects move. These theoretical principles, which we will encounter in Chapter 4, work equally well on Earth as in the most distant corner of the universe. They represent our first complete, coherent description of the behavior of the physical universe. **Newtonian mechanics** had an immediate practical application in the construction of machines, buildings,

and bridges. It is no coincidence that the Industrial Revolution followed hard on the heels of these theoretical and mathematical advances inspired by astronomy.

Newtonian mechanics and other physical theories have stood the test of time and been shown to have great and general validity. Proven theories of this kind are collectively referred to as the **laws of physics**. Astronomers use these laws to interpret and understand their observations of the universe. The laws governing light and its relationship to matter are of particular importance, because the only information we can gather about distant stars and galaxies is in the light that we receive from them. Using the physical laws that describe how objects absorb and emit light, astronomers have measured the temperature of the Sun and even learned what the Sun is made of. By analyzing starlight in the same way, they have discovered that our own Sun is a rather ordinary star and that the observable universe may contain billions of stars just like the Sun.

Technology in Science

An important part of science is the development of new tools for research and new techniques of observation. As an example, until fairly recently everything we knew about the distant universe was based on visible light. Astronomers would peer through telescopes to observe and analyze visible starlight. By the end of the nineteenth century, however, scientists had begun discovering forms of light invisible to the human eye: X rays, gamma rays, radio waves, microwaves, and ultraviolet and infrared radiation.

As we will see in Chapter 6, in recent years astronomers have constructed telescopes that can detect such nonvisible forms of light (**Figure 1-3**). These instruments give us views of the universe vastly different from anything our eyes can see. These new views have allowed us to see through the atmospheres of distant planets,

Figure 1-3   R I $\boxed{V}$ U X G

A Telescope in Space Because it orbits outside the Earth's atmosphere in the vacuum of space, the Hubble Space Telescope (HST) can detect not only visible light but also ultraviolet and near-infrared light coming from distant stars and galaxies. These forms of nonvisible light are absorbed by our atmosphere and hence are difficult or impossible to detect with a telescope on the Earth's surface. This photo of HST was taken by the crew of the space shuttle *Columbia* after a servicing mission in 2002. (NASA)

to study the thin but incredibly violent gas that surrounds our Sun, and even to observe new solar systems being formed around distant stars. Aided by high-technology telescopes, today's astronomers carry on the program of careful observation and logical analysis begun thousands of years ago by their ancient Greek predecessors.

1-2 By exploring the planets, astronomers uncover clues about the formation of the solar system

The science of astronomy allows our intellects to voyage across the cosmos. We can think of three stages in this voyage: from the Earth to the solar system, from the solar system to the stars, and from stars to galaxies and the grand scheme of the universe.

> **Studying planetary science gives us a better perspective on our own unique Earth**

The star we call the Sun and all the celestial bodies that orbit the Sun—including Earth, the other eight planets, all their various moons, and smaller bodies such as asteroids and comets—make up the **solar system.** Since the 1960s a series of unmanned spacecraft has been sent to explore each of the planets (**Figure 1-4**). Using the remote "eyes" of such spacecraft, we have flown over Mercury's cratered surface, peered beneath Venus's poisonous cloud cover, and discovered enormous canyons and extinct volcanoes on Mars. We have found active volcanoes on a moon of Jupiter, probed the atmosphere of Saturn's moon Titan, seen the rings of Uranus up close, and looked down on the active atmosphere of Neptune.

Along with rocks brought back by the Apollo astronauts from the Moon (the only world beyond Earth yet visited by humans), new information from spacecraft has revolutionized our understanding of the origin and evolution of the solar system. We have come to realize that many of the planets and their satellites were shaped by collisions with other objects. Craters on the Moon and on many other worlds are the relics of innumerable impacts by bits of interplanetary rock. The Moon may itself be the result of a catastrophic collision between the Earth and a planet-sized object shortly after the solar system was formed. Such a collision could have torn sufficient material from the primordial Earth to create the Moon.

The oldest objects found on Earth are **meteorites,** bits of interplanetary debris that sometimes fall to our planet's surface. By using radioactive age-dating techniques, scientists have found that all meteorites are 4.56 billion years old—older than any other rocks found on Earth or the Moon. The conclusion is that our entire solar system, including the Sun and planets, formed 4.56 billion years ago. The few thousand years of recorded human history is no more than the twinkling of an eye compared to the long history of our solar system.

The discoveries that we have made in our journeys across the solar system are directly relevant to the quality of human life on our own planet. Until recently, our understanding of geology, weather, and climate was based solely on data from the Earth. Since the advent of space exploration, however, we have been able

Figure 1-4

The Sun and Planets to Scale This montage of images from various spacecraft and ground-based telescopes shows the relative sizes of the planets and the Sun. The Sun is so large compared to the planets that only a portion of it fits into this illustration. The distances from the Sun to each planet are *not* shown to scale; the actual distance from the Sun to the Earth, for instance, is 12,000 times greater than the Earth's diameter. (Calvin J. Hamilton and NASA/JPL) to compare and contrast other worlds with our own. This new knowledge gives us valuable insight into our origins, the nature of our planetary home, and the limits of our natural resources.

1-3 By studying stars and nebulae, astronomers discover how stars are born, grow old, and die

The nearest of all stars to Earth is the Sun. Although humans have used the Sun's warmth since the dawn of our species, it was only in the 1920s and 1930s that physicists figured out how the Sun shines. At the center of the Sun, thermonuclear reactions—so called because they require extremely high temperatures—convert hydrogen (the Sun's primary constituent) into helium. This violent process releases a vast amount of energy, which eventually makes its way to the Sun's surface and escapes as light (see Figure 1-1). All the stars you can see in the nighttime sky also shine by thermonuclear reactions (Figure 1-5). By 1950 physicists could reproduce such thermonuclear reactions here on Earth in the form of a hydrogen bomb (Figure 1-6). While such weapons are capable of destroying life on our planet, peaceful applications of this same process may provide a clean source of energy sometime in the next several decades.

Figure 1-6   R I ☑ U X G

A Thermonuclear Explosion A hydrogen bomb uses the same physical principle as the thermonuclear reactions at the Sun's center: the conversion of matter into energy by nuclear reactions. This thermonuclear detonation on October 31, 1952, had an energy output equivalent to 10.4 million tons of TNT. This is a mere ten-billionth of the amount of energy released by the Sun in one second. (Defense Nuclear Agency)

Thermonuclear reactions consume the material of which stars are made, which means that stars cannot last forever. Rather, they must form, evolve, and eventually die.

CAUTION! Astronomers often use biological terms such as "birth" and "death" to describe stages in the evolution of inanimate objects like stars. Keep in mind that such terms are used only as *analogies*, which help us visualize these stages. They are not to be taken literally!

The Life Stories of Stars

The rate at which stars emit energy in the form of light tells us how rapidly they are consuming their thermonuclear "fuel," and hence how long they can continue to shine before reaching the end of their life spans. More massive stars have more thermonuclear "fuel," but consume it at such a prodigious rate that they live out their lives in just a few million years. Less massive stars have less material to consume, but their thermonuclear reactions proceed so slowly that their life spans are measured in billions of years. (Our own star, the Sun, is in early middle age: it is 4.56 billion years old, with a lifetime of $12\frac{1}{2}$ billion years.)

While no astronomer can watch a single star go through all of its life stages, we have been able to piece together the life stories of stars by observing them at different points in their life cycles. Important pieces of the puzzle have been discovered by studying huge clouds of interstellar gas, called **nebulae** (singular **nebula**), which are found scattered across the sky. Within some

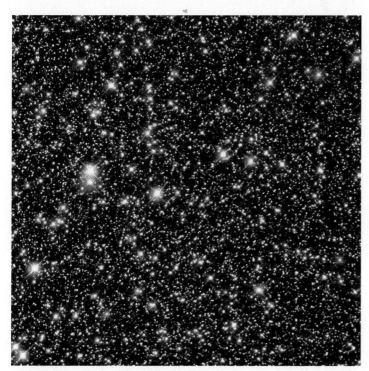

Figure 1-5   R I ☑ U X G

Stars like Grains of Sand This Hubble Space Telescope image shows thousands of stars in the constellation Sagittarius. Each star shines because of thermonuclear reactions that release energy in its interior. Different colors indicate stars with different surface temperatures: stars with the hottest surfaces appear blue, while those with the coolest surfaces appear red. (The Hubble Heritage Team, AURA/STScI/NASA) nebulae, such as the Orion Nebula shown in **Figure 1-7**, stars are born from the material of the nebula itself. Other nebulae reveal what happens when thermonuclear reactions stop and a star dies. Some stars that

are far more massive than the Sun end their lives with a spectacular detonation called a **supernova** (plural **supernovae**) that blows the star apart. The Crab Nebula (**Figure 1-8**) is a striking example of a remnant left behind by a supernova.

Dying stars can produce some of the strangest objects in the sky. Some dead stars become **pulsars,** which spin dizzily at rates of tens or hundreds of rotations per second. And some stars end their lives as almost inconceivably dense objects called **black holes,** whose gravity is so powerful that nothing—not even light—can escape. Even though a black hole itself emits essentially no radiation, a number of black holes have been discovered by Earth-orbiting telescopes. This is done by detecting the X rays emitted by gases falling toward a black hole.

During their death throes, stars return the gas of which they are made to interstellar space. (Figure 1-8 shows these expelled gases expanding away from the site of a supernova explosion.)

Figure 1-8   R I V U X G

The Crab Nebula—Wreckage of an Exploded Star When a dying star exploded in a supernova, it left behind this elegant funeral shroud of glowing gases blasted violently into space. A thousand years after the explosion these gases are still moving outward at about 1800 km/s (roughly 4 million miles per hour). The Crab Nebula is 6500 light-years from Earth and about 13 light-years across. (NASA, ESA, J. Hester and A. Loll/Arizona State University)

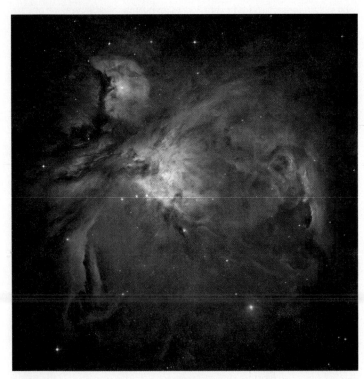

Figure 1-7   R I V U X G

The Orion Nebula—Birthplace of Stars This beautiful nebula is a stellar "nursery" where stars are formed out of the nebula's gas. Intense ultraviolet light from newborn stars excites the surrounding gas and causes it to glow. Many of the stars embedded in this nebula are less than a million years old, a brief interval in the lifetime of a typical star. The Orion Nebula is some 1500 light-years from Earth and is about 30 light-years across. (NASA, ESA, M. Robberto/STScI/ESA, and the Hubble Space Telescope Orion Treasury Project Team)

This gas contains heavy elements—that is, elements heavier than hydrogen and helium—that were created during the star's lifetime by thermonuclear reactions in its interior. Interstellar space thus becomes enriched with newly manufactured atoms and molecules. The Sun and its planets were formed from interstellar material that was enriched in this way. This means that the atoms of iron and nickel that make up the body of the Earth, as well as the carbon in our bodies and the oxygen we breathe, were created deep inside ancient stars. By studying stars and their evolution, we are really studying our own origins.

1-4 By observing galaxies, astronomers learn about the origin and fate of the universe

Stars are not spread uniformly across the universe but are grouped together in huge assemblages called **galaxies.** Galaxies come in a wide range of shapes and sizes. A typical galaxy, like the Milky Way, of which our Sun is part, contains several hundred billion stars. Some galaxies are much smaller, containing only a few million stars. Others are monstrosities that devour neighboring galaxies in a process called "galactic cannibalism."

Figure 1-9   R I [V] U X G

A Galaxy This spectacular galaxy, called M63, contains about a hundred billion stars. M63 has a diameter of about 60,000 light-years and is located about 35 million light-years from Earth. Along this galaxy's spiral arms you can see a number of glowing clumps. Like the Orion Nebula in our own Milky Way Galaxy (see Figure 1-7), these are sites of active star formation. (Subaru Telescope, National Astronomical Observatory of Japan)

Our Milky Way Galaxy has arching spiral arms like those of the galaxy shown in Figure 1-9. These arms are particularly active sites of star formation. In recent years, astronomers have discovered a mysterious object at the center of the Milky Way with a mass millions of times greater than that of our Sun. It now seems certain that this curious object is an enormous black hole.

Some of the most intriguing galaxies appear to be in the throes of violent convulsions and are rapidly expelling matter. The centers of these strange galaxies, which may harbor even more massive black holes, are often powerful sources of X rays and radio waves.

Even more awesome sources of energy are found still deeper in space. At distances so great that their light takes billions of years to reach Earth, we find the mysterious **quasars**. Although quasars look like nearby stars (Figure 1-10), they are among the most distant and most luminous objects in the sky. A typical quasar shines with the brilliance of a hundred galaxies. Detailed observations of quasars imply that they draw their energy from material falling into enormous black holes.

Galaxies and the Expanding Universe

The motions of distant galaxies reveal that they are moving away from us and from each other. In other words, the universe is *expanding*. Extrapolating into the past, we learn that the universe must have been born from an incredibly dense state (perhaps infinitely dense) some 13.7 billion years ago. A variety of evidence indicates that at that moment—the beginning of time—the universe began with a cosmic explosion, known as the **Big Bang**, which occurred throughout all space. Events shortly after the Big Bang dictated the present nature of the universe.

The motions of distant galaxies motivate the ideas of the expanding universe and the Big Bang

Thanks to the combined efforts of astronomers and physicists, we are making steady advances in understanding these cosmic events. This understanding may reveal the origin of some of the most basic properties of physical reality. Studying the most remote galaxies is also helping to answer questions about the ultimate fate of the universe. Such studies suggest that the expansion of the universe will continue forever, and is actually gaining speed.

The work of unraveling the deepest mysteries of the universe requires specialized tools, including telescopes, spacecraft, and computers. But for many purposes the most useful device for studying the universe is the human brain itself. Our goal in this book is to help you use *your* brain to share in the excitement of scientific discovery.

In the remainder of this chapter we introduce some of the key concepts and mathematics that we will use in subsequent

Figure 1-10   R I [V] U X G

A Quasar The two bright starlike objects in this image look almost identical, but they are dramatically different. The object on the left is indeed a star that lies a few hundred light-years from Earth. But the "star" on the right is actually a quasar about 9 billion light-years away. To appear so bright even though they are so distant, quasars like this one must be some of the most luminous objects in the universe. The other objects in this image are galaxies like that in Figure 1-9. (Charles Steidel, California Institute of Technology; and NASA)

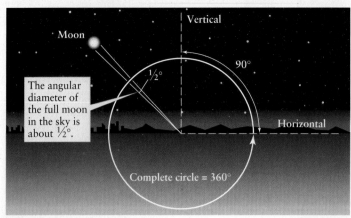

(a) Measuring angles in the sky

(b) Angular distances in the
northern hemisphere

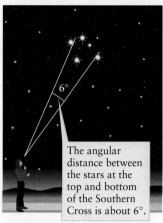

(c) Angular distances in the
southern hemisphere

Figure 1-11

Angles and Angular Measure (a) Angles are measured in degrees (°). There are 360° in a complete circle and 90° in a right angle. For example, the angle between the vertical direction (directly above you) and the horizontal direction (toward the horizon) is 90°. The angular diameter of the full moon in the sky is about $\frac{1}{2}$°. **(b)** The seven bright stars that make up the Big Dipper can be seen from anywhere in the northern hemisphere. The angular distance between the two "pointer stars" at the front of the Big Dipper is about 5°. **(c)** The four bright stars that make up the Southern Cross can be seen from anywhere in the southern hemisphere. The angular distance between the stars at the top and bottom of the cross is about 6°.

chapters. Study these carefully, for you will use them over and over again throughout your own study of astronomy.

1-5 Astronomers use angles to denote the positions and apparent sizes of objects in the sky

Whether they study planets, stars, galaxies, or the very origins of the universe, astronomers must know where to point their telescopes. For this reason, an important part of astronomy is keeping track of the positions of objects in the sky. Angles and a system of angular measure are essential parts for this aspect of astronomy (**Figure 1-11**).

Angular measure is a tool that we will use throughout our study of astronomy

An **angle** is the opening between two lines that meet at a point. **Angular measure** describes the size of an angle exactly. The basic unit of angular measure is the **degree**, designated by the symbol °. A full circle is divided into 360°, and a right angle measures 90° (Figure 1-11*a*). As Figure 1-11*b* shows, if you draw lines from your eye to each of the two "pointer stars" in the Big Dipper, the angle between these lines—that is, the **angular distance** between these two stars—is about 5°. (In Chapter 2 we will see that these two stars "point" to Polaris, the North Star.) The angular distance between the stars that make up the top and bottom of the Southern Cross, which is visible from south of the equator, is about 6° (Figure 1-11*c*).

Astronomers also use angular measure to describe the apparent size of a celestial object—that is, what fraction of the sky that object seems to cover. For example, the angle covered by the diameter of the full moon is about $\frac{1}{2}$° (Figure 1-11*a*). We therefore say that the **angular diameter** (or **angular size**) of the Moon is $\frac{1}{2}$°. Alternatively, astronomers say that the Moon **subtends** an angle of $\frac{1}{2}$°. Ten full moons could fit side by side between the two pointer stars in the Big Dipper.

The adult human hand held at arm's length provides a means of estimating angles, as **Figure 1-12** shows. For example, your fist covers an angle of 10°, whereas the tip of your finger is about 1° wide. You can use various segments of your index finger extended to arm's length to estimate angles a few degrees across.

Figure 1-12

Estimating Angles with Your Hand The adult human hand extended to arm's length can be used to estimate angular distances and angular sizes in the sky.

To talk about smaller angles, we subdivide the degree into 60 **arcminutes** (also called minutes of arc), which is commonly abbreviated as 60 arcmin or 60′. An arcminute is further subdivided into 60 **arcseconds** (or seconds of arc), usually written as 60 arcsec or 60″. Thus,

$$1° = 60 \text{ arcmin} = 60'$$

$$1' = 60 \text{ arcsec} = 60''$$

For example, on January 1, 2007, the planet Saturn had an angular diameter of 19.6 arcsec as viewed from Earth. That is a convenient, precise statement of how big the planet appeared in Earth's sky on that date. (Because this angular diameter is so small, to the naked eye Saturn appears simply as a point of light. To see any detail on Saturn, such as the planet's rings, requires a telescope.)

If we know the angular size of an object as well as the distance to that object, we can determine the actual linear size of the object (measured in kilometers or miles, for example). **Box 1-1** describes how this is done.

BOX 1-1   **Tools of the Astronomer's Trade**

The Small-Angle Formula

You can estimate the angular sizes of objects in the sky with your hand and fingers (see Figure 1-12). Using rather more sophisticated equipment, astronomers can measure angular sizes to a fraction of an arcsecond. Keep in mind, however, that *angular* size is not the same as *actual* size. As an example, if you extend your arm while looking at a full moon, you can completely cover the Moon with your thumb. That's because from your perspective, your thumb has a larger angular size (that is, it subtends a larger angle) than the Moon. But the actual size of your thumb (about 2 centimeters) is much less than the actual diameter of the Moon (more than 3000 kilometers).

The accompanying figure shows how the angular size of an object is related to its linear size. Part **a** of the figure shows that for a given angular size, the more distant the object, the larger its actual size. For example, your thumb held at arm's length just covers the full moon; the angular size of your thumb and the Moon are about the same, but the Moon is much farther away and is far larger in linear size. Part **b** shows that for a given linear size, the angular size decreases the farther away the object. This is why a car looks smaller and smaller as it drives away from you.

We can put these relationships together into a single mathematical expression called the small-angle formula. Suppose that an object subtends an angle α (the Greek letter alpha) and is at a distance *d* from the observer, as in part **c** of the figure. If the angle α is small, as is almost always the case for objects in the sky, the linear size (*D*) of the object is given by the following expression:

The small-angle formula

$$D = \frac{\alpha d}{206{,}265}$$

D = linear size of an object

α = angular size of the object, in arcsec

d = distance to the object

The number 206,265 is required in the formula. Mathematically, it is equal to the number of arcseconds in a complete

(a) For a given angular size α, the more distant the object, the greater its actual (linear) size

(b) For a given linear size, the more distant the object, the smaller its angular size

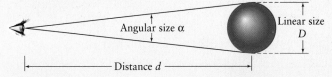

(c) Relating an object's linear size *D*, angular size α, and distance *d*

(a) Two objects that have the same angular size may have different linear sizes if they are at different distances from the observer. **(b)** For an object of a given linear size, the angular size is smaller the farther the object is from the observer. **(c)** The small-angle formula relates the linear size *D* of an object to its angular size α and its distance *d* from the observer.

circle (that is, 360°) divided by the number 2π (the ratio of the circumference of a circle to that circle's radius).

The following examples show two different ways to use the small-angle formula. In both examples we follow a four-step process: Evaluate the **situation** given in the example, decide which **tools** are needed to solve the problem, use those tools to find the **answer** to the problem, and **review** the result to see what it tells you. Throughout this book, we'll use these same four steps in *all* examples that require the use of formulas. We encourage you to follow this four-step process when solving problems for homework or exams. You can remember these steps by their acronym: *S.T.A.R.*

EXAMPLE: On December 11, 2006, Jupiter was 944 million kilometers from Earth and had an angular diameter of 31.2 arcseconds. From this information, calculate the actual diameter of Jupiter in kilometers.

(continued on the next page)

BOX 1-1 *(continued)*

Situation: The astronomical object in this example is Jupiter, and we are given its distance d and its angular size α (the same as angular diameter). Our goal is to find Jupiter's diameter D.

Tools: The equation to use is the small-angle formula, which relates the quantities d, α, and D. Note that when using this formula, the angular size α *must* be expressed in arcseconds.

Answer: The small-angle formula as given is an equation for D. Plugging in the given values $\alpha = 31.2$ arcsec and $d = 944$ million km,

$$D = \frac{31.2 \times 944,000,000 \text{ km}}{206,265} = 143,000 \text{ km}$$

Because the distance d to Jupiter is given in kilometers, the diameter D is also in kilometers.

Review: Does our answer make sense? From Appendix 2 at the back of this book, the equatorial diameter of Jupiter measured by spacecraft flybys is 142,984 km, so our calculated answer is very close.

EXAMPLE: Under excellent conditions, a telescope on Earth can see details with an angular size as small as 1 arcsec. What is the greatest distance at which you could see details as small as 1.7 m (the height of a typical person) under these conditions?

Situation: Now the object in question is a person, whose linear size D we know. Our goal is to find the distance d at which the person has an angular size α equal to 1 arcsec.

Tools: Again we use the small-angle formula to relate d, α, and D.

Answer: We first rewrite the formula to solve for the distance d, then plug in the given values $D = 1.7$ m and $\alpha = 1$ arcsec:

$$d = \frac{206,265 D}{\alpha} = \frac{206,265 \times 1.7 \text{ m}}{1} = 350,000 \text{ m} = 350 \text{ km}$$

Review: This is much less than the distance to the Moon, which is 384,000 km. Thus, even the best telescope on Earth could not be used to see an astronaut walking on the surface of the Moon.

1-6 Powers-of-ten notation is a useful shorthand system for writing numbers

Astronomy is a subject of extremes. Astronomers investigate the largest structures in the universe, including galaxies and clusters of galaxies. But they must also study atoms and atomic nuclei, among the smallest objects in the universe, in order to explain how and why stars shine. They also study conditions ranging from the incredibly hot and dense centers of stars to the frigid near-vacuum of interstellar space. To describe such a wide range of phenomena, we need an equally wide range of both large and small numbers.

> **Learning powers-of-ten notation will help you deal with very large and very small numbers**

Powers-of-Ten Notation: Large Numbers

Astronomers avoid such confusing terms as "a million billion billion" by using a standard shorthand system called **powers-of-ten notation.** All the cumbersome zeros that accompany a large number are consolidated into one term consisting of 10 followed by an **exponent,** which is written as a superscript. The exponent indicates how many zeros you would need to write out the long form of the number. Thus,

$$10^0 = 1 \text{ (one)}$$
$$10^1 = 10 \text{ (ten)}$$
$$10^2 = 100 \text{ (one hundred)}$$
$$10^3 = 1000 \text{ (one thousand)}$$
$$10^4 = 10,000 \text{ (ten thousand)}$$
$$10^6 = 1,000,000 \text{ (one million)}$$
$$10^9 = 1,000,000,000 \text{ (one billion)}$$
$$10^{12} = 1,000,000,000,000 \text{ (one trillion)}$$

and so forth. The exponent also tells you how many tens must be multiplied together to give the desired number, which is why the exponent is also called the **power of ten.** For example, ten thousand can be written as 10^4 ("ten to the fourth" or "ten to the fourth power") because $10^4 = 10 \times 10 \times 10 \times 10 = 10,000$.

In powers-of-ten notation, numbers are written as a figure between one and ten multiplied by the appropriate power of ten. The approximate distance between Earth and the Sun, for example, can be written as 1.5×10^8 kilometers (or 1.5×10^8 km for short). Once you get used to it, this is more convenient than writing "150,000,000 kilometers" or "one hundred and fifty million kilometers." (The same number could also be written as 15×10^7 or 0.15×10^9, but the preferred form is *always* to have the first figure be between 1 and 10.)

Calculators and Powers-of-Ten Notation

Most electronic calculators use a shorthand for powers-of-ten notation. To enter the number 1.5×10^8, you first enter 1.5, then press a key labeled "EXP" or "EE," then enter the exponent 8. (The EXP or EE key takes care of the "$\times 10$" part of the expression.) The number will then appear on your calculator's display as "1.5 E 8," "1.5 8," or some variation of this; typically the "$\times 10$" is not displayed as such. There are some variations from one kind of calculator to another, so you should spend a few minutes reading over your calculator's instruction manual to make sure you know the correct procedure for working with numbers in powers-of-ten notation. You will be using this notation continually in your study of astronomy, so this is time well spent.

CAUTION! Confusion can result from the way that calculators display powers-of-ten notation. Since 1.5×10^8 is displayed as "1.5 8" or "1.5 E 8," it is not uncommon to think that 1.5×10^8 is the same as 1.5^8. That is not correct, however; 1.5^8 is equal to 1.5 multiplied by itself 8 times, or 25.63, which is not even close to $150,000,000 = 1.5 \times 10^8$. Another, not uncommon, mistake is to write 1.5×10^8 as 15^8. If you are inclined to do this, perhaps you are thinking that you can multiply 1.5 by 10, then tack on the exponent later. This also does not work; 15^8 is equal to 15 multiplied by itself 8 times, or 2,562,890,625, which again is nowhere near 1.5×10^8. Reading over the manual for your calculator will help you to avoid these common errors.

Powers-of-Ten Notation: Small Numbers

You can use powers-of-ten notation for numbers that are less than one by using a minus sign in front of the exponent. A negative exponent tells you to *divide* by the appropriate number of tens. For example, 10^{-2} ("ten to the minus two") means to divide by 10 twice, so $10^{-2} = 1/10 \times 1/10 = 1/100 = 0.01$. This same idea tells us how to interpret other negative powers of ten:

$10^0 = 1$ (one)

$10^{-1} = 1/10 = 0.1$ (one tenth)

$10^{-2} = 1/10 \times 1/10 = 1/10^2 = 0.01$ (one hundredth)

$10^{-3} = 1/10 \times 1/10 \times 1/10 = 1/10^3 = 0.001$ (one thousandth)

$10^{-4} = 1/10 \times 1/10 \times 1/10 \times 1/10 = 1/10^4 = 0.0001$
    (one ten-thousandth)

$10^{-6} = 1/10 \times 1/10 \times 1/10 \times 1/10 \times 1/10 \times 1/10 = 1/10^6$
    $= 0.000001$ (one millionth)

$10^{-12} = 1/10 \times 1/10 \times 1/10 \times 1/10 \times 1/10 \times 1/10 \times 1/10$
    $\times 1/10 \times 1/10 \times 1/10 \times 1/10 \times 1/10 = 1/10^{12}$
    $= 0.000000000001$ (one trillionth)

and so forth.

As these examples show, negative exponents tell you how many tenths must be multiplied together to give the desired number. For example, one ten-thousandth, or 0.0001, can be written as 10^{-4} ("ten to the minus four") because $10^{-4} = 1/10 \times 1/10 \times 1/10 \times 1/10 = 0.0001$.

A useful shortcut in converting a decimal to powers-of-ten notation is to notice where the decimal point is. For example, the decimal point in 0.0001 is four places to the left of the "1," so the exponent is -4, that is, $0.0001 = 10^{-4}$.

You can also use powers-of-ten notation to express a number like 0.00245, which is not a multiple of 1/10. For example, $0.00245 = 2.45 \times 0.001 = 2.45 \times 10^{-3}$. (Again, the standard for powers-of-ten notation is that the first figure is a number between one and ten.) This notation is particularly useful when dealing with very small numbers. A good example is the diameter of a hydrogen atom, which is much more convenient to state in powers-of-ten notation (1.1×10^{-10} meter, or 1.1×10^{-10} m) than as a decimal (0.00000000011 m) or a fraction (110 trillionths of a meter.)

Because it bypasses all the awkward zeros, powers-of-ten notation is ideal for describing the size of objects as small as atoms or as big as galaxies (Figure 1-13). Box 1-2 explains how powers-of-ten notation also makes it easy to multiply and divide numbers that are very large or very small.

10^{-15} — Size of a proton
10^{-10} — Size of an atom
— Size of a virus
10^{-5}
1 — Size of a human
10^5
— Diameter of the Earth
— Diameter of the Sun
10^{10} — Distance from Earth to Sun
10^{15}
— Distance to the nearest star beyond the Sun
10^{20} — Diameter of the Galaxy
10^{25} — Size of the observable universe

Figure 1-13

Examples of Powers-of-Ten Notation The scale gives the sizes of objects in meters, ranging from subatomic particles at the left to the entire observable universe on the right. The photograph at the left shows tungsten atoms, 10^{-10} meter in diameter. Second from left is the crystalline skeleton of a diatom (a single-celled organism), 10^{-4} meter (0.1 millimeter) in size. At the center is the Taj Mahal, about 60 meters tall and within reach of our unaided senses. On the right, looking across the Indian Ocean toward the south pole, we see the curvature of the Earth, about 10^7 meters in diameter. At the far right is a galaxy, 10^{21} meters (100,000 light-years) in diameter. (Courtesy of Scientific American Books; NASA; and photograph by David Malin from the Anglo-Australian Observatory)

BOX 1-2   Tools of the Astronomer's Trade

Arithmetic with Powers-of-Ten Notation

Using powers-of-ten notation makes it easy to multiply numbers. For example, suppose you want to multiply 100 by 1000. If you use ordinary notation, you have to write a lot of zeros:

$$100 \times 1000 = 100{,}000 \text{ (one hundred thousand)}$$

By converting these numbers to powers-of-ten notation, we can write this same multiplication more compactly as

$$10^2 \times 10^3 = 10^5$$

Because $2 + 3 = 5$, we are led to the following general rule for *multiplying* numbers expressed in terms of powers of ten: Simply *add* the exponents.

EXAMPLE: $10^4 \times 10^3 = 10^{4+3} = 10^7$

To *divide* numbers expressed in terms of powers of ten, remember that $10^{-1} = 1/10$, $10^{-2} = 1/100$, and so on. The general rule for any exponent n is

$$10^{-n} = \frac{1}{10^n}$$

In other words, dividing by 10^n is the same as multiplying by 10^{-n}. To carry out a division, you first transform it into multiplication by changing the sign of the exponent, and then carry out the multiplication by adding the exponents.

EXAMPLE: $\dfrac{10^4}{10^6} = 10^4 \times 10^{-6} = 10^{4+(-6)} = 10^{4-6} = 10^{-2}$

Usually a computation involves numbers like 3.0×10^{10}, that is, an ordinary number multiplied by a factor of 10 with an exponent. In such cases, to perform multiplication or division, you can treat the numbers separately from the factors of 10^n.

EXAMPLE: We can redo the first numerical example from Box 1-1 in a straightforward manner by using exponents:

$$D = \frac{31.2 \times 944{,}000{,}000 \text{ km}}{206{,}265}$$

$$= \frac{3.12 \times 10 \times 9.44 \times 10^8}{2.06265 \times 10^5} \text{ km}$$

$$= \frac{3.12 \times 9.44 \times 10^{1+8-5}}{2.06265} \text{ km} = 14.3 \times 10^4 \text{ km}$$

$$= 1.43 \times 10 \times 10^4 \text{ km} = 1.43 \times 10^5 \text{ km}$$

1-7 Astronomical distances are often measured in astronomical units, light-years, or parsecs

Astronomers use many of the same units of measurement as do other scientists. They often measure lengths in meters (abbreviated m), masses in kilograms (kg), and time in seconds (s). (You can read more about these units of measurement, as well as techniques for converting between different sets of units, in Box 1-3.)

Specialized units make it easier to comprehend immense cosmic distances

Like other scientists, astronomers often find it useful to combine these units with powers of ten and create new units using prefixes. As an example, the number 1000 (= 10^3) is represented by the prefix "kilo," and so a distance of 1000 meters is the same as 1 kilometer (1 km). Here are some of the most common prefixes, with examples of how they are used:

one-billionth meter	= 10^{-9} m	= 1 nanometer
one-millionth second	= 10^{-6} s	= 1 microsecond
one-thousandth arcsecond	= 10^{-3} arcsec	= 1 milliarcsecond
one-hundredth meter	= 10^{-2} m	= 1 centimeter
one thousand meters	= 10^3 m	= 1 kilometer
one million tons	= 10^6 tons	= 1 megaton

In principle, we could express all sizes and distances in astronomy using units based on the meter. Indeed, we will use kilometers to give the diameters of the Earth and Moon, as well as the Earth-Moon distance. But, while a kilometer (roughly equal to three-fifths of a mile) is an easy distance for humans to visualize, a megameter (10^6 m) is not. For this reason, astronomers have devised units of measure that are more appropriate for the tremendous distances between the planets and the far greater distances between the stars.

When discussing distances across the solar system, astronomers use a unit of length called the **astronomical unit** (abbreviated **AU**). This is the average distance between Earth and the Sun:

$$1 \text{ AU} = 1.496 \times 10^8 \text{ km} = 92.96 \text{ million miles}$$

BOX 1-3   **Tools of the Astronomer's Trade**

Units of Length, Time, and Mass

To understand and appreciate the universe, we need to describe phenomena not only on the large scales of galaxies but also on the submicroscopic scale of the atom. Astronomers generally use units that are best suited to the topic at hand. For example, interstellar distances are conveniently expressed in either light-years or parsecs, whereas the diameters of the planets are more comfortably presented in kilometers.

Most scientists prefer to use a version of the metric system called the International System of Units, abbreviated **SI** (after the French name Système International). In SI units, length is measured in meters (m), time is measured in seconds (s), and mass (a measure of the amount of material in an object) is measured in kilograms (kg). How are these basic units related to other measures?

When discussing objects on a human scale, sizes and distances are usually expressed in millimeters (mm), centimeters (cm), and kilometers (km). These units of length are related to the meter as follows:

$$1 \text{ millimeter} = 0.001 \text{ m} = 10^{-3} \text{ m}$$

$$1 \text{ centimeter} = 0.01 \text{ m} = 10^{-2} \text{ m}$$

$$1 \text{ kilometer} = 1000 \text{ m} = 10^3 \text{ m}$$

Although the English system of inches (in.), feet (ft), and miles (mi) is much older than SI, today the English system is actually based on the SI system: The inch is defined to be exactly 2.54 cm. A useful set of conversions is

$$1 \text{ in.} = 2.54 \text{ cm}$$

$$1 \text{ ft} = 0.3048 \text{ m}$$

$$1 \text{ mi} = 1.609 \text{ km}$$

Each of these equalities can also be written as a fraction equal to 1. For example, you can write

$$\frac{0.3048 \text{ m}}{1 \text{ ft}} = 1$$

Fractions like this are useful for converting a quantity from one set of units to another. For example, the *Saturn V* rocket used to send astronauts to the Moon stands about 363 feet tall. How can we convert this height to meters? The trick is to remember that a quantity does not change if you multiply it by 1. Expressing the number 1 by the fraction (0.3048 m)/(1 ft), we can write the height of the rocket as

$$363 \text{ ft} \times 1 = 363 \text{ ft} \times \frac{0.3048 \text{ m}}{1 \text{ ft}} = 111 \frac{\text{ft} \times \text{m}}{\text{ft}} = 111 \text{ m}$$

EXAMPLE: The diameter of Mars is 6794 km. Let's try expressing this in miles.

CAUTION! You can get into trouble if you are careless in applying the trick of taking the number whose units are to be converted and multiplying it by 1. For example, if we multiply the diameter by 1 expressed as (1.609 km)/(1 mi), we get

$$6794 \text{ km} \times 1 = 6794 \text{ km} \times \frac{1.609 \text{ km}}{1 \text{ mi}} = 10{,}930 \frac{\text{km}^2}{\text{mi}}$$

The unwanted units of km did not cancel, so this cannot be right. Furthermore, a mile is larger than a kilometer, so the diameter expressed in miles should be a smaller number than when expressed in kilometers.

The correct approach is to write the number 1 so that the unwanted units *will* cancel. The number we are starting with is in kilometers, so we must write the number 1 with kilometers in the denominator ("downstairs" in the fraction). Thus, we express 1 as (1 mi)/(1.609 km):

$$6794 \text{ km} \times 1 = 6794 \text{ km} \times 1 \text{ mi}/1.609 \text{ km}$$

$$= 4222 \text{ km} \times \text{mi}/\text{km} = 4222 \text{ mi}$$

Now the units of km cancel as they should, and the distance in miles is a smaller number than in kilometers (as it must be).

When discussing very small distances such as the size of an atom, astronomers often use the micrometer (μm) or the nanometer (nm). These are related to the meter as follows:

$$1 \text{ micrometer} = 1 \text{ μm} = 10^{-6} \text{ m}$$

$$1 \text{ nanometer} = 1 \text{ nm} = 10^{-9} \text{ m}$$

Thus, 1 μm = 10^3 nm. (Note that the micrometer is often called the micron.)

The basic unit of time is the second (s). It is related to other units of time as follows:

$$1 \text{ minute (min)} = 60 \text{ s}$$

$$1 \text{ hour (h)} = 3600 \text{ s}$$

$$1 \text{ day (d)} = 86{,}400 \text{ s}$$

$$1 \text{ year (y)} = 3.156 \times 10^7 \text{ s}$$

(continued on the next page)

BOX 1-3 *(continued)*

In the SI system, speed is properly measured in meters per second (m/s). Quite commonly, however, speed is also expressed in km/s and mi/h:

$$1 \text{ km/s} = 10^3 \text{ m/s}$$

$$1 \text{ km/s} = 2237 \text{ mi/h}$$

$$1 \text{ mi/h} = 0.447 \text{ m/s}$$

$$1 \text{ mi/h} = 1.47 \text{ ft/s}$$

In addition to using kilograms, astronomers sometimes express mass in grams (g) and in solar masses ($M_\odot$), where the subscript $\odot$ is the symbol denoting the Sun. It is especially convenient to use solar masses when discussing the masses of stars and galaxies. These units are related to each other as follows:

$$1 \text{ kg} = 1000 \text{ g}$$

$$1 \text{ M}_\odot = 1.99 \times 10^{30} \text{ kg}$$

CAUTION! You may be wondering why we have not given a conversion between kilograms and pounds. The reason is that these units do not refer to the same physical quantity! A kilogram is a unit of *mass*, which is a measure of the amount of material in an object. By contrast, a pound is a unit of *weight*, which tells you how strongly gravity pulls on that object's material. Consider a person who weighs 110 pounds on Earth, corresponding to a mass of 50 kg. Gravity is only about one-sixth as strong on the Moon as it is on Earth, so on the Moon this person would weigh only one-sixth of 110 pounds, or about 18 pounds. But that person's mass of 50 kg is the same on the Moon; wherever you go in the universe, you take all of your material along with you. We will explore the relationship between mass and weight in Chapter 4.

Thus, the average distance between the Sun and Jupiter can be conveniently stated as 5.2 AU.

 To talk about distances to the stars, astronomers use two different units of length. The **light-year** (abbreviated **ly**) is the distance that light travels in one year. This is a useful concept because the speed of light in empty space always has the same value, 3.00×10^5 km/s (kilometers per second) or 1.86×10^5 mi/s (miles per second). In terms of kilometers or astronomical units, one light-year is given by

$$1 \text{ ly} = 9.46 \times 10^{12} \text{ km} = 63{,}240 \text{ AU}$$

This distance is roughly equal to 6 trillion miles.

CAUTION! Keep in mind that despite its name, the light-year is a unit of distance and *not* a unit of time. As an example, Proxima Centauri, the nearest star other than the Sun, is a distance of 4.2 light-years from Earth. This means that light takes 4.2 years to travel to us from Proxima Centauri.

Physicists often measure interstellar distances in light-years because the speed of light is one of nature's most important numbers. But many astronomers prefer to use another unit of length, the *parsec,* because its definition is closely related to a method of measuring distances to the stars.

Imagine taking a journey far into space, beyond the orbits of the outer planets. As you look back toward the Sun, Earth's orbit subtends a smaller angle in the sky the farther you are from the Sun. As **Figure 1-14** shows, the distance at which 1 AU subtends an angle of 1 arcsec is defined as 1 **parsec** (abbreviated **pc**):

$$1 \text{ pc} = 3.09 \times 10^{13} \text{ km} = 3.26 \text{ ly}$$

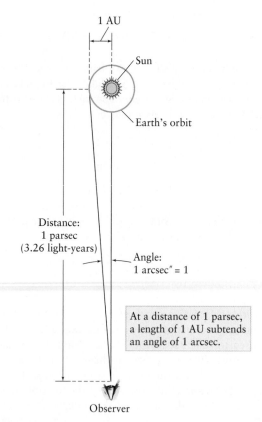

Figure 1-14

A Parsec The parsec, a unit of length commonly used by astronomers, is equal to 3.26 light-years. The parsec is defined as the distance at which 1 AU perpendicular to the observer's line of sight subtends an angle of 1 arcsec.

The distance from Earth to Proxima Centauri can be stated as 1.3 pc or as 4.2 ly. Whether you choose to use parsecs or light-years is a matter of personal taste.

For even greater distances, astronomers commonly use **kiloparsecs** and **megaparsecs** (abbreviated **kpc** and **Mpc**). As we saw before, these prefixes simply mean "thousand" and "million," respectively:

$$1 \text{ kiloparsec} = 1 \text{ kpc} = 1000 \text{ pc} = 10^3 \text{ pc}$$

$$1 \text{ megaparsec} = 1 \text{ Mpc} = 1{,}000{,}000 \text{ pc} = 10^6 \text{ pc}$$

For example, the distance from Earth to the center of our Milky Way Galaxy is about 8 kpc, and the galaxy shown in Figure 1-9 is about 11 Mpc away.

Some astronomers prefer to talk about thousands or millions of light-years rather than kiloparsecs and megaparsecs. Once again, the choice is a matter of personal taste. As a general rule, astronomers use whatever yardsticks seem best suited for the issue at hand and do not restrict themselves to one system of measurement. For example, an astronomer might say that the supergiant star Antares has a diameter of 860 million kilometers and is located at a distance of 185 parsecs from Earth. The *Cosmic Connections* figure on page 16 shows where these different systems are useful.

1-8 Astronomy is an adventure of the human mind

An underlying theme of this book is that the universe is rational. It is not a hodgepodge of unrelated things behaving in unpre-

Studying the universe benefits our lives on Earth

dictable ways. Rather, we find strong evidence that fundamental laws of physics govern the nature of the universe and the behavior of everything in it. These unifying concepts enable us to explore realms far removed from our earthly experience. Thus, a scientist can do experiments in a laboratory to determine the properties of light or the behavior of atoms and then use this knowledge to investigate the structure of the universe.

The discovery of fundamental laws of nature has had a profound influence on humanity. These laws have led to an immense number of practical applications that have fundamentally transformed commerce, medicine, entertainment, transportation, and other aspects of our lives. In particular, space technology has given us instant contact with any point on the globe through communication satellites, precise navigation to any point on Earth using signals from the satellites of the Global Positioning System (GPS), and accurate weather forecasts from meteorological satellites (**Figure 1-15**).

As important as the applications of science are, the pursuit of scientific knowledge for its own sake is no less important. We are fortunate to live in an age in which this pursuit is in full flower. Just as explorers such as Columbus and Magellan discovered the true size of our planet in the fifteenth and sixteenth centuries, astronomers of the twenty-first century are exploring the universe to an extent that is unparalleled in human history. Indeed, even the voyages into space imagined by such great sci-

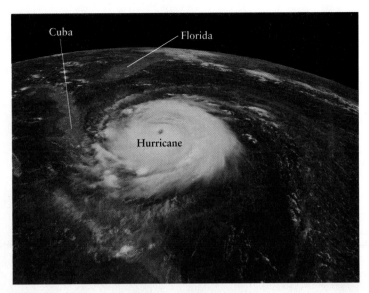

Figure 1-15   R I V U X G

A Hurricane Seen from Space This image of Hurricane Frances was made on September 2, 2004, by *GOES-12* (*Geostationary Operational Environmental Satellite 12*). A geostationary satellite like *GOES-12* orbits around Earth's equator every 24 hours, the same length of time it takes the planet to make a complete rotation. Hence, this satellite remains over the same spot on Earth, from which it can monitor the weather continuously. By tracking hurricanes from orbit, *GOES-12* makes it much easier for meteorologists to give early warning of these immense storms. The resulting savings in lives and property more than pay for the cost of the satellite. (NASA, NOAA)

ence fiction writers as Jules Verne and H. G. Wells pale in comparison to today's reality. Over a few short decades, humans have walked on the Moon, sent robot spacecraft to dig into the Martian soil and explore the satellites of Saturn, and used the most powerful telescopes ever built to probe the limits of the observable universe. Never before has so much been revealed in so short a time.

As you proceed through this book, you will learn about the tools that scientists use to explore the natural world, as well as what they observe with these tools. But, most important, you will see how astronomers build from their observations an understanding of the universe in which we live. It is this search for understanding that makes science more than merely a collection of data and elevates it to one of the great adventures of the human mind. It is an adventure that will continue as long as there are mysteries in the universe—an adventure we hope you will come to appreciate and share.

Key Words

Terms preceded by an asterisk (*) are discussed in the Boxes.

angle, p. 8
angular diameter (angular size), p. 8
angular distance, p. 8
angular measure, p. 8

arcminute (′, minute of arc), p. 9
arcsecond (″, second of arc), p. 9
astronomical unit (AU), p. 12

COSMIC CONNECTIONS — Sizes in the Universe

Powers-of-ten notation provides a convenient way to express the sizes of astronomical objects and distances in space. This illustration suggests the immense distances within our solar system, the far greater distances between the stars within our Milky Way Galaxy, and the truly cosmic distances between galaxies.

Sun: diameter = 1.39×10^6 km

The Sun, Earth, and other planets are members of our solar system

Diameter of Neptune's orbit: 60 AU
1 AU (astronomical unit) = 1.50×10^8 km = average Earth-Sun distance

The Sun is a typical star. Typical distances between our neighboring stars = 1 to 5 ly
1 ly = distance that light travels in one year = 6.32×10^4 AU

Earth: diameter = 6.38×10^3 km

Galaxies are grouped into clusters, which can be up to 10^7 ly across.

Our Sun is one of more than 10^{11} stars in the Milky Way Galaxy.
Distance from the center of the Milky Way to the Sun = 2.8×10^4 ly

Each of the 1.6×10^6 dots in this map of the entire sky represents a relatively nearby galaxy. This is a tiny fraction of the number of galaxies in the observable universe.

Key Ideas

Astronomy, Science, and the Nature of the Universe: The universe is comprehensible. The scientific method is a procedure for formulating hypotheses about the universe. These are tested by observation or experimentation in order to build consistent models or theories that accurately describe phenomena in nature.

• Observations of the heavens have helped scientists discover some of the fundamental laws of physics. The laws of physics are in turn used by astronomers to interpret their observations.

The Solar System: Exploration of the planets provides information about the origin and evolution of the solar system, as well as about the history and resources of Earth.

Stars and Nebulae: Studying the stars and nebulae helps us learn about the origin and history of the Sun and the solar system.

Galaxies: Observations of galaxies tell us about the origin and history of the universe.

Angular Measure: Astronomers use angles to denote the positions and sizes of objects in the sky. The size of an angle is measured in degrees, arcminutes, and arcseconds.

Powers-of-Ten Notation is a convenient shorthand system for writing numbers. It allows very large and very small numbers to be expressed in a compact form.

Units of Distance: Astronomers use a variety of distance units. These include the astronomical unit (the average distance from Earth to the Sun), the light-year (the distance that light travels in one year), and the parsec.

Questions

Review Questions

1. What is the difference between a hypothesis and a theory?
2. What is the difference between a theory and a law of physics?
3. How are scientific theories tested?
4. Describe the role that skepticism plays in science.
5. Describe one reason why it is useful to have telescopes in space.
6. What caused the craters on the Moon?
7. What are meteorites? Why are they important for understanding the history of the solar system?

8. What makes the Sun and stars shine?
9. What role do nebulae like the Orion Nebula play in the life stories of stars?
10. What is the difference between a solar system and a galaxy?
11. What are degrees, arcminutes, and arcseconds used for? What are the relationships among these units of measure?
12. How many arcseconds equal 1°?
13. With the aid of a diagram, explain what it means to say that the Moon subtends an angle of ½°.
14. What is an exponent? How are exponents used in powers-of-ten notation?
15. What are the advantages of using powers-of-ten notation?
16. Write the following numbers using powers-of-ten notation: (**a**) ten million, (**b**) sixty thousand, (**c**) four one-thousandths, (**d**) thirty-eight billion, (**e**) your age in months.
17. How is an astronomical unit (AU) defined? Give an example of a situation in which this unit of measure would be convenient to use.
18. What is the advantage to the astronomer of using the light-year as a unit of distance?
19. What is a parsec? How is it related to a kiloparsec and to a megaparsec?
20. Give the word or phrase that corresponds to the following standard abbreviations: (**a**) km, (**b**) cm, (**c**) s, (**d**) km/s, (**e**) mi/h, (**f**) m, (**g**) m/s, (**h**) h, (**i**) y, (**j**) g, (**k**) kg. Which of these are units of speed? (*Hint:* You may have to refer to a dictionary. All of these abbreviations should be part of your working vocabulary.)
21. In the original (1977) *Star Wars* movie, Han Solo praises the speed of his spaceship by saying, "It's the ship that made the Kessel run in less than 12 parsecs!" Explain why this statement is obvious misinformation.
22. A reporter once described a light-year as "the time it takes light to reach us traveling at the speed of light." How would you correct this statement?

Advanced Questions

Questions preceded by an asterisk (*) involve topics discussed in the Boxes.

> **Problem-solving tips and tools**
>
> The small-angle formula, given in Box 1-1, relates the size of an astronomical object to the angle it subtends. Box 1-3 illustrates how to convert from one unit of measure to another. An object traveling at speed v for a time t covers a distance d given by $d = vt$; for example, a car traveling at 90 km/h (v) for 3 hours (t) covers a distance $d = (90 \text{ km/h})(3 \text{ h}) = 270$ km. Similarly, the time t required to cover a given distance d at speed v is $t = d/v$; for example, if $d = 270$ km and $v = 90$ km/h, then $t = (270 \text{ km})/(90 \text{ km/h}) = 3$ hours.

23. What is the meaning of the letters **R I V U X G** that appear under some of the figures in this chapter? Why in each case is one of the letters highlighted? (*Hint:* See the Preface that precedes Chapter 1.)

24. The diameter of the Sun is 1.4×10^{11} cm, and the distance to the nearest star, Proxima Centauri, is 4.2 ly. Suppose you want to build an exact scale model of the Sun and Proxima Centauri, and you are using a ball 30 cm in diameter to represent the Sun. In your scale model, how far away would Proxima Centauri be from the Sun? Give your answer in kilometers, using powers-of-ten notation.

25. How many Suns would it take, laid side by side, to reach the nearest star? Use powers-of-ten notation. (*Hint:* See the preceding question.)

26. A hydrogen atom has a radius of about 5×10^{-9} cm. The radius of the observable universe is about 14 billion light-years. How many times larger than a hydrogen atom is the observable universe? Use powers-of-ten notation.

27. The Sun's mass is 1.99×10^{30} kg, three-quarters of which is hydrogen. The mass of a hydrogen atom is 1.67×10^{-27} kg. How many hydrogen atoms does the Sun contain? Use powers-of-ten notation.

28. The average distance from the Earth to the Sun is 1.496×10^{8} km. Express this distance (a) in light-years and (b) in parsecs. Use powers-of-ten notation. (c) Are light-years or parsecs useful units for describing distances of this size? Explain.

29. The speed of light is 3.00×10^{8} m/s. How long does it take light to travel from the Sun to Earth? Give your answer in seconds, using powers-of-ten notation. (*Hint:* See the preceding question.)

30. When the *Voyager 2* spacecraft sent back pictures of Neptune during its flyby of that planet in 1989, the spacecraft's radio signals traveled for 4 hours at the speed of light to reach Earth. How far away was the spacecraft? Give your answer in kilometers, using powers-of-ten notation. (*Hint:* See the preceding question.)

31. The star Altair is 5.15 pc from Earth. (a) What is the distance to Altair in kilometers? Use powers-of-ten notation. (b) How long does it take for light emanating from Altair to reach Earth? Give your answer in years. (*Hint:* You do not need to know the value of the speed of light.)

32. The age of the universe is about 13.7 billion years. What is this age in seconds? Use powers-of-ten notation.

*33. Explain where the number 206,265 in the small-angle formula comes from.

*34. At what distance would a person have to hold a European 2-euro coin (which has a diameter of about 2.6 cm) in order for the coin to subtend an angle of (a) 1°? (b) 1 arcmin? (c) 1 arcsec? Give your answers in meters.

*35. A person with good vision can see details that subtend an angle of as small as 1 arcminute. If two dark lines on an eye chart are 2 millimeters apart, how far can such a person be from the chart and still be able to tell that there are two distinct lines? Give your answer in meters.

*36. The average distance to the Moon is 384,000 km, and the Moon subtends an angle of ½°. Use this information to calculate the diameter of the Moon in kilometers.

*37. Suppose your telescope can give you a clear view of objects and features that subtend angles of at least 2 arcsec. What is the diameter in kilometers of the smallest crater you can see on the Moon? (*Hint:* See the preceding question.)

*38. On April 18, 2006, the planet Venus was a distance of 0.869 AU from Earth. The diameter of Venus is 12,104 km. What was the angular size of Venus as seen from Earth on April 18, 2006? Give your answer in arcminutes.

*39. (a) Use the information given in the caption to Figure 1-7 to determine the angular size of the Orion Nebula. Give your answer in degrees. (b) How does the angular diameter of the Orion Nebula compare to the angular diameter of the Moon?

Discussion Questions

40. Scientists assume that "reality is rational." Discuss what this means and the thinking behind it.

41. All scientific knowledge is inherently provisional. Discuss whether this is a weakness or a strength of the scientific method.

42. How do astronomical observations differ from those of other sciences?

Web/eBook Questions

43. Use the links given in the *Universe* Web site or eBook, Chapter 1, to learn about the Orion Nebula (Figure 1-7). Can the nebula be seen with the naked eye? Does the nebula stand alone, or is it part of a larger cloud of interstellar material? What has been learned by examining the Orion Nebula with telescopes sensitive to infrared light?

44. Use the links given in the *Universe* Web site or eBook, Chapter 1, to learn more about the Crab Nebula (Figure 1-8). When did observers on Earth see the supernova that created this nebula? Does the nebula emit any radiation other than visible light? What kind of object is at the center of the nebula?

45. Access the AIMM (Active Integrated Media Module) called "Small-Angle Toolbox" in Chapter 1 of the *Universe* Web site or eBook. Use this to determine the diameters in kilometers of the Sun, Saturn, and Pluto given the following distances and angular sizes:

Object	Distance (km)	Angular size (″)
Sun	1.5×10^{8}	1800
Saturn	1.5×10^{9}	16.5
Pluto	6.3×10^{9}	0.06

Activities

Observing Projects

46. On a dark, clear, moonless night, can you see the Milky Way from where you live? If so, briefly describe its appearance. If not, what seems to be interfering with your ability to see the Milky Way?

47. Look up at the sky on a clear, cloud-free night. Is the Moon in the sky? If so, does it interfere with your ability to see the fainter stars? Why do you suppose astronomers prefer to schedule their observations on nights when the Moon is not in the sky?

48. Look up at the sky on a clear, cloud-free night and note the positions of a few prominent stars relative to such reference markers as rooftops, telephone poles, and treetops. Also note the location from where you make your observations. A few hours later, return to that location and again note the positions of the same bright stars that you observed earlier. How have their positions changed? From these changes, can you deduce the general direction in which the stars appear to be moving?

49. If your book comes with a CD-ROM, use it to install the *Starry Night Enthusiast*™ planetarium software on your computer and run this program to determine when the Moon is visible today from your location. You will see that the Moon can be seen in the daytime as well as at night. Note that the **Time Flow Rate** is set to **1x**, indicating that time is running forward at the normal rate. Set the **Time Flow Rate** to **1 minute** and find the time of moonset at your location. Determine which, if any, of the following planets are visible tonight: Mercury, Venus, Mars, Jupiter, and Saturn. Feel free to experiment with *Starry Night Enthusiast*™.

Operating Hints: (1) To see the sky from your actual location on Earth, select **Set Home Location . . .** in the **File** menu (on a Macintosh, this command is found under the **Starry Night Enthusiast 5.0** menu). Click the **List** tab in the Home Location dialog box; then select the name of your city or town and click the **Save As Home Location** button. You can always return to this starting screen by clicking the **Home** button in the toolbar. (2) To change your viewing direction, move the mouse in the view. When the mouse cursor appears as a little hand, hold down the mouse button (on a Windows computer, the left button) and drag the mouse; this action will move the sky and change the gaze direction. (3) Use the toolbar at the top of the main window to change the time and date appropriate to the display and to adjust the time flow rate, as explained above. (4) You can use the **Find . . .** command in the **Edit** menu or click the **Find** tab on the left side of the main view window to open the Find pane to locate specific planets or stars by name. (5) To learn about any object in the sky, point the cursor at the object. A panel of information about the chosen object will appear on the display. The position of this information panel and its content will depend upon the selected options. These options can be altered from the **Preferences** item in the **File** menu (on a Macintosh, the **Preferences** item is under the **Starry Night Enthusiast 5.0** menu). Select **Cursor Tracking (HUD)** in the left-hand edit box in the Preferences dialog. You can then select the information to be displayed and the position of this information panel. Another way to get information about an object is to position the mouse cursor over the object in the main view window and double-click. This will open the **Info** side-pane which includes specific information about the chosen object under several headings. You can expand and collapse the layers beneath these headings by clicking the + or − icons to the left of the heading label (▶ and ▼ icons on a Macintosh).

You can find further information about the program and its many modes of operation in the *Starry Night*

Enthusiast™ manual (in *Starry Night Enthusiast*™, select **User's Guide** in the **Help** menu).

50. Use the *Starry Night Enthusiast*™ program to investigate the Milky Way Galaxy. Select **Deep Space > Local Universe** in the **Favourites** menu. Click and hold the **minus (−)** symbol in the **Zoom** control on the toolbar to adjust the field of view to an appropriate value for this image of the Milky Way. A foreground image of astronaut's feet may be superimposed upon this view. Click on **View > Feet** to remove this foreground, if desired. The view shows the Milky Way near the center of the window against a background of distant galaxies as seen from a point in space 282,000 light-years from the Sun. To center the Milky Way in the view, position the mouse cursor over the Milky Way and click and hold the mouse button (on a two-button mouse, click the right button). Select **Centre** from the drop down menu that appears. (a) To view the Milky Way from different angles, hold down the **SHIFT** key on the keyboard. The cursor should turn into a small square surrounded by four small triangles. Then hold down the mouse button (the left button on a two-button mouse) and drag the mouse in order to change the viewing angle. This will be equivalent to your moving around the galaxy, viewing it against different backgrounds of both near and far galaxies. How would you describe the shape of the Milky Way? (b) You can zoom in and zoom out on this galaxy using the Zoom buttons at the right side of the toolbar. (c) Another way to zoom in on the view is to change the elevation of the viewing location. Click the **Find** tab at the left of the view window and double-click the entry for the Sun. This will center the view on the Sun. Now use the elevation buttons to the left of the **Home** button in the toolbar to move closer to the Sun. Move in until you can see the planets in their orbits around the Sun, then zoom back out until you can see the entire Milky Way Galaxy again. Are the Sun and planets located at the center of the Milky Way? How would you describe their location?

Collaborative Exercises

51. A scientific theory is fundamentally different than the everyday use of the word "theory." List and describe any three scientific theories of your choice and creatively imagine an additional three hypothetical theories that are not scientific. Briefly describe what is scientific and what is nonscientific about each of these theories.

52. Angular measure describes how far apart two objects appear to an observer. From where you are currently sitting, estimate the angular distance between the floor and the ceiling at the front of the room you are sitting in, the angular distance between the two people sitting closest to you, and the angular size of a clock or an exit sign on the wall. Draw sketches to illustrate each answer and describe how each of your answers would change if you were standing in the very center of the room.

53. Astronomers use powers of ten to describe the distances to objects. List an object or place that is located at very roughly each of the following distances from you: 10^{-2} m, 10^0 m, 10^1 m, 10^3 m, 10^7 m, 10^{10} m, and 10^{20} m.

Why Astronomy?

by Sandra M. Faber

As you study astronomy, you may ask, "Why am I studying this subject? What good is it for people in general and for me in particular?" Admittedly, astronomy does not offer the same practical benefits as other sciences, so how can it be important to your life?

On the most basic level, I think of astronomy as providing the ultimate background for human history. Recorded history goes back about 3000 years. For knowledge of the time before that, we consult archeologists and anthropologists about early human history and paleontologists, biologists, and geologists about the evolution of life and of our planet—altogether going back some five billion years. Astronomy tells us about the time before that, the ten billion years or so when the Sun, solar system, and Milky Way Galaxy formed, and even about the origin of the universe in the Big Bang. Knowledge of astronomy is part of a well-educated person's view of history.

Astronomy challenges our belief system and impels us to put our "philosophical house" in order. For example, the Bible says the world and everything in it were created in six days by the hand of God. However, according to the ancient Egyptians, the Earth arose spontaneously from the infinite waters of the eternal universe, called Nun. Alaskan legends teach that the world was created by the conscious imaginings of a deity named Father Raven.

Modern astronomy, supported by physics and observations, differs from these stories of the creation of Earth. Astronomers believe that the Sun formed about five billion years ago by gravitational collapse from a dense cloud of interstellar gas and dust. At the same time, and over a period of several hundred thousand years, the planets condensed within the swirling solar nebula. Astronomers have actually seen young stars form in this way.

At issue here, really, is the question of how we are to gain information about the nature of the physical world—whether by revelation and intuition or by logic and observation. Where science stops and faith begins is a thorny issue for everyone, but particularly for astronomers—and for astronomy students.

Astronomy cultivates our notions about cosmic time and cosmic evolution. Given the short span of human life, it is all too easy to overlook that the universe is a dynamic place. This idea implies fragility—if something can change, it might even some day disappear. For instance, in another five billion years or so, the Sun will swell up and brighten to 1000 times its present luminosity, incinerating the Earth in the process. This is far enough in the future that neither you nor I need to feel any personal responsibility for preparing to meet this challenge. However, other cosmic catastrophes will inevitably occur before then. The Earth will be hit by a sizable piece of space debris—craters show that this happens every few million years or so. Enormous volcanic eruptions have occurred in the past and will certainly occur again. Another Ice Age is virtually certain to begin within the next 20,000 years, unless we first cook the Earth ourselves by burning too much fossil fuel.

Such common notions as the inevitability of human progress, the desirability of endless economic growth, and the Earth's ability to support its human population are all based on limited experience—they will probably not prove viable in the long run. Consequently, we must rethink who we are as a species and what is our proper activity on Earth. These long-term problems involve the whole human race and are vital to our survival and well-being. Astronomy is essential to developing a perspective on human existence and its relation to the cosmos.

Many astronomers believe that the ultimate, proper concept of "home" for the human race is our universe. It seems increasingly likely that a large number of other universes exist, with the vast majority incapable of harboring intelligent life as we know it. The parallel with Earth is striking. Among the solar system planets, only Earth can support human life. Among the great number of planets in our galaxy, only a small fraction may be such that we can call them home. The fraction of hospitable universes is likely to be smaller still. It seems, then, that our universe is the ultimate "home," a sanctuary in a vast sea of inhospitable universes.

I began this essay by talking about history and ended with issues that border on the ethical and religious. Astronomy is like that: It offers a modern-day version of Genesis—and of the Apocalypse, too. I hope that during this course you will be able to take time out to contemplate the broader implications of what you are studying. This is one of the rare opportunities in life to think about who you are and where you and the human race are going. Don't miss it.

Sandra M. Faber is professor of astronomy at the University of California, Santa Cruz, and astronomer at Lick Observatory. She chaired the now legendary group of astronomers called the Seven Samurai, who surveyed the nearest 400 elliptical galaxies and discovered a new mass concentration, called the Great Attractor. Dr. Faber received the Bok Prize from Harvard University in 1978, was elected to the National Academy of Sciences in 1985, and in 1986 won the coveted Dannie Heineman Prize from the American Astronomical Society, in recognition of her sustained body of especially influential astronomical research.

2

Knowing the Heavens

It is a clear night at the Gemini North Observatory atop Mauna Kea, a dormant volcano on the island of Hawaii. As you gaze toward the north, as in this time-exposure photograph, you find that the stars are not motionless. Rather, they move in counterclockwise circles around a fixed point above the northern horizon. Stars close to this point never dip below the horizon, while stars farther from the fixed point rise in the east and set in the west. These motions fade from view when the Sun rises in the east and illuminates the sky. The Sun, too, arcs across the sky in the same manner as the stars. At day's end, when the Sun sets in the west, the panorama of stars is revealed for yet another night.

These observations are at the heart of *naked-eye astronomy*—the sort that requires no equipment but human vision. Naked-eye astronomy cannot tell us what the Sun is made of or how far away the stars are. For such purposes we need tools such as the Gemini Telescope, housed within the dome shown in the photograph. But by studying naked-eye astronomy, you will learn the answers to equally profound questions such as why there are seasons, why the night sky is different at different times of year, and why the night sky looks different in Australia than in North America. In discovering the answers to these questions, you will learn how the Earth moves through space and will begin to understand our true place in the cosmos.

R I V U X G

The Earth's rotation makes stars appear to trace out circles in the sky. (Gemini Observatory)

2-1 Naked-eye astronomy had an important place in civilizations of the past

Positional astronomy—the study of the positions of objects in the sky and how these positions change—has roots that extend far back in time. Four to five thousand years ago, the inhabitants of the British Isles erected stone structures, such as Stonehenge, that suggest a preoccupation with the motions of the sky. Alignments of these stones appear to show where the Sun rose and set at key times during the year. Stone works of a different sort but with a similar astronomical purpose

Learning Goals

By reading the sections of this chapter, you will learn

2-1 The importance of astronomy in ancient civilizations around the world

2-2 That regions of the sky are divided around groups of stars called constellations

2-3 How the sky changes from night to night

2-4 How astronomers locate objects in the sky

2-5 What causes the seasons

2-6 The effect of changes in the direction of Earth's axis of rotation

2-7 The role of astronomy in measuring time

2-8 How the modern calendar developed

Figure 2-1 R I V U X G

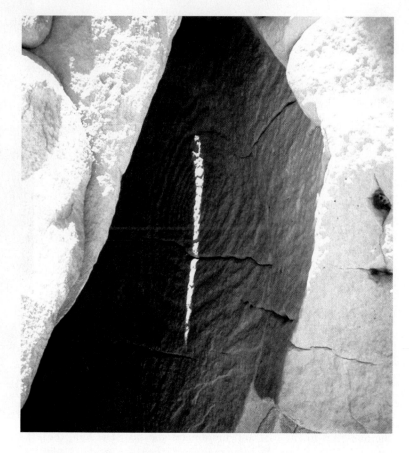

The Sun Dagger at Chaco Canyon On the first day of winter, rays of sunlight passing between stone slabs bracket a spiral stone carving, or petroglyph, at Chaco Canyon in New Mexico. A single band of light strikes the center of the spiral on the first day of summer. This and other astronomically aligned petroglyphs were carved by the ancestral Puebloan culture between 850 and 1250 A.D. (Courtesy Karl Kernberger)

are found in the New World from the American Southwest to the Andes of Peru. The ancestral Puebloans (also called Anasazi) of modern-day New Mexico, Arizona, Utah, and Colorado created stone carvings that were illuminated by the Sun on the first days of summer or winter (Figure 2-1). At the Incan city of Machu Picchu in Peru, a narrow window carved in a rock 2 meters thick looks out on the sunrise on December 21 each year.

> The astronomical knowledge of ancient peoples is the foundation of modern astronomy

Other ancient peoples designed buildings with astronomical orientations. The great Egyptian pyramids, built around 3000 B.C., are oriented north-south and east-west with an accuracy of far less than 1 degree. Similar alignments are found in the grand tomb of Shih Huang Ti (259 B.C.–210 B.C.), the first emperor of China.

Evidence of a highly sophisticated understanding of astronomy can be found in the written records of the Mayan civilization of central America. Mayan astronomers deduced by observation that the apparent motions of the planet Venus follow a cycle that repeats every 584 days. They also developed a technique for calculating the position of Venus on different dates. The Maya believed that Venus was associated with war, so such calculations were important for choosing the most promising dates on which to attack an enemy.

These archaeological discoveries bear witness to an awareness of naked-eye astronomy by the peoples of many cultures. Many of the concepts of modern positional astronomy come to us from these ancients, including the idea of dividing the sky into constellations.

2-2 Eighty-eight constellations cover the entire sky

 Looking at the sky on a clear, dark night, you might think that you can see millions of stars. Actually, the unaided human eye can detect only about 6000 stars. Because half of the sky is below the horizon at any one time, you can see at most about 3000 stars. When ancient peoples looked at these thousands of stars, they imagined that groupings of stars traced out pictures in the sky. Astronomers still refer to these groupings, called **constellations** (from the Latin for "group of stars").

> The constellations provide a convenient framework for stating the position of an object in the heavens

You may already be familiar with some of these pictures or patterns in the sky, such as the Big Dipper, which is actually part of the large constellation Ursa Major (the Great Bear). Many constellations, such as Orion in Figure 2-2, have names derived from the myths and legends of antiquity. Although some star groupings vaguely resemble the figures they are supposed to represent (see Figure 2-2c), most do not.

The term "constellation" has a broader definition in present-day astronomy. On modern star charts, the entire sky is divided into 88 regions, each of which is called a constellation. For example, the constellation Orion is now defined to be an irregular patch of sky whose borders are shown in Figure 2-2b. When astronomers refer to the "Great Nebula" M42 in Orion, they mean that as seen from Earth this nebula appears to be within Orion's

(a)   (b)   (c)

Figure 2-2   RI V UXG

Three Views of Orion The constellation Orion is easily seen on nights from December through March. **(a)** This photograph of Orion shows many more stars than can be seen with the naked eye. **(b)** A portion of a modern star atlas shows the distances in light-years (ly) to some of the stars in Orion. The yellow lines show the borders between Orion and its neighboring constellations (labeled in capitals). **(c)** This fanciful drawing from a star atlas published in 1835 shows Orion the Hunter as well as other celestial creatures. (a: Luke Dodd/Science Photo Library/Photo Researchers; c: Courtesy of Janus Publications)

patch of sky. Some constellations cover large areas of the sky (Ursa Major being one of the biggest) and others very small areas (Crux, the Southern Cross, being the smallest). But because the modern constellations cover the entire sky, every star lies in one constellation or another.

CAUTION! When you look at a constellation's star pattern, it is tempting to conclude that you are seeing a group of stars that are all relatively close together. In fact, most of these stars are nowhere near one another. As an example, Figure 2-2*b* shows the distances in light-years to four stars in Orion. Although Bellatrix (Arabic for "the Amazon") and Mintaka ("the belt") appear to be close to each other, Mintaka is actually more than 600 light-years farther away from us. The two stars only *appear* to be close because they are in nearly the same direction as seen from Earth. The same illusion often appears when you see an airliner's lights at night. It is very difficult to tell how far away a single bright light is, which is why you can mistake an airliner a few kilometers away for a star trillions of times more distant.

The star names shown in Figure 2-2*b* are from the Arabic. For example, Betelgeuse means "armpit," which makes sense when you look at the star atlas drawing in Figure 2-2*c*. Other types of names are also used for stars. For example, Betelgeuse is also known as α Orionis because

it is the brightest star in Orion (α, or alpha, is the first letter in the Greek alphabet).

CAUTION! A number of unscrupulous commercial firms offer to name a star for you for a fee. The money that they charge you for this "service" is real, but the star names are not; none of these names are recognized by astronomers. If you want to use astronomy to commemorate your name or the name of a friend or relative, consider making a donation to your local planetarium or science museum. The money will be put to much better use!

2-3 The appearance of the sky changes during the course of the night and from one night to the next

Go outdoors soon after dark, find a spot away from bright lights, and note the patterns of stars in the sky. Do the same a few hours later. You will find that the entire pattern of stars (as well as the Moon, if it is visible) has shifted its position. New constellations will have risen above the eastern horizon, and some will have disappeared below the western horizon. If you look again before dawn, you will see that the stars that were just rising in the east

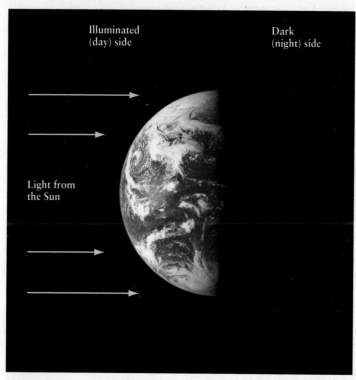

Figure 2-3   R I [V] U X G

Day and Night on the Earth At any moment, half of the Earth is illuminated by the Sun. As the Earth rotates from west to east, your location moves from the dark (night) hemisphere into the illuminated (day) hemisphere and back again. This image was recorded in 1992 by the *Galileo* spacecraft as it was en route to Jupiter. (JPL/NASA)

when the night began are now low in the western sky. This daily motion, or **diurnal motion**, of the stars is apparent in time-exposure photographs (see the photograph that opens this chapter).

If you repeat your observations on the following night, you will find that the motions of the sky are almost but not quite the same. The same constellations rise in the east and set in the west, but a few minutes earlier than on the previous night. If you look again after a month, the constellations visible at a given time of night (say, midnight) will be noticeably different, and after six months you will see an almost totally different set of constellations. Only after a year has passed will the night sky have the same appearance as when you began.

Why does the sky go through diurnal motion? Why do the constellations slowly shift from one night to the next? As we will see, the answer to the first question is that the Earth *rotates* once a day around an axis from the north pole to the south pole, while the answer to the second question is that the Earth also *revolves* once a year around the Sun.

> By understanding the motions of the Earth through space, we can understand why the Sun and stars appear to move in the sky

Diurnal Motion and the Earth's Rotation

To understand diurnal motion, note that at any given moment it is daytime on the half of the Earth illuminated by the Sun and nighttime on the other half (**Figure 2-3**). The Earth rotates from west to east, making one complete rotation every 24 hours, which is why there is a daily cycle of day and night. Because of this rotation, stars appear to us to rise in the east and set in the west, as do the Sun and Moon.

Figure 2-4 helps to further explain diurnal motion. It shows two views of the Earth as seen from a point above the north pole.

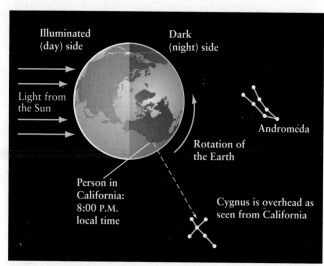

(a) Earth as seen from above the north pole

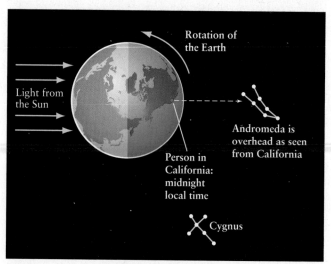

(b) 4 hours (one-sixth of a complete rotation) later

Figure 2-4

Why Diurnal Motion Happens The diurnal (daily) motion of the stars, the Sun, and the Moon is a consequence of the Earth's rotation. **(a)** This drawing shows the Earth from a vantage point above the north pole. In this drawing, for a person in California the local time is 8:00 P.M. and the constellation Cygnus is directly overhead. **(b)** Four hours later, the Earth

has made one-sixth of a complete rotation to the east. As seen from Earth, the entire sky appears to have rotated to the west by one-sixth of a complete rotation. It is now midnight in California, and the constellation directly over California is Andromeda.

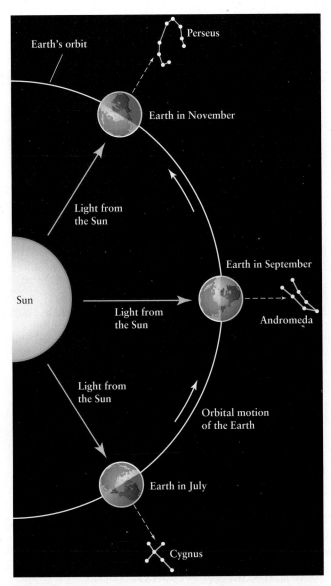

Figure 2-5

Why the Night Sky Changes During the Year As the Earth orbits around the Sun, the nighttime side of the Earth gradually turns toward different parts of the sky. Hence, the particular stars that you see in the night sky are different at different times of the year. This figure shows which constellation is overhead at midnight local time—when the Sun is on the opposite side of the Earth from your location—during different months for observers at midnorthern latitudes (including the United States). If you want to view the constellation Andromeda, the best time of the year to do it is in late September, when Andromeda is nearly overhead at midnight.

At the instant shown in Figure 2-4*a*, it is day in Asia but night in most of North America and Europe. Figure 2-4*b* shows the Earth 4 hours later. Four hours is one-sixth of a complete 24-hour day, so the Earth has made one-sixth of a rotation between Figures 2-4*a* and 2-4*b*. Europe is now in the illuminated half of the Earth (the Sun has risen in Europe), while Alaska has moved from the illuminated to the dark half of the Earth (the Sun has set in Alaska). For a person in California, in Figure 2-4*a* the time is

8:00 P.M. and the constellation Cygnus (the Swan) is directly overhead. Four hours later, the constellation over California is Andromeda (named for a mythological princess). Because the Earth rotates from west to east, it appears to us on Earth that the entire sky rotates around us in the opposite direction, from east to west.

Yearly Motion and the Earth's Orbit

We described earlier that in addition to the diurnal motion of the sky, the constellations visible in the night sky also change slowly over the course of a year. This happens because the Earth orbits, or revolves around, the Sun (Figure 2-5). Over the course of a year, the Earth makes one complete orbit, and the darkened, nighttime side of the Earth gradually turns toward different parts of the heavens. For example, as seen from the northern hemisphere, at midnight in late July the constellation Cygnus is close to overhead; at midnight in late September the constellation Andromeda is close to overhead; and at midnight in late November the constellation Perseus (commemorating a mythological hero) is close to overhead. If you follow a particular star on successive evenings, you will find that it rises approximately 4 minutes earlier each night, or 2 hours earlier each month.

Constellations and the Night Sky

Constellations can help you find your way around the sky. For example, if you live in the northern hemisphere, you can use the Big Dipper in Ursa Major to find the north direction by drawing a straight line through the two stars at the front of the Big Dipper's bowl (Figure 2-6). The first moderately bright star you come to is Polaris, also called the North Star because it is located almost directly over the Earth's north pole. If you draw a line from

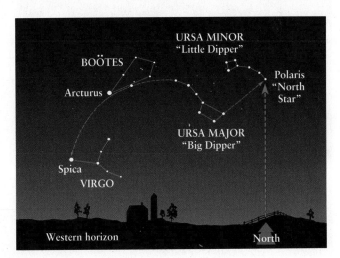

Figure 2-6

The Big Dipper as a Guide The North Star can be seen from anywhere in the northern hemisphere on any night of the year. This star chart shows how the Big Dipper can be used to point out the North Star as well as the brightest stars in two other constellations. The chart shows the sky at around 11 P.M. (daylight savings time) on August 1. Due to the Earth's orbital motion around the Sun, you will see this same view at 1 A.M. on July 1 and at 9 P.M. on September 1. The angular distance from Polaris to Spica is 102°.

Polaris straight down to the horizon, you will find the north direction.

As Figure 2-6 shows, by following the handle of the Big Dipper you can locate the bright reddish star Arcturus in Boötes (the Shepherd) and the prominent bluish star Spica in Virgo (the Virgin). The saying "Follow the arc to Arcturus and speed to Spica" may help you remember these stars, which are conspicuous in the evening sky during the spring and summer.

During winter in the northern hemisphere, you can see some of the brightest stars in the sky. Many of them are in the vicinity of the "winter triangle" (Figure 2-7), which connects bright stars in the constellations of Orion (the Hunter), Canis Major (the Large Dog), and Canis Minor (the Small Dog).

A similar feature, the "summer triangle," graces the summer sky in the northern hemisphere. This triangle connects the brightest stars in Lyra (the Harp), Cygnus (the Swan), and Aquila (the Eagle) (Figure 2-8). A conspicuous portion of the Milky Way forms a beautiful background for these constellations, which are nearly overhead during the middle of summer at midnight.

A wonderful tool to help you find your way around the night sky is the planetarium program *Starry Night Enthusiast*™, which is on the CD-ROM that accompanies certain print copies of this book. (You can also obtain *Starry Night Enthusiast*™ separately.) In addition, at the end of this book you will find a set of selected star charts for the evening

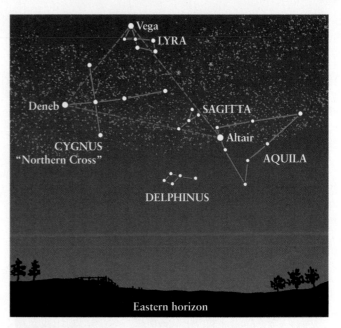

Figure 2-8

The "Summer Triangle" This star chart shows the eastern sky as it appears in the evening during spring and summer in the northern hemisphere (around 1 A.M. daylight savings time on June 1, around 11 P.M. on July 1, and around 9 P.M. on August 1). The angular distance from Deneb to Altair is about 38°. The constellations Sagitta (the Arrow) and Delphinus (the Dolphin) are much fainter than the three constellations that make up the triangle.

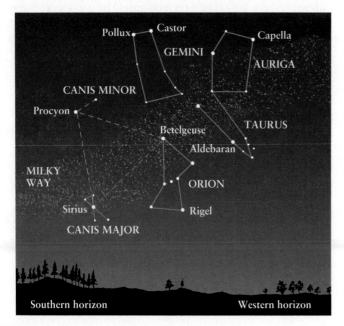

Figure 2-7

The "Winter Triangle" This star chart shows the view toward the southwest on a winter evening in the northern hemisphere (around midnight on January 1, 10 P.M. on February 1, or 8 P.M. on March 1). Three of the brightest stars in the sky make up the "winter triangle," which is about 26° on a side. In addition to the constellations involved in the triangle, the chart shows the prominent constellations Gemini (the Twins), Auriga (the Charioteer), and Taurus (the Bull).

hours of all 12 months of the year. You may find stargazing an enjoyable experience, and *Starry Night Enthusiast*™ and the star charts will help you identify many well-known constellations.

Note that all the star charts in this section and at the end of this book are drawn for an observer in the northern hemisphere. If you live in the southern hemisphere, you can see constellations that are not visible from the northern hemisphere, and vice versa. In the next section we will see why this is so.

2-4 It is convenient to imagine that the stars are located on a celestial sphere

Many ancient societies believed that all the stars are the same distance from the Earth. They imagined the stars to be bits of fire imbedded in the inner surface of an immense hollow sphere, called the **celestial sphere,** with the Earth at its center. In this picture of the universe, the Earth was fixed and did not rotate. Instead, the entire celestial sphere rotated once a day around the Earth from east to west, thereby causing the diurnal motion of the sky. The picture of a rotating celestial sphere fit well with naked-eye observations, and for its time was a useful model of how the

> **The concept of the imaginary celestial sphere helps us visualize the motions of stars in the sky**

universe works. (We discussed the role of models in science in Section 1-1).

Today's astronomers know that this simple model of the universe is not correct. Diurnal motion is due to the rotation of the Earth, not the rest of the universe. Furthermore, as we learned when discussing the constellations in Section 2-2, the stars are not all at the same distance from Earth. Indeed, the stars that you can see with the naked eye range from 4.2 to more than 1000 light-years away, and telescopes allow us to see objects at distances of billions of light-years.

Thus, astronomers now recognize that the celestial sphere is an *imaginary* object that has no basis in physical reality. Nonetheless, the celestial sphere model remains a useful tool of positional astronomy. If we imagine, as did the ancients, that the Earth is stationary and that the celestial sphere rotates around us, it is relatively easy to specify the directions to different objects in the sky and to visualize the motions of these objects.

Figure 2-9 depicts the celestial sphere, with the Earth at its center. (A truly proportional drawing would show the celestial sphere as being millions of times larger than the Earth.) We picture the stars as points of light that are fixed on the inner surface of the celestial sphere. If we project the Earth's equator out into space, we obtain the **celestial equator**. The celestial equator divides the sky into northern and southern hemispheres, just as the Earth's equator divides the Earth into two hemispheres.

If we project the Earth's north and south poles into space, we obtain the **north celestial pole** and the **south celestial pole**. The two celestial poles are where the Earth's axis of rotation (extended out into space) intersects the celestial sphere (see Figure 2-9). The star Polaris is less than 1° away from the north celestial pole, which is why it is called the North Star or the Pole Star.

The point in the sky directly overhead an observer anywhere on Earth is called that observer's **zenith**. The zenith and celestial sphere are shown in Figure 2-10 for an observer located at 35° north latitude (that is, at a location on the Earth's surface 35° north of the equator). The zenith is shown at the top of Figure 2-10, so the Earth and the celestial sphere appear "tipped" compared to Figure 2-9. At any time, an observer can see only half of the celestial sphere; the other half is below the horizon, hidden by the body of the Earth. The hidden half of the celestial sphere is darkly shaded in Figure 2-10.

Motions of the Celestial Sphere

For an observer anywhere in the northern hemisphere, including the observer in Figure 2-10, the north celestial pole is always

Figure 2-9

The Celestial Sphere The celestial sphere is the apparent sphere of the sky. The view in this figure is from the outside of this (wholly imaginary) sphere. The Earth is at the center of the celestial sphere, so our view is always of the *inside* of the sphere. The celestial equator and poles are the projections of the Earth's equator and axis of rotation out into space. The celestial poles are therefore located directly over the Earth's poles.

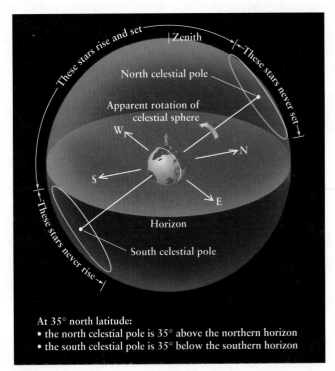

Figure 2-10

The View from 35° North Latitude To an observer at 35° north latitude (roughly the latitude of Los Angeles, Atlanta, Tel Aviv, and Tokyo), the north celestial pole is always 35° above the horizon. Stars within 35° of the *north* celestial pole are circumpolar; they trace out circles around the north celestial pole during the course of the night, and are always above the horizon on any night of the year. Stars within 35° of the *south* celestial pole are always below the horizon and can never be seen from this latitude. Stars that lie between these two extremes rise in the east and set in the west.

(a) At middle northern latitudes

(b) At the north pole

(c) At the equator

Figure 2-11   R I V U X G

The Apparent Motion of Stars at Different Latitudes
As the Earth rotates, stars appear to rotate around us along paths that are parallel to the celestial equator. **(a)** As shown in this long time exposure, at most locations on Earth the rising and setting motions are at an angle to the horizon that depends on the latitude. **(b)** At the north pole (latitude 90° north) the stars appear to move parallel to the horizon. **(c)** At the equator (latitude 0°) the stars rise and set along vertical paths. (a: David Miller/DMI)

above the horizon. As the Earth turns from west to east, it appears to the observer that the celestial sphere turns from east to west. Stars sufficiently near the north celestial pole revolve around the pole, never rising or setting. Such stars are called **circumpolar.** For example, as seen from North America or Europe, Polaris is a circumpolar star and can be seen at any time of night on any night of the year. The photograph that opens this chapter shows the circular trails of stars around the north celestial pole as seen from Hawaii (at 20° north latitude). Stars near the south celestial pole revolve around that pole but always remain below the horizon of an observer in the northern hemisphere. Hence, these stars can never be seen by the observer in Figure 2-10. Stars between those two limits rise in the east and set in the west.

CAUTION! Keep in mind that which stars are circumpolar, which stars never rise, and which stars rise and set depends on the latitude from which you view the heavens. As an example, for an observer at 35° south latitude (roughly the latitude of Sydney, Cape Town, and Buenos Aires), the roles of the north and south celestial poles are the opposite of those shown in Figure 2-10: Objects close to the *south* celestial pole are circumpolar, that is, they revolve around that pole and never rise or set. For an observer in the southern hemisphere, stars close to the *north* celestial pole are always below the horizon and can never be seen. Hence, astronomers in Australia, South Africa, and Argentina never see the North Star but are able to see other stars that are forever hidden from North American or European observers.

For observers at most locations on Earth, stars rise in the east and set in the west at an angle to the horizon (**Figure 2-11a**). To see why this is so, notice that the rotation of the celestial sphere carries stars across the sky in paths that are parallel to the celestial equator. If you stand at the north pole, the north celestial pole is directly above you at the zenith (see Figure 2-9) and the celestial equator lies all around you at the horizon. Hence, as the ce-

lestial sphere rotates, the stars appear to move parallel to the horizon (Figure 2-11b). If instead you stand on the equator, the celestial equator passes from the eastern horizon through the zenith to the western horizon. The north and south celestial poles are 90° away from the celestial equator, so they lie on the northern and southern horizons, respectively. As the celestial sphere rotates around an axis from pole to pole, the stars rise and set straight up and down—that is, in a direction perpendicular to the horizon (Figure 2-11c). At any location on Earth between the equator and either pole, the rising and setting motions of the stars are at an angle intermediate between Figures 2-11b and 2-11c. The particular angle depends on the latitude.

Using the celestial equator and poles, we can define a coordinate system to specify the position of any star on the celestial sphere. As **Box 2-1** describes, the most commonly used coordinate system uses two angles, *right ascension* and *declination,* that are analogous to longitude and latitude on Earth. These coordinates tell us in what direction we should look to see the star. To locate the star's true position in three-dimensional space, we must also know the distance to the star.

2-5 The seasons are caused by the tilt of Earth's axis of rotation

In addition to rotating on its axis every 24 hours, the Earth revolves around the Sun—that is, it *orbits* the Sun—in about 365¼ days (see Figure 2-5). As we travel with the Earth around its orbit, we experience the annual cycle of seasons. But why *are* there seasons? Furthermore, the seasons are opposite in the northern and southern hemispheres. For example, February is midwinter in North America but midsummer in Australia. Why should this be?

The celestial sphere concept helps us visualize how the Sun appears to move relative to the stars

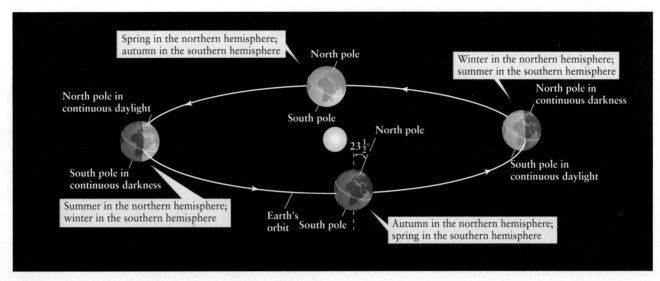

Figure 2-12

The Seasons The Earth's axis of rotation is inclined 23½° away from the perpendicular to the plane of the Earth's orbit. The north pole is aimed at the north celestial pole, near the star Polaris. The Earth maintains this orientation as it orbits the Sun. Consequently, the amount of solar illumination and the number of daylight hours at any location on Earth vary in a regular pattern throughout the year. This is the origin of the seasons.

The Origin of the Seasons

The reason why we have seasons, and why they are different in different hemispheres, is that the Earth's axis of rotation is not perpendicular to the plane of the Earth's orbit. Instead, as **Figure 2-12** shows, the axis is tilted about 23½° away from the perpendicular. The Earth maintains this tilt as it orbits the Sun, with the Earth's north pole pointing toward the north celestial pole. (This stability is a hallmark of all rotating objects. A top will not fall over as long as it is spinning, and the rotating wheels of a motorcycle help to keep the rider upright.)

During part of the year, when the Earth is in the part of its orbit shown on the left side of Figure 2-12, the northern hemisphere is tilted toward the Sun. As the Earth spins on its axis, a point in the northern hemisphere spends more than 12 hours in the sunlight. Thus, the days there are long and the nights are short, and it is summer in the northern hemisphere. The summer is hot not only because of the extended daylight hours but also because the Sun is high in the northern hemisphere's sky. As a result, sunlight strikes the ground at a nearly perpendicular angle that heats the ground efficiently (**Figure 2-13a**). During this same time of year in the southern hemisphere, the days are short and the nights

(a) The Sun in summer

(b) The Sun in winter

Figure 2-13

Solar Energy in Summer and Winter At different times of the year, sunlight strikes the ground at different angles. **(a)** In summer, sunlight is concentrated and the days are also longer, which further increases the heating. **(b)** In winter the sunlight is less concentrated, the days are short, and little heating of the ground takes place. This accounts for the low temperatures in winter.

BOX 2-1

Tools of the Astronomer's Trade

Celestial Coordinates

Your *latitude* and *longitude* describe where on the Earth's surface you are located. The latitude of your location denotes how far north or south of the equator you are, and the longitude of your location denotes how far west or east you are of an imaginary circle that runs from the north pole to the south pole through the Royal Observatory in Greenwich, England. In an analogous way, astronomers use coordinates called *declination* and *right ascension* to describe the position of a planet, star, or galaxy on the celestial sphere.

Declination is analogous to latitude. As the illustration shows, the **declination** of an object is its angular distance north or south of the celestial equator, measured along a circle passing through both celestial poles. Like latitude, it is measured in degrees, arcminutes, and arcseconds (see Section 1-5).

Right ascension is analogous to longitude. It is measured from a line that runs between the north and south celestial poles and passes through a point on the celestial equator called the *vernal equinox* (shown as a blue dot in the illustration). This point is one of two locations where the Sun crosses the celestial equator during its apparent annual motion, as we discuss in Section 2-5. In the Earth's northern hemisphere, spring officially begins when the Sun reaches the vernal equinox in late March. The **right ascension** of an object is the angular distance from the vernal equinox eastward along the celestial equator to the circle used in measuring its declination (see illustration). Astronomers measure right ascension in *time* units (hours, minutes, and seconds), corresponding to the time required for the celestial sphere to rotate through this angle. For

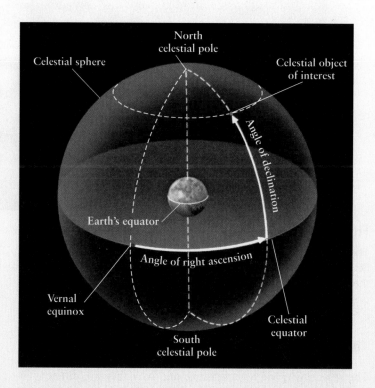

example, suppose there is a star at your zenith right now with right ascension 6^h 0^m 0^s. Two hours and 30 minutes from now, there will be a different object at your zenith with right ascension 8^h 30^m 0^s.

are long, because a point in this hemisphere spends fewer than 12 hours a day in the sunlight. The Sun is low in the sky, so sunlight strikes the surface at a grazing angle that causes little heating (Figure 2-13b), and it is winter in the southern hemisphere.

Half a year later, the Earth is in the part of its orbit shown on the right side of Figure 2-12. Now the situation is reversed, with winter in the northern hemisphere (which is now tilted away from the Sun) and summer in the southern hemisphere. During spring and autumn, the two hemispheres receive roughly equal amounts of illumination from the Sun, and daytime and nighttime are of roughly equal length everywhere on Earth.

CAUTION! A common misconception is that the seasons are caused by variations in the distance from the Earth to the Sun. According to this idea, the Earth is closer to the Sun in summer and farther away in winter. But in fact, the Earth's orbit around the Sun is very nearly circular, and the Earth-Sun distance varies only about 3% over the course of a year. (The Earth's orbit only *looks* elongated in Figure 2-12 because this illustration shows an oblique view.) We are slightly closer to the Sun in January than

in July, but this small variation has little influence on the cycle of the seasons. Also, if the seasons were really caused by variations in the Earth-Sun distance, the seasons would be the same in both hemispheres!

How the Sun Moves on the Celestial Sphere

The plane of the Earth's orbit around the Sun is called the **ecliptic plane** (Figure 2-14a). As a result of the Earth's annual motion around the Sun, it appears to us that the Sun slowly changes its position on the celestial sphere over the course of a year. The circular path that the Sun appears to trace out against the background of stars is called the **ecliptic** (Figure 2-14b). The plane of this path is the same as the ecliptic plane. (The name *ecliptic* suggests that the path traced out by the Sun has something to do with eclipses. We will discuss the connection in Chapter 3.) Because there are 365¼ days in a year and 360° in a circle, the Sun appears to move along the ecliptic at a rate of about 1° per day. This motion is from west to east, that is, in the direction opposite to the apparent motion of the celestial sphere.

The coordinates of the bright star Rigel for the year 2000 are R.A. = 5^h 14^m 32.2^s, Decl. = $-8°$ 12′ 06″. (R.A. and Decl. are abbreviations for right ascension and declination.) A minus sign on the declination indicates that the star is south of the celestial equator; a plus sign (or no sign at all) indicates that an object is north of the celestial equator. As we discuss in Box 2-2, right ascension helps determine the best time to observe a particular object.

It is important to state the year for which a star's right ascension and declination are valid. This is so because of precession, which we discuss in Section 2-6.

EXAMPLE: What are the coordinates of a star that lies exactly halfway between the vernal equinox and the south celestial pole?

Situation: Our goal is to find the right ascension and declination of the star in question.

Tools: We use the definitions depicted in the figure.

Answer: Since the circle used to measure this star's declination passes through the vernal equinox, this star's right ascension is R.A. = 0^h 0^m 0^s. The angle between the celestial equator and south celestial pole is 90° 0′ 0″, so the declination of this star is Decl. = −45° 0′ 0″.

Review: The declination in this example is negative because the star is in the southern half of the celestial sphere.

EXAMPLE: At midnight local time you see a star with R.A. = 2^h 30^m 0^s at your zenith. When will you see a star at your zenith with R.A. = 21^h 0^m 0^s?

Situation: If you held your finger stationary over a globe of the Earth, the longitude of the point directly under your finger would change as you rotated the globe. In the same way, the right ascension of the point directly over your head (the zenith) changes as the celestial sphere rotates. We use this concept to determine the time in question.

Tools: We use the idea that a change in right ascension of 24^h corresponds to an elapsed time of 24 hours and a complete rotation of the celestial sphere.

Answer: The time required for the sky to rotate through the angle between the stars is the difference in their right ascensions: 21^h 0^m 0^s − 2^h 30^m 0^s = 18^h 30^m 0^s. So the second star will be at your zenith $18\frac{1}{2}$ hours after the first one, or at 6:30 P.M. the following evening.

Review: Our answer was based on the idea that the celestial sphere makes *exactly* one complete rotation in 24 hours. If this were so, from one night to the next each star would be in exactly the same position at a given time. But because of the way that we customarily measure time, the celestial sphere makes slightly more than one complete rotation in 24 hours. (We explore the reasons for this in Box 2-2.) As a result, our answer is in error by about 3 minutes. For our purposes, this is a small enough error that we can ignore it.

ANALOGY Note that at the same time that the Sun is making its yearlong trip around the ecliptic, the entire celestial sphere is rotating around us once per day. You can envision the celestial sphere as a merry-go-round rotating clockwise, and the Sun as a restless child who is walking slowly around the merry-go-round's rim in the counterclockwise direction. During the time it takes the child to make a round trip, the merry-go-round rotates $365\frac{1}{4}$ times.

The ecliptic plane is *not* the same as the plane of the Earth's equator, thanks to the $23\frac{1}{2}°$ tilt of the Earth's rotation axis shown in Figure 2-12. As a result, the ecliptic and the celestial equator are inclined to each other by that same $23\frac{1}{2}°$ angle (Figure 2-15).

Equinoxes and Solstices

The ecliptic and the celestial equator intersect at only two points, which are exactly opposite each other on the celestial sphere. Each point is called an **equinox** (from the Latin for "equal night"),

because when the Sun appears at either of these points, day and night are each about 12 hours long at all locations on Earth. The term "equinox" is also used to refer to the date on which the Sun passes through one of these special points on the ecliptic.

On about March 21 of each year, the Sun passes northward across the celestial equator at the **vernal equinox**. This marks the beginning of spring in the northern hemisphere ("vernal" is from the Latin for "spring"). On about September 22 the Sun moves southward across the celestial equator at the **autumnal equinox**, marking the moment when fall begins in the northern hemisphere. Since the seasons are opposite in the northern and southern hemispheres, for Australians and South Africans the vernal equinox actually marks the beginning of autumn. The names of the equinoxes come from astronomers of the past who lived north of the equator.

Between the vernal and autumnal equinoxes lie two other significant locations along the ecliptic. The point on the ecliptic farthest north of the celestial equator is called the **summer solstice**. "Solstice" is from the Latin for "solar standstill," and it is at the summer solstice that the Sun stops moving northward on the

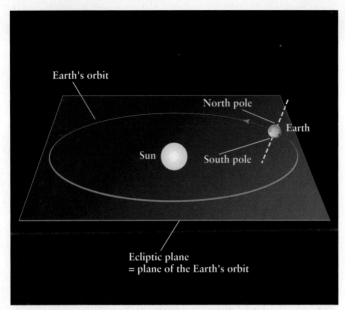

(a) In reality the Earth orbits the Sun once a year

Figure 2-14

The Ecliptic Plane and the Ecliptic (a) The ecliptic plane is the plane in which the Earth moves around the Sun. **(b)** As seen from Earth, the Sun appears to move around the celestial sphere along a circular path called

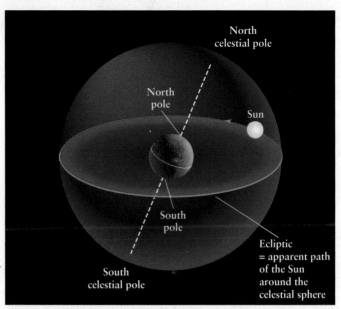

(b) It appears to us that the Sun travels around the celestial sphere once a year

the ecliptic. The Earth takes a year to complete one orbit around the Sun, so as seen by us the Sun takes a year to make a complete trip around the ecliptic.

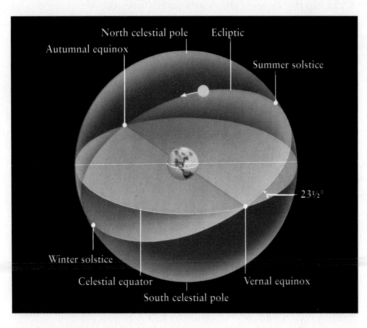

Figure 2-15

The Ecliptic, Equinoxes, and Solstices This illustration of the celestial sphere is similar to Figure 2-14b, but is drawn with the north celestial pole at the top and the celestial equator running through the middle. The ecliptic is inclined to the celestial equator by $23\frac{1}{2}°$ because of the tilt of the Earth's axis of rotation. It intersects the celestial equator at two points, called equinoxes. The northernmost point on the ecliptic is the summer solstice, and the southernmost point is the winter solstice. The Sun is shown in its approximate position for August 1.

celestial sphere. At this point, the Sun is as far north of the celestial equator as it can get. It marks the location of the Sun at the moment summer begins in the northern hemisphere (about June 21). At the beginning of the northern hemisphere's winter (about December 21), the Sun is farthest south of the celestial equator at a point called the **winter solstice.**

Because the Sun's position on the celestial sphere varies slowly over the course of a year, its daily path across the sky (due to the Earth's rotation) also varies with the seasons (**Figure 2-16**). On the first day of spring or the first day of fall, when the Sun is at one of the equinoxes, the Sun rises directly in the east and sets directly in the west.

When the northern hemisphere is tilted away from the Sun and it is winter in the northern hemisphere, the Sun rises in the southeast. Daylight lasts for fewer than 12 hours as the Sun skims low over the southern horizon and sets in the southwest. Northern hemisphere nights are longest when the Sun is at the winter solstice.

The closer you get to the north pole, the shorter the winter days and the longer the winter nights. In fact, anywhere within $23\frac{1}{2}°$ of the north pole (that is, north of latitude $90° - 23\frac{1}{2}° = 66\frac{1}{2}°$ N) the Sun is below the horizon for 24 continuous hours at least one day of the year. The circle around the Earth at $66\frac{1}{2}°$ north latitude is called the **Arctic Circle** (**Figure 2-17**). The corresponding region around the south pole is bounded by the **Antarctic Circle** at $66\frac{1}{2}°$ south latitude. At the time of the winter solstice, explorers south of the Antarctic Circle enjoy "the midnight sun," or 24 hours of continuous daylight.

During summer in the northern hemisphere, when the northern hemisphere is tilted toward the Sun, the Sun rises in the north-

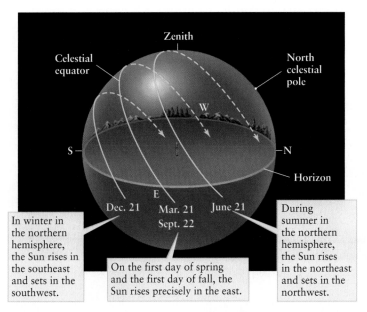

Figure 2-16

The Sun's Daily Path Across the Sky This drawing shows the apparent path of the Sun during the course of a day on four different dates. Like Figure 2-10, this drawing is for an observer at 35° north latitude.

east and sets in the northwest. The Sun is at its northernmost position at the summer solstice, giving the northern hemisphere the greatest number of daylight hours. At the summer solstice the Sun does not set at all north of the Arctic Circle (**Figure 2-18**) and does not rise at all south of the Antarctic Circle.

The variations of the seasons are much less pronounced close to the equator. Between the **Tropic of Capricorn** at 23½° south latitude and the **Tropic of Cancer** at 23½° north latitude, the Sun is directly overhead—that is, at the zenith—at high noon at least one day a year. Outside of the tropics, the Sun is never directly overhead, but is always either south of the zenith (as seen from locations north of the Tropic of Cancer) or north of the zenith (as seen from south of the Tropic of Capricorn).

2-6 The Moon helps to cause precession, a slow, conical motion of Earth's axis of rotation

The Moon is by far the brightest and most obvious naked-eye object in the nighttime sky. Like the Sun, the Moon slowly changes its position relative to the background stars; unlike the Sun, the Moon makes a complete trip around the celestial sphere in only about 4 weeks, or about a month. (The word "month" comes from the same Old English root as the word "moon.") Ancient astronomers realized that this motion occurs because the Moon orbits the Earth in roughly 4 weeks. In 1 hour the Moon moves on the celestial sphere by about ½°, or roughly its own angular size.

> **Precession causes the apparent positions of the stars to slowly change over the centuries**

The Moon's path on the celestial sphere is never far from the Sun's path (that is, the ecliptic). This is because the plane of the Moon's orbit around the Earth is inclined only slightly from the plane of the Earth's orbit around the Sun (the ecliptic plane shown in Figure 2-14a). The Moon's path varies somewhat from one

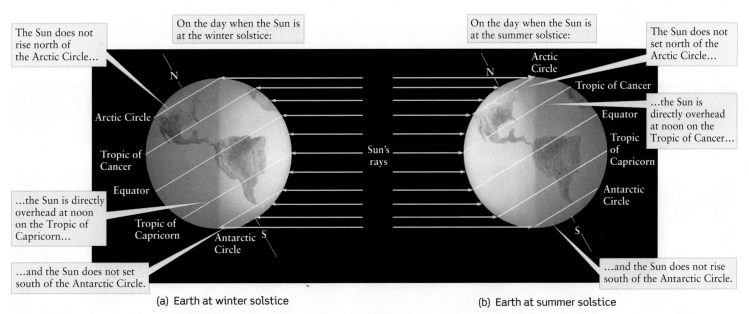

(a) Earth at winter solstice

(b) Earth at summer solstice

Figure 2-17

Tropics and Circles Four important latitudes on Earth are the Arctic Circle (66½° north latitude), Tropic of Cancer (23½° north latitude), Tropic of Capricorn (23½° south latitude), and Antarctic Circle (66½° south latitude). These drawings show the significance of these latitudes when the Sun is (a) at the winter solstice and (b) at the summer solstice.

Figure 2-18 R I V U X G

The Midnight Sun This time-lapse photograph was taken on July 19, 1985, at 69° north latitude in northeast Alaska. At this latitude, the Sun is above the horizon continuously (that is, it is circumpolar) from mid-May to the end of July. (Doug Plummer/Science Photo Library)

11:40 P.M.   12:40 A.M.   1:40 A.M.   2:40 A.M.   3:40 A.M.

month to the next, but always remains within a band called the **zodiac** that extends about 8° on either side of the ecliptic. Twelve famous constellations—Aries, Taurus, Gemini, Cancer, Leo, Virgo, Libra, Scorpius, Sagittarius, Capricornus, Aquarius, and Pisces—lie along the zodiac. The Moon is generally found in one of these 12 constellations. (Thanks to a redrawing of constellation boundaries in the mid-twentieth century, the zodiac actually passes through a thirteenth constellation—Ophiuchus, the Serpent Bearer—between Scorpius and Sagittarius.) As it moves along its orbit, the Moon appears north of the celestial equator for about two weeks and then south of the celestial equator for about the next two weeks. We will learn more about the Moon's motion, as well as why the Moon goes through phases, in Chapter 3.

The Moon not only moves around the Earth but, in concert with the Sun, also causes a slow change in the Earth's rotation. This is because both the Sun and the Moon exert a gravitational pull on the Earth. We will learn much more about gravity in Chapter 4; for now, all we need is the idea that gravity is a universal attraction of matter for other matter.

The gravitational pull of the Sun and the Moon affects the Earth's rotation because the Earth is slightly fatter across the equator than it is from pole to pole: Its equatorial diameter is 43 kilometers (27 miles) larger than the diameter measured from pole to pole. The Earth is therefore said to have an "equatorial bulge." Because of the gravitational pull of the Moon and the Sun on this bulge, the orientation of the Earth's axis of rotation gradually changes.

The Earth behaves somewhat like a spinning top, as illustrated in **Figure 2-19**. If the top were not spinning, gravity would pull the top over on its side. When the top is spinning, gravity causes the top's axis of rotation to trace out a circle, producing a motion called **precession.**

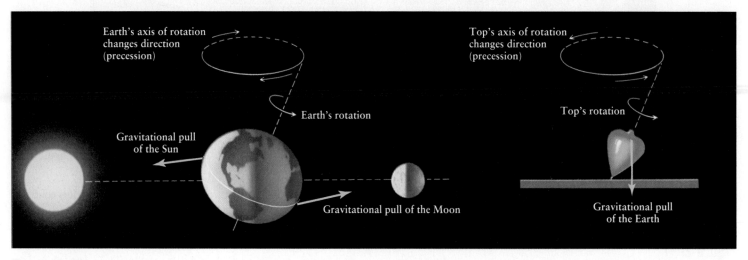

Earth's axis of rotation changes direction (precession)

Top's axis of rotation changes direction (precession)

Earth's rotation

Top's rotation

Gravitational pull of the Sun

Gravitational pull of the Moon

Gravitational pull of the Earth

Figure 2-19

Precession Because the Earth's rotation axis is tilted, the gravitational pull of the Moon and the Sun on the Earth's equatorial bulge together cause the Earth to precess. As the Earth precesses, its axis of rotation slowly traces out a circle in the sky, like the shifting axis of a spinning top.

As the Sun and Moon move along the zodiac, each spends half its time north of the Earth's equatorial bulge and half its time south of it. The gravitational pull of the Sun and Moon tugging on the equatorial bulge tries to twist the Earth's axis of rotation to be perpendicular to the plane of the ecliptic. But because the Earth is spinning, the combined actions of gravity and rotation cause the Earth's axis to trace out a circle in the sky, much like what happens to the toy top. As the axis precesses, it remains tilted about 23½° to the perpendicular.

As the Earth's axis of rotation slowly changes its orientation, the north and south celestial poles—which are the projections of that axis onto the celestial sphere—change their positions relative to the stars. At present, the north celestial pole lies within 1° of the star Polaris, which is why Polaris is the North Star. But 5000 years ago, the north celestial pole was closest to the star Thuban in the constellation of Draco (the Dragon). Thus, that star and not Polaris was the North Star. And 12,000 years from now, the North Star will be the bright star Vega in Lyra (the Harp). It takes 26,000 years for the north celestial pole to complete one full precessional circle around the sky (Figure 2-20). The south celestial pole executes a similar circle in the southern sky.

Precession also causes the Earth's equatorial plane to change its orientation. Because this plane defines the location of the celestial equator in the sky, the celestial equator precesses as well.

The intersections of the celestial equator and the ecliptic define the equinoxes (see Figure 2-15), so these key locations in the sky also shift slowly from year to year. For this reason, the precession of the Earth is also called the **precession of the equinoxes.** The first person to detect the precession of the equinoxes, in the second century B.C., was the Greek astronomer Hipparchus, who compared his own observations with those of Babylonian astronomers three centuries earlier. Today, the vernal equinox is located in the constellation Pisces (the Fishes). Two thousand years ago, it was in Aries (the Ram). Around the year A.D. 2600, the vernal equinox will move into Aquarius (the Water Bearer).

CAUTION! Astrological terms like the "Age of Aquarius" involve boundaries in the sky that are not recognized by astronomers and are generally not even related to the positions of the constellations. For example, most astrologers would call a person born on March 21, 1988, an "Aries" because the Sun was supposedly in the direction of that constellation on March 21. But due to precession, the Sun was actually in the constellation Pisces on that date! Indeed, astrology is *not* a science at all, but merely a collection of superstitions and hokum. Its practitioners use some of the terminology of astronomy but reject the logical thinking that is at the heart of science. James Randi has more to say about astrology and other pseudosciences in his essay "Why Astrology Is Not Science" at the end of this chapter.

The astronomer's system of locating heavenly bodies by their right ascension and declination, discussed in Box 2-1, is tied to the positions of the celestial equator and the vernal equinox. Because of precession, these positions are changing, and thus the coordinates of stars in the sky are also constantly changing. These changes are very small and gradual, but they add up over the years. To cope with this difficulty, astronomers always make note of the date (called the **epoch**) for which a particular set of coordinates is precisely correct. Consequently, star catalogs and star charts are periodically updated. Most current catalogs and star charts are prepared for the epoch 2000. The coordinates in these reference books, which are precise for January 1, 2000, will require very little correction over the next few decades.

2-7 Positional astronomy plays an important role in keeping track of time

Astronomers have traditionally been responsible for telling time. This is because we want the system of timekeeping used in everyday life to reflect the position of the Sun in the sky. Thousands of years ago, the sundial was invented to keep track of **apparent solar time.** To obtain more accurate measurements astronomers use the **meridian.** As Figure 2-21 shows, this is a north-south circle on the celestial sphere that passes through the zenith (the point directly overhead) and both celestial poles. *Local noon* is defined to be when the Sun crosses the **upper meridian,** which is the half of the meridian above the horizon. At *local midnight*, the Sun crosses the **lower**

Ancient scholars developed a system of timekeeping based on the Sun

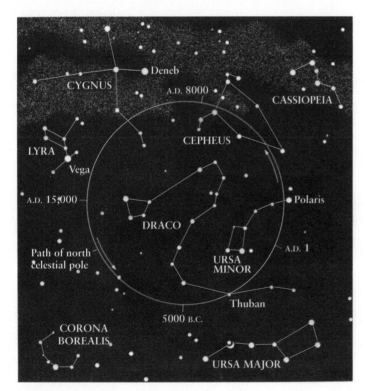

Figure 2-20

Precession and the Path of the North Celestial Pole As the Earth precesses, the north celestial pole slowly traces out a circle among the northern constellations. At present, the north celestial pole is near the moderately bright star Polaris, which serves as the North Star. Twelve thousand years from now the bright star Vega will be the North Star.

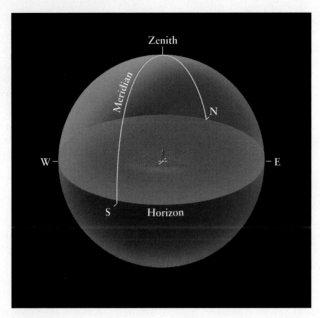

Figure 2-21

The Meridian The meridian is a circle on the celestial sphere that passes through the observer's zenith (the point directly overhead) and the north and south points on the observer's horizon. The passing of celestial objects across the meridian can be used to measure time. The upper meridian is the part above the horizon, and the lower meridian (not shown) is the part below the horizon.

meridian, the half of the meridian below the horizon; this crossing cannot be observed directly.

The crossing of the meridian by any object in the sky is called a **meridian transit** of that object. If the crossing occurs above the horizon, it is an *upper* meridian transit. An **apparent solar day** is the interval between two successive upper meridian transits of the Sun as observed from any fixed spot on the Earth. Stated less formally, an apparent solar day is the time from one local noon to the next local noon, or from when the Sun is highest in the sky to when it is again highest in the sky.

The Sun as a Timekeeper

Unfortunately, the Sun is not a good timekeeper (Figure 2-22). The length of an apparent solar day (as measured by a device such as an hourglass) varies from one time of year to another. There are two main reasons why this is so, both having to do with the way in which the Earth orbits the Sun.

The first reason is that the Earth's orbit is not a perfect circle; rather, it is an ellipse, as Figure 2-22a shows in exaggerated form. As we will learn in Chapter 4, the Earth moves more rapidly along its orbit when it is near the Sun than when it is farther away. Hence, the Sun appears to us to move more than 1° per day along the ecliptic in January, when the Earth is nearest the Sun, and less than 1° per day in July, when the Earth is farthest from the Sun. By itself, this effect would cause the apparent solar day to be longer in January than in July.

The second reason why the Sun is not a good timekeeper is the 23½° angle between the ecliptic and the celestial equator (see

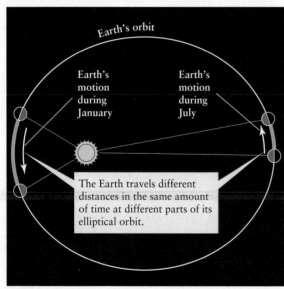

(a) **A month's motion of the Earth along its orbit**

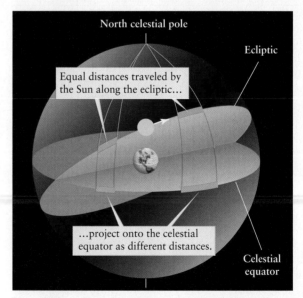

(b) **A day's motion of the Sun along the ecliptic**

Figure 2-22

Why the Sun Is a Poor Timekeeper There are two main reasons that the Sun is a poor timekeeper. **(a)** The Earth's speed along its orbit varies during the year. It moves fastest when closest to the Sun in January and slowest when farthest from the Sun in July. Hence, the apparent speed of the Sun along the ecliptic is not constant. **(b)** Because of the tilt of the

Earth's rotation axis, the ecliptic is inclined with respect to the celestial equator. Therefore, the projection of the Sun's daily progress along the ecliptic onto the celestial equator (shown in blue) varies during the year. This causes further variations in the length of the apparent solar day.

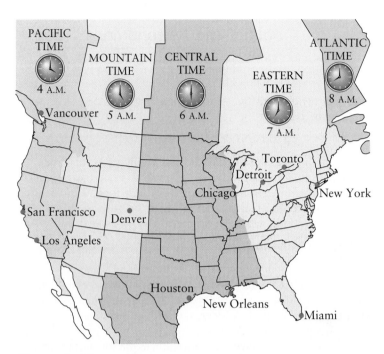

Figure 2-23

Time Zones in North America For convenience, the Earth is divided into 24 time zones, generally centered on 15° intervals of longitude around the globe. There are four time zones across the continental United States, making for a 3-hour time difference between New York and California.

Figure 2-15). As Figure 2-22*b* shows, this causes a significant part of the Sun's apparent motion when near the equinoxes to be in a north-south direction. The net daily eastward progress in the sky is then somewhat foreshortened. At the summer and winter solstices, by contrast, the Sun's motion is parallel to the celestial equator. Thus, there is no comparable foreshortening around the beginning of summer or winter. This effect by itself would make the apparent solar day shorter in March and September than in June or December. Combining these effects with those due to the Earth's noncircular orbit, we find that the length of the apparent solar day varies in a complicated fashion over the course of a year.

To avoid these difficulties, astronomers invented an imaginary object called the **mean sun** that moves along the celestial equator at a uniform rate. (In science and mathematics, "mean" is a synonym for "average.") The mean sun is sometimes slightly ahead of the real Sun in the sky, sometimes behind. As a result, mean solar time and apparent solar time can differ by as much as a quarter of an hour at certain times of the year.

Because the mean sun moves at a constant rate, it serves as a fine timekeeper. A **mean solar day** is the interval between successive upper meridian transits of the mean sun. It is exactly 24 hours long, the average length of an apparent solar day. One 24-hour day as measured by your alarm clock or wristwatch is a mean solar day.

 Time zones were invented for convenience in commerce, transportation, and communication. In a time zone, all clocks and watches are set to the mean solar

time for a meridian of longitude that runs approximately through the center of the zone. Time zones around the world are generally centered on meridians of longitude at 15° intervals. In most cases, going from one time zone to the next requires you to change the time on your wristwatch by exactly 1 hour. The time zones for most of North America are shown in Figure 2-23.

In order to coordinate their observations with colleagues elsewhere around the globe, astronomers often keep track of time using Coordinated Universal Time, somewhat confusingly abbreviated UTC or UT. This is the time in a zone that includes Greenwich, England, a seaport just outside of London where the first internationally accepted time standard was kept. (UTC was formerly known as Greenwich Mean Time.) In North America, Eastern Standard Time (EST) is 5 hours different from UTC; 9:00 A.M. EST is 14:00 UTC. Coordinated Universal Time is also used by aviators and sailors, who regularly travel from one time zone to another.

Although it is natural to want our clocks and method of timekeeping to be related to the Sun, astronomers often use a system that is based on the apparent motion of the stars. This system, called **sidereal time**, is useful when aiming a telescope. Most observatories are therefore equipped with a clock that measures sidereal time, as discussed in Box 2-2.

2-8 Astronomical observations led to the development of the modern calendar

Just as the day is a natural unit of time based on the Earth's rotation, the year is a natural unit of time based on the Earth's revolution about the Sun. However, nature has not arranged things for our convenience. The year does not divide into exactly 365 whole days. Ancient astronomers realized that the length of a year is approximately $365\frac{1}{4}$ days, so the Roman emperor Julius Caesar established the system of "leap years" to account for this extra quarter of a day. By adding an extra day to the calendar every four years, he hoped to ensure that seasonal astronomical events, such as the beginning of spring, would occur on the same date year after year.

> **Our calendar is complex because a year does not contain a whole number of days**

Caesar's system would have been perfect if the year were exactly $365\frac{1}{4}$ days long and if there were no precession. Unfortunately, this is not the case. To be more accurate, astronomers now use several different types of years. For example, the **sidereal year** is defined to be the time required for the Sun to return to the same position with respect to the stars. It is equal to 365.2564 mean solar days, or $365^{\mathrm{d}}\ 6^{\mathrm{h}}\ 9^{\mathrm{m}}\ 10^{\mathrm{s}}$.

The sidereal year is the orbital period of the Earth around the Sun, but it is *not* the year on which we base our calendar. Like Caesar, most people want annual events—in particular, the first days of the seasons—to fall on the same date each year. For example, we want the first day of spring to occur on March 21. But spring begins when the Sun is at the vernal equinox, and the vernal equinox moves slowly against the background stars because of precession. Therefore, to set up a calendar we use the **tropical**

BOX 2-2   Tools of the Astronomer's Trade

Sidereal Time

If you want to observe a particular object in the heavens, the ideal time to do so is when the object is high in the sky, on or close to the upper meridian. This minimizes the distorting effects of the Earth's atmosphere, which increase as you view closer to the horizon. For astronomers who study the Sun, this means making observations at local noon, which is not too different from noon as determined using mean solar time. For astronomers who observe planets, stars, or galaxies, however, the optimum time to observe depends on the particular object to be studied. The problem is this: Given the location of a given object on the celestial sphere, when will that object be on the upper meridian?

To answer this question, astronomers use *sidereal time* rather than solar time. It is different from the time on your wristwatch. In fact, a *sidereal clock* and an ordinary clock even tick at different rates, because they are based on different astronomical objects. Ordinary clocks are related to the position of the Sun, while sidereal clocks are based on the position of the vernal equinox, the location from which right ascension is measured. (See Box 2-1 for a discussion of right ascension.)

Regardless of where the Sun is, midnight sidereal time at your location is defined to be when the vernal equinox crosses your upper meridian. (Like solar time, sidereal time depends on where you are on Earth.) A **sidereal day** is the time between two successive upper meridian passages of the vernal equinox. By contrast, an apparent solar day is the time between two successive upper meridian crossings of the Sun. The illustration shows why these two kinds of day are not equal. Because the Earth orbits the Sun, the Earth must make one complete rotation plus about 1° to get from one local solar noon to the next. This extra 1° of rotation corresponds to 4 minutes of time,

which is the amount by which a solar day exceeds a sidereal day. To be precise:

$$1 \text{ sidereal day} = 23^{\text{h}} \ 56^{\text{m}} \ 4.091^{\text{s}}$$

where the hours, minutes, and seconds are in mean solar time.

One day according to your wristwatch is one mean solar day, which is exactly 24 hours of solar time long. A **sidereal clock** measures sidereal time in terms of sidereal hours, minutes, and seconds, where one sidereal day is divided into 24 sidereal hours.

This explains why a sidereal clock ticks at a slightly different rate than your wristwatch. As a result, at some times of the year a sidereal clock will show a very different time than an ordinary clock. (At local noon on March 21, when the Sun is at the vernal equinox, a sidereal clock will say that it is midnight. Do you see why?)

We can now answer the question in the opening paragraph. The vernal equinox, whose celestial coordinates are R.A. = $0^{\text{h}} \ 0^{\text{m}} \ 0^{\text{s}}$, Decl. = $0° \ 0' \ 0''$, crosses the upper meridian at midnight sidereal time (0:00). The autumnal equinox, which is on the opposite side of the celestial sphere at R.A. = $12^{\text{h}} \ 0^{\text{m}} \ 0^{\text{s}}$, Decl. = $0° \ 0' \ 0''$, crosses the upper meridian 12 sidereal hours later at noon sidereal time (12:00). As these examples illustrate, *any* object crosses the upper meridian when the sidereal time is equal to the object's right ascension. That is why astronomers measure right ascension in units of time rather than degrees, and why right ascension is always given in sidereal hours, minutes, and seconds.

EXAMPLE: Suppose you want to observe the bright star Spica (Figure 2-6), which has epoch 2000 coordinates R.A. = $13^{\text{h}} \ 25^{\text{m}} \ 11.6^{\text{s}}$, Decl. = $-11° \ 9' \ 41''$. What is the best time to do this?

Situation: Our goal is to find the time when Spica passes through your upper meridian, where it can best be observed.

Tools: We use the idea that a celestial object is on your upper meridian when the sidereal time equals the object's right ascension.

Answer: Based on the given right ascension of Spica, it will be best placed for observation when the sidereal time at your location is about 13:25. (Note that sidereal time is measured using a 24-hour clock.)

Review: By itself, our answer doesn't tell you what time on your wristwatch (which measures mean solar time) is best for observing Spica. That's why most observatories are equipped with a sidereal clock.

While sidereal time is extremely useful in astronomy, mean solar time is still the best method of timekeeping for most earthbound purposes. All time measurements in this book are expressed in mean solar time unless otherwise stated.

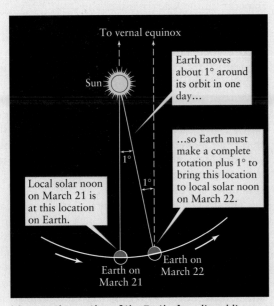

To vernal equinox

Sun

Earth moves about 1° around its orbit in one day...

...so Earth must make a complete rotation plus 1° to bring this location to local solar noon on March 22.

Local solar noon on March 21 is at this location on Earth.

1°

1°

Earth on March 21

Earth on March 22

(a) A month's motion of the Earth along its orbit

year, which is equal to the time needed for the Sun to return to the vernal equinox. This period is equal to 365.2422 mean solar days, or $365^d\ 5^h\ 48^m\ 46^s$. Because of precession, the tropical year is 20 minutes and 24 seconds shorter than the sidereal year.

Caesar's assumption that the tropical year equals $365\frac{1}{4}$ days was off by 11 minutes and 14 seconds. This tiny error adds up to about three days every four centuries. Although Caesar's astronomical advisers were aware of the discrepancy, they felt that it was too small to matter. However, by the sixteenth century the first day of spring was occurring on March 11.

The Roman Catholic Church became concerned because Easter kept shifting to progressively earlier dates. To straighten things out, Pope Gregory XIII instituted a calendar reform in 1582. He began by dropping ten days (October 4, 1582, was followed by October 15, 1582), which brought the first day of spring back to March 21. Next, he modified Caesar's system of leap years.

Caesar had added February 29 to every calendar year that is evenly divisible by four. Thus, for example, 2004, 2008, and 2012 are all leap years with 366 days. But we have seen that this system produces an error of about three days every four centuries. To solve the problem, Pope Gregory decreed that only the century years evenly divisible by 400 should be leap years. For example, the years 1700, 1800, and 1900 (which would have been leap years according to Caesar) were not leap years in the improved Gregorian system, but the year 2000, which can be divided evenly by 400, *was* a leap year.

We use the Gregorian system today. It assumes that the year is 365.2425 mean solar days long, which is very close to the true length of the tropical year. In fact, the error is only one day in every 3300 years. That won't cause any problems for a long time.

Key Words

Terms preceded by an asterisk (*) are discussed in the Boxes.

Antarctic Circle, p. 32	positional astronomy, p. 21
apparent solar day, p. 36	precession, p. 34
apparent solar time, p. 35	precession of the equinoxes,
Arctic Circle, p. 32	p. 35
autumnal equinox, p. 31	*right ascension, p. 29
celestial equator, p. 27	*sidereal clock, p. 38
celestial sphere, p. 26	*sidereal day, p. 38
circumpolar, p. 28	sidereal time, p. 37
constellation, p. 22	sidereal year, p. 37
*declination, p. 30	south celestial pole, p. 27
diurnal motion, p. 24	summer solstice, p. 31
ecliptic, p. 30	time zone, p. 37
ecliptic plane, p. 30	tropical year, p. 37
epoch, p. 35	Tropic of Cancer, p. 33
equinox, p. 31	Tropic of Capricorn, p. 33
lower meridian, p. 35	upper meridian, p. 35
mean solar day, p. 37	vernal equinox, p. 31
mean sun, p. 37	winter solstice, p. 32
meridian, p. 35	zenith, p. 27
meridian transit, p. 36	zodiac, p. 34
north celestial pole, p. 27	

Key Ideas

Ideas preceded by an asterisk (*) are discussed in the Boxes.

Constellations and the Celestial Sphere: It is convenient to imagine the stars fixed to the celestial sphere with the Earth at its center.

• The surface of the celestial sphere is divided into 88 regions called constellations.

Diurnal (Daily) Motion of the Celestial Sphere: The celestial sphere appears to rotate around the Earth once in each 24-hour period. In fact, it is actually the Earth that is rotating.

• The poles and equator of the celestial sphere are determined by extending the axis of rotation and the equatorial plane of the Earth out to the celestial sphere.

• *The positions of objects on the celestial sphere are described by specifying their right ascension (in time units) and declination (in angular measure).

Seasons and the Tilt of the Earth's Axis: The Earth's axis of rotation is tilted at an angle of about $23\frac{1}{2}°$ from the perpendicular to the plane of the Earth's orbit.

• The seasons are caused by the tilt of the Earth's axis.

• Over the course of a year, the Sun appears to move around the celestial sphere along a path called the ecliptic. The ecliptic is inclined to the celestial equator by about $23\frac{1}{2}°$.

• The ecliptic crosses the celestial equator at two points in the sky, the vernal and autumnal equinoxes. The northernmost point that the Sun reaches on the celestial sphere is the summer solstice, and the southernmost point is the winter solstice.

Precession: The orientation of the Earth's axis of rotation changes slowly, a phenomenon called precession.

• Precession is caused by the gravitational pull of the Sun and Moon on the Earth's equatorial bulge.

• Precession of the Earth's axis causes the positions of the equinoxes and celestial poles to shift slowly.

• *Because the system of right ascension and declination is tied to the position of the vernal equinox, the date (or epoch) of observation must be specified when giving the position of an object in the sky.

Timekeeping: Astronomers use several different means of keeping time.

• Apparent solar time is based on the apparent motion of the Sun across the celestial sphere, which varies over the course of the year.

• Mean solar time is based on the motion of an imaginary mean sun along the celestial equator, which produces a uniform mean solar day of 24 hours. Ordinary watches and clocks measure mean solar time.

• *Sidereal time is based on the apparent motion of the celestial sphere.

The Calendar: The tropical year is the period between two passages of the Sun across the vernal equinox. Leap year corrections are needed because the tropical year is not exactly 365 days. The sidereal year is the actual orbital period of the Earth.

Questions

Review Questions

1. Describe three structures or carvings made by past civilizations that show an understanding of astronomy.

2. How are constellations useful to astronomers? How many stars are not part of any constellation?

3. A fellow student tells you that only those stars in Figure 2-2*b* that are connected by blue lines are part of the constellation Orion. How would you respond?

4. Why are different stars overhead at 10:00 P.M. on a given night than two hours later at midnight? Why are different stars overhead at midnight on June 1 than at midnight on December 1?

5. What is the celestial sphere? Why is this ancient concept still useful today?

6. Imagine that someone suggests sending a spacecraft to land on the surface of the celestial sphere. How would you respond to such a suggestion?

7. What is the celestial equator? How is it related to the Earth's equator? How are the north and south celestial poles related to the Earth's axis of rotation?

8. Where would you have to look to see your zenith? Where on Earth would you have to be for the celestial equator to pass through your zenith? Where on Earth would you have to be for the south celestial pole to be at your zenith?

9. How many degrees is the angle from the horizon to the zenith? Does your answer depend on what point on the horizon you choose?

10. Why can't a person in Antarctica use the Big Dipper to find the north direction?

11. Is there any place on Earth where you could see the north celestial pole on the northern horizon? If so, where? Is there any place on Earth where you could see the north celestial pole on the western horizon? If so, where? Explain your answers.

12. How do the stars appear to move over the course of the night as seen from the north pole? As seen from the equator? Why are these two motions different?

13. Using a diagram, explain why the tilt of the Earth's axis relative to the Earth's orbit causes the seasons as we orbit the Sun.

14. Give two reasons why it's warmer in summer than in winter.

15. What is the ecliptic plane? What is the ecliptic?

16. Why is the ecliptic tilted with respect to the celestial equator? Does the Sun appear to move along the ecliptic, the celestial equator, or neither? By about how many degrees does the Sun appear to move on the celestial sphere each day?

17. Where on Earth do you have to be in order to see the north celestial pole directly overhead? What is the maximum possible elevation of the Sun above the horizon at that location? On what date can this maximum elevation be observed?

18. What are the vernal and the autumnal equinoxes? What are the summer and winter solstices? How are these four points related to the ecliptic and the celestial equator?

19. At what point on the horizon does the vernal equinox rise? Where on the horizon does it set? (*Hint:* See Figure 2-16.)

20. How does the daily path of the Sun across the sky change with the seasons? Why does it change?

21. Where on Earth do you have to be in order to see the Sun at the zenith? As seen from such a location, will the Sun be at the zenith every day? Explain.

22. What is precession of the equinoxes? What causes it? How long does it take for the vernal equinox to move 1° along the ecliptic?

23. What is the (fictitious) mean sun? What path does it follow on the celestial sphere? Why is it a better timekeeper than the actual Sun in the sky?

24. Why is it convenient to divide the Earth into time zones?

25. Why is the time given by a sundial not necessarily the same as the time on your wristwatch?

26. What is the difference between the sidereal year and the tropical year? Why are these two kinds of year slightly different in length? Why are calendars based on the tropical year?

27. When is the next leap year? Was 2000 a leap year? Will 2100 be a leap year?

Advanced Questions

Questions preceded by an asterisk (*) involve topics discussed in the Boxes.

Problem-solving tips and tools

 To help you visualize the heavens, it is worth taking the time to become familiar with various types of star charts. These include the simple star charts at the end of this book, the monthly star charts published in such magazines as *Sky & Telescope* and *Astronomy*, and the more detailed maps of the heavens found in star atlases.

 One of the best ways to understand the sky and its motions is to use the *Starry Night Enthusiast™* computer program on the CD-ROM that accompanies certain printed copies of this book. This easy-to-use program allows you to view the sky on any date and at any time, as seen from any point on Earth, and to animate the sky to visualize its diurnal and annual motions.

You may also find it useful to examine a planisphere, a device consisting of two rotatable disks. The bottom disk shows all the stars in the sky (for a particular latitude), and the top one is opaque with a transparent oval window through which only some of the stars can be seen. By rotating the top disk, you can immediately see which constellations are above the horizon at any time of the year. A planisphere is a convenient tool to carry with you when you are out observing the night sky.

---

28. On November 1 at 8:30 P.M. you look toward the eastern horizon and see the bright star Bellatrix (shown in Figure 2-2*b*) rising. At approximately what time will Bellatrix rise one week later, on November 8?

29. Figure 2-4 shows the situation on September 21, when Cygnus is highest in the sky at 8:00 P.M. local time and Andromeda is highest in the sky at midnight. But as Figure 2-5 shows, on

July 21 Cygnus is highest in the sky at midnight. On July 21, at approximately what local time is Andromeda highest in the sky? Explain your reasoning.

30. Figure 2-5 shows which constellations are high in the sky (for observers in the northern hemisphere) in the months of July, September, and November. From this figure, would you be able to see Perseus at midnight on May 15? Draw a picture to justify your answer.

31. Figure 2-6 shows the appearance of Polaris, the Little Dipper, and the Big Dipper at 11 P.M. (daylight savings time) on August 1. Sketch how these objects would appear on this same date at (**a**) 8 P.M. and (**b**) 2 A.M. Include the horizon in your sketches, and indicate the north direction.

32. Figure 2-6 shows the appearance of the sky near the North Star at 11 P.M. (daylight savings time) on August 1. Explain why the sky has this same appearance at 1 A.M. on July 1 and at 9 P.M. on September 1.

33. The time-exposure photograph that opens this chapter shows the trails made by individual stars as the celestial sphere appears to rotate around the Earth. (**a**) For approximately what length of time was the camera shutter left open to take this photograph? (**b**) The stars in this photograph (taken in Hawaii, at roughly 20° north latitude) appear to rotate around one of the celestial poles. Which celestial pole is it? As seen from this location, do the stars move clockwise or counterclockwise around this celestial pole? (**c**) If you were at 20° south latitude, which celestial pole could you see? In which direction would you look to see it? As seen from this location, do the stars move clockwise or counterclockwise around this celestial pole?

34. (**a**) Redraw Figure 2-10 for an observer at the north pole. (*Hint:* The north celestial pole is directly above this observer.) (**b**) Redraw Figure 2-10 for an observer at the equator. (*Hint:* The celestial equator passes through this observer's zenith.) (**c**) Using Figure 2-10 and your drawings from (a) and (b), justify the following rule, long used by navigators: The latitude of an observer in the northern hemisphere is equal to the angle in the sky between that observer's horizon and the north celestial pole. (**d**) State the rule that corresponds to (c) for an observer in the southern hemisphere.

35. The photograph that opens this chapter was taken next to the Gemini North Observatory atop Mauna Kea in Hawaii. The telescope is at longitude 155° 28′ 09″ west and latitude 19° 49′ 26″ north. (**a**) By making measurements on the photograph, find the approximate angular width and angular height of the photo. (**b**) How far (in degrees, arcminutes, and arcseconds) from the south celestial pole can a star be and still be circumpolar as seen from the Gemini North Observatory?

36. The Gemini North Observatory shown in the photograph that opens this chapter is located in Hawaii, roughly 20° north of the equator. Its near-twin, the Gemini South Observatory, is located roughly 30° south of the equator in Chile. Why is it useful to have telescopes in both the northern and southern hemispheres?

37. Is there any place on Earth where all the visible stars are circumpolar? If so, where? Is there any place on Earth where none of the visible stars is circumpolar? If so, where? Explain your answers.

38. The above image of the Earth was made by the *Galileo* spacecraft while en route to Jupiter. South America is at the center of the image and Antarctica is at the bottom of the image. (**a**) In which month of the year was this image made? Explain your reasoning. (**b**) When this image was made, was the Earth relatively close to the Sun or relatively distant from the Sun? Explain your reasoning.

39. Figure 2-16 shows the daily path of the Sun across the sky on March 21, June 21, September 22, and December 21 for an observer at 35° north latitude. Sketch drawings of this kind for (**a**) an observer at 35° south latitude; (**b**) an observer at the equator; and (**c**) an observer at the north pole.

40. Suppose that you live at a latitude of 40° N. What is the elevation (angle) of the Sun above the southern horizon at noon (**a**) at the time of the vernal equinox? (**b**) at the time of the winter solstice? Explain your reasoning. Include a drawing as part of your explanation.

41. In the northern hemisphere, houses are designed to have "southern exposure," that is, with the largest windows on the southern side of the house. But in the southern hemisphere houses are designed to have "northern exposure." Why are houses designed this way, and why is there a difference between the hemispheres?

42. The city of Mumbai (formerly Bombay) in India is 19° north of the equator. On how many days of the year, if any, is the Sun at the zenith at midday as seen from Mumbai? Explain your answer.

43. Ancient records show that 2000 years ago, the stars of the constellation Crux (the Southern Cross) were visible in the southern sky from Greece. Today, however, these stars cannot be seen from Greece. What accounts for this change?

44. The Great Pyramid at Giza has a tunnel that points toward the north celestial pole. At the time the pyramid was built, around 2600 B.C., toward which star did it point? Toward which star does this same tunnel point today? (See Figure 2-20.)

45. The photo on the next page shows a statue of the Greek god Atlas. The globe that Atlas is holding represents the celestial sphere, with depictions of several important constellations and the celestial equator. Although the statue dates from around 150 A.D., it has been proposed that the arrangement

R I [V] U X G

(Scala/Art Resource, NY)

of constellations depicts the sky as it was mapped in an early star atlas that dates from 129 B.C. Explain the reasoning that could lead to such a proposal.

46. Unlike western Europe, Imperial Russia did not use the revised calendar instituted by Pope Gregory XIII. Explain why the Russian Revolution, which started on November 7, 1917, according to the modern calendar, is called "the October revolution" in Russia. What was this date according to the Russian calendar at the time? Explain.

*47. What is the right ascension of a star that is on the meridian at midnight at the time of the autumnal equinox? Explain.

*48. The coordinates on the celestial sphere of the summer solstice are R.A. = 6^h 0^m 0^s, Decl. = $+23°$ $27'$. What are the right ascension and declination of the winter solstice? Explain your answer.

*49. Because 24 hours of right ascension takes you all the way around the celestial equator, $24^h = 360°$, what is the angle in the sky (measured in degrees) between a star with R.A. = 8^h 0^m 0^s, Decl. = $0°$ $0'$ $0''$ and a second star with R.A. = 11^h 20^m 0^s, Decl. = $0°$ $0'$ $0''$? Explain your answer.

*50. On a certain night, the first star in Advanced Question 49 passes through the zenith at 12:30 A.M. local time. At what time will the second star pass through the zenith? Explain your answer.

*51. At local noon on March 21, when the Sun is at the vernal equinox, a sidereal clock will say that it is midnight. Explain why.

*52. (a) What is the sidereal time when the vernal equinox rises? (b) On what date is the sidereal time nearly equal to the solar time? Explain.

*53. How would the sidereal and solar days change (a) if the Earth's rate of rotation increased, (b) if the Earth's rate of rotation decreased, and (c) if the Earth's rotation were retrograde (that is, if the Earth rotated about its axis opposite to the direction in which it revolves about the Sun)?

Discussion Questions

54. Examine a list of the 88 constellations. Are there any constellations whose names obviously date from modern times? Where are these constellations located? Why do you suppose they do not have archaic names?

55. Describe how the seasons would be different if the Earth's axis of rotation, rather than having its present $23\frac{1}{2}°$ tilt, were tilted (a) by 0° or (b) by 90°.

56. In William Shakespeare's *Julius Caesar* (act 3, scene 1), Caesar says:

> *But I am constant as the northern star,*
> *Of whose true-fix'd and resting quality*
> *There is no fellow in the firmament.*

Translate Caesar's statement about the "northern star" into modern astronomical language. Is the northern star truly "constant"? Was the northern star the same in Shakespeare's time (1564–1616) as it is today?

Web/eBook Questions

57. Search the World Wide Web for information about the national flags of Australia, New Zealand, and Brazil and the state flag of Alaska. Which stars are depicted on these flags? Explain any similarities or differences among these flags.

58. Some people say that on the date that the Sun is at the vernal equinox, and only on this date, you can stand a raw egg on end. Others say that there is nothing special about the vernal equinox, and that with patience you can stand a raw egg on end on any day of the year. Search the World Wide Web for information about this story and for hints about how to stand an egg on end. Use these hints to try the experiment yourself on a day when the Sun is *not* at the vernal equinox. What do you conclude about the connection between eggs and equinoxes?

 59. Use the U.S. Naval Observatory Web site to find the times of sunset and sunrise on (a) your next birthday and (b) the date this assignment is due. (c) Are the times the same for the two dates? Explain why or why not.

Activities

Observing Projects

> **Observing tips and tools**
>
> Moonlight is so bright that it interferes with seeing the stars. For the best view of the constellations, do your observing when the Moon is below the horizon. You can find the times of moonrise and moonset in your local newspaper or on the World Wide Web. Each monthly issue of the magazines *Sky & Telescope* and *Astronomy* includes much additional observing information.

60. On a clear, cloud-free night, use the star charts at the end of this book to see how many constellations of the zodiac you can identify. Which ones were easy to find? Which were difficult? Are the zodiacal constellations the most prominent ones in the sky?

61. Examine the star charts that are published monthly in such popular astronomy magazines as *Sky & Telescope* and *As-*

tronomy. How do they differ from the star charts at the end of this book? On a clear, cloud-free night, use one of these star charts to locate the celestial equator and the ecliptic. Note the inclination of the Milky Way to the ecliptic and celestial equator. The Milky Way traces out the plane of our galaxy. What do your observations tell you about the orientation of the Earth and its orbit relative to the galaxy's plane?

62. Suppose you wake up before dawn and want to see which constellations are in the sky. Explain how the star charts at the end of this book can be quite useful, even though chart times are given only for the evening hours. Which chart most closely depicts the sky at 4:00 A.M. on the morning that this assignment is due? Set your alarm clock for 4:00 A.M. to see if you are correct.

63. Use the *Starry Night Enthusiast*™ program to observe the diurnal motion of the sky. Select **Viewing Location . . .** in the **Options** menu, click on the **List** tab, highlight the name of your town or city and click the **Set Location** button to display the sky where you live. a) In the northern hemisphere, press the "N" key to set the gaze direction to the northern sky. (If your location is in the southern hemisphere, press the "S" key to select your southern sky.) Select **Hide Daylight** under the **View** menu. In the toolbar, click on the **Time Flow Rate** control and set the discrete time step to **1 minute**. Then click on the **Play** button (a triangle that points to the right) to run time forward. (To reduce confusion, remove rapidly moving artificial Earth-orbiting satellites from this view by clicking on **View/Solar System** and turning off **Satellites**). Do the background stars appear to rotate clockwise or counterclockwise? Explain this observation in terms of the Earth's rotation. Are any of the stars circumpolar, that is, do they stay above your horizon for the full 24 hours of a day? b) Now center your field of view on the southern horizon if you live in the northern hemisphere) (or the northern horizon, if you live in the southern hemisphere). Describe what you see. Are any of these stars circumpolar?

64. Use the *Starry Night Enthusiast*™ program to observe the Sun's motion on the celestial sphere. Select **Guides > Atlas** in the **Favourites** menu to see the entire celestial sphere as if you were at the center of a transparent Earth. Open the **Find pane** and double-click on the entry for the **Sun** to center it in the view. a) In the toolbar at the top of the main window, set the **Time Flow Rate** to **1 day** and set time to move forward, using the **Play** button (a triangle that points to the right). Observe the Sun for a full year of simulated time. How does the Sun appear to move against the background stars? What path does it follow? Does it ever change direction? b) Open the **Options** pane, expand the **Constellations** layer and select the **Auto Identify, Boundaries and Labels** checkboxes to turn these constellation options on. In the toolbar, click on the **Now** button to return to your present time. Again, open the **Find** pane and double-click on the Sun to center it in your view. In which constellation is the Sun located today? Is this the same as the astrological sign for today's date? Explain your answer in terms of precession. (c) Set the discrete **Time Flow Rate** to **1 day** and click the **Play** button. Through which constellations does the Sun appear to pass over the course of a year?

65. Use the *Starry Night Enthusiast*™ program to investigate the orbits of the Earth and some of the other planets around the Sun from a position 4 AU above the North Pole of the Sun. Select **Viewing Location . . .** in the **Options** menu. In the Viewing Location dialog box, select **stationary location** in the **View from** drop box. In the edit boxes beneath the **Cartesian coordinates** label, enter "0" for the **X** and **Y** values and "4 AU" for the **Z** value. Then click the **Set Location** button. You are now at the above position in space. Open the **Find** pane and center the view upon the Sun from this point in space, 4 AU above the North Pole of the Sun. You will now be able to see the orbits of the four inner planets of the Solar System nearly face-on. To display the Earth's orbit, click the checkbox to the right of the label **Earth** in the **Find** pane. a) Can you tell from this view that the Earth's orbit is not a perfect circle, as we learned in Section 2-7? b) In the **Find** pane, click in the checkbox to the right of the labels for the other three inner planets (Mercury, Venus, and Mars) to display their orbits in the view. Of these four inner planets, which have orbits that are clearly noncircular? (You may want to use a ruler to measure the distance from the Sun to various points on each planet's orbit. If the distance is the same at all points then the orbit is circular; otherwise, it is noncircular.) c) Now change your view so that it is in the plane of the Earth's orbit. Select **Viewing Location . . .** from the **Options** menu. In the Viewing Location dialog box, enter "0" for the **X** and **Z** Cartesian coordinates and enter "4 AU" for the **Y** coordinate. This view shows an edge-on view of the orbits of the four inner planets. Do the orbits of the other planets lie in the same plane as the plane of the Earth's orbit (the ecliptic plane)? Which planet's orbit appears to be tilted the most from the Earth's orbit? (You may need to use the **Zoom** controls at the right side of the toolbar to investigate this.) If you were on any of these other planets in the solar system, would the Sun appear to follow exactly the same path across the celestial sphere as is seen from Earth?

Why Astrology Is Not Science

by James Randi

I'm involved in the strange business of telling folks what they should already know. I meet audiences who believe in all sorts of impossible things, often despite their education and intelligence. My job is to explain how science differs from the unproven, illogical assumptions of pseudoscience—and why it matters. Perhaps my best example is the difference between astronomy and astrology.

Both astrology and astronomy arose from the wonders of the night sky, from the stars to comets, planets, the Sun, and the Moon. Surely, humans have long reasoned, there must be some meaning in their motions. Surely the Moon's effect on tides hints at hidden "causes" for strange events. *Judiciary* (literally "judging") astrology therefore attempted to foretell the future—our earthly future. To serve it, *horary* (literally "hourly") astrology carefully tracked the heavens.

It is the latter that has become astronomy. Thanks to its process of careful measurement and testing, we now understand more about the true nature of the starry universe than astrologers could ever have imagined. With the birth of a new science, astronomers had a logical framework based on physical causes and systematic observations.

Astrology remains a popular delusion. Far too many believe today that patterns in the sky govern our lives. They accept the vague tendencies and portents of seers who cast horoscopes. They shouldn't. Just a glance at the tenets of astrology provides ample evidence of its absurdity.

An individual is said to be born under a sign. To the astrologer, the Sun was located "in" that sign at the moment of birth. (Stars are not seen in the daytime, but no matter—a calculation tells where the Sun is.) Each sign takes its name from a constellation, a totally imaginary figure invented for our convenience in referring to stars. Different cultures have different mythical figures up there, and so different schools of astrology assign different meanings to the signs they use.

In the spirit of equal-opportunity swindling, astrologers divide up the year fairly, ignoring variations in the size of constellations. Since Libra is tiny, while Virgo is huge, they chop some of the sky off Virgo and add it—along with bits of Scorpio—to bring Libra up to size. The Sun could well be declared "in" Libra when it is actually outside that constellation.

It gets worse. Science constantly challenges itself and changes. The rules of astrology could not, although they were made up thousands of years ago, and since then the "fixed" stars have moved. In particular, precession of the equinoxes has shifted objects in the sky relative to our calendar. The constellations have changed but astrology has not. If you were born August 7, you are said to be a Leo, but the Sun that day was really in the same part of the sky as the constellation Cancer.

With a theory like this to back it up, we should not be surprised at the bottom line: *A pseudoscience does not work.* Test after test has checked its predictions, and the result is always the same. One such investigator is Shawn Carlson of the University of California, San Diego. As he put it in *Nature* magazine, astrology is "a hopeless cause." Johannes Kepler, the pioneering astronomer, himself cast horoscopes, but they are little remembered today. Owen Gingerich, a historian of science at Harvard, puts it well: Kepler was the astrologer who destroyed astrology.

Astronomy works, and it works very well indeed. That isn't easy. Because we humans tend to find what we want in any body of data, it takes science's careful process of observation, creative insight, and critical thinking to understand and predict changes in nature. As I write, a transit of Ganymede is due next Thursday at 21:47:20. At exactly that time, the satellite of Jupiter will cross in front of its planet as seen from Earth, and yet most of us will never know it. Still other moons of Jupiter may hold fresh clues to the formation of our entire solar system and the conditions for life elsewhere.

For most people, astronomy has too little fantasy or money in it, and they will never experience the beauty in its predictions. The dedicated labors of generations of scientists have enabled us to perform a genuine wonder.

James ("the Amazing") Randi works tirelessly to expose trickery so that others can relish the greater wonder of science. As a magician, he has had his own television show and an enormous public following. As a lecturer, he addresses teachers, students, and others worldwide. His newsletter and column for *The Skeptic* are key resources for educators. His many books include *Flim-Flam!, The Faith Healers,* and *The Mask of Nostradamus,* about a legendary con man with secrets of his own.

Mr. Randi helped found the Committee for the Scientific Investigation of Claims of the Paranormal, and his $10,000 prize for "the performance of any paranormal event . . . under proper observing conditions" has gone unclaimed for more than 25 years. An amateur archeologist and astronomer as well, he lives in Florida with several untalented parrots and the occasional visiting magus.

3

Eclipses and the Motion of the Moon

The Sun in total eclipse, March 29, 2006. (Stefan Seip)

O n March 29, 2006, a rare cosmic spectacle—a total solar eclipse—was visible along a narrow corridor that extended from the coast of Brazil through equatorial Africa and into central Asia. As shown in this digital composite taken in Turkey, the Moon slowly moved over the disk of the Sun. (Time flows from left to right in this image.) For a few brief minutes the Sun was totally covered, darkening the sky and revealing the Sun's thin outer atmosphere, or corona, which glows with an unearthly pearlescent light.

Such eclipses can be seen only on specific dates from special locations on Earth, so not everyone will ever see the Moon cover the Sun in this way. But anyone can find the Moon in the sky and observe how its appearance changes from night to night, from new moon to full moon and back again, and how the times when the Moon rises and sets differ noticeably from one night to the next.

In this chapter our subject is how the Moon moves as seen from the Earth. We will explore why the Moon goes through a regular cycle of phases, and how the Moon's orbit around the Earth leads to solar eclipses as well as lunar eclipses. We will also see how ancient astronomers used their observations of the Moon to determine the size and shape of the Earth, as well as other features of the solar system. Thus, the Moon—which has always loomed large in the minds of poets, lovers, and dreamers—has also played a key role in the development of our modern picture of the universe.

3-1 The phases of the Moon are caused by its orbital motion

As seen from Earth, both the Sun and the Moon appear to move from west to east on the celestial sphere—that is, relative to the background of stars. They move at very different rates, however. The Sun takes one year to make a complete trip around the imaginary celestial sphere along the path we call the *ecliptic* (Section 2-5). By comparison, the Moon takes only about four weeks. In the past, these similar motions led people to believe that both the Sun and the

> **You can tell the Moon's position relative to the Earth and Sun by observing its phase**

Learning Goals

By reading the sections of this chapter, you will learn

3-1 Why we see the Moon go through phases

3-2 Why we always see the same side of the Moon

3-3 The differences between lunar and solar eclipses

3-4 Why not all lunar eclipses are total eclipses

3-5 Why solar eclipses are visible only from certain special locations on Earth

3-6 How ancient Greek astronomers deduced the sizes of the Earth, the Moon, and the Sun

Figure 3-1   R I V U X G

The Earth and the Moon This picture of the Earth and the Moon was taken in 1992 by the *Galileo* spacecraft on its way toward Jupiter. The Sun, which provides the illumination for both the Earth and the Moon, was far to the right and out of the camera's field of view when this photograph was taken. (NASA/JPL)

Moon orbit around the Earth. We now know that only the Moon orbits the Earth, while the Earth-Moon system as a whole (Figure 3-1) orbits the Sun. (In Chapter 4 we will learn how this was discovered.)

One key difference between the Sun and the Moon is the nature of the light that we receive from them. The Sun emits its own light. So do the stars, which are objects like the Sun but much farther away, and so does an ordinary light bulb. By contrast, the light that we see from the Moon is reflected light. This is sunlight that has struck the Moon's surface, bounced off, and ended up in our eyes here on Earth.

CAUTION! You probably associate *reflection* with shiny objects like a mirror or the surface of a still lake. In science, however, the term refers to light bouncing off any object. You see most objects around you by reflected light. When you look at your hand, for example, you are seeing light from the Sun (or from a light fixture) that has been reflected from the skin of your hand and into your eye. In the same way, moonlight is really sunlight that has been reflected by the Moon's surface.

Understanding the Moon's Phases

Figure 3-1 shows both the Moon and the Earth as seen from a spacecraft. When this image was recorded, the Sun was far off to the right. Hence, only the right-hand hemispheres of both worlds were illuminated by the Sun; the left-hand hemispheres were in darkness and are not visible in the picture. In the same way, when we view the Moon from the Earth, we see only the half of the Moon that faces the Sun and is illuminated. However, not all of the illuminated half of the Moon is necessarily facing us. As the Moon moves around the Earth, from one night to the next we see different amounts of the illuminated half of the Moon. These different appearances of the Moon are called **lunar phases.**

Figure 3-2 shows the relationship between the lunar phase visible from Earth and the position of the Moon in its orbit. For example, when the Moon is at position A, we see it in roughly the same direction in the sky as the Sun. Hence, the dark hemisphere of the Moon faces the Earth. This phase, in which the Moon is barely visible, is called **new moon.** Since a new moon is near the Sun in the sky, it rises around sunrise and sets around sunset.

As the Moon continues around its orbit from position A in Figure 3-2, more of its illuminated half becomes exposed to our view. The result, shown at position B, is a phase called **waxing crescent moon** ("waxing" is a synonym for "increasing"). About a week after new moon, the Moon is at position C; we then see half of the Moon's illuminated hemisphere and half of the dark hemisphere. This phase is called **first quarter moon.**

As seen from Earth, a first quarter moon is one-quarter of the way around the celestial sphere from the Sun. It rises and sets about one-quarter of an Earth rotation, or six hours, after the Sun does: moonrise occurs around noon, and moonset occurs around midnight.

CAUTION! Despite the name, a first quarter moon appears to be *half* illuminated, not one-quarter illuminated! The name means that this phase is one-quarter of the way through the complete cycle of lunar phases.

About four days later, the Moon reaches position D in Figure 3-2. Still more of the illuminated hemisphere can now be seen from Earth, giving us the phase called **waxing gibbous moon** ("gibbous" is another word for "swollen"). When you look at the Moon in this phase, as in the waxing crescent and first quarter phases, the illuminated part of the Moon is toward the west. Two weeks after new moon, when the Moon stands opposite the Sun in the sky (position E), we see the fully illuminated hemisphere. This phase is called **full moon.** Because a full moon is opposite the Sun on the celestial sphere, it rises at sunset and sets at sunrise.

Over the following two weeks, we see less and less of the Moon's illuminated hemisphere as it continues along its orbit, and the Moon is said to be *waning* ("decreasing"). While the Moon is waning, its illuminated side is toward the east. The phases are called **waning gibbous moon** (position F), **third quarter moon** (position G, also called *last quarter moon*), and **waning crescent moon** (position H). A third quarter moon appears one-quarter of the way around the celestial sphere from the Sun, but on the opposite side of the celestial sphere from a first quarter moon. Hence, a third quarter moon rises and sets about one-

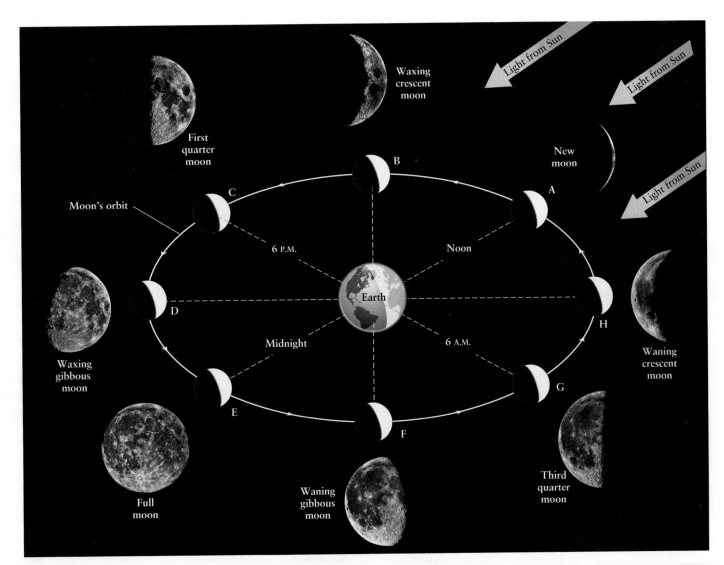

Figure 3-2

Why the Moon Goes Through Phases This figure shows the Moon at eight positions on its orbit, along with photographs of what the Moon looks like at each position as seen from Earth. The changes in phase occur because light from the Sun illuminates one half of the Moon, and as the Moon orbits the Earth we see varying amounts of the Moon's illuminated half. It takes about $29\frac{1}{2}$ days for the Moon to go through a complete cycle of phases. (Photographs from Yerkes Observatory and Lick Observatory)

quarter Earth rotation, or 6 hours, *before* the Sun: moonrise is around midnight and moonset is around noon.

The Moon takes about four weeks to complete one orbit around the Earth, so it likewise takes about four weeks for a complete cycle of phases from new moon to full moon and back to new moon. Since the Moon's position relative to the Sun on the celestial sphere is constantly changing, and since our system of timekeeping is based on the Sun (see Section 2-7), the times of moonrise and moonset are different on different nights. On average the Moon rises and sets about an hour later each night.

Figure 3-2 also explains why the Moon is often visible in the daytime, as shown in **Figure 3-3**. From any location on Earth, about half of the Moon's orbit is visible at any time. For example, if it is midnight at your location, you are in the middle of the dark side of the Earth that faces away from the Sun. At that time you can easily see the Moon if it is at position C, D, E, F, or G. If it is midday at your location, you are in the middle of the Earth's illuminated side, and the Moon will be easily visible if it is at position A, B, C, G, or H. (The Moon is so bright that it can be seen even against the bright blue sky.) You can see that the Moon is prominent in the midnight sky for about half of its orbit, and prominent in the midday sky for the other half.

CAUTION! A very common misconception about lunar phases is that they are caused by the shadow of the *Earth* falling on the Moon. As Figure 3-2 shows, this is not the case at all. Instead, phases are simply the result of our seeing the illuminated half of the Moon at different angles as the Moon moves around its orbit. To help you better visualize how this works, **Box 3-1** describes how you can simulate the cycle shown in Figure 3-2 using ordinary objects on Earth. (As we will learn in Section 3-3, the Earth's shadow does indeed fall on the Moon on rare occasions. When this happens, we see a lunar eclipse.)

Figure 3-3   R I V U X G

The Moon During the Day The Moon can be seen during the daytime as well as at night. The time of day or night when it is visible depends on its phase. [Karl Beath/Gallo Images/Getty Images]

BOX 3-1   The Heavens on the Earth

Phases and Shadows

Figure 3-2 shows how the relative positions of the Earth, Moon, and Sun explain the phases of the Moon. You can visualize lunar phases more clearly by doing a simple experiment here on Earth. All you need are a small round object, such as an orange or a baseball, and a bright source of light, such as a street lamp or the Sun.

In this experiment, you play the role of an observer on the Earth looking at the Moon, and the round object plays the role of the Moon. The light source plays the role of the Sun. Hold the object in your right hand with your right arm stretched straight out in front of you, with the object directly between you and the light source (position A in the accompanying illustration). In this orientation the illuminated half of the object faces away from you, like the Moon when it is in its new phase (position A in Figure 3-2).

Now, slowly turn your body to the left so that the object in your hand "orbits" around you (toward positions C, E, and G in the illustration). As you turn, more and more of the illuminated side of the "moon" in your hand becomes visible, and it goes through the same cycle of phases—waxing crescent, first quarter, and waxing gibbous—as does the real Moon. When you have rotated through half a turn so that the light source is directly behind you, you will be looking face on at the illuminated side of the object in your hand. This corresponds to a full moon (position E in Figure 3-2). Make sure your body does not cast a shadow on the "moon" in your hand—that would correspond to a lunar eclipse!

As you continue turning to the left, more of the unilluminated half of the object becomes visible as its phase moves through waning gibbous, third quarter, and waning crescent. When your body has rotated back to the same orientation that you were in originally, the unilluminated half of your hand-held "moon" is again facing toward you, and its phase is again new. If you continue to rotate, the object in your hand repeats the cycle of "phases," just as the Moon does as it orbits around the Earth.

The experiment works best when there is just one light source around. If there are several light sources, such as in a room with several lamps turned on, the different sources will create multiple shadows and it will be difficult to see the phases of your hand-held "moon." If you do the experiment outdoors using sunlight, you may find that it is best to perform it in the early morning or late afternoon when shadows are most pronounced and the Sun's rays are nearly horizontal.

3-2 The Moon always keeps the same face toward the Earth

Although the phase of the Moon is constantly changing, one aspect of its appearance remains the same: it always keeps essentially the same hemisphere, or face, toward the Earth. Thus, you will always see the same craters and mountains on the Moon, no matter when you look at it; the only difference will be the angle at which these surface features are illuminated by the Sun. (You can verify this by carefully examining the photographs of the Moon in Figure 3-2.)

> **The Moon rotates in a special way: it spins exactly once per orbit**

The Moon's Synchronous Rotation

Why is it that we only ever see one face of the Moon? You might think that it is because the Moon does not rotate (unlike the Earth, which rotates around an axis that passes from its north pole to its south pole). To see that this *cannot* be the case, consider **Figure 3-4**. This figure shows the Earth and the orbiting Moon from a vantage point far above the Earth's north pole. In this figure two craters on the lunar surface have been colored, one in red and one in blue. If the Moon did not rotate on its axis, as in Figure 3-4*a*, at some times the red crater would be visible from Earth, while at other times the blue crater would be visible. Thus, we would see different parts of the lunar surface over time, which does not happen in reality.

In fact, the Moon always keeps the same face toward us because it *is* rotating, but in a very special way: It takes exactly as long to rotate on its axis as it does to make one orbit around the Earth. This situation is called **synchronous rotation**. As Figure 3-4*b* shows, this keeps the crater shown in red always facing the Earth, so that we always see the same face of the Moon. In Chapter 4 we will learn why the Moon's rotation and orbital motion are in step with each other.

An astronaut standing at the spot shown in red in Figure 3-4*b* would spend two weeks (half of a lunar orbit) in darkness, or lunar nighttime, and the next two weeks in sunlight, or lunar daytime. Thus, as seen from the Moon, the Sun rises and sets, and no part of the Moon is perpetually in darkness. This means that there really is no "dark side of the Moon." The side of the Moon that constantly faces away from the Earth is properly called the *far* side. The Sun rises and sets on the far side just as on the side toward the Earth. Hence, the blue crater on the far side of the Moon in Figure 3-4*b* is in sunlight for half of each lunar orbit.

Sidereal and Synodic Months

The time for a complete lunar "day"—the same as the time that it takes the Moon to rotate once on its axis—is about four weeks. (Because the Moon's rotation is synchronous, it takes the same time for one complete lunar orbit.) It also takes about four weeks for the Moon to complete one cycle of its phases as seen from Earth. This regular cycle of phases inspired our ancestors to invent the concept of a month. For historical reasons, none of which

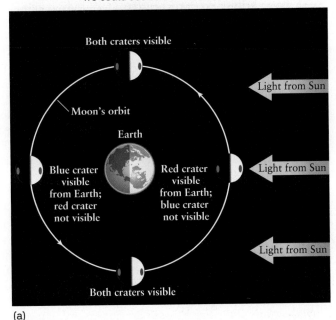

If the Moon did not rotate, we could see all sides of the Moon

(a)

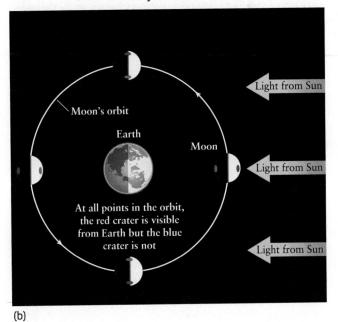

In fact, the Moon does rotate, and we see only one face of the Moon

(b)

Figure 3-4

The Moon's Rotation These diagrams show the Moon at four points in its orbit as viewed from high above the Earth's north pole. **(a)** If the Moon did not rotate, then at various times the red crater would be visible from Earth while at other times the blue crater would be visible. Over a complete orbit, the entire surface of the Moon would be visible. **(b)** In reality, the Moon rotates on its north-south axis. Because the Moon makes one rotation in exactly the same time that it makes one orbit around the Earth, we see only one face of the Moon.

Figure 3-5

The Sidereal and Synodic Months The sidereal month is the time the Moon takes to complete one full revolution around the Earth with respect to the background stars. However, because the Earth is constantly moving along its orbit about the Sun, the Moon must travel through slightly more than 360° of its orbit to get from one new moon to the next. Thus, the synodic month—the time from one new moon to the next—is longer than the sidereal month.

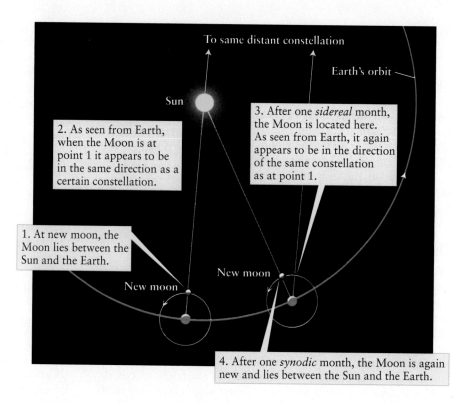

To same distant constellation

Earth's orbit

Sun

2. As seen from Earth, when the Moon is at point 1 it appears to be in the same direction as a certain constellation.

3. After one *sidereal* month, the Moon is located here. As seen from Earth, it again appears to be in the direction of the same constellation as at point 1.

1. At new moon, the Moon lies between the Sun and the Earth.

New moon

New moon

4. After one *synodic* month, the Moon is again new and lies between the Sun and the Earth.

has much to do with the heavens, the calendar we use today has months of differing lengths. Astronomers find it useful to define two other types of months, depending on whether the Moon's motion is measured relative to the stars or to the Sun. Neither corresponds exactly to the familiar months of the calendar.

The **sidereal month** is the time it takes the Moon to complete one full orbit of the Earth, as measured with respect to the stars. This true orbital period is equal to about 27.32 days. The **synodic month**, or *lunar month*, is the time it takes the Moon to complete one cycle of phases (that is, from new moon to new moon or from full moon to full moon) and thus is measured with respect to the Sun rather than the stars. The length of the "day" on the Moon is a synodic month, not a sidereal month.

The synodic month is longer than the sidereal month because the Earth is orbiting the Sun while the Moon goes through its phases. As Figure 3-5 shows, the Moon must travel *more* than 360° along its orbit to complete a cycle of phases (for example, from one new moon to the next). Because of this extra distance, the synodic month is equal to about 29.53 days, about two days longer than the sidereal month.

Both the sidereal month and synodic month vary somewhat from one orbit to another, the latter by as much as half a day. The reason is that the Sun's gravity sometimes causes the Moon to speed up or slow down slightly in its orbit, depending on the relative positions of the Sun, Moon, and Earth. Furthermore, the Moon's orbit changes slightly from one month to the next.

3-3 Eclipses occur only when the Sun and Moon are both on the line of nodes

From time to time the Sun, Earth, and Moon all happen to lie along a straight line. When this occurs, the shadow of the Earth can fall on the Moon or the shadow of the Moon can fall on the Earth. Such phenomena are called **eclipses**. They are perhaps the most dramatic astronomical events that can be seen with the naked eye.

A **lunar eclipse** occurs when the Moon passes through the Earth's shadow. This occurs when the Sun, Earth, and Moon are in a straight line, with the Earth between the Sun and Moon so that the Moon is at full phase (position E in Figure 3-2). At this point in the Moon's orbit, the face of the Moon seen from Earth would normally be fully illuminated by the Sun. Instead, it appears quite dim because the Earth casts a shadow on the Moon.

A **solar eclipse** occurs when the Earth passes through the Moon's shadow. As seen from Earth, the Moon moves in front of the Sun. Once again, this can happen only when the Sun, Moon, and Earth are in a straight line. However, for a solar eclipse to occur, the Moon must be between the Earth and the Sun. Therefore, a solar eclipse can occur only at new moon (position A in Figure 3-2).

CAUTION! Both new moon and full moon occur at intervals of 29½ days. Hence, you might expect that there would be a solar eclipse every 29½ days, followed by a lunar eclipse about two weeks (half a lunar orbit) later. But in fact, there are only a few solar eclipses and lunar eclipses per year. Solar and lunar eclipses are so infrequent because the plane of the Moon's orbit and the plane of the Earth's orbit are not exactly aligned, as Figure 3-6 shows. The angle between the plane of the Earth's orbit and the plane of the Moon's orbit is about 5°. Because of this tilt, new moon and full moon usually occur when the Moon is either above or below the plane of the Earth's orbit. When the Moon is not in the plane of the Earth's orbit, the Sun, Moon, and Earth cannot align perfectly, and an eclipse cannot occur.

The tilt of the Moon's orbit makes lunar and solar eclipses rare events

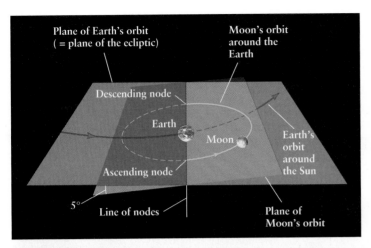

Figure 3-6

The Inclination of the Moon's Orbit This drawing shows the Moon's orbit around the Earth (in yellow) and part of the Earth's orbit around the Sun (in red). The plane of the Moon's orbit (shown in brown) is tilted by about 5° with respect to the plane of the Earth's orbit, also called the plane of the ecliptic (shown in blue). These two planes intersect along a line called the line of nodes.

In order for the Sun, Earth, and Moon to be lined up for an eclipse, the Moon must lie in the same plane as the Earth's orbit around the Sun. As we saw in Section 2-5, this plane is called the *ecliptic plane* because it is the same as the plane of the Sun's apparent path around the sky, or ecliptic (see Figure 2-14). Thus, when an eclipse occurs, the Moon appears from Earth to be on the ecliptic—which is how the ecliptic gets its name.

The planes of the Earth's orbit and the Moon's orbit intersect along a line called the **line of nodes**, shown in Figure 3-6. The line of nodes passes through the Earth and is pointed in a particular direction in space. Eclipses can occur only if the line of nodes is pointed toward the Sun—that is, if the Sun lies on or near the line of nodes—and if, at the same time, the Moon lies on or very near the line of nodes. Only then do the Sun, Earth, and Moon lie in a line straight enough for an eclipse to occur (Figure 3-7).

Anyone who wants to predict eclipses must know the orientation of the line of nodes. But the line of nodes is gradually shifting because of the gravitational pull of the Sun on the Moon. As a result, the line of nodes rotates slowly westward. Astronomers calculate such details to fix the dates and times of upcoming eclipses.

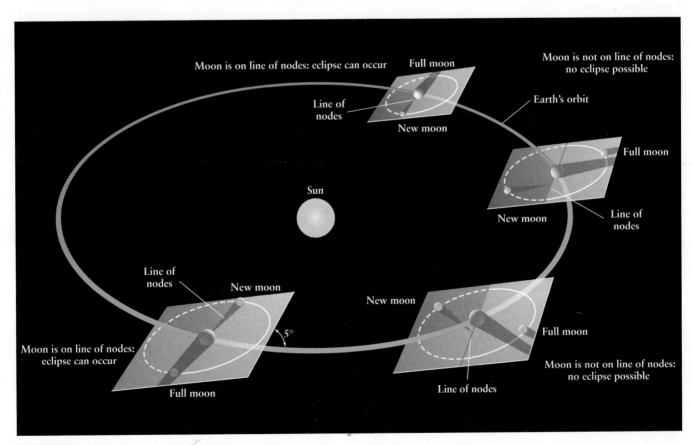

Figure 3-7

Conditions for Eclipses Eclipses can take place only if the Sun and Moon are both very near to or on the line of nodes. Only then can the Sun, Earth, and Moon all lie along a straight line. A solar eclipse occurs only if the Moon is very near the line of nodes at new moon; a lunar eclipse occurs only if the Moon is very near the line of nodes at full moon. If the Sun and Moon are not near the line of nodes, the Moon's shadow cannot fall on the Earth and the Earth's shadow cannot fall on the Moon.

Figure 3-8

Three Types of Lunar Eclipse People on the nighttime side of the Earth see a lunar eclipse when the Moon moves through the Earth's shadow. In the umbra, the darkest part of the shadow, the Sun is completely covered by the Earth. The penumbra is less dark because only part of the Sun is covered by the Earth. The three paths show the motion of the Moon if the lunar eclipse is penumbral (Path 1, in yellow), total (Path 2, in red), or partial (Path 3, in blue). The inset shows these same paths, along with the umbra and penumbra, as viewed from the Earth.

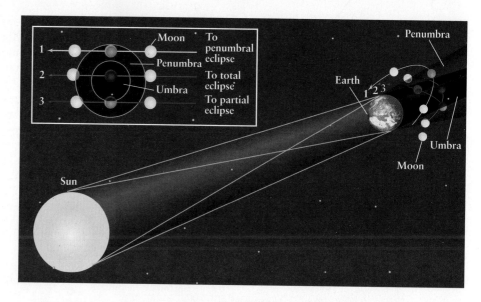

There are at least two—but never more than five—solar eclipses each year. The last year in which five solar eclipses occurred was 1935. The least number of eclipses possible (two solar, zero lunar) happened in 1969. Lunar eclipses occur just about as frequently as solar eclipses, but the maximum possible number of eclipses (lunar and solar combined) in a single year is seven.

3-4 The character of a lunar eclipse depends on the alignment of the Sun, Earth, and Moon

The character of a lunar eclipse depends on exactly how the Moon travels through the Earth's shadow. As **Figure 3-8** shows, the shadow of the Earth has two distinct parts. In the **umbra**, the darkest part of the shadow, no portion of the Sun's surface can be seen. A portion of the Sun's surface is visible in the **penumbra**, which therefore is not quite as dark. Most people notice a lunar eclipse only if the Moon passes into the Earth's umbra. As this umbral phase of the eclipse begins, a bite seems to be taken out of the Moon.

> A lunar eclipse is most impressive when total, but can also be partial or penumbral

The inset in Figure 3-8 shows the different ways in which the Moon can pass into the Earth's shadow. When the Moon passes through only the Earth's penumbra (Path 1), we see a **penumbral eclipse**. During a penumbral eclipse, the Earth blocks only part of the Sun's light and so none of the lunar surface is completely shaded. Because the Moon still looks full but only a little dimmer than usual, penumbral eclipses are easy to miss. If the Moon travels completely into the umbra (Path 2), a **total lunar eclipse** occurs. If only part of the Moon passes through the umbra (Path 3), we see a **partial lunar eclipse**.

If you were on the Moon during a total lunar eclipse, the Sun would be hidden behind the Earth. But some sunlight would be visible through the thin ring of atmosphere around the Earth, just as you can see sunlight through a person's hair if they stand with their head between your eyes and the Sun. As a result, a small amount of light reaches the Moon during a total lunar eclipse, and so the Moon does not completely disappear from the sky as seen from Earth. Most of the sunlight that passes through the Earth's atmosphere is red, and thus the eclipsed Moon glows faintly in reddish hues, as **Figure 3-9** shows. (We'll see in Chapter 5 how our atmosphere causes this reddish color.)

Lunar eclipses occur at full moon, when the Moon is directly opposite the Sun in the sky. Hence, a lunar eclipse can be seen at any place on Earth where the Sun is below the horizon (that is, where it is nighttime). A lunar eclipse has the maximum possible duration if the Moon travels directly through the center of the umbra. The Moon's speed through the Earth's shadow is roughly 1 kilometer per second (3600 kilometers per hour, or 2280 miles per hour), which means that **totality**—the period when the Moon is completely within the Earth's umbra—can last for as long as 1 hour and 42 minutes.

On average, two or three lunar eclipses occur in a year. **Table 3-1** lists all 12 lunar eclipses from 2007 to 2011. Of all lunar eclipses, roughly one-third are total, one-third are partial, and one-third are penumbral.

3-5 Solar eclipses also depend on the alignment of the Sun, Earth, and Moon

As seen from Earth, the angular diameter of the Moon is almost exactly the same as the angular diameter of the far larger but more distant Sun—about 0.5°. Thanks to this coincidence of nature, the Moon just "fits" over the Sun during a **total solar eclipse.**

Total Solar Eclipses

A total solar eclipse is a dramatic event. The sky begins to darken, the air temperature falls, and winds increase as the Moon gradually covers more and more of the Sun's disk. All nature responds: Birds go to roost,

> The Sun's tenuous outer atmosphere is revealed during a total solar eclipse

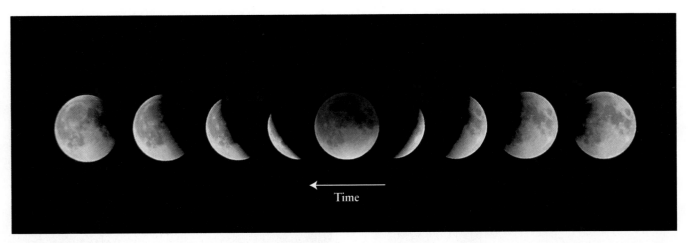

Time

Figure 3-9   R I V UX G

A Total Lunar Eclipse This sequence of nine photographs was taken over a 3-hour period during the lunar eclipse of January 20, 2000. The sequence, which runs from right to left, shows the Moon moving through the Earth's umbra. During the total phase of the eclipse (shown in the center), the Moon has a distinct reddish color. (Fred Espenak, NASA/Goddard Space Flight Center; ©2000 Fred Espenak, MrEclipse.com)

flowers close their petals, and crickets begin to chirp as if evening had arrived. As the last few rays of sunlight peek out from behind the edge of the Moon and the eclipse becomes total, the landscape around you is bathed in an eerie gray or, less frequently, in shimmering bands of light and dark. Finally, for a few minutes the Moon completely blocks out the dazzling solar disk and not much else (**Figure 3-10a**). The **solar corona**—the Sun's thin, hot outer atmosphere, which is normally too dim to be seen—blazes forth in the darkened daytime sky (Figure 3-10b). It is an awe-inspiring sight.

CAUTION! If you are fortunate enough to see a solar eclipse, keep in mind that the only time when it is safe to look at the Sun is during **totality**, when the solar disk is blocked by the Moon and only the solar corona is visible. Viewing this magnificent spectacle cannot harm you in any way. But you must *never* look directly at the Sun when even a portion of its intensely brilliant disk is exposed. *If you look directly at the Sun at any time without a special filter approved for solar viewing, you will suffer permanent eye damage or blindness.*

To see the remarkable spectacle of a total solar eclipse, you must be inside the darkest part of the Moon's shadow, also called the umbra, where the Moon completely blocks the Sun. Because the Sun and the Moon have nearly the same angular diameter as seen from Earth, only the tip of the Moon's umbra reaches the

Table 3-1 Lunar Eclipses, 2007–2011

Date	Type	Where visible	Duration of totality (h = hours, m = minutes)
2007 March 3	Total	Americas, Europe, Africa, Asia	1h 14m
2007 August 28	Total	Eastern Asia, Australia, Pacific, Americas	1h 31m
2008 February 21	Total	Central Pacific, Americas, Europe, Africa	51m
2008 August 16	Partial	South America, Europe, Africa, Asia, Australia	—
2009 February 9	Penumbral	Europe, Asia, Australia, Pacific, North America	—
2009 July 7	Penumbral	Australia, Pacific, Americas	—
2009 August 6	Penumbral	Americas, Europe, Africa, Asia	—
2009 December 31	Partial	Europe, Africa, Asia, Australia	—
2010 June 26	Partial	Asia, Australia, Pacific, Americas	—
2010 December 21	Total	Asia, Australia, Pacific, Americas, Europe	1h 13m
2011 June 15	Total	South America, Europe, Africa, Asia, Australia	1h 41m
2011 December 10	Total	Europe, Africa, Asia, Australia, Pacific, North America	52m

Eclipse predictions by Fred Espenak, NASA/Goddard Space Flight Center. All dates are given in standard astronomical format: year, month, day.

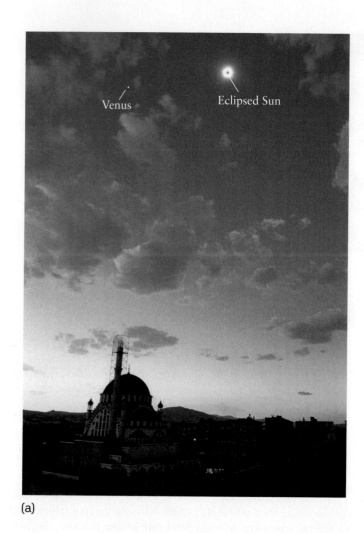

Venus   Eclipsed Sun

(a)

Corona

(b)

Figure 3-10   R I V U X G

A Total Solar Eclipse (a) This photograph shows the total solar eclipse of August 11, 1999, as seen from Elâziğ, Turkey. The sky is so dark that the planet Venus can be seen to the left of the eclipsed Sun. **(b)** When the Moon completely covers the Sun's disk during a total eclipse, the faint solar corona is revealed. (Fred Espenak, MrEclipse.com)

Earth's surface (Figure 3-11). As the Earth rotates, the tip of the umbra traces an **eclipse path** across the Earth's surface. Only those locations within the eclipse path are treated to the spectacle of a total solar eclipse. The inset in Figure 3-11 shows the dark spot on the Earth's surface produced by the Moon's umbra.

Partial Solar Eclipses

Immediately surrounding the Moon's umbra is the region of partial shadow called the penumbra. As seen from this area, the Sun's surface appears only partially covered by the Moon. During a solar eclipse, the Moon's penumbra covers a large portion of the Earth's surface, and anyone standing inside the penumbra sees a **partial solar eclipse.** Such eclipses are much less interesting events than total solar eclipses, which is why astronomy enthusiasts strive to be inside the eclipse path. If you are within the eclipse path, you will see a partial eclipse before and after the brief period of totality (see the photograph that opens this chapter).

The width of the eclipse path depends primarily on the Earth-Moon distance during totality. The eclipse path is widest if the Moon happens to be at **perigee,** the point in its orbit nearest the Earth. In this case the width of the eclipse path can be as great as 270 kilometers (170 miles). In most eclipses, however, the path is much narrower.

Annular Solar Eclipses

In some eclipses the Moon's umbra does not reach all the way to the Earth's surface. This can happen if the Moon is at or near **apogee,** its farthest position from Earth. In this case, the Moon appears too small to cover the Sun completely. The result is a third type of solar eclipse, called an **annular eclipse.** During an annular eclipse, a thin ring of the Sun is seen around the edge of the Moon (Figure 3-12). The length of the Moon's umbra is nearly 5000 kilometers (3100 miles) less than the average distance between the Moon and the Earth's surface. Thus, the Moon's shadow often fails to reach the Earth even when the Sun, Moon, and Earth are properly aligned for an eclipse. Hence, annular eclipses are slightly more common—as well as far less dramatic—than total eclipses.

Even during a total eclipse, most people along the eclipse path observe totality for only a few moments. The Earth's rotation, coupled with the orbital motion of the Moon, causes the umbra to race eastward along the eclipse path at speeds in excess of 1700 kilometers per hour (1060 miles per hour). Because of the umbra's high speed, totality never lasts for more than $7\frac{1}{2}$ minutes. In a typical total solar eclipse, the Sun-Moon-Earth alignment and the Earth-Moon distance are such that totality lasts much less than this maximum.

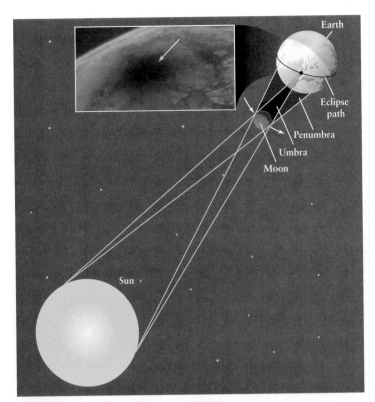

Figure 3-11   R I **V** U X G

The Geometry of a Total Solar Eclipse During a total solar eclipse, the tip of the Moon's umbra reaches the Earth's surface. As the Earth and Moon move along their orbits, this tip traces an eclipse path across the Earth's surface. People within the eclipse path see a total solar eclipse as the tip moves over them. Anyone within the penumbra sees only a partial eclipse. The inset photograph was taken from the *Mir* space station during the August 11, 1999, total solar eclipse (the same eclipse shown in Figure 3-10). The tip of the umbra appears as a black spot on the Earth's surface. At the time the photograph was taken, this spot was 105 km (65 mi) wide and was crossing the English Channel at 3000 km/h (1900 mi/h). (Photograph by Jean-Pierre Haigneré, Centre National d'Etudes Spatiales, France/GSFS)

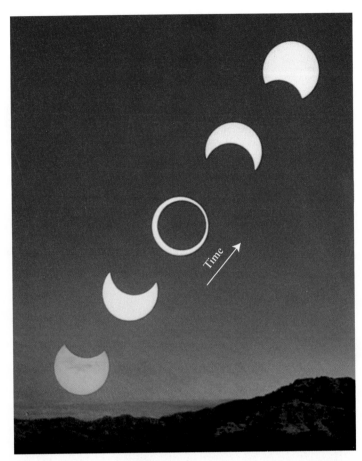

Figure 3-12   R I **V** U X G

An Annular Solar Eclipse This composite of six photographs taken at sunrise in Costa Rica shows the progress of an annular eclipse of the Sun on December 24, 1973. (Five photographs were made of the Sun, plus one of the hills and sky.) Note that at mideclipse the limb, or outer edge, of the Sun is visible around the Moon. (Courtesy of Dennis di Cicco)

The details of solar eclipses are calculated well in advance. They are published in such reference books as the *Astronomical Almanac* and are available on the World Wide Web. **Figure 3-13** shows the eclipse paths for all total solar eclipses from 1997 to 2020. **Table 3-2** lists all the total, annular, and partial eclipses from 2007 to 2011, including the maximum duration of totality for total eclipses.

Ancient astronomers achieved a limited ability to predict eclipses. In those times, religious and political leaders who were able to predict such awe-inspiring events as eclipses must have made a tremendous impression on their followers. One of three priceless manuscripts to survive the devastating Spanish Conquest shows that the Mayan astronomers of Mexico and Guatemala had a fairly reliable method for predicting eclipses. The great Greek astronomer Thales of Miletus is said to have predicted the famous eclipse of 585 B.C., which occurred during the middle of

a war. The sight was so unnerving that the soldiers put down their arms and declared peace.

In retrospect, it seems that what ancient astronomers actually produced were eclipse "warnings" of various degrees of reliability rather than true predictions. Working with historical records, these astronomers generally sought to discover cycles and regularities from which future eclipses could be anticipated. **Box 3-2** describes how you might produce eclipse warnings yourself.

3-6 Ancient astronomers measured the size of the Earth and attempted to determine distances to the Sun and Moon

The prediction of eclipses was not the only problem attacked by ancient astronomers. More than 2000 years ago, centuries before sailors of Columbus's era crossed the oceans, Greek astronomers were fully aware that the Earth is not flat. They had come to this conclusion using a combination of observation and logical

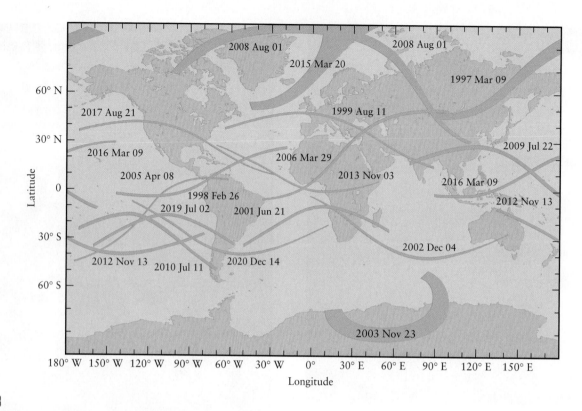

Figure 3-13

Eclipse Paths for Total Eclipses, 1997–2020 This map shows the eclipse paths for all 18 total solar eclipses occurring from 1997 through 2020. In each eclipse, the Moon's shadow travels along the eclipse path in a generally eastward direction across the Earth's surface. (Courtesy of Fred Espenak, NASA/Goddard Space Flight Center)

deduction, much like modern scientists. The Greeks noted that during lunar eclipses, when the Moon passes through the Earth's shadow, the edge of the shadow is always circular. Because a sphere is the only shape that always casts a circular shadow from any angle, they concluded that the Earth is spherical.

> The ideas of geometry made it possible for Greek scholars to estimate cosmic distances

Eratosthenes and the Size of the Earth

Around 200 B.C., the Greek astronomer Eratosthenes devised a way to measure the circumference of the spherical Earth. It was known that on the date of the summer solstice (the first day of summer; see Section 2-5) in the town of Syene in Egypt, near present-day Aswan, the Sun shone directly down the vertical shafts of water wells. Hence, at local noon on that day, the Sun was at

Table 3-2 Solar Eclipses, 2007–2011

Date	Type	Where visible	Notes
2007 March 19	Partial	Asia, Alaska	87% eclipsed
2007 September 11	Partial	South America, Antarctica	75% eclipsed
2008 February 7	Annular	Antarctica, eastern Australia, New Zealand	—
2008 August 1	Total	Northeast North America, Europe, Asia	Maximum duration of totality 2m 27s
2009 January 26	Annular	Southern Africa, Antarctica, southeast Asia, Australia	—
2009 July 22	Total	Eastern Asia, Pacific Ocean, Hawaii	Maximum duration of totality 6m 39s
2010 January 15	Annular	Africa, Asia	—
2010 July 11	Total	Pacific Ocean, South America	Maximum duration of totality 5m 20s
2011 January 4	Partial	Europe, Africa, central Asia	86% eclipsed
2011 June 1	Partial	Eastern Asia, northern North America, Iceland	60% eclipsed
2011 July 1	Partial	Indian Ocean	10% eclipsed
2011 November 25	Partial	Southern Africa, Antarctica, Australia, New Zealand	91% eclipsed

Eclipse predictions by Fred Espenak, NASA/Goddard Space Flight Center. All dates are given in standard astronomical format: year, month, day.

BOX 3-2

Predicting Solar Eclipses

Suppose that you observe a solar eclipse in your home-town and want to figure out when you and your neighbors might see another eclipse. How would you begin?

First, remember that a solar eclipse can occur only if the line of nodes points toward the Sun at the same time that there is a new moon (see Figure 3-7). Second, you must know that it takes 29.53 days (one synodic month) to go from one new moon to the next. Because solar eclipses occur only during new moon, you must wait several whole lunar months for the proper alignment to occur again.

However, there is a complication: The line of nodes gradually shifts its position with respect to the background stars. It takes 346.6 days to move from one alignment of the line of nodes pointing toward the Sun to the next identical alignment. This period is called the **eclipse year.**

Therefore, to predict when you will see another solar eclipse, you need to know how many whole lunar months equal some whole number of eclipse years. This will tell you how long you will have to wait for the next virtually identical alignment of the Sun, the Moon, and the line of nodes. By trial and error, you find that 223 lunar months is the same length of time as 19 eclipse years, because

$$223 \times 29.53 \text{ days} = 19 \times 346.6 \text{ days} = 6585 \text{ days}$$

This calculation is accurate to within a few hours. A more accurate calculation gives an interval, called the **saros,** that is about one-third day longer, or 6585.3 days (18 years, 11.3 days). Eclipses separated by the saros interval are said to form an *eclipse series.*

You might think that you and your neighbors would simply have to wait one full saros interval to go from one solar eclipse to the next. However, because of the extra one-third

day, the Earth will have rotated by an extra 120° (one-third of a complete rotation) when the next solar eclipse of a particular series occurs. The eclipse path will thus be one-third of the way around the world from you. Therefore, you must wait three full saros intervals (54 years, 34 days) before the eclipse path comes back around to your part of the Earth. The illustration shows a series of six solar eclipse paths, each separated from the next by one saros interval.

There is evidence that ancient Babylonian astronomers knew about the saros interval. However, the discovery of the saros is more likely to have come from lunar eclipses than solar eclipses. If you are far from the eclipse path, there is a good chance that you could fail to notice a solar eclipse. Even if half the Sun is covered by the Moon, the remaining solar surface provides enough sunlight for the outdoor illumination not to be greatly diminished. By contrast, anyone on the nighttime side of the Earth can see an eclipse of the Moon unless clouds block the view.

the zenith (see Section 2-4) as seen from Syene. Eratosthenes knew that the Sun never appeared at the zenith at his home in the Egyptian city of Alexandria, which is on the Mediterranean Sea almost due north of Syene. Rather, on the summer solstice in Alexandria, the position of the Sun at local noon was about 7° south of the zenith (Figure 3-14). This angle is about one-fiftieth of a complete circle, so he concluded that the distance from Alexandria to Syene must be about one-fiftieth of the Earth's circumference.

In Eratosthenes's day, the distance from Alexandria to Syene was said to be 5000 stades. Therefore, Eratosthenes found the Earth's circumference to be

$$50 \times 5000 \text{ stades} = 250,000 \text{ stades}$$

Unfortunately, no one today is sure of the exact length of the Greek unit called the stade. One guess is that the stade was about one-sixth of a kilometer, which would mean that Eratosthenes obtained a circumference for the Earth of about 42,000 kilome-

ters. This is remarkably close to the modern value of 40,000 kilometers.

Aristarchus and Distances in the Solar System

Eratosthenes was only one of several brilliant astronomers to emerge from the so-called Alexandrian school, which by his time had a distinguished tradition. One of the first Alexandrian astronomers, Aristarchus of Samos, had proposed a method of determining the relative distances to the Sun and Moon, perhaps as long ago as 280 B.C.

Aristarchus knew that the Sun, Moon, and Earth form a right triangle at the moment of first or third quarter moon, with the right angle at the location of the Moon (Figure 3-15). He estimated that, as seen from Earth, the angle between the Moon and the Sun at first and third quarters is 87°, or 3° less than a right angle. Using the rules of geometry, Aristarchus concluded that the Sun is about 20 times farther from us than is the Moon. We now know that Aristarchus erred in measuring angles and that the

Figure 3-14

Eratosthenes's Method of Determining the Diameter of the Earth
Around 200 B.C., Eratosthenes used observations of the Sun's position at noon on the summer solstice to show that Alexandria and Syene were about 7° apart on the surface of the Earth. This angle is about one-fiftieth of a circle, so the distance between Alexandria and Syene must be about one-fiftieth of the Earth's circumference.

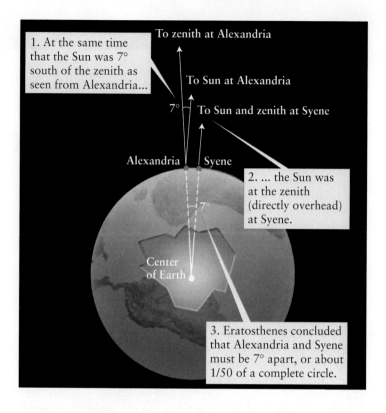

1. At the same time that the Sun was 7° south of the zenith as seen from Alexandria...

To zenith at Alexandria

To Sun at Alexandria

To Sun and zenith at Syene

2. ... the Sun was at the zenith (directly overhead) at Syene.

3. Eratosthenes concluded that Alexandria and Syene must be 7° apart, or about 1/50 of a complete circle.

average distance to the Sun is about 390 times larger than the average distance to the Moon. It is nevertheless impressive that people were trying to measure distances across the solar system more than 2000 years ago.

Aristarchus also made an equally bold attempt to determine the relative sizes of the Earth, Moon, and Sun. From his observations of how long the Moon takes to move through the Earth's shadow during a lunar eclipse, Aristarchus estimated the diame-

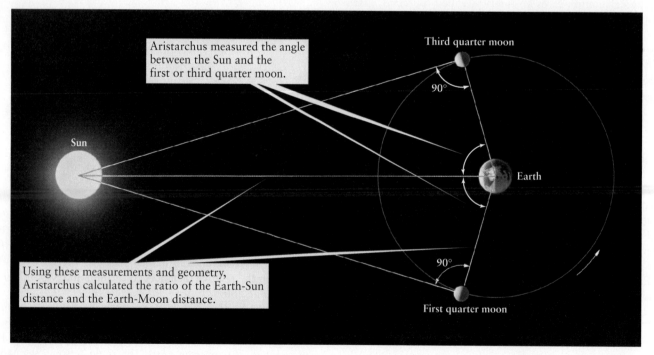

Aristarchus measured the angle between the Sun and the first or third quarter moon.

Using these measurements and geometry, Aristarchus calculated the ratio of the Earth-Sun distance and the Earth-Moon distance.

Third quarter moon

90°

Sun

Earth

90°

First quarter moon

Figure 3-15

Aristarchus's Method of Determining Distances to the Sun and Moon Aristarchus knew that the Sun, Moon, and Earth form a right triangle at first and third quarter phases. Using geometrical arguments, he calculated the relative lengths of the sides of these triangles, thereby obtaining the distances to the Sun and Moon.

Table 3-3 Comparison of Ancient and Modern Astronomical Measurements

	Ancient (km)	Modern (km)
Earth's diameter	13,000	12,756
Moon's diameter	4,300	3,476
Sun's diameter	9×10^4	1.39×10^6
Earth-Moon distance	4×10^5	3.84×10^5
Earth-Sun distance	10^7	1.50×10^8

ter of the Earth to be about 3 times larger than the diameter of the Moon. To determine the diameter of the Sun, Aristarchus simply pointed out that the Sun and the Moon have the same angular size in the sky. Therefore, their diameters must be in the same proportion as their distances (see part *a* of the figure in Box 1-1). In other words, because Aristarchus thought the Sun to be 20 times farther from the Earth than the Moon, he concluded that the Sun must be 20 times larger than the Moon. Once Eratosthenes had measured the Earth's circumference, astronomers of the Alexandrian school could estimate the diameters of the Sun and Moon as well as their distances from Earth.

Table 3-3 summarizes some ancient and modern measurements of the sizes of Earth, the Moon, and the Sun and the distances between them. Some of these ancient measurements are far from the modern values. Yet the achievements of our ancestors still stand as impressive applications of observation and reasoning and important steps toward the development of the scientific method.

Key Words

Terms preceded by an asterisk (*) are discussed in the Boxes.

annular eclipse, p. 54
apogee, p. 54
eclipse, p. 50
eclipse path, p. 54
*eclipse year, p. 57
first quarter moon, p. 46
full moon, p. 46
line of nodes, p. 51
lunar eclipse, p. 50
lunar phases, p. 46
new moon, p. 46
partial lunar eclipse, p. 52
partial solar eclipse, p. 54
penumbra (*plural* penumbrae), p. 52
penumbral eclipse, p. 52
perigee, p. 54

*saros, p. 57
sidereal month, p. 50
solar corona, p. 53
solar eclipse, p. 50
synchronous rotation, p. 49
synodic month, p. 50
third quarter moon, p. 46
totality (lunar eclipse), p. 52
totality (solar eclipse), p. 53
total lunar eclipse, p. 52
total solar eclipse, p. 52
umbra (*plural* umbrae), p. 52
waning crescent moon, p. 46
waning gibbous moon, p. 46
waxing crescent moon, p. 46
waxing gibbous moon, p. 46

Key Ideas

Lunar Phases: The phases of the Moon occur because light from the Moon is actually reflected sunlight. As the relative positions of the Earth, the Moon, and the Sun change, we see more or less of the illuminated half of the Moon.

Length of the Month: Two types of months are used in describing the motion of the Moon.

• With respect to the stars, the Moon completes one orbit around the Earth in a sidereal month, averaging 27.32 days.

• The Moon completes one cycle of phases (one orbit around the Earth with respect to the Sun) in a synodic month, averaging 29.53 days.

The Moon's Orbit: The plane of the Moon's orbit is tilted by about 5° from the plane of the Earth's orbit, or ecliptic.

• The line of nodes is the line where the planes of the Moon's orbit and the Earth's orbit intersect. The gravitational pull of the Sun gradually shifts the orientation of the line of nodes with respect to the stars.

Conditions for Eclipses: During a lunar eclipse, the Moon passes through the Earth's shadow. During a solar eclipse, the Earth passes through the Moon's shadow.

• Lunar eclipses occur at full moon, while solar eclipses occur at new moon.

• Either type of eclipse can occur only when the Sun and Moon are both on or very near the line of nodes. If this condition is not met, the Earth's shadow cannot fall on the Moon and the Moon's shadow cannot fall on the Earth.

Umbra and Penumbra: The shadow of an object has two parts: the umbra, within which the light source is completely blocked, and the penumbra, where the light source is only partially blocked.

Lunar Eclipses: Depending on the relative positions of the Sun, Moon, and Earth, lunar eclipses may be total (the Moon passes completely into the Earth's umbra), partial (only part of the Moon passes into the Earth's umbra), or penumbral (the Moon passes only into the Earth's penumbra).

Solar Eclipses: Solar eclipses may be total, partial, or annular.

• During a total solar eclipse, the Moon's umbra traces out an eclipse path over the Earth's surface as the Earth rotates. Observers outside the eclipse path but within the penumbra see only a partial solar eclipse.

• During an annular eclipse, the umbra falls short of the Earth, and the outer edge of the Sun's disk is visible around the Moon at mideclipse.

The Moon and Ancient Astronomers: Ancient astronomers such as Aristarchus and Eratosthenes made great progress in determining the sizes and relative distances of the Earth, the Moon, and the Sun.

Questions

Review Questions

Questions preceded by an asterisk (*) involve topics discussed in the Boxes.

1. Explain the difference between sunlight and moonlight.
2. (a) Explain why the Moon exhibits phases. (b) A common misconception about the Moon's phases is that they are caused by the Earth's shadow. Use Figure 3-2 to explain why this not correct.

3. How would the sequence and timing of lunar phases be affected if the Moon moved around its orbit (**a**) in the same direction, but at twice the speed; (**b**) at the same speed, but in the opposite direction? Explain your answers.

4. At approximately what time does the Moon rise when it is (**a**) a new moon; (**b**) a first quarter moon; (**c**) a full moon; (**d**) a third quarter moon?

5. Astronomers sometimes refer to lunar phases in terms of the *age* of the Moon. This is the time that has elapsed since new moon phase. Thus, the age of a full moon is half of a $29\frac{1}{2}$-day synodic period, or approximately 15 days. Find the approximate age of (**a**) a waxing crescent moon; (**b**) a third quarter moon; (**c**) a waning gibbous moon.

6. If you lived on the Moon, would you see the Earth go through phases? If so, would the sequence of phases be the same as those of the Moon as seen from Earth, or would the sequence be reversed? Explain using Figure 3-2.

7. Is the far side of the Moon (the side that can never be seen from Earth) the same as the dark side of the Moon? Explain.

8. (**a**) If you lived on the Moon, would you see the Sun rise and set, or would it always be in the same place in the sky? Explain. (**b**) Would you see the Earth rise and set, or would it always be in the same place in the sky? Explain using Figure 3-4.

9. What is the difference between a sidereal month and a synodic month? Which is longer? Why?

10. On a certain date the Moon is in the direction of the constellation Gemini as seen from Earth. When will the Moon next be in the direction of Gemini: one sidereal month later, or one synodic month later? Explain.

11. What is the difference between the umbra and the penumbra of a shadow?

12. Why doesn't a lunar eclipse occur at every full moon and a solar eclipse at every new moon?

13. What is the line of nodes? Why is it important to the subject of eclipses?

14. What is a penumbral eclipse of the Moon? Why do you suppose that it is easy to overlook such an eclipse?

15. Why is the duration of totality different for different total lunar eclipses, as shown in Table 3-1?

16. Can one ever observe an annular eclipse of the Moon? Why or why not?

17. If you were looking at the Earth from the side of the Moon that faces the Earth, what would you see during (**a**) a total lunar eclipse? (**b**) a total solar eclipse? Explain your answers.

18. If there is a total eclipse of the Sun in April, can there be a lunar eclipse three months later in July? Why or why not?

19. Which type of eclipse—lunar or solar—do you think most people on Earth have seen? Why?

20. How is an annular eclipse of the Sun different from a total eclipse of the Sun? What causes this difference?

*21. What is the saros? How did ancient astronomers use it to predict eclipses?

22. How did Eratosthenes measure the size of the Earth?

23. How did Aristarchus try to estimate the distance from the Earth to the Sun and Moon?

24. How did Aristarchus try to estimate the diameters of the Sun and Moon?

Advanced Questions

> **Problem-solving tips and tools**
>
> To estimate the average angular speed of the Moon along its orbit (that is, how many degrees around its orbit the Moon travels per day), divide 360° by the length of a sidereal month. It is helpful to know that the saros interval of 6585.3 days equals 18 years and $11\frac{1}{3}$ days if the interval includes 4 leap years, but is 18 years and $10\frac{1}{3}$ days if it includes 5 leap years.

25. The dividing line between the illuminated and unilluminated halves of the Moon is called the *terminator*. The terminator appears curved when there is a crescent or gibbous moon, but appears straight when there is a first quarter or third quarter moon (see Figure 3-2). Describe how you could use these facts to explain to a friend why lunar phases cannot be caused by the Earth's shadow falling on the Moon.

26. What is the phase of the Moon if it rises at (**a**) midnight; (**b**) sunrise; (**c**) halfway between sunset and midnight; (**d**) halfway between noon and sunset? Explain your answers.

27. The Moon is highest in the sky when it crosses the meridian (see Figure 2-21), halfway between the time of moonrise and the time of moonset. At approximately what time does the Moon cross the meridian if it is (**a**) a new moon; (**b**) a first quarter moon; (**c**) a full moon; (**d**) a third quarter moon? Explain your answers.

28. The Moon is highest in the sky when it crosses the meridian (see Figure 2-21), halfway between the time of moonrise and the time of moonset. What is the phase of the Moon if it is highest in the sky at (**a**) midnight; (**b**) sunrise; (**c**) noon; (**d**) sunset? Explain your answers.

29. Suppose it is the first day of autumn in the northern hemisphere. What is the phase of the Moon if the Moon is located at (**a**) the vernal equinox? (**b**) the summer solstice? (**c**) the autumnal equinox? (**d**) the winter solstice? Explain your answers. (*Hint:* Make a drawing showing the relative positions of the Sun, Earth, and Moon. Compare with Figure 3-2.)

R I [V] U X G

(NASA/JSC)

30. The above photograph of the Earth was taken by the crew of the *Apollo 8* spacecraft as they orbited the Moon. A portion

of the lunar surface is visible at the right-hand side of the photo. In this photo, the Earth is oriented with its north pole approximately at the top. When this photo was taken, was the Moon waxing or waning as seen from Earth? Explain your answer with a diagram.

31. **(a)** The Moon moves noticeably on the celestial sphere over the space of a single night. To show this, calculate how long it takes the Moon to move through an angle equal to its own angular diameter ($\frac{1}{2}°$) against the background of stars. Give your answer in hours. **(b)** Through what angle (in degrees) does the Moon move during a 12-hour night? Can you notice an angle of this size? (*Hint:* See Figure 1-10.)

32. During an occultation, or "covering up," of Jupiter by the Moon, an astronomer notices that it takes the Moon's edge 90 seconds to cover Jupiter's disk completely. If the Moon's motion is assumed to be uniform and the occultation was "central" (that is, center over center), find the angular diameter of Jupiter. (*Hint:* Assume that Jupiter does not appear to move against the background of stars during this brief 90-second interval. You will need to convert the Moon's angular speed from degrees per day to arcseconds per second.)

33. The plane of the Moon's orbit is inclined at a 5° angle from the ecliptic, and the ecliptic is inclined at a $23\frac{1}{2}°$ angle from the celestial equator. Could the Moon ever appear at your zenith if you lived at **(a)** the equator; **(b)** the south pole? Explain your answers.

34. How many more sidereal months than synodic months are there in a year? Explain.

35. Suppose the Earth moved a little faster around the Sun, so that it took a bit less than one year to make a complete orbit. If the speed of the Moon's orbit around the Earth were unchanged, would the length of the sidereal month be the same, longer, or shorter than it is now? What about the synodic month? Explain your answers.

36. If the Moon revolved about the Earth in the same orbit but in the opposite direction, would the synodic month be longer or shorter than the sidereal month? Explain your reasoning.

37. One definition of a "blue moon" is the second full moon within the same calendar month. There is usually only one full moon within a calendar month, so the phrase "once in a blue moon" means "hardly ever." Why are blue moons so rare? Are there any months of the year in which it would be impossible to have two full moons? Explain your answer.

38. You are watching a lunar eclipse from some place on the Earth's night side. Will you see the Moon enter the Earth's shadow from the east or from the west? Explain your reasoning.

39. The total lunar eclipse of October 28, 2004, was visible from South America. The duration of totality was 1 hour, 21 minutes. Was this total eclipse also visible from Australia, on the opposite side of the Earth? Explain your reasoning.

40. During a total solar eclipse, the Moon's umbra moves in a generally eastward direction across the Earth's surface. Use a drawing like Figure 3-11 to explain why the motion is eastward, not westward.

41. A total solar eclipse was visible from Africa on March 29, 2006 (see Figure 3-10). Draw what the eclipse would have looked like as seen from France, to the north of the path of totality. Explain the reasoning behind your drawing.

42. Figures 3-11 and 3-13 show that the path of a total eclipse is quite narrow. Use this to explain why a glow is visible all around the horizon when you are viewing a solar eclipse during totality (see Figure 3-10*a*).

43. **(a)** Suppose the diameter of the Moon were doubled, but the orbit of the Moon remained the same. Would total solar eclipses be more common, less common, or just as common as they are now? Explain. **(b)** Suppose the diameter of the Moon were halved, but the orbit of the Moon remained the same. Explain why there would be *no* total solar eclipses.

44. Just as the distance from the Earth to the Moon varies somewhat as the Moon orbits the Earth, the distance from the Sun to the Earth changes as the Earth orbits the Sun. The Earth is closest to the Sun at its *perihelion*; it is farthest from the Sun at its *aphelion*. In order for a total solar eclipse to have the maximum duration of totality, should the Earth be at perihelion or aphelion? Assume that the Earth-Moon distance is the same in both situations. As part of your explanation, draw two pictures like Figure 3-11, one with the Earth relatively close to the Sun and one with the Earth relatively far from the Sun.

*45. On March 29, 2006, residents of northern Africa were treated to a total solar eclipse. **(a)** On what date and over what part of the world will the next total eclipse of that series occur? Explain. **(b)** On what date might you next expect a total eclipse of that series to be visible from northern Africa? Explain.

Discussion Questions

46. Describe the cycle of lunar phases that would be observed if the Moon moved around the Earth in an orbit perpendicular to the plane of the Earth's orbit. Would it be possible for both solar and lunar eclipses to occur under these circumstances? Explain your reasoning.

47. How would a lunar eclipse look if the Earth had no atmosphere? Explain your reasoning.

48. In his 1885 novel *King Solomon's Mines*, H. Rider Haggard described a total solar eclipse that was seen in both South Africa and in the British Isles. Is such an eclipse possible? Why or why not?

49. Why do you suppose that total solar eclipse paths fall more frequently on oceans than on land? (You may find it useful to look at Figure 3-13.)

50. Examine Figure 3-13, which shows all of the total solar eclipses from 1997 to 2020. What are the chances that you might be able to travel to one of the eclipse paths? Do you think you might go through your entire life without ever seeing a total eclipse of the Sun?

Web/eBook Questions

51. Access the animation "The Moon's Phases" in Chapter 3 of the *Universe* Web site or eBook. This shows the Earth-Moon system as seen from a vantage point looking down onto the north pole. **(a)** Describe where you would be on the diagram if you are on the equator and the time is 6:00 P.M. **(b)** If it is 6:00 P.M. and you are standing on Earth's equator, would a third quarter moon be visible? Why or why not? If it would be visible, describe its appearance.

52. Search the World Wide Web for information about the next total lunar eclipse. Will the total phase of the eclipse be visible from your location? If not, will the penumbral phase be visible? Draw a picture showing the Sun, Earth, and Moon when the totality is at its maximum duration, and indicate your location on the drawing of the Earth.

53. Search the World Wide Web for information about the next total solar eclipse. Through which major cities, if any, does the path of totality pass? What is the maximum duration of totality? At what location is this maximum duration observed? Will this eclipse be visible (even as a partial eclipse) from your location? Draw a picture showing the Sun, Earth, and Moon when the totality is at its maximum duration, and indicate your location on the drawing of the Earth.

54. Access the animation "A Solar Eclipse Viewed from the Moon" in Chapter 3 of the *Universe* Web site or eBook. This shows the solar eclipse of August 11, 1999, as viewed from the Moon. Using a diagram, explain why the stars and the Moon's shadow move in the directions shown in this animation.

Activities

Observing Projects

55. Observe the Moon on each clear night over the course of a month. On each night, note the Moon's location among the constellations and record that location on a star chart that also shows the ecliptic. After a few weeks, your observations will begin to trace the Moon's orbit. Identify the orientation of the line of nodes by marking the points where the Moon's orbit and the ecliptic intersect. On what dates is the Sun near the nodes marked on your star chart? Compare these dates with the dates of the next solar and lunar eclipses.

56. It is quite possible that a lunar eclipse will occur while you are taking this course. Look up the date of the next lunar eclipse in Table 3-1, on the World Wide Web, or in the current issue of a reference such as the *Astronomical Almanac* or *Astronomical Phenomena*. Then make arrangements to observe this lunar eclipse. You can observe the eclipse with the naked eye, but binoculars or a small telescope will enhance your viewing experience. If the eclipse is partial or total, note the times at which the Moon enters and exits the Earth's umbra. If the eclipse is penumbral, can you see any changes in the Moon's brightness as the eclipse progresses?

57. Use the *Starry Night Enthusiast*™ program to observe the motion of the Moon. (a) Display the entire celestial sphere, including the part below the horizon, by moving to the Atlas mode. You do this by selecting **Favourites > Guides > Atlas.** Here, you will see the sky, containing the background stars and the planets, overlaid by a coordinate grid. One axis, the **Right Ascension** axis, is the extension of the Earth's equator on to the sky and is marked in hours along the **Celestial Equator.** At right angles to this equator are the **Declination** lines at constant Right Ascension, converging upon the North and South Celestial Poles. These poles are the extensions of the two ends of the Earth's spin axis. You can use the Hand Tool to explore this coordinate system by moving your viewpoint around the sky. (Move the mouse while holding down the mouse button to

achieve this motion.) Across this sky, inclined at an angle to the celestial equator, is the **Ecliptic,** or the path along which the Sun appears to move across our sky. This is the plane of the Earth's orbit. (If this green line does not appear, open the **Options** pane and check that the **Ecliptic** is selected in the **Guides** layer.) Use the Hand Tool to move the sky around to find the **Moon,** which will be close to, but not on, the ecliptic plane. Once you have found the Moon, use the Hand tool to move the Moon to the right-hand side of the main window. On the toolbar across the top of the main window, click on the **Time Flow Rate** control (immediately to the right of the date and time display) and set the discrete time step to **1 sidereal day.** Then advance time in one-sidereal-day intervals by clicking on the **Step Time Forward** button (the icon consisting of a black vertical line and right-pointing triangle to the far right of the time controls). You will note that the background sky remains fixed, as expected when time moves ahead in sidereal- day intervals. How does the Moon appear to move against the background of stars? Does it ever change direction? (b) Use this **Step Time Forward** button to determine how many days elapse between successive times when the Moon is on the ecliptic. Then move forward in time to a date when the Moon is on the ecliptic and either full or new. What type of eclipse will occur on that date? Confirm your answer by comparing with Tables 3-1 and 3-2 or with lists of eclipses on the World Wide Web.

58. Use the *Starry Night Enthusiast*™ program to examine the Moon as seen from space. Select **Solar System > Inner Solar System** in the **Favourites** menu. Click the **Stop** button in the toolbar to stop time flow. Then, click on the **Find** tab and double-click on the entry for the Moon in the Find pane in order to center the view on the Moon. Close the Find pane and zoom in on the Moon by clicking and holding the mouse cursor on the **Decrease current elevation** button (the downward-pointing arrow to the left of the Home button in the toolbar) to approach the Moon until detail is visible on the lunar surface. You can now view the Moon from any angle by holding down the **Shift** key while holding down the mouse button (the left button on a two-button mouse) and dragging the mouse. This is equivalent to flying a spaceship around the Moon at a constant distance. (a) Use this technique to rotate the Moon and view it from different perspectives. How does the phase of the Moon change as you rotate it around? (*Hint:* Compare with Box 3-1.) (b) Rotate the Moon until you can also see the Sun and note particularly the Moon's phase when it is in front of the Sun. Explain how your observations show that the phases of the Moon cannot be caused by the Earth's shadow falling on the Moon.

Collaborative Exercise

59. Using a bright light source at the center of a darkened room, or a flashlight, use your fist held at arm's length to demonstrate the difference between a full moon and a lunar eclipse. (Use yourself or a classmate as the Earth.) How must your fist "orbit" the Earth so that lunar eclipses do not happen at every full moon? Create a simple sketch to illustrate your answers.

Archaeoastronomy and Ethnoastronomy

by Mark Hollabaugh

Many years ago I read astronomer John Eddy's *National Geographic* article about the Wyoming Medicine Wheel. A few years later I visited this archaeological site in the Bighorn Mountains and started on the major preoccupation of my career.

Archaeoastronomy combines astronomy and archaeology. You may be familiar with sites such as Stonehenge or the Mayan ruins of the Yucatán. You may not know that there are many archaeological sites in the United States that demonstrate the remarkable understanding a people, usually known as the Anasazi, had about celestial motions (see Figure 2-1). Chaco Canyon, Hovenweep, and Chimney Rock in the Four Corners area of the Southwest preserve ruins from this ancient Pueblo culture.

Moonrise at Chimney Rock in southern Colorado provides a good example of the Anasazi's knowledge of the lunar cycles. If you watched the rising of the Moon for many, many years, you would discover that the Moon's northernmost rising point undergoes an 18.6-year cycle. Dr. McKim Malville of the University of Colorado discovered that the Anasazi who lived there knew of the lunar standstill cycle and watched the northernmost rising of the Moon between the twin rock pillars of Chimney Rock.

My own specialty is the ethnoastronomy of the Lakota, or Teton Sioux, who flourished on the Great Plains of what are now Nebraska, the Dakotas, and Wyoming. Ethnoastronomy combines ethnography with astronomy. As an ethnoastronomer, I am less concerned with physical evidence in the form of ruins and more interested in myths, legends, religious belief, and current practices. In my quest to understand the astronomical thinking and customs of the nineteenth-century Lakota, I have traveled to museums, archives, and libraries in Nebraska, Colorado, Wyoming, South Dakota, and North Dakota. I frequently visit the Pine Ridge and Rosebud Reservations in South Dakota. The Lakota, and other Plains Indians, had a rich tradition of understanding celestial motions and developed an even richer explanation of why things appear the way they do in the sky.

My first professional contribution to ethnoastronomy was in 1996 at the Fifth Oxford International Conference on Astronomy in Culture held in Santa Fe, New Mexico. I had noticed that images of eclipses often appeared in Lakota winter counts—their method of making a historic record of events. The great Leonid meteor shower of 1833 appears in almost every Plains Indian winter count. As I looked at the hides or in ledger books recording these winter counts, I wondered why the Sun, the Moon, and the stars are so common among the Lakota of 150 years ago. My curiosity led me to look deeper: What did the Lakota think about eclipses? Why does their central ritual, the Sun Dance, focus on the Sun? Why do so many legends involve the stars?

The Lakota observed lunar and solar eclipses. Perhaps they felt they had the power to restore the eclipsed Sun or Moon. In August 1869, an Indian agency physician in South Dakota told the Lakota there would be an eclipse. When the Sun disappeared from view, the Lakota began firing their guns in the air. In their minds, they were more powerful than the white doctor because the result of their action was to restore the Sun.

The Lakota used a lunar calendar. Their names for the months came from the world around them. October was the moon of falling leaves. In some years, there actually are 13 new moons, and they often called this extra month the "Lost Moon." The lunar calendar often dictated the timing of their sacred rites. Although the Lakota were never dogmatic about it, they preferred to hold their most important ceremony, the Sun Dance, at the time of the full moon in June, which is when the summer solstice occurs.

Why did the Lakota pay attention to the night sky? Lakota elder Ringing Shield's statement about Polaris, recorded in the late nineteenth century, provides a clue: "One star never moves and it is *wakan*. Other stars move in a circle about it. They are dancing in the dance circle." For the Lakota, a driving force in their culture was a quest to understand the nature of the sacred, or *wakan*. Anything hard to understand or different from the ordinary was *wakan*.

Reaching for the stars, as far away as they are, was a means for the Lakota to bring the incomprehensible universe a bit closer to the earth. Their goal was the same as what Dr. Sandra M. Faber says in her essay "Why Astronomy?" at the end of Chapter 1, "a perspective on human existence and its relation to the cosmos."

Mark Hollabaugh teaches physics and astronomy at Normandale Community College, in Bloomington, Minnesota. A graduate of St. Olaf College, he earned his Ph.D. in science education at the University of Minnesota. He has also taught at St. Olaf College, Augsburg College, and the U.S. Air Force Academy. As a boy he watched the dance of the northern lights and the flash of meteors. Meeting *Apollo 13* commander Jim Lovell and going for a ride in Roger Freedman's airplane are as close as he came to his dream of being an astronaut.

4

Gravitation and the Waltz of the Planets

R I **V** U X G

An astronaut orbits the Earth attached to the International Space Station's manipulator arm. (NASA)

Sixty years ago the idea of humans orbiting the Earth or sending spacecraft to other worlds was regarded as science fiction. Today science fiction has become commonplace reality. Literally thousands of artificial satellites orbit our planet to track weather, relay signals for communications and entertainment, and collect scientific data about the Earth and the universe. Humans live and work in Earth orbit (as in the accompanying photograph), have ventured as far as the Moon, and have sent dozens of robotic spacecraft to explore all the planets of the solar system.

While we think of spaceflight as an innovation of the twentieth century, we can trace its origins to a series of scientific revolutions that began in the 1500s. The first of these revolutions overthrew the ancient idea that the Earth is an immovable object at the center of the universe, around which moved the Sun, the Moon, and the planets. In this chapter we will learn how Nicolaus Copernicus, Tycho Brahe, Johannes Kepler, and Galileo Galilei helped us understand that the Earth is itself one of several planets orbiting the Sun.

We will learn, too, about Isaac Newton's revolutionary discovery of why the planets move in the way that they do. This was just one aspect of Newton's immense body of work, which included formulating the fundamental laws of physics and developing a precise mathematical description of the force of gravitation—the force that holds the planets in their orbits.

Gravitation proves to be a truly universal force: it guides spacecraft as they journey across the solar system, and keeps satellites and astronauts in their orbits. In this chapter you will learn more about this important force of nature, which plays a central role in all parts of astronomy.

4-1 Ancient astronomers invented geocentric models to explain planetary motions

Since the dawn of civilization, scholars have attempted to explain the nature of the universe. The ancient Greeks were the first to use the principle that still guides scientists today: *The universe can be described and understood logically.* For example, more than 2500 years ago Pythagoras and his followers put forth the idea that nature can be described with mathematics. About 200 years later, Aristotle asserted that the universe is governed by physical laws. One of the most important tasks before the scholars of ancient Greece was to create a model (see Section 1-1) to explain the motions of objects in the heavens.

Learning Goals

By reading the sections of this chapter, you will learn

4-1 How ancient astronomers attempted to explain the motions of the planets

4-2 What led Copernicus to a Sun-centered model of planetary motion

4-3 How Tycho's naked-eye observations of the sky revolutionized ideas about the heavens

4-4 How Kepler deduced the shapes of the orbits of the planets

4-5 How Galileo's pioneering observations with a telescope supported a Sun-centered model

4-6 The ideas behind Newton's laws, which govern the motion of all physical objects, including the planets

4-7 Why planets stay in their orbits and don't fall into the Sun

4-8 What causes ocean tides on Earth

The Greek Geocentric Model

Most Greek scholars assumed that the Sun, the Moon, the stars, and the planets revolve about a stationary Earth. A model of this kind, in which the Earth is at the center of the universe, is called a **geocentric model**. Similar ideas were held by the scholars of ancient China.

> **The scholars of ancient Greece imagined that the planets followed an ornate combination of circular paths**

Today we recognize that the stars are not merely points of light on an immense celestial sphere. But in fact this is how the ancient Greeks regarded the stars in their geocentric model of the universe. To explain the diurnal motions of the stars, they assumed that the celestial sphere was *real,* and that it rotated around the stationary Earth once a day.

The Sun and Moon both participated in this daily rotation of the sky, which explained their rising and setting motions. To explain why the Sun and Moon both move slowly with respect to the stars, the ancient Greeks imagined that both of these objects orbit around the Earth.

ANALOGY Imagine a merry-go-round that rotates clockwise as seen from above, as in **Figure 4-1a**. As it rotates, two children walk slowly counterclockwise at different speeds around the merry-go-round's platform. Thus, the children rotate along with the merry-go-round and also change their positions with respect to the merry-go-round's wooden horses. This scene is analogous to the way the ancient Greeks pictured the motions of the stars, Sun, and Moon. In their model, the celestial sphere rotated to the west around a stationary Earth (Figure 4-1b). The stars rotate along with the celestial sphere just as the wooden horses rotate along with the merry-go-round in Figure 4-1a. The Sun and Moon are analogous to the two children; they both turn westward with the celestial sphere, making one complete turn each day, and also move slowly eastward at different speeds with respect to the stars.

The geocentric model of the heavens also had to explain the motions of the planets. The ancient Greeks and other cultures of that time knew of five planets: Mercury, Venus, Mars, Jupiter, and Saturn, each of which is a bright object in the night sky. For example, when Venus is at its maximum brilliancy, it is 16 times brighter than the brightest star. (By contrast, Uranus and Neptune are quite dim and were not discovered until after the invention of the telescope.)

Like the Sun and Moon, all of the planets rise in the east and set in the west once a day. And like the Sun and Moon, from night to night the planets slowly move on the celestial sphere, that is, with respect to the background of stars. However, the character of this motion on the celestial sphere is quite different for the planets. Both the Sun and the Moon always move from west to east on the celestial sphere, that is, opposite the direction in which the celestial sphere appears to rotate. The Sun follows the path called the ecliptic (see Section 2-5), while the Moon follows a path that is slightly inclined to the ecliptic (see Section 3-3). Furthermore, the Sun and the Moon each move at relatively constant speeds around the celestial sphere. (The Moon's speed is faster than that of the Sun: It travels all the way around the celestial sphere in about a month while the Sun takes an entire year.) The planets, too, appear to move along paths that are close to the ecliptic. The difference is that each of the planets appears to wander back and forth on the celestial sphere with varying speed. As an example, **Figure 4-2** shows the wandering motion of Mars with respect to the background of stars during 2011 and 2012. (This figure shows that the name *planet* is well deserved; it comes from a Greek word meaning "wanderer.")

CAUTION! On a map of the Earth with north at the top, west is to the left and east is to the right. Why, then, is *east* on the left and *west* on the right in Figure 4-2? The answer is that a map of the Earth is a view looking downward at the ground from above, while a star map like Figure 4-2 is a view looking upward at the sky. If the constellation Leo in Figure 4-2 were directly overhead, Virgo would be toward the eastern horizon and Cancer would be toward the western horizon.

Merry-go-round rotates clockwise

Celestial sphere rotates to the west

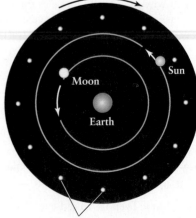

Figure 4-1

A Merry-Go-Round Analogy **(a)** Two children walk at different speeds around a rotating merry-go-round with its wooden horses. **(b)** In an analogous way, the ancient Greeks imagined that the Sun and Moon move around the rotating celestial sphere with its fixed stars. Thus, the Sun and Moon move from east to west across the sky every day and also move slowly eastward from one night to the next relative to the background of stars.

Wooden horses fixed on merry-go-round

Stars fixed on celestial sphere

(a) A rotating merry-go-round

(b) The Greek geocentric model

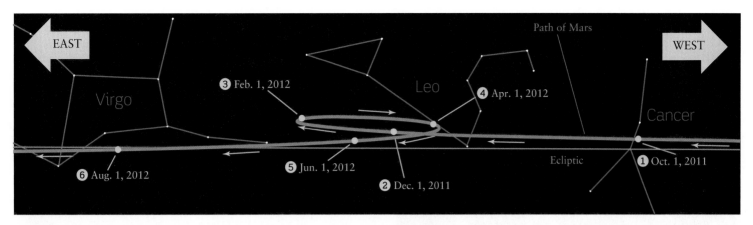

Figure 4-2

The Path of Mars in 2011–2012 From October 2011 through August 2012, Mars will move across the zodiacal constellations Cancer, Leo, and Virgo. Mars's motion will be direct (from west to east, or from right to left in this figure) most of the time but will be retrograde (from east to west, or from left to right in this figure) during February and March 2011. Notice that the speed of Mars relative to the stars is not constant: The planet travels farther across the sky from October 1 to December 1 than it does from December 1 to February 1.

Most of the time planets move slowly eastward relative to the stars, just as the Sun and Moon do. This eastward progress is called **direct motion.** For example, Figure 4-2 shows that Mars will be in direct motion from October 2011 through January 2012 and from April through August 2012. Occasionally, however, the planet will seem to stop and then back up for several weeks or months. This occasional westward movement is called **retrograde motion.** Mars will undergo retrograde motion during February and March 2012 (see Figure 4-2), and will do so again about every 22½ months. All the other planets go through retrograde motion, but at different intervals. In the language of the merry-go-round analogy in Figure 4-1, the Greeks imagined the planets as children walking around the rotating merry-go-round but who keep changing their minds about which direction to walk!

CAUTION! Whether a planet is in direct or retrograde motion, over the course of a single night you will see it rise in the east and set in the west. That's because both direct and retrograde motions are much slower than the apparent daily rotation of the sky. Hence, they are best detected by mapping the position of a planet against the background stars from night to night over a long period. Figure 4-2 is a map of just this sort.

The Ptolemaic System

Explaining the nonuniform motions of the five planets was one of the main challenges facing the astronomers of antiquity. The Greeks developed many theories to account for retrograde motion and the loops that the planets trace out against the background stars. One of the most successful and enduring models was originated by Apollonius of Perga and by Hipparchus in the second century B.C. and expanded upon by Ptolemy, the last of the great Greek astronomers, during the second century A.D. **Figure 4-3a** sketches the basic concept, usually called the **Ptolemaic system.** Each planet is assumed to move in a small circle called an **epicycle,** whose center in turn moves in a larger circle, called a **deferent,** which is centered approximately on the Earth. Both the epicycle and deferent rotate in the same direction, shown as counterclockwise in Figure 4-3a.

As viewed from Earth, the epicycle moves eastward along the deferent. Most of the time the eastward motion of the planet on its epicycle adds to the eastward motion of the epicycle on the deferent (Figure 4-3b). Then the planet is seen to be in direct (eastward) motion against the background stars. However, when the planet is on the part of its epicycle nearest Earth, the motion of the planet along the epicycle is opposite to the motion of the epicycle along the deferent. The planet therefore appears to slow down and halt its usual eastward movement among the constellations, and actually goes backward in retrograde (westward) motion for a few weeks or months (Figure 4-3c). Thus, the concept of epicycles and deferents enabled Greek astronomers to explain the retrograde loops of the planets.

Using the wealth of astronomical data in the library at Alexandria, including records of planetary positions for hundreds of years, Ptolemy deduced the sizes and rotation rates of the epicycles and deferents needed to reproduce the recorded paths of the planets. After years of tedious work, Ptolemy assembled his calculations into 13 volumes, collectively called the *Almagest.* His work was used to predict the positions and paths of the Sun, Moon, and planets with unprecedented accuracy. In fact, the *Almagest* was so successful that it became the astronomer's bible, and for more than 1000 years, the Ptolemaic system endured as a useful description of the workings of the heavens.

A major problem posed by the Ptolemaic system was a philosophical one: It treated each planet independent of the others. There was no rule in the *Almagest* that related the size and rotation speed of one planet's epicycle and deferent to the corresponding sizes and speeds for other planets. This problem made the Ptolemaic system very unsatisfying to many Islamic and European astronomers of the Middle Ages. They felt that a correct model of the universe should be based on a simple set of underlying principles that applied to all of the planets.

The idea that simple, straightforward explanations of phenomena are most likely to be correct is called **Occam's razor,** after William of Occam (or Ockham), the fourteenth-century English philosopher who first expressed it.

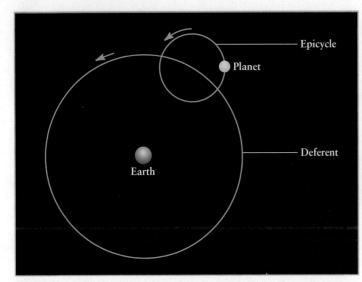

(a) Planetary motion modeled as a combination of circular motions

- Epicycle
- Planet
- Deferent
- Earth

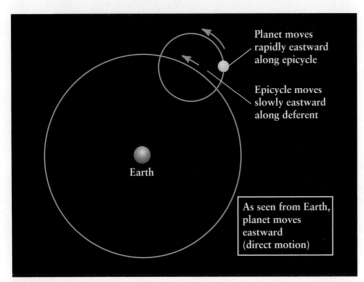

(b) Modeling direct motion

Planet moves rapidly eastward along epicycle

Epicycle moves slowly eastward along deferent

Earth

As seen from Earth, planet moves eastward (direct motion)

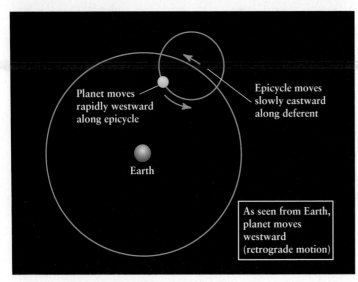

(c) Modeling retrograde motion

Planet moves rapidly westward along epicycle

Epicycle moves slowly eastward along deferent

Earth

As seen from Earth, planet moves westward (retrograde motion)

Figure 4-3

A Geocentric Explanation of Retrograde Motion (a) The ancient Greeks imagined that each planet moves along an epicycle, which in turn moves along a deferent centered approximately on the Earth. The planet moves along the epicycle more rapidly than the epicycle moves along the deferent. (b) At most times the eastward motion of the planet on the epicycle adds to the eastward motion of the epicycle on the deferent. Then the planet moves eastward in direct motion as seen from Earth. (c) When the planet is on the inside of the deferent, its motion along the epicycle is westward. Because this motion is faster than the eastward motion of the epicycle on the deferent, the planet appears from Earth to be moving westward in retrograde motion.

(The "razor" refers to shaving extraneous details from an argument or explanation.) Although Occam's razor has no proof or verification, it appeals to the scientist's sense of beauty and elegance, and it has helped lead to the simple and powerful laws of nature that scientists use today. In the centuries that followed, Occam's razor would help motivate a new and revolutionary view of the universe.

4-2 Nicolaus Copernicus devised the first comprehensive heliocentric model

During the first half of the sixteenth century, a Polish lawyer, physician, canon of the church, and gifted mathematician named Nicolaus Copernicus (Figure 4-4) began to construct a new model of the universe. His model, which placed the Sun at the center, explained the motions of the planets in a more natural way than the Ptolemaic system. As we will see, it also helped lay the foundations of modern physical science. But Copernicus was not the first to conceive a Sun-centered model of planetary motion. In the third century B.C., the Greek astronomer Aristarchus suggested such a model as a way to explain retrograde motion.

A Heliocentric Model Explains Retrograde Motion

Imagine riding on a fast racehorse. As you pass a slowly walking pedestrian, he appears to move backward, even though he is traveling in the same direction as you and your horse. This sort of simple observation inspired Aristarchus to formulate a **heliocentric** (Sun-centered) **model** in which all the planets, including Earth, revolve about the Sun. Different planets take different lengths of time to complete an orbit, so from time to time one planet will overtake another, just as a fast-moving horse overtakes a person on foot. When the Earth overtakes Mars, for example, Mars appears to move backward in retrograde motion, as **Figure 4-5** shows. Thus, in the heliocentric picture, the occasional retrograde motion of a planet is merely the result of the Earth's motion.

> The retrograde motion of the planets is a result of our viewing the universe from a moving Earth

As we saw in Section 3-6, Aristarchus demonstrated that the Sun is bigger than the Earth (see Table 3-3). This made it sensible to imagine the Earth orbiting the larger Sun. He also imagined that the Earth rotated on its axis once a day, which explained the daily rising and setting of the Sun, Moon, and planets and the diurnal motions of the stars. To explain why the apparent motions of the planets never take them far from the ecliptic, Aristarchus proposed that the orbits of the Earth and all the planets must lie in nearly the same plane. (Recall from Section 2-5 that the ecliptic is the projection onto the celestial sphere of the plane of the Earth's orbit.)

This heliocentric model is conceptually much simpler than an Earth-centered system, such as that of Ptolemy, with all its "circles upon circles." In Aristarchus's day, however, the idea of an orbiting, rotating Earth seemed inconceivable, given the Earth's apparent stillness and immobility. Nearly 2000 years would pass before a heliocentric model found broad acceptance.

Copernicus and the Arrangement of the Planets

In the years after 1500, Copernicus came to realize that a heliocentric model has several advantages beyond providing a natural explanation of retrograde motion. In the Ptolemaic system, the arrangement of the planets—that is, which are close to the Earth and which are far away—was chosen in large part by guesswork. But using a heliocentric model, Copernicus could determine the arrangement of the planets without ambiguity.

Copernicus realized that because Mercury and Venus are always observed fairly near the Sun in the sky, their orbits must be smaller than the Earth's. Planets in such orbits are called **inferior planets** (Figure 4-6). The other visible planets—Mars, Jupiter, and Saturn—are sometimes seen on the side of the celestial sphere opposite the Sun, so these planets appear high above the horizon at midnight (when the Sun is far below the horizon). When this

Figure 4-4

Nicolaus Copernicus (1473–1543) Copernicus was the first person to work out the details of a heliocentric system in which the planets, including the Earth, orbit the Sun. (E. Lessing/Magnum)

3. From point 6 to point 9, Mars again appears to move eastward against the background of stars as seen from Earth (direct motion).

2. As Earth passes Mars in its orbit from point 4 to point 6, Mars appears to move westward against the background of stars (retrograde motion).

1. From point 1 to point 4, Mars appears to move eastward against the background of stars as seen from Earth (direct motion).

Figure 4-5

A Heliocentric Explanation of Retrograde Motion In the heliocentric model of Aristarchus, the Earth and the other planets orbit the Sun. The Earth travels around the Sun more rapidly than Mars. Consequently, as the Earth overtakes and passes this slower-moving planet, Mars appears for a few months (from points 4 through 6) to fall behind and move backward with respect to the background of stars.

Figure 4-6

Planetary Orbits and Configurations When and where in the sky a planet can be seen from Earth depends on the size of its orbit and its location on that orbit. Inferior planets have orbits smaller than the Earth's, while superior planets have orbits larger than the Earth's. (Note that in this figure you are looking down onto the solar system from a point far above the Earth's northern hemisphere.)

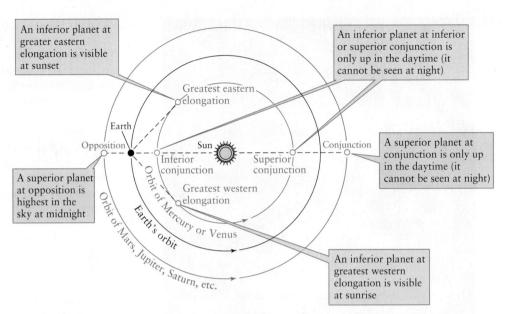

happens, the Earth must lie between the Sun and these planets. Copernicus therefore concluded that the orbits of Mars, Jupiter, and Saturn must be larger than the Earth's orbit. Hence, these planets are called **superior planets.**

Uranus and Neptune, as well as Pluto and a number of other small bodies called asteroids that also orbit the Sun, were discovered after the telescope was invented (and after the death of Copernicus). All of these can be seen at times in the midnight sky, so these also have orbits larger than the Earth's.

The heliocentric model also explains why planets appear in different parts of the sky on different dates. Both inferior planets (Mercury and Venus) go through cycles: The planet is seen in the west after sunset for several weeks or months, then for several weeks or months in the east before sunrise, and then in the west after sunset again.

Figure 4-6 shows the reason for this cycle. When Mercury or Venus is visible after sunset, it is near **greatest eastern elongation.** (The angle between the Sun and a planet as viewed from Earth is called the planet's **elongation.**) The planet's position in the sky is as far east of the Sun as possible, so it appears above the western horizon after sunset (that is, to the east of the Sun) and is often called an "evening star." At **greatest western elongation,** Mercury or Venus is as far west of the Sun as it can possibly be. It then rises before the Sun, gracing the predawn sky as a "morning star" in the east. When Mercury or Venus is at **inferior conjunction,** it is between us and the Sun, and it is moving from the evening sky into the morning sky. At **superior conjunction,** when the planet is on the opposite side of the Sun, it is moving back into the evening sky.

A superior planet such as Mars, whose orbit is larger than the Earth's, is best seen in the night sky when it is at **opposition.** At this point the planet is in the part of the sky opposite the Sun and is highest in the sky at midnight. This is also when the planet appears brightest, because it is closest to us. But when a superior planet like Mars is located behind the Sun at **conjunction,** it is above the horizon during the daytime and thus is not well placed for nighttime viewing.

Planetary Periods and Orbit Sizes

The Ptolemaic system has no simple rules relating the motion of one planet to another. But Copernicus showed that there *are* such rules in a heliocentric model. In particular, he found a correspondence between the time a planet takes to complete one orbit—that is, its **period**—and the size of the orbit.

Determining the period of a planet takes some care, because the Earth, from which we must make the observations, is also moving. Realizing this, Copernicus was careful to distinguish between two different periods of each planet. The **synodic period** is the time that elapses between two successive identical configurations as seen from Earth—from one opposition to the next, for example, or from one conjunction to the next. The **sidereal period** is the true orbital period of a planet, the time it takes the planet to complete one full orbit of the Sun relative to the stars.

The synodic period of a planet can be determined by observing the sky, but the sidereal period has to be found by calculation. Copernicus figured out how to do this (Box 4-1). Table 4-1 shows the results for all of the planets.

Table 4-1   Synodic and Sidereal Periods of the Planets

Planet	Synodic period	Sidereal period
Mercury	116 days	88 days
Venus	584 days	225 days
Earth	—	1.0 year
Mars	780 days	1.9 years
Jupiter	399 days	11.9 years
Saturn	378 days	29.5 years
Uranus	370 days	84.1 years
Neptune	368 days	164.9 years

BOX **4-1**

Tools of the Astronomer's Trade

Relating Synodic and Sidereal Periods

We can derive a mathematical formula that relates a planet's sidereal period (the time required for the planet to complete one orbit) to its synodic period (the time between two successive identical configurations). To start with, let's consider an inferior planet (Mercury or Venus) orbiting the Sun as shown in the accompanying figure. Let P be the planet's sidereal period, S the planet's synodic period, and E the Earth's sidereal period or sidereal year (see Section 2-8), which Copernicus knew to be nearly $365\frac{1}{4}$ days.

The rate at which the Earth moves around its orbit is the number of degrees around the orbit divided by the time to complete the orbit, or $360°/E$ (equal to a little less than 1° per day). Similarly, the rate at which the inferior planet moves along its orbit is $360°/P$.

During a given time interval, the angular distance that the Earth moves around its orbit is its rate, $(360°/E)$, multiplied by the length of the time interval. Thus, during a time S, or one synodic period of the inferior planet, the Earth covers an angular distance of $(360°/E)S$ around its orbit. In that same time, the inferior planet covers an angular distance of $(360°/P)S$. Note, however, that the inferior planet has gained one full lap on the Earth, and hence has covered 360° more than the Earth has (see the figure). Thus, $(360°/P)S = (360°/E)S + 360°$. Dividing each term of this equation by $360°S$ gives

For an inferior planet:

$$\frac{1}{P} = \frac{1}{E} + \frac{1}{S}$$

P = inferior planet's sidereal period

E = Earth's sidereal period = 1 year

S = inferior planet's synodic period

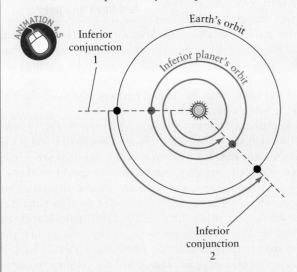

Inferior conjunction 1

Earth's orbit

Inferior planet's orbit

Inferior conjunction 2

A similar analysis for a superior planet (for example, Mars, Jupiter, or Saturn) yields

For a superior planet:

$$\frac{1}{P} = \frac{1}{E} - \frac{1}{S}$$

P = superior planet's sidereal period

E = Earth's sidereal period = 1 year

S = superior planet's synodic period

Using these formulas, we can calculate a planet's sidereal period P from its synodic period S. Often astronomers express P, E, and S in terms of years, by which they mean Earth years of approximately 365.26 days.

EXAMPLE: Jupiter has an observed synodic period of 398.9 days, or 1.092 years. What is its sidereal period?

Situation: Our goal is to find the sidereal period of Jupiter, a superior planet.

Tools: Since Jupiter is a superior planet, we use the second of the two equations given above to determine the sidereal period P.

Answer: We are given the Earth's sidereal period $E = 1$ year and Jupiter's synodic period $S = 1.092$ years. Using the equation $1/P = 1/E - 1/S$,

$$\frac{1}{P} = \frac{1}{1} - \frac{1}{1.092} = 0.08425,$$

so

$$P = \frac{1}{0.08425} = 11.87 \text{ years}$$

Review: Our answer means that it takes 11.87 years for Jupiter to complete one full orbit of the Sun. This is greater than the Earth's 1-year sidereal period because Jupiter's orbit is larger than the Earth's orbit. Jupiter's synodic period of 1.092 years is the time from one opposition to the next, or the time that elapses from when the Earth overtakes Jupiter to when it next overtakes Jupiter. This is so much shorter than the sidereal period because Jupiter moves quite slowly around its orbit. The Earth overtakes it a little less often than once per Earth orbit, that is, at intervals of a little bit more than a year.

Table 4-2   Average Distances of the Planets from the Sun

Planet	Copernican value (AU*)	Modern value (AU)
Mercury	0.38	0.39
Venus	0.72	0.72
Earth	1.00	1.00
Mars	1.52	1.52
Jupiter	5.22	5.20
Saturn	9.07	9.55
Uranus	—	19.19
Neptune	—	30.07

*1 AU = 1 astronomical unit = average distance from the Earth to the Sun.

To find a relationship between the sidereal period of a planet and the size of its orbit, Copernicus still had to determine the relative distances of the planets from the Sun. He devised a straightforward geometric method of determining the relative distances of the planets from the Sun using trigonometry. His answers turned out to be remarkably close to the modern values, as shown in Table 4-2.

The distances in Table 4-2 are given in terms of the astronomical unit, which is the average distance from Earth to the Sun (Section 1-7). Copernicus did not know the precise value of this distance, so he could only determine the *relative* sizes of the orbits of the planets. One method used by modern astronomers to determine the astronomical unit is to measure the Earth-Venus distance very accurately using radar. At the same time, they measure the angle in the sky between Venus and the Sun, and then calculate the Earth-Sun distance using trigonometry. In this way, the astronomical unit is found to be 1.496×10^8 km (92.96 million miles). Once the astronomical unit is known, the average distances from the Sun to each of the planets in kilometers can be determined from Table 4-2. A table of these distances is given in Appendix 2.

By comparing Tables 4-1 and 4-2, you can see the unifying relationship between planetary orbits in the Copernican model: The farther a planet is from the Sun, the longer it takes to travel around its orbit (that is, the longer its sidereal period). That is so for *two* reasons: (1) the larger the orbit, the farther a planet must travel to complete an orbit; and (2) the larger the orbit, the slower a planet moves. For example, Mercury, with its small orbit, moves at an average speed of 47.9 km/s (107,000 mi/h). Saturn travels around its large orbit much more slowly, at an average speed of 9.64 km/s (21,600 mi/h). The older Ptolemaic model offers no such simple relations between the motions of different planets.

The Shapes of Orbits in the Copernican Model

At first, Copernicus assumed that Earth travels around the Sun along a circular path. He found that perfectly circular orbits could not accurately describe the paths of the other planets, so he had to add an epicycle to each planet. This was *not* to explain retrograde motion, which Copernicus realized was because of the differences in orbital speeds of different planets, as shown in Figure 4-5. Rather, the small epicycles helped Copernicus account for slight variations in each planet's speed along its orbit.

Even though he clung to the old notion that orbits must be made up of circles, Copernicus had shown that a heliocentric model could explain the motions of the planets. He compiled his ideas and calculations into a book entitled *De revolutionibus orbium coelestium* (On the Revolutions of the Celestial Spheres), which was published in 1543, the year of his death.

For several decades after Copernicus, most astronomers saw little reason to change their allegiance from the older geocentric model of Ptolemy. The predictions that the Copernican model makes for the apparent positions of the planets are, on average, no better or worse than those of the Ptolemaic model. The test of Occam's razor does not really favor either model, because both use a combination of circles to describe each planet's motion.

More concrete evidence was needed to convince scholars to abandon the old, comfortable idea of a stationary Earth at the center of the universe. The story of how this evidence was accumulated begins nearly 30 years after the death of Copernicus, when a young Danish astronomer pondered the nature of a new star in the heavens.

4-3   Tycho Brahe's astronomical observations disproved ancient ideas about the heavens

On November 11, 1572, a bright star suddenly appeared in the constellation of Cassiopeia. At first, it was even brighter than Venus, but then it began to grow dim. After 18 months, it faded from view. Modern astronomers recognize this event as a supernova explosion, the violent death of a massive star.

> A supernova explosion and a comet revealed to Tycho that our universe is more dynamic than had been imagined

In the sixteenth century, however, the vast majority of scholars held with the ancient teachings of Aristotle and Plato, who had argued that the heavens are permanent and unalterable. Consequently, the "new star" of 1572 could not really be a star at all, because the heavens do not change; it must instead be some sort of bright object quite near the Earth, perhaps not much farther away than the clouds overhead.

The 25-year-old Danish astronomer Tycho Brahe (1546–1601) realized that straightforward observations might reveal the distance to the new star. It is common experience that when you walk from one place to another, nearby objects appear to change position against the background of more distant objects. This phenomenon, whereby the apparent position of an object changes because of the motion of the observer, is called **parallax**. If the new star was nearby, then its position should shift against the background stars over the course of a single night because Earth's rotation changes our viewpoint. Figure 4-7 shows this predicted shift. (Actually, Tycho believed that the heavens rotate about the Earth, as in the Ptolemaic model, but the net effect is the same.)

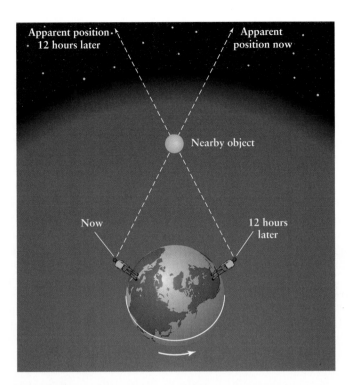

Figure 4-7

A Nearby Object Shows a Parallax Shift Tycho Brahe argued that if an object is near the Earth, an observer would have to look in different directions to see that object over the course of a night and its position relative to the background stars would change. Tycho failed to measure such changes for a supernova in 1572 and a comet in 1577. He therefore concluded that these objects were far from the Earth.

Tycho's careful observations failed to disclose any parallax. The farther away an object is, the less it appears to shift against the background as we change our viewpoint, and the smaller the parallax. Hence, the new star had to be quite far away, farther from Earth than anyone had imagined. This was the first evidence that the "unchanging" stars were in fact changeable.

Tycho also attempted to measure the parallax of a bright comet that appeared in 1577 and, again, found it too small to measure. Thus, the comet also had to be far beyond the Earth. Furthermore, because the comet's position relative to the stars changed from night to night, its motion was more like a planet than a star. But most scholars of Tycho's time taught that the motions of the planets had existed in unchanging form since the beginning of the universe. If new objects such as comets could appear and disappear within the realm of the planets, the conventional notions of planetary motions needed to be revised.

Tycho's observations showed that the heavens are by no means pristine and unchanging. This discovery flew in the face of nearly 2000 years of astronomical thought. In support of these revolutionary observations, the king of Denmark financed the construction of two magnificent observatories for Tycho's use on the island of Hven, just off the Danish coast. The bequest for these two observatories—Uraniborg ("heavenly castle") and Stjerneborg ("star castle")—allowed Tycho to design and have built a set of astronomical instruments vastly superior in quality to any earlier instruments (**Figure 4-8**).

Tycho and the Positions of the Planets

With this state-of-the-art equipment, Tycho proceeded to measure the positions of stars and planets with unprecedented accuracy. In addition, he and his assistants were careful to make several observations of the same star or planet with different instruments, in order to identify any errors that might be caused by the instruments themselves. This painstaking approach to mapping the heavens revolutionized the practice of astronomy and is used by astronomers today.

A key goal of Tycho's observations during this period was to test the ideas Copernicus had proposed decades earlier about the Earth going around the Sun. Tycho argued that if the Earth was in motion, then nearby stars should appear to shift their positions with respect to background stars as we orbit the Sun. Tycho failed to detect any such parallax, and he concluded that the Earth was at rest and the Copernican system was wrong.

On this point Tycho was in error, for nearby stars do in fact shift their positions as he had suggested. But even the nearest stars are so far away that the shifts in their positions are less than an arcsecond, too small to be seen with the naked eye. Tycho would have needed a telescope to detect the parallax that he was looking for, but the telescope was not invented until after his death in

Figure 4-8

Tycho Brahe (1546–1601) Observing This contemporary illustration shows Tycho Brahe with some of the state-of-the-art measuring apparatus at Uraniborg, one of the two observatories that he built under the patronage of Frederik II of Denmark. (This magnificent observatory lacked a telescope, which had not yet been invented.) The data that Tycho collected were crucial to the development of astronomy in the years after his death. (Photo Researchers, Inc.)

WEB LINK 4.5

Figure 4-9

Johannes Kepler (1571–1630) By analyzing Tycho Brahe's detailed records of planetary positions, Kepler developed three general principles, called Kepler's laws, that describe how the planets move about the Sun. Kepler was the first to realize that the orbits of the planets are ellipses and not circles. (E. Lessing/Magnum)

1601. Indeed, the first accurate determination of stellar parallax was not made until 1838.

Although he remained convinced that the Earth was at the center of the universe, Tycho nonetheless made a tremendous contribution toward putting the heliocentric model on a solid foundation. From 1576 to 1597, he used his instruments to make comprehensive measurements of the positions of the planets with an accuracy of 1 arcminute. This is as well as can be done with the naked eye and was far superior to any earlier measurements. Within the reams of data that Tycho compiled lay the truth about the motions of the planets. The person who would extract this truth was a German mathematician who became Tycho's assistant in 1600, a year before the great astronomer's death. His name was Johannes Kepler (**Figure 4-9**).

4-4 Johannes Kepler proposed elliptical paths for the planets about the Sun

The task that Johannes Kepler took on at the beginning of the seventeenth century was to find a model of planetary motion that agreed completely with Tycho's extensive and very accurate observations of planetary positions. To do this, Kepler found that he had to break with an ancient prejudice about planetary motions.

> Kepler's ideas apply not just to the planets, but to all orbiting celestial objects

Elliptical Orbits and Kepler's First Law

Astronomers had long assumed that heavenly objects move in circles, which were considered the most perfect and harmonious of all geometric shapes. They believed that if a perfect God resided in heaven along with the stars and planets, then the motions of these objects must be perfect too. Against this context, Kepler dared to try to explain planetary motions with noncircular curves. In particular, he found that he had the best success with a particular kind of curve called an **ellipse.**

You can draw an ellipse by using a loop of string, two thumbtacks, and a pencil, as shown in **Figure 4-10a**. Each thumbtack in the figure is at a **focus** (plural **foci**) of the ellipse; an ellipse has two foci. The longest diameter of an ellipse, called the **major axis,** passes through both foci. Half of that distance is called the **semimajor axis** and is usually designated by the letter a. A circle is a special case of an ellipse in which the two foci are at the same point (this corresponds to using only a single thumbtack in Figure 4-10a). The semimajor axis of a circle is equal to its radius.

By assuming that planetary orbits were ellipses, Kepler found, to his delight, that he could make his theoretical calculations match precisely to Tycho's observations. This important discovery, first published in 1609, is now called **Kepler's first law:**

The orbit of a planet about the Sun is an ellipse with the Sun at one focus.

The semimajor axis a of a planet's orbit is the average distance between the planet and the Sun.

CAUTION! The Sun is at one focus of a planet's elliptical orbit, but there is *nothing* at the other focus. This "empty focus" has geometrical significance, because it helps to define the shape of the ellipse, but plays no other role.

Ellipses come in different shapes, depending on the elongation of the ellipse. The shape of an ellipse is described by its **eccentricity,** designated by the letter e. The value of e can range from 0 (a circle) to just under 1 (nearly a straight line). The greater the eccentricity, the more elongated the ellipse. Figure 4-10b shows a few examples of ellipses with different eccentricities. Because a circle is a special case of an ellipse, it is possible to have a perfectly circular orbit. But all of the objects that orbit the Sun have orbits that are at least slightly elliptical. The most circular of any planetary orbit is that of Venus, with an eccentricity of just 0.007; Mercury's orbit has an eccentricity of 0.206, and a number of small bodies called comets move in very elongated orbits with eccentricities just less than 1.

Orbital Speeds and Kepler's Second Law

Once he knew the shape of a planet's orbit, Kepler was ready to describe exactly *how* it moves on that orbit. As a planet travels in an elliptical orbit, its distance from the Sun varies. Kepler realized that the speed of a planet also varies along its orbit. A planet moves most rapidly when it is nearest the Sun, at a point on its orbit called **perihelion.** Conversely, a planet moves most slowly when it is farthest from the Sun, at a point called **aphelion** (Figure 4-11).

(a) The geometry of an ellipse

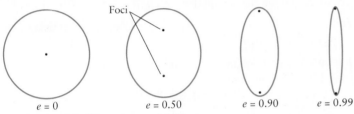

(b) Ellipses with different eccentricities

Figure 4-10

Ellipses (a) To draw an ellipse, use two thumbtacks to secure the ends of a piece of string, then use a pencil to pull the string taut. If you move the pencil while keeping the string taut, the pencil traces out an ellipse. The thumbtacks are located at the two foci of the ellipse. The major axis is the greatest distance across the ellipse; the semimajor axis is half of this distance. **(b)** A series of ellipses with the same major axis but different eccentricities. An ellipse can have any eccentricity from $e = 0$ (a circle) to just under $e = 1$ (virtually a straight line).

After much trial and error, Kepler found a way to describe just how a planet's speed varies as it moves along its orbit. Figure 4-11 illustrates this discovery, also published in 1609. Suppose that it takes 30 days for a planet to go from point *A* to point *B*. During that time, an imaginary line joining the Sun and the planet sweeps out a nearly triangular area. Kepler discovered that a line joining the Sun and the planet also sweeps out exactly the same area during any other 30-day interval. In other words, if the planet also takes 30 days to go from point *C* to point *D*, then the two shaded segments in Figure 4-11 are equal in area. **Kepler's second law** can be stated thus:

A line joining a planet and the Sun sweeps out equal areas in equal intervals of time.

This relationship is also called the **law of equal areas.** In the idealized case of a circular orbit, a planet would have to move at a constant speed around the orbit in order to satisfy Kepler's second law.

ANALOGY An analogy for Kepler's second law is a twirling ice skater holding weights in each hand. If the skater moves the weights closer to her body by pulling her arms straight in, her rate of spin increases and the weights move faster; if she extends her arms so the weights move away from her body, her rate of spin decreases and the weights slow down. Just like the weights, a planet in an elliptical orbit travels at a higher speed when it moves closer to the Sun (toward perihelion) and travels at a lower speed when it moves away from the Sun (toward aphelion).

Orbital Periods and Kepler's Third Law

Kepler's second law describes how the speed of a given planet changes as it orbits the Sun. Kepler also deduced from Tycho's data a relationship that can be used to compare the motions of *different* planets. Published in 1618 and now called **Kepler's third law**, it states a relationship between the size of a planet's orbit and the time the planet takes to go once around the Sun:

Figure 4-11

Kepler's First and Second Laws According to Kepler's first law, a planet travels around the Sun along an elliptical orbit with the Sun at one focus. According to his second law, a planet moves fastest when closest to the Sun (at perihelion) and slowest when farthest from the Sun (at aphelion). As the planet moves, an imaginary line joining the planet and the Sun sweeps out equal areas in equal intervals of time (from *A* to *B* or from *C* to *D*). By using these laws in his calculations, Kepler found a perfect fit to the apparent motions of the planets.

Using Kepler's Third Law

Kepler's third law relates the sidereal period P of an object orbiting the Sun to the semimajor axis a of its orbit:

$$P^2 = a^3$$

You must keep two essential points in mind when working with this equation:

1. The period P *must* be measured in years, and the semimajor axis a *must* be measured in astronomical units (AU). Otherwise you will get nonsensical results.

2. This equation applies *only* to the special case of an object, like a planet, that orbits the Sun. If you want to analyze the orbit of the Moon around the Earth, of a spacecraft around Mars, or of a planet around a distant star, you must use a different, generalized form of Kepler's third law. We discuss this alternative equation in Section 4-7 and Box 4-4.

EXAMPLE: The average distance from Venus to the Sun is 0.72 AU. Use this to determine the sidereal period of Venus.

Situation: The average distance from the Venus to the Sun is the semimajor axis a of the planet's orbit. Our goal is to calculate the planet's sidereal period P.

Tools: To relate a and P we use Kepler's third law, $P^2 = a^3$.

Answer: We first cube the semimajor axis (multiply it by itself twice):

$$a^3 = (0.72)^3 = 0.72 \times 0.72 \times 0.72 = 0.373$$

According to Kepler's third law this is also equal to P^2, the square of the sidereal period. So, to find P, we have to "undo" the square, that is, take the square root. Using a calculator, we find

$$P = \sqrt{P^2} = \sqrt{0.373} = 0.61$$

Review: The sidereal period of Venus is 0.61 years, or a bit more than seven Earth months. This makes sense: A planet with a smaller orbit than the Earth's (an inferior planet) must have a shorter sidereal period than the Earth.

EXAMPLE: A certain small asteroid (a rocky body a few tens of kilometers across) takes eight years to complete one orbit around the Sun. Find the semimajor axis of the asteroid's orbit.

Situation: We are given the sidereal period $P = 8$ years, and are to determine the semimajor axis a.

Tools: As in the preceding example, we relate a and P using Kepler's third law, $P^2 = a^3$.

Answer: We first square the period:

$$P^2 = 8^2 = 8 \times 8 = 64$$

From Kepler's third law, this is also equal to a^3. To determine a, we must take the *cube root* of a^3, that is, find the number whose cube is 64. If your calculator has a cube root function, denoted by the symbol $\sqrt[3]{\ }$, you can use it to find that the cube root of 64 is 4: $\sqrt[3]{64} = 4$. Otherwise, you can determine by trial and error that the cube of 4 is 64:

$$4^3 = 4 \times 4 \times 4 = 64$$

Because the cube of 4 is 64, it follows that the cube root of 64 is 4 (taking the cube root "undoes" the cube).

With either technique you find that the orbit of this asteroid has semimajor axis $a = 4$ AU.

Review: The period is greater than 1 year, so the semimajor axis is greater than 1 AU. Note that $a = 4$ AU is intermediate between the orbits of Mars and Jupiter (see Table 4-3). Many asteroids are known with semimajor axes in this range, forming a region in the solar system called the asteroid belt.

The square of the sidereal period of a planet is directly proportional to the cube of the semimajor axis of the orbit.

Kepler's third law says that the larger a planet's orbit—that is, the larger the semimajor axis, or average distance from the planet to the Sun—the longer the sidereal period, which is the time it takes the planet to complete an orbit. From Kepler's third law one can show that the larger the semimajor axis, the slower the average speed at which the planet moves around its orbit. (By contrast, Kepler's *second* law describes how the speed of a given planet is sometimes faster and sometimes slower than its average speed.) This qualitative relationship between orbital size and orbital speed is just what Aristarchus and Copernicus used to ex-

plain retrograde motion, as we saw in Section 4-2. Kepler's great contribution was to make this relationship a quantitative one.

It is useful to restate Kepler's third law as an equation. If a planet's sidereal period P is measured in years and the length of its semimajor axis a is measured in astronomical units (AU), where 1 AU is the average distance from the Earth to the Sun (see Section 1-7), then Kepler's third law is

Kepler's third law

$$P^2 = a^3$$

P = planet's sidereal period, in years

a = planet's semimajor axis, in AU

Table 4-3 A Demonstration of Kepler's Third Law ($P^2 = a^3$)

Planet	Sidereal period P (years)	Semimajor axis a (AU)	P^2	a^3
Mercury	0.24	0.39	0.06	0.06
Venus	0.61	0.72	0.37	0.37
Earth	1.00	1.00	1.00	1.00
Mars	1.88	1.52	3.53	3.51
Jupiter	11.86	5.20	140.7	140.6
Saturn	29.46	9.55	867.9	871.0
Uranus	84.10	19.19	7,072	7,067
Neptune	164.86	30.07	27,180	27,190

Kepler's third law states that $P^2 = a^3$ for each of the planets. The last two columns of this table demonstrate that this relationship holds true to a very high level of accuracy.

If you know either the sidereal period of a planet or the semimajor axis of its orbit, you can find the other quantity using this equation. **Box 4-2** gives some examples of how this is done.

We can verify Kepler's third law for all of the planets, including those that were discovered after Kepler's death, using data from Tables 4-1 and 4-2. If Kepler's third law is correct, for each planet the numerical values of P^2 and a^3 should be equal. This is indeed true to very high accuracy, as **Table 4-3** shows.

The Significance of Kepler's Laws

Kepler's laws are a landmark in the history of astronomy. They made it possible to calculate the motions of the planets with better accuracy than any geocentric model ever had, and they helped to justify the idea of a heliocentric model. Kepler's laws also pass the test of Occam's razor, for they are simpler in every way than the schemes of Ptolemy or Copernicus, both of which used a complicated combination of circles.

But the significance of Kepler's laws goes beyond understanding planetary orbits. These same laws are also obeyed by spacecraft orbiting the Earth, by two stars revolving about each other in a binary star system, and even by galaxies in their orbits about each other. Throughout this book, we shall use Kepler's laws in a wide range of situations.

As impressive as Kepler's accomplishments were, he did not prove that the planets orbit the Sun, nor was he able to explain why planets move in accordance with his three laws. These advances were made by two other figures who loom large in the history of astronomy: Galileo Galilei and Isaac Newton.

4-5 Galileo's discoveries with a telescope strongly supported a heliocentric model

When Dutch opticians invented the telescope during the first decade of the seventeenth century, astronomy was changed forever. The scholar who used this new tool to amass convincing evidence that the planets orbit the Sun, not the Earth, was the Italian mathematician and physical scientist Galileo Galilei (**Figure 4-12**).

While Galileo did not invent the telescope, he was the first to point one of these new devices toward the sky and to publish his observations. Beginning in 1610, he saw sights of which no one had ever dreamed. He discovered mountains on the Moon, sunspots on the Sun, and the rings of Saturn, and he was the first to see that the Milky Way is not a featureless band of light but rather "a mass of innumerable stars."

> Galileo used the cutting-edge technology of the 1600s to radically transform our picture of the universe

The Phases of Venus

One of Galileo's most important discoveries with the telescope was that Venus exhibits phases like those of the Moon (**Figure 4-13**). Galileo also noticed that the apparent size of Venus as seen through his telescope was related to the planet's phase. Venus appears small at gibbous phase and largest at crescent phase. There is also a correlation between the phases of Venus and the planet's angular distance from the Sun.

Figure 4-14 shows that these relationships are entirely compatible with a heliocentric model in which the Earth and Venus both go around the Sun. They are also completely *incompatible* with the Ptolemaic system, in which the Sun and Venus both orbit

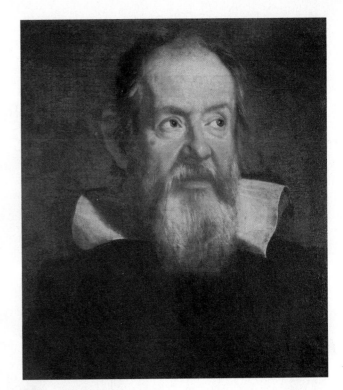

Figure 4-12

Galileo Galilei (1564–1642) Galileo was one of the first people to use a telescope to observe the heavens. He discovered craters on the Moon, sunspots on the Sun, the phases of Venus, and four moons orbiting Jupiter. His observations strongly suggested that the Earth orbits the Sun, not vice versa. (Eric Lessing/Art Resource)

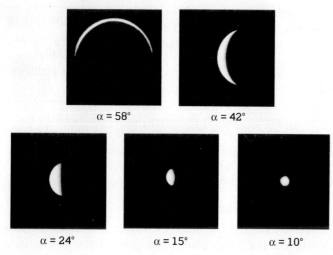

$\alpha = 58°$ $\alpha = 42°$

$\alpha = 24°$ $\alpha = 15°$ $\alpha = 10°$

Figure 4-13   R I V U X G

The Phases of Venus This series of photographs shows how the appearance of Venus changes as it moves along its orbit. The number below each view is the angular diameter α of the planet in arcseconds. Venus has the largest angular diameter when it is a crescent, and the smallest angular diameter when it is gibbous (nearly full). (New Mexico State University Observatory)

the Earth. To explain why Venus is never seen very far from the Sun, the Ptolemaic model had to assume that the deferents of Venus and of the Sun move together in lockstep, with the epicycle of Venus centered on a straight line between the Earth and the Sun (**Figure 4-15**). In this model, Venus was never on the opposite side of the Sun from the Earth, and so it could never have shown the gibbous phases that Galileo observed.

The Moons of Jupiter

Galileo also found more unexpected evidence for the ideas of Copernicus. In 1610 Galileo discovered four moons, now called the Galilean satellites, orbiting Jupiter (**Figure 4-16**). He realized that they were orbiting Jupiter because they appeared to move back and forth from one side of the planet to the other. **Figure 4-17** shows confirming observations made by Jesuit observers in 1620. Astronomers soon realized that the larger the orbit of one of the moons around Jupiter, the slower that moon moves and the longer it takes that moon to travel around its orbit. These are the same relationships that Copernicus deduced for the motions of the planets around the Sun. Thus, the moons of Jupiter behave like a Copernican system in miniature.

Galileo's telescopic observations constituted the first fundamentally new astronomical data in almost 2000 years. Contradicting prevailing opinion and religious belief, his discoveries strongly suggested a heliocentric structure of the universe. The Roman Catholic Church, which was a powerful political force in Italy and whose

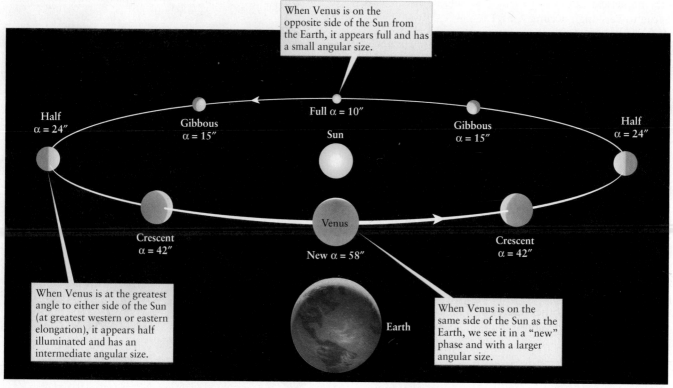

Figure 4-14

The Changing Appearance of Venus Explained in a Heliocentric Model A heliocentric model, in which the Earth and Venus both orbit the Sun, provides a natural explanation for the changing appearance of Venus shown in Figure 4-13.

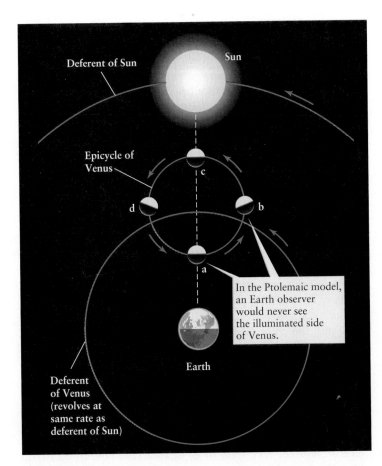

Figure 4-15

The Appearance of Venus in the Ptolemaic Model In the geocentric Ptolemaic model the deferents of Venus and the Sun rotate together, with the epicycle of Venus centered on a line (shown dashed) that connects the Sun and the Earth. In this model an Earth observer would never see Venus as more than half illuminated. (At positions *a* and *c*, Venus appears in a "new" phase; at positions *b* and *d*, it appears as a crescent. Compare with Figure 3-2, which shows the phases of the Moon.) Because Galileo saw Venus in nearly fully illuminated phases, he concluded that the Ptolemaic model must be incorrect.

Figure 4-16   R I V U X G

Jupiter and Its Largest Moons This photograph, taken by an amateur astronomer with a small telescope, shows the four Galilean satellites alongside an overexposed image of Jupiter. Each satellite is bright enough to be seen with the unaided eye, were it not overwhelmed by the glare of Jupiter. (Courtesy of C. Holmes)

Figure 4-17

Early Observations of Jupiter's Moons In 1610 Galileo discovered four "stars" that move back and forth across Jupiter from one night to the next. He concluded that these are four moons that orbit Jupiter, much as our Moon orbits the Earth. This drawing shows notations made by Jesuit observers on successive nights in 1620. The circle represents Jupiter and the stars its moons. Compare the drawing numbered 13 with the photograph in Figure 4-16. (Yerkes Observatory)

doctrine at the time placed the Earth at the center of the universe, cautioned Galileo not to advocate a heliocentric model. He nonetheless persisted and was sentenced to spend the last years of his life under house arrest "for vehement suspicion of heresy." Nevertheless, there was no turning back. (The Roman Catholic Church lifted its ban against Galileo's heliocentric ideas in the 1700s.)

While Galileo's observations showed convincingly that the Ptolemaic model was entirely wrong and that a heliocentric model is the more nearly correct one, he was unable to provide a complete explanation of why the Earth should orbit the Sun and not vice versa. The first person who was able to provide such an explanation was the Englishman Isaac Newton, born on Christmas Day of 1642, a dozen years after the death of Kepler and the same year that Galileo died. While Kepler and Galileo revolutionized our understanding of planetary motions, Newton's contribution was far greater: He deduced the basic laws that govern all motions on Earth as well as in the heavens.

4-6   Isaac Newton formulated three laws that describe fundamental properties of physical reality

Until the mid-seventeenth century, virtually all attempts to describe the motions of the heavens were *empirical,* or based directly on data and observations. From Ptolemy to Kepler, astronomers would adjust their ideas and calculations by trial and error until they ended up with answers that agreed with observation.

> **The same laws of motion that hold sway on Earth apply throughout the universe**

Isaac Newton (Figure 4-18) introduced a new approach. He began with three quite general statements, now called **Newton's laws of motion.** These laws, deduced from experimental observation, apply to all forces and all objects. Newton then showed that Kepler's three laws follow logically from these laws of motion and from a formula for the force of gravity that he derived from observation.

In other words, Kepler's laws are not just an empirical description of the motions of the planets, but a direct consequence of the fundamental laws of physical matter. Using this deeper insight into the nature of motions in the heavens, Newton and his

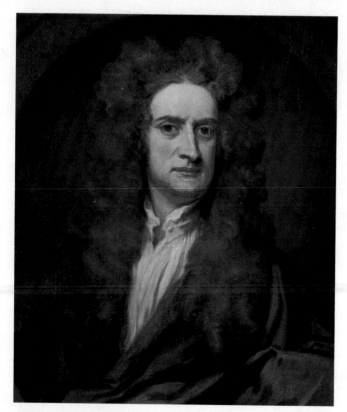

Figure 4-18

Isaac Newton (1642–1727) Using mathematical techniques that he devised, Isaac Newton formulated the law of universal gravitation and demonstrated that the planets orbit the Sun according to simple mechanical rules. (National Portrait Gallery, London)

successors were able to describe accurately not just the orbits of the planets but also the orbits of the Moon and comets.

Newton's First Law

Newton's laws of motion describe objects on Earth as well as in the heavens. Thus, we can understand each of these laws by considering the motions of objects around us. We begin with **Newton's first law of motion:**

An object remains at rest, or moves in a straight line at a constant speed, unless acted upon by a net outside force.

By **force** we mean any push or pull that acts on the object. An *outside* force is one that is exerted on the object by something other than the object itself. The net, or total, outside force is the combined effect of all of the individual outside forces that act on the object.

Right now, you are demonstrating the first part of Newton's first law. As you sit in your chair reading this, there are two outside forces acting on you: The force of gravity pulls you downward, and the chair pushes up on you. These two forces are of equal strength but of opposite direction, so their effects cancel—there is no *net* outside force. Hence, your body remains at rest as stated in Newton's first law. If you try to lift yourself out of your chair by grabbing your knees and pulling up, you will remain at rest because this force is not an outside force.

CAUTION! The second part of Newton's first law, about objects in motion, may seem to go against common sense. If you want to make this book move across the floor in a straight line at a constant speed, you must continually push on it. You might therefore think that there *is* a net outside force, the force of your push. But another force also acts on the book—the force of friction as the book rubs across the floor. As you push the book across the floor, the force of your push exactly balances the force of friction, so again there is no net outside force. The effect is to make the book move in a straight line at constant speed, just as Newton's first law says. If you stop pushing, there will be nothing to balance the effects of friction. Then there will be a net outside force and the book will slow to a stop.

Newton's first law tells us that if no net outside force acts on a moving object, there can be no net outside force acting on that object. This means that a net outside force *must* be acting on the planets. To see why, note that a planet moving through empty space encounters no friction, so the planet would tend to fly off into space at a steady speed along a straight line if there were no other outside force acting on it. Because this does not happen, Newton concluded that a force *must* act continuously on the planets to keep them in their elliptical orbits.

Newton's Second Law

Newton's second law describes how the motion of an object *changes* if there is a net outside force acting on it. To appreciate Newton's second law, we must first understand three quantities that describe motion—speed, velocity, and acceleration.

Speed is a measure of how fast an object is moving. Speed and direction of motion together constitute an object's **velocity.**

Compared with a car driving north at 100 km/h (62 mi/h), a car driving east at 100 km/h has the same speed but a different velocity. We can restate Newton's first law to say that an object has a constant velocity (its speed and direction of motion do not change) if no net outside force acts on the object.

Acceleration is the rate at which velocity changes. Because velocity involves both speed and direction, acceleration can result from changes in either. Contrary to popular use of the term, acceleration does not simply mean speeding up. A car is accelerating if it is speeding up, and it is also accelerating if it is slowing down or turning (that is, changing the direction in which it is moving).

You can verify these statements about acceleration if you think about the sensations of riding in a car. If the car is moving with a constant velocity (in a straight line at a constant speed), you feel the same as if the car were not moving at all. But you can feel it when the car accelerates in any way: You feel thrown back in your seat if the car speeds up, thrown forward if the car

slows down, and thrown sideways if the car changes direction in a tight turn. In Box 4-3 we discuss the reasons for these sensations, along with other applications of Newton's laws to everyday life.

An apple falling from a tree is a good example of acceleration that involves only an increase in speed. Initially, at the moment the stem breaks, the apple's speed is zero. After 1 second, its downward speed is 9.8 meters per second, or 9.8 m/s (32 feet per second, or 32 ft/s). After 2 seconds, the apple's speed is twice this, or 19.6 m/s. After 3 seconds, the speed is 29.4 m/s. Because the apple's speed increases by 9.8 m/s for each second of free fall, the rate of acceleration is 9.8 meters per second per second, or 9.8 m/s^2 (32 ft/s^2). Thus, the Earth's gravity gives the apple a constant acceleration of 9.8 m/s^2 downward, toward the center of the Earth.

A planet revolving about the Sun along a perfectly circular orbit is an example of acceleration that involves change of direction only. As the planet moves along its orbit, its speed remains

---

BOX 4-3   The Heavens on the Earth

Newton's Laws in Everyday Life

In our study of astronomy, we use Newton's three laws of motion to help us understand the motions of objects in the heavens. But you can see applications of Newton's laws every day in the world around you. By considering these everyday applications, we can gain insight into how Newton's laws apply to celestial events that are far removed from ordinary human experience.

Newton's *first* law, or principle of inertia, says that an object at rest naturally tends to remain at rest and that an object in motion naturally tends to remain in motion. This explains the sensations that you feel when riding in an automobile. When you are waiting at a red light, your car and your body are both at rest. When the light turns green and you press on the gas pedal, the car accelerates forward but your body attempts to stay where it was. Hence, the seat of the accelerating car pushes forward into your body, and it feels as though you are being pushed back in your seat.

Once the car is up to cruising speed, your body wants to keep moving in a straight line at this cruising speed. If the car makes a sharp turn to the left, the right side of the car will move toward you. Thus, you will feel as though you are being thrown to the car's right side (the side on the outside of the turn). If you bring the car to a sudden stop by pressing on the brakes, your body will continue moving forward until the seat belt stops you. In this case, it feels as though you are being thrown toward the front of the car.

Newton's *second* law states that the net outside force on an object equals the product of the object's mass and its acceleration. You can accelerate a crumpled-up piece of paper to a pretty good speed by throwing it with a moderate force. But if you try to throw a heavy rock by using the same force, the acceleration will be much less because the rock has much more

mass than the crumpled paper. Because of the smaller acceleration, the rock will leave your hand moving at only a slow speed.

Automobile airbags are based on the relationship between force and acceleration. It takes a large force to bring a fast-moving object suddenly to rest because this requires a large acceleration. In a collision, the driver of a car not equipped with airbags is jerked to a sudden stop and the large forces that act can cause major injuries. But if the car has airbags that deploy in an accident, the driver's body will slow down more gradually as it contacts the airbag, and the driver's acceleration will be less. (Remember that *acceleration* can refer to slowing down as well as to speeding up.) Hence, the force on the driver and the chance of injury will both be greatly reduced.

Newton's *third* law, the principle of action and reaction, explains how a car can accelerate at all. It is not correct to say that the engine pushes the car forward, because Newton's second law tells us that it takes a force acting from outside the car to make the car accelerate. Rather, the engine makes the wheels and tires turn, and the tires push backward on the ground. (You can see this backward force in action when a car drives through wet ground and sprays mud backward from the tires.) From Newton's third law, the ground must exert an equally large forward force on the car, and this is the force that pushes the car forward.

You use the same principles when you walk: You push backward on the ground with your foot, and the ground pushes forward on you. Icy pavement or a freshly waxed floor have greatly reduced friction. In these situations, your feet and the surface under you can exert only weak forces on each other, and it is much harder to walk.

constant. Nevertheless, the planet is continuously being accelerated because its direction of motion is continuously changing.

Newton's second law of motion says that in order to give an object an acceleration (that is, to change its velocity), a net outside force *must* act on the object. To be specific, this law says that the acceleration of an object is proportional to the net outside force acting on the object. That is, the harder you push on an object, the greater the resulting acceleration. This law can be succinctly stated as an equation. If a net outside force F acts on an object of mass m, the object will experience an acceleration a such that

Newton's second law

$$F = ma$$

F = net outside force on an object

m = mass of object

a = acceleration of object

The **mass** of an object is a measure of the total amount of material in the object. It is usually expressed in kilograms (kg) or grams (g). For example, the mass of the Sun is 2×10^{30} kg, the mass of a hydrogen atom is 1.7×10^{-27} kg, and the mass of an average adult is 75 kg. The Sun, a hydrogen atom, and a person have these masses regardless of where they happen to be in the universe.

CAUTION! It is important not to confuse the concepts of mass and weight. **Weight** is the force of gravity that acts on an object and, like any force, is usually expressed in pounds or newtons (1 newton = 0.225 pound).

We can use Newton's second law to relate mass and weight. We have seen that the acceleration caused by the Earth's gravity is 9.8 m/s². When a 50-kg swimmer falls from a diving board, the only outside force acting on her as she falls is her weight. Thus, from Newton's second law ($F = ma$), her weight is equal to her mass multiplied by the acceleration due to gravity:

50 kg $\times$ 9.8 m/s² = 490 newtons = 110 pounds

Note that this answer is correct only when the swimmer is on Earth. She would weigh less on the Moon, where the pull of gravity is weaker, and more on Jupiter, where the gravitational pull is stronger. Floating deep in space, she would have no weight at all; she would be "weightless." Nevertheless, in all these circumstances, she would always have exactly the same mass, because mass is an inherent property of matter unaffected by details of the environment. Whenever we describe the properties of planets, stars, or galaxies, we speak of their masses, never of their weights.

We have seen that a planet is continually accelerating as it orbits the Sun. From Newton's second law, this means that there must be a net outside force that acts continually on each of the planets. As we will see in the next section, this force is the gravitational attraction of the Sun.

Newton's Third Law

The last of Newton's general laws of motion, called **Newton's third law of motion,** is the famous statement about action and reaction:

Whenever one object exerts a force on a second object, the second object exerts an equal and opposite force on the first object.

For example, if you weigh 110 pounds, when you are standing up you are pressing down on the floor with a force of 110 pounds. Newton's third law tells us that the floor is also pushing up against your feet with an equal force of 110 pounds. You can think of each of these forces as a reaction to the other force, which is the origin of the phrase "action and reaction."

Newton realized that because the Sun is exerting a force on each planet to keep it in orbit, each planet must also be exerting an equal and opposite force on the Sun. However, the planets are much less massive than the Sun (for example, the Earth has only 1/300,000 of the Sun's mass). Therefore, although the Sun's force on a planet is the same as the planet's force on the Sun, the planet's much smaller mass gives it a much larger acceleration, according to Newton's second law. This is why the planets circle the Sun instead of vice versa. Thus, Newton's laws reveal the reason for our heliocentric solar system.

4-7 Newton's description of gravity accounts for Kepler's laws and explains the motions of the planets

Tie a ball to one end of a piece of string, hold the other end of the string in your hand, and whirl the ball around in a circle. As the ball "orbits" your hand, it is continuously accelerating because its velocity is changing. (Even if its speed is constant, its direction of motion is changing.) In accordance with Newton's second law, this can happen only if the ball is continuously acted on by an outside force—the pull of the string. The pull is directed along the string toward your hand. In the same way, Newton saw, the force that keeps a planet in orbit around the Sun is a pull that always acts toward the Sun. That pull is **gravity,** or **gravitational force.**

Newton's discovery about the forces that act on planets led him to suspect that the force of gravity pulling a falling apple straight down to the ground is fundamentally the same as the force on a planet that is always directed straight at the Sun.

> Newton's law of gravitation is truly universal: it applies to falling apples as well as to planets and galaxies

In other words, gravity is the force that shapes the orbits of the planets. What is more, he was able to determine how the force of gravity depends on the distance between the Sun and the planet. His result was a law of gravitation that could apply to the motion of distant planets as well as to the flight of a football on Earth. Using this law, Newton achieved the remarkable goal of deducing Kepler's laws from fundamental principles of nature.

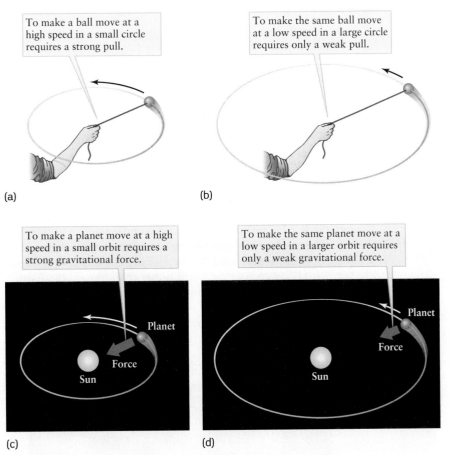

To make a ball move at a high speed in a small circle requires a strong pull.

To make the same ball move at a low speed in a large circle requires only a weak pull.

(a)

(b)

To make a planet move at a high speed in a small orbit requires a strong gravitational force.

To make the same planet move at a low speed in a larger orbit requires only a weak gravitational force.

Planet

Force

Sun

Planet

Force

Sun

(c)

(d)

Figure 4-19

An Orbit Analogy (a) To make a ball on a string move at high speed around a small circle, you have to exert a substantial pull on the string. (b) If you lengthen the string and make the same ball move at low speed around a large circle, much less pull is required. (c), (d) Similarly, a planet that orbits close to the Sun moves at high speed and requires a substantial gravitational force from the Sun, while a planet in a large orbit moves at low speed and requires less gravitational force to stay in orbit.

To see how Newton reasoned, think again about a ball attached to a string. If you use a short string, so that the ball orbits in a small circle, and whirl the ball around your hand at a high speed, you will find that you have to pull fairly hard on the string (Figure 4-19a). But if you use a longer string, so that the ball moves in a larger orbit, and if you make the ball orbit your hand at a slow speed, you only have to exert a light tug on the string (Figure 4-19b). The orbits of the planets behave in the same way; the larger the size of the orbit, the slower the planet's speed (Figure 4-19c, d). By analogy to the force of the string on the orbiting ball, Newton concluded that the force that attracts a planet toward the Sun must decrease with increasing distance between the Sun and the planet.

The Law of Universal Gravitation

Using his own three laws and Kepler's three laws, Newton succeeded in formulating a general statement that describes the nature of the gravitational force. Newton's **law of universal gravitation** is as follows:

Two objects attract each other with a force that is directly proportional to the mass of each object and inversely proportional to the square of the distance between them.

This law states that *any* two objects exert gravitational pulls on each other. Normally, you notice only the gravitational force that the Earth exerts on you, otherwise known as your weight. In fact, you feel gravitational attractions to *all* the objects around you. For example, this book is exerting a gravitational force on you as you read it. But because the force exerted on you by this book is proportional to the book's mass, which is very small compared to the Earth's mass, the force is too small to notice. (It can actually be measured with sensitive equipment.)

Consider two 1-kg objects separated by a distance of 1 meter. Newton's law of universal gravitation says that the force is directly proportional to the mass, so if we double the mass of one object to 2 kg, the force between the objects will double. If we double both masses so that we have two 2-kg objects separated by 1 meter, the force will be $2 \times 2 = 4$ times what it was originally (the force is directly proportional to the mass of *each* object). If we go back to two 1-kg masses, but double their separation to 2 meters, the force will be only one-quarter its original value. This is because the force is inversely proportional to the square of the distance: If we double the distance, the force is multiplied by a factor of

$$\frac{1}{2^2} = \frac{1}{4}$$

Newton's law of universal gravitation can be stated more succinctly as an equation. If two objects have masses m_1 and m_2 and are separated by a distance r, then the gravitational force F between these two objects is given by the following equation:

Newton's law of universal gravitation

$$F = G\left(\frac{m_1 m_2}{r^2}\right)$$

F = gravitational force between two objects

m_1 = mass of first object

m_2 = mass of second object

r = distance between objects

G = universal constant of gravitation

If the masses are measured in kilograms and the distance between them in meters, then the force is measured in newtons. In this formula, G is a number called the **universal constant of gravitation**. Laboratory experiments have yielded a value for G of

$$G = 6.67 \times 10^{-11} \text{ newton} \cdot \text{m}^2/\text{kg}^2$$

We can use Newton's law of universal gravitation to calculate the force with which any two objects attract each other. For example, to compute the gravitational force that the Sun exerts on the Earth, we substitute values for the Earth's mass ($m_1 = 5.98 \times 10^{24}$ kg), the Sun's mass ($m_2 = 1.99 \times 10^{30}$ kg), the distance between them ($r = 1$ AU $= 1.5 \times 10^{11}$ m), and the value of G into Newton's equation. We get

$$F_{\text{Sun-Earth}} = 6.67 \times 10^{-11}\left[\frac{(5.98 \times 10^{24}) \times (1.99 \times 10^{30})}{(1.50 \times 10^{11})^2}\right]$$

$$= 3.53 \times 10^{22} \text{ newtons}$$

If we calculate the force that the Earth exerts on the Sun, we get exactly the same result. (Mathematically, we just let m_1 be the Sun's mass and m_2 be the Earth's mass instead of the other way around. The product of the two numbers is the same, so the force is the same.) This is in accordance with Newton's third law: Any two objects exert *equal* gravitational forces on each other.

Your weight is just the gravitational force that the Earth exerts on you, so we can calculate it using Newton's law of universal gravitation. The Earth's mass is $m_1 = 5.98 \times 10^{24}$ kg, and the distance r to use is the distance between the *centers* of the Earth and you. This is just the radius of the Earth, which is $r = 6378$ km $= 6.378 \times 10^6$ m. If your mass is $m_2 = 50$ kg, your weight is

$$F_{\text{Earth-you}} = 6.67 \times 10^{-11}\left[\frac{(5.98 \times 10^{24}) \times (50)}{(6.378 \times 10^6)^2}\right]$$

$$= 490 \text{ newtons}$$

This is the same as the weight of a 50-kg person that we calculated in Section 4-6. This example shows that your weight would have a different value on a planet with a different mass m_1 and a different radius r.

Gravitational Force and Orbits

Because there is a gravitational force between any two objects, Newton concluded that gravity is also the force that keeps the Moon in orbit around the Earth. It is also the force that keeps artificial satellites in orbit. But if the force of gravity attracts two objects to each other, why don't satellites immediately fall to Earth? Why doesn't the Moon fall into the Earth? And, for that matter, why don't the planets fall into the Sun?

To see the answer, imagine (as Newton did) dropping a ball from a great height above the Earth's surface, as in **Figure 4-20**. After you drop the ball, it, of course, falls straight down (path A in Figure 4-20). But if you *throw* the ball horizontally, it travels some distance across the Earth's surface before hitting the ground (path B). If you throw the ball harder, it travels a greater distance (path C). If you could throw at just the right speed, the curvature of the ball's path will exactly match the curvature of the Earth's surface (path E). Although the Earth's gravity is making the ball fall, the Earth's surface is falling away under the ball at the same rate. Hence, the ball does not get any closer to the surface, and the ball is in circular orbit. So the ball in path E is in fact falling, but it is falling *around* the Earth rather than *toward* the Earth.

A spacecraft is launched into orbit in just this way—by throwing it fast enough. The thrust of a rocket is used to give the spacecraft the necessary orbital speed. Once the spacecraft is in orbit, the rocket turns off and the spacecraft falls continually around the Earth.

Figure 4-20

An Explanation of Orbits If a ball is dropped from a great height above the Earth's surface, it falls straight down (A). If the ball is thrown with some horizontal speed, it follows a curved path before hitting the ground (B, C). If thrown with just the right speed (E), the ball goes into circular orbit; the ball's path curves but it never gets any closer to the Earth's surface. If the ball is thrown with a speed that is slightly less (D) or slightly more (F) than the speed for a circular orbit, the ball's orbit is an ellipse.

CAUTION! An astronaut on board an orbiting spacecraft (like the one shown in the photograph that opens this chapter) feels "weightless." However, this is *not* because she is "beyond the pull of gravity." The astronaut is herself an independent satellite of the Earth, and the Earth's gravitational pull is what holds her in orbit. She feels "weightless" because she and her spacecraft are falling *together* around the Earth, so there is nothing pushing her against any of the spacecraft walls. You feel the same "weightless" sensation whenever you are falling, such as when you jump off a diving board or ride the free-fall ride at an amusement park.

If the ball in Figure 4-20 is thrown with a slightly slower speed than that required for a circular orbit, its orbit will be an ellipse (path D). An elliptical orbit also results if instead the ball is thrown a bit too fast (path F). In this way, spacecraft can be placed into any desired orbit around the Earth by adjusting the thrust of the rockets.

Just as the ball in Figure 4-20 will not fall to Earth if given enough speed, the Moon does not fall to Earth and the planets do not fall into the Sun. The planets acquired their initial speeds around the Sun when the solar system first formed 4.56 billion years ago. Figure 4-20 shows that a circular orbit is a very special case, so it is no surprise that the orbits of the planets are not precisely circular.

CAUTION! Orbiting satellites do sometimes fall out of orbit and crash back to Earth. When this happens, however, the real culprit is not gravity but air resistance. A satellite in a relatively low orbit is actually flying through the tenuous outer wisps of the Earth's atmosphere. The resistance of the atmosphere slows the satellite and changes a circular orbit like E in Figure 4-20 to an elliptical one like D. As the satellite sinks to lower altitude, it encounters more air resistance and sinks even lower. Eventually, it either strikes the Earth or burns up in flight due to air friction. By contrast, the Moon and planets orbit in the near-vacuum of interplanetary space. Hence, they are unaffected by this kind of air resistance, and their orbits are much more long-lasting.

Gravitation and Kepler's Laws

Using his three laws of motion and his law of gravity, Newton found that he could derive Kepler's three laws mathematically. Kepler's first law, concerning the elliptical shape of planetary orbits, proved to be a direct consequence of the $1/r^2$ factor in the law of universal gravitation. (Had the nature of gravity in our universe been different, so that this factor was given by a different function such as $1/r$ or $1/r^3$, elliptical orbits would not have been possible.) The law of equal areas, or Kepler's second law, turns out to be a consequence of the Sun's gravitational force on a planet being directed straight toward the Sun.

Newton also demonstrated that Kepler's third law follows logically from his law of gravity. Specifically, he proved that if two objects with masses m_1 and m_2 orbit each other, the period P of their orbit and the semimajor axis a of their orbit (that is,

the average distance between the two objects) are related by an equation that we call **Newton's form of Kepler's third law:**

$$P^2 = \left[\frac{4\pi^2}{G(m_1 + m_2)} \right] a^3$$

Newton's form of Kepler's third law is valid whenever two objects orbit each other because of their mutual gravitational attraction (**Figure 4-21**). It is invaluable in the study of binary star systems, in which two stars orbit each other. If the orbital period

Figure 4-21   R I V U X G

In Orbit Around the Moon This photograph taken from the spacecraft *Columbia* shows the lunar lander *Eagle* after returning from the first human landing on the Moon in July 1969. Newton's form of Kepler's third law describes the orbit of a spacecraft around the Moon, as well as the orbit of the Moon around the Earth (visible in the distance). (Michael Collins, *Apollo 11*, NASA)

WEB LINK 4.9

P and semimajor axis a of the two stars in a binary system are known, astronomers can use this formula to calculate the sum $m_1 + m_2$ of the masses of the two stars. Within our own solar system, Newton's form of Kepler's third law makes it possible to learn about the masses of planets. By measuring the period and semimajor axis for a satellite, astronomers can determine the sum of the masses of the planet and the satellite. (The satellite can be a moon of the planet or a spacecraft that we place in orbit around the planet. Newton's laws apply in either case.) Box 4-4 gives an example of using Newton's form of Kepler's third law.

Newton also discovered new features of orbits around the Sun. For example, his equations soon led him to conclude that the orbit of an object around the Sun need not be an ellipse. It could be any one of a family of curves called conic sections.

A **conic section** is any curve that you get by cutting a cone with a plane, as shown in Figure 4-22. You can get circles and ellipses by slicing all the way through the cone. You can also get two types of open curves called **parabolas** and **hyperbolas**. If you were to throw the ball in Figure 4-20 with a fast enough speed, it would follow a parabolic or hyperbolic orbit and would fly off into space, never to return. Comets hurtling toward the Sun from the depths of space sometimes follow hyperbolic orbits.

BOX 4-4   Tools of the Astronomer's Trade

Newton's Form of Kepler's Third Law

Kepler's original statement of his third law, $P^2 = a^3$, is valid only for objects that orbit the Sun. (Box 4-2 shows how to use this equation.) But Newton's form of Kepler's third law is much more general: It can be used in *any* situation where two objects of masses m_1 and m_2 orbit each other. For example, Newton's form is the equation to use for a moon orbiting a planet or a satellite orbiting the Earth. This equation is

Newton's form of Kepler's third law

$$P^2 = \left[\frac{4\pi^2}{G(m_1 + m_2)} \right] a^3$$

P = sidereal period of orbit, in seconds

a = semimajor axis of orbit, in meters

m_1 = mass of first object, in kilograms

m_2 = mass of second object, in kilograms

G = universal constant of gravitation = 6.67×10^{-11}

Notice that P, a, m_1, and m_2 *must* be expressed in these particular units. If you fail to use the correct units, your answer will be incorrect.

EXAMPLE: Io (pronounced "eye-oh") is one of the four large moons of Jupiter discovered by Galileo and shown in Figure 4-16. It orbits at a distance of 421,600 km from the center of Jupiter and has an orbital period of 1.77 days. Determine the combined mass of Jupiter and Io.

Situation: We are given Io's orbital period P and semimajor axis a (the distance from Io to the center of its circular orbit, which is at the center of Jupiter). Our goal is to find the sum of the masses of Jupiter (m_1) and Io (m_2).

Tools: Because this is not an orbit around the Sun, we must use Newton's form of Kepler's third law to relate P and a. This relationship also involves m_1 and m_2, whose sum $(m_1 + m_2)$ we are asked to find.

Answer: To solve for $m_1 + m_2$, we rewrite the equation in the form

$$m_1 + m_2 = \frac{4\pi^2 a^3}{GP^2}$$

To use this equation, we have to convert the distance a from kilometers to meters and convert the period P from days to seconds. There are 1000 meters in 1 kilometer and 86,400 seconds in 1 day, so

$$a = (421{,}600 \text{ km}) \times \frac{1000 \text{ m}}{1 \text{ km}} = 4.216 \times 10^8 \text{ m}$$

$$P = (1.77 \text{ days}) \times \frac{86{,}400 \text{ s}}{1 \text{ day}} = 1.529 \times 10^5 \text{ s}$$

We can now put these values and the value of G into the above equation:

$$m_1 + m_2 = \frac{4\pi^2 (4.216 \times 10^8)^3}{(6.67 \times 10^{-11})(1.529 \times 10^5)^2} = 1.90 \times 10^{27} \text{ kg}$$

Review: Io is very much smaller than Jupiter, so its mass is only a small fraction of the mass of Jupiter. Thus, $m_1 + m_2$ is very nearly the mass of Jupiter alone. We conclude that Jupiter has a mass of 1.90×10^{27} kg, or about 300 times the mass of Earth. This technique can be used to determine the mass of any object that has a second, much smaller object orbiting it. Astronomers use this technique to find the masses of stars, black holes, and entire galaxies of stars.

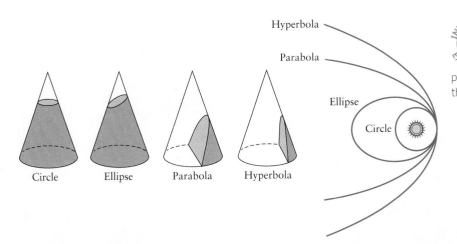

Figure 4-22

Conic Sections A conic section is any one of a family of curves obtained by slicing a cone with a plane. The orbit of one object about another can be any one of these curves: a circle, an ellipse, a parabola, or a hyperbola.

The Triumph of Newtonian Mechanics

Newton's ideas turned out to be applicable to an incredibly wide range of situations. Using his laws of motion, Newton himself proved that the Earth's axis of rotation must precess because of the gravitational pull of the Moon and the Sun on the Earth's equatorial bulge (see Figure 2-19). In fact, all the details of the orbits of the planets and their satellites could be explained mathematically with a body of knowledge built on Newton's work that is today called **Newtonian mechanics.**

Not only could Newtonian mechanics explain a variety of known phenomena in detail, but it could also predict new phenomena. For example, one of Newton's friends, Edmund Halley, was intrigued by three similar historical records of a comet that had been sighted at intervals of 76 years. Assuming these records to be accounts of the same comet, Halley used Newton's methods to work out the details of the comet's orbit and predicted its return in 1758. It was first sighted on Christmas night of 1757, a fitting memorial to Newton's birthday. To this day the comet bears Halley's name (Figure 4-23).

Another dramatic success of Newton's ideas was their role in the discovery of the eighth planet from the Sun. The seventh planet, Uranus, was discovered accidentally by William Herschel in 1781 during a telescopic survey of the sky. Fifty years later, however, it was clear that Uranus was not following its predicted orbit. John Couch Adams in England and Urbain Le Verrier in France independently calculated that the gravitational pull of a yet unknown, more distant planet could explain the deviations of Uranus from its orbit. Le Verrier predicted that the planet would be found at a certain location in the constellation of Aquarius. A brief telescopic search on September 23, 1846, revealed the planet Neptune within 1° of the calculated position. Before it was sighted with a telescope, Neptune was actually predicted with pencil and paper.

Because it has been so successful in explaining and predicting many important phenomena, Newtonian mechanics has become the cornerstone of modern physical science. Even today, as we send astronauts into Earth orbit and spacecraft to the outer planets, Newton's equations are used to calculate the orbits and trajectories of these spacecraft. The *Cosmic Connections* figure on the next page summarizes some of the ways that gravity plays an important role on scales from apples to galaxies.

In the twentieth century, scientists found that Newton's laws do not apply in all situations. A new theory called *quantum mechanics* had to be developed to explain the behavior of matter on the very smallest of scales, such as within the atom and within the atomic nucleus. Albert Einstein developed the *theory of relativity* to explain what happens at very high speeds approaching the speed of light and in places where gravitational forces are very strong. For many purposes in astronomy, however, Newton's laws are as useful today as when Newton formulated them more than three centuries ago.

Figure 4-23

Comet Halley This most famous of all comets orbits the Sun with an average period of about 76 years. During the twentieth century, the comet passed near the Sun in 1910 and again in 1986 (when this photograph was taken). It will next be prominent in the sky in 2061. (Harvard College Observatory/Photo Researchers, Inc.)

COSMIC CONNECTIONS   Universal Gravitation

Gravity is one of the fundamental forces of nature. We can see its effects here on Earth as well as in the farthest regions of the observable universe.[2]

The weight of an ordinary object is just the gravitational force exerted on that object by the Earth.

The Earth is held together by the mutual gravitational attraction of its parts.

Gravitational forces exerted by the Sun keep the planets in their orbits. The farther a planet is from the Sun, the weaker the gravitational force that acts on the planet.

Milky Way Galaxy

Sun

26,000 light-years

The mutual gravitational attraction of all the matter in the Milky Way Galaxy holds it together. The gravitational force of the Galaxy on our Sun and solar system holds us in an immense orbit around the galactic center.

4-8 Gravitational forces between the Earth and Moon produce tides

We have seen how Newtonian mechanics explains why the Moon stays in orbit around the Earth. It also explains why there are ocean tides, as well as why the Moon always keeps the same face toward the Earth. Both of these are the result of *tidal forces*—a consequence of gravity that deforms planets and reshapes galaxies.

Tidal forces are differences in the gravitational pull at different points in an object. As an illustration, imagine that three billiard balls are lined up in space at some distance from a planet, as in **Figure 4-24a**. According to Newton's law of universal gravitation, the force of attraction between two objects is greater the closer the two objects are to each other. Thus, the planet exerts more force on the 3-ball (in red) than on the 2-ball (in blue), and

> **Tidal forces reveal how gravitation can pull objects apart rather than drawing them together**

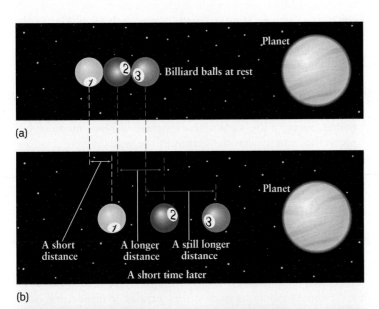

(a)

(b)

(c)

![Animation 4.8]

Figure 4-24

The Origin of Tidal Forces (a) Imagine three identical billiard balls placed some distance from a planet and released. **(b)** The closer a ball is to the planet, the more gravitational force the planet exerts on it. Thus, a short time after the balls are released, the blue 2-ball has moved farther toward the planet than the yellow 1-ball, and the red 3-ball has moved farther still. **(c)** From the perspective of the 2-ball in the center, it appears that forces have pushed the 1-ball away from the planet and pulled the 3-ball toward the planet. These forces are called tidal forces.

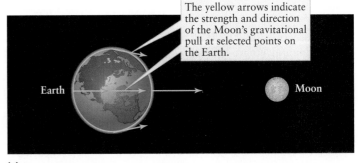

The yellow arrows indicate the strength and direction of the Moon's gravitational pull at selected points on the Earth.

(a)

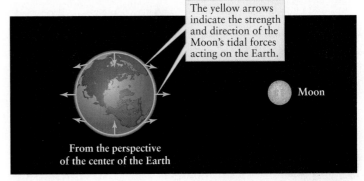

The yellow arrows indicate the strength and direction of the Moon's tidal forces acting on the Earth.

From the perspective of the center of the Earth

(b)

Figure 4-25

Tidal Forces on the Earth (a) The Moon exerts different gravitational pulls at different locations on the Earth. **(b)** At any location, the tidal force equals the Moon's gravitational pull at that location minus the gravitational pull of the Moon at the center of the Earth. These tidal forces tend to deform the Earth into a nonspherical shape.

exerts more force on the 2-ball than on the 1-ball (in yellow). Now, imagine that the three balls are released and allowed to fall toward the planet. Figure 4-24b shows the situation a short time later. Because of the differences in gravitational pull, a short time later the 3-ball will have moved farther than the 2-ball, which will in turn have moved farther than the 1-ball. But now imagine that same motion from the perspective of the 2-ball. From this perspective, it appears as though the 3-ball is pulled toward the planet while the 1-ball is pushed away (Figure 4-24c). These apparent pushes and pulls are called tidal forces.

Tidal Forces on the Earth

The Moon has a similar effect on the Earth as the planet in Figure 4-24 has on the three billiard balls. The arrows in **Figure 4-25a** indicate the strength and direction of the gravitational force of the Moon at several locations on the Earth. The side of the Earth closest to the Moon feels a greater gravitational pull than the Earth's center does, and the side of the Earth that faces away from the Moon feels less gravitational pull than does the Earth's center. This means that just as for the billiard balls in Figure 4-24, there are tidal forces acting on the Earth (Figure 4-25b). These tidal forces exerted by the Moon try to elongate the Earth along a line connecting the centers of the Earth and the Moon and

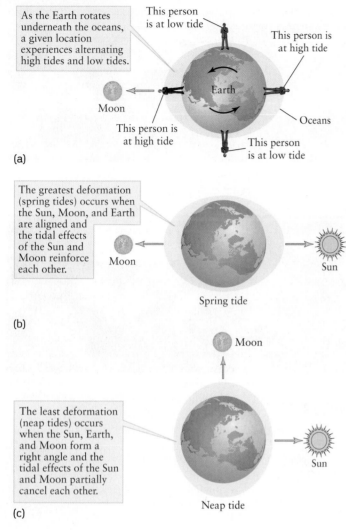

As the Earth rotates underneath the oceans, a given location experiences alternating high tides and low tides.

This person is at low tide

This person is at high tide

Moon

Earth

This person is at high tide

Oceans

This person is at low tide

(a)

The greatest deformation (spring tides) occurs when the Sun, Moon, and Earth are aligned and the tidal effects of the Sun and Moon reinforce each other.

Moon

Sun

Spring tide

(b)

Moon

The least deformation (neap tides) occurs when the Sun, Earth, and Moon form a right angle and the tidal effects of the Sun and Moon partially cancel each other.

Sun

Neap tide

(c)

Figure 4-26

High and Low Tides (a) The gravitational forces of the Moon and Sun deform the Earth's oceans, giving rise to low and high tides. **(b), (c)** The strength of the tides depends on the relative positions of the Sun, Moon, and Earth.

try to squeeze the Earth inward in the direction perpendicular to that line.

Because the body of the Earth is largely rigid, it cannot deform very much in response to the tidal forces of the Moon. But the water in the oceans can and does deform into a football shape, as Figure 4-26a shows. As the Earth rotates, a point on its surface goes from where the water is shallow to where the water is deep and back again. This is the origin of low and high ocean tides. (In this simplified description we have assumed that the Earth is completely covered with water. The full story of the tides is much more complex, because the shapes of the continents and the effects of winds must also be taken into account.)

The Sun also exerts tidal forces on the Earth's oceans. (The tidal effects of the Sun are about half as great as those of the Moon.) When the Sun, Moon, and the Earth are aligned, which happens at either new moon or full moon, the tidal effects of the Sun and Moon reinforce each other and the tidal distortion of the oceans is greatest. This produces large shifts in water level called **spring tides** (Figure 4-26b). At first quarter and last quarter, when the Sun and Moon form a right angle with the Earth, the tidal effects of the Sun and Moon partially cancel each other. Hence, the tidal distortion of the oceans is the least pronounced, producing smaller tidal shifts called **neap tides** (Figure 4-26c).

CAUTION! Note that spring tides have nothing to do with the season of the year called spring. Instead, the name refers to the way that the ocean level "springs up" to a greater than normal height. Spring tides occur whenever there is a new moon or full moon, no matter what the season of the year.

Tidal Forces on the Moon and Beyond

Just as the Moon exerts tidal forces on the Earth, the Earth exerts tidal forces on the Moon. Soon after the Moon formed some 4.56 billion years ago, it was molten throughout its volume. The Earth's tidal forces deformed the molten Moon into a slightly elongated shape, with the long axis of the Moon pointed toward the Earth. The Moon retained this shape and orientation when it cooled and solidified. To keep its long axis pointed toward the Earth, the Moon spins once on its axis as it makes one orbit around the Earth—that is, it is in synchronous rotation (Section 3-2). Hence, the same side of the Moon always faces the Earth,

Figure 4-27   R I **V** U X G

Tidal Forces on a Galaxy For millions of years the galaxies NGC 2207 and IC 2163 have been moving ponderously past each other. The larger galaxy's tremendous tidal forces have drawn a streamer of material a hundred thousand light-years long out of IC 2163. If you lived on a planet orbiting a star within this streamer, you would have a magnificent view of both galaxies. NGC 2207 and IC 2163 are respectively 143,000 light-years and 101,000 light-years in diameter. Both galaxies are 114 million light-years away in the constellation Canis Major. (NASA and the Hubble Heritage Team, AURA/STScI)

IC 2163

NGC 2207

Stars and gas pulled out of IC 2163 by tidal forces from NGC 2207 and this is the side that we see. For the same reason, most of the satellites in the solar system are in synchronous rotation, and thus always keep the same side facing their planet.

Tidal forces are also important on scales much larger than the solar system. Figure 4-27 shows two spiral galaxies, like the one in Figure 1-7, undergoing a near-collision. During the millions of years that this close encounter has been taking place, the tidal forces of the larger galaxy have pulled an immense streamer of stars and interstellar gas out of the smaller galaxy.

Many galaxies, including our own Milky Way Galaxy, show signs of having been disturbed at some time by tidal interactions with other galaxies. By their effect on the interstellar gas from which stars are formed, tidal interactions can actually trigger the birth of new stars. Our own Sun and solar system may have been formed as a result of tidal interactions of this kind. Hence, we may owe our very existence to tidal forces. In this and many other ways, the laws of motion and of universal gravitation shape our universe and our destinies.

Key Words

acceleration, p. 81
aphelion, p. 74
conic section, p. 86
conjunction, p. 70
deferent, p. 67
direct motion, p. 67
eccentricity, p. 74
ellipse, p. 74
elongation, p. 70
epicycle, p. 67
focus (of an ellipse; *plural* foci), p. 74
force, p. 80
geocentric model, p. 66
gravitational force, p. 82
gravity, p. 82
greatest eastern elongation, p. 70
greatest western elongation, p. 70
heliocentric model, p. 68
hyperbola, p. 86
inferior conjunction, p. 70
inferior planet, p. 69
Kepler's first law, p. 74
Kepler's second law, p. 75
Kepler's third law, p. 75
law of equal areas, p. 75
law of universal gravitation, p. 83
major axis (of an ellipse), p. 74

mass, p. 82
neap tides, p. 90
Newtonian mechanics, p. 87
Newton's first law of motion, p. 80
Newton's form of Kepler's third law, p. 85
Newton's second law of motion, p. 82
Newton's third law of motion, p. 82
Occam's razor, p. 67
opposition, p. 70
parabola, p. 86
parallax, p. 72
perihelion, p. 74
period (of a planet), p. 70
Ptolemaic system, p. 67
retrograde motion, p. 67
semimajor axis (of an ellipse), p. 74
sidereal period, p. 70
speed, p. 80
spring tides, p. 90
superior conjunction, p. 70
superior planet, p. 70
synodic period, p. 70
tidal forces, p. 89
universal constant of gravitation, p. 84
velocity, p. 80
weight, p. 82

Key Ideas

Apparent Motions of the Planets: Like the Sun and Moon, the planets move on the celestial sphere with respect to the back-

ground of stars. Most of the time a planet moves eastward in direct motion, in the same direction as the Sun and the Moon, but from time to time it moves westward in retrograde motion.

The Ancient Geocentric Model: Ancient astronomers believed the Earth to be at the center of the universe. They invented a complex system of epicycles and deferents to explain the direct and retrograde motions of the planets on the celestial sphere.

Copernicus's Heliocentric Model: Copernicus's heliocentric (Sun-centered) theory simplified the general explanation of planetary motions.

• In a heliocentric system, the Earth is one of the planets orbiting the Sun.

• A planet undergoes retrograde motion as seen from Earth when the Earth and the planet pass each other.

• The sidereal period of a planet, its true orbital period, is measured with respect to the stars. Its synodic period is measured with respect to the Earth and the Sun (for example, from one opposition to the next).

Kepler's Improved Heliocentric Model and Elliptical Orbits: Copernicus thought that the orbits of the planets were combinations of circles. Using data collected by Tycho Brahe, Kepler deduced three laws of planetary motion: (1) the orbits are in fact ellipses; (2) a planet's speed varies as it moves around its elliptical orbit; and (3) the orbital period of a planet is related to the size of its orbit.

Evidence for the Heliocentric Model: The invention of the telescope led Galileo to new discoveries that supported a heliocentric model. These included his observations of the phases of Venus and of the motions of four moons around Jupiter.

Newton's Laws of Motion: Isaac Newton developed three principles, called the laws of motion, that apply to the motions of objects on Earth as well as in space. These are (1) the tendency of an object to maintain a constant velocity, (2) the relationship between the net outside force on an object and the object's acceleration, and (3) the principle of action and reaction. These laws and Newton's law of universal gravitation can be used to deduce Kepler's laws. They lead to extremely accurate descriptions of planetary motions.

• The mass of an object is a measure of the amount of matter in the object. Its weight is a measure of the force with which the gravity of some other object pulls on it.

• In general, the path of one object about another, such as that of a planet or comet about the Sun, is one of the curves called conic sections: circle, ellipse, parabola, or hyperbola.

Tidal Forces: Tidal forces are caused by differences in the gravitational pull that one object exerts on different parts of a second object.

• The tidal forces of the Moon and Sun produce tides in the Earth's oceans.

• The tidal forces of the Earth have locked the Moon into synchronous rotation.

Questions

Review Questions

1. How did the ancient Greeks explain why the Sun and Moon slowly change their positions relative to the background stars?

2. In what direction does a planet move relative to the stars when it is in direct motion? When it is in retrograde motion? How do these compare with the direction in which we see the Sun move relative to the stars?

3. (a) In what direction does a planet move relative to the horizon over the course of one night? (b) The answer to (a) is the same whether the planet is in direct motion or retrograde motion. What does this tell you about the speed at which planets move on the celestial sphere?

4. What is an epicycle? How is it important in Ptolemy's explanation of the retrograde motions of the planets?

5. What is the significance of Occam's razor as a tool for analyzing theories?

6. How did the models of Aristarchus and Copernicus explain the retrograde motion of the planets?

7. How did Copernicus determine that the orbits of Mercury and Venus must be smaller than the Earth's orbit? How did he determine that the orbits of Mars, Jupiter, and Saturn must be larger than the Earth's orbit?

8. At what configuration (for example, superior conjunction, greatest eastern elongation, and so on) would it be best to observe Mercury or Venus with an Earth-based telescope? At what configuration would it be best to observe Mars, Jupiter, or Saturn? Explain your answers.

9. Is it ever possible to see Mercury at midnight? Explain your answer.

10. Which planets can never be seen at opposition? Which planets can never be seen at inferior conjunction? Explain your answers.

11. What is the difference between the synodic period and the sidereal period of a planet?

12. What is parallax? What did Tycho Brahe conclude from his attempt to measure the parallax of a supernova and a comet?

13. What observations did Tycho Brahe make in an attempt to test the heliocentric model? What were his results? Explain why modern astronomers get different results.

14. What are the foci of an ellipse? If the Sun is at one focus of a planet's orbit, what is at the other focus?

15. What are Kepler's three laws? Why are they important?

16. At what point in a planet's elliptical orbit does it move fastest? At what point does it move slowest? At what point does it sweep out an area at the fastest rate?

17. A line joining the Sun and an asteroid is found to sweep out an area of 6.3 AU^2 during 2010. How much area is swept out during 2011? Over a period of five years?

18. The orbit of a spacecraft about the Sun has a perihelion distance of 0.1 AU and an aphelion distance of 0.4 AU. What is the semimajor axis of the spacecraft's orbit? What is its orbital period?

19. A comet with a period of 125 years moves in a highly elongated orbit about the Sun. At perihelion, the comet comes very close to the Sun's surface. What is the comet's average

distance from the Sun? What is the farthest it can get from the Sun?

20. What observations did Galileo make that reinforced the heliocentric model? Why did these observations contradict the older model of Ptolemy? Why could these observations not have been made before Galileo's time?

21. Why does Venus have its largest angular diameter when it is new and its smallest angular diameter when it is full?

22. What are Newton's three laws? Give an everyday example of each law.

23. How much force do you have to exert on a 3-kg brick to give it an acceleration of 2 m/s^2? If you double this force, what is the brick's acceleration? Explain.

24. What is the difference between weight and mass?

25. What is your weight in pounds and in newtons? What is your mass in kilograms?

26. Suppose that the Earth were moved to a distance of 3.0 AU from the Sun. How much stronger or weaker would the Sun's gravitational pull be on the Earth? Explain.

27. How far would you have to go from Earth to be completely beyond the pull of its gravity? Explain.

28. What are conic sections? In what way are they related to the orbits of planets in the solar system?

29. Why was the discovery of Neptune an important confirmation of Newton's law of universal gravitation?

30. What is a tidal force? How do tidal forces produce tides in the Earth's oceans?

31. What is the difference between spring tides and neap tides?

Advanced Questions

Questions preceded by an asterisk (*) involve topics discussed in the Boxes.

32. Figure 4-2 shows the retrograde motion of Mars as seen from Earth. Sketch a similar figure that shows how Earth would appear to move against the background of stars during this same time period as seen by an observer on Mars.

*33. The synodic period of Mercury (an inferior planet) is 115.88 days. Calculate its sidereal period in days.

*34. Table 4-1 shows that the synodic period is *greater* than the sidereal period for Mercury, but the synodic period is *less* than the sidereal period for Jupiter. Draw diagrams like the one in Box 4-1 to explain why this is so.

*35. A general rule for superior planets is that the greater the average distance from the planet to the Sun, the more frequently that planet will be at opposition. Explain how this rule comes about.

36. In 2006, Mercury was at greatest western elongation on April 8, August 7, and November 25. It was at greatest eastern elongation on February 24, June 20, and October 17. Does Mercury take longer to go from eastern to western elongation, or vice versa? Explain why, using Figure 4-6.

37. Explain why the semimajor axis of a planet's orbit is equal to the average of the distance from the Sun to the planet at perihelion (the *perihelion distance*) and the distance from the Sun to the planet at aphelion (the *aphelion distance*).

38. A certain comet is 2 AU from the Sun at perihelion and 16 AU from the Sun at aphelion. (a) Find the semimajor axis of the comet's orbit. (b) Find the sidereal period of the orbit.

39. A comet orbits the Sun with a sidereal period of 64.0 years. (a) Find the semimajor axis of the orbit. (b) At aphelion, the comet is 31.5 AU from the Sun. How far is it from the Sun at perihelion?

40. One trajectory that can be used to send spacecraft from the Earth to Mars is an elliptical orbit that has the Sun at one focus, its perihelion at the Earth, and its aphelion at Mars. The spacecraft is launched from Earth and coasts along this ellipse until it reaches Mars, when a rocket is fired to either put the spacecraft into orbit around Mars or cause it to land on Mars. (a) Find the semimajor axis of the ellipse. (*Hint:* Draw a picture showing the Sun and the orbits of the Earth, Mars, and the spacecraft. Treat the orbits of the Earth and Mars as circles.) (b) Calculate how long (in days) such a one-way trip to Mars would take.

41. The mass of the Moon is 7.35×10^{22} kg, while that of the Earth is 5.98×10^{24} kg. The average distance from the center of the Moon to the center of the Earth is 384,400 km. What is the size of the gravitational force that the Earth exerts on the Moon? What is the size of the gravitational force that the Moon exerts on the Earth? How do your answers compare with the force between the Sun and the Earth calculated in the text?

42. The mass of Saturn is approximately 100 times that of Earth, and the semimajor axis of Saturn's orbit is approximately 10 AU. To this approximation, how does the gravitational force that the Sun exerts on Saturn compare to the gravitational force that the Sun exerts on the Earth? How do the accelerations of Saturn and the Earth compare?

43. Suppose that you traveled to a planet with 4 times the mass and 4 times the diameter of the Earth. Would you weigh more or less on that planet than on Earth? By what factor?

44. On Earth, a 50-kg astronaut weighs 490 newtons. What would she weigh if she landed on Jupiter's moon Callisto? What fraction is this of her weight on Earth? See Appendix 3 for relevant data about Callisto.

45. Imagine a planet like the Earth orbiting a star with 4 times the mass of the Sun. If the semimajor axis of the planet's orbit is 1 AU, what would be the planet's sidereal period? (*Hint:* Use Newton's form of Kepler's third law. Compared with the case of the Earth orbiting the Sun, by what factor has the quantity $m_1 + m_2$ changed? Has a changed? By what factor must P^2 change?)

46. A satellite is said to be in a "geosynchronous" orbit if it appears always to remain over the exact same spot on the rotating Earth. (a) What is the period of this orbit? (b) At what distance from the center of the Earth must such a satellite be placed into orbit? (*Hint:* Use Newton's form of Kepler's third law.) (c) Explain why the orbit must be in the plane of the Earth's equator.

47. Figure 4-21 shows the lunar module *Eagle* in orbit around the Moon after completing the first successful lunar landing in July 1969. (The photograph was taken from the command module *Columbia*, in which the astronauts returned to Earth.) The spacecraft orbited 111 km above the surface of the Moon. Calculate the period of the spacecraft's orbit. See Appendix 3 for relevant data about the Moon.

*48. In Box 4-4 we analyze the orbit of Jupiter's moon Io. Look up information about the orbits of Jupiter's three other large moons (Europa, Ganymede, and Callisto) in Appendix 3. Demonstrate that these data are in agreement with Newton's form of Kepler's third law.

*49. Suppose a newly discovered asteroid is in a circular orbit with synodic period 1.25 years. The asteroid lies between the orbits of Mars and Jupiter. (a) Find the sidereal period of the orbit. (b) Find the distance from the asteroid to the Sun.

50. The average distance from the Moon to the center of the Earth is 384,400 km, and the diameter of the Earth is 12,756 km. Calculate the gravitational force that the Moon exerts (a) on a 1-kg rock at the point on the Earth's surface closest to the Moon and (b) on a 1-kg rock at the point on the Earth's surface farthest from the Moon. (c) Find the difference between the two forces you calculated in parts (a) and (b). This difference is the tidal force pulling these two rocks away from each other, like the 1-ball and 3-ball in Figure 4-24. Explain why tidal forces cause only a very small deformation of the Earth.

Discussion Questions

51. Which planet would you expect to exhibit the greatest variation in apparent brightness as seen from Earth? Which planet would you expect to exhibit the greatest variation in angular diameter? Explain your answers.

52. Use two thumbtacks, a loop of string, and a pencil to draw several ellipses. Describe how the shape of an ellipse varies as the distance between the thumbtacks changes.

Web/eBook Questions

53. (a) Search the World Wide Web for information about Kepler. Before he realized that the planets move on elliptical paths, what other models of planetary motion did he consider? What was Kepler's idea of "the music of the spheres"? (b) Search the World Wide Web for information about Galileo. What were his contributions to physics? Which of Galileo's new ideas were later used by Newton to construct his laws of motion? (c) Search the World Wide Web for information about Newton. What were some of the contributions that he made to physics other than developing his laws of motion? What contributions did he make to mathematics?

54. Monitoring the Retrograde Motion of Mars. Watching Mars night after night reveals that it changes its position with respect to the background stars. To track its motion, access and view the animation "The Path of Mars in 2011–2012" in Chapter 4 of the *Universe* Web site or eBook. (a) Through which constellations does Mars move? (b) On approximately what date does Mars stop its direct (west-to-east) motion and begin its retrograde motion? (*Hint:* Use the "Stop" function on your animation controls.) (c) Over how many days does Mars move retrograde?

Activities

Observing Projects

55. It is quite probable that within a few weeks of your reading this chapter one of the planets will be near opposition or greatest eastern elongation, making it readily visible in the evening sky. Select a planet that is at or near such a configuration by searching the World Wide Web or by consulting a reference book, such as the current issue of the *Astronomical Almanac* or the pamphlet entitled *Astronomical Phenomena* (both published by the U.S. government). At that configuration, would you expect the planet to be moving rapidly or slowly from night to night against the background stars? Verify your expectations by observing the planet once a week for a month, recording your observations on a star chart.

56. If Jupiter happens to be visible in the evening sky, observe the planet with a small telescope on five consecutive clear nights. Record the positions of the four Galilean satellites by making nightly drawings, just as the Jesuit priests did in 1620 (see Figure 4-17). From your drawings, can you tell which moon orbits closest to Jupiter and which orbits farthest? Was there a night when you could see only three of the moons? What do you suppose happened to the fourth moon on that night?

57. If Venus happens to be visible in the evening sky, observe the planet with a small telescope once a week for a month. On each night, make a drawing of the crescent that you see. From your drawings, can you determine if the planet is nearer or farther from the Earth than the Sun is? Do your drawings show any changes in the shape of the crescent from one week to the next? If so, can you deduce if Venus is coming toward us or moving away from us?

58. Use the *Starry Night Enthusiast™* program to observe the moons of Jupiter. Display the entire celestial sphere (select **Guides > Atlas** in the **Favourites** menu) and center on the planet Jupiter (double-click the entry for Jupiter in the **Find** pane). Then use the zoom controls at the right-hand end of the toolbar (at the top of the main window) to adjust your view to about **20′ × 14′** so that you can see the planet and its four Galilean satellites. (Click on the + button or press the + key on the keyboard to zoom in and click on the − button or press the − key on the keyboard to zoom out.) This is equivalent to increasing the magnification of a telescope while looking at Jupiter. (a) Click on the **Time Flow Rate** control (immedi-

ately to the right of the date and time display) and set the discrete time step to **2 hours.** Using the **Step Forward** button (just to the right of the **Forward** button), observe and draw the positions of the moons relative to Jupiter at 2-hour intervals. (b) From your drawings, can you tell which moon orbits closest to Jupiter and which orbits farthest away? Explain your reasoning. (c) Are there times when only three of the satellites are visible? What happens to the fourth moon at those times?

59. Use the *Starry Night Enthusiast™* program to observe the changing appearance of Mercury. Display the entire celestial sphere (select **Guides > Atlas** in the **Favourites** menu) and center on Mercury (double-click the entry for Mercury in the **Find** pane); then use the zoom controls at the right-hand end of the toolbar (at the top of the main window) to adjust your view so that you can clearly see details on the planet's surface. (Click on the + button to zoom in and on the − button to zoom out.) (a) Click on the **Time Flow Rate** control (immediately to the right of the date and time display) and set the discrete time step to **1 day.** Using the **Step Forward** button, observe and record the changes in Mercury's phase and apparent size from one day to the next. Run time forward for some time to see these changes more graphically. (b) Explain why the phase and apparent size change in the way that you observe.

60. Use the *Starry Night Enthusiast™* program to observe retrograde motion. Display the entire celestial sphere (select **Guides > Atlas** in the **Favourites** menu) and center on Mars (double-click the entry for Mars in the **Find** pane). Set the **Date** (in the time and date display in the toolbar) to **January 1, 2005.** Then click on the **Time Flow Rate** control (immediately to the right of the date and time display) and set the discrete time step to **1 day.** Press and hold down the − key on the keyboard to zoom out to the maximum field of view. Then click on the **Forward** button (a triangle that points to the right) and watch the motion of Mars for two years of simulated time. Since you have centered Mars in the view, the sky appears to move but the relative motion of Mars against this sky is obvious. (a) For most of the two-year period, does Mars move generally to the left (eastward) or to the right (westward) on the celestial sphere? On what date during this period does this *direct* motion end, so that Mars appears to come to a momentary halt, and retrograde motion begins? On what date does *retrograde* motion end and direct motion resume? You may want to use the **Stop** button and the **Back** button (a triangle that points to the left) to help you to pin down the exact dates of these events. (b) You have been observing the motion of Mars as seen from Earth. To observe the motion of Earth as seen from Mars, locate yourself on the north pole of Mars. To do this, select **Viewing Location . . .** in the **Options** menu. In the Viewing Location dialog box that appears on the screen, change the pull-down menus next to the words "View from:" so that they read "the surface of" and "Mars." Click on **Latitude/Longitude** tab in the **Viewing Location** dialog box and change the Latitude to **90° N.** Then click on the **Set Location** button. You will now see a Martian horizon and a pink Martian sky, both of which will interfere with your view of the Earth as

seen from Mars. Remove the horizon by clicking the **Options** tab on the left of the main view to open the **Options** pane. Under the **Local View** heading in the **Options** pane, click the checkbox to the left of the item labeled **Local Horizon.** Then, in the View menu, select **Hide Daylight** to remove the Martian sky. Now center the field of view on the **Earth** (double-click the entry for Earth in the **Find** pane). Set the date to **January 1, 2005,** set the discrete time step to **1 day,** and click on the **Forward** button. As before, watch the motion for two years of simulated time. In which direction does the Earth appear to move most of the time? On what date does its motion change from direct to retrograde? On what date does its motion change from retrograde back to direct? Are these roughly the same dates as you found in part (a)? (c) To understand the motions of Mars as seen from Earth and vice versa, observe the motion of the planets from a point above the solar system. Again, select **Viewing Location . . .** in the Options menu. In the Viewing Location dialog box that appears on the screen, select "stationary location" from the pull-down menu immediately to the right of the words "View from." Then type **0 au** in the **X:** box, **0 au** in the **Y:** box, and **5 au** in the **Z:** box, and click on the **Set Location** button. This will place you at a position 5 AU above the plane of the solar system. Then center the field of view on the Sun (double-click the entry for the Sun in the **Find** pane). Once again, set the date to **January 1, 2005,** set the discrete time step to **1 day,** and click on the **Forward** button. Watch the motions of the planets for two years of simulated time. On what date during this two-year period is Earth directly between Mars and the Sun? How does this date compare to the two dates you recorded in part (a) and the two dates you recorded in part (b)? Explain the significance of this. (*Hint:* see See Figure 4-5.)

5

The Nature of Light

R I **V** U X G

The Ring Nebula is a shell of glowing gases surrounding a dying star. The spectrum of the emitted light reveals which gases are present. (Hubble Heritage Team, AURA/STScI/NASA)

In the early 1800s, the French philosopher Auguste Comte argued that because the stars are so far away, humanity would never know their nature and composition. But the means to learn about the stars was already there for anyone to see—starlight. Just a few years after Comte's bold pronouncement, scientists began analyzing starlight to learn the very things that he had deemed unknowable.

We now know that atoms of each chemical element emit and absorb light at a unique set of wavelengths characteristic of that element alone. The red light in the accompanying image of a gas cloud in space is of a wavelength emitted by nitrogen and no other element; the particular green light in this image is unique to oxygen, and the particular blue light is unique to helium. The light from nearby planets, distant stars, and remote galaxies also has characteristic "fingerprints" that reveal the chemical composition of these celestial objects.

In this chapter we learn about the basic properties of light. Light has a dual nature: it has the properties of both waves and particles. The light emitted by an object depends upon the object's temperature; we can use this to determine the surface temperatures of stars. By studying the structure of atoms, we will learn why each element emits and absorbs light only at specific wavelengths and will see how astronomers determine what the atmospheres of planets and stars are made of. The motion of a light source also affects wavelengths, permitting us to deduce how fast stars and other objects are approaching or receding. These are but a few of the reasons why understanding light is a prerequisite to understanding the universe.

5-1 Light travels through empty space at a speed of 300,000 km/s

Galileo Galilei and Isaac Newton were among the first to ask basic questions about light. Does light travel instantaneously from one place to another, or does it move with a measurable speed? Whatever the nature of light, it does seem to travel swiftly from a source to our eyes. We see a distant event before we hear the accompanying sound. (For example, we see a flash of lightning before we hear the thunderclap.)

In the early 1600s, Galileo tried to measure the speed of light. He and an assistant stood at night on two hilltops a known distance apart, each holding a shuttered lantern. First, Galileo opened the shutter of

The speed of light in a vacuum is a universal constant: It has the same value everywhere in the cosmos

Learning Goals

By reading the sections of this chapter, you will learn

5-1 How we measure the speed of light

5-2 How we know that light is an electromagnetic wave

5-3 How an object's temperature is related to the radiation it emits

5-4 The relationship between an object's temperature and the amount of energy it emits

5-5 The evidence that light has both particle and wave aspects

5-6 How astronomers can detect an object's chemical composition by studying the light it emits

5-7 The quantum rules that govern the structure of an atom

5-8 The relationship between atomic structure and the light emitted by objects

5-9 How an object's motion affects the light we receive from that object his lantern; as soon as his assistant saw the flash of light, he opened his own. Galileo used his pulse as a timer to try to measure the time between opening his lantern and seeing the light from his assistant's lantern. From the distance and time, he hoped to compute the speed at which the light had traveled to the distant hilltop and back.

Galileo found that the measured time failed to increase noticeably, no matter how distant the assistant was stationed. Galileo therefore concluded that the speed of light is too high to be measured by slow human reactions. Thus, he was unable to tell whether or not light travels instantaneously.

The Speed of Light: Astronomical Measurements

The first evidence that light does *not* travel instantaneously was presented in 1676 by Olaus Rømer, a Danish astronomer. Rømer had been studying the orbits of the moons of Jupiter by carefully timing the moments when they passed into or out of Jupiter's shadow. To Rømer's surprise, the timing of these eclipses of Jupiter's moons seemed to depend on the relative positions of Jupiter and the Earth. When the Earth was far from Jupiter (that is, near conjunction; see Figure 4-6), the eclipses occurred several minutes later than when the Earth was close to Jupiter (near opposition).

Rømer realized that this puzzling effect could be explained if light needs time to travel from Jupiter to the Earth. When the Earth is closest to Jupiter, the image of a satellite disappearing into Jupiter's shadow arrives at our telescopes a little sooner than it does when Jupiter and the Earth are farther apart (Figure 5-1). The range of variation in the times at which such eclipses are observed is about 16.6 minutes, which Rømer interpreted as the length of time required for light to travel across the diameter of the Earth's orbit (a distance of 2 AU). The size of the Earth's orbit was not accurately known in Rømer's day, and he never actually calculated the speed of light. Today, using the modern value of 150 million kilometers for the astronomical unit, Rømer's

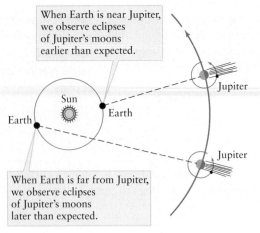

When Earth is near Jupiter, we observe eclipses of Jupiter's moons earlier than expected.

When Earth is far from Jupiter, we observe eclipses of Jupiter's moons later than expected.

Figure 5-1

Rømer's Proof That Light Does Not Travel Instantaneously The timing of eclipses of Jupiter's moons as seen from Earth depends on the Earth-Jupiter distance. Rømer correctly attributed this effect to variations in the time required for light to travel from Jupiter to the Earth.

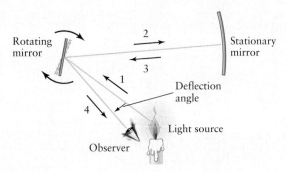

Figure 5-2

The Fizeau-Foucault Method of Measuring the Speed of Light Light from a light source (1) is reflected off a rotating mirror to a stationary mirror (2) and from there back to the rotating mirror (3). The ray that reaches the observer (4) is deflected away from the path of the initial beam because the rotating mirror has moved slightly while the light was making the round trip. The speed of light is calculated from the deflection angle and the dimensions of the apparatus.

method yields a value for the speed of light equal to roughly 300,000 km/s (186,000 mi/s).

The Speed of Light: Measurements on Earth

Almost two centuries after Rømer, the speed of light was measured very precisely in an experiment carried out on Earth. In 1850, the French physicists Armand-Hippolyte Fizeau and Jean Foucault built the apparatus sketched in Figure 5-2. Light from a light source reflects from a rotating mirror toward a stationary mirror 20 meters away. The rotating mirror moves slightly while the light is making the round trip, so the returning light ray is deflected away from the source by a small angle. By measuring this angle and knowing the dimensions of their apparatus, Fizeau and Foucault could deduce the speed of light. Once again, the answer was very nearly 300,000 km/s.

The speed of light in a vacuum is usually designated by the letter *c* (from the Latin *celeritas*, meaning "speed"). The modern value is $c = 299{,}792.458$ km/s (186,282.397 mi/s). In most calculations you can use

$$c = 3.00 \times 10^5 \text{ km/s} = 3.00 \times 10^8 \text{ m/s}$$

The most convenient set of units to use for *c* is different in different situations. The value in kilometers per second (km/s) is often most useful when comparing *c* to the speeds of objects in space, while the value in meters per second (m/s) is preferred when doing calculations involving the wave nature of light (which we will discuss in Section 5-2).

CAUTION! Note that the quantity *c* is the speed of light *in a vacuum*. Light travels more slowly through air, water, glass, or any other transparent substance than it does in a vacuum. In our study of astronomy, however, we will almost always consider light traveling through the vacuum (or near-vacuum) of space.

The speed of light in empty space is one of the most important numbers in modern physical science. This value appears in many equations that describe atoms, gravity, electricity, and mag-

Figure 5-3

A Prism and a Spectrum When a beam of sunlight passes through a glass prism, the light is broken into a rainbow-colored band called a spectrum. The wavelengths of different colors of light are shown on the right (1 nm = 1 nanometer = 10^{-9} m).

netism. According to Einstein's special theory of relativity, nothing can travel faster than the speed of light.

5-2   Light is electromagnetic radiation and is characterized by its wavelength

Light is energy. This fact is apparent to anyone who has felt the warmth of the sunshine on a summer's day. But what exactly is light? How is it produced? What is it made of? How does it move through space? Scholars have struggled with these questions throughout history.

> **Visible light, radio waves, and X rays are all the same type of wave: They differ only in their wavelength**

Newton and the Nature of Color

The first major breakthrough in understanding light came from a simple experiment performed by Isaac Newton around 1670. Newton was familiar with what he called the "celebrated Phenomenon of Colours," in which a beam of sunlight passing through a glass prism spreads out into the colors of the rainbow (**Figure 5-3**). This rainbow is called a **spectrum** (plural **spectra**).

Until Newton's time, it was thought that a prism somehow added colors to white light. To test this idea, Newton placed a second prism so that just one color of the spectrum passed through it (**Figure 5-4**). According to the old theory, this should have caused a further change in the color of the light. But Newton found that each color of the spectrum was unchanged by the second prism; red remained red, blue remained

Figure 5-4

Newton's Experiment on the Nature of Light In a crucial experiment, Newton took sunlight that had passed through a prism and sent it through a second prism. Between the two prisms was a screen with a hole in it that allowed only one color of the spectrum to pass through. This same color emerged from the second prism. Newton's experiment proved that prisms do not add color to light but merely bend different colors through different angles. It also proved that white light, such as sunlight, is actually a combination of all the colors that appear in its spectrum.

blue, and so on. He concluded that a prism merely separates colors and does not add color. Hence, the spectrum produced by the first prism shows that sunlight is a mixture of all the colors of the rainbow.

Newton suggested that light is composed of particles too small to detect individually. In 1678, however, the Dutch physicist and astronomer Christiaan Huygens proposed a rival explanation. He suggested that light travels in the form of waves rather than particles.

Young and the Wave Nature of Light

Around 1801, Thomas Young in England carried out an experiment that convincingly demonstrated the wavelike aspect of light. He passed a beam of light through two thin, parallel slits in an opaque screen, as shown in Figure 5-5a. On a white surface some distance beyond the slits, the light formed a pattern of alternating bright and dark bands. Young reasoned that if a beam of light was a stream of particles (as Newton had suggested), the two beams of light from the slits should simply form bright images of the slits on the white surface. The pattern of bright and dark bands he observed is just what would be expected, however, if light had wavelike properties. An analogy with water waves demonstrates why.

ANALOGY Imagine ocean waves pounding against a reef or breakwater that has two openings (Figure 5-5b). A pattern of ripples is formed on the other side of the barrier as the waves come through the two openings and interfere with each other. At certain points, wave crests arrive simultaneously from the two openings. These reinforce each other and produce high waves. At other points, a crest from one opening meets a trough from the other opening. These cancel each other out, leaving areas of still water. This process of combining two waves also takes place in Young's double-slit experiment: The bright bands are regions where waves from the two slits reinforce each other, while the dark bands appear where waves from the two slits cancel each other.

Maxwell and Light as an Electromagnetic Wave

The discovery of the wave nature of light posed some obvious questions. What exactly is "waving" in light? That is, what is it about light that goes up and down like water waves on the ocean? Because we can see light from the Sun, planets, and stars, light waves must be able to travel across empty space. Hence, whatever is "waving" cannot be any material substance. What, then, is it?

The answer came from a seemingly unlikely source—a comprehensive theory that described electricity and magnetism. Numerous experiments during the first half of the nineteenth century demonstrated an intimate connection between electric and magnetic forces. A central idea to emerge from these experiments is the concept of a *field,* an immaterial yet measurable disturbance of any region of space in which electric or magnetic forces are felt. Thus, an electric charge is surrounded by an electric field, and a magnet is surrounded by a magnetic field. Experiments in the early 1800s demonstrated that moving an electric charge produces a magnetic field; conversely, moving a magnet gives rise to an electric field.

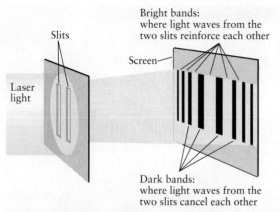

(a) An experiment with light

(b) An analogous experiment with water waves

Figure 5-5

Young's Double-Slit Experiment (a) Thomas Young's classic double-slit experiment can easily be repeated in the modern laboratory by shining light from a laser onto two closely spaced parallel slits. Alternating dark and bright bands appear on a screen beyond the slits. (b) The intensity of light on the screen in (a) is analogous to the height of water waves that pass through a barrier with two openings. (The photograph shows this experiment with water waves in a small tank.) In certain locations, wave crests from both openings reinforce each other to produce extra high waves. At other locations a crest from one opening meets a trough from the other. The crest and trough cancel each other, producing still water. (Eric Schrempp/Photo Researchers)

In the 1860s, the Scottish mathematician and physicist James Clerk Maxwell succeeded in describing all the basic properties of electricity and magnetism in four equations. This mathematical achievement demonstrated that electric and magnetic forces are really two aspects of the same phenomenon, which we now call **electromagnetism.**

By combining his four equations, Maxwell showed that electric and magnetic fields should travel through space in the form of waves at a speed of 3.0×10^5 km/s—a value exactly equal to the best available value for the speed of light. Maxwell's sugges-

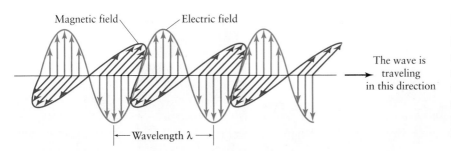

Magnetic field   Electric field

The wave is traveling in this direction

Wavelength λ

Figure 5-6

Electromagnetic Radiation All forms of light consist of oscillating electric and magnetic fields that move through space at a speed of 3.00×10^5 km/s = 3.00×10^8 m/s. This figure shows a "snapshot" of these fields at one instant. The distance between two successive crests, called the wavelength of the light, is usually designated by the Greek letter λ (lambda).

tion that these waves do exist and are observed as light was soon confirmed by experiments. Because of its electric and magnetic properties, light is also called **electromagnetic radiation.**

CAUTION! You may associate the term *radiation* with radioactive materials like uranium, but this term refers to anything that radiates, or spreads away, from its source. For example, scientists sometimes refer to sound waves as "acoustic radiation." Radiation does not have to be related to radioactivity!

Electromagnetic radiation consists of oscillating electric and magnetic fields, as shown in **Figure 5-6**. The distance between two successive wave crests is called the **wavelength** of the light, usually designated by the Greek letter λ (lambda). No matter what the wavelength, electromagnetic radiation always travels at the same speed $c = 3.0 \times 10^5$ km/s = 3.0×10^8 m/s in a vacuum.

More than a century elapsed between Newton's experiments with a prism and the confirmation of the wave nature of light. One reason for this delay is that **visible light,** the light to which the human eye is sensitive, has extremely short wavelengths—less than a thousandth of a millimeter—that are not easily detectable. To express such tiny distances conveniently, scientists use a unit of length called the **nanometer** (abbreviated nm), where 1 nm = 10^{-9} m. Experiments demonstrated that visible light has wavelengths covering the range from about 400 nm for violet light to about 700 nm for red light. Intermediate colors of the rainbow like yellow (550 nm) have intermediate wavelengths, as shown in **Figure 5-7**. (Some astronomers prefer to measure wavelengths in *angstroms*. One angstrom, abbreviated Å, is one-tenth of a nanometer: 1 Å = 0.1 nm = 10^{-10} m. In these units, the wavelengths of visible light extend from about 4000 Å to about 7000 Å. We will not use these units in this book, however.)

Visible and Nonvisible Light

Maxwell's equations place no restrictions on the wavelength of electromagnetic radiation. Hence, electromagnetic waves could and should exist with wavelengths both longer and shorter than the 400–700 nm range of visible light. Consequently, researchers began to look for *invisible* forms of light. These are forms of electromagnetic radiation to which the cells of the human retina do not respond.

The first kind of invisible radiation to be discovered actually preceded Maxwell's work by more than a half century. Around 1800 the British astronomer William Herschel passed sunlight through a prism and held a thermometer just beyond the red end of the visible spectrum. The thermometer registered a temperature increase, indicating that it was being exposed to an invisible form

of energy. This invisible energy, now called **infrared radiation,** was later realized to be electromagnetic radiation with wavelengths somewhat longer than those of visible light.

In experiments with electric sparks in 1888, the German physicist Heinrich Hertz succeeded in producing electromagnetic radiation with even longer wavelengths of a few centimeters or more. These are now known as **radio waves.** In 1895 another German physicist, Wilhelm Röntgen, invented a machine that produces electromagnetic radiation with wavelengths shorter than

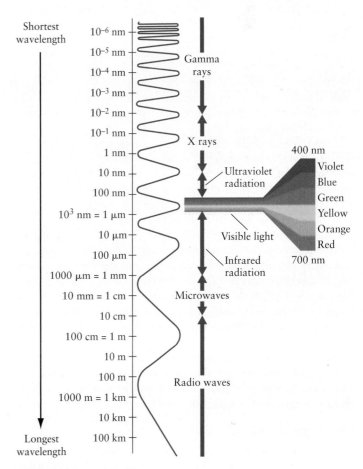

Shortest wavelength

10^{-6} nm
10^{-5} nm
10^{-4} nm
10^{-3} nm
10^{-2} nm
10^{-1} nm
1 nm
10 nm
100 nm
10^3 nm = 1 μm
10 μm
100 μm
1000 μm = 1 mm
10 mm = 1 cm
10 cm
100 cm = 1 m
10 m
100 m
1000 m = 1 km
10 km
100 km

Longest wavelength

Gamma rays

X rays

Ultraviolet radiation

Visible light

Infrared radiation

Microwaves

Radio waves

400 nm
Violet
Blue
Green
Yellow
Orange
Red
700 nm

Figure 5-7

WEB LINK 5.3

The Electromagnetic Spectrum The full array of all types of electromagnetic radiation is called the electromagnetic spectrum. It extends from the longest-wavelength radio waves to the shortest-wavelength gamma rays. Visible light occupies only a tiny portion of the full electromagnetic spectrum.

(a) Mobile phone: radio waves

(b) Microwave oven: microwaves

(c) TV remote: infrared light

(d) Tanning booth: ultraviolet light

(e) Medical imaging: X rays.

(f) Cancer radiotherapy: gamma rays

Figure 5-8

Uses of Nonvisible Electromagnetic Radiation (a) A mobile phone is actually a radio transmitter and receiver. The wavelengths used are in the range 16 to 36 cm. (b) A microwave oven produces radiation with a wavelength near 10 cm. The water in food absorbs this radiation, thus heating the food. (c) A remote control sends commands to a television using a beam of infrared light. (d) Ultraviolet radiation in moderation gives you a suntan, but in excess can cause sunburn or skin cancer.

(e) X rays can penetrate through soft tissue but not through bone, which makes them useful for medical imaging. (f) Gamma rays destroy cancer cells by breaking their DNA molecules, making them unable to multiply. (Ian Britton, Royalty-Free/Corbis, Michael Porsche/Corbis, Bill Lush/Taxi/Getty, Neil McAllister/Alamy, Edward Kinsman/Photo Researchers, Inc., Will and Deni McIntyre/Science Photo Library)

10 nm, now known as **X rays.** The X-ray machines in modern medical and dental offices are direct descendants of Röntgen's invention. Over the years radiation has been discovered with many other wavelengths.

Thus, visible light occupies only a tiny fraction of the full range of possible wavelengths, collectively called the **electromagnetic spectrum.** As Figure 5-7 shows, the electromagnetic spectrum stretches from the longest-wavelength radio waves to the shortest-wavelength gamma rays. Figure 5-8 shows some applications of nonvisible light in modern technology.

On the long-wavelength side of the visible spectrum, infrared radiation covers the range from about 700 nm to 1 mm. Astronomers interested in infrared radiation often express wavelength in *micrometers* or *microns*, abbreviated μm, where 1 μm = 10^{-3} mm = 10^{-6} m. **Microwaves** have wavelengths from roughly 1 mm to 10 cm, while radio waves have even longer wavelengths.

At wavelengths shorter than those of visible light, **ultraviolet radiation** extends from about 400 nm down to 10 nm. Next are X rays, which have wavelengths between about 10 and 0.01 nm, and beyond them at even shorter wavelengths are **gamma rays.** Note that the rough boundaries between different types of radiation are simply arbitrary divisions in the electromagnetic spectrum.

Frequency and Wavelength

Astronomers who work with radio telescopes often prefer to speak of *frequency* rather than wavelength. The **frequency** of a wave is the number of wave crests that pass a given point in one second. Equivalently, it is the number of complete *cycles* of the wave that pass per second (a complete cycle is from one crest to the next). Frequency is usually denoted by the Greek letter ν (nu). The unit of frequency is the cycle per second, also called the *hertz* (abbreviated Hz) in honor of Heinrich Hertz, the physicist who

first produced radio waves. For example, if 500 crests of a wave pass you in one second, the frequency of the wave is 500 cycles per second or 500 Hz.

In working with frequencies, it is often convenient to use the prefix *mega-* (meaning "million," or 10^6, and abbreviated M) or *kilo-* (meaning "thousand," or 10^3, and abbreviated k). For example, AM radio stations broadcast at frequencies between 535 and 1605 kHz (kilohertz), while FM radio stations broadcast at frequencies in the range from 88 to 108 MHz (megahertz).

The relationship between the frequency and wavelength of an electromagnetic wave is a simple one. Because light moves at a constant speed $c = 3 \times 10^8$ m/s, if the wavelength (distance from one crest to the next) is made shorter, the frequency must increase (more of those closely spaced crests pass you each second). Mathematically, the frequency ν of light is related to its wavelength λ by

Frequency and wavelength of an electromagnetic wave

$$\nu = \frac{c}{\lambda}$$

ν = frequency of an electromagnetic wave (in Hz)

c = speed of light = 3×10^8 m/s

λ = wavelength of the wave (in meters)

That is, the frequency of a wave equals the wave speed divided by the wavelength.

For example, hydrogen atoms in space emit radio waves with a wavelength of 21.12 cm. To calculate the frequency of this radiation, we must first express the wavelength in meters rather than centimeters: $\lambda = 0.2112$ m. Then we can use the above formula to find the frequency ν:

$$\nu = \frac{c}{\lambda} = \frac{3 \times 10^8 \text{ m/s}}{0.2112 \text{ m}} = 1.42 \times 10^9 \text{ Hz} = 1420 \text{ MHz}$$

Visible light has a much shorter wavelength and higher frequency than radio waves. You can use the above formula to show that for yellow-orange light of wavelength 600 nm, the frequency is 5×10^{14} Hz or 500 *million* megahertz!

While Young's experiment (Figure 5-5) showed convincingly that light has wavelike aspects, it was discovered in the early 1900s that light *also* has some of the characteristics of a stream of particles and waves. We will explore light's dual nature in Section 5-5.

5-3   An opaque object emits electromagnetic radiation according to its temperature

To learn about objects in the heavens, astronomers study the character of the electromagnetic radiation coming from those objects. Such studies can be very revealing because different kinds of electromagnetic radiation are typically produced in different ways. As an example, on Earth the most common way to generate radio waves is to make an electric current oscillate back and forth (as is done in the broadcast antenna of a radio station). By contrast, X rays for medical and dental purposes are usually produced by bombarding atoms in a piece of metal with fast-moving particles extracted from within other atoms. Our own Sun emits radio waves from near its glowing surface and X rays

> As an object is heated, it glows more brightly and its peak color shifts to shorter wavelengths

from its corona (see the photo that opens Chapter 3). Hence, these observations indicate the presence of electric currents near the Sun's surface and of fast-moving particles in the Sun's outermost regions. (We will discuss the Sun at length in Chapter 18.)

Radiation from Heated Objects

The simplest and most common way to produce electromagnetic radiation, either on or off the Earth, is to heat an object. The hot filament of wire inside an ordinary lightbulb emits white light, and a neon sign has a characteristic red glow because neon gas within the tube is heated by an electric current. In like fashion, almost all the visible light that we receive from space comes from hot objects like the Sun and the stars. The kind and amount of light emitted by a hot object tell us not only how hot it is but also about other properties of the object.

We can tell whether the hot object is made of relatively dense or relatively thin material. Consider the difference between a lightbulb and a neon sign. The dense, solid filament of a lightbulb makes white light, which is a mixture of all different visible wavelengths, while the thin, transparent neon gas produces light of a rather definite red color and, hence, a rather definite wavelength. For now we will concentrate our attention on the light produced by dense, opaque objects. (We will return to the light produced by gases in Section 5-6.) Even though the Sun and stars are gaseous, not solid, it turns out that they emit light with many of the same properties as light emitted by a hot, glowing, solid object.

Imagine a welder or blacksmith heating a bar of iron. As the bar becomes hot, it begins to glow deep red, as shown in Figure 5-9a. (You can see this same glow from the coils of a toaster or from an electric range turned on "high.") As the temperature rises further, the bar begins to give off a brighter orange light (Figure 5-9b). At still higher temperatures, it shines with a brilliant yellow light (Figure 5-9c). If the bar could be prevented from melting and vaporizing, at extremely high temperatures it would emit a dazzling blue-white light.

(a) Hot: glows deep red   (b) Hotter: glows orange   (c) Even hotter: glows yellow

Figure 5-9   R I [V] U X G

Heating a Bar of Iron This sequence of photographs shows how the appearance of a heated bar of iron changes with temperature. As the temperature increases, the bar glows more brightly because it radiates more energy. The color of the bar also changes because as the temperature goes up, the dominant wavelength of light emitted by the bar decreases. (©1984 Richard Megna/Fundamental Photographs)

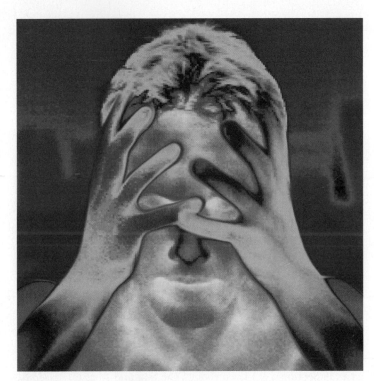

Figure 5-10 R I V U X G

An Infrared Portrait In this image made with a camera sensitive to infrared radiation, the different colors represent regions of different temperature. Red areas (like the man's face) are the warmest and emit the most infrared light, while blue-green areas (including the man's hands and hair) are at the lowest temperatures and emit the least radiation. (Dr. Arthur Tucker/Photo Researchers)

As this example shows, the amount of energy emitted by the hot, dense object and the dominant wavelength of the emitted radiation both depend on the temperature of the object. The hotter the object, the more energy it emits and the shorter the wavelength at which most of the energy is emitted. Colder objects emit relatively little energy, and this emission is primarily at long wavelengths.

These observations explain why you can't see in the dark. The temperatures of people, animals, and furniture are rather less than even that of the iron bar in Figure 5-9a. So, while these objects emit radiation even in a darkened room, most of this emission is at wavelengths greater than those of red light, in the infrared part of the spectrum (see Figure 5-7). Your eye is not sensitive to infrared, and you thus cannot see ordinary objects in a darkened room. But you can detect this radiation by using a camera that is sensitive to infrared light (Figure 5-10).

To better understand the relationship between the temperature of a dense object and the radiation it emits, it is helpful to know just what "temperature" means. The temperature of a substance is directly related to the average speed of the tiny **atoms**—the building blocks that come in distinct forms for each distinct chemical element—that make up the substance. (Typical atoms are about 10^{-10} m = 0.1 nm in diameter, or about 1/5000 as large as a typical wavelength of visible light.)

If something is hot, its atoms are moving at high speeds; if it is cold, its atoms are moving slowly. Scientists usually prefer to use the Kelvin temperature scale, on which temperature is measured in **kelvins** (K) upward from **absolute zero.** This is the coldest possible temperature, at which atoms move as slowly as possible (they can never quite stop completely). On the more familiar Celsius and Fahrenheit temperature scales, absolute zero (0 K) is $-273°$C and $-460°$F. Ordinary room temperature is 293 K, 20°C, or 68°F. **Box 5-1** discusses the relationships among the Kelvin, Celsius, and Fahrenheit temperature scales.

Figure 5-11 depicts quantitatively how the radiation from a dense object depends on its Kelvin temperature. Each curve in this figure shows the intensity of light emitted at each wavelength by a dense object at a given temperature: 3000 K (the temperature at which molten gold boils), 6000K (the temperature of an iron-welding arc), and 12,000 K (a temperature found in special industrial furnaces). In other words, the curves show the spectrum of light emitted by such an object. At any temperature, a hot, dense object emits at all wavelengths, so its spectrum is a smooth, continuous curve with no gaps in it.

The higher the temperature of a blackbody, the shorter the wavelength of maximum emission (the wavelength at which the curve peaks).

Visible light

The higher the temperature of a blackbody, the more light is emitted at *all* wavelengths.

12,000 K

6000 K

3000 K

Intensity

0   500   1000   2000   3000

Wavelength (nm)

Figure 5-11

Blackbody Curves Each of these curves shows the intensity of light at every wavelength that is emitted by a blackbody (an idealized case of a dense object) at a particular temperature. The rainbow-colored band shows the range of visible wavelengths. The vertical scale has been compressed so that all three curves can be seen; the peak intensity for the 12,000-K curve is actually about 1000 times greater than the peak intensity for the 3000-K curve.

BOX 5-1

Temperatures and Temperature Scales

Three temperature scales are in common use. Throughout most of the world, temperatures are expressed in **degrees Celsius** (°C). The Celsius temperature scale is based on the behavior of water, which freezes at 0°C and boils at 100°C at sea level on Earth. This scale is named after the Swedish astronomer Anders Celsius, who proposed it in 1742.

Astronomers usually prefer the Kelvin temperature scale. This is named after the nineteenth-century British physicist Lord Kelvin, who made many important contributions to our understanding of heat and temperature. Absolute zero, the temperature at which atomic motion is at the absolute minimum, is –273°C in the Celsius scale but 0 K in the Kelvin scale. Atomic motion cannot be any less than the minimum, so nothing can be colder than 0 K; hence, there are no negative temperatures on the Kelvin scale. Note that we do *not* use degree (°) with the Kelvin temperature scale.

A temperature expressed in kelvins is always equal to the temperature in degrees Celsius plus 273. On the Kelvin scale, water freezes at 273 K and boils at 373 K. Water must be heated through a change of 100 K or 100°C to go from its freezing point to its boiling point. Thus, the "size" of a kelvin is the same as the "size" of a Celsius degree. When considering temperature changes, measurements in kelvins and Celsius degrees are the same. For extremely high temperatures the Kelvin and Celsius scales are essentially the same: for example, the Sun's core temperature is either 1.55×10^7 K or 1.55×10^7 °C.

The now-archaic Fahrenheit scale, which expresses temperature in **degrees Fahrenheit** (°F), is used only in the United States. When the German physicist Gabriel Fahrenheit introduced this scale in the early 1700s, he intended 100°F to represent the temperature of a healthy human body. On the Fahrenheit scale, water freezes at 32°F and boils at 212°F. There are 180 Fahrenheit degrees between the freezing and boiling points of water, so a degree Fahrenheit is only 100/180 = 5/9 as large as either a Celsius degree or a kelvin.

Two simple equations allow you to convert a temperature from the Celsius scale to the Fahrenheit scale and from Fahrenheit to Celsius:

$$T_F = \frac{9}{5} T_C + 32$$

$$T_C = \frac{5}{9} (T_F - 32)$$

T_F = temperature in degrees Fahrenheit

T_C = temperature in degrees Celsius

EXAMPLE: A typical room temperature is 68°F. We can convert this to the Celsius scale using the second equation:

$$T_C = \frac{5}{9} (68 - 32) = 20°C$$

To convert this to the Kelvin scale, we simply add 273 to the Celsius temperature. Thus,

$$68° = 20°C = 293 \text{ K}$$

The diagram displays the relationships among these three temperature scales.

Kelvin	Celsius	Fahrenheit	
1.55×10^7	1.55×10^7	2.79×10^7	Sun's core temperature
5800	5527	9981	Sun's surface temperature
373	100	212	Boiling point of water
273	0	32	Freezing point of water
0	–273	–460	Absolute zero

The shape of the spectrum depends on temperature, however. An object at relatively low temperature (say, 3000 K) has a low curve, indicating a low intensity of radiation. The **wavelength of maximum emission,** at which the curve has its peak and the emission of energy is strongest, is at a long wavelength. The higher the temperature, the higher the curve (indicating greater intensity) and the shorter the wavelength of maximum emission.

Figure 5-11 shows that for a dense object at a temperature of 3000 K, the wavelength of maximum emission is around 1000 nm (1 μm). Because this is an infrared wavelength well outside the visible range, you might think that you cannot see the radiation from an object at this temperature. In fact, the glow from such an object *is* visible; the curve shows that this object emits plenty of light within the visible range, as well as at even shorter wavelengths.

The 3000-K curve is quite a bit higher at the red end of the visible spectrum than at the violet end, so a dense object at this temperature will appear red in color. Similarly, the 12,000-K curve has its wavelength of maximum emission in the ultraviolet part of the spectrum, at a wavelength shorter than visible light. But such a hot, dense object also emits copious amounts of visible light (much more than at 6000 K or 3000 K, for which the curves are lower) and thus will have a very visible glow. The curve for this temperature is higher for blue light than for red light, and so the color of a dense object at 12,000 K is a brilliant blue or blue-white. These conclusions agree with the color changes of a heated rod shown in Figure 5-9. The same principles apply to stars: A star that looks blue, such as Bellatrix in the constellation Orion (see Figure 2-2a), has a high surface temperature, while a red star such as Betelgeuse (Figure 2-2a) has a relatively cool surface.

These observations lead to a general rule:

The higher an object's temperature, the more intensely the object emits electromagnetic radiation and the shorter the wavelength at which it emits most strongly.

We will make frequent use of this general rule to analyze the temperatures of celestial objects such as planets and stars.

The curves in Figure 5-11 are drawn for an idealized type of dense object called a **blackbody.** A perfect blackbody does not reflect any light at all; instead, it absorbs all radiation falling on it. Because it reflects no electromagnetic radiation, the radiation that it does emit is entirely the result of its temperature. Ordinary objects, like tables, textbooks, and people, are not perfect blackbodies; they reflect light, which is why they are visible. A star such as the Sun, however, behaves very much like a perfect blackbody, because it absorbs almost completely any radiation falling on it from outside. The light emitted by a blackbody is called **blackbody radiation,** and the curves in Figure 5-11 are often called **blackbody curves.**

CAUTION! Despite its name, a blackbody does not necessarily *look* black. The Sun, for instance, does not look black because its temperature is high (around 5800 K), and so it glows brightly. But a room-temperature (around 300 K) blackbody would appear very black indeed. Even if it were as large as the Sun, it would emit only about 1/100,000 as much energy. (Its blackbody curve is far too low to graph in Figure 5-11.) Furthermore, most of this radiation would be at wavelengths that are too long for our eyes to perceive.

Figure 5-12 shows the blackbody curve for a temperature of 5800 K. It also shows the intensity curve for light from the Sun, as measured from above the Earth's atmosphere. (This is necessary because the Earth's atmosphere absorbs certain wavelengths.) The peak of both curves is at a wavelength of about 500 nm, near the middle of the visible spectrum. Note how closely the observed intensity curve for the Sun matches the blackbody curve. This is a strong indication that the temperature of the Sun's glowing surface is about 5800 K—a temperature that we can measure across a distance of 150 million kilometers! The close correlation between blackbody curves and the observed intensity curves for

Figure 5-12

The Sun as a Blackbody This graph shows that the intensity of sunlight over a wide range of wavelengths (solid curve) is a remarkably close match to the intensity of radiation coming from a blackbody at a temperature of 5800 K (dashed curve). The measurements of the Sun's intensity were made above the Earth's atmosphere (which absorbs and scatters certain wavelengths of sunlight). It's not surprising that the range of visible wavelengths includes the peak of the Sun's spectrum; the human eye evolved to take advantage of the most plentiful light available.

most stars is a key reason why astronomers are interested in the physics of blackbody radiation.

Blackbody radiation depends *only* on the temperature of the object emitting the radiation, not on the chemical composition of the object. The light emitted by molten gold at 2000 K is very nearly the same as that emitted by molten lead at 2000 K. Therefore, it might seem that analyzing the light from the Sun or from a star can tell astronomers the object's temperature but not what the star is made of. As Figure 5-12 shows, however, the intensity curve for the Sun (a typical star) is not precisely that of a blackbody. We will see later in this chapter that the *differences* between a star's spectrum and that of a blackbody allow us to determine the chemical composition of the star.

5-4 Wien's law and the Stefan-Boltzmann law are useful tools for analyzing glowing objects like stars

The mathematical formula that describes the blackbody curves in Figure 5-11 is a rather complicated one. But there are two simpler formulas for blackbody radiation that prove to be very useful in many branches of astronomy. They are used by astronomers who investigate the stars as well as by those who study the planets (which are dense, relatively cool objects that emit infrared radiation). One of these formulas relates the temperature of a blackbody to its wavelength of maximum emission,

Two simple mathematical formulas describing blackbodies are essential tools for studying the universe

and the other relates the temperature to the amount of energy that the blackbody emits. These formulas, which we will use throughout this book, restate in precise mathematical terms the qualitative relationships that we described in Section 5-3.

Wien's Law

Figure 5-11 shows that the higher the temperature (T) of a blackbody, the shorter its wavelength of maximum emission (λ_{max}). In 1893 the German physicist Wilhelm Wien used ideas about both heat and electromagnetism to make this relationship quantitative. The formula that he derived, which today is called **Wien's law**, is

Wien's law for a blackbody

$$\lambda_{max} = \frac{0.0029 \text{ K m}}{T}$$

λ_{max} = wavelength of maximum emission of the object (in meters)

T = temperature of the object (in kelvins)

According to Wien's law, the wavelength of maximum emission of a blackbody is inversely proportional to its temperature in kelvins. In other words, if the temperature of the blackbody doubles, its wavelength of maximum emission is halved, and vice versa. For example, Figure 5-11 shows blackbody curves for temperatures of 3000 K, 6000 K, and 12,000 K. From Wien's law, a blackbody with a temperature of 6000 K has a wavelength of maximum emission $\lambda_{max} = (0.0029 \text{ K m})/(6000 \text{ K}) = 4.8 \times 10^{-7}$ m = 480 nm, in the visible part of the electromagnetic spectrum. At 12,000 K, or twice the temperature, the blackbody has a wavelength of maximum emission half as great, or λ_{max} = 240 nm; this is in the ultraviolet. At 3000 K, just half our original temperature, the value of λ_{max} is twice the original value—960 nm, which is an infrared wavelength. You can see that these wavelengths agree with the peaks of the curves in Figure 5-11.

CAUTION! Remember that Wien's law involves the wavelength of maximum emission in *meters*. If you want to convert the wavelength to nanometers, you must multiply the wavelength in meters by $(10^9 \text{ nm})/(1 \text{ m})$.

Wien's law is very useful for determining the surface temperatures of stars. It is not necessary to know how far away the star is, how large it is, or how much energy it radiates into space. All we need to know is the dominant wavelength of the star's electromagnetic radiation.

The Stefan-Boltzmann Law

The other useful formula for the radiation from a blackbody involves the total amount of energy the blackbody radiates at all wavelengths. (By contrast, the curves in Figure 5-11 show how much energy a blackbody radiates at each individual wavelength.)

Energy is usually measured in **joules** (J), named after the nineteenth-century English physicist James Joule. A joule is the amount of energy contained in the motion of a 2-kilogram mass moving at a speed of 1 meter per second. The joule is a convenient unit of energy because it is closely related to the familiar **watt** (W): 1 watt is 1 joule per second, or 1 W = 1 J/s = 1 J s^{-1}. (The superscript -1 means you are dividing by that quantity.) For example, a 100-watt lightbulb uses energy at a rate of 100 joules per second, or 100 J/s. The energy content of food is also often measured in joules; in most of the world, diet soft drinks are labeled as "low joule" rather than "low calorie."

The amount of energy emitted by a blackbody depends both on its temperature and on its surface area. This makes sense: A large burning log radiates much more heat than a burning match, even though the temperatures are the same. To consider the effects of temperature alone, it is convenient to look at the amount of energy emitted from each square meter of an object's surface in a second. This quantity is called the **energy flux** (F). Flux means "rate of flow," and thus F is a measure of how rapidly energy is flowing out of the object. It is measured in joules per square meter per second, usually written as J/m^2/s or J m^{-2} s^{-1}. Alternatively, because 1 watt equals 1 joule per second, we can express flux in watts per square meter (W/m^2, or W m^{-2}).

The nineteenth-century Irish physicist David Tyndall performed the first careful measurements of the amount of radiation emitted by a blackbody. (He studied the light from a heated platinum wire, which behaves approximately like a blackbody.) By analyzing Tyndall's results, the Slovenian physicist Josef Stefan deduced in 1879 that the flux from a blackbody is proportional to the fourth power of the object's temperature (measured in kelvins). Five years after Stefan announced his law, Austrian physicist Ludwig Boltzmann showed how it could be derived mathematically from basic assumptions about atoms and molecules. For this reason, Stefan's law is commonly known as the **Stefan-Boltzmann law**. Written as an equation, the Stefan-Boltzmann law is

Stefan-Boltzmann law for a blackbody

$$F = \sigma T^4$$

F = energy flux, in joules per square meter of surface per second

σ = a constant = 5.67×10^{-8} W m^{-2} K^{-4}

T = object's temperature, in kelvins

The value of the constant σ (the Greek letter sigma) is known from laboratory experiments.

The Stefan-Boltzmann law says that if you double the temperature of an object (for example, from 300 K to 600 K), then the energy emitted from the object's surface each second increases by a factor of 2^4 = 16. If you increase the temperature by a factor of 10 (for example, from 300 K to 3000 K), the rate of energy emission increases by a factor of 10^4 = 10,000. Thus, a chunk of iron at room temperature (around 300 K) emits very little electromagnetic radiation (and essentially no visible light), but an iron bar heated to 3000 K glows quite intensely.

Box 5-2 gives several examples of applying Wien's law and the Stefan-Boltzmann law to typical astronomical problems.

BOX 5-2   Tools of the Astronomer's Trade

Using the Laws of Blackbody Radiation

The Sun and stars behave like nearly perfect blackbodies. Wien's law and the Stefan-Boltzmann law can therefore be used to relate the surface temperature of the Sun or a distant star to the energy flux and wavelength of maximum emission of its radiation. The following examples show how to do this.

EXAMPLE: The maximum intensity of sunlight is at a wavelength of roughly 500 nm = 5.0×10^{-7} m. Use this information to determine the surface temperature of the Sun.

Situation: We are given the Sun's wavelength of maximum emission λ_{max}, and our goal is to find the Sun's surface temperature, denoted by $T_{\odot}$. (The symbol $\odot$ is the standard astronomical symbol for the Sun.)

Tools: We use Wien's law to relate the values of λ_{max} and $T_{\odot}$.

Answer: As written, Wien's law tells how to find λ_{max} if we know the surface temperature. To find the surface temperature from λ_{max}, we first rearrange the formula, then substitute the value of λ_{max}:

$$T_{\odot} = \frac{0.0029 \text{ K m}}{\lambda_{max}} = \frac{0.0029 \text{ K m}}{5.0 \times 10^{-7} \text{ m}} = 5800 \text{ K}$$

Review: This is a very high temperature by Earth standards, about the same as an iron-welding arc.

EXAMPLE: Using detectors above the Earth's atmosphere, astronomers have measured the average flux of solar energy arriving at Earth. This value, called the **solar constant,** is equal to 1370 W m^{-2}. Use this information to calculate the Sun's surface temperature. (This provides a check on our result from the preceding example.)

Situation: The solar constant is the flux of sunlight as measured at the Earth. We want to use the value of the solar constant to calculate $T_{\odot}$.

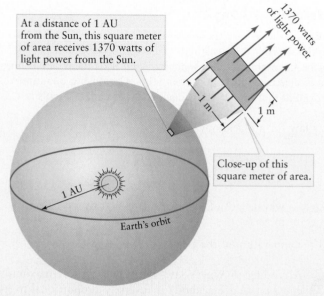

At a distance of 1 AU from the Sun, this square meter of area receives 1370 watts of light power from the Sun.

1370 watts of light power

1 m   1 m

Close-up of this square meter of area.

1 AU

Earth's orbit

Tools: It may seem that all we need is the Stefan-Boltzmann law, which relates flux to surface temperature. However, the quantity F in this law refers to the flux measured at the Sun's surface, *not* at the Earth. Hence, we will first need to calculate F from the given information.

Answer: To determine the value of F, we first imagine a huge sphere of radius 1 AU with the Sun at its center, as shown in the figure. Each square meter of that sphere receives 1370 watts of power from the Sun, so the total energy radiated by the Sun per second is equal to the solar constant multiplied by the sphere's surface area. The result, called the **luminosity** of the Sun and denoted by the symbol $L_{\odot}$, is $L_{\odot} = 3.90 \times 10^{26}$ W. That is, in 1 second the Sun radiates 3.90×10^{26} joules of energy into space. Because we know the size of the Sun, we can compute the energy flux (energy emitted per square meter per second) at its surface. The radius of the Sun is $R_{\odot} = 6.96 \times 10^{8}$ m, and the Sun's surface area is $4\pi R_{\odot}^{2}$.

5-5   Light has properties of both waves and particles

At the end of the nineteenth century, physicists mounted a valiant effort to explain all the characteristics of blackbody radiation. To this end they constructed theories based on Maxwell's description of light as electromagnetic waves. But all such theories failed to explain the characteristic shapes of blackbody curves shown in Figure 5-11.

The Photon Hypothesis

In 1900, however, the German physicist Max Planck discovered that he could derive a formula that correctly described blackbody

curves if he made certain assumptions. In 1905 the great German-born physicist Albert Einstein realized that these assumptions implied a radical new view of the nature of light. One tenet of this new view is that electromagnetic energy is emitted in discrete, particle-like packets, or light *quanta* (the plural of *quantum*, from a Latin word meaning "how much"). The second tenet is that the energy of each light quantum—today called a **photon**—is related to the wavelength of light: the greater the wavelength,

The revolutionary photon concept is necessary to explain blackbody radiation and the photoelectric effect

the lower the energy of a photon associated with that wavelength. Thus, a photon of red light (wavelength $\lambda = 700$ nm) has less en-

Therefore, its energy flux $F_\odot$ is the Sun's luminosity (total energy emitted by the Sun per second) divided by the Sun's surface area (the number of square meters of surface):

$$F_\odot = \frac{L_\odot}{4\pi R_\odot{}^2} = \frac{3.90 \times 10^{26}\ W}{4\pi(6.96 \times 10^8\ m)^2} = 6.41 \times 10^7\ W\ m^{-2}$$

Once we have the Sun's energy flux $F_\odot$ we can use the Stefan-Boltzmann law to find the Sun's surface temperature $T_\odot$:

$$T_\odot{}^4 = F_\odot/\sigma = 1.13 \times 10^{15}\ K^4$$

Taking the fourth root (the square root of the square root) of this value, we find the surface temperature of the Sun to be $T_\odot = 5800$ K.

Review: Our result for $T_\odot$ agrees with the value we computed in the previous example using Wien's law. Notice that the solar constant of 1370 W m^{-2} is very much less than $F_\odot$, the flux at the Sun's surface. By the time the Sun's radiation reaches Earth, it is spread over a greatly increased area.

EXAMPLE: Sirius, the brightest star in the night sky, has a surface temperature of about 10,000 K. Find the wavelength at which Sirius emits most intensely.

Situation: Our goal is to calculate the wavelength of maximum emission of Sirius (λ_{max}) from its surface temperature T.

Tools: We use Wien's law to relate the values of λ_{max} and T.

Answer: Using Wien's law,

$$\lambda_{max} = \frac{0.0029\ K\ m}{T} = \frac{0.0029\ K\ m}{10{,}000\ K}$$

$$= 2.9 \times 10^{-7}\ m = 290\ nm$$

Review: Our result shows that Sirius emits light most intensely in the ultraviolet. In the visible part of the spectrum, it emits more blue light than red light (like the curve for 12,000 K in Figure 5-11), so Sirius has a distinct blue color.

EXAMPLE: How does the energy flux from Sirius compare to the Sun's energy flux?

Situation: To compare the energy fluxes from the two stars, we want to find the *ratio* of the flux from Sirius to the flux from the Sun.

Tools: We use the Stefan-Boltzmann law to find the flux from Sirius and from the Sun, which from the preceding examples have surface temperatures 10,000 K and 5800 K, respectively.

Answer: For the Sun, the Stefan-Boltzmann law is $F_\odot = \sigma T_\odot{}^4$, and for Sirius we can likewise write $F_* = \sigma T_*{}^4$, where the subscripts $\odot$ and $*$ refer to the Sun and Sirius, respectively. If we divide one equation by the other to find the ratio of fluxes, the Stefan-Boltzmann constants cancel out and we get

$$\frac{F_*}{F_\odot} = \frac{T_*{}^4}{T_\odot{}^4} = \frac{(10{,}000\ K)^4}{(5800\ K)^4} = \left(\frac{10{,}000}{5800}\right)^4 = 8.8$$

Review: Because Sirius has such a high surface temperature, each square meter of its surface emits 8.8 times more energy per second than a square meter of the Sun's relatively cool surface. Sirius is actually a larger star than the Sun, so it has more square meters of surface area and, hence, its *total* energy output is *more* than 8.8 times that of the Sun.

ergy than a photon of violet light ($\lambda = 400$ nm). In this picture, light has a dual personality; it behaves as a stream of particlelike photons, but each photon has wavelike properties. In this sense, the best answer to the question "Is light a wave or a stream of particles?" is "Yes!"

It was soon realized that the photon hypothesis explains more than just the detailed shape of blackbody curves. For example, it explains why only ultraviolet light causes suntans and sunburns. The reason is that tanning or burning involves a chemical reaction in the skin. High-energy, short-wavelength ultraviolet photons can trigger these reactions, but the lower-energy, longer-wavelength photons of visible light cannot. Similarly, normal photographic film is sensitive to visible light but not to infrared light; a long-wavelength infrared photon does not have enough

energy to cause the chemical change that occurs when film is exposed to the higher-energy photons of visible light.

 Another phenomenon explained by the photon hypothesis is the **photoelectric effect.** In this effect, a metal plate is illuminated by a light beam. If ultraviolet light is used, tiny negatively charged particles called **electrons** are emitted from the metal plate. (We will see in Section 5-7 that the electron is one of the basic particles of the atom.) But if visible light is used, no matter how bright, no electrons are emitted.

 Einstein explained this behavior by noting that a certain minimum amount of energy is required to remove an electron from the metal plate. The energy of a short-wavelength ultraviolet photon is greater than this minimum value, so an electron that absorbs a photon of ultraviolet light will have enough energy to escape from the plate. But an electron that absorbs a photon of visible light, with its longer wavelength and lower energy, does not gain enough energy to escape and so remains within the metal. Einstein and Planck both won Nobel prizes for their contributions to understanding the nature of light.

The Energy of a Photon

The relationship between the energy E of a single photon and the wavelength of the electromagnetic radiation can be expressed in a simple equation:

Energy of a photon (in terms of wavelength)

$$E = \frac{hc}{\lambda}$$

E = energy of a photon

h = Planck's constant

c = speed of light

λ = wavelength of light

The value of the constant h in this equation, now called *Planck's constant*, has been shown in laboratory experiments to be

$$h = 6.625 \times 10^{-34} \text{ J s}$$

The units of h are joules multiplied by seconds, called "joule-seconds" and abbreviated J s.

Because the value of h is so tiny, a single photon carries a very small amount of energy. For example, a photon of red light

with wavelength 633 nm has an energy of only 3.14×10^{-19} J (Box 5-3). This is why we ordinarily do not notice that light comes in the form of photons; even a dim light source emits so many photons per second that it seems to be radiating a continuous stream of energy.

The energies of photons are sometimes expressed in terms of a small unit of energy called the **electron volt** (eV). One electron volt is equal to 1.602×10^{-19} J, so a 633-nm photon has an energy of 1.96 eV. If energy is expressed in electron volts, Planck's constant is best expressed in electron volts multiplied by seconds, abbreviated eV s:

$$h = 4.135 \times 10^{-15} \text{ eV s}$$

Because the frequency ν of light is related to the wavelength λ by $\nu = c/\lambda$, we can rewrite the equation for the energy of a photon as

Energy of a photon (in terms of frequency)

$$E = h\nu$$

E = energy of a photon

h = Planck's constant

ν = frequency of light

The equations $E = hc/\lambda$ and $E = h\nu$ are together called **Planck's law.** Both equations express a relationship between a particlelike property of light (the energy E of a photon) and a wavelike property (the wavelength λ or frequency ν).

The photon picture of light is essential for understanding the detailed shapes of blackbody curves. As we will see, it also helps to explain how and why the spectra of the Sun and stars differ from those of perfect blackbodies.

BOX 5-3   The Heavens on the Earth

Photons at the Supermarket

A beam of light can be regarded as a stream of tiny packets of energy called photons. The Planck relationships $E = hc/\lambda$ and $E = h\nu$ can be used to relate the energy E carried by a photon to its wavelength λ and frequency ν.

As an example, the laser bar-code scanners used at stores and supermarkets emit orange-red light of wavelength 633 nm. To calculate the energy of a single photon of this light, we must first express the wavelength in meters. A nanometer (nm) is equal to 10^{-9} m, so the wavelength is

$$\lambda = (633 \text{ nm})\left(\frac{10^{-9} \text{ m}}{1 \text{ nm}}\right) = 633 \times 10^{-9} \text{ m} = 6.33 \times 10^{-7} \text{ m}$$

Then, using the Planck formula $E = hc/\lambda$, we find that the energy of a single photon is

$$E = \frac{hc}{\lambda} = \frac{(6.625 \times 10^{-34} \text{ J s})(3 \times 10^8 \text{ m/s})}{6.33 \times 10^{-7} \text{ m}} = 3.14 \times 10^{-19} \text{ J}$$

This is a very small amount of energy. The laser in a typical bar-code scanner emits 10^{-3} joule of light energy per second, so the number of photons emitted per second is

$$\frac{10^{-3} \text{ joule per second}}{3.14 \times 10^{-19} \text{ joule per photon}} = 3.2 \times 10^{15} \text{ photons per second}$$

This is such a large number that the laser beam seems like a continuous flow of energy rather than a stream of little energy packets.

5-6   Each chemical element produces its own unique set of spectral lines

In 1814 the German master optician Joseph von Fraunhofer repeated the classic experiment of shining a beam of sunlight through a prism (see Figure 5-3). But this time Fraunhofer subjected the resulting rainbow-colored spectrum to intense magnification. To his surprise, he discovered that the solar spectrum contains hundreds of fine, dark lines, now called **spectral lines.** By contrast, if the light from a perfect blackbody were sent through a prism, it would produce a smooth, continuous spectrum with no dark lines. Fraunhofer counted more than 600 dark lines in the Sun's spectrum; today we know of more than 30,000. The photograph of the Sun's spectrum in **Figure 5-13** shows hundreds of these spectral lines.

> Spectroscopy is the key to determining the chemical composition of planets and stars

Spectral Analysis

Half a century later, chemists discovered that they could produce spectral lines in the laboratory and use these spectral lines to analyze what kinds of atoms different substances are made of. Chemists had long known that many substances emit distinctive colors when sprinkled into a flame. To facilitate study of these colors, around 1857 the German chemist Robert Bunsen invented a gas burner (today called a Bunsen burner) that produces a clean flame with no color of its own. Bunsen's colleague, the Prussian-

1. Add a chemical substance to a flame

2. Send light from the flame through a narrow slit, then through a prism

3. Bright lines in the spectrum show that the substance emits light at specific wavelengths only

Figure 5-14

The Kirchhoff-Bunsen Experiment In the mid-1850s, Gustav Kirchhoff and Robert Bunsen discovered that when a chemical substance is heated and vaporized, the spectrum of the emitted light exhibits a series of bright spectral lines. They also found that each chemical element produces its own characteristic pattern of spectral lines. (In an actual laboratory experiment, lenses would be needed to focus the image of the slit onto the screen.)

Figure 5-13   R I **V** U X G

The Sun's Spectrum Numerous dark spectral lines are seen in this image of the Sun's spectrum. The spectrum is spread out so much that it had to be cut into segments to fit on this page. (N. A. Sharp, NOAO/NSO/Kitt Peak FTS/AURA/NSF)

born physicist Gustav Kirchhoff, suggested that the colored light produced by substances in a flame might best be studied by passing the light through a prism (**Figure 5-14**). The two scientists promptly discovered that the spectrum from a flame consists of a pattern of thin, bright spectral lines against a dark background. The same kind of spectrum is produced by heated gases such as neon or argon.

Kirchhoff and Bunsen then found that each chemical element produces its own unique pattern of spectral lines. Thus was born in 1859 the technique of **spectral analysis,** the identification of chemical substances by their unique patterns of spectral lines.

A chemical **element** is a fundamental substance that cannot be broken down into more basic chemicals. Some examples are hydrogen, oxygen, carbon, iron, gold, and silver. After Kirchhoff and Bunsen had recorded the prominent spectral lines of all the then-known elements, they soon began to discover other spectral lines in the spectra of vaporized mineral samples. In this way they discovered elements whose presence had never before been suspected. In 1860, Kirchhoff and Bunsen found a new line in the blue portion of the spectrum of a sample of mineral water. After isolating the previously unknown element responsible for making the line, they named it cesium (from the Latin *caesium,* "gray-blue"). The next year, a new line in the red portion of the spectrum of a mineral sample led them to discover the element rubidium (Latin *rubidium,* "red").

Spectral analysis even allowed the discovery of new elements outside Earth. During the solar eclipse of 1868, astronomers found a new spectral line in light coming from the hot gases at the upper surface of the Sun while the main body of the Sun was hidden by the Moon. This line was attributed to a new element

Figure 5-15   R I <u>V</u> U X G

Various Spectra These photographs show the spectra of different types of gases as measured in a laboratory on Earth. Each type of gas has a unique spectrum that is the same wherever in the universe the gas is found. Water vapor (H_2O) is a compound whose molecules are made up of hydrogen and oxygen atoms; the hydrogen molecule (H_2) is made up of two hydrogen atoms. (Ted Kinsman/Science Photo Library)

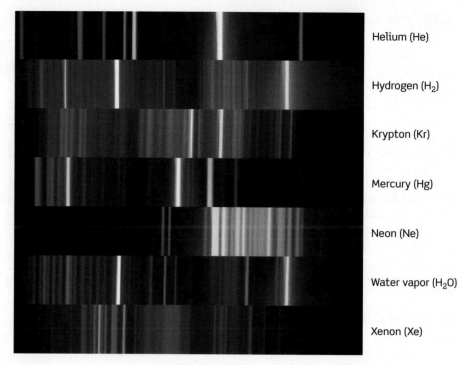

Helium (He)

Hydrogen (H_2)

Krypton (Kr)

Mercury (Hg)

Neon (Ne)

Water vapor (H_2O)

Xenon (Xe)

that was named helium (from the Greek *helios,* "sun"). Helium was not discovered on Earth until 1895, when it was found in gases obtained from a uranium mineral.

A sample of an element contains only a single type of atom; carbon contains only carbon atoms, helium contains only helium atoms, and so on. Atoms of the same or different elements can combine to form **molecules.** For example, two hydrogen atoms (symbol H) can combine with an oxygen atom (symbol O) to form a water molecule (symbol H_2O). Substances like water whose molecules include atoms of different elements are called **compounds.** Just as each type of atom has its own unique spectrum, so does each type of molecule. **Figure 5-15** shows the spectra of several types of atoms and molecules.

Kirchhoff's Laws

The spectrum of the Sun, with its dark spectral lines superimposed on a bright background (see Figure 5-13), may seem to be unrelated to the spectra of bright lines against a dark background produced by substances in a flame (see Figure 5-14). But by the early 1860s, Kirchhoff's experiments had revealed a direct connection between these two types of spectra. His conclusions are summarized in three important statements about spectra that are today called **Kirchhoff's laws.** These laws, which are illustrated in **Figure 5-16,** are as follows:

Law 1 A hot opaque body, such as a perfect blackbody, or a hot, dense gas produces a **continuous spectrum**—a complete rainbow of colors without any spectral lines.

Law 2 A hot, transparent gas produces an **emission line spectrum**—a series of bright spectral lines against a dark background.

Law 3 A cool, transparent gas in front of a source of a continuous spectrum produces an **absorption line spectrum**—a

series of dark spectral lines among the colors of the continuous spectrum. Furthermore, the dark lines in the absorption spectrum of a particular gas occur at exactly the *same* wavelengths as the bright lines in the emission spectrum of that same gas.

Kirchhoff's laws imply that if a beam of white light is passed through a gas, the atoms of the gas somehow extract light of very specific wavelengths from the white light. Hence, an observer who looks straight through the gas at the white-light source (the blackbody in Figure 5-16) will receive light whose spectrum has dark absorption lines superimposed on the continuous spectrum of the white light. The gas atoms then radiate light of precisely these same wavelengths in all directions. An observer at an oblique angle (that is, one who is not sighting directly through the cloud toward the blackbody) will receive only this light radiated by the gas cloud; the spectrum of this light is bright emission lines on a dark background.

CAUTION! Figure 5-16 shows that light can either pass through a cloud of gas or be absorbed by the gas. But there is also a third possibility: The light can simply bounce off the atoms or molecules that make up the gas, a phenomenon called **light scattering.** In other words, photons passing through a gas cloud can miss the gas atoms altogether, be swallowed whole by the atoms (absorption), or bounce off the atoms like billiard balls colliding (scattering). **Box 5-4** describes how light scattering explains the blue color of the sky and the red color of sunsets.

 Whether an emission line spectrum or an absorption line spectrum is observed from a gas cloud depends on the relative temperatures of the gas cloud and its background. Absorption lines are seen if the background is hotter than the gas, and emission lines are seen if the background is cooler.

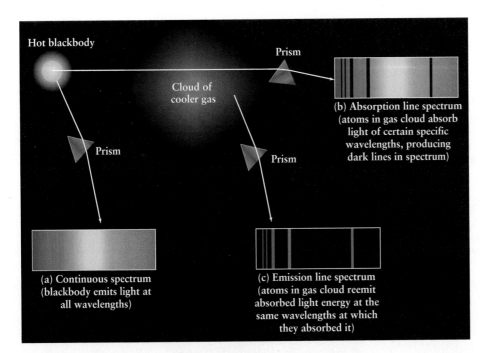

Hot blackbody

Prism

Cloud of cooler gas

Prism

Prism

(b) Absorption line spectrum (atoms in gas cloud absorb light of certain specific wavelengths, producing dark lines in spectrum)

(a) Continuous spectrum (blackbody emits light at all wavelengths)

(c) Emission line spectrum (atoms in gas cloud reemit absorbed light energy at the same wavelengths at which they absorbed it)

Figure 5-16

Continuous, Absorption Line, and Emission Line Spectra A hot, opaque body (like a blackbody) emits a continuous spectrum of light (spectrum *a*). If this light is passed through a cloud of a cooler gas, the cloud absorbs light of certain specific wavelengths, and the spectrum of light that passes directly through the cloud has dark absorption lines (spectrum *b*). The cloud does not retain all the light energy that it absorbs but radiates it outward in all directions. The spectrum of this reradiated light contains bright emission lines (spectrum *c*) with exactly the same wavelengths as the dark absorption lines in spectrum *b*. The specific wavelengths observed depend on the chemical composition of the cloud.

For example, if sodium is placed in the flame of a Bunsen burner in a darkened room, the flame will emit a characteristic orange-yellow glow. (This same glow is produced if we use ordinary table salt, which is a compound of sodium and chlorine.) If we pass the light from the flame through a prism, it displays an emission line spectrum with two closely spaced spectral lines at wavelengths of 588.99 and 589.59 nm, in the orange-yellow part of the spectrum. We now turn on a lightbulb whose filament is hotter than the flame and shine the bulb's white light through the flame. The spectrum of this light after it passes through the flame's sodium vapor is the continuous spectrum from the lightbulb, but with two closely spaced *dark* lines at 588.99 and 589.59 nm. Thus, the chemical composition of the gas is revealed by either bright emission lines or dark absorption lines.

Spectroscopy

Spectroscopy is the systematic study of spectra and spectral lines. Spectral lines are tremendously important in astronomy, because

they provide reliable evidence about the chemical composition of distant objects. As an example, the spectrum of the Sun shown in Figure 5-13 is an absorption line spectrum. The continuous spectrum comes from the hot surface of the Sun, which acts like a blackbody. The dark absorption lines are caused by this light passing through a cooler gas; this gas is the atmosphere that surrounds the Sun. Therefore, by identifying the spectral lines present in the solar spectrum, we can determine the chemical composition of the Sun's atmosphere.

Figure 5-17 shows both a portion of the Sun's absorption line spectrum and the emission line spectrum of iron vapor over the same wavelength range. This pattern of bright spectral lines in the lower spectrum is iron's own distinctive "fingerprint," which no other substance can imitate. Because some absorption lines in the Sun's spectrum coincide with the iron lines, some vaporized iron must exist in the Sun's atmosphere.

Spectroscopy can also help us analyze gas clouds in space, such as the nebula surrounding the star cluster NGC 346 shown

Absorption spectrum of the Sun →

Emission spectrum of iron (in the laboratory on Earth) →

For each emission line of iron, there is a corresponding absorption line in the solar spectrum; hence there must be iron in the Sun's atmosphere

Figure 5-17   R I V U X G

Iron in the Sun The upper part of this figure is a portion of the Sun's spectrum at violet wavelengths, showing numerous dark absorption lines. The lower part of the figure is a corresponding portion of the emission

line spectrum of vaporized iron. The iron lines coincide with some of the solar lines, which proves that there is some iron (albeit a relatively small amount) in the Sun's atmosphere. (Carnegie Observatories)

BOX 5-4   The Heavens on the Earth

Light Scattering

Light scattering is the process whereby photons bounce off particles in their path. These particles can be atoms, molecules, or clumps of molecules. You are reading these words using photons from the Sun or a lamp that bounced off the page—that is, were scattered by the particles that make up the page.

An important fact about light scattering is that very small particles—ones that are smaller than a wavelength of visible light—are quite effective at scattering short-wavelength photons of blue light, but less effective at scattering long-wavelength photons of red light. This fact explains a number of phenomena that you can see here on Earth.

The light that comes from the daytime sky is sunlight that has been scattered by the molecules that make up our atmosphere (see part *a* of the accompanying figure). Air molecules are less than 1 nm across, far smaller than the wavelength of visible light, so they scatter blue light more than red light—which is why the sky looks blue. Smoke particles are also quite small, which explains why the smoke from a cigarette or a fire has a bluish color.

Distant mountains often appear blue thanks to sunlight being scattered from the atmosphere between the mountains and your eyes. (The Blue Ridge Mountains, which extend from Pennsylvania to Georgia, and Australia's Blue Mountains derive their names from this effect.) Sunglasses often have a red or orange tint, which blocks out blue light. This cuts down on the amount of scattered light from the sky reaching your eyes and allows you to see distant objects more clearly.

Light scattering also explains why sunsets are red. The light from the Sun contains photons of all visible wavelengths, but as this light passes through our atmosphere the blue photons are scattered away from the straight-line path from the Sun to your eye. Red photons undergo relatively little scattering, so the Sun always looks a bit redder than it really is. When you look toward the setting sun, the sunlight that reaches your eye has had to pass through a relatively thick layer of atmosphere (part *b* of the accompanying figure). Hence, a large fraction of the blue light from the Sun has been scattered, and the Sun appears quite red.

The same effect also applies to sunrises, but sunrises seldom look as red as sunsets do. The reason is that dust is lifted into the atmosphere during the day by the wind (which is typically stronger in the daytime than at night), and dust particles in the atmosphere help to scatter even more blue light.

If the small particles that scatter light are sufficiently concentrated, there will be almost as much scattering of red light as of blue light, and the scattered light will appear white. This explains the white color of clouds, fog, and haze, in which the scattering particles are ice crystals or water droplets. Whole milk looks white because of light scattering from tiny fat globules; nonfat milk has only a very few of these globules and so has a slight bluish cast.

Light scattering has many applications to astronomy. For example, it explains why very distant stars in our Galaxy appear surprisingly red. The reason is that there are tiny dust particles in the space between the stars, and this dust scatters blue photons. By studying how much scattering takes place, astronomers have learned about the tenuous material that fills interstellar space.

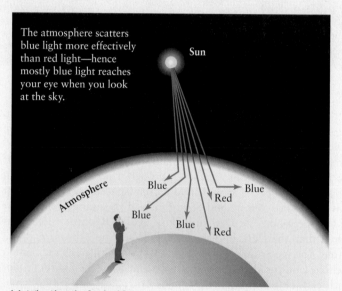

The atmosphere scatters blue light more effectively than red light—hence mostly blue light reaches your eye when you look at the sky.

(a) Why the sky looks blue

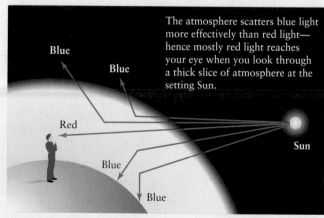

The atmosphere scatters blue light more effectively than red light—hence mostly red light reaches your eye when you look through a thick slice of atmosphere at the setting Sun.

(b) Why the setting Sun looks red

make chemical assays of objects that are almost inconceivably distant. Throughout this book we will see many examples of how astronomers use Kirchhoff's laws to determine the nature of celestial objects.

To make full use of Kirchhoff's laws, it is helpful to understand why they work. Why does an atom absorb light of only particular wavelengths? And why does it then emit light of only these same wavelengths? Maxwell's theory of electromagnetism (see Section 5-2) could not answer these questions. The answers did not come until early in the twentieth century, when scientists began to discover the structure and properties of atoms.

5-7 An atom consists of a small, dense nucleus surrounded by electrons

The first important clue about the internal structure of atoms came from an experiment conducted in 1910 by Ernest Rutherford, a gifted physicist from New Zealand. Rutherford and his colleagues at the University of Manchester in England had been investigating the recently discovered phenomenon of radioactivity. Certain radioactive elements, such as uranium and radium, were known to emit particles of various types. One type, the alpha particle, has about the same mass as a helium atom and is emitted from some radioactive substances with considerable speed.

To decode the information in the light from immense objects like stars and galaxies, we must understand the structure of atoms

In one series of experiments, Rutherford and his colleagues were using alpha particles as projectiles to probe the structure of solid matter. They directed a beam of these particles at a thin sheet of metal (**Figure 5-19**). Almost all the alpha particles passed through the metal sheet with little or no deflection from their straight-line paths. To the surprise of the experimenters, however, an occasional alpha particle bounced back from the metal sheet as though it had struck something quite dense. Rutherford later remarked, "It was almost as incredible as

Figure 5-18   R I **V** U X G

Analyzing the Composition of a Distant Nebula The glowing gas cloud in this Hubble Space Telescope image lies 210,000 light-years away in the constellation Tucana (the Toucan). Hot stars within the nebula emit high-energy, ultraviolet photons, which are absorbed by the surrounding gas and heat the gas to high temperature. This heated gas produces light with an emission line spectrum. The particular wavelength of red light emitted by the nebula is 656 nm, characteristic of hydrogen gas. (NASA, ESA, and A. Nota, STScI/ESA)

in **Figure 5-18**. Such glowing clouds have emission line spectra, because we see them against the black background of space. The particular shade of red that dominates the color of this nebula is due to an emission line at a wavelength near 656 nm. This is one of the characteristic wavelengths emitted by hydrogen gas, so we can conclude that this nebula contains hydrogen. More detailed analyses of this kind show that hydrogen is the most common element in gaseous nebulae, and indeed in the universe as a whole. The spectra of other nebulae, such as the Ring Nebula shown in the image that opens this chapter, also reveal the presence of nitrogen, oxygen, helium, and other gases.

What is truly remarkable about spectroscopy is that it can determine chemical composition at any distance. The 656-nm red light produced by a sample of heated hydrogen gas on Earth (the bright red line in the hydrogen spectrum in Figure 5-15) is the same as that observed coming from the nebula shown in Figure 5-18, located about 210,000 light-years away. By using the basic principles outlined by Kirchhoff, astronomers have the tools to

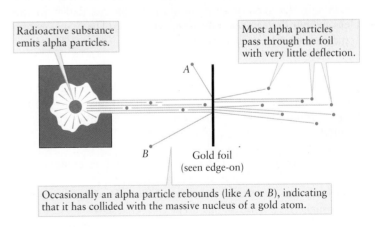

Radioactive substance emits alpha particles.

Most alpha particles pass through the foil with very little deflection.

A

B   Gold foil (seen edge-on)

Occasionally an alpha particle rebounds (like *A* or *B*), indicating that it has collided with the massive nucleus of a gold atom.

Figure 5-19

Rutherford's Experiment Alpha particles from a radioactive source are directed at a thin metal foil. This experiment provided the first evidence that the nuclei of atoms are relatively massive and compact.

Figure 5-20

Rutherford's Model of the Atom Electrons orbit the atom's nucleus, which contains most of the atom's mass. The nucleus contains two types of particles, protons and neutrons.

if you fired a fifteen-inch shell at a piece of tissue paper and it came back and hit you."

Rutherford concluded from this experiment that most of the mass of an atom is concentrated in a compact, massive lump of matter that occupies only a small part of the atom's volume. Most of the alpha particles pass freely through the nearly empty space that makes up most of the atom, but a few particles happen to strike the dense mass at the center of the atom and bounce back.

The Nucleus of an Atom

Rutherford proposed a new model for the structure of an atom, shown in **Figure 5-20**. According to this model, a massive, positively charged **nucleus** at the center of the atom is orbited by tiny, negatively charged electrons. Rutherford concluded that at least 99.98% of the mass of an atom must be concentrated in its nucleus, whose diameter is only about 10^{-14} m. (The diameter of a typical atom is far larger, about 10^{-10} m.)

ANALOGY To appreciate just how tiny the nucleus is, imagine expanding an atom by a factor of 10^{12} to a diameter of 100 meters, about the length of a football field. To this scale, the nucleus would be just a centimeter across—no larger than your thumbnail.

We know today that the nucleus of an atom contains two types of particles, **protons** and **neutrons**. A proton has a positive electric charge, equal and opposite to that of an electron. As its name suggests, a neutron has no electric charge—it is electrically neutral. As an example, an alpha particle (such as those Rutherford's team used) is actually a nucleus of the helium atom, with two protons and two neutrons. Protons and neutrons are held together in a nucleus by the so-called strong nuclear force, whose great strength overcomes the electric repulsion between the posi-

tively charged protons. A proton and a neutron have almost the same mass, 1.7×10^{-27} kg, and each has about 2000 times as much mass as an electron (9.1×10^{-31} kg). In an ordinary atom there are as many positive protons as there are negative electrons, so the atom has no net electric charge. Because the mass of the electron is so small, the mass of an atom is not much greater than the mass of its nucleus. That is why an alpha particle has nearly the same mass as an atom of helium.

While the solar system is held together by gravitational forces, atoms are held together by electrical forces. The electric forces attracting the positively charged protons and the negatively charged electrons keep the atom from coming apart. **Box 5-5** on page 118 describes more about the connection between the structure of atoms and the chemical and physical properties of substances made of those atoms.

Rutherford's experiments clarified the structure of the atom, but they did not explain how these tiny particles within the atom give rise to spectral lines. The task of reconciling Rutherford's atomic model with Kirchhoff's laws of spectral analysis was undertaken by the young Danish physicist Niels Bohr, who joined Rutherford's group at Manchester in 1912.

5-8 Spectral lines are produced when an electron jumps from one energy level to another within an atom

 Niels Bohr began his study of the connection between atomic spectra and atomic structure by trying to understand the structure of hydrogen, the simplest and lightest of the elements. (As we discussed in Section 5-6, hydrogen is also the most common element in the universe.) When Bohr was done, he had not only found a way to explain this

> **Niels Bohr explained spectral lines with a radical new model of the atom**

atom's spectrum but had also found a justification for Kirchhoff's laws in terms of atomic physics.

Hydrogen and the Balmer Series

The most common type of hydrogen atom consists of a single electron and a single proton. Hydrogen atoms have a simple visible-light spectrum consisting of a pattern of lines that begins at a wavelength of 656.3 nm and ends at 364.6 nm. The first spectral line is called H_α (H-alpha), the second spectral line is called H_β (H-beta), the third is H_γ (H-gamma), and so forth. (These are the bright lines in the spectrum of hydrogen shown in Figure 5-15. The fainter lines between these appear when hydrogen atoms form into molecules.) The closer you get to the short-wavelength end of the spectrum at 364.6 nm, the more spectral lines you see.

The regularity in this spectral pattern was described mathematically in 1885 by Johann Jakob Balmer, a Swiss schoolteacher. The spectral lines of hydrogen at visible wavelengths are today called **Balmer lines,** and the entire pattern from H_α onward is called the **Balmer series.** Eight Balmer lines are seen in the spectrum of the star shown in **Figure 5-21**. Stars in general, including

Shorter wavelength

H_θ H_η H_ζ H_ε H_δ H_γ   H_β   H_α

Figure 5-21   R I V U X G

Balmer Lines in the Spectrum of a Star This portion of the spectrum of the star Vega in the constellation Lyra (the Harp) shows eight Balmer lines, from H_α at 656.3 nm through H_θ (H-theta) at 388.9 nm. The series converges at 364.6 nm, slightly to the left of H_θ. (NOAO)

the Sun, have Balmer absorption lines in their spectra, which shows they have atmospheres that contain hydrogen.

Using trial and error, Balmer discovered a formula from which the wavelengths (λ) of hydrogen's spectral lines can be calculated. Balmer's formula is usually written

$$\frac{1}{\lambda} = R\left(\frac{1}{4} - \frac{1}{n^2}\right)$$

In this formula R is the *Rydberg constant* ($R = 1.097 \times 10^7$ m^{-1}), named in honor of the Swedish spectroscopist Johannes Rydberg, and n can be any integer (whole number) greater than 2. To get the wavelength λ_α of the spectral line H_α, you first put $n = 3$ into Balmer's formula:

$$\frac{1}{\lambda_\alpha} = (1.097 \times 10^7 \text{ m}^{-1})\left(\frac{1}{4} - \frac{1}{3^2}\right) = 1.524 \times 10^6 \text{ m}^{-1}$$

Then take the reciprocal:

$$\lambda_\alpha = \frac{1}{1.524 \times 10^6 \text{ m}^{-1}} = 6.563 \times 10^{-7} \text{ m} = 656.3 \text{ nm}$$

To get the wavelength of H_β, use $n = 4$, and to get the wavelength of H_γ, use $n = 5$. If you use $n = \infty$ (the symbol ∞ stands for infinity), you get the short-wavelength end of the hydrogen spectrum at 364.6 nm. (Note that 1 divided by infinity equals zero.)

Bohr's Model of Hydrogen

Bohr realized that to fully understand the structure of the hydrogen atom, he had to be able to derive Balmer's formula using the laws of physics. He first made the rather wild assumption that the electron in a hydrogen atom can orbit the nucleus only in certain specific orbits. (This was a significant break with the ideas of Newton, in whose mechanics any orbit should be possible.) Figure 5-22 shows the four smallest of these **Bohr orbits,** labeled by the numbers $n = 1$, $n = 2$, $n = 3$, and so on.

Although confined to one of these allowed orbits while circling the nucleus, an electron can jump from one Bohr orbit to another. For an electron to do this, the hydrogen atom must gain or lose a specific amount of energy. The atom must absorb energy for the electron to go from an inner to an outer orbit; the atom must release energy for the electron to go from an outer to an inner orbit. As an example, Figure 5-23 shows an electron

jumping between the $n = 2$ and $n = 3$ orbits of a hydrogen atom as the atom absorbs or emits an H_α photon.

When the electron jumps from one orbit to another, the energy of the photon that is emitted or absorbed equals the difference in energy between these two orbits. This energy difference, and hence the photon energy, is the same whether the jump is from a low orbit to a high orbit (Figure 5-23a) or from the high orbit back to the low one (Figure 5-23b). According to Planck and Einstein, if two photons have the same energy E, the relationship $E = hc/\lambda$ tells us that they must also have the same wavelength λ. It follows that if an atom can emit photons of a given energy and wavelength, it can also absorb photons of precisely the same energy and wavelength. Thus, Bohr's picture explains Kirchhoff's observation that atoms emit and absorb the same wavelengths of light.

The Bohr picture also helps us visualize what happens to produce an emission line spectrum. When a gas is heated, its atoms move around rapidly and can collide forcefully with each other. These energetic collisions excite the atoms' electrons into high orbits. The electrons then cascade back down to the innermost

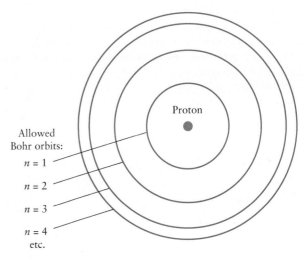

Proton

Allowed Bohr orbits:

$n = 1$
$n = 2$
$n = 3$
$n = 4$
etc.

Figure 5-22

The Bohr Model of the Hydrogen Atom In this model, an electron circles the hydrogen nucleus (a proton) only in allowed orbits $n = 1, 2, 3,$ and so forth. The first four Bohr orbits are shown here. This figure is *not* drawn to scale; in the Bohr model, the $n = 2, 3,$ and 4 orbits are respectively 4, 9, and 16 times larger than the $n = 1$ orbit.

BOX 5-5

Atoms, the Periodic Table, and Isotopes

WEB LINK 5.9

Each different chemical element is made of a specific type of atom. Each specific atom has a characteristic number of protons in its nucleus. For example, a hydrogen atom has 1 proton in its nucleus, an oxygen atom has 8 protons in its nucleus, and so on.

The number of protons in an atom's nucleus is the **atomic number** for that particular element. The chemical elements are most conveniently listed in the form of a **periodic table** (shown in the figure). Elements are arranged in the periodic table in order of increasing atomic number. With only a few exceptions, this sequence also corresponds to increasing average mass of the atoms of the elements. Thus, hydrogen (symbol H), with atomic number 1, is the lightest element. Iron (symbol Fe) has atomic number 26 and is a relatively heavy element.

All the elements listed in a single vertical column of the periodic table have similar chemical properties. For example, the elements in the far right column are all gases under the conditions of temperature and pressure found at the Earth's surface, and they are all very reluctant to react chemically with other elements.

In addition to nearly 100 naturally occurring elements, the periodic table includes a number of artificially produced elements. Most of these elements are heavier than uranium (symbol U) and are highly radioactive, which means that they decay into lighter elements within a short time of being created in laboratory experiments. Scientists have succeeded in creating only a few atoms of elements 104 and above.

The number of protons in the nucleus of an atom determines which element that atom is. Nevertheless, the same element may have different numbers of neutrons in its nucleus. For example, oxygen (O) has atomic number 8, so every oxygen nucleus has exactly 8 protons. But oxygen nuclei can have 8, 9, or 10 neutrons. These three slightly different kinds of oxygen are called **isotopes.** The isotope with 8 neutrons is by far the most abundant variety. It is written as ^{16}O, or oxygen-16. The rarer isotopes with 9 and 10 neutrons are designated as ^{17}O and ^{18}O, respectively.

The superscript that precedes the chemical symbol for an element equals the total number of protons and neutrons in a nucleus of that particular isotope. For example, a nucleus of the most common isotope of iron, ^{56}Fe or iron-56, contains a total of 56 protons and neutrons. From the periodic table, the atomic number of iron is 26, so every iron atom has 26 protons in its nucleus. Therefore, the number of neutrons in an iron-56 nucleus is $56 - 26 = 30$. (Most nuclei have more neutrons than protons, especially in the case of the heaviest elements.)

It is extremely difficult to distinguish chemically between the various isotopes of a particular element. Ordinary chemical reactions involve only the electrons that orbit the atom, never the neutrons buried in its nucleus. But there are small differences in the wavelengths of the spectral lines for different isotopes of the same element. For example, the spectral line wavelengths of the hydrogen isotope ^2H are about 0.03% greater than the wavelengths for the most common hydrogen isotope, ^1H. Thus, different isotopes can be distinguished by careful spectroscopic analysis.

Isotopes are important in astronomy for a variety of reasons. By measuring the relative amounts of different isotopes of a given element in a Moon rock or meteorite, the age of that sample can be determined. The mixture of isotopes left behind when a star explodes into a supernova (see Section 1-3) tells astronomers about the processes that led to the explosion. And knowing the properties of different isotopes of hydrogen and helium is crucial to understanding the nuclear reactions that make the Sun shine. Look for these and other applications of the idea of isotopes in later chapters.

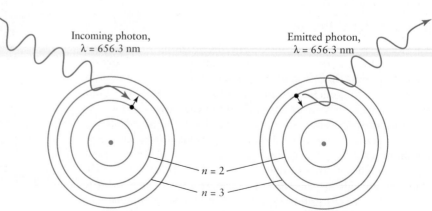

AIMM 5.2

Figure 5-23

The Absorption and Emission of an H$_\alpha$ Photon This schematic diagram, drawn according to the Bohr model, shows what happens when a hydrogen atom **(a)** absorbs or **(b)** emits a photon whose wavelength is 656.3 nm.

Incoming photon,
λ = 656.3 nm

Emitted photon,
λ = 656.3 nm

n = 2

n = 3

(a) Atom absorbs a 656.3-nm photon; absorbed energy causes electron to jump from the n = 2 orbit up the n = 3 orbit

(b) Electron falls from the n = 3 orbit to the n = 2 orbit; energy lost by atom goes into emitting a 656.3-nm photon

Periodic Table of the Elements

1 H Hydrogen																	2 He Helium
3 Li Lithium	4 Be Beryllium											5 B Boron	6 C Carbon	7 N Nitrogen	8 O Oxygen	9 F Fluorine	10 Ne Neon
11 Na Sodium	12 Mg Magnesium											13 Al Aluminum	14 Si Silicon	15 P Phosphorus	16 S Sulfur	17 Cl Chlorine	18 Ar Argon
19 K Potassium	20 Ca Calcium	21 Sc Scandium	22 Ti Titanium	23 V Vanadium	24 Cr Chromium	25 Mn Manganese	26 Fe Iron	27 Co Cobalt	28 Ni Nickel	29 Cu Copper	30 Zn Zinc	31 Ga Gallium	32 Ge Germanium	33 As Arsenic	34 Se Selenium	35 Br Bromine	36 Kr Kryton
37 Rb Rubidium	38 Sr Strontium	39 Y Yttrium	40 Zr Zirconium	41 Nb Niobium	42 Mo Molybdenum	43 Tc Technetium	44 Ru Ruthenium	45 Rh Rhodium	46 Pd Palladium	47 Ag Silver	48 Cd Cadmium	49 In Indium	50 Sn Tin	51 Sb Antimony	52 Te Tellurium	53 I Iodine	54 Xe Xenon
55 Cs Cesium	56 Ba Barium	57 La Lanthanum	72 Hf Hafnium	73 Ta Tantaium	74 W Tungsten	75 Re Rhenium	76 Os Osmium	77 Ir Iridium	78 Pt Platinum	79 Au Gold	80 Hg Mercury	81 Tl Thallium	82 Pb Lead	83 Bi Bismuth	84 Po Polonium	85 At Astatine	86 Rn Radon
87 Fr Francium	88 Ra Radium	89 Ac Actinium	104 Rf Rutherfordium	105 Db Dubnium	106 Sg Seaborgium	107 Bh Bohrium	108 Hs Hassium	109 Mt Meitnerium	110 Ds Darmstadium	111 Rg Roentgenium	112 Uub Ununbium	113 Uut Ununtium	114 Uug Ununguadium	115 Uup Ununpentium	116 Uuh Ununhexium		

58 Ce Cerium	59 Pr Praseodymium	60 Nd Neodymium	61 Pm Promethium	62 Sm Samarium	63 Eu Europium	64 Gd Gadolinium	65 Tb Terbium	66 Dy Dysprosium	67 Ho Holmium	68 Er Erbium	69 Tm Thulium	70 Yb Ytterbium	71 Lu Lutetium
90 Th Thorium	91 Pa Protactinium	92 U Uranium	93 Np Neptunium	94 Pu Plutonium	95 Am Americium	96 Cm Curium	97 Bk Berkelium	98 Cf Californium	99 Es Einsteinium	100 Fm Fermium	101 Md Mendelevium	102 No Nobelium	103 Lr Lawrencium

possible orbit, emitting photons whose energies are equal to the energy differences between different Bohr orbits. In this fashion, a hot gas produces an emission line spectrum with a variety of different wavelengths.

To produce an absorption line spectrum, begin with a relatively cool gas, so that the electrons in most of the atoms are in inner, low-energy orbits. If a beam of light with a continuous spectrum is shone through the gas, most wavelengths will pass through undisturbed. Only those photons will be absorbed whose energies are just right to excite an electron to an allowed outer orbit. Hence, only certain wavelengths will be absorbed, and dark lines will appear in the spectrum at those wavelengths.

Using his picture of allowed orbits and the formula $E = hc/\lambda$, Bohr was able to prove mathematically that the wavelength λ of the photon emitted or absorbed as an electron jumps between an inner orbit N and an outer orbit n is

Bohr formula for hydrogen wavelengths

$$\frac{1}{\lambda} = R\left(\frac{1}{N^2} - \frac{1}{n^2}\right)$$

N = number of inner orbit

n = number of outer orbit

R = Rydberg constant = 1.097×10^7 m^{-1}

λ = wavelength (in meters) of emitted or absorbed photon

If Bohr let $N = 2$ in this formula, he got back the formula that Balmer discovered by trial and error. Hence, Bohr deduced the meaning of the Balmer series: All the Balmer lines are produced by electrons jumping between the second Bohr orbit

Figure 5-24

Electron Transitions in the Hydrogen Atom This diagram shows the photon wavelengths associated with different electron transitions in hydrogen. In each case, the same wavelength occurs whether a photon is emitted (when the electron drops from a high orbit to a low one) or absorbed (when the electron jumps from a low orbit to a high one). The orbits are not shown to scale.

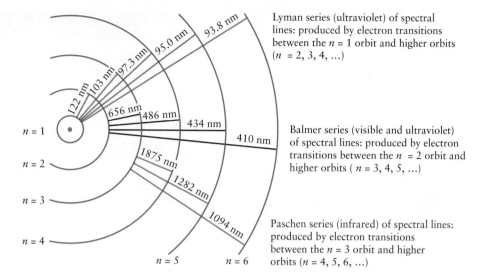

Lyman series (ultraviolet) of spectral lines: produced by electron transitions between the $n = 1$ orbit and higher orbits ($n = 2, 3, 4, ...$)

Balmer series (visible and ultraviolet) of spectral lines: produced by electron transitions between the $n = 2$ orbit and higher orbits ($n = 3, 4, 5, ...$)

Paschen series (infrared) of spectral lines: produced by electron transitions between the $n = 3$ orbit and higher orbits ($n = 4, 5, 6, ...$)

($N = 2$) and higher orbits ($n = 3, 4, 5,$ and so on). Remarkably, as part of his calculation Bohr was able to *derive* the value of the Rydberg constant in terms of Planck's constant, the speed of light, and the mass and electric charge of the electron. This gave particular credence to his radical model.

Bohr's formula also correctly predicts the wavelengths of other series of spectral lines that occur at nonvisible wavelengths (**Figure 5-24**). Using $N = 1$ gives the **Lyman series,** which is entirely in the ultraviolet. All the spectral lines in this series involve electron transitions between the lowest Bohr orbit and all higher orbits ($n = 2, 3, 4,$ and so on). This pattern of spectral lines begins with L_α (Lyman alpha) at 122 nm and converges on L_∞ at 91.2 nm. Using $N = 3$ gives a series of infrared wavelengths called the **Paschen series.** This series, which involves transitions between the third Bohr orbit and all higher orbits, begins with P_α (Paschen alpha) at 1875 nm and converges on P_∞ at 822 nm. Additional series exist at still longer wavelengths.

Atomic Energy Levels

Today's view of the atom owes much to the Bohr model, but is different in certain ways. The modern picture is based on **quantum mechanics,** a branch of physics developed during the 1920s that deals with photons and subatomic particles. As a result of this work, physicists no longer picture electrons as moving in specific orbits about the nucleus. Instead, electrons are now known to have both particle and wave properties and are said to occupy only certain **energy levels** in the atom.

An extremely useful way of displaying the structure of an atom is with an **energy-level diagram.** Figure 5-25 shows such a diagram for hydrogen. The lowest energy level, called the **ground state,** corresponds to the $n = 1$ Bohr orbit. Higher energy levels, called **excited states,** correspond to successively larger Bohr orbits.

An electron can jump from the ground state up to the $n = 2$ level if the atom absorbs a Lyman-alpha photon with a wavelength of 122 nm. Such a photon has energy $E = hc/\lambda = 10.2$ eV (electron volts; see Section 5-5). That's why the energy level of $n = 2$ is shown in Figure 5-25 as having an energy 10.2 eV above that of the ground state (which is usually assigned a value of 0 eV). Similarly, the $n = 3$ level is 12.1 eV above the ground state,

and so forth. Electrons can make transitions to higher energy levels by absorbing a photon or in a collision between atoms; they can make transitions to lower energy levels by emitting a photon.

On the energy-level diagram for hydrogen, the $n = \infty$ level has an energy of 13.6 eV. (This corresponds to an infinitely large orbit in the Bohr model.) If the electron is initially in the ground state and the atom absorbs a photon of any energy greater than 13.6 eV, the electron will be removed completely from the atom. This process is called **ionization.** A 13.6-eV photon has a wave-

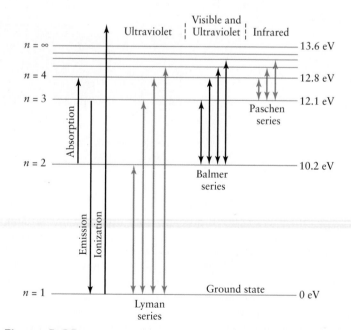

Figure 5-25

Energy-Level Diagram of Hydrogen A convenient way to display the structure of the hydrogen atom is in a diagram like this, which shows the allowed energy levels. The diagram shows a number of possible electron jumps, or transitions, between energy levels. An upward transition occurs when the atom absorbs a photon; a downward transition occurs when the atom emits a photon. (Compare with Figure 5-24.)

length of 91.2 nm, equal to the shortest wavelength in the ultraviolet Lyman series (L$_\infty$). So any photon with a wavelength of 91.2 nm or less can ionize hydrogen. (The Planck formula $E = hc/\lambda$ tells us that the higher the photon energy, the shorter the wavelength.)

As an example, the gaseous nebula shown in Figure 5-18 surrounds a cluster of hot stars which produce copious amounts of ultraviolet photons with wavelengths less than 91.2 nm. Hydrogen atoms in the nebula that absorb these photons become ionized and lose their electrons. When the electrons recombine with the nuclei, they cascade down the energy levels to the ground state and emit visible light in the process. This is what makes the nebula glow.

The Spectra of Other Elements

The same basic principles that explain the hydrogen spectrum also apply to the atoms of other elements. Electrons in each kind of atom can be only in certain energy levels, so only photons of certain wavelengths can be emitted or absorbed. Because each kind of atom has its own unique arrangement of electron levels, the pattern of spectral lines is likewise unique to that particular type of atom (see the photograph that opens this chapter). These patterns are in general much more complicated than for the hydrogen atom. Hence, there is no simple relationship analogous to the Bohr formula that applies to the spectra of all atoms.

The idea of energy levels explains the emission line spectra and absorption line spectra of gases. But what about the continuous spectra produced by dense objects like the filament of a lightbulb or the coils of a toaster? These objects are made of atoms, so why don't they emit light with an emission line spectrum characteristic of the particular atoms of which they are made?

The reason is directly related to the difference between a gas on the one hand and a liquid or solid on the other. In a gas, atoms are widely separated and can emit photons without interference from other atoms. But in a liquid or a solid, atoms are so close that they almost touch, and thus these atoms interact strongly with each other. These interactions interfere with the process of emitting photons. As a result, the pattern of distinctive bright spectral lines that the atoms would emit in isolation becomes "smeared out" into a continuous spectrum.

ANALOGY Think of atoms as being like tuning forks. If you strike a single tuning fork, it produces a sound wave with a single clear frequency and wavelength, just as an isolated atom emits light of definite wavelengths. But if you shake a box packed full of tuning forks, you will hear a clanging noise that is a mixture of sounds of all different frequencies and wavelengths. This is directly analogous to the continuous spectrum of light emitted by a dense object with closely packed atoms.

With the work of such people as Planck, Einstein, Rutherford, and Bohr, the interchange between astronomy and physics came full circle. Modern physics was born when Newton set out to understand the motions of the planets. Two and a half centuries later, physicists in their laboratories probed the properties of light and the structure of atoms. Their labors had immediate applica-

tions in astronomy. Armed with this new understanding of light and matter, astronomers were able to probe in detail the chemical and physical properties of planets, stars, and galaxies.

5-9   The wavelength of a spectral line is affected by the relative motion between the source and the observer

In addition to telling us about temperature and chemical composition, the spectrum of a planet, star, or galaxy can also reveal something about that object's motion through space. This idea dates from 1842, when Christian Doppler, a professor of mathematics in Prague, pointed out that the observed wavelength of light must be affected by motion.

> **The Doppler effect makes it possible to tell whether astronomical objects are moving toward us or away from us**

The Doppler Effect

In **Figure 5-26** a light source is moving from right to left; the circles represent the crests of waves emitted from the moving source at various positions. Each successive wave crest is emitted from a position slightly closer to the observer on the left, so she sees a shorter wavelength—the distance from one crest to the next—than she would if the source were stationary. All the lines in the spectrum

Wave crest 1: emitted when light source was at S$_1$

Wave crest 2: emitted when light source was at S$_2$

Wave crests 3 and 4: emitted when light source was at S$_3$ and S$_4$, respectively

Motion of light source

S$_4$ S$_3$ S$_2$ S$_1$

This observer sees **blueshift**

This observer sees **redshift**

Figure 5-26

The Doppler Effect The wavelength of light is affected by motion between the light source and an observer. The light source shown here is moving, so wave crests 1, 2, etc., emitted when the source was at points S$_1$, S$_2$, etc., are crowded together in front of the source but are spread out behind it. Consequently, wavelengths are shortened (blueshifted) if the source is moving toward the observer and lengthened (redshifted) if the source is moving away from the observer. Motion perpendicular to an observer's line of sight does not affect wavelength.

BOX 5-6

Applications of the Doppler Effect

Doppler's formula relates the radial velocity of an astronomical object to the wavelength shift of its spectral lines. Here are two examples that show how to use this remarkably powerful formula.

EXAMPLE: As measured in the laboratory, the prominent H_α spectral line of hydrogen has a wavelength $\lambda_0 = 656.285$ nm. But in the spectrum of the star Vega (Figure 5-21), this line has a wavelength $\lambda = 656.255$ nm. What can we conclude about the motion of Vega?

Situation: Our goal is to use the ideas of the Doppler effect to find the velocity of Vega toward or away from Earth.

Tools: We use the Doppler shift formula, $\lambda/\lambda_0 = v/c$, to determine Vega's velocity v.

Answer: The wavelength shift is

$$\Delta\lambda = \lambda - \lambda_0 = 656.255 \text{ nm} - 656.285 \text{ nm} = -0.030 \text{ nm}$$

The negative value means that we see the light from Vega shifted to shorter wavelengths—that is, there is a blueshift. (Note that the shift is very tiny and can be measured only using specialized equipment.) From the Doppler shift formula, the star's radial velocity is

$$v = c\frac{\Delta\lambda}{\lambda_0} = (3.00 \times 10^5 \text{ km/s})\left(\frac{-0.030 \text{ nm}}{656.285 \text{ nm}}\right) = -14 \text{ km/s}$$

Review: The minus sign indicates that Vega is coming toward us at 14 km/s. The star may also be moving perpendicular to the line from Earth to Vega, but such motion produces no Doppler shift.

By plotting the motions of stars such as Vega toward and away from us, astronomers have been able to learn how the Milky Way Galaxy (of which our Sun is a part) is rotating. From this knowledge, and aided by Newton's

universal law of gravitation (see Section 4-7), they have made the surprising discovery that the Milky Way contains roughly 10 times more matter than had once been thought! The nature of this unseen *dark matter* is still a subject of debate.

EXAMPLE: In the radio region of the electromagnetic spectrum, hydrogen atoms emit and absorb photons with a wavelength of 21.12 cm, giving rise to a spectral feature commonly called the *21-centimeter line*. The galaxy NGC 3840 in the constellation Leo (the Lion) is receding from us at a speed of 7370 km/s, or about 2.5% of the speed of light. At what wavelength do we expect to detect the 21-cm line from this galaxy?

Situation: Given the velocity of NGC 3840 away from us, our goal is to find the wavelength as measured on Earth of the 21-centimeter line from this galaxy.

Tools: We use the Doppler shift formula to calculate the wavelength shift $\Delta\lambda$, then use this to find the wavelength λ measured on Earth.

Answer: The wavelength shift is

$$\Delta\lambda = \lambda_0\left(\frac{v}{c}\right) = (21.12 \text{ cm})\left(\frac{7370 \text{ km/s}}{3.00 \times 10^5 \text{ km/s}}\right) = 0.52 \text{ cm}$$

Therefore, we will detect the 21-cm line of hydrogen from this galaxy at a wavelength of

$$\lambda = \lambda_0 + \Delta\lambda = 21.12 \text{ cm} + 0.52 \text{ cm} = 21.64 \text{ cm}$$

Review: The 21-cm line has been redshifted to a longer wavelength because the galaxy is receding from us. In fact, most galaxies are receding from us. This observation is one of the key pieces of evidence that the universe is expanding, and has been doing so since the Big Bang that took place almost 14 billion years ago.

---

of an approaching source are shifted toward the short-wavelength (blue) end of the spectrum. This phenomenon is called a **blueshift.**

The source is receding from the observer on the right in Figure 5-26. The wave crests that reach him are stretched apart, so that he sees a longer wavelength than he would if the source were stationary. All the lines in the spectrum of a receding source are shifted toward the longer-wavelength (red) end of the spectrum, producing a **redshift.** In general, the effect of relative motion on wavelength is called the **Doppler effect.** Police radar guns use the Doppler effect to check for cars exceeding the speed limit: the radar gun sends a radio wave toward the car, and measures the wavelength shift of the reflected wave (and thus the speed of the car).

ANALOGY You have probably noticed a similar Doppler effect for sound waves. When a police car is approaching, the sound waves from its siren have a shorter wavelength and higher frequency than if the siren were at rest, and hence you hear a higher pitch. After the police car passes you and is moving away, you hear a lower pitch from the siren because the sound waves have a longer wavelength and a lower frequency.

Suppose that λ_0 is the wavelength of a particular spectral line from a light source that is not moving. It is the wavelength that you might look up in a reference book or determine in a laboratory experiment for this spectral line. If the source is moving, this particular spectral line is shifted to a different wavelength λ. The size of the wavelength shift is usually written as $\Delta\lambda$, where $\Delta\lambda = \lambda - \lambda_0$. Thus, $\Delta\lambda$ is the difference between the wavelength listed in reference books and the wavelength that you actually observe in the spectrum of a star or galaxy.

Doppler proved that the wavelength shift ($\Delta\lambda$) is governed by the following simple equation:

Doppler shift equation

$$\frac{\Delta\lambda}{\lambda_0} = \frac{v}{c}$$

$\Delta\lambda$ = wavelength shift

λ_0 = wavelength if source is not moving

v = velocity of the source measured along the line of sight

c = speed of light = 3.0×10^5 km/s

CAUTION! The capital Greek letter Δ (delta) is commonly used to denote a change in the value of a quantity. Thus, $\Delta\lambda$ is the change in the wavelength λ due to the Doppler effect. It is *not* equal to a quantity Δ multiplied by a second quantity λ!

Interpreting the Doppler Effect

The velocity determined from the Doppler effect is called **radial velocity**, because v is the component of the star's motion parallel to our line of sight, or along the "radius" drawn from Earth to the star. Of course, a sizable fraction of a star's motion may be perpendicular to our line of sight. The speed of this transverse movement across the sky does not affect wavelengths if the speed is small compared with c. Box 5-6 includes two examples of calculations with radial velocity using the Doppler formula.

CAUTION! The redshifts and blueshifts of stars visible to the naked eye, or even through a small telescope, are only a small fraction of a nanometer. These tiny wavelength changes are far too small to detect visually. (Astronomers were able to detect the tiny Doppler shifts of starlight only after they had developed highly sensitive equipment for measuring wavelengths. This was done around 1890, a half-century after Doppler's original proposal.) So, if you see a star with a red color, it means that the star really is red; it does *not* mean that it is moving rapidly away from us.

The Doppler effect is an important tool in astronomy because it uncovers basic information about the motions of planets, stars, and galaxies. For example, the rotation of the planet Venus was deduced from the Doppler shift of radar waves reflected from its surface. Small Doppler shifts in the spectrum of sunlight have shown that the entire Sun is vibrating like an immense gong. The back-and-forth Doppler shifting of the spectral lines of certain stars reveals that these stars are being orbited by unseen companions; from this astronomers have discovered planets around other stars and massive objects that may be black holes. Astronomers also use the Doppler effect along with Kepler's third law to measure the masses of galaxies. These are but a few examples of how

Doppler's discovery has empowered astronomers in their quest to understand the universe.

In this chapter we have glimpsed how much can be learned by analyzing light from the heavens. To analyze this light, however, it is first necessary to collect as much of it as possible, because most light sources in space are very dim. Collecting the faint light from distant objects is a key purpose of telescopes. In the next chapter we will describe both how telescopes work and how they are used.

Key Words

Terms preceded by an asterisk (*) are discussed in the Boxes.

absolute zero, p. 104	joule, p. 107
absorption line spectrum, p. 112	kelvin, p. 104
atom, p. 104	Kirchhoff's laws, p. 112
*atomic number, p. 118	light scattering, p. 112
Balmer line, p. 116	*luminosity, p. 108
Balmer series, p. 116	Lyman series, p. 120
blackbody, p. 106	microwaves, p. 102
blackbody curve, p. 106	molecule, p. 112
blackbody radiation, p. 106	nanometer, p. 101
blueshift, p. 122	neutron, p. 116
Bohr orbits, p. 117	nucleus, p. 116
compound, p. 112	Paschen series, p. 120
continuous spectrum, p. 112	*periodic table, p. 118
*degrees Celsius, p. 105	photoelectric effect, p. 109
*degrees Fahrenheit, p. 105	photon, p. 108
Doppler effect, p. 122	Planck's law, p. 110
electromagnetic radiation, p. 101	proton, p. 116
	quantum mechanics, p. 120
electromagnetic spectrum, p. 102	radial velocity, p. 123
electromagnetism, p. 100	radio waves, p. 101
electron, p. 109	redshift, p. 122
electron volt, p. 110	*solar constant, p. 108
element, p. 111	spectral analysis, p. 111
emission line spectrum, p. 112	spectral line, p. 111
energy flux, p. 107	spectroscopy, p. 113
energy level, p. 120	spectrum (*plural* spectra), p. 99
energy-level diagram, p. 120	Stefan-Boltzmann law, p. 107
excited state, p. 120	ultraviolet radiation, p. 102
frequency, p. 102	visible light, p. 101
gamma rays, p. 102	watt, p. 107
ground state, p. 120	wavelength, p. 101
infrared radiation, p. 101	wavelength of maximum emission, p. 105
ionization, p. 120	Wien's law, p. 107
*isotope, p. 118	X rays, p. 102

Key Ideas

The Nature of Light: Light is electromagnetic radiation. It has wavelike properties described by its wavelength λ and frequency v, and travels through empty space at the constant speed $c = 3.0 \times 10^8$ m/s $= 3.0 \times 10^5$ km/s.

Blackbody Radiation: A blackbody is a hypothetical object that is a perfect absorber of electromagnetic radiation at all wavelengths. Stars closely approximate the behavior of blackbodies, as do other hot, dense objects.

• The intensities of radiation emitted at various wavelengths by a blackbody at a given temperature are shown by a blackbody curve.

• Wien's law states that the dominant wavelength at which a blackbody emits electromagnetic radiation is inversely proportional to the Kelvin temperature of the object: λ_{max} (in meters) = (0.0029 K m)/T.

• The Stefan-Boltzmann law states that a blackbody radiates electromagnetic waves with a total energy flux F directly proportional to the fourth power of the Kelvin temperature T of the object: $F = \sigma T^4$.

Photons: An explanation of blackbody curves led to the discovery that light has particlelike properties. The particles of light are called photons.

• Planck's law relates the energy E of a photon to its frequency v or wavelength λ: $E = hv = hc/\lambda$, where h is Planck's constant.

Kirchhoff's Laws: Kirchhoff's three laws of spectral analysis describe conditions under which different kinds of spectra are produced.

• A hot, dense object such as a blackbody emits a continuous spectrum covering all wavelengths.

• A hot, transparent gas produces a spectrum that contains bright (emission) lines.

• A cool, transparent gas in front of a light source that itself has a continuous spectrum produces dark (absorption) lines in the continuous spectrum.

Atomic Structure: An atom has a small dense nucleus composed of protons and neutrons. The nucleus is surrounded by electrons that occupy only certain orbits or energy levels.

• When an electron jumps from one energy level to another, it emits or absorbs a photon of appropriate energy (and hence of a specific wavelength).

• The spectral lines of a particular element correspond to the various electron transitions between energy levels in atoms of that element.

• Bohr's model of the atom correctly predicts the wavelengths of hydrogen's spectral lines.

The Doppler Shift: The Doppler shift enables us to determine the radial velocity of a light source from the displacement of its spectral lines.

• The spectral lines of an approaching light source are shifted toward short wavelengths (a blueshift); the spectral lines of a receding light source are shifted toward long wavelengths (a redshift).

• The size of a wavelength shift is proportional to the radial velocity of the light source relative to the observer.

Questions

Review Questions

1. When Jupiter is undergoing retrograde motion as seen from Earth, would you expect the eclipses of Jupiter's moons to occur several minutes early, several minutes late, or neither? Explain.

2. Approximately how many times around the Earth could a beam of light travel in one second?

3. How long does it take light to travel from the Sun to the Earth, a distance of 1.50×10^8 km?

4. How did Newton show that a prism breaks white light into its component colors, but does not add any color to the light?

5. For each of the following wavelengths, state whether it is in the radio, microwave, infrared, visible, ultraviolet, X-ray, or gamma-ray portion of the electromagnetic spectrum. Explain your reasoning. (a) 2.6 μm, (b) 34 m, (c) 0.54 nm, (d) 0.0032 nm, (e) 0.620 μm, (f) 310 nm, (g) 0.012 m.

6. What is meant by the frequency of light? How is frequency related to wavelength?

7. A cellular phone is actually a radio transmitter and receiver. You receive an incoming call in the form of a radio wave of frequency 880.65 MHz. What is the wavelength (in meters) of this wave?

8. A light source emits infrared radiation at a wavelength of 1150 nm. What is the frequency of this radiation?

9. (a) What is a blackbody? (b) In what way is a blackbody black? (c) If a blackbody is black, how can it emit light? (d) If you were to shine a flashlight beam on a perfect blackbody, what would happen to the light?

10. Why do astronomers find it convenient to use the Kelvin temperature scale in their work rather than the Celsius or Fahrenheit scale?

11. Explain why astronomers are interested in blackbody radiation.

12. Using Wien's law and the Stefan-Boltzmann law, explain the color and intensity changes that are observed as the temperature of a hot, glowing object increases.

13. If you double the Kelvin temperature of a hot piece of steel, how much more energy will it radiate per second?

14. The bright star Bellatrix in the constellation Orion has a surface temperature of 21,500 K. What is its wavelength of maximum emission in nanometers? What color is this star?

15. The bright star Antares in the constellation Scorpius (the Scorpion) emits the greatest intensity of radiation at a wavelength of 853 nm. What is the surface temperature of Antares? What color is this star?

16. (a) Describe an experiment in which light behaves like a wave. (b) Describe an experiment in which light behaves like a particle.

17. How is the energy of a photon related to its wavelength? What kind of photons carry the most energy? What kind of photons carry the least energy?

18. To emit the same amount of light energy per second, which must emit more photons per second: a source of red light, or a source of blue light? Explain.

19. Explain how we know that atoms have massive, compact nuclei.

20. (a) Describe the spectrum of hydrogen at visible wavelengths. (b) Explain how Bohr's model of the atom accounts for the Balmer lines.

21. Why do different elements display different patterns of lines in their spectra?

22. What is the Doppler effect? Why is it important to astronomers?

23. If you see a blue star, what does its color tell you about how the star is moving through space? Explain your answer.

Advanced Questions

Questions preceded by an asterisk (*) involve topics discussed in the Boxes.

> **Problem-solving tips and tools**
>
> You can find formulas in Box 5-1 for converting between temperature scales. Box 5-2 discusses how a star's radius, luminosity, and surface temperature are related. Box 5-3 shows how to use Planck's law to calculate the energy of a photon. To learn how to do calculations using the Doppler effect, see Box 5-6.

24. Your normal body temperature is 98.6°F. What kind of radiation do you predominantly emit? At what wavelength (in nm) do you emit the most radiation?

25. What is the temperature of the Sun's surface in degrees Fahrenheit?

26. What wavelength of electromagnetic radiation is emitted with greatest intensity by this book? To what region of the electromagnetic spectrum does this wavelength correspond?

27. Black holes are objects whose gravity is so strong that not even an object moving at the speed of light can escape from their surface. Hence, black holes do not themselves emit light. But it is possible to detect radiation from material falling *toward* a black hole. Calculations suggest that as this matter falls, it is compressed and heated to temperatures around 10^6 K. Calculate the wavelength of maximum emission for this temperature. In what part of the electromagnetic spectrum does this wavelength lie?

*28. Use the value of the solar constant given in Box 5-2 and the distance from the Earth to the Sun to calculate the luminosity of the Sun.

*29. The star Alpha Lupi (the brightest in the constellation Lupus, the Wolf) has a surface temperature of 21,600 K. How much more energy is emitted each second from each square meter of the surface of Alpha Lupi than from each square meter of the Sun's surface?

*30. Jupiter's moon Io has an active volcano named Pele whose temperature can be as high as 320°C. (a) What is the wavelength of maximum emission for the volcano at this temperature? In what part of the electromagnetic spectrum is this? (b) The average temperature of Io's surface is −150°C. Compared with a square meter of surface at this temperature, how much more energy is emitted per second from each square meter of Pele's surface?

*31. The bright star Sirius in the constellation of Canis Major (the Large Dog) has a radius of 1.67 $R_\odot$ and a luminosity of 25 $L_\odot$. (a) Use this information to calculate the energy flux at the surface of Sirius. (b) Use your answer in part (a) to calculate the surface temperature of Sirius. How does your answer compare to the value given in Box 5-2?

32. In Figure 5-13 you can see two distinct dark lines at the boundary between the orange and yellow parts of the Sun's spectrum (in the center of the third colored band from the top of the figure). The wavelengths of these dark lines are 588.99 and 589.59 nm. What do you conclude from this about the chemical composition of the Sun's atmosphere? (*Hint:* See Section 5-6.)

33. Instruments on board balloons and spacecraft detect 511-keV photons coming from the direction of the center of our Galaxy. (The prefix k means *kilo*, or thousand, so 1 keV = 10^3 eV.) What is the wavelength of these photons? To what part of the electromagnetic spectrum do these photons belong?

34. (a) Calculate the wavelength of P_δ (P-delta), the fourth wavelength in the Paschen series. (b) Draw a schematic diagram of the hydrogen atom and indicate the electron transition that gives rise to this spectral line. (c) In what part of the electromagnetic spectrum does this wavelength lie?

35. (a) Calculate the wavelength of H_η (H-eta), the spectral line for an electron transition between the $n = 7$ and $n = 2$ orbits of hydrogen. (b) In what part of the electromagnetic spectrum does this wavelength lie? Use this to explain why Figure 5-21 is labeled R I ▣V▣ ▣U▣ X G.

36. Certain interstellar clouds contain a very cold, very thin gas of hydrogen atoms. Ultraviolet radiation with any wavelength shorter than 91.2 nm cannot pass through this gas; instead, it is absorbed. Explain why.

37. (a) Can a hydrogen atom in the ground state absorb an H-alpha (H_α) photon? Explain why or why not. (b) Can a hydrogen atom in the $n = 2$ state absorb a Lyman-alpha (L_α) photon? Explain why or why not.

38. An imaginary atom has just 3 energy levels: 0 eV, 1 eV, and 3 eV. Draw an energy-level diagram for this atom. Show all possible transitions between these energy levels. For each transition, determine the photon energy and the photon wavelength. Which transitions involve the emission or absorption of visible light?

39. The star cluster NGC 346 and nebula shown in Figure 5-18 are located within the Small Magellanic Cloud (SMC), a small galaxy that orbits our Milky Way Galaxy. The SMC and the stars and gas within it are moving away from us at 158 km/s. At what wavelength does the red H_α line of hydrogen (which causes the color of the nebula) appear in the nebula's spectrum?

40. The wavelength of H_β in the spectrum of the star Megrez in the Big Dipper (part of the constellation Ursa Major, the Great Bear) is 486.112 nm. Laboratory measurements demonstrate that the normal wavelength of this spectral line is 486.133 nm. Is the star coming toward us or moving away from us? At what speed?

41. You are given a traffic ticket for going through a red light (wavelength 700 nm). You tell the police officer that because you were approaching the light, the Doppler effect caused a blueshift that made the light appear green (wavelength 500 nm). How fast would you have had to be going for this to be true? Would the speeding ticket be justified? Explain.

Discussion Questions

42. The equation that relates the frequency, wavelength, and speed of a light wave, $v = c/\lambda$, can be rewritten as $c = v\lambda$. A friend who has studied mathematics but not much astronomy or physics might look at this equation and say: "This equation tells me that the higher the frequency v, the greater the wave speed c. Since visible light has a higher frequency than radio waves, this means that visible light goes faster than radio waves." How would you respond to your friend?

43. (a) If you could see ultraviolet radiation, how might the night sky appear different? Would ordinary objects appear different in the daytime? (b) What differences might there be in the appearance of the night sky and in the appearance of ordinary objects in the daytime if you could see infrared radiation?

44. The accompanying visible-light image shows the star cluster NGC 3293 in the constellation Carina (the Ship's Keel). What can you say about the surface temperatures of most of the bright stars in this cluster? In what part of the electromagnetic spectrum do these stars emit most intensely? Are your eyes sensitive to this type of radiation? If not, how is it possible to see these stars at all? There is at least one bright star in this cluster with a distinctly different color from the others; what can you conclude about its surface temperature?

R I V U X G

(David Malin/Anglo-Australian Observatory)

45. The human eye is most sensitive over the same wavelength range at which the Sun emits the greatest intensity of radiation. Suppose creatures were to evolve on a planet orbiting a star somewhat hotter than the Sun. To what wavelengths would their vision most likely be sensitive?

46. Why do you suppose that ultraviolet light can cause skin cancer but ordinary visible light does not?

Web/eBook Questions

47. Search the World Wide Web for information about rainbows. Why do rainbows form? Why do they appear as circular arcs? Why can you see different colors?

48. Measuring Stellar Temperatures. Access the Active Integrated Media Module "Blackbody Curves" in Chapter 5 of the *Universe* Web site or eBook. (a) Use the module to determine the range of temperatures over which a star's peak wavelength is in the visible spectrum. (b) Determine if any of the following stars have a peak wavelength in the visible spectrum: Rigel, $T = 11{,}000$ K; Deneb, $T = 8400$ K; Arcturus, $T = 4290$ K; Vega, $T = 9500$ K; Mira, $T = 2000$ K.

Activities

Observing Projects

49. Turn on an electric stove or toaster oven and carefully observe the heating elements as they warm up. Relate your observations to Wien's law and the Stefan-Boltzmann law.

50. Obtain a glass prism (or a diffraction grating, which is probably more readily available and is discussed in the next chapter) and look through it at various light sources, such as an ordinary incandescent light, a neon sign, and a mercury vapor street lamp. **Do not look at the sun! Looking directly at the Sun causes permanent eye damage or blindness.** Do you have any trouble seeing spectra? What do you have to do to see a spectrum? Describe the differences in the spectra of the various light sources you observed.

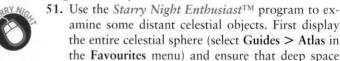

51. Use the *Starry Night Enthusiast*™ program to examine some distant celestial objects. First display the entire celestial sphere (select **Guides > Atlas** in the **Favourites** menu) and ensure that deep space objects are displayed by opening **View > Deep Space** and clicking on **Messier Objects** and **Bright NGC Objects.** You can now search for objects (i), (ii), and (iii) listed below. Click the **Find** tab at the left of the main view window to open the Find pane, click on the magnifying glass icon at the left of the edit box at the top of the Find pane and select **Search All** from the menu, and then type the name of the object in the edit box followed by the **Enter (Return)** key. The object will be centered in the view. For each object, use the zoom controls at the right-hand end of the Toolbar (at the top of the main window) to adjust your view until you can see the object in detail. For each object, state whether it has a continuous spectrum, an absorption line spectrum, or an emission line spectrum, and explain your reasoning. (i) The Lagoon Nebula in Sagittarius. (*Hint:* See Figure 5-18.) (With a field of view of about 6° × 4°, you can compare and contrast the appearance of the Lagoon Nebula with the Trifid Nebula just to the north of it.) (ii) M31, the great galaxy in the constellation Andromeda. (*Hint:* The light coming from this galaxy is the combined light of hundreds of billions of individual stars.) (ii) The Moon. (*Hint:* Recall from Section 3-1 that moonlight is simply reflected sunlight.)

52. Use the *Starry Night Enthusiast*™ program to examine the temperatures of several relatively nearby stars. First display the entire celestial sphere (select **Guides > Atlas** in the **Favourites** menu). You can now search for each of the stars listed below. Open the **Find** pane, click on the magnifying glass icon at the left side of the edit box at the top of the Find pane, select Star from the menu that appears, type the name of the star in the edit box and click the **Enter (Return)** key. (i) Altair; (ii) Procyon; (iii) Epsilon Indi; (iv) Tau Ceti; (v) Epsilon Eridani; (vi) Lalande 21185. Information for each star can then be found by clicking on the **Info** tab at the far left of the *Starry Night Enthusiast*™ window. For each star, record its temperature (listed in the **Info** pane under **Other Data**). Then answer the following questions. (**a**) Which of the stars have a longer wavelength of maximum emission λ_{max} than the Sun? Which of the stars have a shorter λ_{max} than the Sun? (**b**) Which of the stars has a reddish color?

53. Use the *Starry Night Enthusiast*™ program to explore the colors of galaxies. Select **Deep Space > Virgo Cluster** in the **Favourites** menu. The Virgo cluster is a grouping of more than 2000 galaxies extending across about 9 million light-years, located about 50 million light-years from Earth. Move the cursor over one of the larger galaxies in the view (for example, the galaxy named "The Eyes" near the center of the screen) and double-click over the image to center that galaxy in the view. You can zoom in and zoom out on the cluster using the elevation buttons (an upward-pointing triangle and a downward-pointing triangle) to the left of the Home button in the toolbar. You can also rotate the cluster by moving the mouse while holding down both the mouse button (left button on a two-button mouse) and the **Shift** key on the keyboard. (**a**) Zoom in and rotate the image to examine several different galaxies. Are all galaxies the same color? Do all galaxies contain stars of the same surface temperature? (Recall from Section 1-4 that galaxies are assemblages of stars.) (**b**) Examine several spiral galaxies like the one shown in Figure 1-7. What are the dominant colors of the inner regions and outer regions of these galaxies? What can you conclude about the surface temperatures of the brightest stars in the inner regions of spiral galaxies compared to the surface temperatures of those in the outer regions?

Collaborative Exercise

54. The Doppler effect describes how relative motion impacts wavelength. With a classmate, stand up and demonstrate each of the following: (**a**) a blueshifted source for a stationary observer; (**b**) a stationary source and an observer detecting a redshift; and (**c**) a source and an observer both moving in the same direction, but the observer is detecting a redshift. Create simple sketches to illustrate what you and your classmate did.

6

Optics and Telescopes

Visible-light image of galaxy M82

R I V U X G

Infrared image of galaxy M82

R I V U X G

WEB LINK 6.1

Modern telescopes can view the universe in every range of electromagnetic radiation, although some must be placed above Earth's atmosphere. (Visible: NOAO; Infrared: NASA/JPL-Caltech/C. Engelbracht, University of Arizona)

There is literally more to the universe than meets the eye. As seen through a small telescope, the galaxy M82 (shown in the accompanying image) appears as a bright patch that glows with the light of its billions of stars. But when observed with a telescope sensitive to infrared light—with wavelengths longer than your eye can detect—M82 displays an immense halo that extends for tens of thousands of light-years. The spectrum of this halo reveals it to be composed of tiny dust particles, which are ejected by newly formed stars. The halo's tremendous size shows that stars are forming in M82 at a far greater rate than within our own Galaxy.

These observations are just one example of the tremendous importance of telescopes to astronomy. Whether a telescope detects visible or nonvisible light, its fundamental purpose is the same: to gather more light than can the unaided eye. Telescopes are used to gather the feeble light from distant objects to make bright, sharp images. Telescopes also produce finely detailed spectra of objects in space. These spectra reveal the chemical composition of nearby planets as well as of distant galaxies, and help astronomers understand the nature and evolution of astronomical objects of all kinds.

As the accompanying images of the galaxy M82 show, telescopes for nonvisible light reveal otherwise hidden aspects of the universe. In addition to infrared telescopes, radio telescopes have mapped out the structure of our Milky Way Galaxy; ultraviolet telescopes have revealed the workings of the Sun's outer atmosphere; and gamma-ray telescopes have detected the most powerful explosions in the universe. The telescope, in all its variations, is by far astronomers' most useful tool for collecting data about the universe.

6-1 A refracting telescope uses a lens to concentrate incoming light at a focus

The **optical telescope**—that is, a telescope designed for use with visible light—was invented in the Netherlands in the early seventeenth century. Soon after, Galileo used one of these new inventions for his groundbreaking astronomical observations (see Section 4-5). These first telescopes used carefully shaped pieces of glass, or **lenses,** to make distant objects appear larger and brighter. Telescopes of this same basic design are used today by many amateur astronomers. To understand telescopes of this kind, we need to understand how lenses work.

> **Refracting telescopes gave humans the first close-up views of the Moon and planets**

Refraction of Light

All lenses, including those used in telescopes, make use of the same physical principle: *Light travels at a slower speed in a dense substance.* Thus, although the speed of light in a vacuum is 3.00×10^8 m/s, its speed in glass is less than 2×10^8 m/s. Just as a woman's walking pace slows suddenly when she walks from a boardwalk onto a sandy beach, so light slows abruptly as it enters a piece of glass. The same woman easily resumes her original

Learning Goals

By reading the sections of this chapter, you will learn

A car slows down as it moves from pavement to **sand.**

Pavement

Sand

If the car enters the sand at an angle, it slows down and changes direction.

(a) How cars behave

A light beam slows down as it moves from a vacuum into glass.

Vacuum

Glass

If the light beam enters the glass at an angle, it slows down and changes direction.

(b) How light beams behave

Figure 6-1

Refraction **(a)** When a car drives from smooth pavement into soft sand, it slows down. If it enters the sand at an angle, the front wheel on one side feels the drag of the sand before the other wheel, causing the car to veer to the side and change direction. **(b)** Similarly, light slows down when it passes from a vacuum into glass and changes direction if it enters the glass at an angle.

pace when she steps back onto the boardwalk; in the same way, light resumes its original speed upon exiting the glass.

A material through which light travels is called a **medium** (*plural* **media**). As a beam of light passes from one transparent medium into another—say, from air into glass, or from glass back into air—the direction of the light can change. This phenomenon, called **refraction**, is caused by the change in the speed of light.

ANALOGY Imagine driving a car from a smooth pavement onto a sandy beach (Figure 6-1*a*). If the car approaches the beach head-on, it slows down when it enters the sand but keeps moving straight ahead. If the car approaches the beach at an angle, however, one of the front wheels will be slowed by the sand before the other is, and the car will veer from its original direction. In the same way, a beam of light changes direction when it enters a piece of glass at an angle (Figure 6-1*b*).

Figure 6-2*a* shows the refraction of a beam of light passing through a piece of flat glass. As the beam enters the upper surface of the glass, refraction takes place, and the beam is bent to a direction more nearly perpendicular to the surface of the glass. As the beam exits from the glass back into the surrounding air, a second refraction takes place, and the beam bends in the opposite sense. (The amount of bending depends on the speed of light in the glass, so different kinds of glass produce slightly different amounts of refraction.) Because the two surfaces of the glass are parallel, the beam emerges from the glass traveling in the same direction in which it entered.

Lenses and Refracting Telescopes

Something more useful happens if the glass is curved into a convex shape (one that is fatter in the middle than at the edges), like the lens in Figure 6-2*b*. When a beam of light rays passes through the lens, refraction causes all the rays to converge at a point called the **focus.** If the light rays entering the lens are all parallel, the focus occurs at a special point called the **focal point.** The distance from the lens to the focal point is called the **focal length** of the lens.

The case of parallel light rays, shown in Figure 6-2*b*, is not merely a theoretical ideal. The stars are so far away that light rays

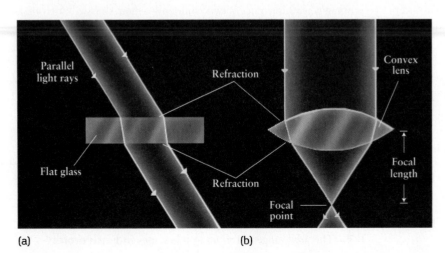

Parallel light rays

Refraction

Flat glass

Refraction

Focal point

Convex lens

Focal length

(a)   **(b)**

Figure 6-2

Refraction and Lenses (a) Refraction is the change in direction of a light ray when it passes into or out of a transparent medium such as glass. When light rays pass through a flat piece of glass, the two refractions bend the rays in opposite directions. There is no overall change in the direction in which the light travels. **(b)** If the glass is in the shape of a convex lens, parallel light rays converge to a focus at a special point called the focal point. The distance from the lens to the focal point is called the focal length of the lens.

from them are essentially parallel, as Figure 6-3 shows. Consequently, a lens always focuses light from an astronomical object to the focal point. If the object has a very small angular size, like a distant star, all the light entering the lens from that object converges onto the focal point. The resulting image is just a single bright dot.

However, if the object is *extended*—that is, has a relatively large angular size, like the Moon or a planet—then light coming from each point on the object is brought to a focus at its own individual point. The result is an extended image that lies in the **focal plane** of the lens (Figure 6-4), which is a plane that includes the focal point. You can use an ordinary magnifying glass in this way to make an image of the Sun on the ground.

To use a lens to make a photograph of an astronomical object, you would place a piece of film or an electronic detector in the focal plane. An ordinary film camera or digital camera works in the same way for taking photographs of relatively close objects here on Earth. Most observations for astronomical research are

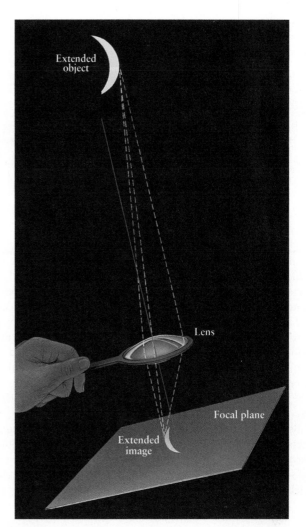

Figure 6-4

A Lens Creates an Extended Image of an Extended Object Light coming from each point on an extended object passes through a lens and produces an image of that point. All of these tiny images put together make an extended image of the entire object. The image of a very distant object is formed in a plane called the focal plane. The distance from the lens to the focal plane is called the focal length.

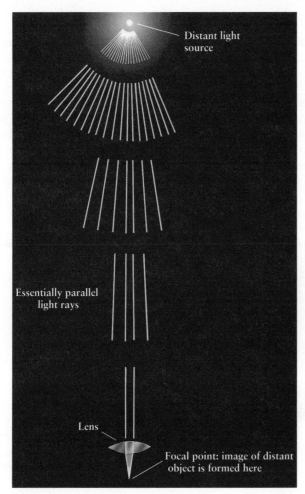

Figure 6-3

Light Rays from Distant Objects Are Parallel Light rays travel away in all directions from an ordinary light source. If a lens is located very far from the light source, only a few of the light rays will enter the lens, and these rays will be essentially parallel. This is why we drew parallel rays entering the lens in Figure 6-2*b*.

done in this way to produce a permanent record of the observation. Furthermore, modern image-recording technology is far more sensitive than the human eye, as we will see in Section 6-4. But many amateur astronomers want to view the image visually, and so they add a second lens to magnify the image formed in the focal plane. Such an arrangement of two lenses is called a **refracting telescope,** or **refractor** (Figure 6-5). The large-diameter, long-focal-length lens at the front of the telescope, called the **objective lens,** forms the image; the smaller, shorter-focal-length lens at the rear of the telescope, called the **eyepiece lens,** magnifies the image for the observer.

Light-Gathering Power

In addition to the focal length, the other important dimension of the objective lens of a refractor is the diameter. Compared with

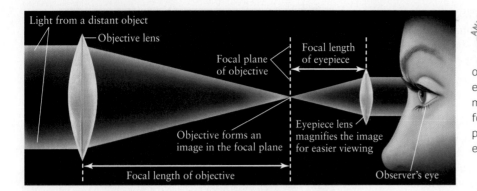

Figure 6-5

A Refracting Telescope A refracting telescope consists of a large-diameter objective lens with a long focal length and a small eyepiece lens of short focal length. The eyepiece lens magnifies the image formed by the objective lens in its focal plane (shown as a dashed line). To take a photograph, the eyepiece is removed and the film or electronic detector is placed in the focal plane.

a small-diameter lens, a large-diameter lens captures more light, produces brighter images, and allows astronomers to detect fainter objects. (For the same reason, the iris of your eye opens when you go into a darkened room to allow you to see dimly lit objects.)

The **light-gathering power** of a telescope is directly proportional to the area of the objective lens, which in turn is proportional to the square of the lens diameter (Figure 6-6). Thus, if you double the diameter of the lens, the light-gathering power increases by a factor of $2^2 = 2 \times 2 = 4$. Box 6-1 describes how to compare the light-gathering power of different telescopes.

Because light-gathering power is so important for seeing faint objects, the lens diameter is almost always given in describing a telescope. For example, the Lick telescope on Mount Hamilton in California is a 90-cm refractor, which means that it is a refracting telescope whose objective lens is 90 cm in diameter. By com-

parison, Galileo's telescope of 1610 was a 3-cm refractor. The Lick telescope has an objective lens 30 times larger in diameter, and so has $30 \times 30 = 900$ times the light-gathering power of Galileo's instrument.

Magnification

In addition to their light-gathering power, telescopes are useful because they magnify distant objects. As an example, the angular diameter of the Moon as viewed with the naked eye is about 0.5°. But when Galileo viewed the Moon through his telescope, its apparent angular diameter was 10°, large enough so that he could identify craters and mountain ranges. The **magnification,** or **magnifying power,** of a telescope is the ratio of an object's angular diameter seen through the telescope to its naked-eye angular diameter. Thus, the magnification of Galileo's telescope was 10°/0.5° = 20 times, usually written as 20×.

Small-diameter objective lens:
dimmer image, less detail

Large-diameter objective lens:
brighter image, more detail

Figure 6-6　R I **V** U X G

Light-Gathering Power These two photographs of the galaxy M31 in Andromeda were taken using the same exposure time and at the same magnification, but with two different telescopes with objective lenses of different diameters. The right-hand photograph is brighter and shows more detail because it was made using the large-diameter lens, which intercepts more starlight than a small-diameter lens. This same principle applies to telescopes that use curved mirrors rather than lenses to collect light (see Section 6-2). (Association of Universities for Research in Astronomy)

BOX 6-1   Tools of the Astronomer's Trade

Magnification and Light-Gathering Power

The magnification of a telescope is equal to the focal length of the objective divided by the focal length of the eyepiece. Telescopic eyepieces are usually interchangeable, so the magnification of a telescope can be changed by using eyepieces of different focal lengths.

EXAMPLE: A small refracting telescope has an objective of focal length 120 cm. If the eyepiece has a focal length of 4.0 cm, what is the magnification of the telescope?

Situation: We are given the focal lengths of the telescope's objective and eyepiece lenses. Our goal is to calculate the magnification provided by this combination of lenses.

Tools: We use the relationship that the magnification equals the focal length of the objective (120 cm) divided by the focal length of the eyepiece (4.0 cm).

Answer: Using this relationship,

$$\text{Magnification} = \frac{120 \text{ cm}}{4.0 \text{ cm}} = 30 \text{ (usually written as } 30\times)$$

Review: A magnification of $30\times$ means that as viewed through this telescope, a large lunar crater that subtends an angle of 1 arcminute to the naked eye will appear to subtend an angle 30 times greater, or 30 arcminutes (one-half of a degree). This makes the details of the crater much easier to see.

If a 2.0-cm-focal-length eyepiece is used instead, the magnification will be $(120 \text{ cm})/(2.0 \text{ cm}) = 60\times$. The shorter the focal length of the eyepiece, the greater the magnification.

The light-gathering power of a telescope depends on the diameter of the objective lens; it does not depend on the focal length. The light-gathering power is proportional to the square of the diameter. As an example, a fully dark adapted human eye has a pupil diameter of about 5 mm. By comparison, a small telescope whose objective lens is 5 cm in diameter has 10 times the diameter and $10^2 = 100$ times the light-gathering

power of the eye. (Recall that there are 10 mm in 1 cm.) Hence, this telescope allows you to see objects 100 times fainter than you can see without a telescope.

EXAMPLE: The same relationships apply to reflecting telescopes, discussed in Section 6-2. Each of the two Keck telescopes on Mauna Kea in Hawaii (discussed in Section 6-3; see Figure 6-16) uses a concave mirror 10 m in diameter to bring starlight to a focus. How many times greater is the light-gathering power of either Keck telescope compared to that of the human eye?

Situation: We are given the diameters of the pupil of the human eye (5 mm) and of the mirror of either Keck telescope (10 m). Our goal is to compare the light-gathering powers of these two optical instruments.

Tools: We use the relationship that light-gathering power is proportional to the square of the diameter of the area that gathers light. Hence, the *ratio* of the light-gathering powers is equal to the square of the ratio of the diameters.

Answer: We first calculate the ratio of the diameter of the Keck mirror to the diameter of the pupil. To do this, we must first express both diameters in the same units. Because there are 1000 mm in 1 meter, the diameter of the Keck mirror can be expressed as

$$10 \text{ m} \times \frac{1000 \text{ mm}}{1 \text{ m}} = 10,000 \text{ mm}$$

Thus, the light-gathering power of either of the Keck telescopes is greater than that of the human eye by a factor of

$$\frac{(10,000 \text{ mm})^2}{(5 \text{ mm})^2} = (2000)^2 = 4 \times 10^6 = 4,000,000$$

Review: Either Keck telescope can gather *4 million* times as much light as a human eye. When it comes to light-gathering power, the bigger the telescope, the better!

 The magnification of a refracting telescope depends on the focal lengths of both of its lenses:

$$\text{Magnification} = \frac{\text{focal length of objective lens}}{\text{focal length of eyepiece lens}}$$

This formula shows that using a long-focal-length objective lens with a short-focal-length eyepiece gives a large magnification. Box 6-1 illustrates how this formula is used.

CAUTION! Many people think that the primary purpose of a telescope is to magnify images. But in fact magnification is *not* the most important aspect of a telescope. The reason is that there is a limit to how sharp any astronomical image can be, due either to the blurring caused by the Earth's atmosphere or to fundamental limitations imposed by the nature of light itself. (We will describe these effects in more detail in Section 6-3.) Magnifying a blurred image may make it look bigger but will not make it any clearer. Thus, beyond a certain point, there is nothing to be gained by further magnification. Astronomers put much more store in the light-gathering power of a telescope than in its magnification. Greater light-gathering power means brighter images, which makes it easier to see faint details.

Disadvantages of Refracting Telescopes

If you were to build a telescope like that in Figure 6-5 using only the instructions given so far, you would probably be disappointed with the results. The problem is that a lens bends different colors of light through different angles, just as a prism does (recall Figure 5-3). As a result, different colors do not focus at the same point, and stars viewed through a telescope that uses a simple lens are surrounded by fuzzy, rainbow-colored halos. Figure 6-7a shows this optical defect, called **chromatic aberration.**

One way to correct for chromatic aberration is to use an objective lens that is not just a single piece of glass. Different types of glass can be manufactured by adding small amounts of chemicals to the glass when it is molten. Because of these chemicals, the speed of light varies slightly from one kind of glass to another, and the refractive properties vary as well. If a thin lens is mounted just behind the main objective lens of a telescope, as shown in Figure 6-7b, and if the telescope designer carefully chooses two different kinds of glass for these two lenses, different colors of light can be brought to a focus at the same point.

Chromatic aberration is only the most severe of a host of optical problems that must be solved in designing a high-quality refracting telescope. Master opticians of the nineteenth century devoted their careers to solving these problems, and several magnificent refractors were constructed in the late 1800s (Figure 6-8).

Unfortunately, there are several negative aspects of refractors that even the finest optician cannot overcome:

1. Because faint light must readily pass through the objective lens, the glass from which the lens is made must be totally free of defects, such as the bubbles that frequently form when molten glass is poured into a mold. Such defect-free glass is extremely expensive.

2. Glass is opaque to certain kinds of light. Ultraviolet light is absorbed almost completely, and even visible light is dimmed substantially as it passes through the thick slab of glass that makes up the objective lens.

3. It is impossible to produce a large lens that is entirely free of chromatic aberration.

4. Because the lens can be supported only around its edges, it tends to sag and distort under its own weight. This has adverse effects on the image clarity.

For these reasons and more, few major refractors have been built since the beginning of the twentieth century. Instead, astronomers have avoided all of the limitations of refractors by building telescopes that use a mirror instead of a lens to form an image.

6-2 A reflecting telescope uses a mirror to concentrate incoming light at a focus

Almost all modern telescopes form an image using the principle of **reflection.** To understand reflection, imagine drawing a dashed line perpendicular to the surface of a flat mirror at the point where a light ray strikes the mirror (Figure 6-9a). The angle i between the *incident* (arriving) light ray and the perpendicular is al-

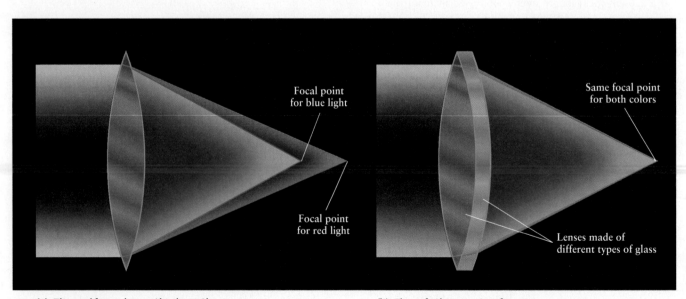

(a) The problem: chromatic aberration

(b) The solution: use two lenses

Figure 6-7

Chromatic Aberration (a) A single lens suffers from a defect called chromatic aberration, in which different colors of light are brought to a focus at different distances from the lens. **(b)** This problem can be corrected by adding a second lens made from a different kind of glass.

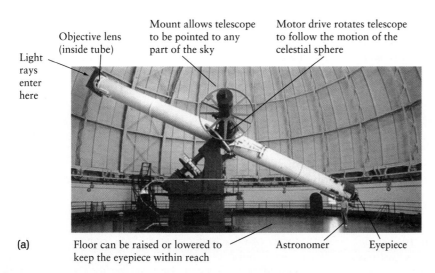

Light rays enter here

Objective lens (inside tube)

Mount allows telescope to be pointed to any part of the sky

Motor drive rotates telescope to follow the motion of the celestial sphere

(a)

Floor can be raised or lowered to keep the eyepiece within reach

Astronomer   Eyepiece

Objective lens made of two different types of glass (see Figure 6-7a)

(b)

Figure 6-8   R I [V] U X G

A Large Refracting Telescope (a) This giant refractor, built in 1897, is housed at Yerkes Observatory near Chicago. The telescope tube is 19.5 m (64 ft) long; it has to be this long because the focal length of the objective is just under 19.5 m (see Figure 6-5). As the Earth rotates, the motor drive rotates the telescope in the opposite direction in order to keep the object being studied within the telescope's field of view. **(b)** This historical photograph shows the astronomer George van Biesbrock with the objective lens of the Yerkes refractor. This lens, the largest ever made, is 102 cm (40 in.) in diameter. (Yerkes Observatory)

ways equal to the angle r between the *reflected* ray and the perpendicular.

In 1663, the Scottish mathematician James Gregory first proposed a telescope using reflection from a concave mirror—one that is fatter at the edges than at the middle. Such a mir-

> **All of the largest professional telescopes—and most amateur telescopes—are reflecting telescopes**

ror makes parallel light rays converge to a focus (Figure 6-9b). The distance between the reflecting surface and the focus is the focal length of the mirror. A telescope that uses a curved mirror to make an image of a distant object is called a **reflecting telescope,** or **reflector.** Using terminology similar to that used for refractors, the mirror that forms the image is called the **objective mirror** or **primary mirror.**

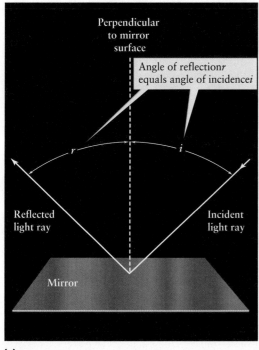

Perpendicular to mirror surface

Angle of reflection *r* equals angle of incidence *i*

r   *i*

Reflected light ray

Incident light ray

Mirror

(a)

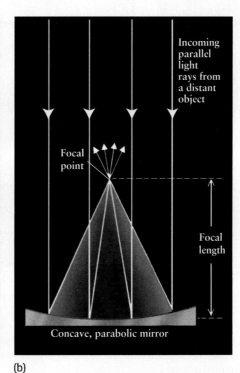

Incoming parallel light rays from a distant object

Focal point

Focal length

Concave, parabolic mirror

(b)

Figure 6-9

Reflection (a) The angle at which a beam of light approaches a mirror, called the angle of incidence (*i*), is always equal to the angle at which the beam is reflected from the mirror, called the angle of reflection (*r*). **(b)** A concave mirror causes parallel light rays to converge to a focus at the focal point. The distance between the mirror and the focal point is the focal length of the mirror.

To make a reflector, an optician grinds and polishes a large slab of glass into the appropriate concave shape. The glass is then coated with silver, aluminum, or a similar highly reflective substance. Because light reflects off the surface of the glass rather than passing through it, defects within the glass—which would have very negative consequences for the objective lens of a refracting telescope—have no effect on the optical quality of a reflecting telescope.

Another advantage of reflectors is that they do not suffer from the chromatic aberration that plagues refractors. This is because reflection is not affected by the wavelength of the incoming light, so all wavelengths are reflected to the same focus. (A small amount of chromatic aberration may arise if the image is viewed using an eyepiece lens.) Furthermore, the mirror can be fully supported by braces on its back, so that a large, heavy mirror can be mounted without much danger of breakage or surface distortion.

Designs for Reflecting Telescopes

Although a reflecting telescope has many advantages over a refractor, the arrangement shown in Figure 6-9a is not ideal. One problem is that the focal point is in front of the objective mirror.

If you try to view the image formed at the focal point, your head will block part or all of the light from reaching the mirror.

To get around this problem, in 1668 Isaac Newton simply placed a small, flat mirror at a 45° angle in front of the focal point, as sketched in Figure 6-10a. This secondary mirror deflects the light rays to one side, where Newton placed an eyepiece lens to magnify the image. A reflecting telescope with this optical design is appropriately called a **Newtonian reflector** (Figure 6-10b). The magnifying power of a Newtonian reflector is calculated in the same way as for a refractor: The focal length of the objective mirror is divided by the focal length of the eyepiece (see Box 6-1).

Later astronomers modified Newton's original design. The objective mirrors of some modern reflectors are so large that an astronomer could actually sit in an "observing cage" at the undeflected focal point directly in front of the objective mirror. (In practice, riding in this cage on a winter's night is a remarkably cold and uncomfortable experience.) This arrangement is called a **prime focus** (Figure 6-11a). It usually provides the highest-quality image, because there is no need for a secondary mirror (which might have imperfections).

(a)

(b)

Figure 6-10   R I **V** U X G

A Newtonian Telescope (a) In a Newtonian telescope, the image made by the objective is moved from point 1 to point 2 by means of a flat mirror called the secondary. An eyepiece magnifies this image, just as for a refracting telescope (Figure 6-5). **(b)** This is a replica of a Newtonian telescope built by Isaac Newton in 1672. The objective mirror is 3 cm (1.3 inches) in diameter and the magnification is 40× (Royal Greenwich Observatory/Science Photo Library)

Incoming parallel rays from a distant object

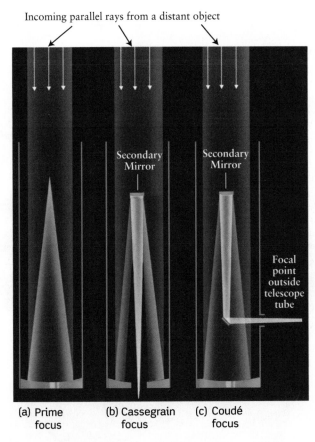

(a) Prime
focus

(b) Cassegrain
focus

(c) Coudé
focus

Figure 6-11

Designs for Reflecting Telescopes Three common optical
designs for reflecting telescopes are shown here. **(a)** Prime
focus is used only on some large telescopes; an observer or instrument is
placed directly at the focal point, within the barrel of the telescope.
(b) The Cassegrain focus is used on reflecting telescopes of all sizes, from
90-mm (3.5-in.) reflectors used by amateur astronomers to giant research
telescopes on Mauna Kea (see Figure 6-14). **(c)** The coudé focus is useful
when large and heavy optical apparatus is to be used at the focal point.
Light reflects off the objective mirror to a secondary mirror, then back down
to a third, angled mirror. With this arrangement, the heavy apparatus at the
focal point does not have to move when the telescope is repositioned.

Another popular optical design, called a **Cassegrain focus** after the French contemporary of Newton who first proposed it,
also has a convenient, accessible focal point. A hole is drilled directly through the center of the primary mirror, and a convex secondary mirror placed in front of the original focal point reflects
the light rays back through the hole (Figure 6-11*b*).

A fourth design is useful when there is optical equipment too
heavy or bulky to mount directly on the telescope. Instead, a series of mirrors channels the light rays away from the telescope to
a remote focal point where the equipment is located. This design
is called a **coudé focus,** from a French word meaning "bent like
an elbow" (Figure 6-11*c*).

CAUTION! You might think that the secondary mirror in the
Newtonian design and the Cassegrain and coudé designs shown

in Figure 6-11 would cause a black spot or hole in the center of
the telescope image. But this does not happen. The reason is that
light from every part of the object lands on every part of the primary, objective mirror. Hence, any portion of the mirror can itself produce an image of the distant object, as Figure 6-12 shows.
The only effect of the secondary mirror is that it prevents part
of the light from reaching the objective mirror, which reduces
somewhat the light-gathering power of the telescope.

Spherical Aberration

A reflecting telescope must be designed to minimize a defect called
spherical aberration (Figure 6-13). At issue is the precise shape of
a mirror's concave surface. A spherical surface is easy to grind
and polish, but different parts of a spherical mirror have slightly
different focal lengths (Figure 6-13*a*). This results in a fuzzy
image.

One common way to eliminate spherical aberration is to polish the mirror's surface to a parabolic shape, because a parabola
reflects parallel light rays to a common focus (Figure 6-13*b*). Unfortunately, the astronomer then no longer has a wide-angle view.
Furthermore, unlike spherical mirrors, parabolic mirrors suffer

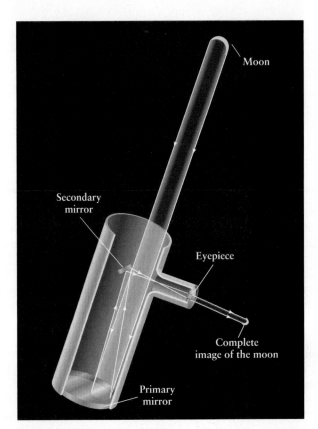

Figure 6-12

The Secondary Mirror Does Not Cause a Hole in the Image This
illustration shows how even a small portion of the primary (objective)
mirror of a reflecting telescope can make a complete image of the Moon.
Thus, the secondary mirror does not cause a black spot or hole in the
image. (It does, however, make the image a bit dimmer by reducing the
total amount of light that reaches the primary mirror.)

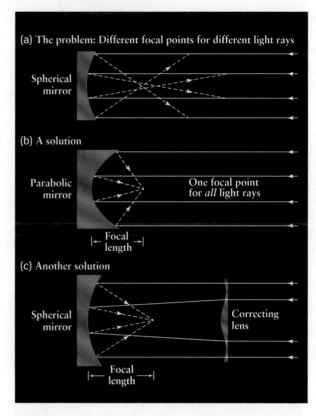

Figure 6-13

Spherical Aberration (a) Different parts of a spherically concave mirror reflect light to slightly different focal points. This effect, called spherical aberration, causes image blurring. This difficulty can be corrected by either **(b)** using a parabolic mirror or **(c)** using a correcting lens in front of the mirror.

from a defect called **coma,** wherein star images far from the center of the field of view are elongated to look like tiny teardrops. A different approach is to use a spherical mirror, thus minimizing coma, and to place a thin correcting lens at the front of the telescope to eliminate spherical aberration (Figure 6-13c). This approach is only used on relatively small reflecting telescopes for amateur astronomers.

The Largest Reflectors

As of 2007, there were 13 optical reflectors in operation with primary mirrors between 8 meters (26.2 feet) and 11 meters (36.1 feet) in diameter. Figure 6-14a shows the objective mirror of one of the four Very Large Telescope (VLT) units in Chile, and Figure 6-14b shows the objective and secondary mirrors of the Gemini North telescope in Hawaii. A near-twin of Gemini North, called Gemini South, is in Cerro Pachón, Chile. These twins allow astronomers to observe both the northern and southern parts of the celestial sphere with essentially the same state-of-the-art instrument. Two other "twins" are the side-by-side 8.4-m objective mirrors of the Large Binocular Telescope in Arizona. Combining the light from these two mirrors gives double the light-gathering power, equivalent to a single 11.8-m mirror.

(a) A large objective mirror

(b) A large Cassegrain telescope

Figure 6-14   R I V U X G

Reflecting Telescopes (a) This photograph shows technicians preparing an objective mirror 8.2 meters in diameter for the European Southern Observatory in Chile. The mirror was ground to a curved shape with a remarkable precision of 8.5 nanometers. **(b)** This view of the Gemini North telescope shows its 8.1-meter objective mirror (1). Light incident on this mirror is reflected toward the 1.0-meter secondary mirror (2), then through the hole in the objective mirror (3) to the Cassegrain focus (see Figure 6-11b). (a: SAGEM; b: NOAO/AURA/NSF)

Several other reflectors around the world have objective mirrors between 3 and 6 meters in diameter, and dozens of smaller but still powerful telescopes have mirrors in the range of 1 to 3 meters. There are thousands of professional astronomers, each of whom has several ongoing research projects, and thus the demand for all of these telescopes is high. On any night of the year, nearly every research telescope in the world is being used to explore the universe.

6-3   Telescope images are degraded by the blurring effects of the atmosphere and by light pollution

In addition to providing a brighter image, a large telescope also helps achieve a second major goal: It produces star images that are sharp and crisp. A quantity called **angular resolution** gauges how well fine details can be seen. Poor angular resolution causes star images to be fuzzy and blurred together.

To determine the angular resolution of a telescope, pick out two adjacent stars whose separate images are just barely discernible (Figure 6-15). The angle θ (the Greek letter theta) between these stars is the telescope's angular resolution; the *smaller* that angle, the finer the details that can be seen and the sharper the image.

When you are asked to read the letters on an eye chart, what's being measured is the angular resolution of your eye. If you have 20/20 vision, the angular resolution θ of your eye is about 1 arcminute, or 60 arcseconds. (You may want to review the definitions of these angular measures in Section 1-5.) Hence, with the naked eye it is impossible to distinguish two stars less than 1 arcminute apart or to see details on the Moon with an angular size smaller than this. All the planets have angular sizes (as seen from Earth) of 1 arcminute or less, which is why they appear as featureless points of light to the naked eye.

> **Adaptive optics produces sharper images by "undoing" atmospheric turbulence**

Limits to Angular Resolution

One factor limiting angular resolution is **diffraction,** which is the tendency of light waves to spread out when they are confined to a small area like the lens or mirror of a telescope. (A rough analogy is the way water exiting a garden hose sprays out in a wider angle when you cover part of the end of the hose with your thumb.) As a result of diffraction, a narrow beam of light tends to spread out within a telescope's optics, thus blurring the image. If diffraction were the only limit, the angular resolution of a telescope would be given by the formula

Diffraction-limited angular resolution

$$\theta = 2.5 \times 10^5 \frac{\lambda}{D}$$

θ = diffraction-limited angular resolution of a telescope, in arcseconds

Two light sources with angular separation greater than angular resolution of telescope: Two sources easily distinguished

(a)

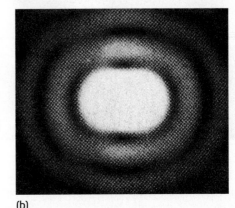

Light sources moved closer so that angular separation equals angular resolution of telescope: Just barely possible to tell that there are two sources

(b)

Figure 6-15   R I V U X G

Angular Resolution The angular resolution of a telescope indicates the sharpness of the telescope's images. **(a)** This telescope view shows two sources of light whose angular separation is greater than the angular resolution. **(b)** The light sources have been moved together so that their angular separation is equal to the angular resolution. If the sources were moved any closer together, the telescope image would show them as a single source.

λ = wavelength of light, in meters

D = diameter of telescope objective, in meters

For a given wavelength of light, using a telescope with an objective of *larger* diameter D *reduces* the amount of diffraction and makes the angular resolution θ *smaller* (and hence better). For example, with red light with wavelength 640 nm, or 6.4×10^{-7} m, the diffraction-limited resolution of an 8-meter telescope (see Figure 6-15) would be

$$\theta = (2.5 \times 10^5) \frac{6.4 \times 10^{-7} \text{ m}}{8 \text{ m}} = 0.02 \text{ arcsec}$$

In practice, however, ordinary optical telescopes cannot achieve such fine angular resolution. The problem is that turbulence in the air causes star images to jiggle around and twinkle. Even through the largest telescopes, a star still looks like a tiny blob rather than a pinpoint of light. A measure of the limit that atmospheric turbulence places on a telescope's resolution is called the **seeing disk.** This disk is the angular diameter of a star's image broadened by turbulence. The size of the seeing disk varies

WEB LINK 6.3 **Figure 6-16   R I V U X G**

The Telescopes of Mauna Kea
The summit of Mauna Kea—an extinct Hawaiian volcano that reaches more than 4100 m (13,400 ft) above the waters of the Pacific—has nighttime skies that are unusually clear, still, and dark. To take advantage of these superb viewing conditions, Mauna Kea has become the home of many powerful telescopes. (Richard J. Wainscoat, University of Hawaii)

from one observatory site to another and from one night to another. At the observatories on Kitt Peak in Arizona and Cerro Tololo in Chile, the seeing disk is typically around 1 arcsec. Some of the very best conditions in the world can be found at the observatories atop Mauna Kea in Hawaii, where the seeing disk is often as small as 0.5 arcsec. This is one reason why so many telescopes have been built there (Figure 6-16).

Active Optics and Adaptive Optics

In many cases the angular resolution of a telescope is even worse than the limit imposed by the seeing disk. This occurs if the objective mirror deforms even slightly due to variations in air temperature or flexing of the telescope mount. To combat this, many large telescopes are equipped with an **active optics** system. Such a system adjusts the mirror shape every few seconds to help keep the telescope in optimum focus and properly aimed at its target.

Changing the mirror shape is also at the heart of a more refined technique called **adaptive optics.** The goal of this technique is to compensate for atmospheric turbulence, so that the angular resolution can be smaller than the size of the seeing disk and can even approach the theoretical limit set by diffraction. Turbulence causes the image of a star to "dance" around erratically. In an adaptive optics system, sensors monitor this dancing motion 10 to 100 times per second, and a powerful computer rapidly calculates the mirror shape needed to compensate. Fast-acting mechanical devices called *actuators* then deform the mirror accordingly, at a much faster rate than in an active optics system. In some adaptive optics systems, the actuators deform a small secondary mirror rather than the large objective mirror.

One difficulty with adaptive optics is that a fairly bright star must be in or near the field of the telescope's view to serve as a "target" for the sensors that track atmospheric turbulence. This is seldom the case, since the field of view of most telescopes is rather narrow. Astronomers get around this limitation by shining a laser beam toward a spot in the sky near the object to be observed (Figure 6-17). The laser beam causes atoms in the upper atmosphere to glow, making an artificial "star." The light that comes down to Earth from this "star" travels through the same part of our atmosphere as the light from the object being observed, so its image in the telescope will "dance" around in the same erratic way as the image of a real star.

Figure 6-18 shows the dramatic improvement in angular resolution possible with adaptive optics. Images made with adaptive optics are nearly as sharp as if the telescope were in the vacuum of space, where there is no atmospheric distortion whatsoever and the only limit on angular resolution is diffraction. A number of large telescopes are now being used with adaptive optics systems.

CAUTION! The images in Figure 6-18 are **false color** images: They do not represent the true color of the stars shown. False color is often used when the image is made using wavelengths that the eye cannot detect, as with the infrared images in Figure 6-18. A different use of false color is to indicate the relative brightness of different parts of the image, as in the infrared image of a person in Figure 5-10. Throughout this book, we'll always point out when false color is used in an image.

Interferometry

Several large observatories are developing a technique called **interferometry** that promises to further improve the angular resolution of telescopes. The idea is to have two widely separated telescopes observe the same object simultaneously, then use fiber optic cables to "pipe" the light signals from each telescope to a central location where they "interfere" or blend together. This makes the combined signal sharp and clear. The effective resolution of such a combination of telescopes is equivalent to that of one giant telescope with a diameter equal to the **baseline,** or distance between the two telescopes. For example, the Keck I and

Figure 6-17 R I V U X G

Creating an Artificial "Star" A laser beam shines upward from Yepun, an 8.2-meter telescope at the European Southern Observatory in the Atacama Desert of Chile. (Figure 6-14a shows the objective mirror for this telescope.) The beam strikes sodium atoms that lie about 90 km (56 miles) above the Earth's surface, causing them to glow and make an artificial "star." Tracking the twinkling of this "star" makes it possible to undo the effects of atmospheric turbulence on telescope images. (European Southern Observatory)

Keck II telescopes atop Mauna Kea (Figure 6-16) are 85 meters apart, so when used as an interferometer the angular resolution is the same as a single 85-meter telescope.

Interferometry has been used for many years with radio telescopes (which we will discuss in Section 6-6), but is still under development with telescopes for visible light or infrared wavelengths. Astronomers are devoting a great deal of effort to this development because the potential rewards are great. For example, the Keck I and II telescopes used together should give an angular resolution as small as 0.005 arcsec, which corresponds to being able to read the bottom row on an eye chart 36 km (22 miles) away!

Light Pollution

 Light from city street lamps and from buildings also degrades telescope images. This **light pollution** illuminates the sky, making it more difficult to see the stars. You can appreciate the problem if you have ever looked at the night sky from a major city. Only a few of the very brightest stars can be seen, as against the thousands that can be seen with the naked eye in the desert or the mountains. To avoid light pollution, observatories are built in remote locations far from any city lights.

Unfortunately, the expansion of cities has brought light pollution to observatories that in former times had none. As an example, the growth of Tucson, Arizona, has had deleterious effects on observations at the nearby Kitt Peak National Observatory. Efforts have been made to have cities adopt light fixtures that provide safe illumination for their citizens but produce little light pollution. These efforts have met with only mixed success.

One factor over which astronomers have absolutely no control is the weather. Optical telescopes cannot see through clouds, so it is important to build observatories where the weather is usually good. One advantage of mountaintop observatories such as Mauna Kea is that most clouds form at altitudes below the observatory, giving astronomers a better chance of having clear skies.

In many ways the best location for a telescope is in orbit around Earth, where it is unaffected by weather, light pollution, or atmospheric turbulence. We will discuss orbiting telescopes in Section 6-7.

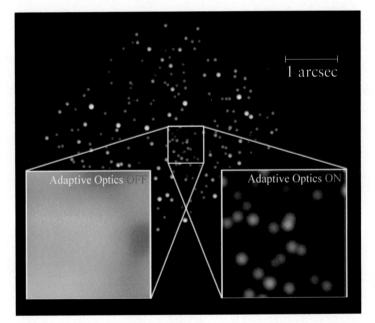

Figure 6-18 R I V U X G

Using Adaptive Optics to "Unblur" Telescope Images The two false-color, inset images show the same 1-arcsecond-wide region of the sky at infrared wavelengths as observed with the 10.0-m Keck II telescope on Mauna Kea (see Figure 6-16). Without adaptive optics, it is impossible to distinguish individual stars in this region. With adaptive optics turned on, more than two dozen stars can be distinguished. (UCLA Galactic Center Group)

6-4 A charge-coupled device is commonly used to record the image at a telescope's focus

Telescopes provide astronomers with detailed pictures of distant objects. The task of recording these pictures is called **imaging.**

Astronomical imaging really began in the nineteenth century with the invention of photography. It was soon realized that this new invention was a boon to astronomy. By taking long exposures with a camera mounted at the focus of a telescope, an astronomer can record features too faint to be seen by simply looking through the telescope. Such long exposures can reveal details in galaxies, star clusters, and nebulae that would not be visible to an astronomer looking through a telescope. Indeed, most large, modern telescopes do not have eyepieces at all.

> **The same technology that makes digital cameras possible has revolutionized astronomy**

Unfortunately, photographic film is not a very efficient light detector. Only about 1 out of every 50 photons striking photographic film triggers the chemical reaction needed to produce an image. Thus, roughly 98% of the light falling onto photographic film is wasted.

Charge-Coupled Devices

The most sensitive light detector currently available to astronomers is the **charge-coupled device (CCD).** At the heart of a CCD is a semiconductor wafer divided into an array of small light-sensitive squares called picture elements or, more commonly, **pixels (Figure 6-19a).** For example, each of the 40 CCDs shown in Figure 6-19a has more than 9.4 million pixels arranged in 2048 rows by 4608 columns. They have about a thousand times more pixels per square centimeter than on a typical computer screen, which means that a CCD of this type can record very fine image details. CCDs with smaller numbers of pixels are used in digital cameras, scanners, and fax machines.

When an image from a telescope is focused on the CCD, an electric charge builds up in each pixel in proportion to the number of photons falling on that pixel. When the exposure is finished, the amount of charge on each pixel is read into a computer, where the resulting image can be stored in digital form and either viewed on a monitor or printed out. Compared with photographic film, CCDs are some 35 times more sensitive to light (they commonly respond to 70% of the light falling on them, versus 2% for film), can record much finer details, and respond more uniformly to light of different colors. Figures 6-19b, 6-19c, and 6-19d show the dramatic difference between photographic and CCD images. The great sensitivity of CCDs also makes them useful for **photometry,** which is the measurement of the brightnesses of stars and other astronomical objects.

In the modern world of CCD astronomy, astronomers need no longer spend the night in the unheated dome of a telescope. Instead, they operate the telescope electronically from a separate control room, where the electronic CCD images can be viewed on a computer monitor. The control room need not even be adjacent

(a) A mosaic of 40 charge-coupled devices (CCDs)

(b) An image made with photographic film

(c) An image of the same region of the sky made with a CCD

(d) Combining several CCD images made with different color filters

Figure 6-19 R I ⬛V⬛ U X G

Charge-Coupled Devices (CCDs) and Imaging (a) The 40 CCDs in this mosaic are used to record the light gathered by the Canada-France-Hawaii Telescope (see Figure 6-16). After an exposure, data from each of the 377 million light-sensitive pixels in the mosaic is transferred to a waiting computer. **(b)** This negative print (black stars and white sky) shows a portion of the sky as imaged with a 4-meter telescope and photographic film. **(c)** This negative image of the same region of the sky was made with the same telescope, but with the photographic film replaced by a CCD. Many more stars and galaxies are visible. **(d)** To produce this color positive view of the same region, a series of CCD images were made using different color filters. These were then combined using a computer image-processing program.

(a: J. C. Cullandre/Canada-France-Hawaii Telescope; b, c, d: Patrick Seitzer, National Optical Astronomy Observatories) to the telescope. Although the Keck I and II telescopes (see Figure 6-16) are at an altitude of 4100 m (13,500 feet), astronomers can now make observations from a facility elsewhere on the island of Hawaii that is much closer to sea level. This saves the laborious drive to the summit of Mauna Kea and eliminates the need for astronomers to acclimate to the high altitude.

Most of the images that you will see in this book were made with CCDs. Because of their extraordinary sensitivity and their ability to be used in conjunction with computers, CCDs have attained a role of central importance in astronomy.

6-5 Spectrographs record the spectra of astronomical objects

We saw in Section 5-6 how the spectrum of an astronomical object provides a tremendous amount of information about that object, including its chemical composition and temperature. This is why measuring spectra, or **spectroscopy**, is one of the most important uses of telescopes. Indeed, some telescopes are designed solely for measuring the spectra of distant, faint objects; they are never used for imaging.

> **The spectrum of a planet, star, or galaxy can reveal more about its nature than an image**

Spectrographs and Diffraction Gratings

An essential tool of spectroscopy is the **spectrograph**, a device that records spectra. This optical device is mounted at the focus of a telescope. Figure 6-20*a* shows one design for a spectrograph, in which a **diffraction grating** is used to form the spectrum of a planet, star, or galaxy. This is a piece of glass on which thousands of very regularly spaced parallel lines have been cut. Some of the finest diffraction gratings have more than 10,000 lines per centimeter, which are usually cut by drawing a diamond back and forth across the glass. When light is shone on a diffraction grating, a spectrum is produced by the way in which light waves leaving different parts of the grating interfere with each other. (This same effect produces the rainbow of colors you see reflected from a compact disc or DVD, as shown in Figure 6-20*b*. Information is stored on the disc in a series of closely spaced pits, which act as a diffraction grating.)

Older types of spectrographs used a prism rather than a diffraction grating to form a spectrum. This had several drawbacks. A prism does not disperse the colors of the rainbow evenly: Blue and violet portions of the spectrum are spread out more than the red portion. In addition, because the blue and violet wavelengths must pass through more of the prism's glass than do the red wavelengths (examine Figure 5-3), light is absorbed unevenly across the spectrum. Indeed, a glass prism is opaque to near-ultraviolet light. For these reasons, diffraction gratings are preferred in modern spectrographs.

In Figure 6-20*a* the spectrum of a planet, star, or galaxy formed by the diffraction grating is recorded on a CCD. Light from a hot gas (such as helium, neon, argon, iron, or a combination of these) is then focused on the spectrograph slit. The result is a *comparison spectrum* alongside the spectrum of the celestial object under study. The wavelengths of the bright spectral lines

(a)

(b)

Figure 6-20

A Grating Spectrograph (a) This optical device uses a diffraction grating to break up the light from a source into a spectrum. The spectrum is then recorded on a CCD. **(b)** A diffraction grating has a large number of parallel lines in its surface that reflect light of different colors in different direction. A compact disc, which stores information in a series of closely spaced pits, reflects light in a similar way. (Dale E. Boyer/Photo Researchers)

(a) Two representations of an absorption line spectrum

(b) Two representations of an emission line spectrum

Figure 6-21

Two Ways to Represent Spectra When a CCD is placed at the focus of a spectrograph, it records the rainbow-colored spectrum. A computer program can be used to convert the recorded data into a graph of intensity versus wavelength. **(a)** Absorption lines appear as dips on such a graph, while **(b)** emission lines appear as peaks. The dark absorption lines and bright emission lines in this example are the Balmer lines of hydrogen (see Section 5-8).

of the comparison spectrum are known from laboratory experiments and can therefore serve as reference markers. (See Figure 5-17, which shows the spectrum of the Sun and a comparison spectrum of iron.)

When the exposure is finished, electronic equipment measures the charge that has accumulated in each pixel. These data are used to graph light intensity against wavelength. Dark absorption lines in the spectrum appear as depressions or valleys on the graph, while bright emission lines appear as peaks. Figure 6-21 compares two ways of exhibiting spectra with absorption lines and emission lines. Later in this book we shall see spectra presented in both these ways.

6-6 A radio telescope uses a large concave dish to reflect radio waves to a focus

For thousands of years, all the information that astronomers gathered about the universe was based on ordinary visible light. In the twentieth century, however, astronomers first began to explore the nonvisible electromagnetic radiation coming from astronomical objects. In this way they have discovered aspects of the cosmos that are forever hidden to optical telescopes.

Astronomers have used ultraviolet light to map the outer regions of the Sun and the clouds of Venus, and used infrared radiation to see new stars and perhaps new planetary sys-

> **Observing at radio wavelengths reveals aspects of the universe hidden from ordinary telescopes**

tems in the process of formation. By detecting radio waves from Jupiter and Saturn, they have mapped the intense magnetic fields that surround those giant planets; by detecting curious bursts of X rays from space, they have learned about the utterly alien conditions in the vicinity of a black hole. It is no exaggeration to say that today's astronomers learn as much about the universe using telescopes for nonvisible wavelengths as they do using visible light.

Radio Astronomy

Radio waves were the first part of the electromagnetic spectrum beyond the visible to be exploited for astronomy. This happened as a result of a research project seemingly unrelated to astronomy. In the early 1930s, Karl Jansky, a young electrical engineer at Bell Telephone Laboratories, was trying to locate what was causing interference with the then-new transatlantic radio link. By 1932, he realized that one kind of radio noise is strongest when the constellation Sagittarius is high in the sky. The center of our Galaxy is located in the direction of Sagittarius, and Jansky concluded that he was detecting radio waves from an astronomical source.

At first only Grote Reber, a radio engineer living in Illinois, took up Jansky's research. In 1936 Reber built in his backyard the first **radio telescope**, a radio-wave detector dedicated to astronomy. He modeled his design after an ordinary reflecting telescope, with a parabolic metal "dish" (reflecting antenna) measuring 31 ft (10 m) in diameter and a radio receiver at the focal point of the dish.

Reber spent the years from 1938 to 1944 mapping radio emissions from the sky at wavelengths of 1.9 m and 0.63 m. He found radio waves coming from the entire Milky Way, with the greatest emission from the center of the Galaxy. These results, together with the development of improved radio technology during World War II, encouraged the growth of radio astronomy and the construction of new radio telescopes around the world. Radio observatories are as common today as major optical observatories.

Modern Radio Telescopes

Like Reber's prototype, a typical modern radio telescope has a large parabolic dish (Figure 6-22). An antenna tuned to the desired frequency is located at the focus (like the prime focus design for optical reflecting telescopes shown in Fig. 6-11*a*). The incoming signal is relayed from the antenna to amplifiers and recording instruments, typically located in a room at the base of the telescope's pier.

CAUTION! The radio telescope in Figure 6-22 looks like a radar dish but is used in a different way. In radar, the dish is used to send out a narrow beam of radio waves. If this beam encounters an object like an airplane, some of the radio waves will be reflected back to the radar dish and detected by a receiver at the focus of the dish. Thus, a radar dish looks for radio waves *reflected* by distant objects. By contrast, a radio telescope is designed to receive radio waves *emitted* by objects in space.

Many radio telescope dishes, like the one in Figure 6-22, have visible gaps in them like a wire mesh. This does not affect their

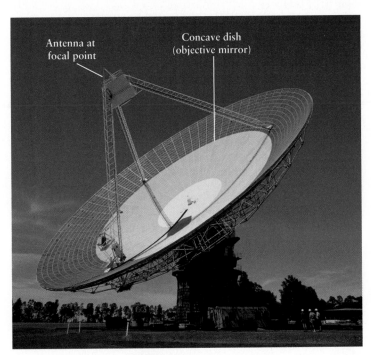

Figure 6-22   R I V U X G

A Radio Telescope The dish of the Parkes radio telescope in New South Wales, Australia, is 64 m (210 ft) in diameter. Radio waves reflected from the dish are brought to a focus and collected by an antenna at the focal point. (David Nunuk/Photo Researchers)

reflecting power because the holes are much smaller than the wavelengths of the radio waves. The same idea is used in the design of microwave ovens. The glass window in the oven door would allow the microwaves to leak out, so the window is covered by a metal screen with small holes. These holes are much smaller than the 12.2-cm (4.8-in.) wavelength of the microwaves, so the screen reflects the microwaves back into the oven.

Radio Telescopes: Limits to Angular Resolution

One great drawback of early radio telescopes was their very poor angular resolution. Recall from Section 6-3 that angular resolution is the smallest angular separation between two stars that can just barely be distinguished as separate objects. Unlike visible light, radio waves are only slightly affected by turbulence in the atmosphere, so the limitation on the angular resolution of a radio telescope is diffraction. The problem is that diffraction-limited angular resolution is directly proportional to the wavelength being observed: The longer the wavelength, the larger (and hence worse) the angular resolution and the fuzzier the image. (See the formula for angular resolution θ in Section 6-3.) As an example, a 1-m radio telescope detecting radio waves of 5-cm wavelength has 100,000 times poorer angular resolution than a 1-m optical telescope. Because radio radiation has very long wavelengths, small radio telescopes can produce only blurry, indistinct images.

A very large radio telescope can produce a somewhat sharper radio image, because as the diameter of the telescope increases, the angular resolution decreases. In other words, the bigger the dish, the better the resolution. For this reason, most modern radio telescopes have dishes more than 30 m (100 ft) in diameter. This is also useful for increasing light-gathering power, because radio signals from astronomical objects are typically very weak in comparison with the intensity of visible light. But even the largest single radio dish in existence, the 305-m (1000-ft) Arecibo radio telescope in Puerto Rico, cannot come close to the resolution of the best optical instruments.

To improve angular resolution at radio wavelengths, astronomers combine observations from two or more widely separated telescopes using the interferometry technique that we described in Section 6-3. This is much easier than for visible or infrared light because radio signals can be carried over electrical wires. Consequently, two radio telescopes observing the same astronomical object can be hooked together, even if they are separated by many kilometers. As for visible light, the resulting angular resolution is that of a single telescope whose diameter is equal to the baseline, or separation between the telescopes.

One of the largest arrangements of radio telescopes for interferometry is the Very Large Array (VLA), located in the desert near Socorro, New Mexico (Figure 6-23). The VLA consists of 27 parabolic dishes, each 25 m (82 ft) in diameter. These 27 telescopes are arranged along the arms of a gigantic Y that covers an area 27 km (17 mi) in diameter. By pointing all 27 telescopes at the same object and combining the 27 radio signals, this system can produce radio views of the sky with an angular resolution as small as 0.05 arcsec, comparable to that of the very best optical telescopes.

WEB LINK 6.6

Figure 6-23   R I V U X G

The Very Large Array (VLA) The 27 radio telescopes of the VLA in central New Mexico are arranged along the arms of a Y. The north arm of the array is 19 km long; the southwest and southeast arms are each 21 km long. By spreading the telescopes out along the legs and combining the signals received, the VLA can give the same angular resolution as a single dish many kilometers in radius. (Courtesy of NRAO/AUI)

WEB LINK 6.7

Dramatically better angular resolution can be obtained by combining the signals from radio telescopes at different observatories thousands of kilometers apart. This technique is called **very-long-baseline interferometry (VLBI)**. VLBI is used by a system called the Very Long Baseline Array (VLBA), which consists of ten 25-meter dishes at different locations between Hawaii and the Caribbean. Although the ten dishes are not physically connected, they are all used to observe the same object at the same time. The data from each telescope are recorded electronically and processed later. By carefully synchronizing the ten recorded signals, they can be combined just as if the telescopes had been linked together during the observation. With the VLBA, features smaller than 0.001 arcsec can be distinguished at radio wavelengths. This angular resolution is 100 times better than a large optical telescope with adaptive optics.

Even better angular resolution can be obtained by adding radio telescopes in space; the baseline is then the distance from the VLBA to the orbiting telescope. The first such space radio telescope, the Japanese HALCA spacecraft, operated from 1997 to 2003. Orbiting at a maximum distance of 21,400 km (13,300 mi) above the Earth's surface, HALCA gave a baseline three times longer—and thus an angular resolution three times better—than can be obtained with Earthbound telescopes alone. Successors to HALCA may be placed in orbit after 2010.

Figure 6-24 shows how optical and radio images of the same object can give different and complementary kinds of information. The visible-light image of Saturn (Figure 6-24*a*) shows clouds in the planet's atmosphere and the structure of the rings.

(a) R I V U X G

(b) R I V U X G

Figure 6-24

Optical and Radio Views of Saturn (a) This picture was taken by a spacecraft 18 million kilometers from Saturn. The view was produced by sunlight reflecting from the planet's cloudtops and rings. **(b)** This VLA image shows radio emission from Saturn at a wavelength of 2 cm. In this false-color image, the most intense radio emission is shown in red, the least intense in blue. Yellow and green represent intermediate levels of radio intensity; black indicates no detectable radio emission. Note the radio "shadow" caused by Saturn's rings where they lie in front of the planet. (a: NASA; b: Image courtesy of NRAO/AUI)

Like the visible light from the Moon, the light used to make this image is just reflected sunlight. By contrast, the false-color radio image of Saturn (Figure 6-24*b*) is a record of waves *emitted* by the planet and its rings. Analyzing such images provides information about the structure of Saturn's atmosphere and rings that could never be obtained from a visible-light image such as Figure 6-24*a*.

6-7 Telescopes in orbit around the Earth detect radiation that does not penetrate the atmosphere

The many successes of radio astronomy show the value of observations at nonvisible wavelengths. But the Earth's atmo-

Figure 6-25

The Transparency of the Earth's Atmosphere This graph shows the percentage of radiation that can penetrate the Earth's atmosphere at different wavelengths. Regions in which the curve is high are called "windows," because the atmosphere is relatively transparent at those wavelengths. There are also three wavelength ranges in which the atmosphere is opaque and the curve is near zero: at wavelengths less than about 290 nm, which are absorbed by atmospheric oxygen and nitrogen; between the optical and radio windows, due to absorption by water vapor and carbon dioxide; and at wavelengths longer than about 20 m, which are reflected back into space by ionized gases in the upper atmosphere.

sphere is opaque to many wavelengths. Other than visible light and radio waves, very little radiation from space manages to penetrate the air we breathe. To overcome this, astronomers have placed a variety of telescopes in orbit.

> **Space telescopes make it possible to study the universe across the entire electromagnetic spectrum**

Figure 6-25 shows the transparency of the Earth's atmosphere to different wavelengths of electromagnetic radiation. The atmosphere is most transparent in two wavelength regions, the **optical window** (which includes the entire visible spectrum) and the **radio window** (which includes part, but not all, of the radio spectrum). There are also several relatively transparent regions at infrared wavelengths between 1 and 40 μm. Infrared radiation within these wavelength intervals can penetrate the Earth's atmosphere somewhat and can be detected with ground-based telescopes. This wavelength range is called the *near-infrared,* because it lies just beyond the red end of the visible spectrum.

Infrared Astronomy

Water vapor is the main absorber of infrared radiation from space, which is why infrared observatories are located at sites with exceptionally low humidity. The site must also be at high altitude to get above as much of the atmosphere's water vapor as possible. One site that meets both criteria is the summit of Mauna Kea in Hawaii, shown in Figure 6-16. (The complete lack of vegetation on the summit attests to its extreme dryness.) Some of the telescopes on Mauna Kea are designed exclusively for detecting infrared radiation. Others, such as the Keck I and Keck II telescopes, are used for both visible and near-infrared observations (see Figure 6-18).

 Even at the elevation of Mauna Kea, water vapor in the atmosphere restricts the kinds of infrared observations that astronomers can make. This situation can be improved by carrying telescopes on board high-altitude balloons or aircraft. But the ultimate solution is to place a telescope in Earth's orbit and radio its data back to astronomers on the ground. The first such orbiting infrared observatory, the Infrared Astronomical Satellite (IRAS), was launched in 1983. During its nine-month mission, IRAS used its 57-cm (22-in.) telescope to map almost the entire sky at wavelengths from 12 to 100 μm.

The IRAS data revealed the presence of dust disks around nearby stars. Planets are thought to coalesce from disks of this kind, so this was the first concrete (if indirect) evidence that there might be planets orbiting other stars. The dust that IRAS detected is warm enough to emit infrared radiation but too cold to emit much visible light, so it remained undetected by ordinary optical telescopes. IRAS also discovered distant, ultraluminous galaxies that emit almost all their radiation at infrared wavelengths.

In 1995 the Infrared Space Observatory (ISO), a more advanced 60-cm reflector with better light detectors, was launched into orbit by the European Space Agency. During its 2½-year mission, ISO made a number of groundbreaking observations of very distant galaxies and of the thin, cold material between the stars of our own Galaxy. Like IRAS, ISO had to be cooled by liquid helium to temperatures just a few degrees above absolute zero. Had this not been done, the infrared blackbody radiation from the telescope itself would have outshone the infrared radiation from astronomical objects. The ISO mission came to an end when the last of the helium evaporated into space.

At the time of this writing the largest orbiting infrared observatory is the Spitzer Space Telescope, an 85-cm infrared telescope designed to survey the infrared sky with unprecedented resolution (Figure 6-26). Placed in orbit in 2003, the Spitzer Space Telescope is being used to study the formation of new stars, probe the inner structure of galaxies, and examine the first galaxies that formed after the Big Bang. The images that open this chapter illustrate how Spitzer can reveal aspects of astronomical objects that are hidden from visible-light telescopes.

Figure 6-26   R I V U X G

The Spitzer Space Telescope Launched in 2003, this is the largest
infrared telescope ever placed in space. Its 85-cm objective mirror and
three science instruments, kept cold by 360 liters of liquid helium, enable
the Spitzer Space Telescope to observe the universe at wavelengths from
3 to 180 μm. The background to this illustration of the telescope is a
false-color Spitzer image of the Milky Way and the constellation Orion at
100 μm, showing emission from warm interstellar dust. (NASA/JPL-Caltech)

Ultraviolet Astronomy

Astronomers are also very interested in observing at ultraviolet
wavelengths. These observations can reveal a great deal about hot
stars, ionized clouds of gas between the stars, and the Sun's high-
temperature corona (see Section 3-5), all of which emit copious
amounts of ultraviolet light. The spectrum of ultraviolet sunlight
reflected from a planet can also reveal the composition of the
planet's atmosphere. However, Earth's atmosphere is opaque to
ultraviolet light except for the narrow *near-ultraviolet* range,
which extends from about 400 nm (the violet end of the visible
spectrum) down to 300 nm.

To see shorter-wavelength *far-ultraviolet* light, astronomers
must again make their observations from space. The first ultraviolet telescope was placed in orbit in 1962, and several others have
since followed it into space. Small rockets have also lifted ultraviolet cameras briefly above the Earth's atmosphere. Figure 6-27
shows an ultraviolet view of the constellation Orion, along with
infrared and visible views.

The Far Ultraviolet Spectroscopic Explorer (FUSE),
which went into orbit in 1999, specializes in measuring spectra at wavelengths from 90 to 120 nm.
Highly ionized oxygen atoms, which can exist only in an extremely high-temperature gas, have a characteristic spectral
line in this range. By looking for this spectral line in various
parts of the sky, FUSE has confirmed that our Milky Way
Galaxy (Section 1-4) is surrounded by an immense "halo" of
gas at temperatures in excess of 200,000 K. Only an ultraviolet telescope could have detected this "halo," which is thought
to have been produced by exploding stars called supernovae
(Section 1-3).

The Hubble Space Telescope

Infrared and ultraviolet satellites give excellent views of the heavens at selected wavelengths. But since the 1940s astronomers had
dreamed of having one large telescope that could be operated at
any wavelength from the near-infrared through the visible range
and out into the ultraviolet. This is the mission of the Hubble
Space Telescope (HST), which was placed in a 600-km-high orbit by the space shuttle *Discovery* in 1990 (Figure 6-28). HST has
a 2.4-meter (7.9-ft) objective mirror and was designed to observe
at wavelengths from 115 nm to 1 μm. Like most ground-based
telescopes, HST uses a CCD to record images. (In fact, the development of HST helped drive advances in CCD technology.) The
images are then radioed back to Earth in digital form.

The great promise of HST was that from its vantage point
high above the atmosphere, its angular resolution would be limited only by diffraction. But soon after HST was placed in orbit,
astronomers discovered that a manufacturing error had caused
the telescope's objective mirror to suffer from spherical aberration. The mirror should have been able to concentrate 70% of a
star's light into an image with an angular diameter of 0.1 arcsec.
Instead, only 20% of the light was focused into this small area.
The remainder was smeared out over an area about 1 arcsec wide,
giving images little better than those achieved at major ground-
based observatories.

On an interim basis, astronomers used only the 20% of incoming starlight that was properly focused and, with computer
processing, discarded the remaining poorly focused 80%. This
was practical only for brighter objects on which astronomers
could afford to waste light. But many of the observing projects
scheduled for HST involved extremely dim galaxies and nebulae.

These problems were resolved by a second space shuttle mission in 1993. Astronauts installed a set of small secondary mirrors whose curvature exactly compensated for the error in
curvature of the primary mirror. Once these were in place, HST
was able to make truly sharp images of extremely faint objects.
Astronomers have used the repaired HST to make discoveries
about the nature of planets, the evolution of stars, the inner

Immense clouds of dust are heated by hot, luminous, newly-formed stars; the clouds glow at (a) ultraviolet and (b) infrared wavelengths

At visible wavelengths the dust is opague, obscuring the newly-formed stars

R I V **U** X G

(a) Ultraviolet Orion

R **I** V U X G

(b) Infrared Orion

R I **V** U X G

(c) Visible Orion

(d) A star chart of Orion

Figure 6-27

Orion Seen at Ultraviolet, Infrared, and Visible Wavelengths (a) An ultraviolet view of the constellation of Orion was obtained during a brief rocket flight in 1975. This 100-s exposure covers the wavelength range 125–200 nm. **(b)** The false-color view from the Infrared Astronomical Satellite displays emission at different wavelengths in different colors:

red for 100-μm radiation, green for 60-μm radiation, and blue for 12-μm radiation. Compare these images with **(c)** an ordinary visible-light photograph and (d) a star chart of Orion. (a: G. R. Carruthers, Naval Research Laboratory; b: NASA; c: R. C. Mitchell, Central Washington University)

Figure 6-28   R I **V** U X G

WEB LINK 6.11

The Hubble Space Telescope The largest telescope yet placed in orbit, HST is a joint project of NASA and the European Space Agency (ESA). HST has helped discover new moons of Pluto, probed the formation of stars, and found evidence that the universe is now expanding at a faster rate than several billion years ago. This photograph was taken from the space shuttle *Discovery* during a 1997 mission to service HST. (NASA) workings of galaxies, and the expansion of the universe. You will see many HST images in later chapters.

The success of HST has inspired plans for its larger successor, the James Webb Space Telescope or JWST (Figure 6-29). Planned for a 2013 launch, JWST will observe at visible and infrared wavelengths from 600 nm to 28 μm. With its 6.5-m objective mirror—2.5 times the diameter of the HST objective mirror, with six times the light-gathering power—JWST will study faint objects such as planetary systems forming around other stars and galaxies near the limit of the observable universe. Unlike HST, which is in a relatively low-altitude orbit around the Earth, JWST will orbit the Sun some 1.5 million km beyond the Earth. In this orbit the telescope's view will not be blocked by the Earth. Furthermore, by remaining far from the radiant heat of the Earth it will be easier to keep JWST at the very cold temperatures required by its infrared detectors.

X-Ray Astronomy

Space telescopes have also made it possible to explore objects whose temperatures reach the almost inconceivable values of 10^6 to 10^8 K. Atoms in such a high-temperature gas move so fast that when they collide, they emit X-ray photons of very high energy and very short wavelengths less than 10 nm. X-ray telescopes designed to detect these photons must be placed in orbit, since Earth's atmosphere is totally opaque at these wavelengths.

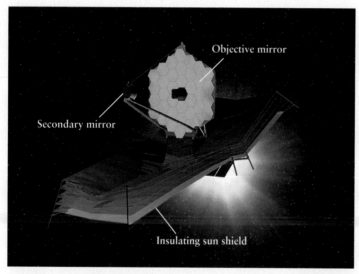

Figure 6-29

The James Webb Space Telescope (JWST) The successor to the Hubble Space Telescope, JWST will have an objective mirror 6.5 m (21 ft) in diameter. Like many Earthbound telescopes, JWST will be of Cassegrain design (see Figures 6-11b and 6-14b). Rather than using liquid helium to keep the telescope and instruments at the low temperatures needed to observe at infrared wavelengths, JWST will keep cool using a multilayer sunshield the size of two tennis courts. (Courtesy of Northrop Grumman Space Technology)

CAUTION! X-ray telescopes work on a very different principle from the X-ray devices used in medicine and dentistry. If you have your foot "X rayed" to check for a broken bone, a piece of photographic film (or an electronic detector) sensitive to X rays is placed under your foot and an X-ray beam is directed at your foot from above. The radiation penetrates through soft tissue but not through bone, so the bones cast an "X-ray shadow" on the film. A fracture will show as a break in the shadow. X-ray telescopes, by contrast, do *not* send beams of X rays toward astronomical objects in an attempt to see inside them. Rather, these telescopes detect X rays that the objects emit on their own.

Astronomers got their first quick look at the X-ray sky from brief rocket flights during the late 1940s. These observations confirmed that the Sun's corona (see Section 3-5) is a source of X rays, and must therefore be at a temperature of millions of kelvins. In 1962 a rocket experiment revealed that objects beyond the solar system also emit X rays.

Since 1970, a series of increasingly sensitive and sophisticated X-ray observatories have been placed in orbit, including NASA's Einstein Observatory, the European Space Agency's Exosat, and the German-British-American ROSAT. These telescopes have shown that other stars also have high-temperature coronae, and have found hot, X-ray-emitting gas clouds so immense that hundreds of galaxies fit inside them. They also discovered unusual stars that emit X rays in erratic bursts. These bursts are now thought to be coming from heated gas swirling around a small but massive object—possibly a black hole.

X-ray astronomy took a quantum leap forward in 1999 with the launch of NASA's Chandra X-ray Observatory and the European Space Agency's XMM-Newton. Named for the Indian-American astrophysicist Subrahmanyan Chandrasekhar, Chandra can view the X-ray sky with an angular resolution of 0.5 arcsec (Figure 6-30a). This is comparable to the best ground-based optical telescopes, and more than a thousand times better than the resolution of the first orbiting X-ray telescope. Chandra can also measure X-ray spectra 100 times more precisely than any previous spacecraft and can detect variations in X-ray emissions on time scales as short as 16 microseconds. This latter capability is essential for understanding how X-ray bursts are produced around black holes.

XMM-Newton (for *X-ray Multi-mirror Mission*) is actually three X-ray telescopes that all point in the same direction (Figure 6-30b). Their combined light-gathering power is five times greater than that of Chandra, which makes XMM-Newton able to observe fainter objects. (For reasons of economy, the mirrors were not ground as precisely as those on Chandra, so the angular resolution of XMM-Newton is only about 6 arcseconds.) It also carries a small but highly capable telescope for ultraviolet and visible observations. Hot X-ray sources are usually accompanied by cooler material that radiates at these longer wavelengths, so XMM-Newton can observe these hot and cool regions simultaneously.

Gamma-Ray Astronomy

Gamma rays, the shortest-wavelength photons of all, help us to understand phenomena even more energetic than those that pro-

(a) Chandra X-ray Observatory

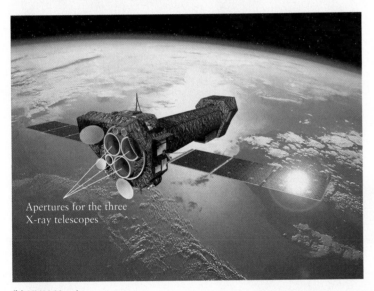

(b) XMM-Newton

Figure 6-30

Two Orbiting X-Ray Observatories (a) X rays are absorbed by ordinary mirrors like those used in optical reflectors, but they can be reflected if they graze the mirror surface at a very shallow angle. In the Chandra X-ray Observatory, X rays are focused in this way onto a focal plane 10 m (33 ft) behind the mirror. (b) XMM-Newton is about the same size as Chandra, and its three X-ray telescopes form images in the same way. (a: NASA/Chandra X-ray Observatory Center/Smithsonian Astrophysical Observatory; b: D. Ducros/European Space Agency)

Like X rays, gamma rays do not penetrate Earth's atmosphere, so space telescopes are required. One of the most important gamma-ray telescopes placed in orbit to date is the Compton Gamma Ray Observatory (CGRO), shown in Figure 6-31. One particularly important task for CGRO was the study of gamma-ray bursts, which are brief, unpredictable, and very intense flashes of gamma rays that are found in all parts of the sky. By analyzing data from CGRO and other orbiting observatories, astronomers have shown that the sources of these gamma-ray bursts are billions of light-years away. For these bursts to be visible across such great distances, their sources must be among the most energetic objects in the universe. By combining these gamma-ray observations with images made by optical telescopes, astronomers have found that at least some of the gamma-ray bursts emanate from stars that explode catastrophically. Since the CGRO mission ended in 2000, more advanced gamma-ray telescopes, including the European Space Agency's INTEGRAL (International Gamma-Ray Astrophysics Laboratory), have taken its place.

The *Cosmic Connections* figure shows the wavelengths at which Earth-orbiting telescopes are particularly useful, as well as summarizing the design of refracting and reflecting Earth-based telescopes. The advantages and benefits of Earth-orbiting observatories cannot be overemphasized. We are no longer limited to the narrow ranges of whatever wavelengths manage to leak through our shimmering, hazy atmosphere (Figure 6-32). For the first time, we are really *seeing* the universe.

Figure 6-31   R I **V** U X G

The Compton Gamma Ray Observatory (CGRO) This photograph shows the Compton Observatory being deployed from the space shuttle *Atlantis* in 1991. Named in honor of Arthur Holly Compton, an American scientist who made important discoveries about gamma rays, CGRO carried four different gamma-ray detectors. When its mission ended in 2000, CGRO disintegrated as it reentered our atmosphere. (NASA)

duce X rays. As an example, when a massive star explodes into a supernova, it produces radioactive atomic nuclei that are strewn across interstellar space. Observing the gamma rays emitted by these nuclei helps astronomers understand the nature of supernova explosions.

(a) R I V U X G

(c) R I V U X G

(b) R I V U X G

(d) R I V U X G

(e) R I V U X G

Figure 6-32

WEB LINK 6.15

The Entire Sky at Five Wavelength Ranges These five views show the entire sky at visible, radio, infrared, X-ray, and gamma-ray wavelengths. The entire celestial sphere is mapped onto an oval, with the Milky Way stretching horizontally across the center. The black crescents in the infrared and X-ray images are where data are missing. (GSFC/NASA) **(a)** In the visible view the constellation Orion is at the right, Sagittarius in the middle, and Cygnus toward the left. Many of the dark areas along the Milky Way are locations where interstellar dust is sufficiently thick to block visible light. **(b)** The radio view shows the sky at a wavelength of 21 cm. This wavelength is emitted by hydrogen atoms in interstellar space. The brightest regions (shown in red) are in the plane of the Milky Way, where the hydrogen is most concentrated. **(c)** The infrared view from IRAS shows emission at 100 μm, 60 μm, and 12 μm. Most of

the emission is from dust particles in the plane of the Milky Way that have been warmed by starlight. **(d)** The X-ray view from ROSAT shows wavelengths of 0.8 nm (blue), 1.7 nm (green), and 5.0 nm (red), corresponding to photon energies of 1500, 750, and 250 eV. Extremely high temperature gas emits these X rays. The white regions, which emit strongly at all X-ray wavelengths, are remnants of supernovae. **(e)** The gamma-ray view from the Compton Gamma Ray Observatory includes all wavelengths less than about 1.2×10^{-5} nm (photon energies greater than 10^8 eV). The diffuse radiation from the Milky Way is emitted when fast-moving subatomic particles collide with the nuclei of atoms in interstellar gas clouds. The bright spots above and below the Milky Way are distant, extremely energetic galaxies.

Today, telescopes can view the universe in every range of electromagnetic radiation, although some must be above Earth's atmosphere to receive radiation without interference.

Light from distant object

Objective lens

A refracting telescope uses a lens to form an image. An eyepiece is used to magnify the image.

Image of distant object

Eyepiece for magnifying image

Light from distant object

Secondary mirror

A reflecting telescope uses a curved mirror to form an image. An eyepiece can be used to magnify the image, or the image can be recorded on a CCD.

Objective mirror

Image

CCD detector

The Earth's atmosphere is transparent to visible wavelengths, so visible-light telescopes (refracting or reflecting) can be used from Earth's surface.

Infrared telescopes are best placed in orbit since most infrared wavelengths do not penetrate the atmosphere.

Infrared telescope

Ultraviolet telescope

Gamma-ray, x-ray, and ultraviolet telescopes must be placed in orbit because these wavelengths do not penetrate the atmosphere.

X-ray telescope

Radio telescope

The atmosphere is transparent to radio waves, so radio telescopes can be used from the Earth's surface.

Gamma-ray telescope

Key Words

Key Ideas

Refracting Telescopes: Refracting telescopes, or refractors, produce images by bending light rays as they pass through glass lenses.

• Chromatic aberration is an optical defect whereby light of different wavelengths is bent in different amounts by a lens.

• Glass impurities, chromatic aberration, opacity to certain wavelengths, and structural difficulties make it inadvisable to build extremely large refractors.

Reflecting Telescopes: Reflecting telescopes, or reflectors, produce images by reflecting light rays to a focus point from curved mirrors.

• Reflectors are not subject to most of the problems that limit the useful size of refractors.

Angular Resolution: A telescope's angular resolution, which indicates ability to see fine details, is limited by two key factors.

• Diffraction is an intrinsic property of light waves. Its effects can be minimized by using a larger objective lens or mirror.

• The blurring effects of atmospheric turbulence can be minimized by placing the telescope atop a tall mountain with very smooth air. They can be dramatically reduced by the use of adaptive optics and can be eliminated entirely by placing the telescope in orbit.

Charge-Coupled Devices: Sensitive light detectors called charge-coupled devices (CCDs) are often used at a telescope's focus to record faint images.

Spectrographs: A spectrograph uses a diffraction grating to form the spectrum of an astronomical object.

Radio Telescopes: Radio telescopes use large reflecting dishes to focus radio waves onto a detector.

• Very large dishes provide reasonably sharp radio images. Higher resolution is achieved with interferometry techniques that link smaller dishes together.

Transparency of the Earth's Atmosphere: The Earth's atmosphere absorbs much of the radiation that arrives from space.

• The atmosphere is transparent chiefly in two wavelength ranges known as the optical window and the radio window. A few wavelengths in the near-infrared also reach the ground.

Telescopes in Space: For observations at wavelengths to which the Earth's atmosphere is opaque, astronomers depend on telescopes carried above the atmosphere by rockets or spacecraft.

• Satellite-based observatories provide new information about the universe and permit coordinated observation of the sky at all wavelengths.

Questions

Review Questions

1. Describe refraction and reflection. Explain how these processes enable astronomers to build telescopes.
2. Explain why a flat piece of glass does not bring light to a focus while a curved piece of glass can.
3. Explain why the light rays that enter a telescope from an astronomical object are essentially parallel.
4. With the aid of a diagram, describe a refracting telescope. Which dimensions of the telescope determine its light-gathering power? Which dimensions determine the magnification?
5. What is the purpose of a telescope eyepiece? What aspect of the eyepiece determines the magnification of the image? In what circumstances would the eyepiece not be used?
6. Do most professional astronomers actually look through their telescopes? Why or why not?
7. Quite often advertisements appear for telescopes that extol their magnifying power. Is this a good criterion for evaluating telescopes? Explain your answer.
8. What is chromatic aberration? For what kinds of telescopes does it occur? How can it be corrected?
9. With the aid of a diagram, describe a reflecting telescope. Describe four different ways in which an astronomer can access the focal plane.
10. Explain some of the disadvantages of refracting telescopes compared to reflecting telescopes.
11. What kind of telescope would you use if you wanted to take a color photograph entirely free of chromatic aberration? Explain your answer.
12. Explain why a Cassegrain reflector can be substantially shorter than a refractor of the same focal length.

13. No major observatory has a Newtonian reflector as its primary instrument, whereas Newtonian reflectors are extremely popular among amateur astronomers. Explain why this is so.

14. What is spherical aberration? How can it be corrected?

15. What is diffraction? Why does it limit the angular resolution of a telescope? What other physical phenomenon is often a more important restriction on angular resolution?

16. What is active optics? What is adaptive optics? Why are they useful? Would either of these be a good feature to include on a telescope to be placed in orbit?

17. Explain why combining the light from two or more optical telescopes can give dramatically improved angular resolution.

18. What is light pollution? What effects does it have on the operation of telescopes? What can be done to minimize these effects?

19. What is a charge-coupled device (CCD)? Why have CCDs replaced photographic film for recording astronomical images?

20. What is a spectrograph? Why do many astronomers regard it as the most important device that can be attached to a telescope?

21. What are the advantages of using a diffraction grating rather than a prism in a spectrograph?

22. Compare an optical reflecting telescope and a radio telescope. What do they have in common? How are they different?

23. Why can radio astronomers make observations at any time during the day, whereas optical astronomers are mostly limited to observing at night? (*Hint:* Does your radio work any better or worse in the daytime than at night?)

24. Why are radio telescopes so large? Why does a single radio telescope have poorer angular resolution than a large optical telescope? How can the resolution be improved by making simultaneous observations with several radio telescopes?

25. What are the optical window and the radio window? Why isn't there an X-ray window or an ultraviolet window?

26. Why is it necessary to keep an infrared telescope at a very low temperature?

27. How are the images made by an X-ray telescope different from those made by a medical X-ray machine?

28. Why must astronomers use satellites and Earth-orbiting observatories to study the heavens at X-ray and gamma-ray wavelengths?

29. Show by means of a diagram why the image formed by a simple refracting telescope is upside down.

30. Ordinary photographs made with a telephoto lens make distant objects appear close. How does the focal length of a telephoto lens compare with that of a normal lens? Explain your reasoning.

31. The observing cage in which an astronomer can sit at the prime focus of the 5-m telescope on Palomar Mountain is about 1 m in diameter. Calculate what fraction of the incoming starlight is blocked by the cage.

32. (a) Compare the light-gathering power of the Keck I 10.0-m telescope with that of the Hubble Space Telescope (HST), which has a 2.4-m objective mirror. (b) What advantages does Keck I have over HST? What advantages does HST have over Keck I?

33. Suppose your Newtonian reflector has an objective mirror 20 cm (8 in.) in diameter with a focal length of 2 m. What magnification do you get with eyepieces whose focal lengths are (a) 9 mm, (b) 20 mm, and (c) 55 mm? (d) What is the telescope's diffraction-limited angular resolution when used with orange light of wavelength 600 nm? (e) Would it be possible to achieve this angular resolution if you took the telescope to the summit of Mauna Kea? Why or why not?

34. Several groups of astronomers are making plans for large ground-based telescopes. (a) What would be the diffraction-limited angular resolution of a telescope with a 40-meter objective mirror? Assume that yellow light with wavelength 550 nm is used. (b) Suppose this telescope is placed atop Mauna Kea. How will the actual angular resolution of the telescope compare to that of the 10-meter Keck I telescope? Assume that adaptive optics is not used.

35. The Hobby-Eberly Telescope (HET) at the McDonald Observatory in Texas has a spherical mirror, which is the least expensive shape to grind. Consequently, the telescope has spherical aberration. Explain why this doesn't affect the usefulness of HET for spectroscopy. (The telescope is not used for imaging.)

36. The four largest moons of Jupiter are roughly the same size as our Moon and are about 628 million (6.28×10^8) kilometers from Earth at opposition. What is the size in kilometers of the smallest surface features that the Hubble Space Telescope (resolution of 0.1 arcsec) can detect? How does this compare with the smallest features that can be seen on the Moon with the unaided human eye (resolution of 1 arcmin)?

37. The Hubble Space Telescope (HST) has been used to observe the galaxy M100, some 70 million light-years from Earth. (a) If the angular resolution of the HST image is 0.1 arcsec, what is the diameter in light-years of the smallest detail that can be discerned in the image? (b) At what distance would a U.S. dime (diameter 1.8 cm) have an angular size of 0.1 arcsec? Give your answer in kilometers.

38. At its closest to Earth, Pluto is 28.6 AU from Earth. Can the Hubble Space Telescope distinguish any features on Pluto? Justify your answer using calculations.

39. The Institute of Space and Astronautical Science in Japan proposes to place a radio telescope into an even higher orbit than the HALCA telescope. Using this telescope in concert with ground-based radio-telescopes, baselines as long as

Advanced Questions

Questions preceded by an asterisk (*) involve topics discussed in the Boxes.

Problem-solving tips and tools

You may find it useful to review the small-angle formula discussed in Box 1-1. The area of a circle is proportional to the square of its diameter. Data on the planets can be found in the appendices at the end of this book. Section 5-2 discusses the relationship between frequency and wavelength. Box 6-1 gives examples of how to calculate magnifying power and light-gathering power.

25,000 km may be obtainable. Astronomers want to use this combination to study radio emission at a frequency of 43 GHz from the molecule silicon monoxide, which is found in the interstellar clouds from which stars form. (1 GHz = 1 gigahertz = 10^9 Hz.) (a) What is the wavelength of this emission? (b) Taking the baseline to be the effective diameter of this radio-telescope array, what angular resolution can be achieved?

40. The mission of the Submillimeter Wave Astronomy Satellite (SWAS), launched in 1998, was to investigate interstellar clouds within which stars form. One of the frequencies at which it observed these clouds is 557 GHz (1 GHz = 1 gigahertz = 10^9 Hz), characteristic of the emission from interstellar water molecules. (a) What is the wavelength (in meters) of this emission? In what part of the electromagnetic spectrum is this? (b) Why was it necessary to use a satellite for these observations? (c) SWAS had an angular resolution of 4 arcminutes. What was the diameter of its primary mirror?

41. To search for ionized oxygen gas surrounding our Milky Way Galaxy, astronomers aimed the ultraviolet telescope of the FUSE spacecraft at a distant galaxy far beyond the Milky Way. They then looked for an ultraviolet spectral line of ionized oxygen in that galaxy's spectrum. Were they looking for an emission line or an absorption line? Explain.

42. A sufficiently thick interstellar cloud of cool gas can absorb low-energy X rays but is transparent to high-energy X rays and gamma rays. Explain why both part *b* and part *d* of Figure 6-32 reveal the presence of cool gas in the Milky Way. Could you infer the presence of this gas from the visible-light image in Figure 6-32*a*? Explain.

Discussion Questions

43. If you were in charge of selecting a site for a new observatory, what factors would you consider important?

44. Discuss the advantages and disadvantages of using a small telescope in Earth's orbit versus a large telescope on a mountaintop.

Web/e-Book Questions

45. Several telescope manufacturers build telescopes with a design called a "Schmidt-Cassegrain." These use a correcting lens in an arrangement like that shown in Figure 6-13*c*. Consult advertisements on the World Wide Web to see the appearance of these telescopes and find out their cost. Why do you suppose they are very popular among amateur astronomers?

46. The Large Zenith Telescope (LZT) in British Columbia, Canada, uses a 5.0-m *liquid* mirror made of mercury. Use the World Wide Web to investigate this technology. How can a liquid metal be formed into the necessary shape for a telescope mirror? What are the advantages of a liquid mirror? What are the disadvantages?

47. Three of the telescopes shown in Figure 6-16—the James Clerk Maxwell Telescope (JCMT), the Caltech Submillimeter Observatory (CSO), and the Submillimeter Array (SMA)—are designed to detect radiation with wavelengths close to 1 mm. Search for current information about JCMT, CSO, and SMA on the World Wide Web. What kinds of celestial objects emit radiation at these wavelengths? What can astronomers see using JCMT, CSO, and SMA that cannot be observed at other wavelengths? Why is it important that they be at high altitude? How large are the primary mirrors used in JCMT, CSO, and SMA? What are the differences among the three telescopes? Which can be used in the daytime? What recent discoveries have been made using JCMT, CSO, or SMA?

48. In 2003 an ultraviolet telescope called GALEX (Galaxy Evolution Explorer) was placed into orbit. Use the World Wide Web to learn about GALEX and its mission. What aspects of galaxies was GALEX designed to investigate? Why is it important to make these observations using ultraviolet wavelengths?

49. At the time of this writing, NASA's plans for the end of the Hubble Space Telescope's mission were uncertain. Consult the Space Telescope Science Institute Web site to learn about plans for HST's final years of operation. Are future space shuttle missions planned to service HST? If so, what changes will be made to HST on such missions? What will become of HST at the end of its mission lifetime?

50. NASA and the European Space Agency (ESA) have plans to launch a number of advanced space telescopes. These include Herschel, the Wide-Field Infrared Survey Explorer (WISE), the Gamma-ray Large Area Space Telescope (GLAST), the Space Interferometry Mission (SIM), and the X-ray Evolving Universe Spectroscopy (XEUS) mission. Search the World Wide Web for information about at least two of these. What are the scientific goals of these projects? What is unique about each telescope? What advantages would they have over existing ground-based or orbiting telescopes? What kind of orbit will each telescope be in? When will each be launched and placed in operation?

51. **Telescope Magnification.** Access the Active Integrated Media Module "Telescope Magnification" in Chapter 6 of the *Universe* Web site or eBook. A common telescope found in department stores is a 3-inch (76-mm) diameter refractor that boasts a magnification of 300 times. Use the magnification calculator to determine the magnifications that are achieved by using each of the following commonly found eyepieces on that telescope: Eyepiece A with focal length 40 mm; Eyepiece B with focal length 25 mm; Eyepiece C with focal length 12 mm; and Eyepiece D with focal length 2.5 mm.

Activities

Observing Projects

52. Obtain a telescope during the daytime along with several eyepieces of various focal lengths. If you can determine the telescope's focal length, calculate the magnifying powers of the eyepieces. Focus the telescope on some familiar object, such as a distant lamppost or tree. **DO NOT FOCUS ON THE SUN! Looking directly at the Sun can cause blindness.** Describe the image you see through the telescope. Is it upside down? How does the image move as you slowly and gently

shift the telescope left and right or up and down? Examine the eyepieces, noting their focal lengths. By changing the eyepieces, examine the distant object under different magnifications. How do the field of view and the quality of the image change as you go from low power to high power?

53. On a clear night, view the Moon, a planet, and a star through a telescope using eyepieces of various focal lengths and known magnifying powers. (To determine the locations in the sky of the Moon and planets, you may want to use the *Starry Night Enthusiast*™ program on the CD-ROM that comes with certain printed copies of this book. You may also want to consult such magazines as *Sky & Telescope* and *Astronomy* or their Web sites.) In what way does the image seem to degrade as you view with increasingly higher magnification? Do you see any chromatic aberration? If so, with which object and which eyepiece is it most noticeable?

54. Many towns and cities have amateur astronomy clubs. If you are so inclined, attend a "star party" hosted by your local club. People who bring their telescopes to such gatherings are delighted to show you their instruments and take you on a telescopic tour of the heavens. Such an experience can lead to a very enjoyable, lifelong hobby.

55. The field of view of a typical small telescope for amateur astronomers is about 30 arcminutes, or 30′. (Many research telescopes have much smaller fields of view. For instance, the widest field of view available to the Hubble Space Telescope is only 3.4 arcminutes.) Use the *Starry Night Enthusiast*™ program to simulate the view that such a telescope provides of various celestial objects. First display the entire celestial sphere (select **Guides > Atlas** in the **Favourites** menu). Open the **Find** pane (select **Find . . .** under the **Edit** menu or click the **Find** tab on the left side of the view window). Click on the magnifying glass icon at the left-hand side of the edit box at the top of the Find pane and select the **Orbiting Objects** item from the drop-down menu that appears. This will bring up a list of Solar System objects in the Find pane. For each of the following objects, (i) the Moon, (ii) Jupiter, and (iii) Saturn, double-click on the name of the object in the Find pane to center it in the view. Use the zoom controls (+ and − buttons) at the right-hand end of the toolbar to adjust the field of view to approximately 30 arcminutes. (You can display an indication of this specific field of view on the sky by opening the **FOV (Field of View)** pane on the left side of the window and click-

ing on **Other > 30 Arcminutes.**) Describe how much detail you can see on each object. Open the Find pane again. Click on the magnifying glass icon on the left-hand side of the edit box at the top of the Find pane and select Messier Objects from the drop-down menu that appears. From the list of Messier Objects that populates the Find pane, double-click the entry for "M31 (Andromeda Galaxy)" to center this object in the view. Again, use the zoom controls to adjust the field of view to approximately 30 arcminutes. Describe how much detail you can see for this object.

56. Use the *Starry Night Enthusiast*™ program to explore the concept of angular resolution. Click the **Find** tab to the left of the main view window to open the Find pane. Click on the magnifying glass icon at the left-hand side of the edit box at the top of the Find pane and select the **Orbiting Objects** item from the drop-down menu that appears. This will bring up a list of Solar System objects in the Find pane. Double-click the entry labeled **The Moon.** You can zoom in and zoom out using the **Zoom** buttons at the right side of the toolbar. You can also rotate the Moon by putting the mouse cursor over the image, holding down the mouse button and the **Shift** key on the keyboard, and moving the mouse. (On a two-button mouse, hold down the left mouse button.) (a) What is the size of the smallest detail that you can see? (You will have to make measurements on the screen using a ruler and compare it to the diameter of the Moon, which is 3476 km.) (b) The angular resolution of the Hubble Space Telescope (HST) is 0.1 arcsec. How far away from the Moon could HST be and still be able to resolve details as small as you determined in part (a)? Give your answer in kilometers and in astronomical units (AU).

Collaborative Exercises

57. Stand up and have everyone in your group join hands, making as large a circle as possible. If a telescope mirror were built as big as your circle, what would be its diameter? What would be your telescope's diffraction-limited angular resolution for blue light? Would atmospheric turbulence have a noticeable effect on the angular resolution?

58. Are there enough students in your class to stand and join hands and make two large circles that recreate the sizes of the two Keck telescopes? Explain how you determined your answer.

7

Comparative Planetology I: Our Solar System

WEB LINK 7.1 March 10, 2006: The *Mars Reconnaissance Orbiter* spacecraft arrives at Mars (artist's impression). (JPL/NASA)

Fifty years ago, astronomers knew precious little about the worlds that orbit the Sun. Even the best telescopes provided images of the planets that were frustratingly hazy and indistinct. Of asteroids, comets, and the satellites of the planets, we knew even less.

Today, our knowledge of the solar system has grown exponentially, due almost entirely to robotic spacecraft. (The illustration shows an artist's impression of one such robotic explorer, the *Mars Reconnaissance Orbiter* spacecraft, as it approached Mars in 2006.) Spacecraft have been sent to fly past all the planets at close range, revealing details unimagined by astronomers of an earlier generation. We have landed spacecraft on the Moon, Venus, and Mars and dropped a probe into the immense atmosphere of Jupiter. This is truly the golden age of solar system exploration.

In this chapter we paint in broad outline our present understanding of the solar system. We will see that the planets come in a variety of sizes and chemical compositions. There is also rich variety among the moons of the planets and among smaller bodies we call asteroids, comets, and trans-Neptunian objects. We will investigate the nature of craters on the Moon and other worlds of the solar system. And by exploring the magnetic fields of planets, we will be able to peer inside the Earth and other worlds and learn about their interior compositions.

An important reason to study the solar system is to search for our own origins. In Chapter 8 we will see how astronomers have used evidence from the present-day solar system to understand how the Sun, the Earth, and the other planets formed some four and a half billion years ago, and how they have evolved since then. But for now, we invite you to join us on a guided tour of the worlds that orbit our Sun.

7-1 The solar system has two broad categories of planets: Earthlike and Jupiterlike

Each of the planets that orbit the Sun is unique. Only Earth has liquid water and an atmosphere breathable by humans; only Venus has a perpetual cloud layer made of sulfuric acid droplets; and only Jupiter has immense storm systems that persist for centuries. But there are also striking similarities among planets.

Learning Goals

By reading the sections of this chapter, you will learn

7-1 The important differences between the two broad categories of planets: terrestrial and Jovian

7-2 The similarities and differences among the large planetary satellites, including Earth's Moon

7-3 How the spectrum of sunlight reflected from a planet reveals the composition of its atmosphere and surface

7-4 Why some planets have atmospheres and others do not

7-5 The categories of the many small bodies that also orbit the Sun

7-6 How craters on a planet or satellite reveal the age of its surface and the nature of its interior

7-7 Why a planet's magnetic field indicates a fluid interior in motion

7-8 How the diversity of the solar system is a result of its origin and evolution

Volcanoes are found not only on Earth but also on Venus and Mars; rings encircle Jupiter, Saturn, Uranus, and Neptune; and impact craters dot the surfaces of Mercury, Venus, Earth, and Mars, showing that all of these planets have been bombarded by interplanetary debris.

We can understand the most important similarities and differences among the planets by comparing their orbits, masses, and diameters

How can we make sense of the many similarities and differences among the planets? An important step is to organize our knowledge of the planets in a systematic way. There are two useful ways to do this. First, we can contrast the orbits of different planets around the Sun; and second, we can compare the planets' physical properties such as diameter, mass, average density, and chemical composition.

Comparing the Planets: Orbits

The planets fall naturally into two classes according to the sizes of their orbits. As Figure 7-1 shows, the orbits of the four inner planets (Mercury, Venus, Earth, and Mars) are crowded in close to the Sun. In contrast, the orbits of the next four planets (Jupiter, Saturn, Uranus, and Neptune) are widely spaced at great distances from the Sun. Table 7-1 lists the orbital characteristics of these eight planets.

CAUTION! While Figure 7-1 shows the orbits of the planets, it does not show the planets themselves. The reason is simple: If Jupiter, the largest of the planets, were to be drawn to the same scale as the rest of this figure, it would be a dot just 0.0002 cm across—about $1/300$ of the width of a human hair and far too small to be seen without a microscope. The planets themselves are *very* small compared to the distances between them. Indeed, while an airliner traveling at 1000 km/h (620 mi/h) can fly around the Earth in less than two days, at this speed it would take 17 *years* to fly from the Earth to the Sun. The solar system is a very large and very empty place!

Most of the planets have orbits that are nearly circular. As we learned in Section 4-4, Kepler discovered in the seventeenth

Figure 7-1

The Solar System to Scale This scale drawing shows the orbits of the planets around the Sun. The four inner planets are crowded in close to the Sun, while the four outer planets orbit the Sun at much greater distances. On the scale of this drawing, the planets themselves would be much smaller than the diameter of a human hair and too small to see.

Table 7-1 Characteristics of the Planets

	The Inner (Terrestrial) Planets			
	Mercury	**Venus**	**Earth**	**Mars**
Average distance from Sun (10^6 km)	57.9	108.2	149.6	227.9
Average distance from Sun (AU)	0.387	0.723	1.000	1.524
Orbital period (years)	0.241	0.615	1.000	1.88
Orbital eccentricity	0.206	0.007	0.017	0.093
Inclination of orbit to the ecliptic	7.00°	3.39°	0.00°	1.85°
Equatorial diameter (km)	4880	12,104	12,756	6794
Equatorial diameter (Earth = 1)	0.383	0.949	1.000	0.533
Mass (kg)	3.302×10^{23}	4.868×10^{24}	5.974×10^{24}	6.418×10^{23}
Mass (Earth = 1)	0.0553	0.8150	1.0000	0.1074
Average density (kg/m³)	5430	5243	5515	3934

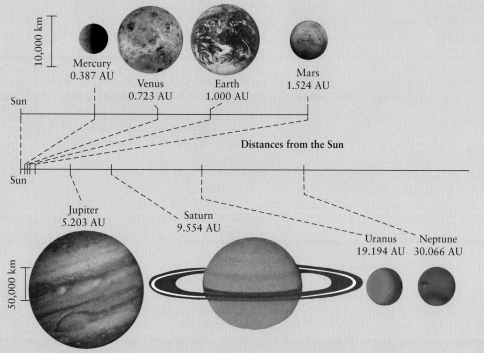

R I V U X G
(NASA)

	The Outer (Jovian) Planets			
	Jupiter	**Saturn**	**Uranus**	**Neptune**
Average distance from Sun (10^6 km)	778.3	1429	2871	4498
Average distance from Sun (AU)	5.203	9.554	19.194	30.066
Orbital period (years)	11.86	29.46	84.10	164.86
Orbital eccentricity	0.048	0.053	0.043	0.010
Inclination of orbit to the ecliptic	1.30°	2.48°	0.77°	1.77°
Equatorial diameter (km)	142,984	120,536	51,118	49,528
Equatorial diameter (Earth = 1)	11.209	9.449	4.007	3.883
Mass (kg)	1.899×10^{27}	5.685×10^{26}	8.682×10^{25}	1.024×10^{26}
Mass (Earth = 1)	317.8	95.16	14.53	17.15
Average density (kg/m³)	1326	687	1318	1638

century that these orbits are actually ellipses. Astronomers denote the elongation of an ellipse by its *eccentricity* (see Figure 4-10*b*). The eccentricity of a circle is zero, and indeed most of the eight planets (with the notable exception of Mercury) have orbital eccentricities that are very close to zero.

If you could observe the solar system from a point several astronomical units (AU) above the Earth's north pole, you would see that all the planets orbit the Sun in the same counterclockwise direction. Furthermore, the orbits of the eight planets all lie in nearly the same plane. In other words, these orbits are inclined at only slight angles to the plane of the ecliptic, which is the plane of the Earth's orbit around the Sun (see Section 2-5). What is more, the plane of the Sun's equator is very closely aligned with the orbital planes of the planets. As we will see in Chapter 8, these near-alignments are not a coincidence. They provide important clues about the origin of the solar system.

Not included in Figure 7-1 or Table 7-1 is Pluto, which is in an even larger orbit than Neptune. Until the late 1990s, Pluto was generally regarded as the ninth planet. But in light of recent discoveries many astronomers now consider Pluto to be simply one member of a large collection of *trans-Neptunian objects* that orbit far from the Sun. These all orbit the Sun in the same direction as the planets, though many of them have orbits that are steeply inclined to the plane of the ecliptic and have high eccentricities (that is, the orbits are quite elongated and noncircular). We will discuss trans-Neptunian objects, along with other small bodies that orbit the Sun, in Section 7-5.

Comparing the Planets: Physical Properties

When we compare the physical properties of the planets, we again find that they fall naturally into two classes—four small inner planets and four large outer ones. The four small inner planets are called **terrestrial planets** because they resemble the Earth (in Latin, *terra*). They all have hard, rocky surfaces with mountains, craters, valleys, and volcanoes. You could stand on the surface of any one of them, although you would need a protective spacesuit on Mercury, Venus, or Mars. The four large outer planets are called **Jovian planets** because they resemble Jupiter. (Jove was an-

other name for the Roman god Jupiter.) An attempt to land a spacecraft on the surface of any of the Jovian planets would be futile, because the materials of which these planets are made are mostly gaseous or liquid. The visible "surface" features of a Jovian planet are actually cloud formations in the planet's atmosphere. The photographs in **Figure 7-2** show the distinctive appearances of the two classes of planets.

The most apparent difference between the terrestrial and Jovian planets is their *diameters*. You can compute the diameter of a planet if you know its angular diameter as seen from Earth and its distance from Earth. For example, on March 16, 2007, Venus was 1.97×10^8 km from Earth and had an angular diameter of 12.7 arcsec. Using the small-angle formula from Box 1-1, we can calculate the diameter of Venus to be 12,100 km (7520 mi). Similar calculations demonstrate that the Earth, with its diameter of about 12,756 km (7926 mi), is the largest of the four inner, terrestrial planets. In sharp contrast, the four outer, Jovian planets are much larger than the terrestrial planets. First place goes to Jupiter, whose equatorial diameter is more than 11 times that of the Earth. On the other end of the scale, Mercury's diameter is less than two-fifths that of the Earth. Figure 7-2 shows the Sun and the planets drawn to the same scale. The diameters of the planets are given in Table 7-1.

The *masses* of the terrestrial and Jovian planets are also dramatically different. If a planet has a satellite, you can calculate the planet's mass from the satellite's period and semimajor axis by using Newton's form of Kepler's third law (see Section 4-7 and Box 4-4). Astronomers have also measured the mass of each planet by sending a spacecraft to pass near the planet. The planet's gravitational pull (which is proportional to its mass) deflects the spacecraft's path, and the amount of deflection tells us the planet's mass. Using these techniques, astronomers have found that the four Jovian planets have masses that are tens or hundreds of times greater than the mass of any of the terrestrial planets. Again, first place goes to Jupiter, whose mass is 318 times greater than the Earth's.

Once we know the diameter and mass of a planet, we can learn something about what that planet is made of. The trick is to calculate the planet's **average density,** or mass divided by vol-

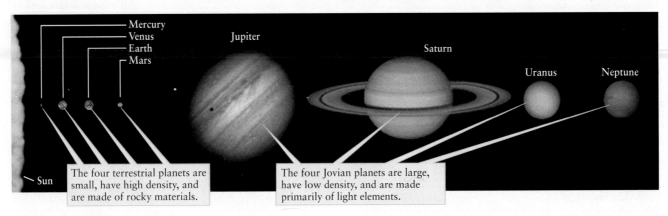

Figure 7-2   R I ⟨V⟩ U X G

The Planets to Scale This figure shows the planets from Mercury to Neptune to the same scale. The four terrestrial planets have orbits nearest the Sun, and the Jovian planets are the next four planets from the Sun. (Calvin J. Hamilton and NASA/JPL)

BOX 7-1

Average Density

Average density—the mass of an object divided by that object's volume—is a useful quantity for describing the differences between planets in our solar system. This same quantity has many applications here on Earth.

A rock tossed into a lake sinks to the bottom, while an air bubble produced at the bottom of a lake (for example, by the air tanks of a scuba diver) rises to the top. These are examples of a general principle: An object sinks in a fluid if its average density is greater than that of the fluid, but rises if its average density is less than that of the fluid. The average density of water is 1000 kg/m³, which is why a typical rock (with an average density of about 3000 kg/m³) sinks, while an air bubble (average density of about 1.2 kg/m³) rises.

At many summer barbecues, cans of soft drinks are kept cold by putting them in a container full of ice. When the ice melts, the cans of diet soda always rise to the top, while the cans of regular soda sink to the bottom. Why is this? The average density of a can of diet soda—which includes water, flavoring, artificial sweetener, and the trapped gas that makes the drink fizzy—is slightly less than the density of water, and so

the can floats. A can of regular soda contains sugar instead of artificial sweetener, and the sugar is a bit heavier than the sweetener. The extra weight is just enough to make the average density of a can of regular soda slightly more than that of water, making the can sink. (You can test these statements for yourself by putting unopened cans of diet soda and regular soda in a sink or bathtub full of water.)

The concept of average density provides geologists with important clues about the early history of the Earth. The average density of surface rocks on Earth, about 3000 kg/m³, is less than the Earth's average density of 5515 kg/m³. The simplest explanation is that in the ancient past, the Earth was completely molten throughout its volume, so that low-density materials rose to the surface and high-density materials sank deep into the Earth's interior in a process called *chemical differentiation*. This series of events also suggests that the Earth's core must be made of relatively dense materials, such as iron and nickel. A tremendous amount of other geological evidence has convinced scientists that this picture is correct.

---

ume, measured in kilograms per cubic meter (kg/m³). The average density of any substance depends in part on that substance's composition. For example, air near sea level on Earth has an average density of 1.2 kg/m³, water's average density is 1000 kg/m³, and a piece of concrete has an average density of 2000 kg/m³. Box 7-1 describes some applications of the idea of average density to everyday phenomena on Earth.

The four inner, terrestrial planets have very high average densities (see Table 7-1); the average density of the Earth, for example, is 5515 kg/m³. By contrast, a typical rock found on the Earth's surface has a lower average density, about 3000 kg/m³. Thus, the Earth must contain a large amount of material that is denser than rock. This information provides our first clue that terrestrial planets have dense iron cores.

In sharp contrast, the outer, Jovian planets have quite low densities. Saturn has an average density less than that of water. This information strongly suggests that the giant outer planets are composed primarily of light elements such as hydrogen and helium. All four Jovian planets probably have large cores of mixed rock and highly compressed water buried beneath low-density outer layers tens of thousands of kilometers thick.

We can conclude that the following general rule applies to the planets:

The terrestrial planets are made of rocky materials and have dense iron cores, which gives these planets high average densities. The Jovian planets are composed primarily of light elements such as hydrogen and helium, which gives these planets low average densities.

7-2 Seven large satellites are almost as big as the terrestrial planets

All the planets except Mercury and Venus have satellites. More than 140 satellites are known: Earth has one (the Moon), Mars has two, Jupiter has at least 62, Saturn at least 43, Uranus at least 24, and Neptune at least 13. Dozens of other small satellites probably remain to be discovered as our telescope technology continues to improve. Like the terrestrial planets, all of the satellites of the planets have solid surfaces.

You can see that there is a striking difference between the terrestrial planets, with few or no satellites, and the Jovian planets, each of which has so many moons that it resembles a solar system in miniature. In Chapter 8 we will explore this and other evidence that the Jovian planets formed in a manner similar to the solar system as a whole, but on a smaller scale.

The various moons of the planets are not simply copies of the Earth's Moon

Of the known satellites, seven are roughly as big as the planet Mercury. Table 7-2 lists these satellites and shows them to the same scale. Note that the Earth's Moon and Jupiter's satellites Io and Europa have relatively high average densities, indicating that these satellites are made primarily of rocky materials. By contrast, the average densities of Ganymede, Callisto, Titan, and Triton are all relatively low. Planetary scientists conclude that the interiors of these four satellites also contain substantial amounts of water ice, which is less dense than rock. (In Section 7-4 we

Table 7-2 The Seven Giant Satellites

	Moon	Io	Europa	Ganymede	Callisto	Titan	Triton
Parent planet	Earth	Jupiter	Jupiter	Jupiter	Jupiter	Saturn	Neptune
Diameter (km)	3476	3642	3130	5268	4806	5150	2706
Mass (kg)	7.35×10^{22}	8.93×10^{22}	4.80×10^{22}	1.48×10^{23}	1.08×10^{23}	1.34×10^{23}	2.15×10^{22}
Average density (kg/m³)	3340	3530	2970	1940	1850	1880	2050
Substantial atmosphere?	No	No	No	No	No	Yes	No

Moon   Io   Europa   Ganymede   Callisto   Titan   Triton

R I V U X G

(NASA/JPL/Space Science Institute)

will learn about types of frozen "ice" made of substances other than water.)

CAUTION! Water ice may seem like a poor material for building a satellite, since the ice you find in your freezer can easily be cracked or crushed. But under high pressure, such as is found in the interior of a large satellite, water ice becomes as rigid as rock. (It also becomes denser than the ice found in ice cubes, although not as dense as rock.) Note that water ice is an important constituent only for satellites in the outer solar system, where the Sun is far away and temperatures are very low. For example, the surface temperature of Titan is a frigid 95 K ($-178°C = -288°F$). In Section 7-4 we will learn more about the importance of temperature in determining the composition of a planet or satellite.

The satellites listed in Table 7-2 are actually unusually large. Most of the known satellites have diameters less than 2000 km, and many are just a few kilometers across.

Interplanetary spacecraft have made many surprising and fascinating discoveries about the satellites of the solar system. We now know that Jupiter's satellite Io is the most geologically active world in the solar system, with numerous geyserlike volcanoes that continually belch forth sulfur-rich compounds. The fractured surface of Europa, another of Jupiter's large satellites, suggests that a worldwide ocean of liquid water may lie beneath its icy surface. Saturn's largest satellite, Titan, has an atmosphere nearly twice as dense as the Earth's atmosphere, with a perpetual haze layer that gives it a featureless appearance (see the photographs in Table 7-2).

7-3 Spectroscopy reveals the chemical composition of the planets

As we have seen, the average densities of the planets and satellites give us a crude measure of their **chemical compositions**—that is, what substances they are made of. For example, the low average density of the Moon (3340 kg/m³) compared with the Earth (5515 kg/m³) tells us that the Moon contains relatively little iron or other dense metals. But to truly understand the nature of the planets and satellites, we need to know their chemical compositions in much greater detail than we can learn from average density alone.

> The light we receive from a planet or satellite is reflected sunlight—but with revealing differences in its spectrum

The most accurate way to determine chemical composition is by directly analyzing samples taken from a planet's atmosphere and soil. Unfortunately, of all the planets and satellites, we have such direct information only for the Earth and the three worlds on which spacecraft have landed—Venus, the Moon, and Mars. In all other cases, astronomers must analyze sunlight reflected from the distant planets and their satellites. To do that, astronomers bring to bear one of their most powerful tools, **spectroscopy,** the systematic study of spectra and spectral lines. (We discussed spectroscopy in Sections 5-6 and 6-5.)

Determining Atmospheric Composition

Spectroscopy is a sensitive probe of the composition of a planet's *atmosphere.* If a planet has an atmosphere, then sunlight reflected from that planet must have passed through its atmosphere. During this passage, some of the wavelengths of sunlight will have been absorbed. Hence, the spectrum of this reflected sunlight will have dark absorption lines. Astronomers look at the particular wavelengths absorbed and the amount of light absorbed at those wavelengths. Both of these depend on the kinds of chemicals present in the planet's atmosphere and the abundance of those chemicals.

For example, astronomers have used spectroscopy to analyze the atmosphere of Saturn's largest satellite, Titan (see **Figure 7-3a** and Table 7-2). The graph in Figure 7-3b shows the spectrum of visible sunlight reflected from Titan. (We first saw this method of displaying spectra in Figure 6-21.) The dips in this curve of intensity versus wavelength represent absorption lines. However, not all of these absorption lines are produced in the atmosphere of Titan (Figure 7-3c). Before reaching Titan, light from the Sun's glowing surface must pass through the Sun's hydrogen-rich atmosphere. This produces the hydrogen absorption line in Figure 7-3b at a wavelength of 656 nm. After being reflected from Titan, the light must pass through the Earth's atmosphere before reaching the telescope; this is where the oxygen absorption line in

Figure 7-3b is produced. Only the two dips near 620 nm and 730 nm are caused by gases in Titan's atmosphere.

These two absorption lines are caused not by individual atoms in the atmosphere of Titan but by atoms combined to form molecules. (We introduced the idea of molecules in Section 5-6.) Molecules, like atoms, also produce unique patterns of lines in the spectra of astronomical objects. The absorption lines in Figure 7-3b indicate the presence in Titan's atmosphere of molecules of methane (CH_4, a molecule made of one carbon atom and four hydrogen atoms). This shows that Titan is a curious place indeed, because on Earth, methane is a rather rare substance that is the primary ingredient in natural gas! When we examine other planets and satellites with atmospheres, we find that all of their spectra have absorption lines of molecules of various types.

In addition to visible-light measurements such as those shown in Figure 7-3b, it is very useful to study the *infrared* and *ultraviolet* spectra of planetary atmospheres. Many molecules have much stronger spectral lines in these nonvisible wavelength bands than in the visible. As an example, the ultraviolet spectrum of Titan shows that nitrogen molecules (N_2) are the dominant constituent of Titan's atmosphere. Furthermore, Titan's infrared spectrum

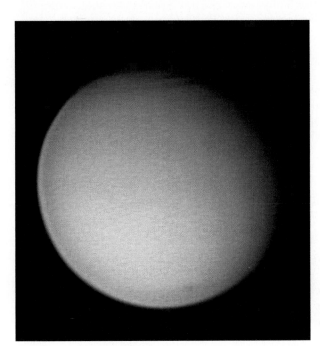

(a) Saturn's satellite Titan

Figure 7-3   R I $\boxed{\text{V}}$ U X G

Analyzing a Satellite's Atmosphere through its Spectrum (a) Titan is the only satellite in the solar system with a substantial atmosphere. **(b)** The dips in the spectrum of sunlight reflected from Titan are due to absorption by hydrogen atoms (H), oxygen molecules (O_2), and methane molecules (CH_4). Of these, only methane is actually present in Titan's atmosphere. **(c)** This illustration shows the path of the light that reaches us from Titan. To interpret the spectrum of this light as shown in (b), astronomers must account for the absorption that takes place in the atmospheres of the Sun and Earth. (a: NASA/JPL/Space Science Institute)

(b) The spectrum of sunlight reflected from Titan

(c) Interpreting Titan's spectrum

(a) Jupiter's moon Europa   R I V U X G

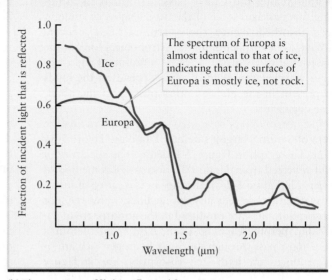

The spectrum of Europa is almost identical to that of ice, indicating that the surface of Europa is mostly ice, not rock.

(b) The spectum of light reflected from Europa

Figure 7-4

Analyzing a Satellite's Surface from Its Spectrum

(a) Unlike Titan (Figure 7-3a), Jupiter's satellite Europa has no atmosphere. **(b)** Infrared light from the Sun that is reflected from the surface of Europa, has almost exactly the same spectrum as sunlight reflected from ordinary water ice.

includes spectral lines of a variety of molecules that contain carbon and hydrogen, indicating that Titan's atmosphere has a very complex chemistry. None of these molecules could have been detected by visible light alone. Since the Earth's atmosphere is largely opaque to infrared and ultraviolet wavelengths (see Section 6-7), telescopes in space are important tools for these spectroscopic studies of the solar system.

It is a testament to the power of spectroscopy that when the robotic spacecraft *Huygens* landed on Titan in 2005, its onboard instruments confirmed the presence of methane and nitrogen in Titan's atmosphere—just as had been predicted years before by spectroscopic observations.

Determining Surface Composition

Spectroscopy can also provide useful information about the *solid surfaces* of planets and satellites without atmospheres. When light shines on a solid surface, some wavelengths are absorbed while others are reflected. (For example, a plant leaf absorbs red and violet light but reflects green light—which is why leaves look green.) Unlike a gas, a solid illuminated by sunlight does not produce sharp, definite spectral lines. Instead, only broad absorption features appear in the spectrum. By comparing such a spectrum with the spectra of samples of different substances on Earth, astronomers can infer the chemical composition of the surface of a planet or satellite.

As an example, Figure 7-4*a* shows Jupiter's satellite Europa (see Table 7-2), and Figure 7-4*b* shows the infrared spectrum of light reflected from the surface of Europa. Because this spectrum is so close to that of water ice—that is, frozen water—astronomers conclude that water ice is the dominant constituent of Europa's surface. (We saw in Section 7-2 that water ice cannot be the dom-

inant constituent of Europa's *interior*, because this satellite's density is too high.)

Unfortunately, spectroscopy tells us little about what the material is like just below the surface of a satellite or planet. For this purpose, there is simply no substitute for sending a spacecraft to a planet and examining its surface directly.

7-4 The Jovian planets are made of lighter elements than the terrestrial planets

Spectroscopic observations from Earth and spacecraft show that the outer layers of the Jovian planets are composed primarily of the lightest gases, hydrogen and helium (see Box 5-5). In contrast, chemical analysis of soil samples from Venus, Earth, and Mars demonstrate that the terrestrial planets are made mostly of heavier elements, such as iron, oxygen, silicon, magnesium, nickel, and sulfur. Spacecraft images such as Figure 7-5 and Figure 7-6 only hint at these striking differences in chemical composition, which are summarized in Table 7-3.

Temperature plays a major role in determining whether the materials of which planets are made exist as solids, liquids, or gases. Hydrogen (H_2) and helium (He) are gaseous except at extremely low temperatures and extraordinarily high pressures. By contrast, rock-forming substances such as iron and silicon are solids except

> **Differences in distance from the Sun, and hence in temperature, explain many distinctions between terrestrial and Jovian planets**

Figure 7-5   R I V U X G

A Jovian Planet This Hubble Space Telescope image gives a detailed view of Jupiter's cloudtops. Jupiter is composed mostly of the lightest elements, hydrogen and helium, which are colorless; the colors in the atmosphere are caused by trace amounts of other substances. The giant storm at lower right, called the Great Red Spot, has been raging for more than 300 years. (Space Telescope Science Institute/JPL/NASA)

 Figure 7-6   R I V U X G

A Terrestrial Planet Mars is composed mostly of heavy elements such as iron, oxygen, silicon, magnesium, nickel, and sulfur. The planet's red surface can be seen clearly in this Hubble Space Telescope image because the Martian atmosphere is thin and nearly cloudless. Olympus Mons, the extinct volcano to the left of center, is nearly 3 times the height of Mount Everest. (Space Telescope Science Institute/JPL/NASA)

at temperatures well above 1000 K. (You may want to review the discussion of temperature scales in Box 5-1.) Between these two extremes are substances such as water (H_2O), carbon dioxide (CO_2), methane (CH_4), and ammonia (NH_3), which solidify at low temperatures (from below 100 to 300 K) into solids called **ices.** (In astronomy, frozen water is just one kind of "ice.") At somewhat higher temperatures, they can exist as liquids or gases. For example, clouds of ammonia ice crystals are found in the cold upper atmosphere of Jupiter, but within Jupiter's warmer interior, ammonia exists primarily as a liquid.

CAUTION! The Jovian planets are sometimes called "gas giants." It is true that their primary constituents, including hydrogen, helium, ammonia, and methane, are gases under normal conditions on Earth. But in the interiors of these planets, pressures are so high that these substances are *liquids*, not gases. The Jovian planets might be better described as "liquid giants"!

As you might expect, a planet's surface temperature is related to its distance from the Sun. The four inner planets are quite warm. For example, midday temperatures on Mercury may climb to 700 K (= 427°C = 801°F), and during midsummer on Mars, it is sometimes as warm as 290 K (= 17°C = 63°F). The outer planets, which receive much less solar radiation, are cooler. Typ-

ical temperatures range from about 125 K (= –148°C = –234°F) in Jupiter's upper atmosphere to about 55 K (= –218°C = –360°F) at the tops of Neptune's clouds.

How Temperature Affects Atmospheres

The higher surface temperatures of the terrestrial planets help to explain the following observation: The atmospheres of the terrestrial planets contain virtually *no* hydrogen molecules or helium atoms.

Table 7-3   Comparing Terrestrial and Jovian Planets

	Terrestrial Planets	Jovian Planets
Distance from the Sun	Less than 2 AU	More than 5 AU
Size	Small	Large
Composition	Mostly rocky materials containing iron, oxygen, silicon, magnesium, nickel, and sulfur	Mostly light elements such as hydrogen and helium
Density	High	Low

BOX 7-2 · · · · · · · · · Tools of the Astronomer's Trade

Kinetic Energy, Temperature, and Whether Planets Have Atmospheres

A moving object possesses energy as a result of its motion. The faster it moves, the more energy it has. Energy of this type is called **kinetic energy**. If an object of mass m is moving with a speed v, its kinetic energy E_k is given by

Kinetic energy

$$E_k = \frac{1}{2} mv^2$$

E_k = kinetic energy of an object

m = mass of object

v = speed of object

This expression for kinetic energy is valid for all objects, both big and small, from atoms and molecules to planets and stars, as long as their speeds are slow in relation to the speed of light. If the mass is expressed in kilograms and the speed in meters per second, the kinetic energy is expressed in joules (J).

EXAMPLE: An automobile of mass 1000 kg driving at a typical freeway speed of 30 m/s (= 108 km/h = 67 mi/h) has a kinetic energy of

$$E_k = \frac{1}{2} (1000 \times 30^2) = 450{,}000 \text{ J} = 4.5 \times 10^5 \text{ J}$$

Consider a gas, such as the atmosphere of a star or planet. Some of the gas atoms or molecules will be moving slowly, with little kinetic energy, while others will be moving faster and have more kinetic energy. The temperature of the gas is a direct measure of the *average* amount of kinetic energy per atom or molecule. The hotter the gas, the faster atoms or molecules move, on average, and the greater the average kinetic energy of an atom or molecule.

If the gas temperature is sufficiently high, typically several thousand kelvins, molecules move so fast that when they collide with one another, the energy of the collision can break the molecules apart into their constituent atoms. Thus, the Sun's atmosphere, where the temperature is 5800 K, consists primarily of individual hydrogen atoms rather than hydrogen molecules. By contrast, the hydrogen atoms in the Earth's atmosphere (temperature 290 K) are combined with oxygen atoms into molecules of water vapor (H_2O).

The physics of gases tells us that in a gas of temperature T (in kelvins), the average kinetic energy of an atom or molecule is

Kinetic energy of a gas atom or molecule

$$E_k = \frac{3}{2} kT$$

E_k = average kinetic energy of a gas atom or molecule, in joules

$k = 1.38 \times 10^{-23}$ J/K

T = temperature of gas, in kelvins

The quantity k is called the Boltzmann constant. Note that the higher the gas temperature, the greater the average kinetic energy of an atom or molecule of the gas. This average kinetic energy becomes zero at absolute zero, or $T = 0$, the temperature at which molecular motion is at a minimum.

At a given temperature, all kinds of atoms and molecules will have the same average kinetic energy. But the average *speed* of a given kind of atom or molecule depends on its mass. To see this, note that the average kinetic energy of a gas atom or molecule can be written in two equivalent ways:

$$E_k = \frac{1}{2} mv^2 = \frac{3}{2} kT$$

where v represents the average speed of an atom or molecule in a gas with temperature T. Rearranging this equation, we obtain

Average speed of a gas atom or molecule

$$v = \sqrt{\frac{3kT}{m}}$$

v = average speed of a gas atom or molecule, in m/s

$k = 1.38 \times 10^{-23}$ J/K

T = temperature of gas, in kelvins

m = mass of the atom or molecule, in kilograms

For a given gas temperature, the greater the mass of a given type of gas atom or molecule, the slower its average

Instead, the atmospheres of Venus, Earth, and Mars are composed of heavier molecules such as nitrogen (N_2, 14 times more massive than a hydrogen molecule), oxygen (O_2, 16 times more massive), and carbon dioxide (22 times more massive). To understand the connection between surface temperature and the absence of hydrogen and helium, we need to know a few basic facts about gases.

The temperature of a gas is directly related to the speeds at which the atoms or molecules of the gas move: The higher the gas temperature, the greater the speed of its atoms or molecules. Furthermore, for a given temperature, lightweight atoms and molecules move more rapidly than heavy ones. On the four inner, terrestrial planets, where atmospheric temperatures are high, low-speed. (The value of v given by this equation is actually slightly higher than the average speed of the atoms or molecules in the gas, but it is close enough for our purposes here. If you are studying physics, you may know that v is actually the root-mean-square speed.)

EXAMPLE: What is the average speed of the oxygen molecules that you breathe at a room temperature of 20°C (= 68°F)?

Situation: We are given the temperature of a gas and are asked to find the average speed of the gas molecules.

Tools: We use the relationship $v = \sqrt{3kT/m}$, where T is the gas temperature in kelvins and m is the mass of a single oxygen molecule in kilograms.

Answer: To use the equation to calculate the average speed v, we must express the temperature T in kelvins (K) rather than degrees Celsius (°C). As we learned in Box 5-1, we do this by adding 273 to the Celsius temperature, so 20°C becomes $(20 + 273) = 293$ K. The mass m of an oxygen molecule is not given, but from a reference book you can find that the mass of an oxygen *atom* is 2.66×10^{-26} kg. The mass of an oxygen molecule (O_2) is twice the mass of an oxygen atom, or $2(2.66 \times 10^{-26}$ kg$) = 5.32 \times 10^{-26}$ kg. Thus, the average speed of an oxygen molecule in 20°C air is

$$v = \sqrt{\frac{3(1.38 \times 10^{-23})(293)}{5.32 \times 10^{-26}}} = 478 \text{ m/s} = 0.478 \text{ km/s}$$

Review: This is about 1700 kilometers per hour, or about 1100 miles per hour. Hence, atoms and molecules move rapidly in even a moderate-temperature gas.

In some situations, atoms and molecules in a gas may be moving so fast that they can overcome the attractive force of a planet's gravity and escape into interplanetary space. The minimum speed that an object at a planet's surface must have in order to permanently leave the planet is called the planet's **escape speed.** The escape speed for a planet of mass M and radius R is given by

$$v_{escape} = \sqrt{\frac{2GM}{R}}$$

where $G = 6.67 \times 10^{-11}$ N m^2/kg^2 is the universal constant of gravitation.

The accompanying table gives the escape speed for the Sun, the planets, and the Moon. For example, to get to another planet, a spacecraft must leave Earth with a speed greater than 11.2 km/s (25,100 mi/h).

A good rule of thumb is that a planet can retain a gas if the escape speed is at least 6 times greater than the average speed of the molecules in the gas. (Some molecules are moving slower than average, and others are moving faster, but very few are moving more than 6 times faster than average.) In such a case, very few molecules will be moving fast enough to escape from the planet's gravity.

EXAMPLE: Consider the Earth's atmosphere. We saw that the average speed of oxygen molecules is 0.478 km/s at room temperature. The escape speed from the Earth (11.2 km/s) is much more than 6 times the average speed of the oxygen molecules, so the Earth has no trouble keeping oxygen in its atmosphere.

A similar calculation for hydrogen molecules (H_2) gives a different result, however. At 293 K, the average speed of a hydrogen molecule is 1.9 km/s. Six times this speed is 11.4 km/s, which is slightly higher than the escape speed from the Earth. Thus, the Earth does not retain hydrogen in its atmosphere. Any hydrogen released into the air slowly leaks away into space. On Jupiter, by contrast, the escape speed is so high that even the lightest gases such as hydrogen are retained in its atmosphere. But on Mercury the escape speed is low and the temperature high (so that gas molecules move faster), so Mercury cannot retain any significant atmosphere at all.

Object	Escape speed (km/s)
Sun	618
Mercury	4.3
Venus	10.4
Earth	11.2
Moon	2.4
Mars	5.0
Jupiter	59.5
Saturn	35.5
Uranus	21.3
Neptune	23.5

mass hydrogen molecules and helium atoms move so swiftly that they can escape from the relatively weak gravity of these planets. Hence, the atmospheres that surround the terrestrial planets are composed primarily of more massive, slower-moving molecules such as CO_2, N_2, O_2, and water vapor (H_2O). On the four Jovian planets, low temperatures and relatively strong gravity prevent even lightweight hydrogen and helium gases from escaping into space, and so their atmospheres are much more extensive. The combined mass of Jupiter's atmosphere, for example, is about a million (10^6) times greater than that of the Earth's atmosphere. This is comparable to the mass of the entire Earth! Box 7-2 describes more about the ability of a planet's gravity to retain gases.

7-5 Small chunks of rock and ice also orbit the Sun

In addition to the eight planets, many smaller objects orbit the Sun. These fall into three broad categories: asteroids, which are rocky objects found in the inner solar system; trans-Neptunian objects, which are found beyond Neptune in the outer solar system and contain both | **The smaller bodies of the solar system contain important clues about its origin and evolution**

rock and ice; and comets, which are mixtures of rock and ice that originate in the outer solar system but can venture close to the Sun.

Asteroids

Within the orbit of Jupiter are hundreds of thousands of rocky objects called **asteroids.** There is no sharp dividing line between planets and asteroids, which is why asteroids are also called **minor planets.** The largest asteroid, Ceres, has a diameter of about 900 km. The next largest, Pallas and Vesta, are each about 500 km in diameter. Still smaller ones, like the asteroid shown in close-up in **Figure 7-7**, are increasingly numerous. Hundreds of thousands of kilometer-sized asteroids are known, and there are probably hundreds of thousands more that are boulder-sized or smaller. All of these objects orbit the Sun in the same direction as the planets.

Most (although not all) asteroids orbit the Sun at distances of 2 to 3.5 AU. This region of the solar system between the orbits of Mars and Jupiter is called the **asteroid belt.**

CAUTION! One common misconception about asteroids is that they are the remnants of an ancient planet that somehow broke apart or exploded, like the fictional planet Krypton in the comic book adventures of Superman. In fact, the combined mass of the asteroids is less than that of the Moon, and they were probably never part of any planet-sized body. The early solar system is thought to have been filled with asteroidlike objects, most of which were incorporated into the planets. The "leftover" objects that missed out on this process make up our present-day population of asteroids.

Trans-Neptunian Objects

While asteroids are the most important small bodies in the inner solar system, the outer solar system is the realm of the **trans-Neptunian objects.** As the name suggests, these are small bodies whose orbits lie beyond the orbit of Neptune. The first of these to be discovered (1930) was Pluto, with a diameter of only 2274 km. This is larger than any asteroid, but smaller than any planet or any of the satellites listed in Table 7-2. The orbit of Pluto has a greater semimajor axis (39.54 AU), is more steeply inclined to the ecliptic (17.15°), and has a greater eccentricity (0.250) than that of any of the planets (**Figure 7-8**). In fact, Pluto's noncircular orbit sometimes takes it nearer the Sun than Neptune. (Happily, the orbits of Neptune and Pluto are such that they will never collide.) Pluto's density is only 2000 kg/m³, about the same as Neptune's moon Triton, shown in Table 7-2. Hence, its composition is thought to be a mixture of about 70% rock and 30% ice.

Figure 7-7 R I V U X G

An Asteroid The asteroid shown in this image, 433 Eros, is only 33 km (21 mi) long and 13 km (8 mi) wide—about the same size as the island of Manhattan. Because Eros is so small, its gravity is too weak to have pulled it into a spherical shape. This image was taken in March 2000 by *NEAR Shoemaker,* the first spacecraft to orbit around and land on an asteroid. (*NEAR* Project, Johns Hopkins University, Applied Physics Laboratory, and NASA)

Since 1992 astronomers have discovered more than 900 other trans-Neptunian objects, and are discovering more each year. Like asteroids, all trans-Neptunian objects orbit the Sun in the same direction as the planets. A handful of trans-Neptunian objects are comparable in size to Pluto; at least one, Eris, is even larger than Pluto, as well as being in an orbit that is much larger, more steeply inclined, and more eccentric (Figure 7-8). Table 7-4 lists the seven largest trans-Neptunian objects known as of this writing. (Note that Charon is actually a satellite of Pluto.)

Just as most asteroids lie in the asteroid belt, most trans-Neptunian objects orbit within a band called the **Kuiper belt** (pronounced "ki-per") that extends from 30 AU to 50 AU from the Sun and is centered on the plane of the ecliptic. Like asteroids, many more trans-Neptunian objects remain to be discovered: astronomers estimate that there are 35,000 or more such objects with diameters greater than 100 km. If so, the combined mass of all trans-Neptunian objects is comparable to the mass of Jupiter, and is several hundred times greater than the combined mass of all the asteroids found in the inner solar system.

Like asteroids, trans-Neptunian objects are thought to be debris left over from the formation of the solar system. In the inner regions of the solar system, rocky fragments have been able to endure continuous exposure to the Sun's heat, but any ice originally present would have evaporated. Far from the Sun, ice has survived for billions of years. Thus, debris in the solar system naturally divides into two families (asteroids and trans-Neptunian objects), which can be arranged according to distance from the Sun, just like the two categories of planets (terrestrial and Jovian).

Comets

Two objects in the Kuiper belt can collide if their orbits cross each other. When this happens, a fragment a few kilometers across can be knocked off one of the colliding objects and be diverted into an elongated orbit that brings it close to the Sun. Such small

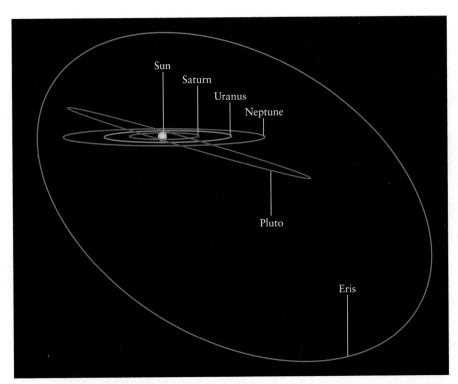

Figure 7-8

Trans-Neptunian Objects Pluto and Eris are the two largest trans-Neptunian objects, small worlds of rock and ice that orbit beyond Neptune. Unlike the orbits of the planets, the orbits of these two objects are steeply inclined to the ecliptic: Pluto's orbit is tilted by about 17°, and that of Eris is tilted by 44°.

Table 7-4   Seven Large Trans-Neptunian Objects

	Quaoar	2003 EL61	Sedna	2005 FY9	Pluto	Charon (satellite of Pluto)	Eris
Average distance from Sun (AU)	43.54	43.34	489	45.71	39.54	39.54	67.67
Orbital period (years)	287	285	10,800	309	248.6	248.6	557
Orbital eccentricity	0.035	0.189	0.844	0.155	0.250	0.250	0.442
Inclination of orbit to the ecliptic	8.0°	28.2°	11.9°	29.0°	17.15°	17.15°	44.2°
Approximate diameter (km)	1250	1500	1600	1800	2274	1190	2900

3476 km = diameter of the Earth's Moon

Charon

Quaoar   2003 EL61   Sedna   2005 FY9   Pluto   Eris

(Note: To date astronomers have been able to make images of the surfaces of Pluto and Charon only.)

R I **V** **U** X G

(Images of Pluto and Charon: Alan Stern, Southwest Research Institute; Marc Buie, Lowell Observatory; NASA; and ESA)

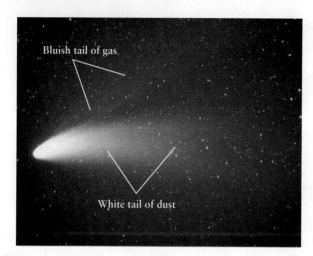

Bluish tail of gas

White tail of dust

Figure 7-9   R I V U X G

A Comet This photograph shows Comet Hale-Bopp as it appeared in April 1997. The solid part of a comet like this is a chunk of dirty ice a few tens of kilometers in diameter. When a comet passes near the Sun, solar radiation vaporizes some of the icy material, forming a bluish tail of gas and a white tail of dust. Both tails can extend for tens of millions of kilometers. (Richard J. Wainscoat, University of Hawaii)

objects, each a combination of rock and ice, are called **comets.** When a comet comes close enough to the Sun, the Sun's radiation vaporizes some of the comet's ices, producing long flowing tails of gas and dust particles (**Figure 7-9**). Astronomers deduce the composition of comets by studying the spectra of these tails.

CAUTION! Science-fiction movies and television programs sometimes show comets tearing across the night sky like a rocket. That would be a pretty impressive sight—but that is not what comets look like. Like the planets, comets orbit the Sun. And like the planets, comets move hardly at all against the background of stars over the course of a single night (see Section 4-1). If you are lucky enough to see a bright comet, it will not zoom dramatically from horizon to horizon. Instead, it will seem to hang majestically among the stars, so you can admire it at your leisure.

Some comets appear to originate from locations far beyond the Kuiper belt. The source of these is thought to be a swarm of comets that forms a spherical "halo" around the solar system called the **Oort cloud.** This hypothesized "halo" extends to 50,000 AU from the Sun (about one-fifth of the way to the nearest other star). Because the Oort cloud is so distant, it has not yet been possible to detect objects in the Oort cloud directly.

7-6 Cratering on planets and satellites is the result of impacts from interplanetary debris

One of the great challenges in studying planets and satellites is how to determine their internal structures. Are they solid or liquid inside? If there is liquid in the interiors, is the liquid calm or in agitated motion? Because planets are opaque, we cannot see directly into their interiors to answer these questions. But we can gather important clues about the interiors of terrestrial planets and satellites by studying the extent to which their

> Scientists study craters on a planet or satellite to learn the age and geologic history of the surface

surfaces are covered with craters (**Figure 7-10**). To see how this is done, we first need to understand where craters come from.

The Origin of Craters

The planets orbit the Sun in roughly circular orbits. But many asteroids and comets are in more elongated orbits. Such an elongated orbit can put these small objects on a collision course with a planet or satellite. If the object collides with a Jovian planet, it is swallowed up by the planet's thick atmosphere. (Astronomers actually saw an event of this kind in 1994, when a comet crashed into Jupiter.) But if the object collides with the solid surface of a terrestrial planet or a satellite, the result is an **impact crater** (see Figure 7-10). Such impact craters, found throughout the solar system, offer stark evidence of these violent collisions.

The easiest way to view impact craters is to examine the Moon through a telescope or binoculars. Some 30,000 lunar craters are visible, with diameters ranging from 1 km to several hundred km. Close-up photographs from lunar orbit have revealed millions of craters too small to be seen from Earth (Figure 7-10a). These smaller craters are thought to have been caused by impacts of relatively small objects called **meteoroids,** which range in size from a few hundred meters across to the size of a pebble or smaller. Meteoroids are the result of collisions between asteroids, whose orbits sometimes cross. The chunks of rock that result from these collisions go into independent orbits around the Sun, which can lead them to collide with the Moon or another world.

When German astronomer Franz Gruithuisen proposed in 1824 that lunar craters were the result of impacts, a major sticking point was the observation that nearly all craters are circular. If craters were merely gouged out by high-speed rocks, a rock striking the Moon in any direction except straight downward would have created a noncircular crater. A century after Gruithuisen, it was realized that a meteoroid colliding with the Moon generates a shock wave in the lunar surface that spreads out from the point of impact. Such a shock wave produces a circular crater no matter what direction the meteoroid was moving. (In a similar way, the craters made by artillery shells are almost always circular.) Many of the larger lunar craters also have a central peak, which is characteristic of a high-speed impact (see Figure 7-10a). Craters made by other processes, such as volcanic action, would not have central peaks of this sort.

Comparing Cratering on Different Worlds

Not all planets and satellites show the same amount of cratering. The Moon is heavily cratered over its entire surface, with craters on top of craters, as shown in Figure 7-10a. On Earth, by contrast, craters are very rare: Geologists have identified fewer than 200 impact craters on our planet (Figure 7-10b). Our understanding is that both the Earth and the Moon formed at nearly the same time and have been bombarded at comparable rates over

Central peak

50 km

(a) A crater on the Moon

50 km

(b) A crater on the Earth

50 km

(c) A crater on Mars

Figure 7-10   R I [V] U X G

Impact Craters These images, all taken from spacecraft, show impact craters on three different worlds. **(a)** The Moon's surface has craters of all sizes. The large crater near the middle of this image is about 80 km (50 miles) in diameter, equal to the length of San Francisco Bay. **(b)** Manicouagan Reservoir in Quebec is the relic of a crater formed by an impact more than 200 million years ago. The crater was eroded over the ages by the advance and retreat of glaciers, leaving a ring lake 100 km (60 miles) across. **(c)** Lowell Crater in the southern highlands of Mars is 201 km (125 miles) across. Like the image of the Moon in part (a), there are craters on top of craters. Note the light-colored frost formed by condensation of carbon dioxide from the Martian atmosphere. (a: NASA; b: JSC/NASA; c: NASA/JPL/MSSS)

their history. Why, then, are craters so much rarer on the Earth than on the Moon?

The answer is that the Earth is a *geologically active* planet: the continents slowly change their positions over eons, new material flows onto the surface from the interior (as occurs in a volcanic eruption), and old surface material is pushed back into the interior (as occurs off the coast of Chile, where the ocean bottom is slowly being pushed beneath the South American continent). These processes, coupled with erosion from wind and water, cause craters on Earth to be erased over time. The few craters found on Earth today must be relatively recent, since there has not yet been time to erase them.

The Moon, by contrast, is geologically *inactive*. There are no volcanoes and no motion of continents (and, indeed, no continents). Furthermore, the Moon has neither oceans nor an atmosphere, so there is no erosion as we know it on Earth. With none of the processes that tend to erase craters on Earth, the Moon's surface remains pockmarked with the scars of billions of years of impacts.

In order for a planet to be geologically active, its interior must be at least partially molten. This is necessary so that continents can slide around on the underlying molten material and so that molten lava can come to the surface, as in a volcanic eruption. Hence, we would expect geologically inactive (and hence heavily cratered) worlds like the Moon to have less molten material in their interiors than does the Earth. Investigations of these inactive worlds bear this out. But *why* is the Moon's interior less molten than the Earth's?

ANALOGY To see one simple answer to this question, notice that a large turkey or roast taken from the oven will stay warm inside for hours, but a single meatball will cool off much more rapidly. The reason is that the meatball has more surface area

relative to its volume, and so it can more easily lose heat to its surroundings. A planet or satellite also tends to cool down as it emits electromagnetic radiation into space (see Section 5-3); the smaller the planet or satellite, the greater its surface area relative to its volume, and the more readily it can radiate away heat (**Figure 7-11**). Both the Earth and Moon were probably completely molten when they first formed, but because the Moon (diameter 3476 km) is so much smaller than the Earth (diameter 12,756 km), it has lost much of its internal heat and has a much more solid interior.

Cratering Measures Geologic Activity

By considering these differences between the Earth and the Moon, we have uncovered a general rule for worlds with solid surfaces:

The smaller the terrestrial world, the less internal heat it is likely to have retained, and, thus, the less geologic activity it will display on its surface. The less geologically active the world, the older and hence more heavily cratered its surface.

This rule means that we can use the amount of cratering visible on a planet or satellite to estimate the age of its surface and how geologically active it is. As an example, Mercury has a heavily cratered surface, which means that the surface is very old. This accords with Mercury being the smallest of the terrestrial planets (see Table 7-1): Due to its small size, it has lost the internal heat required to sustain geologic activity. On Venus, by comparison, there are only about a thousand craters larger than a few kilometers in diameter, many more than have been found on Earth but only a small fraction of the number on the Moon or Mercury. Venus is only slightly smaller than the Earth, and it has enough internal heat to power the geologic activity required to erase most of its impact craters.

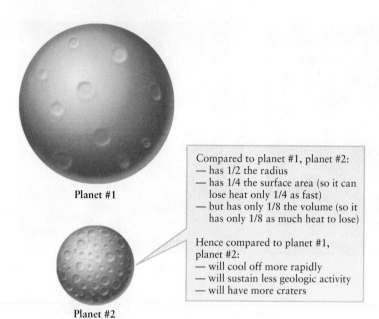

Figure 7-11

Planet Size and Cratering Of these two hypothetical planets, the smaller one (#2) has less volume and less internal heat, as well as less surface area from which to radiate heat into space. But the *ratio* of surface area to volume is greater for the smaller planet. Hence, the smaller planet will lose heat faster, have a colder interior, and be less geologically active. It will also have a more heavily cratered surface, since it takes geologic activity to erase craters.

Mars is an unusual case, in that extensive cratering (Figure 7-10*c*) is found only in the higher terrain; the lowlands of Mars are remarkably smooth and free of craters. Thus, it follows that the Martian highlands are quite old, while the lowlands have a younger surface from which most craters have been erased. Considering the planet as a whole, the amount of cratering on Mars is intermediate between that on Mercury and the Earth. This agrees with our general rule, because Mars is intermediate in size between Mercury and Earth. The interior of Mars was once hotter and more molten than it is now, so that geologic processes were able to erase some of the impact craters. A key piece of evidence that supports this picture is that Mars has a number of immense volcanoes (**Figure 7-12**). These were active when Mars was young, but as this relatively small planet cooled down and its interior solidified, the supply of molten material to the volcanoes from the Martian interior was cut off. As a result, all of the volcanoes of Mars are now inactive.

As for all rules, there are limitations and exceptions to the rule relating a world's size to its geologic activity. One limitation is that the four terrestrial planets all have slightly different compositions, which affects the types and extent of geologic activity that can take place on their surfaces. This also complicates the relationship between the number of craters and the age of the surface. An important exception to our rule is Jupiter's satellite Io, which, despite its small size, is the most volcanic world in the solar system (see Section 7-2). Something must be supplying Io with energy to keep its interior hot; this turns out to be Jupiter, which exerts powerful tides on Io as it moves in a relatively small orbit around its planet. These tides cause Io to flex like a ball of clay

Figure 7-12   R I ☒V U X G

A Martian Volcano Olympus Mons is the largest of the inactive volcanoes of Mars and the largest volcano in the solar system. The base of Olympus Mons measures 600 km (370 mi) in diameter, and the scarps (cliffs) that surround the base are 6 km (4 mi) high. The caldera, or volcanic crater, at the summit is approximately 70 km across, large enough to contain the state of Rhode Island. This view was creating by combining a number of images taken from Mars orbit. (© Calvin J. Hamilton)

being kneaded between your fingers, and this flexing heats up the satellite's interior. But despite these limitations and exceptions, the relationships between a world's size, internal heat, geologic activity, and amount of cratering are powerful tools for understanding the terrestrial planets and satellites.

7-7   A planet with a magnetic field indicates a fluid interior in motion

The amount of impact cratering on a terrestrial planet or satellite provides indirect evidence about whether the planet or satellite has a molten interior. But another, more direct tool for probing the interior of *any* planet or satellite is an ordinary compass, which senses the magnetic field outside the planet or satellite. Magnetic field measurements prove to be an extremely powerful way to investigate the internal structure of a world without having to actually dig into its interior. To illustrate how this works, consider the behavior of a compass on Earth.

> By studying the magnetic field of a planet or satellite, scientists can learn about that world's interior motions

The needle of a compass on Earth points north because it aligns with the Earth's *magnetic field*. Such fields arise whenever electrically charged particles are in motion. For example, a loop of wire carrying an electric current generates a magnetic field in the space around it. The magnetic field that surrounds an ordinary bar magnet (**Figure 7-13*a***) is created by the motions of negatively charged electrons within the iron atoms of which the magnet is made. The Earth's magnetic field is similar to that of a bar magnet, as Figure 7-13*b* shows. The consensus among geolo-

(a)

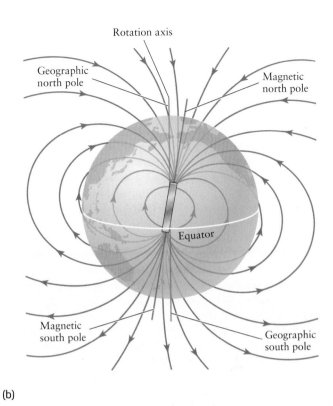

(b)

Figure 7-13

The Magnetic Fields of a Bar Magnet and of the Earth (a) This picture was made by placing a piece of paper on top of a bar magnet, then spreading iron filings on the paper. The pattern of the filings show the magnetic field lines, which appear to stream from one of the magnet's poles to the other. **(b)** The Earth's magnetic field lines have a similar pattern. Although the Earth's field is produced in a different way—by electric currents in the liquid portion of our planet's interior—the field is much the same as if there were a giant bar magnet inside the Earth. This "bar magnet" is not exactly aligned with the Earth's rotation axis, which is why the magnetic north and south poles are not at the same locations as the true, or geographic, poles. A compass needle points toward the north magnetic pole, not the true north pole. (a: Jules Bucher/Photo Researchers)

gists is that this magnetic field is caused by the motion of the liquid portions of the Earth's interior. Because this molten material (mostly iron) conducts electricity, these motions give rise to electric currents, which in turn produce the Earth's magnetic field. Our planet's rotation helps to sustain these motions and hence the magnetic field. This process for producing a magnetic field is called a **dynamo.**

CAUTION! While the Earth's magnetic field is similar to that of a giant bar magnet, you should not take this picture too literally. The Earth is *not* simply a magnetized ball of iron. In an iron bar magnet, the electrons of different atoms orbit their nuclei in the same general direction, so that the magnetic fields generated by individual atoms add together to form a single, strong field. But at temperatures above 770°C (1418°F = 1043 K), the orientations of the electron orbits become randomized. The fields of individual atoms tend to cancel each other out, and the iron loses its magnetism. Geological evidence shows that almost all of the Earth's interior is hotter than 770°C, so the iron there cannot be extensively magnetized. The correct picture is that the Earth acts as a dynamo: The liquid iron carries electric currents, and these currents create the Earth's magnetic field.

If a planet or satellite has a mostly solid interior, then the dynamo mechanism cannot work: Material in the interior cannot flow, there are no electric currents, and the planet or satellite does not generate a magnetic field. One example of this is the Moon. As we saw in Section 7-6, the extensive cratering of the lunar surface indicates that the Moon has no geologic activity and must therefore have a mostly solid interior. Measurements made during the *Apollo* missions, in which 12 humans visited the lunar surface between 1969 and 1972, showed that the present-day Moon indeed has no global magnetic field. However, careful magnetic measurements of lunar rocks returned by the *Apollo* astronauts indicate that the Moon *did* have a weak magnetic field when the rocks solidified. These rocks, like the rest of the lunar surface, are very old. Hence, in the distant past the Moon may have had a small amount of molten iron in its interior that acted as a dynamo. This material presumably solidified at least partially as the Moon cooled, so that the lunar magnetic field disappeared.

We have now identified another general rule about planets and satellites:

A planet or satellite with a global magnetic field has liquid material in its interior that conducts electricity and is in motion, generating the magnetic field.

Thus, by studying the magnetic field of a planet or satellite, we can learn about that world's interior. This explains why many spacecraft carry devices called **magnetometers** to measure magnetic fields. Magnetometers are often placed on a long boom extending outward from the body of the spacecraft (Figure 7-14). This isolates them from the magnetic fields produced by electric currents in the spacecraft's own circuitry.

Measurements made with magnetometers on spacecraft have led to a number of striking discoveries. For example, it has been found that Mercury has a planetwide magnetic field like the Earth's, although it is only about 1% as strong as the Earth's field. Mercury has a heavily cratered surface and hence little or no geologic activity, which by itself would suggest that the planet's interior is mostly solid. The magnetic field measurements show that *some* of Mercury's interior must be in the liquid state to act as a dynamo. By contrast, Venus has no measurable planetwide magnetic field, even though the paucity of craters on its surface indicates the presence of geologic activity and hence a hot interior of the planet. One possible reason for the lack of a magnetic field on Venus is that the planet turns on its axis very slowly, taking 243 days for a complete rotation. Because of this slow rotation, the fluid material within the planet is hardly agitated at all, and so may not move in the fashion that generates a magnetic field.

Like Venus, Mars has no planetwide magnetic field. But the magnetometer aboard the *Mars Global Surveyor* spacecraft, which went into orbit around Mars in 1997, found magnetized regions in the cratered highlands of the Martian southern hemisphere (Figure 7-15). As we saw in Section 7-6, these portions of the Martian surface are very old, and so would have formed when the planet's interior was still hot and molten. Electric currents in the flowing molten material could then produce a planetwide

magnetic field. As surface material cooled and solidified, it became magnetized by the planetwide field, and this material has retained its magnetization over the eons. *Mars Global Surveyor* has not found magnetized areas in the younger terrain of the lowlands. Hence, the planetwide Martian magnetic field must have shut off by the time the lowland terrain formed.

The most intense planetary magnetic field in the solar system is that of Jupiter: at the tops of Jupiter's clouds, the magnetic field is about 14 times stronger than the field at the Earth's surface. It is thought that Jupiter's field, like the Earth's, is produced by a dynamo acting deep within the planet's interior. Unlike the Earth, however, Jupiter is composed primarily of hydrogen and helium, not substances like iron that conduct electricity. How, then, can Jupiter have a dynamo that generates such strong magnetic fields?

To answer this question, recall from Section 5-8 that a hydrogen atom consists of a single proton orbited by a single electron. Deep inside Jupiter, the pressure is so great and hydrogen atoms are squeezed so close together that electrons can hop from one atom to another. This hopping motion creates an electric current, just as the ordered movement of electrons in the copper wires of a flashlight constitutes an electric current. In other words, the highly compressed hydrogen deep inside Jupiter behaves like an electrically conducting metal; thus, it is called **liquid metallic hydrogen.**

Laboratory experiments show that hydrogen becomes a liquid metal when the pressure is more than about 1.4 million times ordinary atmospheric pressure on Earth. Recent calculations suggest that this transition occurs about 7000 km below Jupiter's cloudtops. Most of the planet's enormous bulk lies below this level, so there is a tremendous amount of liquid metallic hydrogen within Jupiter. Since Jupiter rotates rapidly—a "day" on

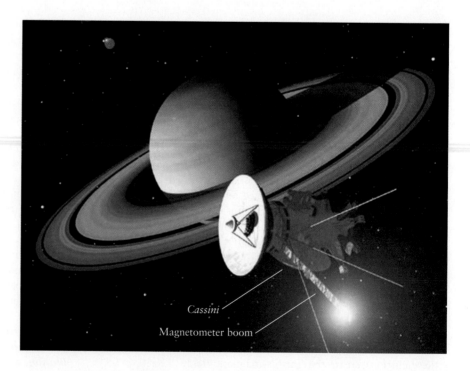

Figure 7-14

Probing the Magnetic Field of Saturn This illustration depicts the *Cassini* spacecraft as it entered orbit around Saturn on July 1, 2004. In addition to telescopes for observing Saturn and its satellites, *Cassini* carries a magnetometer for exploring Saturn's magnetic field. The magnetometer is located on the long boom that extends down and to the right from the body of the spacecraft. (The glow at the end of the boom is the reflection of the Sun.) (JPL/NASA)

Cassini
Magnetometer boom

Figure 7-15

Relic Magnetism on Mars Portions of the highlands of Mars became magnetized during the period early in Martian history when the planet had an extensive magnetic field. The planetwide field has long since disappeared, but certain surface regions remained magnetized. The false colors in this illustration indicate the direction and strength of the field at different locations on Mars superimposed on images of the two Martian hemispheres. (MGS Magnetometer Team, GSFC/NASA)

Jupiter is just less than 10 hours long—this liquid metal moves rapidly, generating the planet's powerful magnetic field. Saturn also has a magnetic field produced by dynamo action in liquid metallic hydrogen. (The field is weaker than Jupiter's because Saturn is a smaller planet with less internal pressure, so there is less of the liquid metal available.)

Uranus and Neptune also have magnetic fields, but they cannot be produced in the same way: Because these planets are relatively small, the internal pressure is not great enough to turn liquid hydrogen into a metal. Instead, it is thought that both Uranus and Neptune have large amounts of liquid water in their interiors and that this water has molecules of ammonia and other substances dissolved in it. (The fluid used for washing windows has a similar chemical composition.) Under the pressures found in this interior water, the dissolved molecules lose one or more electrons and become electrically charged (that is, they become ionized; see Section 5-8). Water is a good conductor of electricity when it has such electrically charged molecules dissolved in it, and electric currents in this fluid are probably the source of the magnetic fields of Uranus and Neptune.

7-8 The diversity of the solar system is a result of its origin and evolution

Our brief tour of the solar system has revealed its almost dizzying variety. No two planets are alike, satellites come in all sizes, the extent of cratering varies from one terrestrial planet to another, and the magnetic fields of different planets vary dramatically in their strength and in how they are produced. (The *Cosmic Connections* figure that closes this chapter summarizes these properties of the planets.) All of this variety leads us to a simple yet profound question: *Why* are the planets and satellites of the solar system so different from each other?

Among humans, the differences from one individual to another result from heredity (the genetic traits passed on from an individual's parents) and environment (the circumstances under which the individual

The similarities and differences among the planets can be logically explained by a model of the solar system's origin and evolution

COSMIC CONNECTIONS
Characteristics of the Planets

The Inner (Terrestrial) Planets — Close to the Sun - Small diameter, small mass - High density

	Mercury	Venus	Earth	Mars
Average distance from Sun (AU)	0.387	0.723	1.000	1.524
Equatorial diameter (Earth = 1)	0.383	0.949	1.000	0.533
Mass (Earth = 1)	0.0553	0.8150	1.0000	0.1074
Average density (kg/m^3)	5430	5243	5515	3934

Mercury **Venus** **Earth** **Mars**

Atmosphere
None

Carbon dioxide
Nitrogen
Other

Atmosphere
Carbon dioxide

Nitrogen
Other Oxygen

Atmosphere
Nitrogen, oxygen

Carbon dioxide
Nitrogen
Other

Atmosphere
Carbon dioxide

Magnetic field
Weak

Magnetic field
None

Magnetic field
Moderate, due to liquid iron core

Magnetic field
None

Interior
Iron-nickel core, rocky shell

Interior
Iron-nickel core, rocky shell

Interior
Iron-nickel core, rocky shell

Interior
Iron-nickel core, rocky shell

The Outer (Jovian) Planets   Far from the Sun - Large diameter, large mass - Low density

	Jupiter	Saturn	Uranus	Neptune
Average distance from Sun (AU)	5.203	9.554	19.194	30.066
Equatorial diameter (Earth = 1)	11.209	9.449	4.007	3.883
Mass (Earth = 1)	317.8	95.16	14.53	17.15
Average density (kg/m^3)	1326	687	1318	1638

Jupiter

Saturn

Uranus

Neptune

Atmosphere
Hydrogen, helium

Atmosphere
Hydrogen, helium

Atmosphere
Hydrogen, helium

Atmosphere
Hydrogen, helium

Magnetic field
Strong, due to
liquid metallic hydrogen

Magnetic field
Strong, due to
liquid metallic hydrogen

Magnetic field
Moderate, due to
dissolved ions

Magnetic field
Moderate, due to
dissolved ions

Interior
Rocky core, liquid
hydrogen and helium

Interior
Rocky core, liquid
hydrogen and helium

Interior
Rocky core, liquid
water and ammonia

Interior
Rocky core, liquid
water and ammonia

matures to an adult). As we will find in the fzollowing chapter, much the same is true for the worlds of the solar system.

In Chapter 8 we will see evidence that the entire solar system shares a common "heredity," in that the planets, satellites, comets, asteroids, and the Sun itself formed from the same cloud of interstellar gas and dust. The composition of this cloud was shaped by cosmic processes, including nuclear reactions that took place within stars that died long before our solar system was formed. We will see how different planets formed in different environments depending on their distance from the Sun and will discover how these environmental variations gave rise to the planets and satellites of our present-day solar system. And we will see how we can test these ideas of solar system origin and evolution by studying planetary systems orbiting other stars.

Our journey through the solar system is just beginning. In this chapter we have explored space to examine the variety of the present-day solar system; in Chapter 8 we will journey through time to see how our solar system came to be.

Key Words

Terms preceded by an asterisk (*) are discussed in the Boxes.

asteroid, p. 170	Kuiper belt, p. 170
asteroid belt, p. 170	liquid metallic hydrogen,
average density, p. 162	p. 176
chemical composition, p. 164	magnetometer, p. 176
comet, p. 172	meteoroid, p. 172
dynamo, p. 175	minor planet, p. 170
*escape speed, p. 169	Oort cloud, p. 172
ices, p. 167	spectroscopy, p. 164
impact crater, p. 172	terrestrial planet, p. 162
Jovian planet, p. 162	trans-Neptunian object,
*kinetic energy, p. 168	p. 170

Key Ideas

Properties of the Planets: All of the planets orbit the Sun in the same direction and in almost the same plane. Most of the planets have nearly circular orbits.

• The four inner planets are called terrestrial planets. They are relatively small (with diameters of 5000 to 13,000 km), have high average densities (4000 to 5500 kg/m^3), and are composed primarily of rocky materials.

• The four giant outer planets are called Jovian planets. They have large diameters (50,000 to 143,000 km) and low average densities (700 to 1700 kg/m^3) and are composed primarily of light elements such as hydrogen and helium.

Satellites and Small Bodies in the Solar System: Besides the planets, the solar system includes satellites of the planets, asteroids, comets, and trans-Neptunian objects.

• Seven large planetary satellites (one of which is the Moon) are comparable in size to the planet Mercury. The remaining satellites of the solar system are much smaller.

• Asteroids are small, rocky objects, while comets and trans-Neptunian objects are made of ice and rock. All are remnants left over from the formation of the planets.

• Most asteroids are found in the asteroid belt between the orbits of Mars and Jupiter, and most trans-Neptunian objects lie in the Kuiper belt outside the orbit of Neptune. Pluto is one of the largest members of the Kuiper belt.

Spectroscopy and the Composition of the Planets: Spectroscopy, the study of spectra, provides information about the chemical composition of objects in the solar system.

• The spectrum of a planet or satellite with an atmosphere reveals the atmosphere's composition. If there is no atmosphere, the spectrum indicates the composition of the surface.

• The substances that make up the planets can be classified as gases, ices, or rock, depending on the temperatures at which they solidify.

Impact Craters: When an asteroid, comet, or meteoroid collides with the surface of a terrestrial planet or satellite, the result is an impact crater.

• Geologic activity renews the surface and erases craters, so a terrestrial world with extensive cratering has an old surface and little or no geologic activity.

• Because geologic activity is powered by internal heat, and smaller worlds lose heat more rapidly, as a general rule smaller terrestrial worlds are more extensively cratered.

Magnetic Fields and Planetary Interiors: Planetary magnetic fields are produced by the motion of electrically conducting liquids inside the planet. This mechanism is called a dynamo. If a planet has no magnetic field, that is evidence that there is little such liquid material in the planet's interior or that the liquid is not in a state of motion.

• The magnetic fields of terrestrial planets are produced by metals such as iron in the liquid state. The stronger fields of the Jovian planets are generated by liquid metallic hydrogen or by water with ionized molecules dissolved in it.

Questions

Review Questions

1. Do all the planets orbit the Sun in the same direction? Are all of the orbits circular?
2. What are the characteristics of a terrestrial planet?
3. What are the characteristics of a Jovian planet?
4. What is meant by the average density of a planet? What does the average density of a planet tell us?
5. In what ways are the largest satellites similar to the terrestrial planets? In what ways are they different? Which satellites are the largest?
6. The absorption lines in the spectrum of a planet or satellite do not necessarily indicate the composition of the planet or satellite's atmosphere. Why not?

7. What are the differences in chemical composition between the terrestrial and Jovian planets?

8. Why are hydrogen and helium abundant in the atmospheres of the Jovian planets but present in only small amounts in the Earth's atmosphere?

9. What is an asteroid? What is a trans-Neptunian object? In what ways are these minor members of the solar system like or unlike the planets?

10. What are the asteroid belt, the Kuiper belt, and the Oort cloud? Where are they located? How do the objects found in these three regions compare?

11. In what ways is Pluto similar to a terrestrial planet? In what ways is it different?

12. What is the connection between comets and the Kuiper belt? Between comets and the Oort cloud?

13. What is one piece of evidence that impact craters are actually caused by impacts?

14. What is the relationship between the extent to which a planet or satellite is cratered and the amount of geologic activity on that planet or satellite?

15. How do we know that the surface of Venus is older than the Earth's surface but younger than the Moon's surface?

16. Why do smaller worlds retain less of their internal heat?

17. How does the size of a terrestrial planet influence the amount of cratering on the planet's surface?

18. How is the magnetic field of a planet different from that of a bar magnet? Why is a large planet more likely to have a magnetic field than a small planet?

19. Could you use a compass to find your way around Venus? Why or why not?

20. If Mars has no planetwide magnetic field, why does it have magnetized regions on its surface?

21. What is liquid metallic hydrogen? Why is it found only in the interiors of certain planets?

Advanced Questions

Questions preceded by an asterisk (*) involve topics discussed in the Boxes.

Problem-solving tips and tools

The volume of a sphere of radius r is $4\pi r^3/3$, and the surface area of a sphere of radius r is $4\pi r^2$. The surface area of a circle of radius r is πr^2. The average density of an object is its mass divided by its volume. To calculate escape speeds, you will need to review Box 7-2. Be sure to use the same system of units (meters, seconds, kilograms) in all your calculations involving escape speeds, orbital speeds, and masses. Appendix 6 gives conversion factors between different sets of units, and Box 5-1 has formulas relating various temperature scales.

---

22. Mars has two small satellites, Phobos and Deimos. Phobos circles Mars once every 0.31891 day at an average altitude of 5980 km above the planet's surface. The diameter of Mars is 6794 km. Using this information, calculate the mass and average density of Mars.

23. Figure 7-3 shows the spectrum of Saturn's largest satellite, Titan. Can you think of a way that astronomers can tell which absorption lines are due to Titan's atmosphere and which are due to the atmospheres of the Sun and Earth? Explain.

*24. (a) Find the mass of a hypothetical spherical asteroid 2 km in diameter and composed of rock with average density 2500 kg/m³. (b) Find the speed required to escape from the surface of this asteroid. (c) A typical jogging speed is 3 m/s. What would happen to an astronaut who decided to go for a jog on this asteroid?

*25. The hypothetical asteroid described in Question 24 strikes the Earth with a speed of 25 km/s. (a) What is the kinetic energy of the asteroid at the moment of impact? (b) How does this energy compare with that released by a 20-kiloton nuclear weapon, like the device that destroyed Hiroshima, Japan, on August 6, 1945? (*Hint:* 1 kiloton of TNT releases 4.2×10^{12} joules of energy.)

*26. Suppose a spacecraft landed on Jupiter's moon Europa (see Table 7-2), which moves around Jupiter in an orbit of radius 670,900 km. After collecting samples from the satellite's surface, the spacecraft prepares to return to Earth. (a) Calculate the escape speed from Europa. (b) Calculate the escape speed from Jupiter at the distance of Europa's orbit. (c) In order to begin its homeward journey, the spacecraft must leave Europa with a speed greater than either your answer to (a) or your answer to (b). Explain why.

*27. A hydrogen atom has a mass of 1.673×10^{-27} kg, and the temperature of the Sun's surface is 5800 K. What is the average speed of hydrogen atoms at the Sun's surface?

*28. The Sun's mass is 1.989×10^{30} kg, and its radius is 6.96×10^8 m. (a) Calculate the escape speed from the Sun's surface. (b) Using your answer to Question 27, explain why the Sun has lost very little hydrogen over its entire 4.56-billion-year history.

*29. Saturn's satellite Titan has an appreciable atmosphere, yet Jupiter's satellite Ganymede—which is about the same size and mass as Titan—has no atmosphere. Explain why there is a difference.

30. The distance from the asteroid 433 Eros (Figure 7-7) to the Sun varies between 1.13 and 1.78 AU. (a) Find the period of Eros's orbit. (b) Does Eros lie in the asteroid belt? How can you tell?

31. Imagine a trans-Neptunian object with roughly the same mass as Earth but located 50 AU from the Sun. (a) What do you think this object would be made of? Explain your reasoning. (b) On the basis of this speculation, assume a reasonable density for this object and calculate its diameter. How many times bigger or smaller than Earth would it be?

32. Consider a hypothetical trans-Neptunian object located 100 AU from the Sun. (a) What would be the orbital period (in years) of this object? (b) There are 360 degrees in a circle, and 60 arcminutes in a degree. How long would it take this object to move 1 arcminute across the sky? (c) Trans-Neptunian objects are discovered by looking for "stars" that move on the celestial sphere. Use your answer from part (b)

to explain why these discoveries require patience. (**d**) Discovering trans-Neptunian objects also requires large telescopes equipped with sensitive detectors. Explain why.

33. The surfaces of Mercury, the Moon, and Mars are riddled with craters formed by the impact of space debris. Many of these craters are billions of years old. By contrast, there are only a few conspicuous craters on the Earth's surface, and these are generally less than 500 million years old. What do you suppose explains the difference?

34. During the period of most intense bombardment by space debris, a new 1-km-radius crater formed somewhere on the Moon about once per century. During this same period, what was the probability that such a crater would be created within 1 km of a certain location on the Moon during a 100-year period? During a 10^6-year period? (*Hint:* If you drop a coin onto a checkerboard, the probability that the coin will land on any particular one of the board's 64 squares is $^1/_{64}$.)

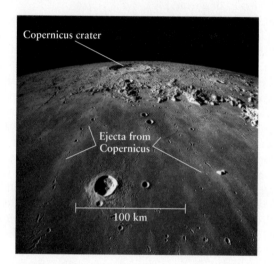

R I V U X G

(Courtesy of USRA)

35. When an impact crater is formed, material (called *ejecta*) is sprayed outward from the impact. (The accompanying photograph of the Moon shows light-colored ejecta extending outward from the crater Copernicus.) While ejecta are found surrounding the craters on Mercury, they do not extend as far from the craters as do ejecta on the Moon. Explain why, using the difference in surface gravity between the Moon (surface gravity = 0.17 that on Earth) and Mercury (surface gravity = 0.38 that on Earth).

36. Mercury rotates once on its axis every 58.646 days, compared to 1 day for the Earth. Use this information to argue why Mercury's magnetic field should be much smaller than the Earth's.

37. Suppose *Mars Global Surveyor* had discovered magnetized regions in the lowlands of Mars. How would this discovery

have affected our understanding of the evolution of the Martian interior?

38. Liquid metallic hydrogen is the source of the magnetic fields of Jupiter and Saturn. Explain why liquid metallic hydrogen cannot be the source of the Earth's magnetic field.

Discussion Questions

*39. There are no asteroids with an atmosphere. Discuss why not.

40. The *Galileo* spacecraft that orbited Jupiter from 1995 to 2003 discovered that Ganymede (Table 7-2) has a magnetic field twice as strong as that of Mercury. Does this discovery surprise you? Why or why not?

Web/eBook Questions

41. Search the World Wide Web for information about impact craters on Earth. Where is the largest crater located? How old is it estimated to be? Which crater is closest to where you live?

42. **Determining Terrestrial Planet Orbital Periods.** Access the animation "Planetary Orbits" in Chapter 7 of the *Universe* Web site or eBook. Focus on the motions of the inner planets at the last half of the animation. Using the stop and start buttons, determine how many days it takes Mars, Venus, and Mercury to orbit the Sun once if Earth takes approximately 365 days.

Activities

Observing Projects

43. Use a telescope or binoculars to observe craters on the Moon. Make a drawing of the Moon, indicating the smallest and largest craters that you can see. Can you estimate their sizes? For comparison, the Moon as a whole has a diameter of 3476 km. *Hint:* You can see craters most distinctly when the Moon is near first quarter or third quarter (see Figure 3-2). At these phases, the Sun casts long shadows across the portion of the Moon in the center of your field of view, making the variations in elevation between the rims and centers of craters easy to identify. You can determine the phase of the Moon by looking at a calendar or the weather page of the newspaper, by using the *Starry Night Enthusiast*™ program, or on the World Wide Web.

44. Use the *Starry Night Enthusiast*™ program to examine magnified images of the terrestrial planets Mercury, Venus, Earth, and Mars and the asteroid Ceres. In the **Favourites** menu, under the **Solar System** submenu, select the desired planet. This will place you in the position of an astronaut orbiting above the surface of the planet. The astronaut's spacesuit and feet are shown in the foreground in this view but can be removed by clicking on **View > Feet**. To show the planet as we see it from Earth, we need to show its atmosphere. Select **Solar System > Planets-Moons . . .** in the **Options** menu. In the Planets-Moons Options dialog that pops up, click on the **Show atmosphere** checkbox to turn this option on and then click the

OK button. The mouse icon will change to the **location scroller** when moved over the planet. You can use this scroller to rotate the image to see different views of the planet. This is equivalent to flying around the planet at a fixed distance. Follow the above steps to examine each planet and asteroid from different viewpoints and describe each planet's appearance. From what you observe in each case, is there any way of knowing whether you are looking at a planet's surface or at complete cloud cover over the planet? Which planet or planets have clouds? Which planet or asteroid shows the heaviest cratering? Which of these planets show evidence of liquid water?

45. Use the *Starry Night Enthusiast*™ program to examine the Jovian planets Jupiter, Saturn, Uranus, and Neptune. Select each of these planets from the **Solar System** submenu in the **Favourites** menu. If you desire, you can remove the image of the astronaut's feet by selecting **Feet** in the **View** menu. Position the mouse cursor over the planet and click and drag the image to examine the planet from different views. Describe each planet's appearance. Which has the greatest color contrast in its cloud tops? Which has the least color contrast? What can you say about the thickness of Saturn's rings compared to their diameter?

AU Microscopii
(disk seen edge-on)

HD 107146
(disk seen face-on)

Size of
Neptune's
Orbit

Size of
Neptune's Orbit

R I **V** U X G

Planets are thought to form within the disks surrounding young stars such as these. Neptune's orbit, shown for scale, is about 60 AU across. (NASA, ESA, D. R. Ardila (JHU), D. A. Golimowski (JHU), J. E. Krist (STScI/JPL), M. Clampin (NASA/GSFC), J. P. Williams (UH/IfA), J. P. Blakeslee (JHU), H. C. Ford (JHU), G. F. Hartig (STScI), G. D. Illingworth (UCO-Lick) and the ACS Science Team)

8

Comparative Planetology II: The Origin of Our Solar System

What did our solar system look like before the planets were fully formed? The answer may lie in these remarkable images from the Hubble Space Telescope. Each image shows an immense disk of gas and dust centered on a young star. (In each image the light from the star itself was blocked out within the telescope to make the rather faint disk more visible.) Astronomers strongly suspect that in its infancy our own Sun was surrounded by a disk of this kind and that the planets coalesced from it.

In this chapter we will examine the evidence that led astronomers to this picture of the origin of the planets. We will see how the abundances of different chemical elements in the solar system indicate that the Sun and planets formed from a thin cloud of interstellar matter some 4.56 billion years ago, an age determined by measuring the radioactivity of meteorites. We will learn how the nature of planetary orbits gives important clues to what happened as this cloud contracted and how evidence from meteorites reveals the chaotic conditions that existed within the cloud. And we will see how this cloud eventually evolved into the solar system that we see today.

In the past few years astronomers have been able to test this picture of planetary formation by examining disks around young stars (like the ones in the accompanying images). Most remarkably of all, they have discovered planets in orbit around dozens of other stars. These recent observations provide valuable information about how our own system of planets came to be.

8-1 Any model of solar system origins must explain the present-day Sun and planets

How did the Sun and planets form? In other words, where did the solar system come from? This question has tantalized astronomers for centuries. Our goal in this chapter is to examine

Learning Goals

By reading the sections of this chapter, you will learn

8-1 The key characteristics of the solar system that must be explained by any theory of its origins

8-2 How the abundances of chemical elements in the solar system and beyond explain the sizes of the planets

8-3 How we can determine the age of the solar system by measuring abundances of radioactive elements

8-4 Why scientists think the Sun and planets all formed from a cloud called the solar nebula

8-5 How the solar nebula model explains the formation of the terrestrial planets

8-6 Two competing models for the origin of the Jovian planets

8-7 How astronomers test the solar nebula model by observing planets around other stars

Table 8-1   Three Key Properties of Our Solar System

Any theory of the origin of the solar system must be able to account for these properties of the planets.	
Property 1: *Sizes and compositions of terrestrial planets versus Jovian planets*	The terrestrial planets, which are composed primarily of rocky substances, are relatively small, while the Jovian planets, which are composed primarily of hydrogen and helium, are relatively large (see Sections 7-1 and 7-4).
Property 2: *Directions and orientations of planetary orbits*	All of the planets orbit the Sun in the same direction, and all of their orbits are in nearly the same plane (see Section 7-1).
Property 3: *Sizes of terrestrial planet orbits versus Jovian planet orbits*	The terrestrial planets orbit close to the Sun, while the Jovian planets orbit far from the Sun (see Section 7-1).

our current understanding of how the solar system came to be—that is, our current best *theory* of the origin of the solar system.

Recall from Section 1-1 that a theory is not merely a set of wild speculations, but a self-consistent collection of ideas that must pass the test of providing an accurate description of the real world. Since no humans were present to witness the formation of the planets, scientists must base their theories of solar system origins on their observations of the present-day solar system. (In an analogous way, paleontologists base their understanding of the lives of dinosaurs on the evidence provided by fossils that have survived to the present day.) In so doing, they are following the steps of the scientific method that we described in Section 1-1.

What key attributes of the solar system should guide us in building a theory of solar system origins? Among the many properties of the planets that we discussed in Chapter 7, three of the most important are listed in Table 8-1.

Any theory that attempts to describe the origin of the solar system must be able to explain how these attributes came to be. We begin by considering what Property 1 tells us; we will return to Properties 2 and 3 and the orbits of the planets later in this chapter.

8-2   The cosmic abundances of the chemical elements are the result of how stars evolve

The small sizes of the terrestrial planets compared to the Jovian planets (Property 1 in Table 8-1) suggest that some chemical elements are quite common in our solar system, while others are quite rare. The tremendous masses of the Jovian planets—Jupiter alone has more mass than all of the other planets combined—means that the elements of which they are made, primarily hydrogen and helium, are very abundant. The Sun, too, is made almost entirely of hydrogen and helium: Its average density of 1410 kg/m³ is in the same range as the densities of the Jovian planets (see Table 7-1), and its absorption spectrum (see Figure 5-12) shows the dominance of hydrogen and helium in the Sun's atmosphere. Hydrogen, the most abundant element, makes up nearly three-quarters of the combined mass of the Sun and planets. Helium is the second most abundant element. Together, hydrogen and helium account for about 98% of the mass of all the material

> The terrestrial planets are small because they are made of less abundant elements

in the solar system. All of the other chemical elements are relatively rare; combined, they make up the remaining 2% (Figure 8-1).

The dominance of hydrogen and helium is not merely a characteristic of our local part of the universe. By analyzing the spectra of stars and galaxies, astronomers have found essentially the same pattern of chemical abundances out to the farthest distance attainable by the most powerful telescopes. Hence, the vast majority of the atoms in the universe are hydrogen and helium atoms. The elements that make up the bulk of the Earth—mostly iron, oxygen, and silicon—are relatively rare in the universe as a whole, as are the elements of which living organisms are made—carbon, oxygen, nitrogen, and phosphorus, among others. (You may find

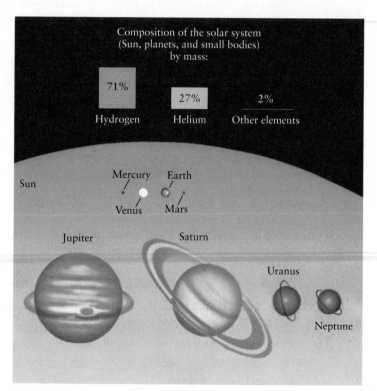

Figure 8-1

Composition of the Solar System Hydrogen and helium make up almost all of the mass of our solar system. Other elements such as carbon, oxygen, nitrogen, iron, gold, and uranium constitute only 2% of the total mass.

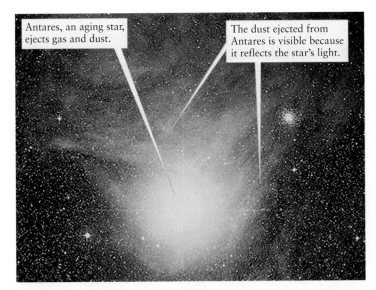

Figure 8-2   R I V U X G

A Mature Star Ejecting Gas and Dust The star Antares is shedding material from its outer layers, forming a thin cloud around the star. We can see the cloud because some of the ejected material has condensed into tiny grains of dust that reflect the star's light. (Dust particles in the air around you reflect light in the same way, which is why you can see them within a shaft of sunlight in a darkened room). Antares lies some 600 light-years from Earth in the constellation Scorpio. (David Malin/Anglo-Australian Observatory)

it useful to review the periodic table of the elements, described in Box 5-5.)

The Origin of the Elements and Cosmic "Recycling"

There is a good reason for this overwhelming abundance of hydrogen and helium. A wealth of evidence has led astronomers to conclude that the universe began some 13.7 billion years ago with a violent event called the Big Bang (see Section 1-4). Only the lightest elements—hydrogen and helium, as well as tiny amounts of lithium and perhaps beryllium—emerged from the enormously high temperatures following this cosmic event. All the heavier elements were manufactured by stars later, either by thermonuclear fusion reactions deep in their interiors or by the violent explosions that mark the end of massive stars. Were it not for these processes that take place only in stars, there would be no heavy elements in the universe, no planet like our Earth, and no humans to contemplate the nature of the cosmos.

Because our solar system contains heavy elements, it must be that at least some of its material was once inside other stars. But how did this material become available to help build our solar system? The answer is that near the ends of their lives, stars cast much of their matter back out into space. For most stars this process is a comparatively gentle one, in which a star's outer layers are gradually expelled. **Figure 8-2** shows a star losing material in this fashion. This ejected material appears as the cloudy region, or **nebulosity** (from *nubes*, Latin for "cloud"), that surrounds the star and is illuminated by it. A few stars eject matter much more dramatically

at the very end of their lives, in a spectacular detonation called a *supernova*, which blows the star apart (see Figure 1-6).

No matter how it escapes, the ejected material contains heavy elements dredged up from the star's interior, where they were formed. This material becomes part of the **interstellar medium,** a tenuous collection of gas and dust that pervades the spaces between the stars. As different stars die, they increasingly enrich the interstellar medium with heavy elements. Observations show that new stars form as condensations in the interstellar medium (Figure 8-3). Thus, these new stars have an adequate supply of heavy elements from which to develop a system of planets, satellites, comets, and asteroids. Our own solar system must have formed from enriched material in just this way. Thus, our solar system contains "recycled" material that was produced long ago inside a now-dead star. This "recycled" material includes all of the carbon in your body, all of the oxygen that you breathe, and all of the iron and silicon in the soil beneath your feet.

The Abundances of the Elements

Stars create different heavy elements in different amounts. For example, oxygen (as well as carbon, silicon, and iron) is readily produced in the interiors of massive stars, whereas gold (as well as silver, platinum, and uranium) is created only under special

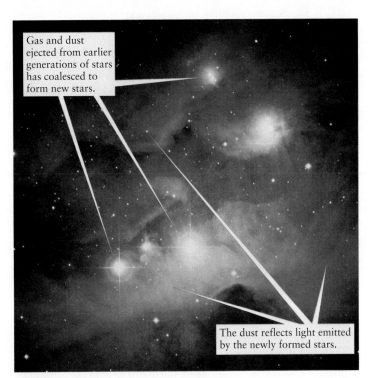

Figure 8-3   R I V U X G

New Stars Forming from Gas and Dust Unlike Figure 8-2, which depicts an old star that is ejecting material into space, this image shows young stars in the constellation Orion (the Hunter) that have only recently formed from a cloud of gas and dust. The bluish, wispy appearance of the cloud (called NGC 1973-1975-1977) is caused by starlight reflecting off interstellar dust grains within the cloud (see Box 5-4). The grains are made of heavy elements produced by earlier generations of stars. (David Malin/Anglo-Australian Observatory)

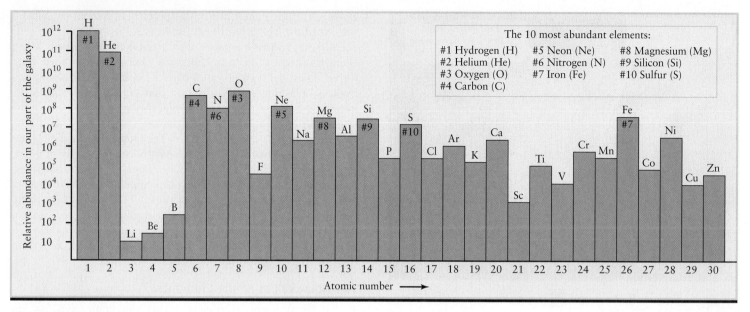

Figure 8-4

Abundances of the Lighter Elements This graph shows the abundances in our part of the Galaxy of the 30 lightest elements (listed in order of increasing atomic number) compared to a value of 10^{12} for hydrogen. The inset lists the 10 most abundant of these elements, which are also indicated in the graph. Notice that the vertical scale is not linear; each division on the scale corresponds to a tenfold increase in abundance. All elements heavier than zinc (Zn) have abundances of fewer than 1000 atoms per 10^{12} atoms of hydrogen.

circumstances. Consequently, gold is rare in our solar system and in the universe as a whole, while oxygen is relatively abundant (although still much less abundant than hydrogen or helium).

A convenient way to express the relative abundances of the various elements is to say how many atoms of a particular element are found for every trillion (10^{12}) hydrogen atoms. For example, for every 10^{12} hydrogen atoms in space, there are about 100 billion (10^{11}) helium atoms. From spectral analysis of stars and chemical analysis of Earth rocks, Moon rocks, and bits of interplanetary debris called meteorites, scientists have determined the relative abundances of the elements in our part of the Milky Way Galaxy today. Figure 8-4 shows the relative abundances of the 30 lightest elements, arranged in order of their **atomic numbers.** An element's atomic number is the number of protons in the nucleus of an atom of that element. It is also equal to the number of electrons orbiting the nucleus (see Box 5-5). In general, the greater the atomic number of an atom, the greater its mass.

CAUTION! Figure 8-1 shows that our solar system has 2.6 times (71% versus 27%) as much hydrogen than helium by *mass*, while Figure 8-4 shows that there is about 10 times as much hydrogen than helium by *number of atoms*. The explanation of this seeming inconsistency is that a helium atom has about 4 times the mass of a hydrogen atom, which makes helium more important on a per-mass basis than on a per-number basis.

The inset in Figure 8-4 lists the 10 most abundant elements. Note that even oxygen (chemical symbol O), the third most abundant element, is quite rare relative to hydrogen (H) and helium (He): There are only 8.5×10^8 oxygen atoms for each 10^{12} hydrogen atoms and each 10^{11} helium atoms. Expressed another way, for each oxygen atom in our region of the Milky Way Galaxy, there are about 1200 hydrogen atoms and 120 helium atoms.

In addition to the 10 most abundant elements listed in Figure 8-3, five elements are moderately abundant: sodium (Na), aluminum (Al), argon (Ar), calcium (Ca), and nickel (Ni). These elements have abundances in the range of 10^6 to 10^7 relative to the standard 10^{12} hydrogen atoms. Most of the other elements are much rarer. For example, for every 10^{12} hydrogen atoms in the solar system, there are only six atoms of gold.

The small cosmic abundances of elements other than hydrogen and helium help to explain why the terrestrial planets are so small (Property 1 in Table 8-1). Because the heavier elements required to make a terrestrial planet are rare, only relatively small planets can form out of them. By contrast, hydrogen and helium are so abundant that it was possible for these elements to form large Jovian planets.

8-3 The abundances of radioactive elements reveal the solar system's age

The heavy elements can tell us even more about the solar system: They also help us determine its age. The particular heavy elements that provide us with this information are *radioactive*. Their atomic nuclei are unstable because they contain too many protons or too many neutrons. A radioactive nucleus therefore ejects particles until it becomes stable. In doing so, a nucleus may change from one element to another. Physicists refer to this transmutation as **radioactive decay.** For example, a radioactive form of the element rubidium (atomic number 37) decays into the element strontium (atomic number 38) when one of the neutrons in the rubidium nucleus decays into a proton and an electron (which is ejected from the nucleus).

> Our solar system, which formed 9 billion years after the Big Bang, is a relative newcomer in the universe

Experiment shows that each type of radioactive nucleus decays at its own characteristic rate, which can be measured in the laboratory. This observation is the key to a technique called **radioactive dating,** which is used to determine the ages of rocks. If a rock contained a certain amount of radioactive rubidium when it first formed, over time more and more of the atoms of rubidium within the rock will decay into strontium atoms. The ratio of the number of strontium atoms the rock contains to the number of rubidium atoms it contains then gives a measure of the age of the rock. **Box 8-1** describes radioactive dating in more detail.

Dating the Solar System

Scientists have applied techniques of radioactive dating to rocks taken from all over the Earth. The results show that most rocks are tens or hundreds of millions of years old, but that some rocks are as much as 4 billion (4×10^9) years old. These results confirm that geologic processes have produced new surface material over the Earth's history, as we concluded from the small number of impact craters found on Earth (see Section 7-6). They also show that the Earth is at least 4×10^9 years old.

Radioactive dating has also been applied to rock samples brought back from the Moon by the *Apollo* astronauts. The oldest *Apollo* specimen, collected from one of the most heavily cratered and hence most ancient regions of the Moon, is 4.5×10^9 years old. But the oldest rocks found anywhere in the solar system are **meteorites,** bits of interplanetary debris that survive passing through the Earth's atmosphere and land on our planet's surface (**Figure 8-5**). Radioactive dating of meteorites reveals that they are all nearly the *same* age, about 4.56 billion years old. The absence of any younger or older meteorites indicates that these are all remnants of objects that formed at the same time in the early solar system. We conclude that the age of the oldest meteorites, about 4.56×10^9 years, is the age of the solar system itself. Note that this almost inconceivably long span of time is only about one-third of the current age of the universe, 13.7×10^9 years.

Thus, by studying the abundances of radioactive elements, we are led to a remarkable insight: Some 4.56 billion years ago, a collection of hydrogen, helium, and heavy elements came together to form the Sun and all of the objects that orbit around it. All of those heavy elements, including the carbon atoms in your body and the oxygen atoms that you breathe, were created and cast off by stars that lived and died long before our solar system formed, during the first 9 billion years of the universe's existence. We are literally made of star dust.

Figure 8-5   R I V U X G

A Meteorite Although it resembles an ordinary Earth rock, this is actually a meteorite that fell from space. The proof of its extraterrestrial origin is the meteorite's surface, which shows evidence of having been melted by air friction as it entered our atmosphere at 40,000 km/h (25,000 mi/h). Meteorites are the oldest objects in the solar system. (Ted Kinsman/Photo Researchers, Inc.)

8-4 The Sun and planets formed from a solar nebula

We have seen how processes in the Big Bang and within ancient stars produced the raw ingredients of our solar system. But given these ingredients, how did they combine to make the Sun and planets? Astronomers have developed a variety of models for the origin of the solar system. The test of these models is whether they explain the properties of the present-day system of Sun and planets.

> Astronomers see young stars that may be forming planets today in the same way that our solar system did billions of years ago

The Failed Tidal Hypothesis

Any model of the origin of the solar system must explain why all the planets orbit the Sun in the same direction and in nearly the same plane (Property 2 in Table 8-1). One model that was devised explicitly to address this issue was the *tidal hypothesis,* proposed in the early 1900s. As we saw in Section 4-8, two nearby planets, stars, or galaxies exert tidal forces on each other that cause the objects to elongate. In the tidal hypothesis, another star happened to pass close by the Sun, and the star's tidal forces drew a long filament out of the Sun. The filament material would then go into orbit around the Sun, and all of it would naturally orbit in the same direction and in the same plane. From this filament the planets would condense. However, it was shown in the 1930s that tidal forces strong enough to pull a filament out of the Sun would cause the filament to disperse before it could condense into planets. Hence, the tidal hypothesis cannot be correct.

BOX 8-1   Tools of the Astronomers Trade

Radioactive Dating

How old are the rocks found on the Earth and other planets? Are rocks found at different locations the same age or of different ages? How old are meteorites? Questions like these are important to scientists who wish to reconstruct the history of our solar system. But simply looking at a rock cannot tell us whether it was formed a thousand years, a million years, or a billion years ago. Fortunately, most rocks contain trace amounts of radioactive elements such as uranium. By measuring the relative abundances of various radioactive isotopes and their decay products within a rock, scientists can determine the rock's age.

As we saw in Box 5-5, every atom of a particular element has the same number of protons in its nucleus. However, different isotopes of the same element have different numbers of neutrons in their nuclei. For example, the common isotopes of uranium are ^{235}U and ^{238}U. Each isotope of uranium has 92 protons in its nucleus (correspondingly, uranium is element 92 in the periodic table; see Box 5-5). However, a ^{235}U nucleus contains 143 neutrons, whereas a ^{238}U nucleus has 146 neutrons.

A radioactive nucleus with too many protons or too many neutrons is unstable; to become stable, it *decays* by ejecting particles until it becomes stable. If the number of protons (the atomic number) changes in this process, the nucleus changes from one element to another.

Some radioactive isotopes decay rapidly, while others decay slowly. Physicists find it convenient to talk about the decay rate in terms of an isotope's **half-life.** The half-life of an isotope is the time interval in which one-half of the nuclei decay. For example, the half-life of ^{238}U is 4.5 billion (4.5×10^9) years. This means that if you start out with 1 kg of ^{238}U, after 4.5 billion years you will have only $\frac{1}{2}$ kg of ^{238}U remain-

ing; the other $\frac{1}{2}$ kg will have turned into other elements. If you want another half-life, so that a total of 9.0 billion years has elapsed, only $\frac{1}{4}$ kg of ^{238}U—one-half of one-half of the original amount—will remain. Several isotopes useful for determining the ages of rocks are listed in the accompanying table.

To see how geologists date rocks, consider the slow conversion of radioactive rubidium (^{87}Rb) into strontium (^{87}Sr). (The periodic table in Box 5-5 shows that the atomic numbers for these elements are 37 for rubidium and 38 for strontium, so in the decay a neutron is transformed into a proton. In this process an electron is ejected from the nucleus.) Over the years, the amount of ^{87}Rb in a rock decreases, while the amount of ^{87}Sr increases. Because the ^{87}Sr appears in the rock due to radioactive decay, this isotope is called *radiogenic.* Dating the rock is not simply a matter of measuring its ratio of rubidium to strontium, however, because the rock already had some strontium in it when it was formed. Geologists must therefore determine how much fresh strontium came from the decay of rubidium after the rock's formation.

To do this, geologists use as a reference another isotope of strontium whose concentration has remained constant. In this case, they use ^{86}Sr, which is stable and is not created by radioactive decay; it is said to be *nonradiogenic.* Dating a rock thus entails comparing the ratio of radiogenic and nonradiogenic strontium (^{87}Sr/^{86}Sr) in the rock to the ratio of radioactive rubidium to nonradiogenic strontium (^{87}Rb/^{86}Sr). Because the half-life for converting ^{87}Rb into ^{87}Sr is known, the rock's age can then be calculated from these ratios (see the table).

Radioactive isotopes decay with the same half-life no matter where in the universe they are found. Hence, scientists have used the same techniques to determine the ages of rocks from the Moon and of meteorites.

Original Radioactive Isotope	Final Stable Isotope	Half-Life (Years)	Range of Ages that Can Be Determined (Years)
Rubidium (^{87}Rb)	Strontium (^{87}Sr)	47.0 billion	10 million–4.56 billion
Uranium (^{238}U)	Lead (^{206}Pb)	4.5 billion	10 million–4.56 billion
Potassium (^{40}K)	Argon (^{40}Ar)	1.3 billion	50,000–4.56 billion
Carbon (^{14}C)	Nitrogen (^{14}N)	5730	100–70,000

The Successful Nebular Hypothesis

An entirely different model is now thought to describe the most likely series of events that led to our present solar system (**Figure 8-6**). The central idea of this model dates to the late 1700s, when the German philosopher Immanuel Kant and the French scientist Pierre-Simon de Laplace turned their attention to the manner in which the planets orbit the Sun. Both concluded that the arrangement of the orbits—all in the same direction and in nearly the same plane—could not be mere coincidence. To explain the or-

bits, Kant and Laplace independently proposed that our entire solar system, including the Sun as well as all of its planets and satellites, formed from a vast, rotating cloud of gas and dust called the **solar nebula** (Figure 8-6*a*). This model is called the **nebular hypothesis.**

The consensus among today's astronomers is that Kant and Laplace were exactly right. In the modern version of the nebular hypothesis, at the outset the solar nebula was similar in character to the nebulosity shown in Figure 8-3 and had a mass somewhat greater than that of our present-day Sun.

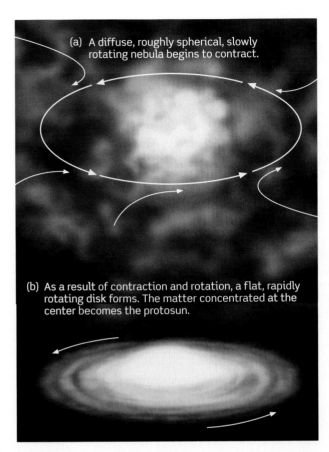

Figure 8-6

The Birth of the Solar System (a) A cloud of interstellar gas and dust begins to contract because of its own gravity. (b) As the cloud flattens and spins more rapidly around its rotation axis, a central condensation develops that evolves into a glowing protosun. The planets will form out of the surrounding disk of gas and dust.

(a) A diffuse, roughly spherical, slowly rotating nebula begins to contract.

(b) As a result of contraction and rotation, a flat, rapidly rotating disk forms. The matter concentrated at the center becomes the protosun.

As the newly created protosun continued to contract and become denser, its temperature continued to climb as well. After about 10^5 (100,000) years, the protosun's surface temperature stabilized at about 6000 K, but the temperature in its interior kept increasing to ever higher values as the central regions of the protosun became denser and denser. Eventually, after perhaps 10^7 (10 million) years had passed since the solar nebula first began to contract, the gas at the center of the protosun reached a density of about 10^5 kg/m^3 (a hundred times denser than water) and a temperature of a few million kelvins (that is, a few times 10^6 K). Under these extreme conditions, nuclear reactions that convert hydrogen into helium began in the protosun's interior. When this happened, the energy released by these reactions stopped the contraction and a true star was born. Nuclear reactions continue to the present day in the interior of the Sun and are the source of all the energy that the Sun radiates into space.

The Protoplanetary Disk

If the solar nebula had not been rotating at all, everything would have fallen directly into the protosun, leaving nothing behind to form the planets. Instead, the solar nebula must have had an overall slight rotation, which caused its evolution to follow a different path. As the slowly rotating nebula collapsed inward, it would naturally have tended to rotate faster. This relationship between the size of an object and its rotation speed is an example of a general principle called the **conservation of angular momentum.**

ANALOGY Figure skaters make use of the conservation of angular momentum. When a spinning skater pulls her arms and legs in close to her body, the rate at which she spins automatically increases (**Figure 8-7**). Even if you are not a figure skater,

Each part of the nebula exerted a gravitational attraction on the other parts, and these mutual gravitational pulls tended to make the nebula contract. As it contracted, the greatest concentration of matter occurred at the center of the nebula, forming a relatively dense region called the **protosun.** As its name suggests, this part of the solar nebula eventually developed into the Sun. The planets formed from the much sparser material in the outer regions of the solar nebula. Indeed, the mass of all the planets together is only 0.1% of the Sun's mass.

Evolution of the Protosun

When you drop a ball, the gravitational attraction of the Earth makes the ball fall faster and faster as it falls; in the same way, material falling inward toward the protosun would have gained speed. As this fast-moving material ran into the protosun, the energy of the collision was converted into thermal energy, causing the temperature deep inside the solar nebula to climb. This process, in which the gravitational energy of a contracting gas cloud is converted into thermal energy, is called **Kelvin-Helmholtz contraction,** after the nineteenth-century physicists who first described it.

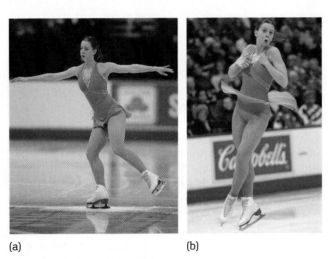

(a)   **(b)**

Figure 8-7   R I V U X G

Conservation of Angular Momentum A figure skater who **(a)** spins slowly with her limbs extended will naturally speed up when **(b)** she pulls her limbs in. In the same way, the solar nebula spun more rapidly as its material contracted toward the center of the nebula. (AP Photo/ Amy Sancetta)

(a)

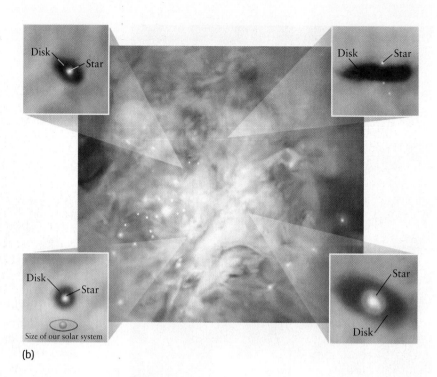

(b)

VIDEO 8.1

Figure 8-8   R I **V** U X G

Protoplanetary Disks (a) The Orion Nebula is a star-forming region located some 1500 light-years from Earth. It is the middle "star" in Orion's "sword" (see Figure 2-2a). The smaller, bluish nebula is the object shown in Figure 8-3. **(b)** This view of the center of the Orion Nebula is a mosaic of Hubble Space Telescope images. The four insets are false-color close-ups of four protoplanetary disks that lie within the nebula. A young, recently formed star is at the center of each disk. (The disk at upper right is seen nearly edge-on.) The inset at the lower left shows the size of our own solar system for comparison. (a: Anglo-Australian Observatory image by David Malin; b: C. R. O'Dell and S. K. Wong, Rice University; NASA)

you can demonstrate this by sitting on an office chair. Sit with your arms outstretched and hold a weight, like a brick or a full water bottle, in either hand. Now use your feet to start your body and the chair rotating, lift your feet off the ground, and then pull your arms inward. Your rotation will speed up quite noticeably.

As the solar nebula began to rotate more rapidly, it also tended to flatten out (Figure 8-6b). From the perspective of a particle rotating along with the nebula, it felt as though there were a force pushing the particle away from the nebula's axis of rotation. (Likewise, passengers on a spinning carnival ride seem to feel a force pushing them outward and away from the ride's axis of rotation.) This apparent force was directed opposite to the inward pull of gravity, and so it tended to slow the contraction of material toward the nebula's rotation axis. But there was no such effect opposing contraction in a direction parallel to the rotation axis. Some 10^5 (100,000) years after the solar nebula first began to contract, it had developed the structure shown in Figure 8-5b, with a rotating, flattened disk surrounding what will become the protosun. This disk is called the **protoplanetary disk** or **proplyd**, since planets formed from its material. This model explains why their orbits all lie in essentially the same plane and why they all orbit the Sun in the same direction.

There were no humans to observe these processes taking place during the formation of the solar system. But Earth astronomers have seen disks of material surrounding other stars that formed only recently. These, too, are called protoplanetary disks, because it is thought that planets can eventually form from these disks around other stars. Hence, these disks are planetary systems that are still "under construction." By studying these disks around other stars, astronomers are able to examine what our solar nebula may have been like some 4.56×10^9 years ago.

WEB LINK 8.5

Figure 8-8 shows a number of protoplanetary disks in the Orion Nebula, a region of active star formation. A star is visible at the center of each disk, which reinforces the idea that our Sun began to shine before the planets were fully formed. (The images that open this chapter show even more detailed views of disks surrounding young stars.) A study of 110 young stars in the Orion Nebula detected protoplanetary disks around 56 of them, which suggests that systems of planets may form around a substantial fraction of stars. Later in this chapter we will see direct evidence for planets that have formed around stars other than the Sun.

8-5 The terrestrial planets formed by the accretion of planetesimals

We have seen how the solar nebula would have contracted to form a young Sun with a protoplanetary disk rotating around it. But how did the material in this disk form into planets? And why are terrestrial planets in the inner solar system, while the giant Jovian planets are in the outer solar system (Property 3 in Table 8-1)? In this section and the next we will see how the nebular hypothesis provides answers to these questions.

> **Rocky planets formed in the inner solar nebula as a consequence of the high temperatures close to the protosun**

Temperatures in the Solar Nebula

To understand how the planets, asteroids, and comets formed, we must look at the conditions that prevailed within the solar nebula. The density of material in the part of the nebula outside the protosun was rather low, as was the pressure of the nebula's gas. If the pressure is sufficiently low, a substance cannot remain in the liquid state, but must end up as either a solid or a gas. For a given pressure, what determines whether a certain substance is a solid or a gas is its **condensation temperature.** If the temperature of a substance is above its condensation temperature, the substance is a gas; if the temperature is below the condensation temperature, the substance solidifies into tiny specks of dust or snowflakes. You can often see similar behavior on a cold morning. The air temperature can be above the condensation temperature of water, while the cold windows of parked cars may have temperatures below the condensation temperature. Thus, water molecules in the air remain as a gas (water vapor) but form solid ice particles (frost) on the car windows.

Substances such as water (H_2O), methane (CH_4), and ammonia (NH_3) have low condensation temperatures, ranging from 100 to 300 K. Rock-forming substances have much higher condensation temperatures, in the range from 1300 to 1600 K. The gas cloud from which the solar system formed had an initial temperature near 50 K, so all of these substances could have existed in solid form. Thus, the solar nebula would have been populated by an abundance of small ice particles and solid dust grains like the one shown in Figure 8-9. (Recall from Section 7-4 that "ice" can refer to frozen CO_2, CH_4, or NH_3, as well as frozen water.) But hydrogen and helium, the most abundant elements in the solar nebula, have condensation temperatures so near absolute zero that these substances always existed as gases during the creation of the solar system. You can best visualize the initial state of the solar nebula as a thin gas of hydrogen and helium strewn with tiny dust particles.

This state of affairs changed as the central part of the solar nebula underwent Kelvin-Helmholtz contraction to form the protosun. During this phase, the protosun was actually quite a bit more luminous than the present-day Sun, and it heated the innermost part of the solar nebula to temperatures above 2000 K. Meanwhile, temperatures in the outermost regions of the solar nebula remained below 50 K.

Figure 8-10 shows the probable temperature distribution in the solar nebula at this stage in the formation of our solar system. In the inner regions of the solar nebula, water, methane, and ammonia were vaporized by the high temperatures. Only materials with high condensation temperatures could have remained solid. Of these materials, iron, silicon, magnesium, and sulfur were particularly abundant, followed closely by aluminum, calcium, and nickel. (Most of these elements were present in the form of oxides, which are chemical compounds containing oxy-

 Figure 8-9

A Grain of Cosmic Dust This highly magnified image shows a microscopic dust grain that came from interplanetary space. It entered Earth's upper atmosphere and was collected by a high-flying aircraft. Dust grains of this sort are abundant in star-forming regions like that shown in Figure 8-3. These tiny grains were also abundant in the solar nebula and served as the building blocks of the planets. (Donald Brownlee, University of Washington)

gen. These compounds also have high condensation temperatures.) In contrast, ice particles and ice-coated dust grains were able to survive in the cooler, outer portions of the solar nebula.

Planetesimals, Protoplanets, and Terrestrial Planets

In the inner part of the solar nebula, the grains of high-condensation-temperature materials would have

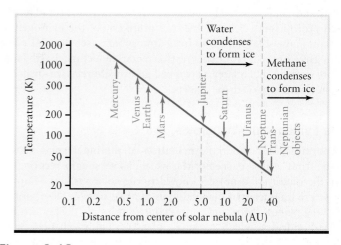

Figure 8-10

Temperature Distribution in the Solar Nebula This graph shows how temperatures probably varied across the solar nebula as the planets were forming. Note the general decline in temperature with increasing distance from the center of the nebula. Beyond 5 AU from the center of the nebula, temperatures were low enough for water to condense and form ice; beyond 30 AU, methane (CH_4) could also condense into ice.

collided and merged into small chunks. Initially, electric forces—that is, chemical bonds—held these chunks together, in the same way that chemical bonds hold an ordinary rock together. Over a few million years, these chunks coalesced into roughly 10^9 asteroidlike objects called **planetesimals,** with diameters of a kilometer or so. These were large enough to be held together by the gravitational attraction of the different parts of the planetesimal for each other. During the next stage, gravitational attraction between the planetesimals caused them to collide and accumulate into still-larger objects called **protoplanets,** which were roughly the size and mass of our Moon. This accumulation of material to form larger and larger objects is called **accretion.** During the final stage, these Moon-sized protoplanets collided to form the inner planets. This final episode must have involved some truly spectacular, world-shattering collisions.

In the inner solar nebula only materials with high condensation temperatures could form dust grains and hence protoplanets, so the result was a set of planets made predominantly of materials such as iron, silicon, magnesium, and nickel, and their oxides. This is just the composition of the present-day terrestrial planets.

At first the material that coalesced to form protoplanets in the inner solar nebula remained largely in solid form, despite the high temperatures close to the protosun. But as the protoplanets grew, they were heated by the violent impacts of planetesimals as well as the decay of radioactive elements such as uranium, and this heat caused melting. Thus, the terrestrial planets began their existence as spheres of at least partially molten rocky materials. Material was free to move within these molten spheres, so the denser, iron-rich minerals sank to the centers of the planets while the less dense silicon-rich minerals floated to their surfaces. This process is called **chemical differentiation** (see Box 7-1). In this way the terrestrial planets developed their dense iron cores. We saw in Section 7-7 that this electrically conductive iron gives rise to the magnetic fields of the terrestrial planets.

Because the materials that went into the terrestrial planets are relatively scarce, these planets ended up having relatively small mass and hence relatively weak gravity. As a result, these terrestrial protoplanets were unable to capture any of the hydrogen or helium in the solar nebula to form atmospheres (see Box 7-2). The thin envelopes of atmosphere that encircle Venus, Earth, and Mars evolved much later as trapped gases were released from the molten interiors of these planets.

The Evidence of Chondrules

Important clues about the evolution of the inner solar system come from studies of meteorites, which, as we saw in Section 8-3, are the oldest solid objects known in the solar system. Many of these are fragments of planetesimals that were never incorporated into the planets. Most meteorites contain not only dust grains but also **chondrules,** which are small, glassy, roughly spherical blobs (**Figure 8-11**). (The "ch" in "chondrule" is pronounced as in "chord.") Liquids tend to form spherical drops (like drops of water), so the shape of chondrules shows that they were once molten.

Attempts to produce chondrules in the laboratory show that the blobs must have melted suddenly, then solidified over the space of only an hour or so. This means they could not have been melted by the temperature of the inner solar nebula, which would have remained high for hundreds of thousands of years after the

Chondrules   1 cm

Figure 8-11   R I **V** U X G

Chondrules This is a cross-section of a meteorite that landed in Chihuahua, Mexico in 1969. The light-colored spots are individual chondrules that range in size from a few millimeters to a few tenths of a millimeter. Many of the chondrules are coated with dust grains like the one shown in Figure 8-8. (The R. N. Hartman Collection)

formation of the protosun. Instead, chondrules must have been produced by sudden, high-energy events that quickly melted material and then permitted it to cool. These events were probably shock waves propagating through the gas of the solar nebula, like the sonic booms produced by an airplane traveling faster than sound. The sheer number of chondrules suggests that such shock waves occurred throughout the inner solar nebula over a period of more than 10^6 years. The words "cloud" and "nebula" may suggest material moving in gentle orbits around the protosun; the existence of chondrules suggests that conditions in the inner solar nebula were at times quite violent.

Simulating the Formation of Planets

Astronomers use computer simulations to learn how the inner planets could have formed from planetesimals. A computer is programmed to simulate a large number of particles circling a newborn sun along orbits dictated by Newtonian mechanics. As the simulation proceeds, the particles coalesce to form larger objects, which in turn collide to form planets. By performing a variety of simulations, each beginning with somewhat different numbers of planetesimals in different orbits, it is possible to see what kinds of planetary systems would have been created under different initial conditions. Such studies demonstrate that a wide range of initial conditions ultimately lead to basically the same result in the

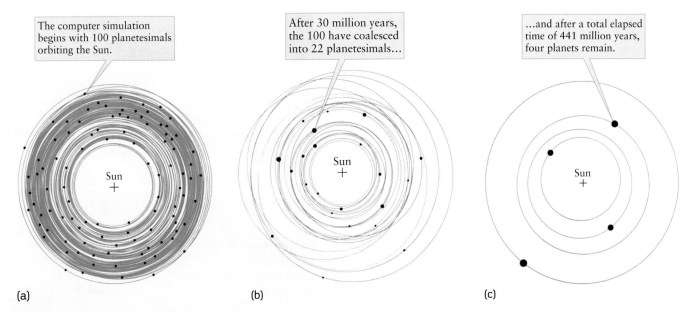

The computer simulation begins with 100 planetesimals orbiting the Sun.

Sun +

(a)

After 30 million years, the 100 have coalesced into 22 planetesimals...

Sun +

(b)

...and after a total elapsed time of 441 million years, four planets remain.

Sun +

(c)

Figure 8-12

Accretion of the Terrestrial Planets These three drawings show the results of a computer simulation of the formation of the inner planets. In this simulation, the inner planets were essentially formed after 150 million years. (Adapted from George W. Wetherill)

inner solar system: Accretion continues for roughly 10^8 (100 million) years and typically forms four or five terrestrial planets with orbits between 0.3 and 1.6 AU from the Sun.

Figure 8-12 shows one particular computer simulation. The calculations began with 100 planetesimals, each having a mass of 1.2×10^{23} kg. This choice ensures that the total mass (1.2×10^{25} kg) equals the combined mass of the four terrestrial planets (Mercury through Mars) plus their satellites. The initial orbits of planetesimals are inclined to each other by angles of less than 5° to simulate a thin layer of particles orbiting the protosun.

After an elapsed time simulating 30 million (3×10^7) years, the 100 original planetesimals have merged into 22 protoplanets (Figure 8-9b). After 79 million (7.9×10^7) years, these have combined into 11 larger protoplanets. Nearly another 100 million (10^8) years elapse before further mergers leave just six growing protoplanets. Finally, Figure 8-9c shows four planets following nearly circular orbits after a total elapsed time of 441 million (4.41×10^8) years. In this particular simulation, the fourth planet from the Sun ends up being the most massive; in our own solar system, the third planet (Earth) is the most massive of the terrestrial planets. Note that due to collisions, the four planets in the simulation end up in orbits that are nearly circular, just like the orbits of most of the planets in our solar system.

8-6 Gases in the outer solar nebula formed the Jovian planets, whose gravitational influence gave rise to the small bodies of the solar system

We have seen how the high temperatures in the inner solar nebula led to the formation of the small, rocky terrestrial planets. To explain the very different properties of the Jovian planets, we need to consider the conditions that prevailed in the relatively cool outer regions of the solar nebula. Two leading models explain how the Jovian planets could have formed there.

Low temperatures in the outer solar nebula made it possible for planets to grow to titanic size

The Core Accretion Model

Like the inner planets, the outer planets may have begun to form by the accretion of planetesimals. The key difference is that ices as well as rocky grains were able to survive in the outer regions of the solar nebula, where temperatures were relatively low (see Figure 8-10). The elements of which ices are made are much more abundant than those that form rocky grains. Thus, much more solid material would have been available to form planetesimals in the outer solar nebula than in the inner part. As a result, solid objects several times larger than any of the terrestrial planets could have formed in the outer solar nebula. Each such object, made up of a mixture of ices and rocky material, could have become the core of a Jovian planet and served as a "seed" around which the rest of the planet eventually grew.

Thanks to the lower temperatures in the outer solar system, gas atoms (principally hydrogen and helium) were moving relatively slowly and so could more easily be captured by the strong gravity of these massive cores (see Box 7-2). Thus, the core of a Jovian protoplanet could have captured an envelope of gas as it continued to grow by accretion. This picture is called the **core accretion model** for the origin of the Jovian planets.

Calculations based on the core accretion model suggest that both rocky materials and gas slowly accumulated for several million years, until the masses of the core and the envelope became equal. From that critical moment on, the envelope pulled in all the gas it could get, dramatically increasing the protoplanet's mass and size. This runaway growth of the protoplanet would have continued until all the available gas was used up. The result was a huge planet with an enormously thick, hydrogen-rich envelope surrounding a rocky core with 5 to 10 times the mass of the Earth. This scenario could have occurred at four different distances from the Sun, thus creating the four Jovian planets.

In the core accretion model, Uranus and Neptune probably did not form at their present locations, respectively about 19 and 30 AU from the Sun. The solar nebula was too sparse at those distances to allow these planets to have grown to the present-day sizes. Instead, it is thought that Uranus and Neptune formed between 4 and 10 AU from the Sun, but were flung into larger orbits by gravitational interactions with Jupiter.

The Disk Instability Model

An alternative model for the origin of the Jovian planets, the **disk instability model,** suggests that they formed directly from the gas of the solar nebula. If the gas of the outer solar nebula was not smooth but clumpy, a sufficiently large and massive clump of gas could begin to collapse inward all on its own, like the protosun but on a smaller scale. Such a large clump would attract other gas, forming a very large planet of hydrogen and helium within just a few hundreds or thousands of years.

In the disk instability model the rocky core of a Jovian planet is *not* the original nucleus around which the planet grew; rather, it is the result of planetesimals and icy dust grains that were drawn gravitationally into the planet and then settled at the planet's center. Another difference is that in the disk instability model it is possible for Uranus and Neptune to have formed in their present-day orbits, since they would have grown much more rapidly than would be possible in the core accretion model.

Planetary scientists are actively researching whether the core accretion model or the disk instability model is the more correct description of the formation of the Jovian planets. In either case, we can explain why the terrestrial planets formed in the inner solar system while the Jovian planets formed in the outer solar system.

Figure 8-13 summarizes our overall picture of the formation of the terrestrial and Jovian planets.

The Origin of Jovian Satellites and Small Bodies

Like the terrestrial planets, the Jovian planets were quite a bit hotter during their formation than they are today. A heated gas expands, so these gas-rich planets must also have been much larger than at present. As each planet cooled and contracted, it would have formed a disk like a solar nebula in miniature (see Figure 8-6). Many of the satellites of the Jovian planets, including those shown in Table 7-2, are thought to have formed from ice particles and dust grains within these disks.

Jupiter, the most massive of the planets, was almost certainly the first to form. This may explain why there are asteroids, trans-Neptunian objects, and comets (see Section 7-5). Jupiter's tremendous mass would have exerted strong gravitational forces on any nearby planetesimals. Some would have been sent crashing into the Sun, while others would have been sent crashing into the terrestrial planets to form impact craters (see Section 7-6). Still others would have been ejected completely from the solar system. Only a relatively few rocky planetesimals survived to produce the present-day population of asteroids. Indeed, the mass of all known asteroids combined is less than the mass of the Moon.

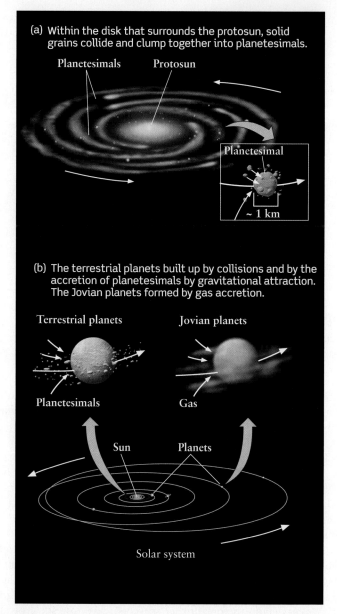

(a) Within the disk that surrounds the protosun, solid grains collide and clump together into planetesimals.

Planetesimals   Protosun

Planetesimal

~ 1 km

(b) The terrestrial planets built up by collisions and by the accretion of planetesimals by gravitational attraction. The Jovian planets formed by gas accretion.

Terrestrial planets   Jovian planets

Planetesimals   Gas

Sun   Planets

Solar system

Figure 8-13

Terrestrial Versus Jovian Planet Formation
(a) Planetesimals about 1 km in size formed in the solar nebula from small dust grains sticking together. **(b)** Planetesimals in the inner solar system grouped together to form the terrestrial planets, as in Figure 8-12. In the outer solar system, the Jovian planets may have begun as terrestrial-like planets that accumulated massive envelopes of hydrogen and helium. Alternatively, the Jovian planets may have formed directly from the gas of the solar nebula.

By contrast, trans-Neptunian objects (including Pluto and the other objects shown in Table 7-4) began as icy planetesimals beyond the orbit of Jupiter. The strong gravitational forces from all of the Jovian planets pushed these into orbits even further from the Sun, forming the immense ring of the Kuiper belt (**Figure 8-14**). The images that open this chapter show two young stars surrounded by dusty disks, one seen

Figure 8-14

The Kuiper Belt: A Dusty Debris Ring The gravitational influence of the Jovian planets pushed small, icy objects to the outer reaches of the solar system beyond Neptune. The result shown in this artist's conception is the Kuiper belt, a ring populated by trans-Neptunian objects like Pluto, icy planetesimals, and dust particles. (NASA/JPL-Caltech/T. Pyle, SSC)

edge-on and the other seen face-on, that resemble the Kuiper belt. As we saw in Section 7-5, the combined mass of all trans-Neptunian objects is hundreds of times larger than that of the asteroid belt, in part because the materials that form ices are much more abundant than those that form rock.

Some of the smaller icy objects are thought to have been pushed as far as 50,000 AU from the Sun, forming a spherical "halo" around the solar system called the Oort cloud. From time to time one of the smaller chunks of ice and rock from either the Kuiper belt or the Oort cloud is deflected into the inner solar system. If one of these chunks comes close enough to the Sun, it be-

gins to evaporate, producing a visible tail and appearing as a comet (see Figure 7-9).

The Final Stages of Solar System Evolution

While the planets, satellites, asteroids, and comets were forming, the protosun was also evolving into a full-fledged star with nuclear reactions occurring in its core (see Section 1-3). The time required for this to occur was about 10^8 years, roughly the same as that required for the formation of the terrestrial planets. Before the onset of nuclear reactions, however, the young Sun probably expelled a substantial portion of its mass into space (**Figure 8-15**).

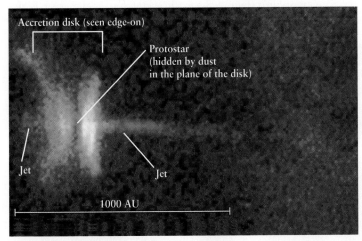

Accretion disk (seen edge-on)

Protostar
(hidden by dust
in the plane of the disk)

Jet

Jet

1000 AU

(a) Jets from a young star

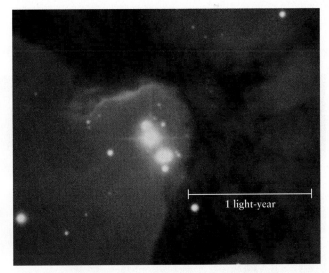

1 light-year

(b) Winds from young stars

Figure 8-15   R I **V** U X G

Jets and Winds from Young Stars (a) This protostar in the constellation Taurus (the Bull) is ejecting matter in two immense jets directed perpendicular to the plane of the accretion disk. Red denotes light emitted by hot ionized gas in the jet, while green denotes starlight scattered by dust particles in the disk. **(b)** An outpouring of particles and radiation from the surfaces of these young stars has

carved out a cavity in the surrounding dusty material. The stars lie within the Trifid Nebula in the constellation Sagittarius (the Archer). The scale of this image is about 100 times greater than that of image (a).
(a: C. Burrows, the WFPC-2 Investigation Definition Team, and NASA;
b: David Malin/Anglo-Australian Observatory)

Magnetic fields within the solar nebula would have funneled a portion of the nebula's mass into oppositely directed **jets** along the rotation axis of the nebula. Figure 8-15*a* shows one such jet emanating from a young star.

In addition, instabilities within the young Sun would have caused it to eject its tenuous outermost layers into space. This brief but intense burst of mass loss, observed in many young stars across the sky, is called a **T Tauri wind,** after the star in the constellation Taurus (the Bull) where it was first identified (see Figure 8-15*b*). Each of the protoplanetary disks in Figure 8-8 has a T Tauri star at its center. (The present-day Sun also loses mass from its outer layers in the form of high-speed electrons and protons, a flow called the **solar wind.** But this is minuscule compared with a T Tauri wind, which causes a star to lose mass 10^6 to 10^7 times faster than in the solar wind.)

With the passage of time, the combined effects of jets, the T Tauri wind, and accretion onto the planets would have swept the solar system nearly clean of gas and dust. With no more interplanetary material to gather up, the planets would have stabilized at roughly their present-day sizes, and the formation of the solar system would have been complete.

The *Cosmic Connections* figure summarizes our modern-day picture of the origin of our solar system. Prior to 1995 the only fully formed planetary system to which we could apply this model was our own. As we will see in the next section, astronomers can now further test this model on an ever-growing number of planets known to orbit other stars.

8-7 A variety of observational techniques reveal planets around other stars

If planets formed around our Sun, have they formed around other stars? That is, are there **extrasolar planets** orbiting stars other than the Sun? Our model for the formation of the planets would seem to suggest so. This model is based on the laws of physics and chemistry, which to the best of our understanding are the same throughout the universe. The discovery of a set of planets orbiting another star, with terrestrial planets in orbit close to the star and Jovian planets orbiting farther away, would be a tremendous vindication of our theory of solar system formation. It would also tell us that our planetary system is not unique in the universe. Because at least one planet in our solar system—the Earth—has the ability to support life, perhaps other planetary systems could also harbor living organisms.

Since 1995 astronomers have in fact discovered about 200 planets orbiting stars other than the Sun. However, none of these is a terrestrial planet. Most of them are more massive than Jupiter, and some are in eccentric, noncircular orbits quite unlike planetary orbits in our solar system. To appreciate how remarkable these discoveries are, we must look at the process that astronomers go through to search for extrasolar planets.

> Several stars are known to have systems of two or more planets, but none of these resemble our own solar system

Searching for Extrasolar Planets

It is very difficult to make direct observations of planets orbiting other stars. The problem is that planets are small and dim compared with stars; at visible wavelengths, the Sun is 10^9 times brighter than Jupiter and 10^{10} times brighter than the Earth. A hypothetical planet orbiting a distant star, even a planet 10 times larger than Jupiter, would be lost in the star's glare as seen through even the largest telescope on Earth.

Instead, indirect methods are used to search for extrasolar planets. One very powerful method is to search for stars that appear to "wobble" (**Figure 8-16**). If a star has a planet, it is not

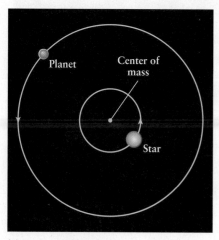

(a) A star and its planet

(b) The astrometric method

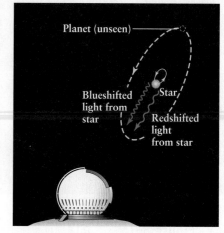

(c) The radial velocity method

Figure 8-16

Detecting a Planet by Measuring Its Parent Star's Motion
(a) A planet and its star both orbit around their common center of mass, always staying on opposite sides of this point. Even if the planet cannot be seen, its presence can be inferred if the star's motion can be detected.
(b) The astrometric method of detecting the unseen planet involves making direct measurements of the star's orbital motion. **(c)** In the radial velocity method, astronomers measure the Doppler shift of the star's spectrum as it moves alternately toward and away from the Earth. The amount of Doppler shift determines the size of the star's orbit, which in turn tells us about the unseen planet's orbit.

An overview of our present-day understanding of how the solar system formed.

Vast, rotating cloud of gas and dust (solar nebula)

Kelvin-Helmholtz contraction

Increased density, temperature

Protosun forms, begins to grow

Tens of millions of years

Protoplanetary disk forms, rock and ice particles collide, start to form planetesimals

Inner disk: rock particles collide, grow

Outer disk: ice particles collide, attract gas

Several hundred million years

Terrestrial planet in molten state

Several hundred million years

Accretion of terrestrial planets, protosun becomes hot enough for nuclear fusion to begin

Jovian planets accrete from gas in outer disk, terrestrial planets heat up, begin chemical differentiation

Several hundred million years

Jovian planet with core of rock, ices

T Tauri wind sweeps away gas and dust, leaving planets, moons, asteroids, Kuiper belt, Oort cloud

quite correct to say that the planet orbits the star. Rather, both the planet and the star move in elliptical orbits around a point called the **center of mass**. Imagine the planet and the star as sitting at opposite ends of a very long seesaw; the center of mass is the point where you would have to place the fulcrum in order to make the seesaw balance. Because of the star's much greater mass, the center of mass is much closer to the star than to the planet. Thus, while the planet may move in an orbit that is hundreds of millions of kilometers across, the star will move in a much smaller orbit (Figure 8-16a).

As an example, the Sun and Jupiter both orbit their common center of mass with an orbital period of 11.86 years. (Jupiter has more mass than the other eight planets put together, so it is a reasonable approximation to consider the Sun's wobble as being due to Jupiter alone.) Jupiter's orbit has a semimajor axis of 7.78×10^8 km, while the Sun's orbit has a much smaller semimajor axis of 742,000 km. The Sun's radius is 696,000 km, so the Sun slowly wobbles around a point not far outside its surface. If astronomers elsewhere in the Galaxy could detect the Sun's wobbling motion, they could tell that there was a large planet (Jupiter) orbiting the Sun. They could even determine the planet's mass and the size of its orbit, even though the planet itself was unseen.

Detecting the wobble of other stars is not an easy task. One approach to the problem, called the **astrometric method**, involves making very precise measurements of a star's position in the sky relative to other stars. The goal is to find stars whose positions change in a cyclic way (Figure 8-16b). The measurements must be made with very high accuracy (0.001 arcsec or better) and, ideally, over a long enough time to span an entire orbital period of the star's motion. Because this technique is so challenging to put into practice, it has not yet led to any confirmed detections of extrasolar planets.

A different approach to the problem is the **radial velocity method** (Figure 8-16c). This is based on the Doppler effect, which we described in Section 5-9. A wobbling star will alternately move away from and toward the Earth. This will cause the dark absorption lines in the star's spectrum (see Figure 5-13) to change their wavelengths in a periodic fashion. When the star is moving away from us, its spectrum will undergo a redshift to longer wavelengths. When the star is approaching, there will be a blueshift of the spectrum to shorter wavelengths. These wavelength shifts are very small because the star's motion around its orbit is quite slow. As an example, the Sun moves around its small orbit at only 12.5 m/s (45 km/h, or 28 mi/h). If the Sun were moving directly toward an observer at this speed, the hydrogen absorption line at a wavelength of 656 nm in the Sun's spectrum would be shifted by only 2.6×10^{-5} nm, or about 1 part in 25 million. Detecting these tiny shifts requires extraordinarily careful measurements and painstaking data analysis.

In 1995, Michel Mayor and Didier Queloz of the Geneva Observatory in Switzerland announced that by using the radial velocity method, they had discovered a planet orbiting the star 51 Pegasi, which is 47.9 light-years from the Sun in the constellation Pegasus (the Winged Horse). Their results were soon confirmed by the team of Geoff Marcy of San Francisco State University and Paul Butler of the University of California, Berkeley, using observations made at the University of California's Lick Observatory. For the first time, solid evidence had been found for a planet orbiting a star like our own Sun. Astronomers have since used the radial velocity method to discover planets orbiting more than 160 other stars at distances up to 600 light-years from the Sun. In at least 20 cases the star's "wobble" is too complex to have been caused by a single planet, so in these cases the star has a multiple-planet system. Figure 8-17 depicts a selection of these extrasolar planetary discoveries.

Extrasolar Planets: Surprising Masses and Orbits

Most of the extrasolar planets discovered by the radial velocity method have masses comparable with or larger than that of Saturn, and thus are presumably Jovian planets made primarily of hydrogen and helium. (This is difficult to confirm directly, because the spectra of the planets are very faint.) According to the picture we presented in Section 8-5, such Jovian planets would be expected to orbit relatively far from their stars, where temperatures were low enough to allow the buildup of a massive envelope of hydrogen and helium gas. But as Figure 8-17 shows, many extrasolar planets are in fact found orbiting very *close* to their stars. For example, the planet orbiting 51 Pegasi has a mass at least 0.46 times that of Jupiter but orbits only 0.052 AU from its star with an orbital period of just 4.23 days. In our own solar system, this orbit would lie well inside the orbit of Mercury!

Another surprising result is that many of the extrasolar planets found so far have orbits with very large eccentricities (see Figure 4-10b). As an example, the planet around the star 16 Cygni B has an orbital eccentricity of 0.67; its distance from the star varies between 0.55 AU and 2.79 AU. This is quite unlike planetary orbits in our own solar system, where no planet has an orbital eccentricity greater than 0.25.

Do these observations mean that our picture of how planets form is incorrect? If Jupiterlike extrasolar planets such as that orbiting 51 Pegasi formed close to their stars, the mechanism of their formation must have been very different from that which operated in our solar system.

But another possibility is that extrasolar planets actually formed at large distances from their stars and have migrated inward since their formation. If enough gas and dust remain in a disk around a star after its planets form, interactions between the disk material and an orbiting planet will cause the planet to lose energy and to spiral inward toward the star around which it orbits. The planet's inward migration can eventually stop because of subtle gravitational effects from the disk or from the star. Gravitational interactions between the planet and the disk, or between planets in the same planetary system, could also have forced an extrasolar planet into a highly eccentric, noncircular orbit.

One piece of evidence that young planets may spiral inward toward their parent stars is the spectrum of the star HD 82943, which has at least two planets orbiting it. The spectrum shows that this star's atmosphere contains a rare form of lithium that is found in planets, but which is destroyed in stars by nuclear reactions within 30 million years. The presence of this exotic form of lithium, known as ^{6}Li, means that in the past HD 82943 was orbited by at least one other planet.

Figure 8-17

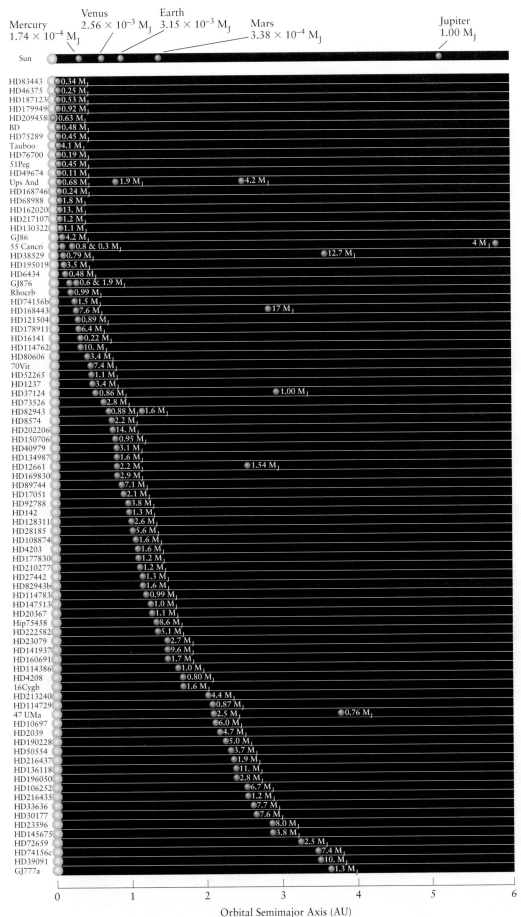

A Selection of Extrasolar Planets This figure summarizes what we know about planets orbiting a number of other stars. The star name is given at the left of each line. Each planet is shown at its average distance from its star (equal to the semimajor axis of its orbit). The mass of each planet—actually a lower limit—is given as a multiple of Jupiter's mass (M_J), equal to 318 Earth masses. Comparison with our own solar system (at the top of the figure) shows how closely many these extrasolar planets orbit their stars. (Adapted from the California and Carnegie Planet Search)

This planet vanished when it spiraled too close to the star, was vaporized, and was swallowed whole.

Are all stars equally likely to have planets? The answer appears to be no: Only about 5% of the stars surveyed with the radial velocity method show evidence of planets. However, the percentage is 10% to 20% for stars in which heavy elements are at least as abundant as in the Sun. The conclusion is that heavy elements appear to play an important role in the evolution of planets from an interstellar cloud. This agrees with the idea discussed in Section 8-5 that planetesimals (which are composed of heavy elements) are the building blocks of planets.

Analyzing Extrasolar Planets with the Transit Method

The radial velocity method gives us only a partial picture of what extrasolar planets are like. It cannot give us precise values for a planet's mass, only a lower limit. (An exact determination of the

mass would require knowing how the plane of the planet's orbit is inclined to our line of sight. Unfortunately, this angle is not known because the planet itself is unseen.) Furthermore, the radial velocity method by itself cannot tell us the diameter of a planet or what the planet is made of.

A technique called the **transit method** makes it possible to fill in these blanks about the properties of certain extrasolar planets. This method looks for the rare situation in which a planet comes between us and its parent star, an event called a **transit** (**Figure 8-18**). As in a partial solar eclipse (Section 3-5), this causes a small but measurable dimming of the star's light. If a transit is seen, the orbit must be nearly edge-on to our line of sight. Knowing the orientation of the orbit, the information obtained by radial velocity measurements of the star tells us the true mass (not just a lower limit) of the orbiting planet. About 5% of all known extrasolar planets undergo transits of their parent

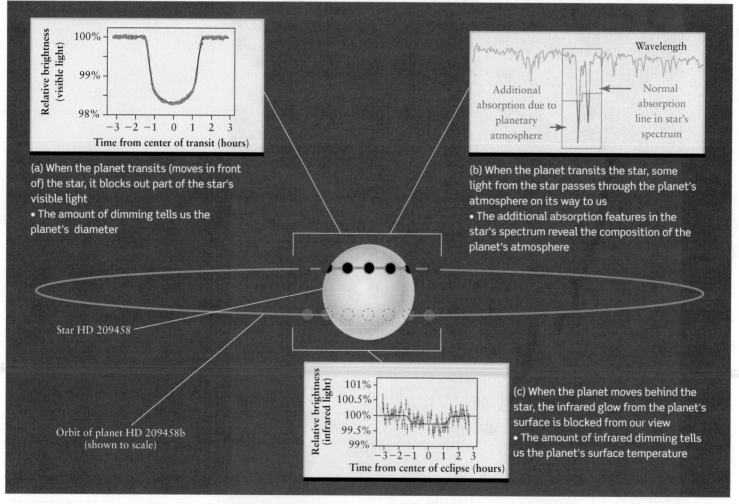

(a) When the planet transits (moves in front of) the star, it blocks out part of the star's visible light
• The amount of dimming tells us the planet's diameter

(b) When the planet transits the star, some light from the star passes through the planet's atmosphere on its way to us
• The additional absorption features in the star's spectrum reveal the composition of the planet's atmosphere

Star HD 209458

Orbit of planet HD 209458b (shown to scale)

(c) When the planet moves behind the star, the infrared glow from the planet's surface is blocked from our view
• The amount of infrared dimming tells us the planet's surface temperature

Figure 8-18

A Transiting Extrasolar Planet If the orbit of an extrasolar planet is nearly edge-on to our line of sight, like the planet that orbits the star HD 209458, we can learn about the planet's **(a)** diameter, **(b)** atmospheric composition, and **(c)** surface temperature. (S. Seager and C. Reed, *Sky and*

Telescope; H. Knutson, D. Charbonneau, R. W. Noyes (Harvard-Smithsonian CfA), T. M. Brown (HAO/NCAR), and R. L. Gilliland (STScI); A. Feild (STScI); NASA/JPL-Caltech/D. Charbonneau, Harvard-Smithsonian CfA)

(a) No microlensing

(b) Microlensing by star

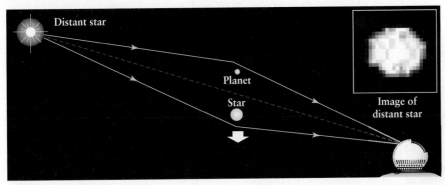

(c) Microlensing by star and planet

Figure 8-19

Microlensing Reveals an Extrasolar Planet
(a) A star with a planet drifts across the line of sight between a more distant star and a telescope on Earth. **(b)** The gravity of the closer star bends the light rays from the distant star, focusing the distant star's light and making it appear brighter. **(c)** The gravity of the planet causes a second increase in the distant star's brightness.

star, and all of these planets turn out to have masses comparable to that of Jupiter. These results reinforce the idea that other stars can end up with planets much like those in our solar system.

There are three other benefits of the transit method. One, the amount by which the star is dimmed during the transit depends on how large the planet is, and so tells us the planet's diameter (Figure 8-18*a*). Two, during the transit the star's light passes through the planet's atmosphere, where certain wavelengths are absorbed by the atmospheric gases. This absorption affects the spectrum of starlight that we measure and so allows us to determine the composition of the planet's atmosphere (Figure 8-18*b*). Three, with an infrared telescope it's possible to detect a slight dimming when the planet goes *behind* the star (Figure 8-18*c*). That is because the planet emits infrared radiation due to its own temperature, and this radiation is blocked

when the planet is behind the star. Measuring the amount of dimming tells us the amount of radiation emitted by the planet, which in turn tells us the planet's surface temperature.

One example of what the transit method can tell us is about the planet that orbits the star HD 209458, which is 153 light-years from the Sun. The mass of this planet is 0.69 that of Jupiter, but its diameter is 1.32 times larger than that of Jupiter. This low-mass, large-volume planet thus has only one-quarter the density of Jupiter. It is also very hot: Because the planet orbits just 0.047 AU from the star, its surface temperature is a torrid 1130 K (860°C, or 1570°F). Observations of the spectrum during a transit reveal the presence of hydrogen, carbon, oxygen, and sodium in a planetary atmosphere that is literally evaporating away in the intense starlight. Discoveries such as these give insight into the exotic circumstances that can be found in other planetary systems.

Other Techniques for Discovering Extrasolar Planets

In 2004 astronomers began to use a property of space discovered by Albert Einstein as a tool for detecting extrasolar planets. Einstein's general theory of relativity makes the remarkable statement that a star's gravity can deflect the path of a light beam just as it deflects the path of a planet or spacecraft. If a star drifts through the line of sight between Earth and a more distant star, the closer star's gravity acts like a lens that focuses the more distant star's light. Such **microlensing** causes the distant star's image as seen in a telescope to become brighter. If the closer star has a planet, the planet's gravity will cause a secondary brightening whose magnitude depends on the planet's mass (**Figure 8-19**). Using this technique astronomers have detected a handful of extrasolar planets at distances up to 21,000 light-years, and expect to find many more.

Since 2004 astronomers have also been able to record images of a few extrasolar planets. At infrared wavelengths a star outshines a Jupiter-sized planet by only about 100 to 1, compared to 10^9 to 1 at visible wavelengths. Hence, by using an infrared telescope aided by adaptive optics (Section 6-3) it is possible to resolve both a star and its planet if the two are sufficiently far apart. **Figure 8-20** shows an infrared image of the star 2M1207 and its planet. Unlike planets detected with the radial velocity or transit method, it is impractical to watch this planet go through a complete orbit: It is so far from its parent star that the orbital period is more than a thousand years. Instead, astronomers verify that it is a planet by confirming that the star and planet move together

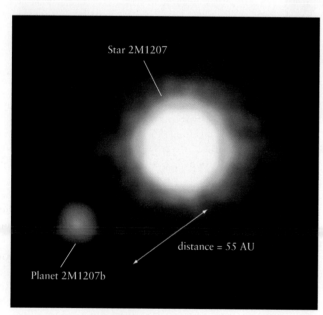

Figure 8-20   R I V U X G

Imaging an Extrasolar Planet This infrared image from the European Southern Observatory shows the star 2M1207 and a planet with about 1.5 times the diameter of Jupiter. First observed in 2004, this extrasolar planet was the first to be visible in a telescopic image. 2M1207 and its planet lie about 170 light-years from the Sun in the constellation Hydra (the Water Snake). (ESO/VLT/NACO)

through space, as they must do if they are held together by their mutual gravitational attraction. While it is not possible to measure the planet's mass directly, the spectrum and brightness of the planet tell astronomers its surface temperature and diameter.

The Search for Earthlike Planets

If there are planets orbiting other stars, are there any that resemble our own Earth? As of this writing we do not know the answer to this question. An Earth-sized planet has such weak gravity that it can produce only a tiny "wobble" in its parent star, making it very difficult to detect such planets using the radial velocity method. If an Earth-sized planet were to transit its parent star, its diameter is so small that it would cause only a miniscule dimming of the star's light. For these reasons, as of this writing, astronomers have been unable to detect extrasolar planets less massive than 7.5 Earth masses.

In the near future, however, orbiting telescopes being planned by both NASA and the European Space Agency (ESA) will search for terrestrial planets around other stars. One such orbiting telescope, scheduled for launch in 2008, is NASA's Kepler, which will use a one-meter telescope and sensitive detectors to look for transits of Earth-sized planets within a narrow patch of sky.

An even broader search for extrasolar planets will be carried out by ESA's Gaia mission, scheduled for launch in 2011. Gaia will survey a billion (10^9) stars—1% of all the stars in the Milky Way Galaxy—and will be able to detect star "wobbles" as small as 2×10^{-6} arcsec. With such high-precision data the astrometric method (Figure 8-16*b*) will finally come into its own. Astronomers estimate that Gaia will find from 10,000 to 50,000 Jupiter-sized planets. If our solar system is fairly typical, a substantial number of these planets should move around their parent stars in orbits like those of the Jovian planets around the Sun, with semimajor axes greater than 3 AU and low eccentricities. If Gaia finds such planetary systems that resemble our own, they will be natural places to look for Earth-sized planets. The next few decades may tell whether our own solar system is an exception or just one of a host of similar systems throughout the Galaxy.

Key Words

Term preceded by an asterisk (*) is discussed in Box 8-1.

accretion, p. 194
astrometric method (for detecting extrasolar planets), p. 200
atomic number, p. 188
center of mass, p. 200
chemical differentiation, p. 194
chondrule, p. 194
condensation temperature, p. 193
conservation of angular momentum, p. 191
core accretion model, p. 195

disk instability model, p. 196
extrasolar planet, p. 198
*half-life, p. 190
interstellar medium, p. 187
jets, p. 198
Kelvin-Helmholtz contraction, p. 191
meteorite, p. 189
microlensing, p. 204
nebulosity, p. 187
nebular hypothesis, p. 190
planetesimal, p. 194
protoplanet, p. 194

protoplanetary disk (proplyd),
  p. 192
protosun, p. 191
radial velocity method (for
  detecting extrasolar planets),
  p. 200
radioactive dating, p. 189

radioactive decay, p. 189
solar nebula, p. 190
solar wind, p. 198
T Tauri wind, p. 198
transit, p. 202
transit method (for detecting
  extrasolar planets), p. 202

Key Ideas

Models of Solar System Formation: The most successful model of the origin of the solar system is called the nebular hypothesis. According to this hypothesis, the solar system formed from a cloud of interstellar material called the solar nebula. This occurred 4.56 billion years ago (as determined by radioactive dating).

The Solar Nebula and Its Evolution: The chemical composition of the solar nebula, by mass, was 98% hydrogen and helium (elements that formed shortly after the beginning of the universe) and 2% heavier elements (produced much later in the centers of stars, and cast into space when the stars died). The heavier elements were in the form of ice and dust particles.

• The nebula flattened into a disk in which all the material orbited the center in the same direction, just as do the present-day planets.

Formation of the Planets and Sun: The terrestrial planets, the Jovian planets, and the Sun followed different pathways to formation.

• The four terrestrial planets formed through the accretion of dust particles into planetesimals, then into larger protoplanets.

• In the core accretion model, the four Jovian planets began as rocky protoplanetary cores, similar in character to the terrestrial planets. Gas then accreted onto these cores in a runaway fashion.

• In the alternative disk instability model, the Jovian planets formed directly from the gases of the solar nebula. In this model the cores formed from planetesimals falling into the planets.

• The Sun formed by gravitational contraction of the center of the nebula. After about 10^8 years, temperatures at the protosun's center became high enough to ignite nuclear reactions that convert hydrogen into helium, thus forming a true star.

Extrasolar Planets: Astronomers have discovered planets orbiting other stars.

• Most of these planets are detected by the "wobble" of the stars around which they orbit.

• A small but growing number of extrasolar planets have been discovered by the transit method, by microlensing, and direct imaging.

• Most of the extrasolar planets discovered to date are quite massive and have orbits that are very different from planets in our solar system.

Questions

Review Questions

1. Describe three properties of the solar system that are thought to be a result of how the solar system formed.
2. The graphite in your pencil is a form of carbon. Where were these carbon atoms formed?
3. What is the interstellar medium? How does it become enriched over time with heavy elements?
4. What is the evidence that other stars existed before our Sun was formed?
5. Why are terrestrial planets smaller than Jovian planets?
6. How do radioactive elements make it possible to determine the age of the solar system? What are the oldest objects that have been found in the solar system?
7. What is the tidal hypothesis? What aspect of the solar system was it designed to explain? Why was this hypothesis rejected?
8. What is the nebular hypothesis? Why is this hypothesis accepted?
9. What was the protosun? What caused it to shine? Into what did it evolve?
10. Why is it thought that a disk appeared in the solar nebula?
11. What are proplyds? What do they tell us about the plausibility of our model of the solar system's origin?
12. What is meant by a substance's condensation temperature? What role did condensation temperatures play in the formation of the planets?
13. What is a planetesimal? How did planetesimals give rise to the terrestrial planets?
14. (a) What is meant by accretion? (b) Why are the terrestrial planets denser at their centers than at their surfaces?
15. If hydrogen and helium account for 98% of the mass of all the atoms in the universe, why aren't the Earth and Moon composed primarily of these two gases?
16. What is a chondrule? How do we know they were not formed by the ambient heat of the solar nebula?
17. Why did the terrestrial planets form close to the Sun while the Jovian planets formed far from the Sun?
18. What are the competing models of how the Jovian planets formed?
19. Explain how our current understanding of the formation of the solar system can account for the following characteristics of the solar system: (a) All planetary orbits lie in nearly the same plane. (b) All planetary orbits are nearly circular. (c) The planets orbit the Sun in the same direction in which the Sun itself rotates.
20. Why is the combined mass of all the asteroids so small? Why is the combined mass of all trans-Neptunian objects so much greater than that for the asteroids?
21. Explain why most of the satellites of Jupiter orbit that planet in the same direction that Jupiter rotates.
22. What is the radial velocity method used to detect planets orbiting other stars? Why is it difficult to use this method to detect planets like Earth?
23. Summarize the differences between the planets of our solar system and those found orbiting other stars.

24. Is there evidence that planets have fallen into their parent stars? Explain.

25. What does it mean for a planet to transit a star? What can we learn from such events?

26. What is microlensing? How does it enable astronomers to discover extrasolar planets?

27. A 1999 news story about the discovery of three planets orbiting the star Upsilon Andromedae ("Ups And" in Figure 8-17) stated that "the newly discovered galaxy, with three large planets orbiting a star known as Upsilon Andromedae, is 44 light-years away from Earth." What is wrong with this statement?

Advanced Questions

Questions preceded by an asterisk (*) involve topics discussed in Box 4-4 or Box 8-1.

> ### Problem-solving tips and tools
>
> The volume of a disk of radius r and thickness t is $\pi r^2 t$. Box 1-1 explains the relationship between the angular size of an object and its actual size. An object moving at speed v for a time t travels a distance $d = vt$; Appendix 6 includes conversion factors between different units of length and time. To calculate the mass of 70 Virginis or the orbital period of the planet around 2M1207, review Box 4-4. Section 5-4 describes the properties of blackbody radiation.

28. Figure 8-4 shows that carbon, nitrogen, and oxygen are among the most abundant elements (after hydrogen and helium). In our solar system, the atoms of these elements are found primarily in the molecules CH_4 (methane), NH_3 (ammonia), and H_2O (water). Explain why you suppose this is.

29. (a) If the Earth had retained hydrogen and helium in the same proportion to the heavier elements that exist elsewhere in the universe, what would its mass be? Give your answer as a multiple of the Earth's actual mass. Explain your reasoning. (b) How does your answer to (a) compare with the mass of Jupiter, which is 318 Earth masses? (c) Based on your answer to (b), would you expect Jupiter's rocky core to be larger, smaller, or the same size as the Earth? Explain your reasoning.

*30. If you start with 0.80 kg of radioactive potassium (^{40}K), how much will remain after 1.3 billion years? After 2.6 billion years? After 3.9 billion years? How long would you have to wait until there was *no* ^{40}K remaining?

*31. Three-quarters of the radioactive potassium (^{40}K) originally contained in a certain volcanic rock has decayed into argon (^{40}Ar). How long ago did this rock form?

32. Suppose you were to use the Hubble Space Telescope to monitor one of the protoplanetary disks shown in Figure 8-8b. Over the course of 10 years, would you expect to see planets forming within the disk? Why or why not?

33. The protoplanetary disk at the upper right of Figure 8-8b is seen edge-on. The diameter of the disk is about 700 AU. (a) Make measurements on this image to determine the thickness of the disk in AU. (b) Explain why the disk will continue to flatten as time goes by.

R **I** V U X G

(Courtesy of D. Padgett and W. Brandner, IPAC/ Caltech; K. Stapelfeldt, JPL; and NASA)

34. The accompanying infrared image shows IRAS 04302+2247, a young star that is still surrounded by a disk of gas and dust. The scale bar at the lower left of the image shows that at the distance of IRAS 04302+2247, an angular size of 2 arcseconds corresponds to a linear size of 280 AU. Use this information to find the distance to IRAS 04302+2247.

35. The image accompanying Question 34 shows a dark, opaque disk of material surrounding the young star IRAS 04302+2247. The disk is edge-on to our line of sight, so it appears as a dark band running vertically across this image. The material to the left and right of this band is still falling onto the disk. (a) Make measurements on this image to determine the diameter of the disk in AU. Use the scale bar at the lower left of this image. (b) If the thickness of the disk is 50 AU, find its volume in cubic meters. (c) The total mass of the disk is perhaps 2×10^{28} kg (0.01 of the mass of the Sun). How many atoms are in the disk? Assume that the disk is all hydrogen. A single hydrogen atom has a mass of 1.673×10^{-27} kg. (d) Find the number of atoms per cubic meter in the disk. Is the disk material thick or thin compared to the air that you breathe, which contains about 5.4×10^{25} atoms per cubic meter?

36. If the material in the jet shown in Figure 8-15a is moving at 200 km/s, how long ago was the material at the far right-hand end of the jet ejected from the star? Give your answer in years. (You will need to make measurements on the image.)

*37. The planet discovered orbiting the star 70 Virginis ("70Vir" in Figure 8-17), 59 light-years from Earth, moves in an orbit with semimajor axis 0.48 AU and eccentricity 0.40. The period of the orbit is 116.7 days. Find the mass of 70 Virginis. Compare your answer with the mass of the Sun. (*Hint:* The planet has far less mass than the star.)

*38. Because of the presence of Jupiter, the Sun moves in a small orbit of radius 742,000 km with a period of 11.86 years. (a) Calculate the Sun's orbital speed in meters per second. (b) An astronomer on a hypothetical planet orbiting the star

Vega, 25 light-years from the Sun, wants to use the astrometric method to search for planets orbiting the Sun. What would be the angular diameter of the Sun's orbit as seen by this alien astronomer? Would the Sun's motion be discernible if the alien astronomer could measure positions to an accuracy of 0.001 arcsec? (c) Repeat part (b), but now let the astronomer be located on a hypothetical planet in the Pleiades star cluster, 360 light-years from the Sun. Would the Sun's motion be discernible to this astronomer?

39. (a) Figure 8-18c shows how astronomers determine that the planet of HD 209458 has a surface temperature of 1130 K. Treating the planet as a blackbody, calculate the wavelength at which it emits most strongly. (b) The star HD 209458 itself has a surface temperature of 6030 K. Calculate its wavelength of maximum emission, assuming it to be a blackbody. (c) If a high-resolution telescope were to be used in an attempt to record an image of the planet orbiting HD 209458, would it be better for the telescope to use visible or infrared light? Explain your reasoning.

*40. (a) The star 2M1207 shown in Figure 8-20 is 170 light-years from Earth. Find the angular distance between this star and its planet as seen from Earth. Express your answer in arcseconds. (b) The mass of 2M1207 is 0.025 that of the Sun; the mass of the planet is very much smaller. Calculate the orbital period of the planet, assuming that the distance between the star and planet shown in Figure 8-20 is the semimajor axis of the orbit. Is it possible that an astronomer could observe a complete orbit in one lifetime?

Discussion Questions

41. Propose an explanation why the Jovian planets are orbited by terrestrial-like satellites.

42. Suppose that a planetary system is now forming around some protostar in the sky. In what ways might this planetary system turn out to be similar to or different from our own solar system? Explain your reasoning.

43. Suppose astronomers discovered a planetary system in which the planets orbit a star along randomly inclined orbits. How might a theory for the formation of that planetary system differ from that for our own?

Web/eBook Questions

44. Search the World Wide Web for information about recent observations of protoplanetary disks. What insights have astronomers gained from these observations? Is there any evidence that planets have formed within these disks?

45. In 2000, extrasolar planets with masses comparable to that of Saturn were first detected around the stars HD 16141 (also called 79 Ceti) and HD 46375. Search the World Wide Web for information about these "lightweight" planets. Do these planets move around their stars in the same kind of orbit as Saturn follows around the Sun? Why do you suppose this is? How does the discovery of these planets reinforce the model of planet formation described in this chapter?

46. In 2006 a planet called XO-1b was discovered using the transit method. Search the World Wide Web for information about this planet and how it was discovered. What unusual kind of telescope was used to make this discovery? Have other extrasolar planets been discovered using the same kind of telescope?

Activities

Observing Projects

47. Use the *Starry Night Enthusiast*™ program to investigate stars that have planets orbiting them. First display the entire celestial sphere (select **Guides > Atlas** in the **Favourites** menu). Then use the **Find** pane to find and center each of the stars listed below. To do this, click the magnifying glass icon on the left side of the edit box at the top of the Find pane and select **Star** from the drop-down menu; then type the name of the star in the edit box and press the Enter or Return key on the keyboard. Click on the **Info** tab on the left-hand side of the *Starry Night Enthusiast*™ window for full information about the star. For each star, record the luminosity of the star (a measure of the star's total light output). How far from Earth is each star? Which stars are more luminous than the Sun? Which are less luminous? How do you think these differences would have affected temperatures in the nebula in which each star's planets formed (see Figure 8-10)? (i) 47 Ursae Majoris; (ii) 51 Pegasi; (iii) 70 Virginis; (iv) Rho Coronae Borealis.

48. Use the *Starry Night Enthusiast*™ program to examine stars that have planets. Select **Stars > Extra-Solar Planets** in the **Favourites** menu. In the star map that appears, each circled star has one or more planets. (You can zoom in and zoom out using the buttons at the right side of the toolbar. Click the **Info** tab at the left of the main window to open the Info pane and open the **Other Data** panel. Then click on a circled star to learn more about its properties. Note that the information given for each star includes the *apparent magnitude*, which is a measure of how bright each star appears as seen from Earth. Apparent magnitude uses a "backwards" scale: The greater the value of the apparent magnitude, the *dimmer* the star. Most of the brighter stars you can see with the naked eye from the Earth have apparent magnitudes between 0 and 1, while the dimmest star you can see from a dark location has apparent magnitude 6. Are most of the circled stars visible to the naked eye? List at least two stars that are visible, and include their apparent magnitudes.

Alien Planets
by Geoff Marcy

Astronomers have now surveyed 2000 nearby stars in the search for orbiting planets. My team is using the 10-meter Keck telescope in Hawaii to measure the Doppler effect in stars that wobble because of planets orbiting around them. So far, we and other teams have found more than 200 planets, some more massive than Jupiter and some less massive than Neptune. We have now found 20 stars that have multiple planets orbiting them—full planetary systems. Often the planets orbit synchronously, with one planet orbiting the star two or three times for every orbit of another planet. This synchronization of the planets allows gravity to shepherd their motion, keeping them synchronized forever.

We hope to compare the architecture of these planetary systems to that of our own solar system. Is our solar system a common type? Do all planetary systems have giant planets orbiting far away from the host star, with small rocky planets orbiting close in?

Remarkably, the planetary systems observed so far seem very different from our own solar system. In some, a giant planet orbits very close to the star. More surprising is that almost every alien planet that orbits beyond 0.1 AU from its star resides in a highly elliptical orbit that resembles eccentric comet orbits. Why are these planets in such elliptical orbits so different from the nearly circular orbits of planets in our own solar system? Astronomers are baffled by this difference. One possibility is that planets usually form as close-knit families around a star. Over time, the planets may gravitationally pull on one

another, yanking themselves out of their original circular orbits. These perturbed planets may then venture near other planets, yanking on them, too. Many of the planets may gravitationally slingshot each other completely out of the planetary system. These far flung planets are destined to roam the darkness of galactic space, cooling to frigid temperatures that are inhospitable to life.

The extrasolar planets cast a mystery back on the Earth. Why is our solar system immune to this chaotic episode that scatters planets into wacky orbits? Imagine the disaster if our Earth were suddenly thrust into a highly elliptical orbit. During half the year, we would be roasting too close to the Sun, with oceans vaporized into an enormous steam bath. The other half of the year our orbit would carry us too far from the Sun, causing all the oceans to freeze over. Life on Earth would be severely challenged to survive in such an elongated orbit.

Why are we humans so lucky as to live in a stable planetary system in which circular orbits ensure that we receive the same warming light intensity from our Sun, all year round? Perhaps the question to ask is: Would life have evolved here if the Earth suffered from wild temperature swings that would make liquid water rare? After all, intelligent life might not have evolved here if the Earth suffered from wild temperature swings that would make liquid water rare. We humans, and species in general, may flourish mostly on worlds that maintain nearly constant temperature. If so, our solar system and the Earth might be a rare galactic quirk that just happens to be suitable for life. We wouldn't be here otherwise. If this picture is correct, our Earth is indeed a precious oasis in our Milky Way Galaxy.

In the next few years, we plan to use the Doppler technique to discover planets having masses only 10 or 20 times that of our Earth. These would likely be rocky planets, allowing liquid water to puddle in ponds and lakes. Who knows what biology might emerge in such extraterrestrial chemistry labs! In the next decade, we will discover more planetary systems by using the Doppler method. We are working night and day to find Jovian planets that orbit 5 AU from their stars, similar to the orbital radius of our Jupiter. Their orbits, circular or elliptical, will shed light on a key question: How common are configurations like our solar system, and how common is advanced life in the universe?

When Geoff Marcy sat down to write this essay, more than 200 planets were known to orbit other stars. He leads the team that has found more than 120 of them. Dr. Marcy became interested in astronomy at the age of 14, when his parents bought him a small reflecting telescope. Since receiving his Ph.D. from the University of California, Santa Cruz, he has studied stars similar to our Sun. He helped show that magnetic regions on their surfaces cause "star spots" and stellar flares. He also showed that brown dwarfs—stars too small to burn hydrogen—rarely orbit other stars. His work may soon reveal whether a planetary system like ours is common or a quirk of the cosmos. Dr. Marcy is an astronomer at the University of California, Berkeley.

9

The Living Earth

R I V U X G

The two hemispheres of Earth. (NASA)

When astronauts made the first journeys from the Earth to the Moon between 1968 and 1972, they often reported that our planet is the most beautiful sight visible from space. Of all the worlds of the solar system, only Earth has the distinctive green color of vegetation, and only Earth has liquid water on its surface. Compared with the Moon, whose lifeless, dry surface has been ravaged by billions of years of impacts by interplanetary debris, the Earth seems an inviting and tranquil place.

Yet the appearance of tranquility is deceiving. The seemingly solid surface of the planet is in a state of slow but constant motion, driven by the flow of molten rock within the Earth's interior. Sunlight provides the Earth's atmosphere with the energy that drives our planet's sometimes violent weather, including all the cloud patterns visible in the images shown here.

Unseen in these images are immense clouds of subatomic particles that wander around the outside of the planet in two giant belts, held in thrall by the Earth's magnetic field. And equally unseen are trace gases in the Earth's atmosphere whose abundance may determine the future of our climate, and on which may depend the survival of entire species.

In this chapter our goal is to learn about the Earth's components—its dynamic oceans and atmosphere, its ever-changing surface, and its hot, active interior—and how they interact to make up our planetary home. In later chapters we will use this knowledge as a point of reference for studying the Moon and the other planets.

9-1 The Earth's atmosphere, oceans, and surface are extraordinarily active

The crew of an alien spacecraft exploring our solar system might overlook the Earth altogether. Although the largest of the terrestrial planets, with a mass greater than that of Mercury, Venus, and Mars put together, Earth is far smaller than any of the giant Jovian planets (see Table 7-1 and Table 9-1). But a more careful inspection would reveal the Earth to be unique among all the planets that orbit the Sun.

Dynamic Earth: Water, Air, and Land

Unlike the arid surfaces of Venus and Mars, the Earth's surface is very wet. Indeed, nearly 71% of the Earth's surface is covered with water (Figure 9-1). Water is also locked into the chemical structure of many Earth rocks, including those found in the driest deserts. Furthermore, the Earth's liquid water is in constant motion. The oceans ebb and flow with the tide, streams and rivers flow down to the sea, and storms whip lake waters into a frenzy.

> To understand our dynamic planet, we must understand the energy sources that power its activity

The Earth's atmosphere is also in a state of perpetual activity. Winds blow at all altitudes, with speeds and directions that change from one hour to the next. The atmosphere is at its most

Learning Goals

By reading the sections of this chapter, you will learn

9-1 What powers the motions of the Earth's atmosphere, oceans, and land surfaces

9-2 How scientists have deduced the layered structure of our planet's interior

9-3 The evidence that the continents are being continuously moved and reshaped

9-4 How life on Earth is protected from subatomic particles emitted by the Sun

9-5 How the evolution of life has transformed the Earth's atmosphere

9-6 What causes the patterns of weather in our atmosphere

9-7 How human civilization is causing dramatic and adverse changes to our planetary habitat

Table 9-1   Earth Data

Average distance from Sun:	1.000 AU $= 1.496 \times 10^8$ km
Maximum distance from Sun:	1.017 AU $= 1.521 \times 10^8$ km
Minimum distance from Sun:	0.983 AU $= 1.471 \times 10^8$ km
Eccentricity of orbit:	0.017
Average orbital speed:	29.79 km/s
Orbital period:	365.256 days
Rotation period:	23.9345 hours
Inclination of equator to orbit:	23.45°
Diameter (equatorial):	12,756 km
Mass:	5.974×10^{24} kg
Average density:	5515 kg/m^3
Escape speed:	11.2 km/s
Albedo:	0.31
Surface temperature range	Maximum: 60°C = 140°F = 333 K Mean: 14°C = 57°F = 287 K Minimum: −90°C = −130°F = 183 K
Atmospheric composition (by number of molecules):	78.08% nitrogen (N_2) 20.95% oxygen (O_2) 0.035% carbon dioxide (CO_2) about 1% water vapor

WEB LINK 9.2   (NASA)

dynamic when water evaporates to form clouds, then returns to the surface as rain or snow (Figure 9-2). While other worlds of the solar system have atmospheres, only the Earth's contains the oxygen that animals (including humans) need to sustain life.

The combined effects of water and wind cause erosion of mountains and beaches. But these are by no means the only forces reshaping the face of our planet. All across the Earth we find evidence that the surface has been twisted, deformed, and folded (Figure 9-3). What is more, new material is continually being added to the Earth's surface as lava pours forth from volcanoes and from immense cracks in the ocean floors. At the same time, other geologic activity slowly pushes old surface material back into our planet's interior. These processes and others work together to renew and refresh our planet's exterior. Thus, although the Earth is some 4.56 billion (4.56×10^9) years old, much of its surface is a few hundred million (a few times 10^8) years old (Figure 9-4). (We saw in Section 8-2 how scientists use radioactive dating to determine the ages of rocks and of the Earth as a whole.)

Earth's Energy Sources

What powers all of this activity in the Earth's oceans, atmosphere, and surface? There are three energy sources: radiation from the Sun, the tidal effects of the Moon, and the Earth's own internal heat (Table 9-2).

Figure 9-1   R I V U X G

The Earth's Dynamic Oceans Nearly three-quarters of the Earth's surface is covered with water, a substance that is essential to the existence of life. In contrast, there is no liquid water at all on Mercury, Venus, Mars, or the Moon. (Farley Lewis/Photo Researchers)

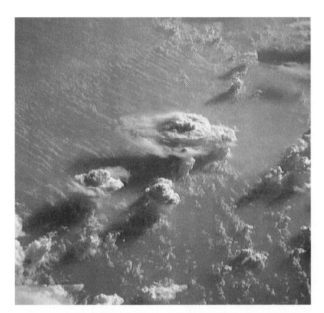

Figure 9-2   R I V U X G

The Earth's Dynamic Atmosphere This space shuttle image shows thunderstorm clouds over the African nation of Zaire. The tops of thunderstorm clouds frequently reach altitudes of 10,000 m (33,000 ft) or higher. At any given time, nearly 2000 thunderstorms are in progress over the Earth's surface. (JSC/NASA)

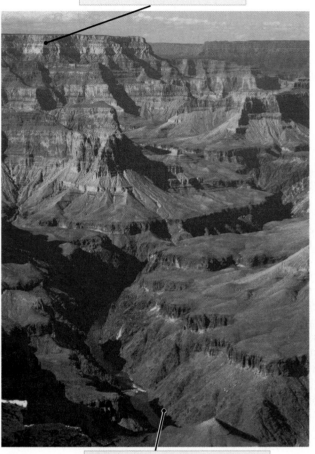

2. Over millions of years, layers of sediments built up over that rock. The most recent layer—the top—is about 250 million years old.

1. The rocks at the bottom of the Grand Canyon are 1.7–2.0 billion years old.

Figure 9-4   R I V U X G

Old and Young Rocks in the Grand Canyon The ages of rocks in Arizona's Grand Canyon demonstrate that geologic processes take place over very long time scales. (John Wang/PhotoDisc/Getty Images)

Figure 9-3   R I V U X G

The Earth's Dynamic Surface These tan-colored ridges, or hogbacks, in Colorado's Rocky Mountains were once layers of sediment at the bottom of an ancient body of water. Forces within the Earth folded this terrain and rotated the layers into a vertical orientation. The layers were revealed when wind and rain eroded away the surrounding material. (Tom Till/DRK)

Table 9-2   The Earth's Energy Sources

Activity	Energy Sources
Motion of water in oceans, lakes, rivers	Solar energy, tidal forces
Motion of the atmosphere	Solar energy
Reshaping of surface	Earth's internal heat
Life	Solar energy (a few species that live on the ocean floor make use of the Earth's internal heat)

The Sun is the principal source of energy for the atmosphere. The Earth's surface is warmed by sunlight, which in turn warms the air next to the surface. Hot air is less dense than cool air and so tends to rise (see the discussion of density in Box 7-1). As the air rises, it transfers heat to its surroundings. As a result, the rising air cools and becomes denser. It then sinks downward to be heated again, and the process starts over. This up-and-down motion is called **convection,** and the overall pattern of circulation is called a **convection current.** (You can see convection currents in action by heating water on a stove, as Figure 9-5 shows.) In Section 9-5 we will see how the Earth's rotation modifies the up-and-down motion of the atmosphere to produce a pattern of global circulation.

Solar energy also powers atmospheric activity by evaporating water from the surface. The energy in the water vapor is released when it condenses to form water droplets, like those that make up clouds. In a typical thunderstorm (see Figure 9-2), some 5×10^8 kg of water vapor is lifted to great heights. The amount of energy released when this water condenses is as much as a city of 100,000 people uses in a month!

Solar energy also helps to power the oceans. Warm water from near the equator moves toward the poles, while cold polar water returns toward the equator. As we saw in Section 4-8, however, there is back-and-forth motion of the oceans due to the tidal forces of the Moon and Sun. Sometimes these two influences can reinforce each other, as when a storm (caused by solar energy) reaches a coastline at high tide (caused by tidal forces), producing waves strong enough to seriously erode beaches and sea cliffs.

Neither solar energy nor tidal forces can explain the reshaping of the Earth's surface suggested by Figure 9-3. Rather, all geologic activity is powered by heat flowing from the interior of the Earth itself. The Earth formed by collisions among planetesimals

(see Section 8-4), and these collisions heated the body of the Earth. The interior remains hot to this day. An additional source of heat is the decay of radioactive elements such as uranium and thorium deep inside the Earth.

The heat flow from the Earth's interior to its surface is minuscule—just 1/6000 as great as the flow of energy we receive from the Sun—but it has a profound effect on the face of our planet. In Sections 9-2 and 9-3 we will explore the interior structure of the Earth and discover the connection between geologic activity on the Earth's surface and heat coming from its interior.

The Earth's Surface Temperature and the Greenhouse Effect

Since so little heat comes from inside our planet, the average surface temperature of the Earth depends almost entirely on the amount of energy that reaches us from the Sun in the form of electromagnetic radiation. (In an analogous way, whether you feel warm or cool outdoors on a summer day depends on whether you are in the sunlight and receiving lots of solar energy or in the shade and receiving little of this energy.)

If the Earth did nothing but *absorb* radiation from the Sun, it would get hotter and hotter until the surface temperature became high enough to melt rock. Happily, there are two reasons why this does not happen. One is that the clouds, snow, ice, and sand *reflect* about 31% of the incoming sunlight back into space. The fraction of incoming sunlight that a planet reflects is called its **albedo** (from the Latin for "whiteness"); thus the Earth's albedo is about 0.31. Thus, only 69% of the incoming solar energy is absorbed by the Earth. A second reason is that the Earth also *emits* radiation into space because of its temperature, in accordance with the laws that describe heated dense objects (see Section 5-3). The Earth's average surface temperature is nearly constant, which means that on the whole it is neither gaining nor losing energy. Thus, the rate at which the Earth emits energy into space must equal the rate at which it absorbs energy from the Sun.

To better understand this balance between absorbed and emitted radiation, remember that Wien's law tells us that the wavelength at which such an object emits most strongly (λ_{max}) is inversely proportional to its temperature (T) on the Kelvin scale (see Section 5-4 and Box 5-2). For example, the Sun's surface temperature is about 5800 K, and sunlight has its greatest intensity at a wavelength λ_{max} of 500 nm, in the middle of the visible spectrum. The Earth's average surface temperature of 287 K is far lower than the Sun's, so the Earth radiates most strongly at longer wavelengths in the infrared portion of the electromagnetic spectrum. The Stefan-Boltzmann law tells us that temperature also determines the *amount* of radiation that the Earth emits: The higher the temperature, the more energy it radiates.

Given the amount of energy reaching us from the Sun each second as well as the Earth's albedo, we can calculate the amount of solar energy that the Earth should *absorb* each second. Since this must equal the amount of electromagnetic energy that the Earth *emits* each second, which in turn depends on the Earth's average surface temperature, we can calculate what the Earth's average surface temperature should be. The result is a very chilly 254 K ($-19°C = -2°F$), so cold that oceans and lakes around the world should be frozen over. In fact, the Earth's actual average surface temperature is 287 K ($14°C = 57°F$). What is wrong with our model? Why is the Earth warmer than we would expect?

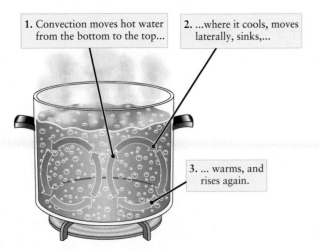

1. Convection moves hot water from the bottom to the top...

2. ...where it cools, moves laterally, sinks,...

3. ... warms, and rises again.

Figure 9-5

Convection in the Kitchen Heat supplied to the bottom of a pot warms the water there, making the water expand and lowering its average density. This low-density water rises and transfers heat to its cooler surroundings. The water that began at the bottom thus cools down, becomes denser, and sinks back to the bottom to repeat the process. (Adapted from F. Press, R. Siever, T. Grotzinger, and T. H. Jordan, *Understanding Earth,* 4th ed., W. H. Freeman, 2004)

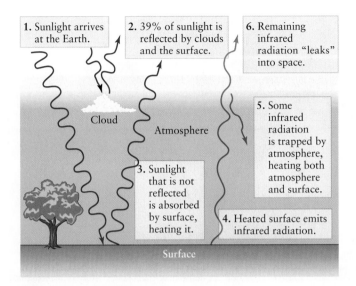

Figure 9-6

The Greenhouse Effect Sunlight warms the Earth's surface, which due to its temperature emits infrared radiation. Much of this radiation is absorbed by atmospheric water vapor and carbon dioxide, helping to raise the average temperature of the surface. Some infrared radiation does penetrate the atmosphere and leaks into space. In a state of equilibrium, the rate at which the Earth loses energy to space in this way is equal to the rate at which it absorbs energy from the Sun.

The explanation for this discrepancy is called the **greenhouse effect:** Our atmosphere prevents some of the radiation emitted by the Earth's surface from escaping into space. Certain gases in our atmosphere called **greenhouse gases,** among them water vapor and carbon dioxide, are transparent to visible light but not to infrared radiation. Consequently, visible sunlight has no trouble entering our atmosphere and warming the surface. But the infrared radiation coming from the heated surface is partially trapped by the atmosphere, thus raising the temperatures of both the atmosphere and the surface. As the surface and atmosphere become hotter, they both emit more infrared radiation, part of which is able to escape into space. The temperature levels off when the amount of infrared energy that escapes just balances the amount of solar energy reaching the surface (**Figure 9-6**). The result is that our planet's surface is some 33°C (59°F) warmer than it would be without the greenhouse effect, and water remains unfrozen over most of the Earth.

The warming caused by the greenhouse effect gives our planet the moderate temperatures needed for the existence of life. For over a century, however, our technological civilization has been adding greenhouse gases to the atmosphere at an unprecedented rate. As we will see in Section 9-7, the likely consequences to our climate are extremely grave.

9-2 Studies of earthquakes reveal the Earth's layered interior structure

The kinds of rocks found on and near the Earth's surface provide an important clue about our planet's interior. The densities of

typical surface rocks are around 3000 kg/m³, but the average density of the Earth as a whole (that is, its mass divided by its volume) is 5515 kg/m³. The interior of the Earth must therefore be composed of substances

> **Like an exotic dessert, our planet's interior has layers of both solid and liquid material**

much denser than those found near the surface. But what is this substance? Why is the Earth's interior more dense than its crust? And is the Earth's interior solid like rock or molten like lava?

An Iron-Rich Planet

Iron (chemical symbol Fe) is a good candidate for the substance that makes up most of the Earth's interior. This is so for two reasons. First, iron atoms are quite massive (a typical iron atom has 56 times the mass of a hydrogen atom), and second, iron is relatively abundant. (Figure 8-4 shows that it is the seventh most abundant element in our part of the Milky Way Galaxy.) Other elements such as lead and uranium have more massive atoms, but these elements are quite rare. Hence, the solar nebula could not have had enough of these massive atoms to create the Earth's dense interior. Furthermore, iron is common in meteoroids that strike the Earth, which suggests that it was abundant in the planetesimals from which the Earth formed.

The Earth was almost certainly molten throughout its volume soon after its formation, about 4.56×10^9 years ago. Energy released by the violent impacts of numerous meteoroids and asteroids and by the decay of radioactive isotopes likely melted the solid material collected from the earlier planetesimals. Gravity caused abundant, dense iron to sink toward the Earth's center, forcing less dense material to the surface. **Figure 9-7a** shows this process of *chemical differentiation*. (We discussed chemical differentiation in Section 8-4.) The result was a planet with the layered structure shown in Figure 9-7b—a central **core** composed of almost pure iron, surrounded by a **mantle** of dense, iron-rich minerals. The mantle, in turn, is surrounded by a thin **crust** of relatively light silicon-rich minerals. We live on the surface of this crust.

Seismic Waves as Earth Probes

How do we know that this layered structure is correct? The challenge in testing this model is that the Earth's interior is as inaccessible as the most distant galaxies in space. The deepest wells go down only a few kilometers, barely penetrating the surface of our planet. Despite these difficulties, geologists have learned basic properties of the Earth's interior by studying earthquakes and the seismic waves that they produce.

 Over the centuries, stresses build up in the Earth's crust. Occasionally, these stresses are relieved with a sudden motion called an **earthquake.** Most earthquakes occur deep within the Earth's crust. The point on the Earth's surface directly over an earthquake's location is called the **epicenter.**

Earthquakes produce three different kinds of **seismic waves,** which travel around or through the Earth in different ways and at different speeds. Geologists use sensitive instruments called **seismographs** to detect and record these vibratory motions. The first type of wave, which is analogous to water waves on the ocean, causes the rolling motion that people feel around an epicenter. These are called **surface waves** because they travel only

Figure 9-7

Chemical Differentiation and the Earth's Internal Structure (a) The newly formed Earth was molten throughout its volume. Dense materials such as iron (shown in orange) sank toward the center, while low-density materials (shown in blue) rose toward the surface. **(b)** The present-day Earth is no longer completely molten inside. A dense, solid iron core is surrounded by a less dense liquid core and an even less dense mantle. The crust, which includes the continents and ocean floors, is the least dense of all; it floats atop the mantle like the skin that forms on the surface of a cooling cup of cocoa.

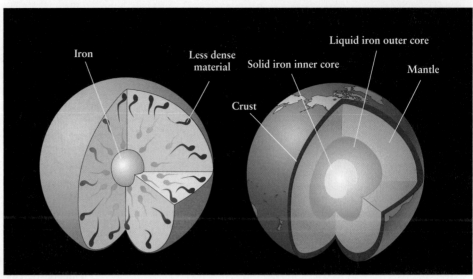

(a) During differentiation, iron sank to the center and less dense material floated upward

(b) As a result of differentiation, the Earth has the layered structure that we see today

over the Earth's surface. The two other kinds of waves, called **P waves** (for "primary") and **S waves** (for "secondary"), travel through the interior of the Earth. P waves are called *longitudinal* waves because their oscillations are parallel to the direction of wave motion, like a spring that is alternately pushed and pulled. In contrast, S waves are called *transverse* waves because their vibrations are perpendicular to the direction in which the waves move. S waves are analogous to waves produced by a person shaking a rope up and down (Figure 9-8).

What makes seismic waves useful for learning about the Earth's interior is that they do not travel in straight lines. Instead, the paths that they follow through the body of the Earth are bent

Direction of vibration

P waves are longitudinal waves

Direction of vibration

S waves are transverse waves

Figure 9-8

Seismic Waves Earthquakes produce two kinds of waves that travel through the body of our planet. One kind, called P waves, are longitudinal waves. They are analogous to those produced by pushing a spring in and out. The other kind, S waves, are transverse waves analogous to the waves produced by shaking a rope up and down.

because of the varying density and composition of the Earth's interior. We saw in Section 6-1 that light waves behave in a very similar way. Just as light waves bend, or refract, when they pass from air into glass or vice versa (see Figure 6-2), seismic waves refract as they pass through different parts of the Earth's interior. By studying how the paths of these waves bend, geologists can map out the general interior structure of the Earth.

One key observation about seismic waves and how they bend has to do with the differences between S and P waves. When an earthquake occurs, seismographs relatively close to the epicenter record both S and P waves, but seismometers on the opposite side of Earth record only P waves. The absence of S waves was first explained in 1906 by British geologist Richard Dixon Oldham, who noted that transverse vibrations such as S waves cannot travel far through liquids. Oldham therefore concluded that our planet has a molten core. Furthermore, there is a region in which neither S waves nor P waves from an earthquake can be detected (Figure 9-9). This "shadow zone" results from the specific way in which P waves are refracted at the boundary between the solid mantle and the molten core. By measuring the size of the shadow zone, geologists have concluded that the radius of the molten core is about 3500 km (2200 mi), about 55% of our planet's overall radius but about double the radius of the Moon (1738 km = 1080 mi).

As the quality and sensitivity of seismographs improved, geologists discovered faint traces of P waves in an earthquake's shadow zone. In 1936, the Danish seismologist Inge Lehmann explained that some of the P waves passing through the Earth are deflected into the shadow zone by a small, solid **inner core** at the center of our planet. The radius of this inner core is about 1300 km (800 mi).

Earth's Major Layers

The seismic evidence reveals that our planet has a curious internal structure—a liquid **outer core** sandwiched between a solid inner core and a solid mantle. Table 9-3 summarizes this structure. To understand

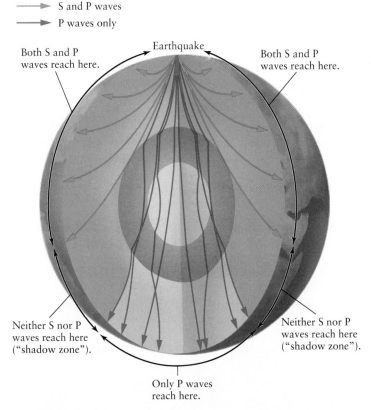

S and P waves

P waves only

Both S and P
waves reach here.

Earthquake

Both S and P
waves reach here.

Neither S nor P
waves reach here
("shadow zone").

Neither S nor P
waves reach here
("shadow zone").

Only P waves
reach here.

Figure 9-9

The Earth's Internal Structure and the Paths of Seismic Waves
Seismic waves follow curved paths because of differences in the density
and composition of the material in the Earth's interior. The paths curve
gradually where there are gradual changes in density and composition.
Sharp bends occur only where there is an abrupt change from one kind of
material to another, such as at the boundary between the outer core and
the mantle. Only P waves can pass through the Earth's liquid outer core.

this arrangement, we must look at how temperature and pressure
inside the Earth affect the melting point of rock.

Both temperature and pressure increase with increasing depth
below the Earth's surface. The temperature of the Earth's interior
rises steadily from about 14°C on the surface to nearly 5000°C
at our planet's center (Figure 9-10).

The Earth's outermost layer, the crust, is only about 5 to
35 km thick. It is composed of rocks for which the **melting point**,
or temperature at which the rock changes from solid to liquid, is
far higher than the temperatures actually found in the crust. Thus,
the crust is solid.

The Earth's mantle, which extends to a depth of about
2900 km (1800 mi), is largely composed of substances rich in
iron and magnesium. On the Earth's surface, specimens of these
substances have melting points slightly over 1000°C. However,
the melting point of a substance depends on the pressure to which
it is subjected—the higher the pressure, the higher the melting
point. As Figure 9-10 shows, the actual temperatures throughout
most of the mantle are less than the melting point of all the substances of which the mantle is made. Hence, the mantle is primarily solid. However, the upper levels of the mantle—called the
asthenosphere (from the Greek *asthenia*, meaning "weakness")—
are able to flow slowly and are therefore referred to as being
plastic.

CAUTION! It may seem strange to think of a plastic material
as one that is able to flow. Normally we think of plastic objects
as being hard and solid, like the plastic out of which compact
discs and video cassettes are made. But to make these objects,
the plastic material is heated so that it can flow into a mold,
then cooled so that it solidifies. Strictly, the material is only
"plastic" when it is able to flow. Maple syrup and tree sap are
two other substances that flow at warm temperatures but become solid when cooled.

At the boundary between the mantle and the outer core, there
is an abrupt change in chemical composition, from silicon-rich
materials to almost-pure iron with a small admixture of nickel.
Because this nearly pure iron has a lower melting point than the
iron-rich materials, the melting curve in Figure 9-10 has a "jog"
as it crosses from the mantle to the outer core and remains. As a
result, the melting point is less than the temperature at depths of
about 2900 to 5100 km (1800 to 3200 mi), and the outer core is
liquid.

At depths greater than about 5100 km, the pressure is more
than 10^{11} newtons per square meter. This is about 10^6 times ordinary atmospheric pressure, or about 10^4 tons per square inch.
Because the melting point of the iron-nickel mixture under this
pressure is higher than the actual temperature (see Figure 9-10),
the Earth's inner core is solid.

Table 9-3   The Earth's Internal Structure

Region	Depth Below Surface (km)	Distance From Center (km)	Average Density (kg/m³)
Crust (solid)	0–5 (under oceans) 0–35 (under continents)	6343–6378	3500
Mantle (plastic, solid)	from bottom of crust to 2900	3500–6343	3500–5500
Outer core (liquid)	2900–5100	1300–3500	10,000–12,000
Inner core (solid)	5100–6400	0–1300	13,000

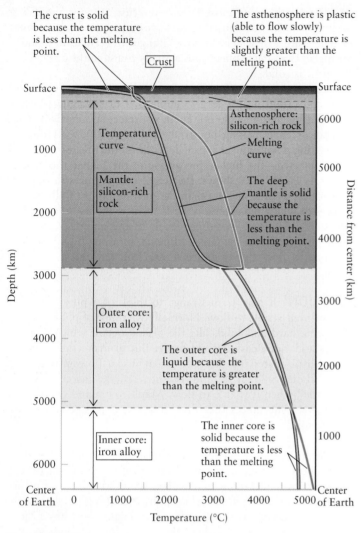

The crust is solid because the temperature is less than the melting point.

The asthenosphere is plastic (able to flow slowly) because the temperature is slightly greater than the melting point.

The deep mantle is solid because the temperature is less than the melting point.

The outer core is liquid because the temperature is greater than the melting point.

The inner core is solid because the temperature is less than the melting point.

Figure 9-10

Temperature and Melting Point of Rock Inside the Earth The temperature (yellow curve) rises steadily from the Earth's surface to its center. The melting point of the Earth's material (red curve) is also shown on this graph. Where the temperature is below the melting point, as in the mantle and inner core, the material is solid; where the temperature is above the melting point, as in the outer core, the material is liquid. (Adapted from T. Grotzinger, T. H. Jordan, F. Press, and R. Siever, *Understanding Earth*, 5th ed., W. H. Freeman, 2007)

 Until the 1980s, geologists knew little more about the inner core than that it is solid and dense. Since then, evidence has accumulated that suggests the inner core is a single **crystal** of iron—that is, iron atoms arranged in orderly rows, like carbon atoms are arranged within a diamond. If so, the inner core is the largest crystalline object anywhere in the solar system. Furthermore, there are strong indications that the inner core is rotating at a slightly faster rate than the rest of the Earth! These remarkable discoveries suggest that even more surprises may lurk deep within the Earth's interior.

Heat naturally flows from where the temperature is high to where it is low. Figure 9-10 thus explains why there is a heat flow from the center of the Earth outward. We will see in the next section how this heat flow acts as the "engine" that powers our planet's geologic activity.

9-3 Plate movement produces earthquakes, mountain ranges, and volcanoes that shape the Earth's surface

 One of the most important geological discoveries of the twentieth century was the realization that the Earth's crust is constantly changing. We have learned that the crust is divided into huge **plates** whose motions produce earthquakes, volcanoes, mountain ranges, and oceanic trenches. This picture of the Earth's crust has come to be the central unifying theory of geology, much as the theory of evolution has become the centerpiece of modern biology.

Evidence from the bottom of the ocean confirmed the theory of moving continents

Continental Drift

Anyone who carefully examines a map of the Earth might come up with the idea of moving continents. South America, for example, would fit snugly against Africa were it not for the Atlantic Ocean. As **Figure 9-11** shows, the fit between land masses on ei-

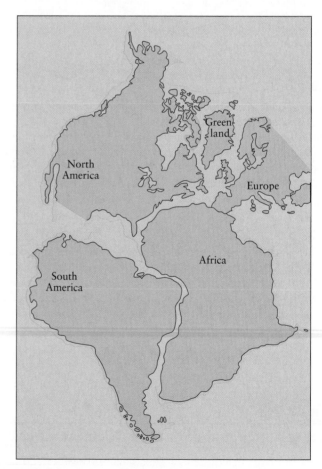

Figure 9-11

Fitting the Continents Together Africa, Europe, Greenland, North America, and South America fit together remarkably well. The fit is especially convincing if the edges of the continental shelves (shown in yellow) are used, rather than today's shorelines. This strongly suggests that these continents were in fact joined together at some point in the past. (Adapted from P. M. Hurley)

(a) 237 million years ago: the supercontinent Pangaea

(b) 152 million years ago: the breakup of Pangaea

(c) The continents today

Figure 9-12

The Breakup of the Supercontinent Pangaea **(a)** The shapes of the continents led Alfred Wegener to conclude that more than 200 million (2×10^8) years ago, the continents were merged into a single supercontinent, which he called Pangaea. **(b)** Pangaea first split into two smaller land masses, Laurasia and Gondwana. **(c)** Over millions of years, the continents moved to their present-day locations. Among the evidence confirming this picture are nearly identical rock formations 200 million years in age that today are thousands of kilometers apart but would have been side by side on Pangaea. (Adapted from F. Press, R. Siever, T. Grotzinger, and T. H. Jordan, *Understanding Earth,* 4th ed., W. H. Freeman, 2004)

ther side of the Atlantic Ocean is quite remarkable. This observation inspired the German meteorologist Alfred Wegener to advocate "continental drift"—the idea that the continents on either side of the Atlantic Ocean have simply drifted apart. After much research, in 1915 Wegener published the theory that there had originally been a single gigantic supercontinent, which he called Pangaea (meaning "all lands"), that began to break up and drift apart roughly 200 million years ago. (No humans were present to witness this; the earliest humans did not appear on Earth until about 1.9 million years ago. At the time that Pangaea broke up, during what geologists refer to as the early Jurassic period, dinosaurs were the dominant land animals.)

Other geologists refined this theory, arguing that Pangaea must have first split into two smaller supercontinents, which they called Laurasia and Gondwana, separated by what they called the Tethys Ocean. Gondwana later split into Africa and South America, with Laurasia dividing to become North America and Eurasia. According to this theory, the Mediterranean Sea is a surviving remnant of the ancient Tethys Ocean (**Figure 9-12**).

Initially, most geologists ridiculed Wegener's ideas. Although it was generally accepted that the continents do "float" on the denser, somewhat plastic mantle beneath them, few geologists could accept the idea that entire continents could move around the Earth at speeds as great as several centimeters per year. The "continental drifters" could not explain what forces could be shoving the massive continents around.

Plate Tectonics

 Beginning in the mid-1950s, however, geologists found evidence that material is being forced upward to the crust from deep within the Earth. Bruce C. Heezen of Columbia University and his colleagues began discovering long mountain ranges on the ocean floors, such as the Mid-Atlantic Ridge, which stretches all the way from Iceland to Antarctica (**Figure 9-13**). During the 1960s, Harry Hess of Princeton University and Robert Dietz of the University of California again carefully examined the floor of the Atlantic Ocean. They concluded that rock from the Earth's mantle is being melted and then forced upward along the Mid-Atlantic Ridge, which is in essence a long chain of underwater volcanoes.

The upwelling of new material from the mantle to the crust forces the existing crusts apart, causing **seafloor spreading.** For example, the floor of the Atlantic Ocean to the east of the Mid-Atlantic Ridge is moving eastward and the floor to the west is moving westward. By explaining what fills in the gap between continents as they move apart, seafloor spreading helps to fill out the

Figure 9-13

The Mid-Atlantic Ridge This artist's rendition shows the Mid-Atlantic Ridge, an immense mountain ridge that rises up from the floor of the North Atlantic Ocean. It is caused by lava seeping up from the Earth's interior along a rift that extends from Iceland to Antarctica. (Courtesy of M. Tharp and B. C. Heezen)

Figure 9-14

The Earth's Major Plates The boundaries of the Earth's major plates are the scenes of violent seismic and geologic activity. Most earthquakes occur where plates separate, collide, or rub together. Plate boundaries are therefore easily identified simply by plotting earthquake epicenters (shown here as dots) on a map. The colors of the dots indicate the depths at which the earthquakes originate. (Data from Harvard CMT catalog; plot by M. Boettcher and T. Jordan.)

theories of continental drift. Because of the seafloor spreading from the Mid-Atlantic Ridge, South America and Africa are moving apart at a speed of roughly 3 cm per year. Working backward, these two continents would have been next to each other some 200 million years ago—just as Wegener suggested (see Figure 9-11).

In the early 1960s, geologists began to find additional evidence supporting the existence of large, moving plates. Thus was born the modern theory of crustal motion, which came to be known as **plate tectonics** (from the Greek *tekton,* meaning "builder").

Geologists today realize that earthquakes tend to occur at the boundaries of the Earth's crustal plates, where the plates are colliding, separating, or rubbing against each other. The boundaries of the plates therefore stand out clearly when the epicenters of earthquakes are plotted on a map, as in Figure 9-14.

The vast majority of volcanoes also occur at plate boundaries. Figure 9-15 shows a volcanic eruption at the boundary between the Pacific and Eurasian plates.

Convection and Plate Motion

What makes the plates move? The answer is twofold. First, heat flows outward from the Earth's hot core to its cool crust, and second, this heat flows through the mantle by convection, the same process that takes place in the Earth's atmosphere or in a pot of boiling water.

Figure 9-5 shows that convection occurs when a fluid is heated from below. The heat that drives convection in the Earth's

outer layers actually comes from very far below, at the boundary between the outer and inner cores. As material deep in the liquid core cools and solidifies to join the solid portion of the core, it releases the energy needed to heat the overlying mantle and cause

Figure 9-15   R I V U X G

An Erupting Volcano On September 30, 1994, a volcano erupted at the boundary of the Pacific and Eurasian plates on Russia's Kamchatka Peninsula. This space shuttle photograph shows an immense cloud of ash belching forth from this volcano. The cloud reached an altitude of 18 km (60,000 ft) above sea level and was carried by the winds for at least 1000 km from the volcano. (*STS-68,* NASA)

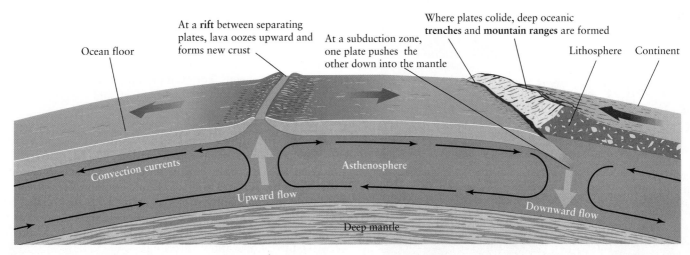

Ocean floor

At a **rift** between separating plates, lava oozes upward and forms new crust

At a subduction zone, one plate pushes the other down into the mantle

Where plates colide, deep oceanic **trenches** and **mountain ranges** are formed

Lithosphere   Continent

Convection currents

Asthenosphere

Upward flow

Downward flow

Deep mantle

Figure 9-16

The Mechanism of Plate Tectonics Convection currents in the asthenosphere, the soft upper layer of the mantle, are responsible for pushing around rigid, low-density crustal plates. New crust forms in oceanic rifts, where lava oozes upward between separating plates. Mountain ranges and deep oceanic trenches are formed where plates collide.

convection. (In Section 9-1 we described a similar process in Earth's atmosphere: When water vapor in the air cools and forms liquid drops, it releases the energy that powers thunderstorms.)

The asthenosphere, or upper level of the mantle, is hot and soft enough to permit an oozing, plastic flow. Atop the asthenosphere is a rigid layer, called the **lithosphere** (from the Greek word for "rock.") The lithosphere is divided into plates that ride along the convection currents of the asthenosphere. The crust is simply the uppermost layer of the lithosphere, with a somewhat different chemical composition than the lithosphere's lower regions.

Figure 9-16 shows how convection causes plate movement. Molten subsurface rock seeps upward along **oceanic rifts,** where plates are separating. The Mid-Atlantic Ridge, shown in Figure 9-13, is an oceanic rift. Where plates collide, cool crustal material from one of the plates sinks back down into the mantle along a **subduction zone.** One such subduction zone is found along the west coast of South America, where the oceanic Nazca plate is being subducted into the mantle under the continental South American plate at a relatively speedy 10 centimeters per year. As the material from the subducted plate sinks, it pulls the rest of its plate along with it, thus helping to keep the plates in motion. New material is added to the crust from the mantle at the oceanic rifts and is "recycled" back into the mantle at the subduction zones. In this way the total amount of crust remains essentially the same.

The boundaries between plates are the sites of some of the most impressive geological activity on our planet. Great mountain ranges, such as the Sierras and Cascades along the western coast of North America and the Andes along South America's west coast, are thrust up by ongoing collisions between continental plates and the plates of the ocean floor. Subduction zones, where old crust is drawn back down into the mantle, are typically the locations of deep oceanic trenches, such as the Peru-Chile Trench off the west coast of South America. **Figure 9-17** and **Figure 9-18**

show two well-known geographic features that resulted from tectonic activity at plate boundaries.

Plate Tectonics and the Varieties of Rocks

Plate tectonics also helps to explain geology on the scale of individual rocks and minerals. These are composed of chemical

The Red Sea is formed by the spreading apart of the African plate (at lower left) and the Arabian plate (at upper right)

Saudi Arabia

Gulf of Aqaba

Sinai Peninsula

Egypt

Gulf of Suez

Red Sea

Figure 9-17   R I ⟨V⟩ U X G

The Separation of Two Plates The plates that carry Africa and Arabia are moving apart, leaving a great rift that has been flooded to form the Red Sea. This view from orbit shows the northern Red Sea, which splits into the Gulf of Suez and the Gulf of Aqaba. (*Gemini 12,* NASA)

The Himalayas are formed by the collision of the Australia-India plate (on the left) and the Eurasian plate (on the right)

Tibet

India

Himalayas

Figure 9-18   R I V U X G

The Collision of Two Plates The plates that carry India and China are colliding. Instead of one plate being subducted beneath the other, both plates are pushed upward, forming the Himalayas. Mount Everest is one of the snow-covered peaks near the center of this photograph taken from orbit. (*Apollo 7*, NASA)

elements (see the periodic table in Box 5-5), but individual elements are rarely found in a pure state. The exceptions include diamonds, which are crystals of carbon, as well as nuggets of gold, silver, and copper. Somewhat more common are solids composed of a particular chemical combination of elements called a *compound*. An example is feldspar, a crystalline material made up of

potassium, aluminum, silicon, and oxygen atoms. Diamond and feldspar are both **minerals**—naturally occurring solids composed of a single element or compound. Most common of all are **rocks,** which are solid parts of the Earth's crust that are composed of one or more minerals. For example, granite contains both feldspar and quartz (a mineral containing silicon and oxygen).

Geologic processes create three major categories of rocks (**Figure 9-19**). **Igneous rocks** result when minerals cool from a molten state. (Molten rock is called **magma** when it is buried below the surface and **lava** when it flows out upon the surface, as in a volcanic eruption.) The ocean floor is made predominantly of a type of igneous rock called basalt (Figure 9-19*a*). This is just what we would expect if the ocean floor is produced by material welling up from the mantle, as predicted by the theory of seafloor spreading.

Sedimentary rocks are produced by the action of wind, water, or ice. For example, winds can pile up layer upon layer of sand grains. Other minerals present amid the sand can gradually cement the grains together to produce sandstone (Figure 9-19*b*). Minerals that precipitate out of the oceans can cover the ocean floor with layers of a sedimentary rock called limestone. The motion of tectonic plates can move such rocks to places far from where they were formed. This explains why sedimentary rock from the floor of an ancient ocean can now be found in central Alaska and as hogback ridges in Colorado (see Figure 9-3).

Sometimes igneous or sedimentary rocks become buried far beneath the surface, where they are subjected to enormous pressure and high temperatures. These severe conditions change the structure of the rocks, producing **metamorphic rock** (Figure 9-19*c*). The presence of metamorphic rocks at the Earth's surface tells us that tectonic activity sometimes lifts up material from deep within the crust. This can happen when two plates collide, as shown in Figure 9-18. The boundaries of the colliding plates are thrust upward to form a new chain of mountains, and buried metamorphic rocks are brought to the surface.

(a) Igneous rock (basalt)   (b) Sedimentary rock (sandstone)

(c) Metamorphic rocks (marble, schist)

Figure 9-19   R I V U X G

Igneous, Sedimentary, and Metamorphic Rocks (a) Igneous rocks such as basalt are created when molten materials solidify. This example contains iron-rich minerals that give it its dark color. **(b)** Sedimentary rocks such as sandstone are typically formed when loose particles of soil or sand are fused into rock by the presence of other minerals, which act

as a cement. **(c)** Metamorphic rocks are produced when igneous or sedimentary rocks are subjected to high temperatures and pressures deep within the Earth's crust. Marble (left) is formed from sedimentary limestone; schist (right) is formed from igneous rock. (W. J. Kaufmann III; specimens courtesy Mineral Museum, California Division of Mines)

The Cycle of Supercontinents

Plate tectonic theory offers insight into geology on the largest of scales, that of an entire supercontinent. In recent years, geologists have uncovered evidence that points to a whole succession of supercontinents that once broke apart and then reassembled. Pangaea is only the most recent supercontinent in this cycle, which repeats about every 500 million (5×10^8) years. As a result, intense episodes of mountain building have occurred at roughly 500-million-year intervals.

Apparently, a supercontinent sows the seeds of its own destruction because it blocks the flow of heat from the Earth's interior. As soon as a supercontinent forms, temperatures beneath it rise, much as they do under a book lying on an electric blanket. As heat accumulates, the lithosphere domes upward and cracks. Molten rock from the overheated asthenosphere wells up to fill the resulting fractures, which continue to widen as pieces of the fragmenting supercontinent move apart.

It can take a very long time for the heat trapped under a supercontinent to escape. Although Pangaea broke apart some 200 million years ago (see Figure 9-11), the mantle under its former location is still hot and is still trying to rise upward. As a result, Africa—which lies close to the center of this mass of rising material—sits several tens of meters higher than the other continents.

The changes wrought by plate tectonics are very slow on the scale of a human lifetime, but they are very rapid in comparison with the age of the Earth. For example, the period over which Pangaea broke into Laurasia and Gondwana was only about 0.4% of the Earth's age of 4.56×10^9 years. (To put this in perspective, 0.4% of a human lifetime is about 4 months.) The lesson of plate tectonics is that the seemingly permanent face of the Earth is in fact dynamic and ever-changing.

9-4   The Earth's magnetic field produces a magnetosphere that traps particles from the solar wind

We saw in Section 7-7 that Earth's magnetic field provides evidence that our planet's interior is partially molten. Our understanding is that the magnetic field is produced by the motion of iron-rich material in the liquid outer core. Because this material is a good conductor of electricity, its motions generate a magnetic field in a mechanism called a *dynamo*.

> **Like the motions of tectonic plates, the Earth's magnetic field results from our planet's internal heat**

The Source of Earth's Magnetic Field

The same energy source that powers plate tectonics is also responsible for the Earth's dynamo. As we saw in Section 9-3, energy is released when material from our planet's liquid core cools and solidifies onto the solid core. This released energy stirs up the motions of the liquid core, and these motions generate a magnetic field. (While there is also fluid motion in the mantle, it does not produce an appreciable magnetic field: The silicon-rich rocks that comprise the mantle are poor conductors of electricity, and their motions are a million times slower than those in the outer core.)

 Studies of ancient rocks reinforce the idea that our planet's magnetism is due to fluid material in motion. When iron-bearing lava cools and solidifies to form igneous rock, it becomes magnetized in the direction of the Earth's magnetic field. By analyzing samples of igneous rock of different ages from around the world, geologists have found that the Earth's magnetic field actually flips over and reverses direction on an irregular schedule ranging from tens of thousands to hundreds of thousands of years. As an example, lava that solidified 30,000 years ago is magnetized in the opposite direction to lava that has solidified recently. Therefore, 30,000 years ago a compass needle would have pointed south, not north! If the Earth was a permanent magnet, like the small magnets used to attach notes to refrigerators, it would be hard to imagine how its magnetic field could spontaneously reverse direction. But computer simulations show that fields produced by moving fluids in the Earth's outer core can indeed change direction from time to time.

The Magnetosphere

The Earth's magnetic field has important effects far above the atmosphere, where it interacts dramatically with charged particles from the Sun. This *solar wind* is a flow of mostly protons and electrons that streams constantly outward from the Sun's upper atmosphere. Near the Earth, the particles in the solar wind move at speeds of roughly 450 km/s, or about a million miles per hour. Because this is considerably faster than sound waves can travel in the very thin gas between the planets, the solar wind is said to be *supersonic*. (Because the gas between the planets is so thin, interplanetary sound waves carry too little energy to be heard by astronauts.)

If the Earth had no magnetic field, we would be continually bombarded by the solar wind. But our planet does have a magnetic field, and the forces that this field can exert on charged particles are strong enough to deflect them away from us. The region of space around a planet in which the motion of charged particles is dominated by the planet's magnetic field is called the planet's **magnetosphere**. **Figure 9-20** is a scale drawing of the Earth's magnetosphere, which was discovered in the late 1950s by the first satellites placed in orbit.

When the supersonic particles in the solar wind first encounter the Earth's magnetic field, they abruptly slow to subsonic speeds. The boundary where this sudden decrease in velocity occurs is called a **shock wave**. Still closer to the Earth lies another boundary, called the **magnetopause**, where the outward magnetic pressure of the Earth's field is exactly counterbalanced by the impinging pressure of the solar wind. Most of the particles of the solar wind are deflected around the magnetopause, just as water is deflected to either side of the bow of a ship.

Some charged particles of the solar wind manage to leak through the magnetopause. When they do, they are trapped by the Earth's magnetic field in two huge, doughnut-shaped rings around Earth called the **Van Allen belts**. These belts were discovered in 1958 during the flight of the first successful U.S. Earth-orbiting satellite. They are named after the physicist James Van Allen, who

Figure 9-20

The Earth's Magnetosphere The Earth's magnetic field carves out a cavity in the solar wind, shown here in cross section. A shock wave marks the boundary where the supersonic solar wind is abruptly slowed to subsonic speeds. Most of the particles of the solar wind are deflected around the Earth in a turbulent region (colored blue in this drawing). The Earth's magnetic field also traps some charged particles in two huge, doughnut-shaped rings called the Van Allen belts (shown in red). This figure shows only a slice through the Van Allen belts.

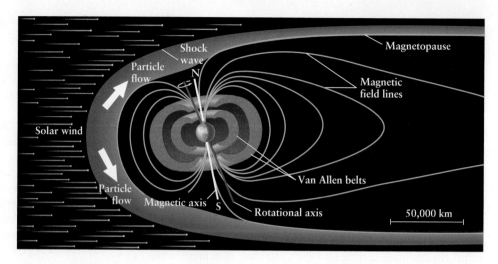

insisted that the satellite carry a Geiger counter to detect charged particles. The inner Van Allen belt, which extends over altitudes of about 2000 to 5000 km, contains mostly protons. The outer Van Allen belt, about 6000 km thick, is centered at an altitude of about 16,000 km above the Earth's surface and contains mostly electrons.

Aurorae

Sometimes the magnetosphere becomes overloaded with particles. The particles then leak through the magnetic fields at their weakest points and cascade down into the Earth's upper atmosphere, usually in a ring-shaped pattern (**Figure 9-21a**). As these high-speed charged particles collide with atoms in the upper atmosphere, they excite the atoms to high energy levels. The atoms then emit visible light as they drop down to their ground states, like the excited gas atoms in a neon light (see Section 5-8). The result is a beautiful, shimmering display called an **aurora** (*plural* **aurorae**). These are also called the **northern lights** (**aurora borealis**) or **southern lights** (**aurora australis**), depending on the hemisphere in which the phenomenon is observed. Figures 9-21*b* and 9-21*c* show the aurorae as seen from orbit and from the Earth's surface.

Occasionally, a violent event on the Sun's surface sends a particularly intense burst of protons and electrons toward Earth. The resulting auroral display can be exceptionally bright and can often be seen over a wide range of latitudes. Such events also disturb radio transmissions and can damage communication satellites and transmission lines.

It is remarkable that the Earth's magnetosphere, including its vast belts of charged particles, was entirely unknown until a few decades ago. Such discoveries remind us of how little we truly understand and how much remains to be learned even about our own planet.

(a) R I V U X G

(b) R I V U X G

(c) R I V U X G

Figure 9-21

The Aurora An increased flow of charged particles from the Sun can overload the Van Allen belts and cascade toward the Earth, producing aurorae. **(a)** This false-color ultraviolet image from the *Dynamics Explorer 1* spacecraft shows a glowing oval of auroral emission about 4500 km in diameter centered on the north magnetic pole. **(b)** This photograph shows the aurora australis over Antarctica as seen from an orbiting space shuttle. **(c)** This view from Alaska shows aurorae at their typical altitudes of 100 to 400 km. The green color shows that the light is emitted by excited oxygen atoms in the upper atmosphere. (a: Courtesy of L. A. Frank and J. D. Craven, University of Iowa; b: MTU/Geological & Mining Engineering & Sciences; c: J. Finch/Photo Researchers, Inc.)

9-5   The Earth's atmosphere has changed substantially over our planet's history

Both plate tectonics and the Earth's magnetic field paint a picture of a planet with a dynamic, evolving surface and interior. Our atmosphere, too, has changed and evolved substantially over the Earth's 4.56-billion-year history. As we will see, this evolution explains why we have an atmosphere totally unlike that of any other world in the solar system.

The composition of our atmosphere has been dramatically altered by the evolution of life

Earth's Early Atmosphere

When the Earth first formed by accretion of planetesimals (see Section 8-4), gases were probably trapped within the Earth's interior in the same proportions that they were present in the solar nebula. But since the early Earth was hot enough to be molten throughout its volume, most of these trapped gases were released. As we discussed in Section 7-4 and Box 7-2, the Earth's gravity was too weak to prevent hydrogen and helium—the two most common kinds of atoms in the universe, but also the least massive—from leaking away into space. The atmosphere that remained still contained substantial amounts of hydrogen, but in the form of relatively massive molecules of water vapor (H_2O) made by combining two atoms of hydrogen with one atom of oxygen, the third most common element in our part of the Milky Way Galaxy (see Figure 8-4). In fact, water vapor was probably the dominant constituent of the early atmosphere, which is thought to have been about 100 times denser than our present-day atmosphere.

Water vapor is one of the greenhouse gases that we discussed in Section 9-1. It traps infrared radiation from the Earth's surface, and its presence in the early atmosphere helped to sustain the high temperature of the surface. But if water vapor had been the *only* constituent of the atmosphere, you would not be reading these words! The reason is that as the molten Earth inevitably cooled, water vapor in the atmosphere condensed into raindrops and fell to Earth to form the oceans. As water vapor was lost from the atmosphere, its contribution to the greenhouse effect weakened, which would have made the surface temperature drop even lower. A further cooling factor is that the young Sun was only about 70% as bright as it is today. The net result is that the entire Earth should have been frozen over! Ice is so reflective that the Earth would have remained frozen even as the Sun aged and became more luminous. Liquid water is absolutely essential to all living creatures, so life as we know it would probably never have evolved on such an icy Earth.

In fact, the first living organisms appeared on Earth within 400 million years after the planet first formed. Hence, the Earth could not have been frozen over for very long, if at all. What saved our planet from a perpetual deep freeze was the presence of carbon, the fourth most common of the elements. When combined with oxygen, carbon forms carbon dioxide (CO_2), a greenhouse gas that remains gaseous at low temperatures. Carbon dioxide would have been released into the atmosphere by volcanic activity, a process called **outgassing** (see Figure 9-15). It would also have been added to the atmosphere by carbon-rich meteors striking the planet's surface. Once CO_2 was in the atmosphere, its greenhouse effect raised the planet's temperature and melted the ice. Some of the water evaporated into the atmosphere, further enhancing the Earth's greenhouse effect.

Fewer than one in 2500 of the molecules in today's atmosphere is carbon dioxide. But CO_2 must have been thousands of times more abundant in the early Earth's atmosphere in order to melt the global ice sheath. Unlike hydrogen or helium, this excess CO_2 was not lost into space; instead, it has been trapped in rocks. Carbon dioxide dissolves in rainwater and falls into the oceans, where it combines with other substances to form a class of minerals called *carbonates*. (Limestone and marble are examples of carbonate-bearing rock.) These form sediments on the ocean floor, which are eventually recycled into the crust by subduction. As an example, marble (Figure 9-19*c*) is a metamorphic rock formed deep within the crust from limestone, a carbonate-rich sedimentary rock.

This process removed most of the CO_2 from the atmosphere within the first billion (10^9) years after the formation of the Earth. Although this weakened the greenhouse effect, temperatures remained warm as the Sun's brightness increased. The small amounts of atmospheric CO_2 that remain today are the result of a balance between volcanic activity (which releases CO_2 into the atmosphere) and the formation of carbonates (which removes CO_2). (As we will see in Section 9-7, human activity is upsetting this balance.)

Life's Impact on Earth's Atmosphere

The appearance of life on Earth set into motion a radical transformation of the atmosphere. Early single-cell organisms converted energy from sunlight into chemical energy using **photosynthesis,** a chemical process that consumes CO_2 and water and releases oxygen (O_2). Oxygen molecules are very reactive, so originally most of the O_2 produced by photosynthesis combined with other substances to form minerals called oxides. (Evidence for this can be found in rock formations of various ages. The oldest rocks have very low oxide content, while oxides are prevalent in rocks that formed after the appearance of organisms that used photosynthesis.) But as life proliferated, the amount of photosynthesis increased dramatically. Eventually so much oxygen was being produced that it could not all be absorbed to form oxides, and O_2 began to accumulate in the atmosphere. Figure 9-22 shows how the amount of O_2 in the atmosphere has increased over the history of the Earth.

About 2 billion (2×10^9) years ago, a new type of life evolved to take advantage of the newly abundant oxygen. These new organisms produce energy by consuming oxygen and releasing carbon dioxide—a process called **respiration** that is used by all modern animals, including humans. Such organisms thrived because photosynthetic plants continued to add even more oxygen to the atmosphere.

Several hundred million years ago, the number of oxygen molecules in the atmosphere stabilized at 21% of the total, the same as the present-day value. This value represents a balance between the release of oxygen from plants by photosynthesis and the intake of oxygen by respiration. Thus, the abundance of oxygen is regulated almost exclusively by the presence of life—a situation that has no parallel anywhere else in the solar system.

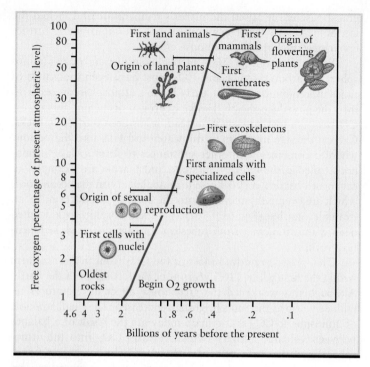

Figure 9-22

The Increase in Atmospheric Oxygen This graph shows how the amount of oxygen in the atmosphere (expressed as a percentage of its present-day value) has evolved with time. Note that the atmosphere contained essentially no oxygen until about 2 billion years ago. (Adapted from Preston Cloud, "The Biosphere," *Scientific American*, September 1983, p. 176)

The most numerous molecules in our atmosphere are nitrogen (N_2), which make up 78% of the total. These, too, are a consequence of the presence of life on Earth. Certain bacteria extract oxygen from minerals called nitrates, and in the process liberate nitrogen into the atmosphere. The amount of atmospheric nitrogen is kept in check by lightning. The energy in a lightning flash causes atmospheric nitrogen and oxygen to combine into nitrogen oxides, which dissolve in rainwater, fall into the oceans, and form nitrates.

Comparing Atmospheres: Earth, Venus, and Mars

Table 9-4 shows the dramatic differences between the Earth's atmosphere and those of Venus and Mars. (Mercury, the other terrestrial planet, is too small and has too little gravity to hold an appreciable atmosphere.) The greater intensity of sunlight on Venus caused higher temperatures, which boiled any liquid water and made it impossible for CO_2 to be taken out of the atmosphere and put back into rocks. The atmosphere thus became far denser than our own and rich in greenhouse gases. The result was a very strong greenhouse effect that raised temperatures on Venus to their present value of about 460°C (733 K = 855°F).

Just the opposite happened on Mars, where sunlight is less than half as intense as on Earth. The lower temperatures drove CO_2 from the atmosphere into Martian rocks, and froze all of

Table 9-4   Chemical Compositions of Three Planetary Atmospheres

	Venus	Earth	Mars
Nitrogen (N_2)	3.5%	78.08%	2.7%
Oxygen (O_2)	almost zero	20.95%	almost zero
Carbon dioxide (CO_2)	96.5%	0.035%	95.3%
Water vapor (H_2O)	0.003%	about 1%	0.03%
Other gases	almost zero	almost zero	2%

the water beneath the planet's surface. The atmosphere that remains on Mars has a similar composition to that of Venus but is less than 1/10,000 as dense. On neither Venus nor Mars was life able to blossom and transform the atmosphere as it did here on Earth.

We have thus uncovered a general rule about the terrestrial planets:

The closer a terrestrial planet is to the Sun, the more important the greenhouse effect in that planet's atmosphere. The stronger the greenhouse effect, the higher the planet's surface temperature.

This rule suggests that if the Earth had formed a bit closer to or farther from the Sun, temperatures on our planet might have been too high or too low for life ever to evolve. Thus, our very existence is a result of the Earth's special position in the solar system.

9-6   Like Earth's interior, our atmosphere has a layered structure

While life has shaped our atmosphere's chemical composition, the structure of the atmosphere is controlled by the influence of sunlight. Scientists describe the structure of any atmosphere in terms of two properties: temperature and atmospheric pressure, which vary with altitude.

> **Circulation in our atmosphere results from convection and the Earth's rotation**

Pressure, Temperature, and Convection in the Atmosphere

Atmospheric pressure at any height in the atmosphere is caused by the weight of all the air above that height.

The average atmospheric pressure at sea level is defined to be one **atmosphere** (1 atm), equal to 1.01×10^5 N/m² or 14.7 pounds per square inch. As you go up in the atmosphere, there is less air above you to weigh down on you. Hence, atmospheric pressure decreases smoothly with increasing altitude.

Unlike atmospheric pressure, temperature varies with altitude in a complex way. Figure 9-23 shows that temperature decreases with increasing altitude in some layers of the atmosphere, but in other layers actually *increases* with increasing altitude. These differences result from the individual ways in which each layer is heated.

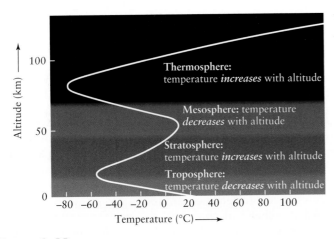

Figure 9-23

Temperature Profile of the Earth's Atmosphere This graph shows how the temperature in the Earth's atmosphere varies with altitude. In the troposphere and mesosphere, temperature decreases with increasing altitude; in the stratosphere and thermosphere, temperature actually increases with increasing altitude.

The lowest layer, called the **troposphere,** extends from the surface to an average altitude of 12 km (roughly 7.5 miles, or 39,000 ft). It is heated only indirectly by the Sun. Sunlight warms the Earth's surface, which heats the lower part of the troposphere. By contrast, the upper part of the troposphere remains at cooler temperatures. This vertical temperature variation causes convection currents that move up and down

through the troposphere (see Section 8-1). All of the Earth's weather is a consequence of this convection.

Convection on a grand scale is caused by the temperature difference between the Earth's equator and its poles. If the Earth did not rotate, heated air near the equator would rise upward and flow at high altitude toward the poles. There it would cool and sink to lower altitudes, at which it would flow back to the equator. However, the Earth's rotation breaks up this simple convection pattern into a series of **convection cells.** In these cells, air flows east and west as well as vertically and in a north-south direction. The structure of these cells explains why the prevailing winds blow in different directions at different latitudes (Figure 9-24).

Upper Layers of the Atmosphere

Almost all the oxygen in the troposphere is in the form of O_2, a molecule made of two oxygen atoms. But in the **stratosphere,** which extends from about 12 to 50 km (about 7.5 to 31 mi) above the surface, an appreciable amount of oxygen is in the form of **ozone,** a molecule made of *three* oxygen atoms (O_3). Unlike O_2, ozone is very efficient at absorbing ultraviolet radiation from the Sun, which means that the stratosphere can directly absorb solar energy. The result is that the temperature actually increases as you move upward in the stratosphere. Convection requires that the temperature must decrease, not increase, with increasing altitude, so there are essentially no convection currents in the stratosphere.

Above the stratosphere lies the **mesosphere.** Very little ozone is found here, so solar ultraviolet radiation is not absorbed within the mesosphere, and atmospheric temperature again declines with increasing altitude.

Figure 9-24

Circulation Patterns in the Earth's Atmosphere The dominant circulation in our atmosphere consists of six convection cells, three in the northern hemisphere and three in the southern hemisphere. In the northern temperate region (including the continental United States), the prevailing winds at the surface are from the southwest toward the northeast. Farther south, within the northern tropical region (for example, Hawaii), the prevailing surface winds are from northeast to southwest.

The temperature of the mesosphere reaches a minimum of about $-75°C$ ($= -103°F = 198$ K) at an altitude of about 80 km (50 mi). This minimum marks the bottom of the atmosphere's thinnest and uppermost layer, the **thermosphere,** in which temperature once again rises with increasing altitude. This is not due to the presence of ozone, because in this very low-density region oxygen and nitrogen are found as individual atoms rather than in molecules. Instead, the thermosphere is heated because these isolated atoms absorb very-short-wavelength solar ultraviolet radiation (which oxygen and nitrogen molecules cannot absorb).

CAUTION! At altitudes near 300 km the temperature of the thermosphere is about 1000°C (1800°F). This is near the altitude at which the space shuttle and satellites orbit the Earth. Nonetheless, a satellite in orbit does *not* risk being burned up as it moves through the thermosphere. The reason is that the thermosphere is far less dense than the atmosphere at sea level. The high temperature simply means that an average atom in the thermosphere is moving very fast (see Box 7-2). But because the thermosphere is so thin (only about 10^{-11} as dense as the air at sea level), these fast-moving atoms are few and far between. Hence, the thermosphere contains very little energy. Nearly all the heat that an orbiting satellite receives is from sunlight, not the thermosphere.

The *Cosmic Connections* figure compares the layered structure of our atmosphere with that of the Earth's interior.

9-7 A burgeoning human population is profoundly altering the Earth's biosphere

One of the extraordinary characteristics of the Earth is that it is covered with life, from the floors of the oceans to the tops of mountains and from frigid polar caps to blistering deserts. So far in this chapter we have hinted at how our planet and living organisms interact: the greenhouse effect has given the Earth a suitable temperature for the evolution of life (Section 9-1), and over billions of years that evolution has transformed the chemical composition of the atmosphere (Section 9-5). Let's explore in greater depth how Earth and the organisms that live on it, especially humans, affect each other.

> **Global warming and its consequences pose a major challenge to our civilization**

The Biosphere and Natural Climate Variation

All life on Earth subsists in a relatively thin layer called the **biosphere,** which includes the oceans, the lowest few kilometers of the troposphere, and the crust to a depth of almost 3 kilometers. **Figure 9-25** is a portrait of the Earth's biosphere based on NASA satellite data. The biosphere, which has taken billions of years to evolve to its present state, is a delicate, highly complex system in which plants and animals depend on each other for their mutual survival.

low   high
Ocean chlorophyll index

low   high
Land vegetation index

Figure 9-25

The Earth's Biosphere This image, based on data from the *SeaWIFS* spacecraft, shows the distribution of plant life over the Earth's surface. The ocean colors show where free-floating microscopic plants called phytoplankton are found. (NASA Visible Earth)

COSMIC CONNECTIONS

Comparing Earth's Atmosphere and Interior

Earth's atmosphere (which is a gas) and Earth's interior (which is partly solid, partly liquid) both decrease in density and pressure as you go farther away from the Earth's center.

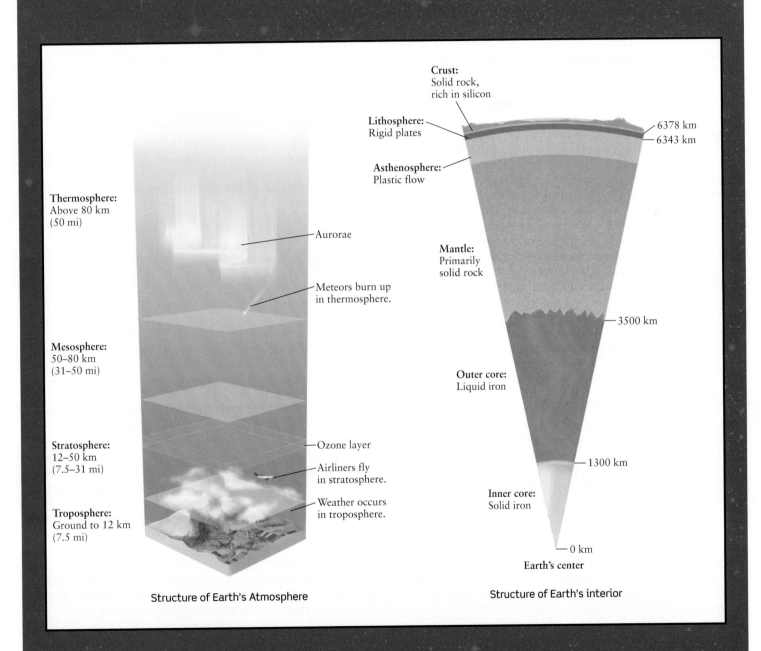

Thermosphere: Above 80 km (50 mi)

Aurorae

Meteors burn up in thermosphere.

Mesosphere: 50–80 km (31–50 mi)

Stratosphere: 12–50 km (7.5–31 mi)

Ozone layer

Airliners fly in stratosphere.

Weather occurs in troposphere.

Troposphere: Ground to 12 km (7.5 mi)

Structure of Earth's Atmosphere

Crust: Solid rock, rich in silicon

Lithosphere: Rigid plates

6378 km
6343 km

Asthenosphere: Plastic flow

Mantle: Primarily solid rock

3500 km

Outer core: Liquid iron

1300 km

Inner core: Solid iron

0 km

Earth's center

Structure of Earth's interior

The state of the biosphere depends crucially on the temperature of the oceans and atmosphere. Even small temperature changes can have dramatic consequences. An example that recurs every three to seven years is the El Niño phenomenon, in which temperatures at the surface of the equatorial Pacific Ocean rise by 2 to 3°C. Ordinarily, water from the cold depths of the ocean is able to well upward, bringing with it nutrients that are used by microscopic marine organisms called phytoplankton that live near the surface (see Figure 9-25). But during an El Niño, the warm surface water suppresses this upwelling, and the phytoplankton starve. This wreaks havoc on organisms such as mollusks that feed on phytoplankton, on the fish that feed on the mollusks, and on the birds and mammals that eat the fish. During the 1982–1983 El Niño, a quarter of the adult sea lions off the Peruvian coast starved, along with all of their pups.

Many different factors can change the surface temperature of our planet. One is that the amount of energy radiated by the Sun can vary up or down by a few tenths of a percent. Reduced solar brightness may explain the period from 1450 to 1850 when European winters were substantially colder than they are today.

Other factors are the gravitational influences of the Moon and the other planets. Thanks to these influences, the eccentricity of the Earth's orbit varies with a period of 90,000 to 100,000 years, the tilt of its rotation axis varies between 22.1° and 24.5° with a 40,000-year period, and the orientation of its rotation axis changes due to precession (see Section 2-6) with a 26,000-year period. These variations can affect climate by altering the amount of solar energy that heats the Earth during different parts of the year. They help explain why the Earth periodically undergoes an extended period of low temperatures called an *ice age,* the last of which ended about 11,000 years ago.

As we saw in Sections 9-1 and 9-5, one of the most important factors affecting global temperatures is the abundance of greenhouse gases such as CO_2. Geologic processes can alter this abundance, either by removing CO_2 from the atmosphere (as happens when fresh rock is uplifted and exposed to the air, where it can absorb atmospheric CO_2 to form carbonates) or by supplying new CO_2 (as in volcanic eruptions like that shown in Figure 9-15). From time to time in our planet's history, natural events have caused dramatic increases in the amount of greenhouse gases in the atmosphere. One such event may have taken place 251 million (2.51×10^8) years ago, when Siberia went through a period of intense volcanic activity. (The evidence for this is a layer of igneous rock in Siberia that covers an area about the size of Europe. Like the rock shown in Figure 9-19a, this layer—which radioactive dating reveals to be 2.51×10^8 years old—is solidified lava.) The tremendous amounts of CO_2 released by this activity would have elevated the global temperature by several degrees. Remarkably, the fossil record reveals that 95% of all species on Earth became extinct at this same time. The coincidence of these two events suggests that greenhouse-induced warming can have catastrophic effects on life.

Human Effects on the Biosphere: Deforestation

Our species is having an increasing effect on the biosphere because our population is skyrocketing. Figure 9-26 shows the sharp rise in the human population that began in the late 1700s with

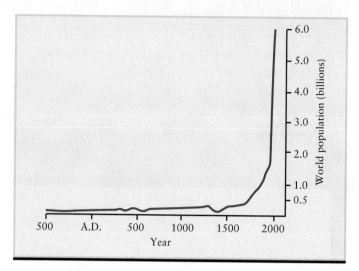

Figure 9-26

The Human Population Data and estimates from the U.S. Bureau of the Census, the Population Reference Bureau, and the United Nations Population Fund were combined to produce this graph showing the human population from 500 B.C. to 2000 A.D. The population began to rise in the eighteenth century and has been increasing at an astonishing rate since 1900.

the Industrial Revolution and the spread of modern ideas about hygiene. This rise accelerated in the twentieth century thanks to medical and technological advances ranging from antibiotics to high-yield grains. In 1960 there were 3 billion people on Earth; in 1975, 4 billion; and in 1999, 6 billion. Projections by the United Nations Population Division show that there will be more than 8 billion people on Earth by the year 2030 and more than 9 billion by 2050.

Every human being has basic requirements: food, clothing, and housing. We all need fuel for cooking and heating. To meet these demands, we cut down forests, cultivate grasslands, and build sprawling cities. A striking example of this activity is occurring in the Amazon rain forest of Brazil. Tropical rain forests are vital to our planet's ecology because they absorb significant amounts of CO_2 and release O_2 through photosynthesis. Although rain forests occupy only 7% of the world's land areas, they are home to at least 50% of all plant and animal species on Earth. Nevertheless, to make way for farms and grazing land, people simply cut down the trees and set them on fire—a process called slash-and-burn. Such deforestation, along with extensive lumbering operations, is occurring in Malaysia, Indonesia, and Papua New Guinea. The rain forests that once thrived in Central America, India, and the western coast of Africa are almost gone (Figure 9-27).

Human Effects on the Biosphere: The Ozone Layer

Human activity is also having potentially disastrous effects on the upper atmosphere. Certain chemicals released into the air—in particular chlorofluorocarbons (CFCs), which have been used in refrigeration and electronics, and methyl bromide, which is used in fumigation—are destroying the ozone in the stratosphere. We saw in Section 9-6 that ozone molecules absorb ultraviolet

Figure 9-27   R I V̄ U X G

The Deforestation of Amazonia The Amazon, the world's largest rain forest, is being destroyed at a rate of 20,000 square kilometers per year in order to provide land for grazing and farming and as a source for lumber. About 80% of the logging is being carried out illegally. (Martin Wendler/Okapia/Photo Researchers)

Figure 9-29

Atmospheric Carbon Dioxide Is Increasing This graph shows measurements of atmospheric carbon dioxide in parts per million (ppm). The sawtooth pattern results from plants absorbing more carbon dioxide during the spring and summer. The CO_2 concentration in the atmosphere has increased by 21% since continuous observations started in 1958. (NOAA/Scripps Institution of Oceanography)

radiation from the Sun. Without this high-altitude **ozone layer**, solar ultraviolet radiation would beat down on the Earth's surface with greatly increased intensity. Such radiation breaks apart most of the delicate molecules that form living tissue. Hence, a complete loss of the ozone layer would lead to a catastrophic ecological disaster.

In 1985 scientists discovered an **ozone hole**, a region with an abnormally low concentration of ozone, over Antarctica. Since then this hole has generally expanded from one year to the next (**Figure 9-28**). Smaller but still serious effects have been observed in the stratosphere above other parts of the Earth. As a result, there has been a worldwide increase in the number of deaths due to skin cancer caused by solar ultraviolet radiation. By international agreement, CFCs are being replaced by compounds that do not deplete stratospheric ozone, and sunlight naturally produces

more ozone in the stratosphere. But even in the best of circumstances, the damage to the ozone layer is not expected to heal for many decades.

Human Effects on the Biosphere: Global Warming and Climate Change

The most troubling influence of human affairs on the biosphere is a consequence of burning fossil fuels (petroleum and coal) in automobiles, airplanes, and power plants as well as burning forests and brushland for agriculture and cooking (Figure 9-27). This burning releases carbon dioxide into the Earth's atmosphere—and we are adding CO_2 to the atmosphere faster than plants and geological processes can extract it. **Figure 9-29** shows how the

Figure 9-28

The Antarctic Ozone Hole These two false-color images show that there was a net decrease of 50% in stratospheric ozone over Antarctica between October 1979 and September 2003. The amount of ozone at midlatitudes, where most of the human population lives, decreased by 10 to 20% over the same period. (GSFC/NASA)

October 1979   September 2003

Figure 9-30

Atmospheric CO₂ and Changes in Global Temperature This figure shows how the carbon dioxide concentration in our atmosphere (upper curve) and the Earth's average surface temperature (lower curve) have changed since 1000 A.D. The increase in CO₂ since 1800 due to burning fossil fuels has strengthened the greenhouse effect and caused a dramatic temperature increase. (Intergovernmental Panel on Climate Change and Hadley Centre for Climate Prediction and Research, U.K. Meteorological Office)

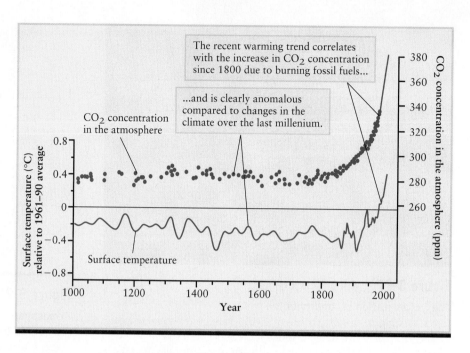

carbon dioxide content of the atmosphere has increased since 1958, when scientists began to measure this quantity on an ongoing basis.

To put the values shown in Figure 9-29 into perspective, we need to know the atmospheric CO₂ concentration in earlier eras. Scientists have learned this by analyzing air bubbles trapped at various depths in the ice that blankets the Antarctic and Greenland. Each winter a new ice layer is deposited, so the depth of a bubble indicates the year in which it was trapped. Figure 9-30 uses data obtained in this way to show how the atmospheric CO₂ concentration has varied since 1000 A.D. While there has been some natural variation in the concentration, its value has skyrocketed since the beginning of the Industrial Revolution around 1800. Data from older, deeper bubbles of trapped air show that in the 650,000 years before the Industrial Revolution, the CO₂ concentration was never greater than 300 parts per million. The present-day CO₂ concentration is greater than this by 25%, and has grown to its present elevated level in just over half a century. If there are no changes in our energy consumption habits, by 2050 the atmospheric CO₂ concentration will be greater than 600 parts per million.

Increasing atmospheric CO₂ is of concern because this strengthens the greenhouse effect and raises the Earth's average surface temperature. Figure 9-30 also shows how this average temperature has varied since 1000 A.D. (Like the CO₂ data, the temperature data from past centuries comes from analyzing trapped air bubbles. Scientists measure the relative amounts of two different types of atmospheric oxygen called ¹⁶O and ¹⁸O, described in Box 5-5, in a bubble. The ratio of these is a sensitive measure of the temperature at that time the air was trapped inside the ice.) The recent dramatic increase in atmospheric CO₂ concentration has produced an equally dramatic increase in the average surface temperature. This temperature increase is called **global warming.** Other explanations for global warming have been proposed, such as changes in the Sun's brightness, but only greenhouse gases produced by human activity can explain the

steep temperature increase shown in Figure 9-30. These gases include methane (CH₄) and nitrous oxide (N₂O), which are released in relatively small amounts by agriculture and industry but which are far more effective greenhouse gases than CO₂.

The effects of global warming can be seen around the world. Nine of the 10 warmest years on record have occurred since 1995, and the years since 2000 have seen increasing numbers of droughts, water shortages, and unprecedented heat waves. Glaciers worldwide are receding, the size of the ice cap around the North Pole has decreased by 20% since 1979, and a portion of the Antarctic ice shelf has broken off (Figure 9-31).

Figure 9-31   R I V U X G

A Melting Antarctic Ice Shelf Global warming caused the Larsen B ice shelf to break up in early 2002. This ice shelf, which was about the size of Rhode Island, is thought to have been part of the Antarctic coast for the past 12,000 years. (NASA/GSFC/LaRC/JPL, MIST Team)

Unfortunately, global warming is predicted to intensify in the decades to come. The UN Intergovernmental Panel on Climate Change predicts that if nothing is done to decrease the rate at which we add greenhouse gases to our atmosphere, the average surface temperature will continue to rise by an additional 1.4 to 5.8°C during the twenty-first century. What is worse, the temperature increase is predicted to be greater at the poles than at the equator. The global pattern of atmospheric circulation (Figure 9-24) depends on the temperature difference between the warm equator and cold poles, so this entire pattern will be affected. The same is true for the circulation patterns in the oceans. As a result, temperatures will rise in some regions and decline in others and the patterns of rainfall will be substantially altered. Agriculture depends on rainfall, so these changes in rainfall patterns can cause major disruptions in the world food supply. Studies suggest that the climate changes caused by a 3°C increase in the average surface temperature would cause a worldwide drop in cereal crops of 20 to 400 million tons, putting 400 million more people at risk of hunger.

The melting of polar ice due to global warming poses an additional risk to our civilization. When floating ice like that found near the North Pole melts, the ocean level remains the same. (You can see this by examining a glass of water with an ice cube floating in it. The water level doesn't change when the ice melts.) But the ocean level rises when ice on land melts and runs off into the sea. The Greenland ice cap has been melting at an accelerating rate since 2000 and has the potential to raise sea levels by half a meter or more. Low-lying coastal communities such as Boston and New Orleans, as well as river cities such as London, will have a greater risk of catastrophic flooding. Some island nations of the Pacific will disappear completely beneath the waves.

While global warming is an unintended consequence of human activity, the solution to global warming will require concerted and thoughtful action. Global warming cannot be stopped completely: Even if we were to immediately halt all production of greenhouse gases, the average surface temperature would increase an additional 2°C by 2100 thanks to the natural inertia of the Earth's climate system. Instead, our goal is to minimize the effects of global warming by changing the ways in which we produce energy, making choices about how to decrease our requirements for energy, and finding ways to remove CO_2 from the atmosphere and trap it in the oceans or beneath our planet's surface. Confronting global warming is perhaps the greatest challenge to face our civilization in the twenty-first century.

Key Words

albedo, p. 212
asthenosphere, p. 215
atmosphere (atm), p. 224
atmospheric pressure, p. 224
aurora (*plural* aurorae), p. 222
aurora australis, p. 222
aurora borealis, p. 222
biosphere, p. 226
convection, p. 210
convection cell, p. 225
convection current, p. 210
core, p. 213
crust (of Earth), p. 213
crystal, p. 216
earthquake, p. 213
epicenter, p. 213
global warming, p. 230
greenhouse effect, p. 213
greenhouse gas, p. 213
igneous rock, p. 220
inner core (of Earth), p. 214
lava, p. 220
lithosphere, p. 219
magma, p. 220
magnetopause, p. 221
magnetosphere, p. 221
mantle, p. 213
melting point, p. 215
mesosphere, p. 225
metamorphic rock, p. 220
mineral, p. 220
northern lights, p. 222
oceanic rift, p. 219
outer core (of Earth), p. 214
outgassing, p. 223
ozone, p. 225
ozone layer, p. 229
ozone hole, p. 229
P wave, p. 214
photosynthesis, p. 223
plastic, p. 215
plate (lithospheric), p. 216
plate tectonics, p. 218
respiration, p. 223
rock, p. 220
S wave, p. 214
seafloor spreading, p. 217
sedimentary rock, p. 220
seismic wave, p. 213
seismograph, p. 213
shock wave, p. 221
southern lights, p. 222
stratosphere, p. 225
subduction zone, p. 219
surface wave, p. 214
thermosphere, p. 226
troposphere, p. 225
Van Allen belts, p. 221

Key Ideas

The Earth's Energy Sources: All activity in the Earth's atmosphere, oceans, and surface is powered by three sources of energy.

• Solar energy is the energy source for the atmosphere. In the greenhouse effect, some of this energy is trapped by infrared-absorbing gases in the atmosphere, raising the Earth's surface temperature.

• Tidal forces from the Moon and Sun help to power the motion of the oceans.

• The internal heat of the Earth is the energy source for geologic activity.

The Earth's Interior: Studies of seismic waves (vibrations produced by earthquakes) show that the Earth has a small, solid inner core surrounded by a liquid outer core. The outer core is surrounded by the dense mantle, which in turn is surrounded by the thin low-density crust.

• Seismologists deduce the Earth's interior structure by studying how longitudinal P waves and transverse S waves travel through the Earth's interior.

• The Earth's inner and outer cores are composed of almost pure iron with some nickel mixed in. The mantle is composed of iron-rich minerals.

• Both temperature and pressure steadily increase with depth inside the Earth.

Plate Tectonics: The Earth's crust and a small part of its upper mantle form a rigid layer called the lithosphere. The lithosphere is divided into huge plates that move about over the plastic layer called the asthenosphere in the upper mantle.

• Plate tectonics, or movement of the plates, is driven by convection within the asthenosphere. Molten material wells up at oceanic rifts, producing seafloor spreading, and is returned to the asthenosphere in subduction zones. As one end of a plate is subducted back into the asthenosphere, it helps to pull the rest of the plate along.

• Plate tectonics is responsible for most of the major features of the Earth's surface, including mountain ranges, volcanoes, and the shapes of the continents and oceans.

• Plate tectonics is involved in the formation of the three major categories of rocks: igneous rocks (cooled from molten material), sedimentary rocks (formed by the action of wind, water, and ice), and metamorphic rocks (altered in the solid state by extreme heat and pressure).

The Earth's Magnetic Field and Magnetosphere: Electric currents in the liquid outer core generate a magnetic field. This magnetic field produces a magnetosphere that surrounds the Earth and blocks the solar wind from hitting the atmosphere.

• A bow-shaped shock wave, where the supersonic solar wind is abruptly slowed to subsonic speeds, marks the outer boundary of the magnetosphere.

• Most of the particles of the solar wind are deflected around the Earth by the magnetosphere.

• Some charged particles from the solar wind are trapped in two huge, doughnut-shaped rings called the Van Allen belts. An excess of these particles can initiate an auroral display.

The Earth's Atmosphere: The Earth's atmosphere differs from those of the other terrestrial planets in its chemical composition, circulation pattern, and temperature profile.

• The Earth's atmosphere evolved from being mostly water vapor to being rich in carbon dioxide. A strong greenhouse effect kept the Earth warm enough for water to remain liquid and to permit the evolution of life.

• The appearance of photosynthetic living organisms led to our present atmospheric composition, about four-fifths nitrogen and one-fifth oxygen.

• The Earth's atmosphere is divided into layers called the troposphere, stratosphere, mesosphere, and thermosphere. Ozone molecules in the stratosphere absorb ultraviolet light.

• Because of the Earth's rapid rotation, the circulation in its atmosphere is complex, with three circulation cells in each hemisphere.

The Biosphere: Human activity is changing the Earth's biosphere, on which all living organisms depend.

• Industrial chemicals released into the atmosphere have damaged the ozone layer in the stratosphere.

• Deforestation and the burning of fossil fuels are increasing the greenhouse effect in our atmosphere and warming the planet. This can lead to destructive changes in the climate.

Questions

Review Questions

1. Describe the various ways in which the Earth is unique among the planets of our solar system.

2. Describe the various ways in which the Earth's surface is reshaped over time.

3. Why are typical rocks found on the Earth's surface much younger than the Earth itself?

4. What is convection? What causes convection in the Earth's atmosphere?

5. Describe how energy is transferred from the Earth's surface to the atmosphere by both convection and radiation.

6. If heat flows to the Earth's surface from both the Sun and the Earth's interior, why do we say that the motions of the atmosphere are powered by the Sun?

7. How does solar energy help power the motion of water in the Earth's oceans?

8. How does the greenhouse effect influence the temperature of the atmosphere? Which properties of greenhouse gases in the atmosphere cause this effect?

9. How do we know that the Earth was once entirely molten?

10. What are the different types of seismic waves? Why are seismic waves useful for probing the Earth's interior structure?

11. Describe the interior structure of the Earth.

12. What is a plastic material? Which parts of the Earth's interior are described as being plastic?

13. The deepest wells and mines go down only a few kilometers. What, then, is the evidence that iron is abundant in the Earth's core? That the Earth's outer core is molten but the inner core is solid?

14. The inner core of the Earth is at a higher temperature than the outer core. Why, then, is the inner core solid and the outer core molten instead of the other way around?

15. Describe the process of plate tectonics. Give specific examples of geographic features created by plate tectonics.

16. Explain how convection in the Earth's interior drives the process of plate tectonics.

17. Why do you suppose that active volcanoes, such as Mount St. Helens in Washington State, are usually located in mountain ranges that border on subduction zones?

18. What is the difference between a rock and a mineral?

19. What are the differences among igneous, sedimentary, and metamorphic rocks? What do these rocks tell us about the sites at which they are found?

20. Why do geologists suspect that Pangaea was the most recent in a succession of supercontinents?

21. Why do geologists think that the Earth's magnetic field is produced in the liquid outer core rather than in the mantle?

22. Describe the Earth's magnetosphere. If the Earth did not have a magnetic field, do you think aurorae would be more common or less common than they are today?

23. What are the Van Allen belts?

24. What phenomena tended to make the young Earth freeze over? What other phenomena prevented this from happening?

25. Summarize the history of the Earth's atmosphere. What role has biological activity played in this evolution?

26. Ozone and carbon dioxide and ozone each make up only a fraction of a percent of our atmosphere. Why, then, should we be concerned about small increases or decreases in the atmospheric abundance of these gases?

27. How are scientists able to measure what the atmospheric CO_2 concentration and average surface temperature were in the distant past?

28. Does global warming increase the surface temperature of all parts of the Earth by equal amounts or by different amounts? What consequences does this have?

Advanced Questions

29. The total power in sunlight that reaches the top of our atmosphere is 1.75×10^{17} W. (a) How many watts of power are reflected back into space due to the Earth's albedo? (b) If the Earth had no atmosphere, all of the solar power that was not reflected would be absorbed by the Earth's surface. In equilibrium, the heated surface would act as a blackbody that radiates as much power as it absorbs from the Sun. How much power would the entire Earth radiate? (c) How much power would one square meter of the surface radiate? (d) What would be the average temperature of the surface? Give your answer in both the Kelvin and Celsius scales. (e) Why is the Earth's actual average temperature higher than the value you calculated in (d)?

30. On average, the temperature beneath the Earth's crust increases at a rate of 20°C per kilometer. At what depth would water boil? (Assume the surface temperature is 20°C and ignore the effect of the pressure of overlying rock on the boiling point of water.)

31. What fractions of Earth's total volume are occupied by the core, the mantle, and the crust?

32. What fraction of the total mass of the Earth lies within the inner core?

33. (a) Using the mass and diameter of the Earth listed in Table 9-1, verify that the average density of the Earth is 5500 kg/m³. (b) Assuming that the average density of material in the Earth's mantle is about 3500 kg/m³, what must the average density of the core be? Is your answer consistent with the values given in Table 9-3?

34. Identical fossils of the reptile *Mesosaurus,* which lived 300 million years ago, are found in eastern South America and western Africa and nowhere else in the world. Explain how these fossils provide evidence for the theory of plate tectonics.

35. Measurements of the sea floor show that the Eurasian and North American plates have moved 60 km apart in the past 3.3 million years. How far apart (in millimeters) do they move in one year? (By comparison, your fingernails grow at a rate of about 50 mm/year.)

36. The oldest rocks found on the continents are about 4 billion years old. By contrast, the oldest rocks found on the ocean floor are only about 200 million years old. Explain why there is such a large difference in ages.

37. It was stated in Section 7-7 that iron loses its magnetism at temperatures above 770°C. Use this fact and Figure 9-10 to explain why the Earth's magnetic field cannot be due to a magnetized solid core, but must instead be caused by the motion of electrically conducting material in the liquid core.

38. Most auroral displays (like those shown in Figure 9-21c) have a green color dominated by emission from oxygen atoms at a wavelength of 557.7 nm. (a) What minimum energy (in electron volts) must be imparted to an oxygen atom to make it emit this wavelength? (b) Why is your answer in (a) a *minimum* value?

39. Describe how the present-day atmosphere and surface temperature of the Earth might be different (a) if carbon dioxide had never been released into the atmosphere; (b) if carbon dioxide had been released, but life had never evolved on Earth.

40. The photograph below shows the soil at Daspoort Tunnel near Pretoria, South Africa. The whitish layer that extends from lower left to upper right is 2.2 billion years old. Its color is due to a lack of iron oxide. More recent soils typically contain iron oxide and have a darker color. Explain what this tells us about the history of the Earth's atmosphere.

R I V̄ U X G

(Courtesy of H. D. Holland)

41. Earth's atmospheric pressure decreases by a factor of one-half for every 5.5-km increase in altitude above sea level. Construct a plot of pressure versus altitude, assuming the pressure at sea level is one atmosphere (1 atm). Discuss the characteristics of your graph. At what altitude is the atmospheric pressure equal to 0.001 atm?

42. The Earth is at perihelion on January 3 and at aphelion on July 4. Because of precession, in 15,000 A.D. the amount of sunlight in summer will be more than at present in the northern hemisphere but less than at present in the southern hemisphere. Explain why.

43. Antarctica has an area of 13 million square kilometers and is covered by an icecap that varies in thickness from 300 meters near the coast to 1800 meters in the interior. Estimate the volume of this icecap. Assuming that water and ice have roughly the same density, estimate the amount by which the water level of the world's oceans would rise if Antarctica's ice were to melt completely (see Figure 9-31). What portions of the Earth's surface would be inundated by such a deluge?

Discussion Questions

44. The human population on Earth is currently doubling about every 30 years. Describe the various pressures placed on the Earth by uncontrolled human population growth. Can such growth continue indefinitely? If not, what natural and human controls might arise to curb this growth? It has been suggested that overpopulation problems could be solved by colonizing the Moon or Mars. Do you think this is a reasonable solution? Explain your answer.

45. In order to alleviate global warming, it will be necessary to dramatically reduce the amount of carbon dioxide that we release into the atmosphere by burning petroleum. What changes in technology and society do you think will be needed to bring this about?

Web/eBook Questions

46. Search the World Wide Web for information about Rodinia, the supercontinent that predated Pangaea. When did this supercontinent form? When did it break apart? Do we have any evidence of supercontinents that predated Rodinia? Why or why not?

47. Because of plate tectonics, the arrangement of continents in the future will be different from today. Search the World Wide Web for information about "Pangaea Ultima," a supercontinent that may form in the distant future. When is it predicted to form? How will it compare to the supercontinent that existed 200 million years ago (see Figure 9-12a)?

48. Use the World Wide Web to investigate the current status of the Antarctic ozone hole. Is the situation getting better or worse? Is there a comparable hole over the North Pole? Why do most scientists blame the chemicals called CFCs for the existence of the ozone hole?

49. Use the World Wide Web to research predictions of the Earth's future surface temperature. What are some predictions for the surface temperature in 2050? In 2100? What effects may the increased temperature have on human health?

50. The Kyoto Protocol is an agreement made by a number of nations around the world to reduce their production of greenhouse gases. Use the World Wide Web to investigate the status of the Kyoto Protocol. How many nations have signed the protocol? What fraction of the world's greenhouse gas production comes from these nations? Have any developed nations failed to sign? How effective has the protocol been?

51. **Observing Mountain Range Formation.** Access the two animations "The Collision of Two Plates: South America" and "The Collision of Two Plates: The Himalayas" in Chapter 9 of the *Universe* Web site or eBook. (a) In which case are the mountains formed by tectonic uplift? In which case are the mountains formed by volcanoes from rising lava? (b) For each animation, describe which plate is moving in which direction.

Activities

Observing Project

52. Use the *Starry Night Enthusiast*™ program to view the Earth from space. In the **Favourites** menu, select **Solar System > Earth** to place yourself in the position of an astronaut flying outside his spaceship at about 5800 km above the Earth's surface on September 26, 2002. [You may need to click once on a **Zoom** button to show the Earth against the background stars. You may also see the astronaut's spacesuit and feet in the foreground. This foreground can be removed by clicking off **Feet** in the **View** menu.] Click and hold down and move the mouse to use the **Location Scroller** to rotate the cloud-enshrouded Earth beneath you, allowing you to view different parts of the Earth. You can change the **Time** to view different parts of this surface in daylight. (a) Is there any evidence for the presence of an atmosphere on Earth? Explain. Remove the cloud cover by opening **Options > Solar System,** clicking on **Planets-Moons** to display the Planets-Moons Options panel and turning off **Show atmosphere.** It is also helpful to remove all the Earth-orbiting satellites by clicking on **View > Solar System > Satellites.** (b) Can you see any evidence of plate tectonics as you rotate the Earth again? Locate the position of your home and zoom in on it, using the Zoom tool on the right of the toolbar to set the Field of View to about 16° × 11°. Again, use the location scroller to move around on the Earth. (c) Can you see any evidence of the presence of life or of man-made objects? Right-click on the Earth and click on Google Map to examine regions of the Earth in great detail on these images from an orbiting satellite. You can attempt to locate your own home on these maps. As an exercise, find the country of Panama, zoom in progressively upon the Panama Canal and search for ships traversing this incredible waterway. (d) What do you conclude about the importance of sending spacecraft to planets to explore their surfaces at close range? Close the internet connection to Google Map. In *Starry Night Enthusiast*™ open the **Favourites** menu, select **Solar System,** and click on **Earth** to return to the initial time and position above the Earth. Rotate the Earth to place eastern Brazil, on the eastern tip of South America, near to the center of the window. At the time of this observation this area is near the terminator (the dividing line between the day and night sides of the Earth). (e) As seen from eastern Brazil, is the sun rising or setting? Explain.

10

Our Barren Moon

Walking on the lunar surface, April 1972. (John Young, *Apollo 16*, NASA/Science Photo Library)

Other than the Earth itself, the most familiar world in the solar system is the Moon. It is so near to us that some of its surface features are visible to the naked eye. But the Moon is also a strange and alien place, with dramatic differences from our own Earth. It has no substantial atmosphere, no global magnetic field, and no liquid water of any kind. On the airless Moon, the sky is black even at midday. And unlike the geologically active Earth, the Moon has a surface that changes very little with time. When *Apollo 16* astronauts Charlie Duke (shown in the photograph) and John Young explored the region of the lunar highlands known as the Descartes Formation, the youngest rocks that they found were virtually the same age as the oldest rocks found on Earth.

Because the Moon has changed so little over the eons, it has preserved the early history of the inner solar system and the terrestrial planets. In this chapter we will see evidence that the Moon's formation and early evolution was violent and chaotic. The *Apollo* missions revealed that the Moon may have formed some 4.5 billion years ago as the result of a titanic collision between the young Earth and a rogue protoplanet the size of Mars. The young Moon was then pelted incessantly for hundreds of millions of years by large chunks of interplanetary debris, producing a cratered landscape. We will discover that the largest of these impacts formed immense circular basins that then flooded with molten lava. You can see the solidified lava as dark patches that cover much of the face of the Moon. In later chapters we will find other evidence that catastrophic collisions have played an important role in shaping the solar system.

10-1 The Moon's airless, dry surface is covered with plains and craters

Compared with the distances between planets, which are measured in tens or hundreds of millions of kilometers, the average distance from the Earth to the Moon is just 384,400 km (238,900 mi). Despite being so close, the Moon has an angular diameter of just ½°. This tells us that it is a rather small world. The diameter of the Moon is 3476 km (2160 mi), just 27% of the Earth's diameter and about the same as the distance from Las Vegas to Philadelphia or from London to Cairo (Figure 10-1). Table 10-1 lists some basic data about the Moon.

Motions of the Earth-Moon System

Although the Moon has a small mass (just 1.23% of the Earth's mass), Newton's third law tells us that the Moon exerts just as much gravitational force on the Earth as the Earth does on the Moon (see Section 4-6). Hence, it is not strictly correct to say that the Moon orbits the Earth; rather, they both orbit around a point between their centers called the **center of mass** of the Earth-Moon

Learning Goals

By reading the sections of this chapter, you will learn

10-1 The nature of the Moon's surface

10-2 The story of human exploration of the Moon

10-3 How we have learned about the Moon's interior

10-4 How Moon rocks differ from rocks on Earth

10-5 Why scientists think the Moon formed as the result of a violent collision between worlds

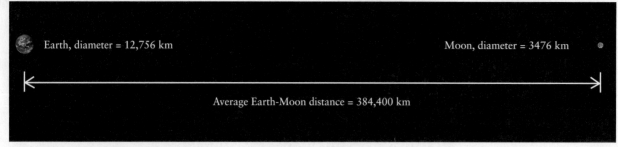

Earth, diameter = 12,756 km

Moon, diameter = 3476 km

Average Earth-Moon distance = 384,400 km

(a) The Earth-Moon system

Earth

Moon

(b) The Earth and Moon to scale, shown 10 times larger than in part (a)

Figure 10-1   R I V U X G

Comparing the Earth and Moon (a) The Moon is a bit more than one-quarter the diameter of the Earth. The average Earth-Moon distance is about 30 times the Earth's diameter. **(b)** The Earth has blue oceans, an atmosphere streaked with white clouds, and continents continually being reshaped by plate tectonics. By contrast, the Moon has no oceans, no atmosphere, and no evidence of plate tectonics. (a: NASA; b: John Sanford, Science Photo Library)

system. (Because the Earth is so much more massive than the Moon, the center of mass is close to the Earth's center: Indeed, it actually lies within the Earth's interior.) The center of mass then follows an elliptical orbit around the Sun (**Figure 10-2***a*). In a similar way, a spinning wrench sliding along a table actually rotates around its center of mass (Figure 10-2*b*).

The Moon's Rotation and Libration

Even without binoculars or a telescope, you can easily see that dark gray and light gray areas cover vast expanses of the Moon's rocky surface. If you observe the Moon over many nights, you will see the Moon go through its phases as the dividing line between the illuminated and darkened portion of the Moon (the **terminator**) shifts. But you will always see the same pattern of dark and light areas, because the Moon keeps the same side turned toward the Earth; you never see the **far side** (or back side) of the Moon. We saw in Section 3-2 that this is the result of the Moon's **synchronous rotation:** It takes precisely the same amount of time for the Moon to rotate on its axis (one lunar "day") as it does to complete one orbit around the Earth.

 With patience, however, you can see slightly more than half the lunar surface. This is because the Moon appears to wobble slightly over the course of an orbit: It seems to rock back and forth around its north-south axis and to nod up and down in a north-south direction. This apparent peri-

odic wobbling, called **libration,** permits us to view 59% of our satellite's surface.

 We stress that the Moon only *appears* to be wobbling. In fact, its rotation axis keeps nearly the same orientation in space as the Moon goes around its orbit. The reason why the Moon seems to rock back and forth is that its orbit around the Earth is slightly elliptical. As a result, the Moon's orbital motion does not keep pace with its rotation at all points around the orbit. Furthermore, because the Moon's rotation axis is not exactly perpendicular to the plane of its orbit, it appears to us to nod up and down.

An Airless World

Clouds always cover some portion of the Earth's surface, but clouds are never seen on the Moon (see Figure 10-1). This is because the Moon is too small a world to have an atmosphere. Its surface gravity is low, only about one-sixth as great as that of the Earth, and thus any gas molecules can easily escape into space (see Box 7-2). Because there is no atmosphere to scatter sunlight, the daytime sky on the Moon appears as black as the nighttime sky on the Earth (see the photograph that opens this chapter).

The absence of an atmosphere means that the Moon can have no liquid water on its surface. Here on Earth, water molecules are kept in the liquid state by the pressure of the atmosphere pushing down on them. To see what happens to liquid water on the

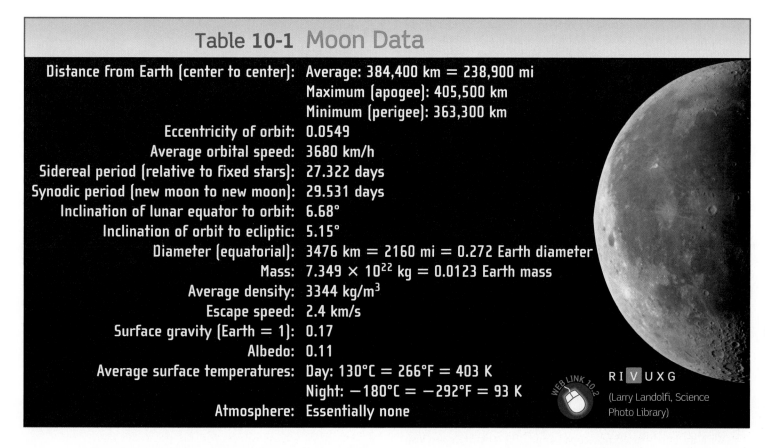

Table 10-1 Moon Data	
Distance from Earth (center to center):	Average: 384,400 km = 238,900 mi
	Maximum (apogee): 405,500 km
	Minimum (perigee): 363,300 km
Eccentricity of orbit:	0.0549
Average orbital speed:	3680 km/h
Sidereal period (relative to fixed stars):	27.322 days
Synodic period (new moon to new moon):	29.531 days
Inclination of lunar equator to orbit:	6.68°
Inclination of orbit to ecliptic:	5.15°
Diameter (equatorial):	3476 km = 2160 mi = 0.272 Earth diameter
Mass:	7.349×10^{22} kg = 0.0123 Earth mass
Average density:	3344 kg/m³
Escape speed:	2.4 km/s
Surface gravity (Earth = 1):	0.17
Albedo:	0.11
Average surface temperatures:	Day: 130°C = 266°F = 403 K
	Night: −180°C = −292°F = 93 K
Atmosphere:	Essentially none

R I V U X G

(Larry Landolfi, Science Photo Library)

airless Moon, consider what happens when the pressure of the atmosphere is reduced. The air pressure in Denver, Colorado (elevation 1650 m, or 5400 ft), is only 83% as great as at sea level. Therefore, in Denver, less energy has to be added to the molecules in a pot of water to make them evaporate, and water boils at only

95°C (= 368 K = 203°F), as compared with 100°C (= 373 K = 212°F) at sea level. (This is why some foods require longer cooking times at high elevations, where the boiling water used for cooking is not as hot as at sea level.) On the Moon, the atmospheric pressure is essentially zero and the boiling temperature is

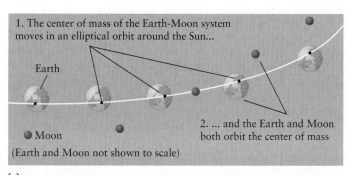

1. The center of mass of the Earth-Moon system moves in an elliptical orbit around the Sun...

Earth

Moon
(Earth and Moon not shown to scale)

2. ... and the Earth and Moon both orbit the center of mass

(a)

Figure 10-2   R I V U X G

Motion of the Earth-Moon System (a) The Earth and Moon both orbit around their center of mass, which in turn follows an elliptical orbit around the Sun. **(b)** The illustration in part (a) is analogous to this time-lapse image of a spinning wrench sliding across a table.
(b: Berenice Abbott/Photo Researchers)

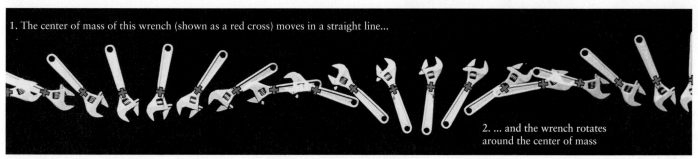

1. The center of mass of this wrench (shown as a red cross) moves in a straight line...

2. ... and the wrench rotates around the center of mass

(b)

lower than the temperature of the lunar surface. Under these conditions, water can exist as a solid (ice) or a vapor but not as a liquid.

CAUTION! If astronauts were to venture out onto the lunar surface without wearing spacesuits, the lack of atmosphere would *not* cause their blood to boil. That's because our skin is strong enough to contain bodily fluids. In the same way, if we place an orange in a sealed container and remove all the air, the orange retains all of its fluid and does not explode. It even tastes the same afterward! The reasons why astronauts wore pressurized spacesuits on the Moon (see the photograph that opens this chapter) were to provide oxygen for breathing, to regulate body temperature, and to protect against ultraviolet radiation from the Sun. Since the Moon has no atmosphere, and thus no ozone layer like that of the Earth to screen out ultraviolet wavelengths (Section 9-7), an unprotected astronaut would suffer a sunburn after just 10 seconds of exposure!

Lunar Craters

Since the Moon has no atmosphere to obscure its surface, you can get a clear view of several different types of lunar features with a small telescope or binoculars. These include *craters*, the dark-colored *maria*, and the light-colored *lunar highlands* or *terrae* (**Figure 10-3**). Taken together, these different features show the striking differences between the surfaces of the Earth and the Moon.

With an Earth-based telescope, some 30,000 **craters** are visible, with diameters ranging from 1 km to several hundred km. Close-up photographs from lunar orbit have revealed millions of

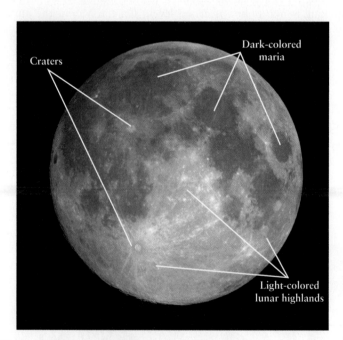

Figure 10-3   R I **V** U X G

The Moon as Seen from Earth The key features of the lunar surface can be seen with binoculars, a small telescope, or even the naked eye. (Larry Landolfi, Science Photo Library)

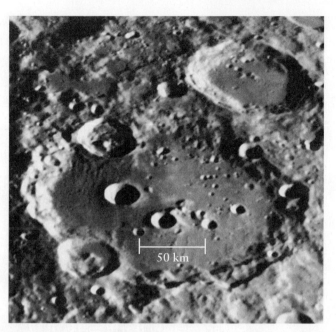

Figure 10-4   R I **V** U X G

The Crater Clavius This photograph from the 5-meter (200-in.) Palomar telescope shows one of the largest craters on the Moon. Clavius has a diameter of 232 km (144 mi) and a depth of 4.9 km (16,000 ft), measured from the crater's floor to the top of the surrounding rim. Later impacts formed the smaller craters inside Clavius. (Palomar Observatory)

craters too small to be seen from Earth. Following a tradition established in the seventeenth century, the most prominent craters are named after famous philosophers, mathematicians, and scientists, such as Plato, Aristotle, Pythagoras, Copernicus, and Kepler. Virtually all lunar craters, both large and small, are the result of the Moon's having been bombarded by meteoritic material. For this reason, they are also called **impact craters** (see Section 7-6). One piece of evidence that lunar craters are caused by impacts is that many large craters have a central peak, with a shape that could only result from high-speed impacts (see Figure 7-10*a*).

A meteoroid striking the lunar surface generates a shock wave over the surface outward from the point of impact. Such shock waves can spread over an area much larger than the size of the meteoroid that generated the wave. Very large craters more than 100 km across, such as the crater Clavius shown in **Figure 10-4**, were probably created by the impact of a fast-moving piece of rock only a few kilometers in radius.

Maria: Dark and Relatively Young

The large dark areas on the lunar surface shown in Figure 10-3 are called **maria** (pronounced "MAR-ee-uh"). They form a pattern that looks crudely like a human face, sometimes called "the man in the Moon." The singular form of maria is **mare** (pronounced "MAR-ay"), which means "sea" in Latin. These terms date from the seventeenth century, when observers using early telescopes thought these dark features were large bodies of water.

They gave the maria fanciful and romantic names such as Mare Tranquillitatis (Sea of Tranquillity), Mare Nubium (Sea of Clouds), Mare Nectaris (Sea of Nectar), Mare Serenitatis (Sea of Serenity), and Mare Imbrium (Sea of Showers). While modern astronomers still use these names, they now understand that the maria are not bodies of water (which, as we have seen, could not exist on the airless Moon) but rather the remains of huge lava flows. The maria get their distinctive appearance from the dark color of the solidified lava.

On the whole, the maria have far fewer craters than the surrounding terrain. Since craters are caused by meteoritic bombardment, this means that the maria have not been exposed to that bombardment for as long as the surrounding terrain. Hence, the maria must be relatively young, and the lava flows that formed them must have occurred at a later stage in the Moon's geologic history (**Figure 10-5**). The craters that are found in the maria were caused by the relatively few impacts that have occurred since the maria solidified (**Figure 10-6**).

Although they are much larger than craters, maria also tend to be circular in shape (see Figure 10-5). This suggests that, like craters, these depressions in the lunar surface were also created by impacts. It is thought that very large meteoroids or asteroids some tens of kilometers across struck the Moon, forming basins. These depressions subsequently flooded with lava that flowed out from the Moon's interior through cracks in the lunar crust. The

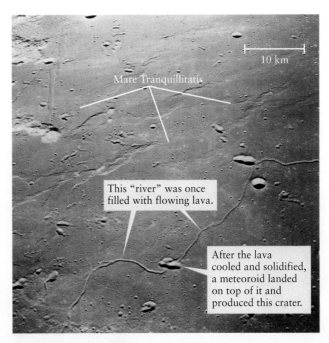

Figure 10-6   R I V U X G

Details of Mare Tranquillitatis This color photograph from lunar orbit reveals numerous tiny craters in the surface of a typical mare. The curving "river," or rille, was carved out by flowing lava that later solidified; similar features are found in volcanic areas on Earth. The *Apollo 11* landing site is near the center of the top edge of this image. (*Apollo 10*, NASA)

solidified lava formed the maria that we see today. The *Cosmic Connections* figure depicts our understanding of how lunar craters and maria formed.

An impact large enough to create a mare basin must have thrust upward tremendous amounts of material around the impact site. This explains the mountain ranges that curve in circular arcs around mare basins. Analysis of lunar rocks as well as observations made from lunar orbit have validated this picture of the origin of lunar mountains.

Lunar Highlands: Light-Colored and Ancient

The light-colored terrain that surrounds the maria is called **terrae**, from the Latin for "lands." (Seventeenth-century astronomers thought that this terrain was the dry land that surrounded lunar oceans.) Detailed measurements by astronauts in lunar orbit demonstrated that the maria on the Moon's Earth-facing side are 2 to 5 km below the average lunar elevation, while the terrae extend to several kilometers above the average elevation. For this reason terrae are often called **lunar highlands**. The flat, low-lying, dark gray maria cover only 15% of the lunar surface; the remaining 85% is light gray lunar highlands. Since the lunar highlands are more heavily cratered than the maria, we can conclude that the highlands are older.

 Remarkably, when the Moon's far side was photographed for the first time by the Soviet *Luna 3* spacecraft in 1959, scientists were surprised to find almost

Figure 10-5   R I V U X G

Mare Imbrium from Earth Mare Imbrium is the largest of 14 dark plains that dominate the Earth-facing side of the Moon. Its diameter of 1100 km (700 mi) is about the same as the distance from Chicago to Philadelphia or from London to Rome. The maria formed after the surrounding light-colored terrain, so they have not been exposed to meteoritic bombardment for as long and hence have fewer craters. (Carnegie Observatories)

COSMIC CONNECTIONS

The Formation of Craters and Maria on the Moon

The physical features of lunar craters and maria show that they were formed by meteoroids impacting the Moon's surface at high speed. The crater labeled in the photograph, called Aristillus after a Greek astronomer who lived around 300 B.C., is 55 km (34 mi) across and 3600 m (12,000 ft) deep. (Photograph: Lunar and Planetary Institute/Universities Space Research Association)

Meteoroid

Lunar surface

1. A meteoroid approaches the lunar surface.

Ejecta

Compressed rock

2. When the meteoroid strikes, it ejects rocks and boulders and produces a shock wave in the lunar surface.

Ejecta blanket

Secondary craters

Central peak

Crater wall

3. What remains is a crater with a central peak. The ejected material forms a blanket around the crater and can cause secondary craters.

Ejecta blanket

Central peak

Crater wall

Mare

R I **V** U X G

Lava from interior flows to surface.

Lava fills crater and solidifies to form the dark mare floor.

4. A very large meteoroid can crack the lunar crust, forming a mare.

no maria there. Thus, lunar highlands predominate on the side of the Moon that faces away from the Earth (Figure 10-7). Presumably, the crust is thicker on the far side, so even the most massive impacts there were unable to crack through the crust and let lava flood onto the surface. The Moon is not unique in having two hemispheres with different crustal thicknesses: The Earth's northern hemisphere has much more continental crust than its southern hemisphere, and the southern hemisphere of Mars has a thicker crust than the northern hemisphere.

Although the Moon's far side is deficient in maria compared to the near side, the highlands on the far side are as heavily pockmarked by impacts as those on the near side. The most prominent feature on the far side is the South Pole–Aitken Basin, the largest known impact crater in the solar system. This basin is about 2500 km (1600 mi) across—about the same size as the continent of Europe—and 12 km (7 mi) deep.

In contrast to the Earth and its active geology, the Moon's surface is geologically dead

A World Without Plate Tectonics

Something is entirely missing from our description of the Moon's surface—any mention of plate tectonics. Indeed, no evidence for plate tectonics can be found on the Moon. Recall from Section 9-3 that the Earth's major mountain ranges were created by collisions between lithospheric plates. In contrast, it appears that the Moon is a *one-plate world*: Its entire lithosphere is a single, solid plate. There are no rifts where plates are moving apart, no mountain ranges of the sort formed by collisions between plates, and no subduction zones. Thus, while lunar mountain ranges bear the names of terrestrial ranges, such as the Alps, Carpathians, and Apennines, they are entirely different in character: they were formed by material ejected from impact sites, not by plate tectonics. Geologically, the Moon is a dead world.

In Section 7-6 we discussed the general rule that a world with abundant craters has an old, geologically inactive surface. The Moon is a prime example of this rule. Without an atmosphere to cause erosion and without plate tectonics, the surface of the Moon changes only very slowly. That is why the Moon's craters, many of which are now known to be billions of years old, remain visible today. On the Earth, by contrast, the continents and seafloor are continually being reshaped by seafloor spreading, mountain building, and subduction. Thus, any craters that were formed by impacts on the ancient Earth's surface have long since been erased. There are fewer than 200 impact craters on the Earth today; these were formed within the last few hundred million years and have not yet been erased.

Near side: highlands and maria

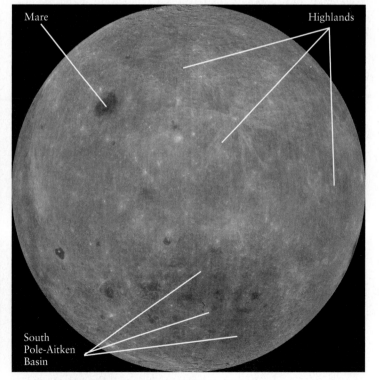

Far side: highlands but almost no maria

Figure 10-7   R I V U X G

The Near and Far Sides of the Moon About 50,000 individual infrared images from the *Clementine* spacecraft were combined to create these views of the near (Earth-facing) side and far side of the Moon. (NASA)

10-2 Human exploration of the Moon in the 1960s and 1970s has been continued by robotic spacecraft

 For thousands of years, storytellers have invented fabulous tales of voyages to the Moon. The seventeenth-century French satirist Cyrano de Bergerac wrote of a spacecraft whose motive power was provided by spheres of morning dew. (After all, reasoned Cyrano, dew rises with the morning sun, so a spacecraft would rise along with it.) In his imaginative tale *Somnium*, Cyrano's contemporary Johannes Kepler (the same Kepler who deduced the laws of planetary motion) used the astronomical knowledge of his time to envision what it would be like to walk on the Moon's surface. In 1865 the French author Jules Verne published *From the Earth to the Moon*, a story of a spacecraft sent to the Moon from a launching site in Florida.

Almost exactly a hundred years later, Florida was indeed the starting point of the first voyages made by humans to other worlds. Between 1969 and 1972, 12 astronauts walked on the Moon, and fantasy became magnificent reality.

> **The Moon is the only world other than Earth to be visited by humans**

The First Spacecraft to the Moon

As in Verne's story, politics had much to do with the motivation for going to the Moon. But in fact, there were excellent scientific reasons to do so. Because the Earth is a geologically active planet, typical terrestrial surface rocks are only a few hundred million years old, which is only a small fraction of the Earth's age. Thus, all traces of the Earth's origins have been erased. By contrast, rocks on the geologically inactive surface of the Moon have been essentially undisturbed for billions of years. Samples of lunar rocks have helped answer many questions about the Moon and have shed light on the origin of the Earth and the birth of the entire solar system.

Lunar missions began in 1959, when the Soviet Union sent three small spacecraft—*Lunas 1, 2,* and *3*—toward the Moon. American attempts to reach the Moon began in the early 1960s with Project Ranger. Equipped with six television cameras, the *Ranger* spacecraft transmitted close-up views of the lunar surface taken in the final few minutes before they crashed onto the Moon. These images showed far more detail than even the best Earth-based telescope.

In 1966 and 1967, five *Lunar Orbiter* spacecraft were placed in orbit around the Moon and made a high-resolution photographic survey of 99.5% of the lunar surface. Some of these photographs show features as small as 1 meter across and were essential in helping NASA scientists choose suitable landing sites for the astronauts.

Even after the *Ranger* missions, it was unclear whether the lunar surface was a safe place to land. One leading theory was that billions of years of impacts by meteoroids had ground the Moon's surface to a fine powder, much like that used by dental hygienists to polish teeth. Perhaps a spacecraft attempting to land would simply sink into a quicksandlike sea of dust! Thus, before a manned landing on the Moon could be attempted, it was necessary to try a soft landing of an unmanned spacecraft. This was the purpose of the *Surveyor* program. Between June 1966 and January 1968, five *Surveyors* successfully touched down on the Moon, sending back pictures and data directly from the lunar surface. These missions demonstrated convincingly that the Moon's surface was as solid as that of the Earth. Figure 10-8 shows an astronaut visiting *Surveyor 3,* two and a half years after it had landed on the Moon.

The *Apollo* and *Luna* Missions

The first of six manned lunar landings took place on July 20, 1969, when the *Apollo 11* lunar module *Eagle* set down in Mare Tranquillitatis (see Figure 10-6). Astronauts Neil Armstrong and Edwin "Buzz" Aldrin were the first humans to set foot on the Moon. *Apollo 12* also landed in a mare, Oceanus Procellarum (Ocean of Storms). The *Apollo 13* mission experienced a nearly catastrophic inflight explosion that made it impossible for it to land on the Moon. Fortunately, titanic efforts by the astronauts and ground personnel brought it safely home. The remaining four missions, *Apollo 14* through *Apollo 17,* were made in progressively more challenging terrain, chosen to permit the exploration of a wide variety of geologically interesting features. The first era of human exploration of the Moon culminated in December 1972, when *Apollo 17* landed in

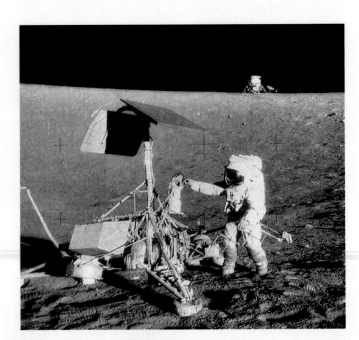

Figure 10-8   R I V U X G

Visiting an Unmanned Pioneer *Surveyor 3* landed on the Moon in April 1967. It was visited in November 1969 by *Apollo 12* astronaut Alan Bean, who took pieces of the spacecraft back to the Earth for analysis. (The *Apollo 12* lunar module is visible in the distance.) The retrieved pieces showed evidence of damage by micrometeoroid impacts, demonstrating that small bits of space debris are still colliding with the Moon. (Pete Conrad, *Apollo 12,* NASA) a rugged region of the lunar highlands. Figure 10-9 shows one of the *Apollo* bases.

In a curious postscript to the *Apollo* missions, in the 1990s an "urban legend" arose that the Moon landings had never taken place and were merely an elaborate hoax by NASA. While amusing, this legend does not hold up under even the slightest scientific scrutiny. For example, believers in the "Moon hoax" cannot explain why the rocks brought back from the Moon are substantially different in their chemistry from Earth rocks, as we will see later in this chapter. It makes as much sense to ask "Do you believe that humans actually went to the Moon?" as it does to ask "Do you believe there was a First World War?" or "Do you believe there is such a place as Antarctica?"

Between 1966 and 1976, a series of unmanned Soviet spacecraft also orbited the Moon and soft-landed on its surface. The first soft landing, by *Luna 9*, came four months before the U.S. *Surveyor 1* arrived. The first lunar satellite, *Luna 10*, orbited the Moon four months before *Lunar Orbiter 1*. In the 1970s, robotic spacecraft named *Luna 17* and *Luna 21* landed vehicles that roamed the lunar surface, and *Lunas 16, 20,* and *24* brought samples of rock back to the Earth. The *Luna* samples complement those returned by the *Apollo* astronauts because they were taken from different parts of the Moon's surface.

Recent Lunar Exploration

Unmanned spacecraft continued the exploration of the Moon in the 1990s and 2000s. In 1994 a small spacecraft named *Clementine* spent more than two months observing the Moon from lunar orbit. Although originally in-

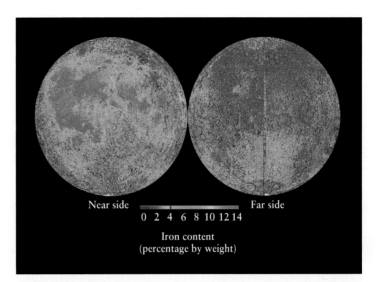

Figure 10-10

Iron on the Moon Images at different wavelengths from the *Clementine* spacecraft were used to make this map of the concentration of iron on the lunar surface. The areas of highest concentration (red) coincide closely with the maria on the near side (compare Figure 10-7), confirming that the maria formed from iron-rich lavas. The lowest iron concentration (blue) is found in the lunar highlands. (No data were collected in the gray areas.) The large green region of intermediate iron concentration on the far side is the South Pole–Aitken Basin (see Figure 10-7). The impact that created this basin may have excavated all the way to the Moon's mantle, in which iron is more abundant than in the crust. (JSC/NASA)

tended to test lightweight imaging systems for future defense satellites, *Clementine* proved to be a remarkable tool for lunar exploration. It carried an array of very high-resolution cameras sensitive to ultraviolet and infrared light as well as the visible spectrum (see Figure 10-7). Different types of materials on the lunar surface reflect different wavelengths, so analysis of *Clementine* images made in different parts of the electromagnetic spectrum has revealed the composition of large areas of the Moon's surface (Figure 10-10). The task of analyzing the Moon's surface composition was carried out in even greater depth by the European Space Agency's *SMART-1* spacecraft, which orbited the Moon from 2004 to 2006 and carried even more sensitive instruments than *Clementine*.

One remarkable result of the *Clementine* mission was evidence for a patch of water ice tens or hundreds of kilometers across at the Moon's south pole, in low-lying locations where sunlight never reaches and where temperatures are low enough to prevent ice from evaporating. Radio waves beamed from *Clementine* toward these areas were detected by radio telescopes on the Earth and showed the characteristic signature of reflection from ice.

Data from the *Lunar Prospector* spacecraft, which went into orbit around the Moon in January 1998, reinforced the *Clementine* findings and indicated that even more ice might lie in deep craters at the Moon's north pole. The ice may have been deposited by comets (see Section 7-5) that crashed into the lunar surface.

Figure 10-9   R I **V** U X G

The *Apollo 15* Base This photograph shows astronaut James Irwin and the *Apollo 15* landing site at the foot of the Apennine Mountains near the eastern edge of Mare Imbrium. The hill in the background is about 5 km away and rises about 3 km high. At the right is a Lunar Rover, an electric-powered vehicle used by astronauts to explore a greater radius around the landing site. The last three *Apollo* missions carried Lunar Rovers. (Dave Scott, *Apollo 15*, NASA)

One complication is that the *Lunar Prospector* observations show only that there is an excess of hydrogen atoms at the lunar poles. These could be trapped within water molecules (H_2O) or within other, more exotic minerals. At the end of the *Lunar Prospector* mission in July 1999, the spacecraft was crashed into a crater at the lunar south pole in the hope that a plume of water vapor might rise from the impact site. Although more than a dozen large Earth-based telescopes watched the Moon during the impact, no such plume was observed, nor was any debris of any kind. Radar observations made from Earth in 2003 suggest that if there is ice at the lunar poles, it is in sheets no more than a few centimeters thick. A new generation of European, Japanese, Indian, Chinese, and U.S. unmanned spacecraft may help provide a definitive answer about ice at the Moon's poles.

10-3 The Moon has no global magnetic field but has a small core beneath a thick mantle

We saw in Section 7-6 and Section 7-7 that a small planet or satellite will have little internal heat, and hence is likely to have a mostly solid interior. Lacking substantial amounts of moving molten material in its interior, a small world is unlikely to generate a global magnetic field. As we will see, these general rules describe the Moon very well.

> Compared to earthquakes, moonquakes are few and feeble—but they reveal key aspects of the Moon's interior

The Moon's Interior: The Core

The Moon landings gave scientists an unprecedented opportunity to explore an alien world. The *Apollo* spacecraft were therefore packed with an assortment of apparatus that the astronauts deployed around the landing sites (see Figure 10-9). For example, all the missions carried magnetometers, which confirmed that the present-day Moon has no global magnetic field. However, careful magnetic measurements of lunar rocks returned by the *Apollo* astronauts indicated that the Moon did have a weak magnetic field when the rocks solidified billions of years ago. Hence, the Moon must originally have had a small, molten, iron-rich core. (We saw in Section 7-7 that the Earth's magnetic field is generated by fluid motions in its interior.) The core presumably solidified at least partially as the Moon cooled, so that the lunar magnetic field disappeared.

More recently, scientists used the *Lunar Prospector* to probe the character of the Moon's core. The spacecraft made extensive measurements of lunar gravity and of how the Moon responds when it passes through the Earth's magnetosphere (Section 9-4). When combined with detailed observations of how the Earth's tidal forces affect the motions of the Moon, the data indicate that the present-day Moon has a partially liquid, iron-rich core about 700 km (435 mi) in diameter. While 32% of the Earth's mass is in its core, the core of the Moon contains only 2% to 3% of the lunar mass.

The Moon's Interior: The Mantle

The *Apollo* astronauts also set up seismometers on the Moon's surface, which made it possible for scientists to investigate the Moon's interior using seismic waves just as is done for the Earth (Section 9-2). It was found that the Moon exhibits far less seismic activity than the Earth. Roughly 3000 **moonquakes** were detected per year, whereas a similar seismometer on the Earth would record hundreds of thousands of earthquakes per year. Furthermore, typical moonquakes are very weak. A major terrestrial earthquake is in the range from 6 to 8 on the Richter scale, while a major moonquake measures 0.5 to 1.5 on that scale and would go unnoticed by a person standing near the epicenter.

Analysis of the feeble moonquakes reveals that most originate 600 to 800 km below the surface, deeper than most earthquakes. On Earth, the deepest earthquakes mark the boundary between the solid lithosphere and the plastic asthenosphere. This is because the lithosphere is brittle enough to fracture and produce seismic waves, whereas the deeper rock of the asthenosphere oozes and flows rather than cracks. We may conclude that the Moon's lithosphere is about 800 km thick (Figure 10-11). By contrast, the Earth's lithosphere is only 50 to 100 km thick.

The lunar asthenosphere extends down to the Moon's relatively small iron-rich core, which is more than 1400 km below the lunar surface. The asthenosphere and the lower part of the lithosphere are presumably composed of relatively dense iron-rich rocks. The uppermost part of the lithosphere is a lower-density crust, with an average thickness of about 60 km on the Earth-facing side and up to 100 km on the far side. For comparison, the thickness of the Earth's crust ranges from 5 km under the oceans to about 30 km under major mountain ranges on the con-

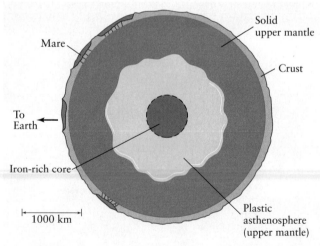

Figure 10-11

The Internal Structure of the Moon Like the Earth, the Moon has a crust, a mantle, and a core. The lunar crust has an average thickness of about 60 km on the Earth-facing side but about 100 km on the far side. The crust and solid upper mantle form a lithosphere about 800 km thick. The plastic (nonrigid) asthenosphere extends all the way to the base of the mantle. The iron-rich core is roughly 700 km in diameter. (Compare Figure 9-9.)

tinents (see Table 9-3). This slightly lopsided distribution of mass between the near and far sides of the Moon helps to keep the Moon in synchronous rotation, so that we always see the same side of the Moon (see Section 4-8).

On the whole, the Moon's interior is much more rigid and less fluid than the interior of the Earth. Plate tectonics requires that there be fluid motion just below a planet's surface, so it is not surprising that there is no evidence for plate tectonics on the lunar surface.

The Origin of Moonquakes

Earthquakes tend to occur at the boundaries between tectonic plates, where plates move past each other, collide, or undergo subduction (see Figure 9-14). But there are no plate motions on the Moon. What, then, causes moonquakes? The answer turns out to be the Earth's tidal forces. Just as the tidal forces of the Sun and Moon deform the Earth's oceans and give rise to ocean tides (see Figure 4-26), the Earth's tidal forces deform the solid body of the Moon. The amount of the deformation changes as the Moon moves around its elliptical orbit and the Earth-Moon distance varies. As a result, the body of the Moon flexes slightly, triggering moonquakes. The tidal stresses are greatest when the Moon is nearest the Earth—that is, at perigee (see Section 3-5)— which is just when the *Apollo* seismometers reported the most moonquakes. Box 10-1 has more information about how the tidal stresses on the Moon depend on its distance from Earth.

Without tectonics, and without the erosion caused by an atmosphere or oceans, the only changes occurring today on the lunar surface are those due to meteoroid impacts. These impacts were also monitored by the seismometers left on the Moon by the *Apollo* astronauts. These devices, which remained in operation for several years, were sensitive enough to detect a hit by a grapefruit-sized meteoroid anywhere on the lunar surface. The data show that every year the Moon is struck by 80 to 150 meteoroids having masses between 0.1 and 1000 kg (roughly $\frac{1}{4}$ lb to 1 ton).

10-4 Lunar rocks reveal a geologic history quite unlike that of Earth

One of the most important scientific goals of the *Apollo* missions was to learn as much as possible about the lunar surface. Although only one of the 12 astronauts who visited the Moon was a professional geologist, all of the astronauts received intensive training in field geology, and they were in constant communication with planetary scientists on Earth as they explored the lunar surface.

Weathering by Meteoroids

The Moon has no atmosphere or oceans to cause "weathering" of the surface. But the *Apollo* astronauts found that the lunar surface has undergone a kind of erosion due to billions of years of relentless mete-

Although no wind or rain disturbs the lunar surface, it is "gardened" over the ages by the impact of tiny meteoroids

oritic bombardment. This bombardment has pulverized the surface rocks into a layer of fine powder and rock fragments called the **regolith**, from the Greek for "blanket of stone" (Figure 10-12). This layer varies in depth from about 2 to 20 m. Just as the rough-surfaced acoustic tile used in ceilings absorbs sound waves, the regolith absorbs most of the sunlight that falls on it. This helps to account for the Moon's very low albedo (see Table 10-1).

Moon Rocks

In addition to their observations of six different landing sites, the *Apollo* astronauts brought back 2415 individual samples of lunar material totaling 382 kg (842 lb). In addition, the unmanned Soviet spacecraft *Luna 16, 20,* and *24* returned 300 g ($\frac{2}{3}$ lb) from three other sites on the Moon. This collection of lunar material has provided important information about the early history of the Moon that could have been obtained in no other way.

All of the lunar samples are igneous rocks; there are no metamorphic or sedimentary rocks (see Section 9-3, especially Figure 9-19). This suggests that most or all of the Moon's surface was once molten. Indeed, Moon rocks are composed mostly of the same minerals that are found in volcanic rocks on Earth.

Astronauts who visited the maria discovered that these dark regions of the Moon are covered with basaltic rock quite similar to the dark-colored rocks formed by lava from volcanoes on

Figure 10-12   R I <u>V</u> U X G

The Regolith Billions of years of bombardment by space debris have pulverized the uppermost layer of the Moon's surface into powdered rock. This layer, called the regolith, is utterly bone-dry. It nonetheless sticks together like wet sand, as shown by the sharp outline of an astronaut's bootprint. (*Apollo 11,* NASA)

BOX 10-1

Tools of the Astronomer's Trade

Calculating Tidal Forces

The gravitational force between any two objects decreases with increasing distance between the objects. This principle leads to a simple formula for estimating the tidal force that the Earth exerts on parts of the Moon.

Consider two small objects, each of mass m, on opposite sides of the Moon. Because the two objects are at different distances from the Earth's center, the Earth exerts different forces F_{near} and F_{far} on them (see the accompanying illustration). The consequence of this difference is that the two objects tend to pull away from each other and away from the center of the Moon (see Section 4-8, especially Figure 4-23). The *tidal force* on these two objects, which is the force that tends to pull them apart, is the difference between the forces on the individual objects:

$$F_{tidal} = F_{near} - F_{far}$$

Both F_{near} and F_{far} are inversely proportional to the square of the distance from the Earth (see Section 4-7). If this distance were doubled, both forces would decrease to $\frac{1}{2}^2 = \frac{1}{4}$ of their original values. But the tidal force F_{tidal}, which is the *difference* between these two forces, decreases even more rapidly with distance: It is inversely proportional to the *cube* of the distance from the Earth to the Moon. That is, doubling the Earth-Moon distance decreases the tidal force to $\frac{1}{2}^3 = \frac{1}{8}$ of its original value. If we use the symbols M_{Earth} for the Earth's mass (5.974×10^{24} kg), d for the Moon's diameter, r for the center-to-center distance from the Earth to the Moon, and G for the universal constant of gravitation (6.67×10^{-11} newton m²/kg²), the tidal force is given approximately by the formula

$$F_{tidal} = \frac{2GM_{Earth}md}{r^3}$$

EXAMPLE: Calculate the tidal force for two 1-kg lunar rocks at the locations shown in the figure.

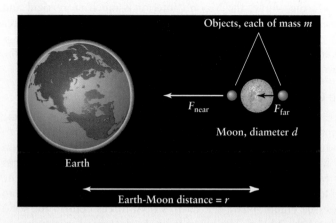

Situation: We are asked to calculate a tidal force. We know the masses of the Earth (M_{Earth}) and of each of the two rocks ($m = 1$ kg). The Earth-Moon distance (r) and the diameter of the Moon (d, equal to the separation between the two rocks) are given in Table 10-1.

Tools: We use the equation given above for the tidal force F_{tidal}.

Answer: To use the given values in the equation, we must first convert the values for the average Earth-Moon distance r and the Moon's diameter d from kilometers to meters:

$$d = 3476 \text{ km} \times \frac{10^3 \text{ m}}{1 \text{ km}} = 3.476 \times 10^6 \text{ m}$$

$$r = 384{,}400 \text{ km} \times \frac{10^3 \text{ m}}{1 \text{ km}} = 3.844 \times 10^8 \text{ m}$$

The tidal force is then

$$F_{tidal} = \frac{2(6.67 \times 10^{-11})(5.974 \times 10^{24})(1)(3.476 \times 10^6)}{(3.844 \times 10^8)^3}$$

$$= 4.88 \times 10^{-5} \text{ newton}$$

Hawaii and Iceland. The rock of these low-lying lunar plains is called **mare basalt** (Figure 10-13).

In contrast to the dark maria, the lunar highlands are composed of a light-colored rock called **anorthosite** (Figure 10-14). On the Earth, anorthositic rock is found only in such very old mountain ranges as the Adirondacks in the eastern United States. In comparison to the mare basalts, which have more of the heavier elements like iron, manganese, and titanium, anorthosite is rich in silicon, calcium, and aluminum. Anorthosite is therefore

less dense than basalt. During the period when the Moon's surface was molten, the less dense anorthosite rose to the top, forming the terrae that make up the majority of the present-day lunar surface.

Although anorthosite is the dominant rock type in the lunar highlands, most of the rock samples collected there were not pure anorthosite but rather **impact breccias**. These are composites of different rock types that were broken apart, mixed, and fused together by a series of meteoritic impacts (Figure 10-15). The

Review: To put our result into perspective, let's compare it to the *weight* of a 1-kg rock on the Moon. The weight of an object is equal to its mass multiplied by the acceleration due to gravity (see Section 4-6). This acceleration is equal to 9.8 m/s² on Earth, but is only 0.17 as great on the Moon. So, the weight of a 1-kg rock on the Moon is

$$0.17 \times 1 \text{ kg} \times 9.8 \text{ m/s}^2 = 1.7 \text{ newtons}$$

Thus, the tidal force on lunar rocks is *much* less than their weight. This is a good thing: If the tidal force (which tries to pull a rock away from the Moon's center) were greater than the weight (which pulls it toward the Moon's center), loose objects would levitate off the Moon's surface and fly into space!

The tidal force of the Earth on the Moon deforms the solid body of the Moon, so that it acquires *tidal bulges* on its near and far sides. The mass of each bulge must be proportional to the tidal force that lifts lunar material away from the Moon's center. So, the above formula tells us that the mass of each bulge is inversely proportional to the cube of the Earth-Moon distance r:

$$m_{\text{bulge}} = \frac{A}{r^3}$$

(Determining the value of the constant A requires knowing how easily the Moon deforms, which is a complicated problem beyond our scope.) We can then use our formula for the tidal force F_{tidal} to tell us the *net* tidal force on the Moon. To do this, we think of the objects of mass m on opposite sides of the Moon as being the tidal bulges. So, we substitute m_{bulge} for m in the tidal force formula:

$$F_{\text{tidal-net}} = \frac{2GM_{\text{Earth}}m_{\text{bulge}}d}{r^3} = \frac{2GM_{\text{Earth}}}{r^3} \frac{A}{r^3}d$$

$$= \frac{2GM_{\text{Earth}}Ad}{r^6}$$

Thus, the net tidal force on the Moon is inversely proportional to the *sixth* power of the Earth-Moon distance. Even a small decrease in this distance can cause a substantial increase in the net tidal force.

EXAMPLE: Compare the net tidal force on the Moon at perigee (when it is closest to the Earth) and at apogee (when it is farthest away).

Situation: We are asked to *compare* the tidal forces on the Moon at two different distances from the Earth. To do this, we will find the *ratio* of the tidal force at perigee to the tidal force at apogee.

Tools: We use the equation given above for the net tidal force on the Moon.

Answer: The Earth-Moon distances at perigee and apogee are given in Table 10-1. If we take the ratio of the tidal forces at these two distances, the constants (including the unknown quantity A) cancel out:

$$\frac{F_{\text{tidal-net-perigee}}}{F_{\text{tidal-net-apogee}}} = \frac{2GM_{\text{Earth}}Ad/r^6_{\text{perigee}}}{2GM_{\text{Earth}}Ad/r^6_{\text{apogee}}} = \left(\frac{r_{\text{apogee}}}{r_{\text{perigee}}}\right)^6$$

$$= \left(\frac{405,500 \text{ km}}{363,000 \text{ km}}\right)^6 = (1.116)^6 = 1.93$$

Review: Our result tells us that the net tidal force at perigee is 1.93 times as great (that is, almost double) as at apogee. Tidal forces on the Moon help create the stresses that generate moonquakes, so it is not surprising that these quakes are more frequent at perigee than at apogee.

prevalence of breccias is evidence that the lunar highlands have undergone eons of bombardment from space.

Every rock found on Earth, even those from the driest desert, has some water locked into its mineral structure. But lunar rocks are totally dry. With the exception of ice at the lunar poles (see Section 10-2), there is no evidence that water has ever existed on the Moon. Because the Moon lacks both an atmosphere and liquid water, it is not surprising that no traces of life have ever been found there.

By carefully measuring the abundances of trace amounts of radioactive elements in lunar samples and applying the principles of radioactive dating (see Section 8-2), geologists have determined that lunar rocks formed more than 3 billion years ago. Of these ancient rocks, however, anorthosite is much older than the mare basalts. Typical anorthositic specimens from the highlands are between 4.0 and 4.3 billion years old. All these extremely ancient specimens are probably samples of the Moon's original crust. In sharp contrast, all the mare basalts are between 3.1 and 3.8 bil-

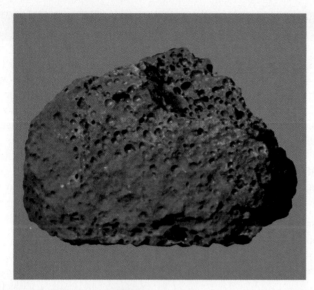

Figure 10-13   R I V U X G

Mare Basalt This 1.53-kg (3.38-lb) specimen of mare basalt was brought back by the crew of *Apollo 15*. Gas must have dissolved under pressure in the lava from which this rock solidified. When the lava reached the airless lunar surface, the pressure on the lava dropped and the gas formed bubbles. Some of the bubbles were frozen in place as the rock cooled, forming the holes on its surface. (Figure 9-19a shows a sample of basalt from Earth.) (NASA)

lion years old. With no plate tectonics to bring fresh material to the surface, no new anorthosites or basalts have formed on the Moon for more than 3 billion years.

The History of Lunar Cratering

By correlating the ages of Moon rocks with the density of craters at the sites where the rocks were collected, geologists have found that the rate of impacts on the Moon has changed over the ages. The ancient, heavily cratered lunar highlands are evidence of an intense bombardment that dominated the Moon's early history. For nearly a billion years, rocky debris left over from the formation of the planets rained down upon the Moon's young surface. As Figure 10-16 shows, this barrage extended from 4.56 billion years ago, when the Moon's surface solidified, until about 3.8 billion years ago. At its peak, this bombardment from space would have produced a new crater of about 1 km radius somewhere on the Moon about once per century. (If this seems like a long time interval, remember that we are talking about a bombardment that lasted hundreds of millions of years and produced millions of craters over that time.)

The rate of impacts should have tapered off as meteoroids and planetesimals were swept up by the newly formed planets. In fact, radioactive dating shows that the impact rate spiked upward between 4.0 and 3.8 billion years ago (see Figure 10-16). One proposed explanation is that the orbits of Jupiter and Saturn changed slightly but suddenly about 4.0 billion years ago during the process of settling into their present-day orbits. This would have changed the gravitational forces that these planets exerted on the asteroid belt and could have disturbed the orbits of many asteroids, sending them careening toward the inner solar system.

Whatever the explanation for the final epoch of heavy bombardment, major impacts during this period gouged out the mare basins. Between 3.8 and 3.1 billion years ago iron-rich magma oozed up from the Moon's still-molten mantle, flooding the mare basins with lava that solidified to form basalt (see the *Cosmic Connections* figure).

The relative absence of craters in the maria (see Figure 10-5) tells us that the impact rate has been quite low since the lava solidified. During these past 3.1 billion years the Moon's surface has changed very little.

Figure 10-14   R I V U X G

Anorthosite The light-colored lunar terrae (highlands) are composed of this ancient type of rock, which is thought to be the material of the original lunar crust. Lunar anorthosites vary in color from dark gray to white; this sample from the *Apollo 16* mission is a medium gray. It measures 18 × 16 × 7 cm. (NASA)

Figure 10-15   R I V U X G

Impact Breccias These rocks are evidence of the Moon's long history of bombardment from space. Parts of the Moon's original crust were shattered and strewn across the surface by meteoritic impacts. Later impacts compressed and heated this debris, welding it into the type of rock shown here. Such impact breccias are rare on Earth but abundant on the Moon. (NASA)

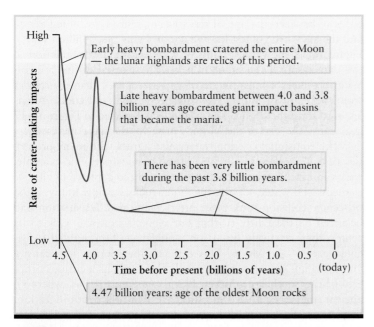

Figure 10-16

The Rate of Crater Formation on the Moon This graph shows the rate at which impact craters formed over the Moon's history. The impact rate today is only about 1/10,000 as great as during the most intense bombardment. (Adapted from T. Grotzinger, T. H. Jordan, F. Press, and R. Siever, *Understanding Earth,* 5th ed., W. H. Freeman, 2007)

10-5 The Moon probably formed from debris cast into space when a huge protoplanet struck the young Earth

The lunar rocks collected by the *Apollo* and *Luna* missions have provided essential information about the history of the Moon. In particular, they have helped astronomers come to a consensus about one of the most important questions in lunar science: Where did the Moon come from?

The Receding Moon

Seventy years before the first spacecraft landed on the Moon, the British astronomer George Darwin (second son of the famous evolutionist Charles Darwin) deduced that the Moon must be slowly moving away from the Earth. He began with the notion that the Moon's tidal forces elongate the oceans into a football shape (see Section 4-8, especially Figure 4-26). However, the long axis of this "football" does not point precisely at the Moon. The Earth spins on its axis more rapidly than the Moon revolves around the Earth, and this rapid rotation carries the tidal bulge about 10° ahead of a line connecting the Earth and Moon (Figure 10-17). This misaligned bulge produces a small but steady gravitational force that tugs the Moon forward and lifts it into an ever larger orbit around the Earth. In other words, Darwin concluded that the Moon must be spiraling away from the Earth.

It became possible to test Darwin's conclusions with the aid of a simple device that the *Apollo 11, 14,* and *15* and *Luna 21*

missions left on the Moon—a set of reflectors, similar to the orange and red ones found in automobile taillights. Since 1969, scientists on Earth have been firing pulses of laser light at these reflectors and measuring very accurately the length of time it takes each pulse to return to Earth. Knowing the speed of light, they can use this data to calculate the distance to the Moon to an accuracy of just 3 centimeters. They have found that the Moon is moving away from the Earth at a very gradual rate of 3.8 cm per year—which is just what Darwin predicted.

As the Moon moves away from the Earth, its sidereal period is getting longer in accordance with Kepler's third law (see Section 4-7). At the same time, friction between the oceans and the body of the Earth is gradually slowing the Earth's rotation. The length of the Earth's day is therefore increasing, by approximately 0.002 second per century.

Theories of the Moon's Origins

These observations mean that in the distant past, the Earth was spinning faster than it is now and the Moon was much closer. Darwin theorized that the early Earth may have been spinning so fast that a large fraction of its mass tore away, and that this fraction coalesced to form the Moon. This is called the **fission theory** of the Moon's origin. Prior to the *Apollo* program, two other theories were in competition with the fission theory. The **capture theory** posits that the Moon was formed elsewhere in the solar system and then drawn into orbit about the Earth by gravitational forces. By contrast, the **co-creation theory** proposes that the Earth and the Moon were formed at the same time but separately.

One fact used to support the fission theory was that the Moon's average density (3344 kg/m^3) is comparable to that of the Earth's outer layers, as would be expected if the Moon had been

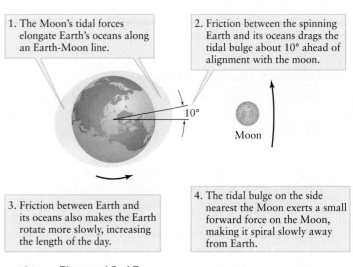

Figure 10-17

The Moon's Tidal Recession The Earth's rapid rotation drags the tidal bulge of the oceans about 10° ahead of a direct alignment with the Moon. The bulge on the side nearest the Moon exerts more gravitational force than the other, more distant bulge. The net effect is a small forward force on the Moon that makes it spiral slowly away from the Earth.

flung out of a rapidly rotating proto-Earth. However, the fission theory predicts that lunar and terrestrial rocks should be similar in composition. In fact, samples returned from the Moon by the *Apollo* and *Luna* missions show that they are significantly different.

Volatile elements (such as potassium and sodium) boil at relatively low temperatures under 900°C, whereas refractory elements (such as titanium, calcium, and aluminum) boil at 1400°C or higher. Compared with terrestrial rocks, the lunar rocks have slightly greater proportions of the refractory elements and slightly lesser proportions of the volatile elements. This suggests that the Moon formed from material that was heated to a higher temperature than that from which the Earth formed. Some of the volatile elements boiled away, leaving the young moon relatively enriched in the refractory elements.

 Proponents of the capture theory used these differences to argue that the Moon formed elsewhere and was later captured by the Earth. They also noted that the plane of the Moon's orbit is close to the plane of the ecliptic, as would be expected if the Moon had originally been in orbit about the Sun but was later captured by the Earth.

There are difficulties with the capture theory, however. If the Earth did capture the Moon intact, then a host of conditions must have been satisfied entirely by chance. The Moon must have coasted to within 50,000 km of the Earth at exactly the right speed to leave a solar orbit and must have gone into orbit around the Earth without actually hitting our planet.

Proponents of the co-creation theory argued that it is easier for a planet to capture swarms of tiny rocks. In this theory the Moon formed from just such rocky debris. Great numbers of rock fragments orbited the newborn Sun in the plane of the ecliptic along with the protoplanets. Heat generated in collisions could have baked the volatile elements out of these smaller rock fragments. Then, just as planetesimals accreted to form the proto-Earth in orbit about the Sun, the fragments in orbit about the Earth accreted to form the Moon.

The Collisional Ejection Theory

The present scientific consensus is that *none* of these three theories—fission, capture, and co-creation—is correct. Instead, the evidence points toward an idea proposed in 1975 by William Hartmann and Donald Davis and independently by Alastair Cameron and William Ward. In this collisional ejection theory, the proto-Earth was struck off-center by a Mars-sized object, and this collision ejected debris from which the Moon formed.

The collisional ejection theory agrees with what we know about the early history of the solar system. We saw in Section 8-4 that smaller objects collided and fused together to form the inner planets. Some of these collisions must have been quite spectacular, especially near the end of planet formation, when most of the mass of the inner solar system was contained in a few dozen large protoplanets. The collisional ejection theory proposes that one such protoplanet struck the proto-Earth about 4.51 billion years ago, about

> The difference in chemical composition between lunar rocks and Earth rocks helps to constrain theories of the Moon's origin

halfway between the age of the solar system (4.56 billion years) and the age of the oldest known Moon rocks (4.47 billion years). Figure 10-18 shows the results of a supercomputer simulation of this cataclysm. In the simulation, energy released during the collision produces a huge plume of vaporized rock that squirts out from the point of impact. As this ejected material cools, it coalesces to form the Moon. The tidal effects depicted in Figure 10-17 then cause the size of the Moon's orbit to gradually increase.

The collisional ejection theory also agrees with many properties of lunar rocks and of the Moon as a whole. For example, rock vaporized by the impact would have been depleted of volatile elements and water, leaving the moon rocks we now know. If the off-center collision took place after chemical differentiation had occurred on the Earth, so that our planet's iron had sunk to its center, then relatively little iron would have been ejected to become part of the Moon. This explains the Moon's low density and the small size of its iron core. It also explains a curious property of the South Pole–Aitken Basin (see Figure 10-7), where an ancient impact excavated the surface down to a depth of 12 km. Data from the *Clementine* spacecraft show that iron concentration in this basin is only 10% (see Figure 10-10), far less than the 20–30% concentration at a corresponding depth below the Earth's surface.

An Overview of the Moon's History

We can now summarize the entire geologic history of the Moon. The surface of the newborn Moon was probably molten for many years, both from heat released during the impact of rock fragments falling into the young satellite and from the decay of short-lived radioactive isotopes. As the Moon gradually cooled, low-density lava floating on the Moon's surface began to solidify into the anorthositic crust that exists today. The barrage of rock fragments that ended about 3.8 billion years ago produced the craters that cover the lunar highlands.

During the relatively brief period of heavy bombardment between 4.0 and 3.8 billion years ago (see Figure 10-16), more than a dozen asteroid-size objects, each measuring at least 100 km across, struck the Moon and blasted out the vast mare basins. Meanwhile, heat from the decay of long-lived radioactive elements like uranium and thorium began to melt the inside of the Moon. From 3.8 to 3.1 billion years ago, great floods of molten rock gushed up from the lunar interior, filling the impact basins and creating the basaltic maria we see today.

Very little has happened on the Moon since those ancient times. A few relatively fresh craters have been formed, but the astronauts visited a world that has remained largely unchanged for more than 3 billion years. During that same period on the Earth, by contrast, the continents have been totally transformed time and time again (see Section 9-3).

Many questions and mysteries still remain. The six American and three Soviet lunar landings have brought back samples from only nine locations, barely scratching the lunar surface. We still know very little about the Moon's far side and poles. Are there really vast ice deposits at the poles? How much of the Moon's interior is molten? Just how large is its iron core? How old are the youngest lunar rocks? Did lava flows occur over western Oceanus Procellarum only 2 billion years ago, as crater densities there sug-

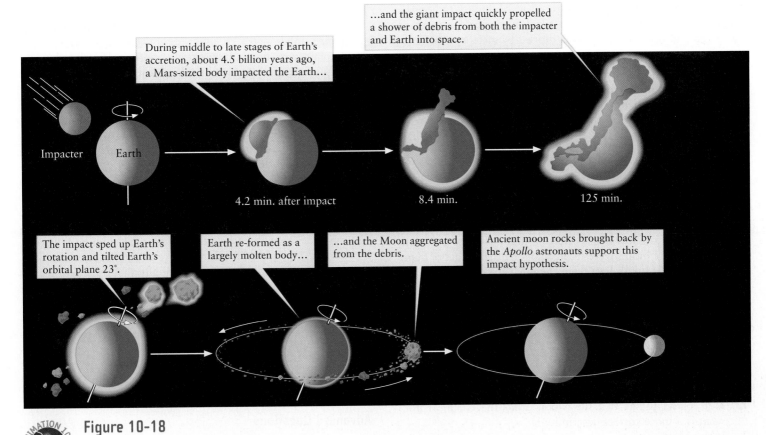

During middle to late stages of Earth's accretion, about 4.5 billion years ago, a Mars-sized body impacted the Earth...

...and the giant impact quickly propelled a shower of debris from both the impacter and Earth into space.

Impacter   Earth

4.2 min. after impact   8.4 min.   125 min.

The impact sped up Earth's rotation and tilted Earth's orbital plane 23°.

Earth re-formed as a largely molten body...

...and the Moon aggregated from the debris.

Ancient moon rocks brought back by the *Apollo* astronauts support this impact hypothesis.

Figure 10-18

The Formation of the Moon This figure shows how the Moon could have formed in the aftermath of a collision between a Mars-sized protoplanet and the proto-Earth. (Adapted from T. Grotzinger, T. H. Jordan, F. Press, and R. Siever, *Understanding Earth*, 5th ed., W. H. Freeman, 2007)

ANIMATION 10.2

gest? Could there still be active volcanoes on the Moon? There is still much to be learned by exploring the Moon.

Key Words

anorthosite, p. 246	impact crater, p. 238
capture theory (of Moon's formation), p. 249	libration, p. 236
	lunar highlands, p. 239
center of mass, p. 235	mare (*plural* maria), p. 238
co-creation theory (of Moon's formation), p. 249	mare basalt, p. 246
	moonquake, p. 244
collisional ejection theory (of Moon's formation), p. 250	refractory element, p. 250
	regolith, p. 245
crater, p. 238	synchronous rotation, p. 238
far side (of the Moon), p. 236	terminator, p. 236
fission theory (of Moon's formation), p. 249	terrae, p. 239
	volatile element, p. 250
impact breccia, p. 246	

Key Ideas

Appearance of the Moon: The Earth-facing side of the Moon displays light-colored, heavily cratered highlands and dark-colored, smooth-surfaced maria. The Moon's far side has almost no maria.

• Virtually all lunar craters were caused by space debris striking the surface. There is no evidence of plate tectonic activity on the Moon.

Internal Structure of the Moon: Much of our knowledge about the Moon has come from human exploration in the 1960s and early 1970s and from more recent observations by unmanned spacecraft.

• Analysis of seismic waves and other data indicates that the Moon has a crust thicker than that of the Earth (and thickest on the far side of the Moon), a mantle with a thickness equal to about 80% of the Moon's radius, and a small iron core.

• The Moon's lithosphere is far thicker than that of the Earth. The lunar asthenosphere probably extends from the base of the lithosphere to the core.

• The Moon has no global magnetic field today, although it had a weak magnetic field billions of years ago.

Geologic History of the Moon: The anorthositic crust exposed in the highlands was formed between 4.3 and 4.0 billion years ago. An era of heavy bombardment formed the maria basins between 4.0 and 3.8 billion years ago, and the mare basalts solidified between 3.8 and 3.1 billion years ago.

• The Moon's surface has undergone very little change over the past 3 billion years.

• Meteoroid impacts have been the only significant "weathering" agent on the Moon. The Moon's regolith, or surface layer of powdered and fractured rock, was formed by meteoritic action.

• All of the lunar rock samples are igneous rocks formed largely of minerals found in terrestrial rocks. The lunar rocks contain no water and also differ from terrestrial rocks in being relatively enriched in the refractory elements and depleted in the volatile elements.

Origin of the Moon: The collisional ejection theory of the Moon's origin holds that the proto-Earth was struck by a Mars-sized protoplanet and that debris from this collision coalesced to form the Moon. This theory successfully explains most properties of the Moon.

• The Moon was molten in its early stages, and the anorthositic crust solidified from low-density magma that floated to the lunar surface. The mare basins were created later by the impact of planetesimals and filled with lava from the lunar interior.

• Tidal interactions between the Earth and Moon are slowing the Earth's rotation and pushing the Moon away from the Earth.

Questions

Review Questions

1. Is it correct to say that the Moon orbits the Earth? If not, what is a more correct description?
2. If the Moon always keeps the same face toward the Earth, how is it possible for Earth observers to see more than half of the Moon's surface?
3. Why does the sky look black on the Moon even during daytime?
4. Why is it impossible for liquid water to exist on the surface of the Moon?
5. Describe two reasons why astronauts needed to wear spacesuits on the lunar surface.
6. Describe the kinds of features that can be seen on the Moon with a small telescope.
7. Are impact craters on the Moon the same size as the meteoroids that made the impact? Explain your answer.
8. Describe the differences between the maria and the lunar highlands. Which kind of terrain covers more of the Moon's surface? Which kind of terrain is more heavily cratered? Which kind of terrain was formed later in the Moon's history? How do we know?
9. Describe the differences between the near and far sides of the Moon. What is thought to be the likely explanation for these differences?
10. What does it mean to say the Moon is a "one-plate world"? What is the evidence for this statement?
11. Why was it necessary to send unmanned spacecraft to land on the Moon before sending humans there?
12. What is the evidence that ice exists at the lunar poles? Is this evidence definitive?
13. Why was it useful for the *Apollo* astronauts to bring magnetometers and seismometers to the Moon?
14. Could you use a magnetic compass to navigate on the Moon? Why or why not?

15. Describe the evidence that (a) the Moon has a more solid interior than the Earth and (b) the Moon's interior is not completely solid.
16. Explain why moonquakes occur more frequently when the Moon is at perigee than at other locations along its orbit.
17. Why is the Earth geologically active while the Moon is not?
18. What is the regolith? What causes its powdery character?
19. Why are there no sedimentary rocks on the Moon?
20. On the basis of moon rocks brought back by the astronauts, explain why the maria are dark-colored but the lunar highlands are light-colored.
21. Briefly describe the main differences and similarities between Moon rocks and Earth rocks.
22. Rocks found on the Moon are between 3.1 and 4.47 billion years old. By contrast, the majority of the Earth's surface is made of oceanic crust that is less than 200 million years old, and the very oldest Earth rocks are about 4 billion years old. If the Earth and Moon are essentially the same age, why is there such a disparity in the ages of rocks on the two worlds?
23. If the Earth's tidal bulge pointed directly toward the Moon, would the Moon still be receding from the Earth? Explain.
24. Why do most scientists favor the collisional ejection theory of the Moon's formation?
25. Some people who supported the fission theory proposed that the Pacific Ocean basin is the scar left when the Moon pulled away from the Earth. Explain why this idea is probably wrong.

Advanced Questions

Questions preceded by an asterisk (*) involve topics discussed in Box 10-1.

Problem-solving tips and tools

Recall that the average density of an object is its mass divided by its volume. The volume of a sphere is $4\pi r^3/3$, where r is the sphere's radius. The surface area of a sphere of radius r is $4\pi r^2$, while the surface area of a circle of radius r is πr^2. Recall also that the acceleration of gravity on the Earth's surface is 9.8 m/s^2. You may find it useful to know that a 1-pound (1-lb) weight presses down on the Earth's surface with a force of 4.448 newtons. You might want to review Newton's universal law of gravitation in Section 4-7. The time to travel a certain distance is equal to the distance traveled divided by the speed of motion. Consult Table 10-1 and the Appendices for any additional data.

---

26. Suppose two worlds (say, a planet and its satellite) have masses m_1 and m_2, and the center-to-center distance between the worlds is r. The distance d_{cm} from the center of world 1 to the center of mass of the system of two worlds is given by the formula

$$d_c = \frac{m_2 r}{m_1 + m_2}$$

(a) Suppose world 1 is the Earth and world 2 is the Moon. If the Earth and Moon are at their average center-to-center distance, find the distance from the center of the Earth to the center of mass of the Earth-Moon system. (b) Is the Earth-

Moon system's center of mass within the Earth? How far below the Earth's surface is it located? (c) Find the distance from the center of the Sun (mass 1.989×10^{30} kg) to the center of mass of the Sun-Earth system. How does this compare to the diameter of the Sun? Is it a good approximation to say that the Earth orbits around the center of the Sun?

27. If you view the Moon through a telescope, you will find that details of its craters and mountains are more visible when the Moon is near first quarter phase or third quarter phase than when it is at full phase. Explain why.

28. In a whimsical moment during the *Apollo 14* mission, astronaut Alan Shepard hit two golf balls over the lunar surface. Give two reasons why they traveled much farther than golf balls do on Earth.

29. Temperature variations between day and night are much more severe on the Moon than on Earth. Explain why.

30. How much would an 80-kg person weigh on the Moon? How much does that person weigh on the Earth?

31. Using the diameter and mass of the Moon given in Table 10-1, verify that the Moon's average density is about 3344 kg/m³. Explain why this average density implies that the Moon's interior contains much less iron than the interior of the Earth.

*32. In Box 10-1 we calculated the tidal force that the Earth exerts on two 1-kg rocks located on the near and far sides of the Moon. We assumed that the Earth-Moon distance was equal to its average value. Repeat this calculation (a) for the Moon at perigee and (b) for the Moon at apogee. (c) What is the ratio of the tidal force on the rocks at perigee to the tidal force at apogee?

33. The youngest lunar anorthosites are 4.0 billion years old, and the youngest mare basalts are 3.1 billion years old. Would you expect to find any impact breccias on the Moon that formed less than 3.1 billion years ago? Explain your answer.

34. In the maria, the lunar regolith is about 2 to 8 meters deep. In the lunar highlands, by contrast, it may be more than 15 meters deep. Explain how the different ages of the maria and highlands can account for these differences.

35. The mare basalts are volcanic rock. Is it likely that active volcanoes exist anywhere on the Moon today? Explain.

36. Calculate the round-trip travel time for a pulse of laser light that is fired from a point on the Earth nearest the Moon, hits a reflector at the point on the Moon nearest the Earth, and returns to its point of origin. Assume that the Earth and Moon are at their average separation from each other.

37. Before the *Apollo* missions to the Moon, there were two diametrically opposite schools of thought about the history of lunar geology. The "cold moon" theory held that all lunar surface features were the result of impacts. The most violent impacts melted the surface rock, which then solidified to form the maria. The opposite "hot moon" theory held that all lunar features, including maria, mountains, and craters, were the result of volcanic activity. Explain how lunar rock samples show that neither of these theories is entirely correct.

*38. When the Moon originally coalesced, it may have been only one-tenth as far from the Earth as it is now. (a) When the Moon first coalesced, was the Earth's tidal force strong enough to lift rocks off the lunar surface? Explain. (b) Compared with the net tidal force that the Earth exerts on the Moon today, how many times larger was the net tidal force

on the newly coalesced Moon? (This strong tidal force kept the one axis of the Moon oriented toward the Earth, and the Moon kept that orientation after it solidified.)

Discussion Questions

39. Comment on the idea that without the presence of the Moon in our sky, astronomy would have developed far more slowly.

40. No *Apollo* mission landed on the far side of the Moon. Why do you suppose this was? What would have been the scientific benefits of a mission to the far side?

41. NASA is planning a new series of manned missions to the Moon. Compare the advantages and disadvantages of exploring the Moon with astronauts as opposed to using mobile, unmanned instrument packages.

42. Describe how you would empirically test the idea that human behavior is related to the phases of the Moon. What problems are inherent in such testing?

43. How would our theories of the Moon's history have been affected if astronauts had discovered sedimentary rock on the Moon?

44. Imagine that you are planning a lunar landing mission. What type of landing site would you select? Where might you land to search for evidence of recent volcanic activity?

Web/eBook Questions

45. In 2005 the *SMART-1* spacecraft detected calcium on the lunar surface. Search the World Wide Web for information about the *SMART-1* mission and this discovery. How was the presence of calcium detected? What does this tell astronomers about the origin of the Moon?

46. **Determining the Size of the Planetesimal That Formed the Moon.** Access the animation "The Formation of the Moon" in Chapter 10 of the *Universe* Web site or eBook. By making measurements on the screen with a ruler, determine how many times larger the proto-Earth is than the impacting planetesimal. If Mars is about 50% the size of Earth, how does the planetesimal compare in size with present-day Mars?

Activities

Observing Projects

Observing tips and tools

If you do not have access to a telescope, you can learn a lot by observing the Moon through binoculars. Note that the Moon will appear right-side-up through binoculars but inverted through a telescope; if you are using a map of the Moon to aid your observations, you will need to take this into account. Inexpensive maps of the Moon can be purchased from most good bookstores or educational supply stores. You can determine the phase of the Moon either by looking at a calendar (most of which tell you the dates of new moon, first quarter, full moon, and third quarter), by checking the weather page of your newspaper, by consulting the current issue of *Sky & Telescope* or *Astronomy* magazine, by using the *Starry Night Enthusiast*™ program, or by using the World Wide Web.

47. Observe the Moon through a telescope every few nights over a period of two weeks between new moon and full moon. Make sketches of various surface features, such as craters, mountain ranges, and maria. How does the appearance of these features change with the Moon's phase? Which features are most easily seen at a low angle of illumination? Which features show up best with the Sun nearly overhead?

48. Use the *Starry Night Enthusiast*™ program to observe the changing appearance of the Moon. Display the entire celestial sphere by selecting **Guides > Atlas** in the **Favourites** menu. Open the **Find** pane by clicking the **Find** tab at the top of the left border of the view window. Click the menu button (the blue colored button with a downward- pointing arrowhead) at the left of the entry in the list for the **Moon** and select the **Magnify** option. This will center a magnified image of the Moon in the view as seen from the center of a transparent Earth. Close the **Find** pane by clicking its tab. Click on the down arrow to the right of the **Time Flow Rate** control and select **hours** from the drop-down menu. Then click on the **Run Time Forward** button (a triangle that points to the right like a Play button). (a) Describe how the phase of the Moon changes over time. (b) Look carefully at features near the left-hand and right-hand limbs (edges) of the Moon. Are these features always at the same position relative to the limb? Explain in terms of libration.

49. Use the *Starry Night Enthusiast*™ program to observe the apparent change in size of the Moon as seen from the surface of the Earth. Click the **Home** button in the toolbar. Stop the time flow by clicking the **Stop** time button (the button with a square icon in the **Time Flow Rate** section of the toolbar). Click each field of the **Time** and **Date** display pane in the toolbar and use the keyboard to set the time to 12:00:00 A.M. and the date to October 16, 2006. You can now set up the view of the Moon as if you were observing from a transparent Earth in continuous darkness. Open the **Options** pane by clicking its tab on the left border of the view window. In the **Local View** layer of the Options pane, uncheck the boxes beside the **Daylight** and **Local Horizon** options. Open the **Find** pane and click the menu button at the left of the entry for the Moon and select **Magnify** from the menu to display a stationary image of the magnified Moon in the center of the view. Notice the Moon's phase. (a) Select **Options > Solar System > Planets-Moons. . .** from the menu. In the Planets-Moons dialog box, click the slide control next to the **Show dark side** label near the top of the window and adjust the control all the way to the right (**Brighter**) side of the scale. Then click the **OK** button to close the dialog box. Note that the image of the Moon now appears full because *Starry Night Enthusiast*™ has artificially brightened the dark side of the image of the Moon, effectively removing the appearance of the Moon's phases. (b) Use the + button in the Zoom section of the toolbar to adjust the field of view to about 55′ × 45′. *Starry Night Enthusiast*™ can display a reference field of view (FOV) upon this sky. Open the **FOV** pane by clicking its tab. Select the **30 Arcminutes** option. Then click the FOV tab to close the pane. (c) The final view is of the Moon, its dark side artificially brightened, as it would be seen from your home location if the Earth were airless and transparent, surrounded by a yellow FOV circle 30 arcminutes in diameter. Note the size of the Moon relative to this reference circle. Set the **Time Flow Rate** in the toolbar to 1 minute. **Run Time Forward** for least 24 hours and observe the apparent size of the Moon relative to the reference circle. (c) Note that the apparent size of the Moon changes somewhat over the course of a day (of simulated time). Explain in terms of the Earth's rotation. (*Hint:* In this view, *Starry Night Enthusiast*™ has made the Earth transparent, so you can see objects that would normally be below the horizon. As the Earth rotates, your observing location on the surface is carried along and your distance from the Moon changes.) (d) Change the **Time Flow Rate** to **1 day** and again click the **Run Time Forward** button. Does the apparent size of the Moon always stay the same, or does it vary? Explain what this tells you about the shape of the Moon's orbit.

50. Use the *Starry Night Enthusiast*™ program to examine the Moon. Select **Solar System > Moon** from the **Favourites** menu. (If desired, remove the image of the astronaut by clicking on **Feet** in the **View** menu) You can rotate the image of the Moon by placing the mouse cursor over the image, holding down the mouse button, and moving the mouse. (On a two-button mouse, hold down the left mouse button.) (a) From what you can see in the image, what evidence can you find that the Moon is geologically inactive? Explain. (b) Spreading outward from some of the largest craters on the Moon are straight lines of light-colored material called rays that were caused by material ejected outward by the impact that caused the crater. Rotate the Moon around to see the entire illuminated surface. Can you find any rays on the Moon? Zoom in on the Moon to examine various surface features such as craters and mountain ranges. Estimate the length of several rays extending from craters by measuring their length on the screen with a ruler and comparing them to the diameter of the Moon, which is about 30 minutes of arc when seen from Earth.

Collaborative Exercise

51. The image of Crater Clavius in Figure 10-4 reveals numerous craters. Using the idea that the Moon's landscape can only be changed by impacts, make a rough sketch showing ten of the largest craters and label them from oldest (those that showed up first) to youngest (the most recent ones). Explain your reasoning and any uncertainties.

11

Mercury, Venus, and Mars: Earthlike yet Unique

Mercury	Venus	Mars
R I **V** U X G	R I **V** U X G	R I **V** U X G

(Mercury: NASA; Venus: NASA/JPL, MIT, and USGS; Mars: NASA, J. Bell/Cornell University, and M. Wolff/SSI)

These images above show the three planets that share the inner solar system with our Earth. All three have solid surfaces, and, in principle, a properly protected astronaut could stand on any of them. The differences between these three worlds, however, are pronounced.

Mercury is small, airless, and extensively cratered. It is also mysterious: Half of its surface has never been viewed at close range, and it has a perplexing and unexpected magnetic field. It also rotates in a manner unique in the solar system, spinning three times on its axis for every two orbits around the Sun.

Venus is nearly the same size as Earth, but it is shrouded by a perpetual cloud cover that hides its surface from view. To penetrate the clouds and learn what lies beneath, two advanced technologies were needed: radar, which allowed astronomers to "see"

through the clouds, and unmanned spacecraft, which orbited Venus at close range and even landed on its surface. The false-color topographic map shown here reveals that Venus has no true continents but merely highlands (shown in red) that rise gently above the planet's lower-lying areas (shown in blue). We have also learned that Venus's atmosphere is thick and searing hot, and that its clouds contain droplets of corrosive sulfuric acid.

Mars has captivated the popular imagination like no other planet. But rather than being the abode of warlike aliens, Mars proves to be an enigmatic world. Some parts of its surface are drier than any desert on Earth, while other locations show evidence of having been underwater for extended periods. Our challenge is to understand how these three nominally Earthlike worlds evolved to be so unique and so different than our own Earth.

Learning Goals

By reading the sections of this chapter, you will learn

11-1 What astronomers have learned by observing the terrestrial planets from Earth

11-2 The radically different ways in which Mercury, Venus, and Mars rotate on their axes

11-3 The outstanding features of Mercury, and why its magnetic field came as a surprise

11-4 How the advent of the space age transformed our understanding of Venus and Mars

11-5 How geologic activity took a very different form on Venus than on Earth, and why it essentially stopped on Mars

11-6 The key differences among the atmospheres of Earth, Venus, and Mars

11-7 How the atmospheres of Earth, Venus, and Mars evolved to their present states

11-8 The evidence that there was once liquid water on Mars

11-9 What we know about the two small satellites of Mars

11-1 Mercury, Venus, and Mars can all be seen with the naked eye

At various times Mercury, Venus, and Mars are among the brightest objects in the sky. At their greatest brilliance, each of these planets appears brighter than any star, which is why they have been known since ancient times and why their motions played a role in the religious beliefs of many ancient cultures.

Naked-Eye Observations of Mercury and Venus

Mercury and Venus are *inferior* planets whose orbits around the Sun are smaller than the Earth's,: that is, each moves in an orbit with a semimajor axis *a* less than 1 astronomical unit (AU). For Mercury, $a = 0.387$ AU (57.9 million kilometers, or 36.0 million miles), and for

> Viewing Mercury is difficult because it is so close to the Sun; Venus is easier to see but covered with clouds; Mars is best seen at opposition

Venus, $a = 0.723$ AU (108 million kilometers, or 67.2 million miles). As a result, Mercury and Venus never appear very far from the Sun in our sky (**Figure 11-1**). Table 11-1 and Table 11-2 summarize some basic information about these planets and their orbits.

We get the best view of either Mercury or Venus when it appears as far from the Sun in the sky as it can be, at its greatest eastern or western elongation (Section 4-2). On dates near its **greatest eastern elongation**, Mercury or Venus appears after sunset over the western horizon as an "evening star." On dates near its **greatest western elongation**, Mercury or Venus appears as a "morning star" that rises before the Sun in the eastern sky.

CAUTION! Remember that "greatest *western* elongation" and "greatest *eastern* elongation" refer to a planet's position in the sky relative to the Sun. When Mercury or Venus is at greatest western elongation, it is west of the Sun in the sky and has already risen at dawn when the Sun is rising in the east. When Mercury or Venus is at greatest eastern elongation, it is east of the Sun in the sky and is still above the horizon when the Sun sets in the west.

Because Mercury's orbit is so close to the Sun, its maximum elongation is only 28° and it is quite difficult to observe. (It is said that Copernicus never saw it.) The celestial sphere appears to rotate at 15° per hour (360° divided by 24 hours), so Mercury never rises more than 2 hours before sunrise or sets more than 2 hours after sunset. Unfortunately, Mercury's orbit is very elliptical (its eccentricity of 0.206 is greater than that of any other planet) and is inclined to the ecliptic—the plane of the Earth's orbit around the Sun—by about 7°. As a result, at greatest elongation Mercury is often much less than 28° from the horizon at sunset or sunrise. Hence, some elongations are more favorable for viewing Mercury than others, as **Figure 11-2** shows. A total of six or seven greatest elongations (both eastern and western) occur each year, but usually only two of these will be favorable for viewing the planet.

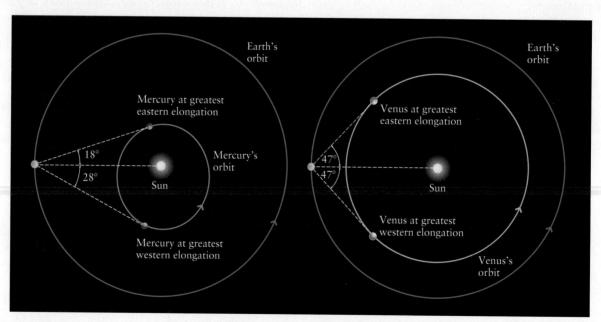

Figure 11-1

The Orbits of Mercury and Venus Mercury moves around the Sun every 88 days in a rather eccentric orbit. As seen from the Earth, the angle between Mercury and the Sun at greatest eastern or western elongation can be as large as 28° (when Mercury is near aphelion) or as small as 18° (near perihelion). By contrast, Venus follows a larger, nearly circular orbit with a 224.7-day period. The angle between Venus and the Sun at eastern or western elongation is 47°.

Table 11-1	Mercury Data
Average distance from Sun:	$0.387 \text{ AU} = 5.79 \times 10^7 \text{ km}$
Maximum distance from Sun:	$0.467 \text{ AU} = 6.98 \times 10^7 \text{ km}$
Minimum distance from Sun:	$0.307 \text{ AU} = 4.60 \times 10^7 \text{ km}$
Eccentricity of orbit:	0.206
Average orbital speed:	47.9 km/s
Orbital period:	87.969 days
Rotation period:	58.646 days
Inclination of equator to orbit:	0.5°
Inclination of orbit to ecliptic:	7° 00′ 16″
Diameter (equatorial):	4880 km = 0.383 Earth diameter
Mass:	$3.302 \times 10^{23} \text{ kg} = 0.0553 \text{ Earth mass}$
Average density:	5430 kg/m^3
Escape speed:	4.3 km/s
Surface gravity (Earth = 1):	0.38
Albedo:	0.12
Average surface temperatures:	Day: 350°C = 662°F = 623 K
	Night: −170°C = 2274°F = 103 K
Atmosphere:	Essentially none

RIVUXG
(Astrogeology Team, U. S. Geological Survey)

Mercury is at its maximum possible elongation (28°)...

28°

...and its orbital plane is inclined at a large angle to the observer's horizon.

Favorable elongation

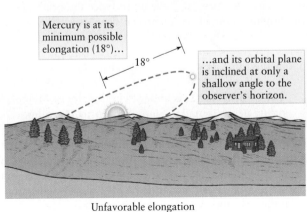

Mercury is at its minimum possible elongation (18°)...

18°

...and its orbital plane is inclined at only a shallow angle to the observer's horizon.

Unfavorable elongation

Figure 11-2

Favorable Versus Unfavorable Elongations The tilt of the Earth's axis, the inclination and eccentricity of Mercury's orbit, and the latitude of the observer combine to make an elongation either favorable (left illustration) or unfavorable (right illustration) for viewing Mercury.

Table 11-2   Venus Data

Average distance from Sun:	$0.723 \text{ AU} = 1.082 \times 10^8 \text{ km}$
Maximum distance from Sun:	$0.728 \text{ AU} = 1.089 \times 10^8 \text{ km}$
Minimum distance from Sun:	$0.718 \text{ AU} = 1.075 \times 10^8 \text{ km}$
Eccentricity of orbit:	0.0068
Average orbital speed:	35.0 km/s
Orbital period:	224.70 days
Rotation period:	243.01 days (retrograde)
Inclination of equator to orbit:	177.4°
Inclination of orbit to ecliptic:	3.39°
Diameter (equatorial):	12,104 km = 0.949 Earth diameter
Mass:	$4.868 \times 10^{24} \text{ kg} = 0.815$ Earth mass
Average density:	5243 kg/m^3
Escape speed:	10.4 km/s
Surface gravity (Earth = 1):	0.91
Albedo:	0.59
Average surface temperature:	460°C = 860°F = 733 K
Atmospheric composition (by number of molecules):	96.5% carbon dioxide (CO_2) 3.5% nitrogen (N_2) 0.003% water vapor (H_2O)

WEB LINK 11.3

R I V U X G

(NASA)

Like all the planets, Mercury shines by reflected sunlight. Because it orbits so close to the Sun, it is exposed to sunlight that on average is about 7 times as intense as on the Earth. Furthermore, Mercury is never more than 1.5 AU from Earth. These factors explain why Mercury can have up to twice the brightness of Sirius, the brightest star in the nighttime sky. Mercury would be even brighter if its surface did not have a rather low albedo of 0.12 (that is, it reflects only about 12% of the sunlight that falls on it—about the same as weathered asphalt).

WEB LINK 11.6   The orbit of Venus is about twice the size of Mercury's orbit, nearly circular, and inclined to the ecliptic by only 3.39°. As a result, Venus can be seen about 47° away from the Sun and all elongations are nearly equally favorable. At greatest eastern elongation, which occurs at intervals of about 19 months, Venus sets in the west about 3 hours after sunset. About 5 months later Venus is at greatest western elongation and rises in the east about 3 hours before sunrise.

At its greatest brilliance, Venus is 16 times brighter than Sirius and is outshone only by the Sun and the Moon. Venus is more brilliant than Mercury because it is relatively large (more than double the diameter of Mercury and almost the same size as the Earth) and has a much higher albedo of 0.59 (it reflects 59% of the sunlight that falls on it).

Naked-Eye Observations of Mars

WEB LINK 11.7   Unlike Mercury and Venus, Mars is a *superior* planet with an orbit larger than the Earth's orbit. As Table 11-3 shows, its orbital semimajor axis is $a = 1.524$ AU (227.9 million kilometers, or 141.6 million miles). Hence, Mars and the Sun are sometimes on opposite sides of the sky as seen from Earth—that is, Mars is at opposition (see Figure 4-6)—and appears high in our night sky. Such oppositions occur at intervals of about 780 days. However, some oppositions provide better views of Mars than others because Mars has a noticeably elongated orbit (Figure 11-3). As a result, we get the best view of Mars when it is simultaneously at opposition and near the perihelion of its elliptical orbit, a configuration called a **favorable opposition**. The Earth-Mars distance can then be as small as 0.37 AU, or 56 million kilometers (35 million miles), and the angular diameter of Mars can be as large as 25.1 arcsec—about the same as a person's head seen at a distance of 1.6 km (1 mile). Under these conditions, Mars appears in the nighttime sky as a brilliant red object $3\frac{1}{2}$ times brighter than Sirius.

Table 11-3	Mars Data
Average distance from Sun:	1.524 AU $= 2.279 \times 10^8$ km
Maximum distance from Sun:	1.666 AU $= 2.492 \times 10^8$ km
Minimum distance from Sun:	1.381 AU $= 2.067 \times 10^8$ km
Eccentricity of orbit:	0.093
Average orbital speed:	24.1 km/s
Orbital period:	686.98 days $= 1.88$ years
Rotation period:	$24^h\ 37^m\ 22^s$
Inclination of equator to orbit:	25.19°
Inclination of orbit to ecliptic:	1.85°
Diameter (equatorial):	6794 km $= 0.533$ Earth diameter
Mass:	6.418×10^{23} kg $= 0.107$ Earth mass
Average density:	3934 kg/m³
Escape speed:	5.0 km/s
Surface gravity (Earth $= 1$):	0.38
Albedo:	0.15
Surface temperatures:	Maximum: 20°C $=$ 70°F $=$ 293 K Mean: -23°C $= -10$°F $=$ 250 K Minimum: -140°C $= -220$°F $=$ 133 K
Atmospheric composition (by number of molecules):	95.3% carbon dioxide (CO_2) 2.7% nitrogen (N_2) 0.03% water vapor (H_2O) 2% other gases

WEB LINK 11.4

R I V U X G

(NASA, USGS)

Unfortunately, the time from one favorable opposition to the next is about 15 years. Hence, astronomers take advantage of all oppositions, even the not-so-favorable ones, to observe Mars. When Mars is at opposition but near aphelion the Earth-Mars distance can be as large as 0.68 AU, or 101 million kilometers (63 million miles).

Telescopic Views of Mercury, Venus, and Mars

Naked-eye observations of Mercury are best made at dusk or dawn, but the best telescopic views are obtained in the daytime when the planet is high in the sky, far above the degrading atmospheric effects near the horizon. A yellow filter can eliminate much of the blue light from our sky. The photograph in Figure 11-4a, which is among the finest Earth-based views of Mercury, was taken during daytime.

While an Earth-based telescope reveals some surface details on Mercury, Venus appears almost completely featureless, with neither continents nor mountains (see Figure 4-13). Astronomers soon realized that they were not seeing the true surface of the planet. Rather, Venus must be covered by a thick, unbroken layer of clouds. A cloud layer would also explain why Venus reflects such a large fraction of the sunlight that falls on it. Direct evidence that Venus has a thick atmosphere came when nineteenth-century astronomers observed Venus near the time of inferior conjunction. This is when Venus lies most nearly between the Earth and the Sun, so that we see the planet lit from behind. As Figure 11-4b shows, sunlight is scattered by Venus's atmosphere, producing a luminescent ring around the planet that would not otherwise be present. No such ring is seen around Mercury at inferior conjunction, which indicates that Mercury has no appreciable atmosphere.

Figure 11-3

The Orbits of Earth and Mars The best times to observe Mars are at opposition, when Mars is on the side of Earth opposite the Sun. The red and blue dots connected by dashed lines show the positions of Mars and Earth at various oppositions. The months shown around the Sun refer to the time of year when the Earth is at each position around its orbit.

(a) Mercury at greatest elongation

(b) Venus at inferior conjunction

Figure 11-4   R I [V] U X G

Earth-Based Views of Mercury, Venus, and Mars
(a) Hazy markings are faintly visible in this photograph of Mercury, one of the best ever produced with an Earth-based telescope. Note that Mercury goes through phases like Venus (see Figure 4-13). **(b)** Venus at inferior conjunction shows its atmosphere as a faint ring surrounding the planet. This photograph was processed to reveal the ring, which is too faint to show in Figure 4-13. **(c)** A 25-cm (10-in.) telescope with a CCD detector (see Section 6-4) was used to make this image of Mars. Compare this with the Hubble Space Telescope image that opens this chapter, which was taken during the same month in 2003 and which shows the same face of Mars. (a: New Mexico State University Observatory; b: Lowell Observatory; c: Courtesy of Sheldon Faworski and Donald Parker)

(c) Mars at opposition

Unlike airless Venus or cloud-shrouded Venus, Mars has a thin, almost cloudless atmosphere that permits a clear view of the surface. Under the right conditions, a relatively small telescope can reveal substantial detail on the Martian surface. Figure 11-4c shows an image of Mars during the very favorable 2003 opposition, when Mars was closer to Earth than it had been in several thousand years. The thin strip of blue and orange around the right-hand edge of Mars in this image is caused by the Martian atmosphere. This image also shows that different parts of the Martian surface have different colors. Most striking are the whitish ice caps at the Martian poles, which bear a striking superficial resemblance to the Arctic and Antarctic regions on Earth.

CAUTION! Although Mars appears bright red to Earth observers, its surface is actually reddish-brown. A sunlit brown surface, if surrounded by the blackness of space, is perceived by the human eye as having a red color.

11-2 While Mars rotates much like the Earth, Mercury's rotation is coupled to its orbital motion and Venus's rotation is slow and retrograde

We have seen that the surface of Mars is relatively easy to observe with visible-light telescopes. By contrast, only a few faint, hazy markings can be seen on Mercury's surface, and the surface of Venus is perpetually hidden by clouds. These differences explain why the rotation of Mars has been well understood since the mid-seventeenth century, while it was not until the 1960s that astronomers accurately measured the rotation of Mercury and Venus. As we will see, Mars rotates in a very Earthlike way, while Mercury and Venus rotate like no other objects in the solar system.

> **Radar technology revealed the curious rotation of Mercury and Venus**

The Rotation of Mars

 The first reliable record of surface features on Mars was made by the Dutch scientist Christiaan Huygens, who observed the planet during the opposition of 1659. He identified a prominent dark feature that we now call Syrtis Major. After observing this feature for several weeks, Huygens concluded that the rotation period of Mars is approximately 24 hours, the same as the Earth.

In 1666, the Italian astronomer Gian Domenico Cassini made more-refined observations which showed that a Martian day is 37.5 minutes longer than on Earth. More than a century later, the German-born English astronomer William Herschel found that Mars's axis of rotation is not perpendicular to the plane of the planet's orbit, but is tilted by about 25° away from the perpendicular. This is very close to the Earth's 23½° tilt (see Figure 2-12). This striking coincidence means that Mars experiences Earthlike seasons, with opposite seasons in the northern and southern Martian hemispheres (see Section 2-5). Because Mars takes nearly 2 (Earth) years to orbit the Sun, the Martian seasons last nearly twice as long as on Earth.

The Challenge of Observing Mercury's Rotation

During the 1880s, the Italian astronomer Giovanni Schiaparelli attempted to make the first map of Mercury. Unfortunately, Schiaparelli's telescopic views of Mercury were so vague and indistinct that he made an understandable but major error, which went uncorrected for more than half a century. He erroneously concluded that Mercury always keeps the same side facing the Sun.

Many objects in our solar system are in synchronous rotation, so that their rotation period equals their period of revolution—a situation also called **1-to-1 spin-orbit coupling.** We saw in Section 4-8 how the Earth's tidal forces keep the Moon in synchronous rotation, so that it always keeps the same side toward the Earth. Tidal forces also keep the two moons of Mars with the same side facing their parent planet, and likewise many of the satellites of Jupiter and Saturn.

It had been suggested as early as 1865 that the Sun's tidal forces would keep Mercury in synchronous rotation. But this is not how Mercury rotates. The first clue about Mercury's true rotation period came in 1962, when astronomers detected radio radiation coming from the innermost planet.

As we learned in Sections 5-3 and 5-4, every dense object (such as a planet) emits electromagnetic radiation with a continuous spectrum. The dominant wavelength of this radiation depends on the object's temperature. Only at absolute zero does a dense object emit no radiation at all.

If Mercury exhibited synchronous rotation, one side of the planet would remain in perpetual, frigid darkness, quite near absolute zero. To test this idea, in 1962 the radio astronomer W. E. Howard and his colleagues at the University of Michigan began monitoring radiation from Mercury. (If the planet had one side that was very cold, most of the emission from that side should be at quite long wavelengths. Hence, a radio telescope was used for these observations.) They were surprised to discover that the temperature on the planet's nighttime side is about 100 K (= −173°C = −280°F), not nearly as cold as expected. These observations were in direct contradiction to the idea of synchronous rotation.

One idea offered to explain Howard's observations was that Mercury rotates synchronously, but that winds in the planet's atmosphere carried warmth from the daytime side of the planet around to the nighttime side. But, in fact, there are no winds on Mercury. Its daytime temperature is too high and its gravity too weak to retain any substantial atmosphere (see Box 7-2).

The definitive observation was made in 1965, when Rolf B. Dyce and Gordon H. Pettengill used the giant 1000-ft radio telescope at the Arecibo Observatory in Puerto Rico to bounce powerful radar pulses off Mercury. The outgoing radiation consisted of microwaves of a very specific wavelength. In the reflected signal echoed back from the planet, the wavelength had shifted as a result of the Doppler effect (see Section 5-9, especially Figure 5-26). As Mercury rotates, one side of the planet approaches the Earth, while the other side recedes from the Earth. Microwaves reflected from the planet's approaching side were shortened in wavelength, whereas those from its receding side were lengthened (**Figure 11-5**). Thus, the radar pulse went out at one specific wavelength, but it came back spread over a small wavelength

Figure 11-5

Measuring Mercury's Rotation Period As Mercury rotates, one side of the planet moves away from the Earth and the other side toward the Earth. If radio waves of a single wavelength are beamed toward Mercury, waves reflected from these two sides will be Doppler shifted to longer and shorter wavelengths, respectively. By measuring the wavelength shift of the reflected radiation, astronomers deduced how rapidly Mercury rotates and thus determined its rotation period.

range. From the width of this wavelength range, Dyce and Pettengill deduced that Mercury's rotation period is approximately 58.6 days.

Spin-Orbit Coupling and the Curious Rotation of Mercury

Giuseppe Colombo, an Italian physicist with a long-standing interest in Mercury, was intrigued by Dyce and Pettingill's results.

Colombo noted that their result for the rotation period is very close to two-thirds of Mercury's sidereal period of 87.969 days:

$$2/3 \ (87.969 \text{ days}) = 58.646 \text{ days}$$

Colombo therefore boldly speculated that Mercury's true rotation period is exactly 58.646 days, or 58 days and $15\frac{1}{2}$ hours. He realized that this figure would mean that Mercury is locked into a **3-to-2 spin-orbit coupling:** The planet makes *three* complete rotations on its axis for every *two* complete orbits around the Sun. No other planet or satellite in the solar system has this curious relationship between its rotation and its orbital motion.

Figure 11-6 shows how gravitational forces from the Sun cause Mercury's 3-to-2 spin-orbit coupling. The force of gravity decreases with increasing distance, which explains how the Moon is able to raise a tidal bulge in the Earth's oceans (see Section 4-8). Mercury has no oceans, but it has a natural bulge of its own; thanks to the Sun's tidal forces, the planet is slightly elongated along one axis. The stronger gravitational force that the Sun exerts on the near side of the planet tends to twist the long axis to point toward the Sun, as Figure 11-6a shows. Indeed, Mercury's long axis would always point toward the Sun if its orbit were circular or nearly so. In this case the same side of Mercury would always face the Sun, and there would be synchronous rotation (Figure 11-6b). But Mercury's orbit has a rather high eccentricity, and the twisting effect on Mercury decreases rapidly as the planet moves away from the Sun. As a result, Mercury's long axis points toward the Sun only at perihelion (Figure 11-6c). From one perihelion to the next, alternate sides of Mercury face the Sun.

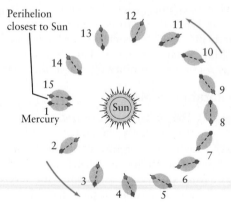

(a) Mercury is slightly elongated: The different gravitational forces on Mercury's two ends tend to twist the long axis to point toward the Sun.

(b) If Mercury were in a circular orbit, its long axis would always point toward the Sun: Mercury would be in synchronous rotation (1-to-1 spin-orbit coupling).

(c) In fact Mercury is in an elliptical orbit, and its long axis only points toward the Sun at perihelion: Mercury spins on its axis 1½ times during each complete orbit (3-to-2 spin-orbit coupling).

Figure 11-6

Mercury's Spin-Orbit Coupling (a) The Sun's gravitational force on the near side of Mercury's long axis is greater than the force on the far side. This tends to twist the long axis to point toward the Sun. **(b)** If Mercury were in a circular orbit, the twisting effect shown in (a) would always keep the same side of Mercury (shown by a red dot) facing the Sun.

(c) Because Mercury's orbit is rather elongated, its rotation is more complex: The end of Mercury's long axis marked by a red dot faces the Sun at one perihelion (point 1), but at the next perihelion (point 15) the opposite end—marked by a blue dot—faces the Sun.

Because of the 3-to-2 spin-orbit coupling, the average time from sunrise to sunset on Mercury is just equal to its orbital period, or just under 88 days. This helps explain the tremendous difference between daytime and nighttime temperatures on Mercury. Not only is sunlight about 7 times more intense on Mercury than on the Earth (because of the smaller size of Mercury's orbit), but that sunlight has almost 3 Earth months to heat up the surface. As a result, daytime temperatures at the equator are high enough— about 430°C (roughly 700 K, or 800°F)—to melt lead. (In comparison, a typical kitchen oven reaches only about 300°C or 550°F.)

The time from sunset to the next sunrise is also 88 days, and the surface thus has almost 3 Earth months of darkness during which to cool down. Therefore, nighttime temperatures reach to below −170°C (about 100 K, or −270°F). This is cold enough to freeze carbon dioxide and methane. The surface of Mercury is truly inhospitable!

While the Sun moves slowly across Mercury's sky, it does not always move from east to west as it does as seen from Earth. The reason is that Mercury's speed along its elliptical orbit varies substantially, in accordance with Kepler's second law (see Section 4-4, especially Figure 4-11). It moves fastest (59 km/s) at perihelion and slowest (39 km/s) at aphelion.

As seen from Mercury's surface, the Sun rises in the east and sets in the west, just as it does on the Earth. When Mercury is near perihelion, however, the planet's rapid motion along its orbit outpaces its leisurely rotation about its axis (one rotation in 58.646 days). Hence, the usual east-west movement of the Sun across Mercury's sky is interrupted, and the Sun actually stops and moves backward (from west to east) for a few Earth days. If you were standing on Mercury watching a sunset occurring at perihelion, the Sun would not simply set. It would dip below the western horizon and then come back up, only to set a second time a day or two later.

Observing Venus's Rotation: Penetrating the Clouds with Radio

Venus's perpetual cloud cover makes it impossible to measure the planet's rotation using visible-light telescopes. (It is possible to use visible-light telescopes to track features in the planet's cloud cover. But winds in the atmosphere cause such features to move relative to Venus's surface, so this approach does not give a direct measurement of how the planet itself rotates.) To penetrate the planet's perpetual cloud cover, astronomers applied to Venus the same technique of using radio waves and the Doppler effect that revealed the rotation of Mercury.

Clouds may contain gas, dust, haze, water droplets, or other small particles. In general, electromagnetic radiation can pass easily through such a cloud only if the wavelength is large compared to the size of the particles. We then say that the cloud is *transparent* to the radiation. For example, clouds in the Earth's atmosphere are made of water droplets with an average diameter of about 20 μm (2×10^{-5} m, or 20,000 nm). Visible light has wavelengths between 400 to 700 nm, which is less than the droplet size. Hence, visible light cannot easily pass through such clouds, which is why cloudy days are darker than sunny days. But radio waves, with wavelengths of 0.1 m or more, and microwaves, with

wavelengths from 10^{-3} m to 0.1 m, can pass through such a cloud with ease. That is why your mobile phone, radio, or television works just as well on a cloudy or foggy day as on a clear one.

Like clouds in the Earth's atmosphere, the clouds of Venus are transparent to radio waves and microwaves. Beginning in the early 1960s, advances in radio telescope technology made it possible to send microwave radiation to Venus and detect the waves reflected from its surface. Just as for Mercury (see Figure 11-5), the Doppler shift of waves reflected from the two sides of Venus reveal the speed and direction of the planet's rotation.

Venus: Where the Sun Rises in the West

These measurements show that Venus rotates very slowly; the sidereal rotation period of the planet is 243.01 days, even longer than the planet's 224.7-day orbital period. To an imaginary inhabitant of Venus who could somehow see through the cloud cover, stars would move across the sky at a rate of only $1\frac{1}{2}°$ per Earth day. As seen from the Earth, by contrast, stars move $1\frac{1}{2}°$ across the sky in just 6 minutes.

Not only does Venus rotate slowly, it also rotates in an unusual direction. Most of the planets in the solar system spin on their axes in **prograde rotation**. Prograde means "forward," and planets with prograde rotation spin on their axes in the same direction in which they orbit the Sun (Figure 11-7a). As seen from the surface of a planet with prograde rotation, such as the Earth, the Sun and stars rise in the east and set in the west.

Venus is an exception to the rule: It spins in **retrograde rotation**. That is, the direction in which it spins on its axis is backward compared to the direction in which it orbits the Sun, as Figure 11-7b shows. (Recall that "retrograde" means "backward.") If an observer on the surface of Venus could see through the thick layer of clouds, the Sun and stars would rise in the west and set in the east!

CAUTION! Be careful not to confuse retrograde *rotation* with retrograde *motion*. When Venus is near inferior conjunction, so that it is passing the Earth on its smaller, faster orbit around the Sun, it appears to us to move from east to west on the celestial sphere from one night to the next. This apparent motion is called retrograde motion, because it is backward to the apparent west-to-east motion that Venus displays most of the time. (You may want to review the discussion of retrograde motion in Sections 4-1 and 4-2.) By contrast, the retrograde rotation of Venus means that it spins on its axis from east to west, rather than west to east like the Earth.

Venus's retrograde rotation adds to the puzzle of our solar system's origins. If you could view our solar system from a great distance above the Earth's north pole, you would see all the planets orbiting the Sun counterclockwise. Closer examination would reveal that the Sun and most of the planets also *rotate* counterclockwise on their axes; the exceptions are Venus and Uranus. Most of the satellites of the planets also move counterclockwise along their orbits and rotate in the same direction. Thus, the rotation of Venus is not just opposite to its orbital motion; it is opposite to most of the orbital and rotational motions in the solar system!

(a) Prograde rotation

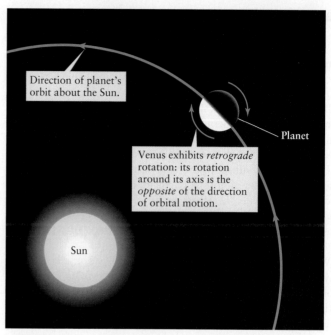

(b) Retrograde rotation

Figure 11-7

Prograde and Retrograde Rotation (a) If you could view the solar system from a point several astronomical units above the Earth's north pole, you would see all of the planets orbiting the Sun in a counterclockwise direction. Most of the planets also rotate counterclockwise on their axes in prograde (forward) rotation. **(b)** Venus is an exception; from your perspective high above the plane of the solar system, you would see Venus rotating clockwise in retrograde (backward) rotation.

As we saw in Section 8-4, the Sun and planets formed from a solar nebula that was initially rotating. (Had it not been rotating, all of the material in the nebula would have fallen inward toward the protosun, leaving nothing to form the planets.) Because the planets formed from the material of this rotating cloud, they tend to rotate on their axes in that same direction. It is difficult to imagine how Venus and Uranus could have bucked this trend. One theory suggests that the tidal forces exerted by the Sun tended to elongate Venus's atmosphere toward the Sun, much as the Moon's tidal forces elongate the Earth's oceans toward the Moon (see Section 4-8, especially Figure 4-25). Friction between the atmosphere and Venus's surface then acted to slow the planet's rotation. By themselves, these tidal effects would put Venus into synchronous prograde rotation. But in addition, the Sun's heat on the illuminated side of Venus generates strong winds in the planet's atmosphere, and the frictional forces of these winds drive Venus into slow retrograde rotation.

Venus's rotation has other puzzling aspects. The length of an apparent solar day on Venus—that is, from when the Sun is highest in the sky to when it is again highest in the sky (see Section 2-7)—is 116.8 Earth days. Remarkably, this is almost exactly one-fifth as long as the planet's 584-day synodic period. (We saw in Section 4-2 that this is the time between successive inferior conjunctions, when Venus and the Earth are closest together.) Is this curious ratio the result of gravitational interactions between Venus and the Earth? Or is it sheer coincidence? No one knows the answer.

11-3 Mercury is cratered like the Moon but has a surprising magnetic field

Imagine an alien astronomer observing Earth from Mercury, from above the clouds of Venus, or from Mars with the same telescope technology available to our own astronomers. Such an astronomer would be unable to resolve any features on Earth less than a few hundred kilometers across. It would be impossible for her to learn about the Earth's mountain ranges, its volcanoes, or even its cities. Earth astronomers face the same limitations in studying the other terrestrial planets. To truly understand these other worlds, it is essential to study them at close range using spacecraft.

Spacecraft observations at close range were necessary to reveal Mercury's unusual properties

In this section we will see what a single spacecraft has been able to teach us about Mercury's surface and interior. Later in this chapter we will see how a small fleet of spacecraft sent to Venus and Mars has revolutionized our knowledge of those planets.

Mariner 10 to Mercury

The first spacecraft to provide detailed knowledge about Mercury was *Mariner 10*. Launched in 1973, this spacecraft went into an orbit around the Sun that brought it close to Mercury on three occasions in 1974 and 1975. The images transmitted back to Earth showed features as small as 1 km, hundreds of times smaller than the finest details made using Earth-based telescopes. **Figure 11-8** shows a mosaic made by combining a number of *Mariner 10* images.

The *Mariner 10* images show only one hemisphere of Mercury's surface, since the time interval between flybys was equal to twice the planet's orbital period and 3 times its rotation period. Hence, the same side of Mercury was illuminated by the Sun during each flyby. The other side of Mercury has yet to be explored at close range.

Mercury's Cratered Surface

As *Mariner 10* closed in on Mercury, scientists were struck by the Moonlike pictures appearing on their television monitors. It was obvious that Mercury, like the Moon, is a barren, heavily cratered world, with no evidence for plate tectonics. This is what we would expect based on Mercury's small size (see Table 7-1 and the figure that opens this chapter). As we saw in Section 7-6, a small world is expected to have little internal heat, and, hence, little or no geologic activity and an ancient, cratered surface (see Figure 7-11). But craters are not the only features of Mercury's surface; there are also gently rolling plains, long, meandering cliffs, and an unusual sort of jumbled terrain. As we will see, Mercury's distinctive surface shows that this planet is *not* merely a somewhat larger version of the Moon.

Figure 11-9 shows a typical close-up view sent back from *Mariner 10*. The consensus among astronomers is that most of the craters on both Mercury and the Moon were produced during the first 700 million years after the planets formed. Debris re-

Figure 11-9   R I V U X G

Mercurian Craters and Plains This view of Mercury's northern hemisphere was taken by *Mariner 10* at a range of 55,000 km (34,000 mi) from the planet's surface. Numerous craters and extensive intercrater plains appear in this photograph, which covers an area 480 km (300 mi) wide. (NASA)

maining after planet formation rained down on these young worlds, gouging out most of the craters we see today. We saw in Section 10-4 that the strongest evidence for this chronology comes from analysis and dating of Moon rocks. *Mariner 10* did not land on Mercury, so we are not able to make the same kind of direct analysis of rocks from the planet's surface.

Unlike the Moon, Mercury's surface has extensive low-lying plains (examine the upper half of Figure 11-9). These large, smooth areas are about 2 km lower than the cratered terrain. These probably have a similar origin to the lunar maria, which were produced by extensive lava flows between 3.1 and 3.8 billion years ago (see Section 10-4). Primordial lava flows probably also explain the plains of Mercury. As large meteoroids punctured the planet's thin, newly formed crust, lava welled up from the molten interior to flood low-lying areas.

From the number of craters that pit them, Mercury's plains appear to have been formed near the end of the era of heavy bombardment, a little more than 3.8 billion years ago. Mercury's plains are therefore older than most of the lunar maria. Analysis of *Mariner 10* images made at different wavelengths also shows that the material in Mercury's plains has a lower iron content than the lunar maria. This is presumably why the plains of Mercury do not have the dark color of the Moon's maria.

Scarps: Evidence of a Wrinkled Planet

Mariner 10 also revealed numerous long cliffs, called **scarps**, meandering across Mercury's surface (**Figure 11-10**). Some scarps rise as much as 3 km (2 mi) above the surrounding plains and are 20 to 500 km long. These cliffs probably formed as the planet cooled and contracted.

Figure 11-8   R I V U X G

Mariner 10 **Mosaic of Mercury** To make this mosaic of the side of Mercury viewed by *Mariner 10,* scientists combined dozens of images recorded by the spacecraft as it flew toward and away from the planet. The blank swatch running down from Mercury's north pole is a region that was not imaged. (Astrogeology Team, U.S. Geological Survey)

1. The floors of these craters were flooded by lava from Mercury's interior.

2. Some time after the lava cooled, Mercury's crust contracted to form this scarp.

3. This crater was distorted when the scarp formed.

50 km

Figure 11-10   R I V U X G

A Scarp A long cliff, or scarp, called Santa Maria Rupes runs across this *Mariner 10* image. This scarp is more than a kilometer high and extends for several hundred kilometers across a region near Mercury's equator. The area shown is approximately 200 km (120 mi) across. (NASA)

ANALOGY The shrinkage of Mercury's crust is probably a result of a general property of most materials: They shrink when cooled and expand when heated. The lid on a jar of jam makes a useful analogy. When the jar is kept cold in the refrigerator, the metal lid contracts more than the glass jar. This makes a tight seal, but also makes the lid hard to unscrew. You can loosen the lid by running hot water over it; this makes the metal lid expand more than the glass jar, helping to break the seal.

If there had still been molten material beneath Mercury's surface when this contraction took place, lava would have leaked onto the surface around the scarps. This does not appear to have happened. Hence, the scarps must have formed relatively late in Mercury's history, after the lava flows had ended and after the planet had solidified to a substantial depth beneath the surface. With the exception of the scarps, there are no features on Mercury that resemble the boundaries of tectonic plates. Thus, we can regard Mercury's crust, like that of the Moon, as a single plate.

Fire and Ice: The Caloris Basin and Mercury's Poles

The most impressive feature discovered by *Mariner 10* was a huge impact scar called the Caloris Basin (*calor* is the Latin word for "heat"). The Sun is directly over the Caloris Basin during alternating perihelion passages, and thus it is the hottest place on the planet once every 176 days. During each of the *Mariner 10* flybys, the Caloris Basin happened to lie on the terminator (the line dividing day from night; see Section 10-1). Hence, in Figure 11-11 slightly more than half of this basin is hidden on the night side of the planet.

The Caloris Basin, which measures 1300 km (810 mi) in diameter, is both filled with and surrounded by smooth lava plains. The basin was probably gouged out by the impact of a large me-

teoroid that penetrated the planet's crust, allowing lava to flood out onto the surface and fill the basin. Because relatively few craters pockmark these lava flows, the Caloris impact must be relatively young. It must have occurred toward the end of the crater-making period that dominated the first 700 million years of our solar system's history.

The impact that created the huge Caloris Basin must have been so violent that it shook the entire planet. On the side of Mercury opposite the Caloris Basin, *Mariner 10* discovered a consequence of this impact—a jumbled, hilly region covering about 500,000 square kilometers, about twice the size of the state of Wyoming. Geologists theorize that seismic waves from the Caloris impact became focused as they passed through Mercury. As this concentrated seismic energy reached the far surface of the planet, it deformed the surface and created jumbled terrain. (This same process may also have taken place on the Moon: There is a region of chaotic hills on the side of the Moon directly opposite from the large impact basin called Mare Orientale.)

In the decades since the *Mariner 10* mission, astronomers have continued to study Mercury with both visible-light and radio telescopes. Radar observations have detected patches at Mercury's north and south poles that are unusually effective at reflecting radio waves. It has been suggested that these patches may be regions of water ice deep within craters where the Sun's rays never reach. (We saw in Section 10-2 that ice may also exist in shadowed regions at the Moon's poles.) If this is correct, then Mercury is truly a world of extremes, with perpetual ice caps at one or both poles but with midday temperatures at the equator that are high enough to melt lead.

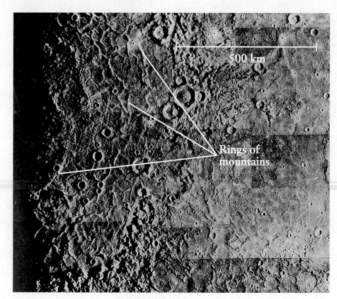

500 km

Rings of mountains

Figure 11-11   R I V U X G

The Caloris Basin An impact created this 1300-km (810-mi) diameter basin on Mercury about 4 billion years ago. Its center is hidden in the shadows just beyond the left side of the image. The impact fractured the surface extensively, forming several concentric chains of mountains. The mountains in the outermost ring are up to 2 km (6500 ft) high. (NASA)

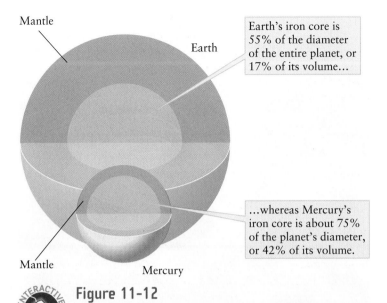

Mantle

Earth

Earth's iron core is 55% of the diameter of the entire planet, or 17% of its volume…

Mantle

Mercury

…whereas Mercury's iron core is about 75% of the planet's diameter, or 42% of its volume.

Figure 11-12

The Internal Structures of Mercury and the Earth
Compared to the Earth's core, Mercury's iron-rich core takes up a much larger fraction of the planet's volume. Indeed, Mercury is the most iron-rich planet in the solar system. Surrounding Mercury's core is a 600-km-thick rocky mantle.

Mercury's Interior and Iron Core

By measuring how much *Mariner 10* was deflected by Mercury's gravity, scientists were able to make a precise determination of Mercury's mass. Given the mass and the diameter of Mercury, they could then calculate that the planet has an average density of 5430 kg/m³. This is quite similar to the Earth's average density of 5515 kg/m³. We saw in Section 9-2 that typical rocks from the Earth's surface have a density of only about 3000 kg/m³ because they are composed primarily of lightweight, mineral-forming elements. The higher average density of our planet is

caused by the Earth's iron core. By studying how the Earth vibrates during earthquakes, geologists have deduced that our planet's iron core occupies about 17% of the Earth's volume.

Pressures inside a planet increase the density of rock by squeezing its atoms closer together. Because the Earth is 18 times more massive than Mercury, the weight of this larger mass compresses the Earth's core much more than Mercury's core. If the Earth were uncompressed, its density would be 4400 kg/m³, whereas Mercury's uncompressed density would still be a hefty 5300 kg/m³. This higher uncompressed density means that Mercury has a *larger* proportion of iron than the Earth: An iron core must occupy about 42% of Mercury's volume. Mercury is therefore the most iron-rich planet in the solar system. Figure 11-12 is a scale drawing of the interior structures of Mercury and the Earth.

Several theories have been proposed to account for Mercury's high iron content. According to one theory, the inner regions of the primordial solar nebula were so warm that only those substances with high condensation temperatures—like iron-rich minerals—could have condensed into solids. Another theory suggests that a brief episode of very powerful solar winds could have stripped Mercury of its low-density mantle shortly after the Sun formed. A third possibility is that during the final stages of planet formation, Mercury was struck by a large planetesimal. Supercomputer simulations show that this cataclysmic collision would have ejected much of the lighter mantle, leaving a disproportionate amount of iron to reaccumulate to form the planet we see today (Figure 11-13).

Clues About the Core: Mercury's Magnetic Field

An important clue to the structure of Mercury's iron core came from the *Mariner 10* magnetometers, which discovered that Mercury has a magnetic field similar to that of the Earth but only about 1% as strong. We saw in Section 7-7 that for a planet to have a global magnetic field, it must have moving liquid material in its interior that conducts electricity. This moving material acts

Proto-Mercury

Planetesimal ⅙ the mass of proto-Mercury

5000 km

Just before contact

Planetesimal plows into proto-Mercury's interior

5000 km

3 minutes after contact

Impact vaporizes the rocky mantles of both worlds

5000 km

6 minutes after contact

Mercury re-forms from iron-rich material within this banana-shaped region

Other material disperses into space

5000 km

30 minutes after contact

Figure 11-13

Stripping Mercury's Mantle by a Collision To account for Mercury's high iron content, one theory proposes that a collision with a planet-sized object stripped Mercury of most of its rocky mantle. These four images

show a supercomputer simulation of a head-on collision between proto-Mercury and a planetesimal one-sixth its mass. (Courtesy of W. Benz, A. G. W. Cameron, and W. Slattery) as a dynamo and generates the magnetic field. Hence, we conclude that (1) at least part of Mercury's core must be in a liquid state, and (2) there must be a source of energy to make material flow within this liquid region in order to produce electric currents. The second requirement is evidence that Mercury, like the Earth, also has a *solid* part of its core: As material deep in the liquid core cools and solidifies to join the solid portion of the core, it releases the energy needed to stir up the motions of the remaining liquid material. Mercury's magnetic field therefore suggests that its core has the same general structure as the Earth's core (recall Figure 9-9).

Mercury's magnetic field comes as something of a surprise. The ancient, cratered surface of Mercury is evidence of a lack of geologic activity, which in turn shows that Mercury must have lost the internal heat that powers such activity. This is as we would expect for such a small planet (see Section 7-6). But the existence of a global magnetic field shows that Mercury is still partially molten inside, which is evidence that the planet has some internal heat. The evidence from cratering and from the global magnetic field are thus in contradiction with each other.

Two other observations further complicate our understanding of Mercury's interior. One is that the amount of electrically conducting molten material that remains within Mercury is thought to be too little to account for even the relatively weak global magnetic field. Another is that a planet must rotate fairly rapidly in order to stir its molten interior into the kind of motion that generates a magnetic field. But as we saw in Section 11-2, Mercury rotates very *slowly*: It takes 58.646 days to spin on its axis. We are forced to conclude that we still have much to learn about the interior of Mercury.

Mercury's Magnetosphere

Whatever the origin of Mercury's magnetic field, we know that it interacts with the solar wind much as does the Earth's field, although on a much smaller scale (**Figure 11-14**). The solar wind is a constant flow of charged particles (mostly protons and electrons) away from the outer layers of the Sun's upper atmosphere. A planet's magnetic field can repel and deflect the impinging particles, thereby forming an elongated "cavity" in the solar wind called a magnetosphere. (We discussed the Earth's magnetosphere in Section 9-4.)

Charged-particle detectors aboard *Mariner 10* mapped the structure of Mercury's magnetosphere. When the particles in the solar wind first encounter Mercury's magnetic field, they are abruptly slowed, producing a shock wave that marks the boundary where this sudden decrease in velocity occurs. Most of the particles from the solar wind are deflected around the planet, just as water is deflected to either side of the bow of a ship. Mercury's magnetosphere thus prevents the solar wind from reaching the planet's surface. Because Mercury's magnetic field is much weaker than that of the Earth, it is not able to capture particles from the solar wind. Hence, it has no structures like the Van Allen belts that surround the Earth (recall Figure 9-20).

Future Exploration of Mercury

Much about Mercury still remains unknown. What is the character of the side of the planet that *Mariner 10* did not see? What is

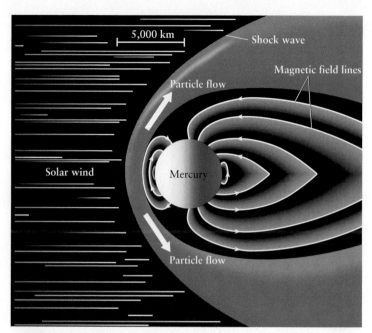

Figure 11-14

Mercury's Magnetosphere Mercury's weak magnetic field has only 1% the strength of the Earth's field. This is just strong enough to carve out a cavity in the solar wind, preventing the impinging particles from striking the planet's surface directly. Most of the particles of the solar wind are deflected around the planet in a turbulent region colored purple in this scale drawing. (Compare this with the Earth's much larger magnetosphere, shown in Figure 9-20.)

the chemical composition of the surface rocks? Was the surface of Mercury once completely molten? What caused Mercury to have such a high iron content? What portion of the planet's core is liquid, and what portion is solid? Is there indeed ice perpetually frozen at the poles? Learning the answers will require extensive observations of Mercury by a spacecraft at close range.

 One such spacecraft, *MESSENGER* (MErcury Surface, Space ENvironment, GEochemistry, and Ranging), was launched in 2004 and is scheduled to go into orbit around Mercury in 2011. Its sophisticated instruments will image the side of the planet that *Mariner 10* did not observe, make detailed measurements of surface composition, and help reveal the nature of Mercury's core. Another Mercury mission, *BepiColombo*, is being planned by the European Space Agency in collaboration with the Japanese Institute of Space and Astronautical Science. Named for the physicist who first proposed Mercury's 3-to-2 spin-orbit coupling, *BepiColombo* is actually two spacecraft: an orbiter that will study the planet itself and a smaller orbiter that will explore Mercury's magnetosphere. If launched as scheduled in 2013, *BepiColombo* will reach Mercury in 2019. Until this new generation of spacecraft begins returning new data, Mercury will remain one of the most enigmatic worlds of the solar system.

11-4 The first missions to Venus and Mars demolished decades of speculations about those planets

Prior to the 1960s, scientists developed a number of intriguing ideas about Venus and Mars. While these ideas were consistent with observations made using Earth-based telescopes, they needed to be radically revised with the advent of spacecraft data.

Speculations About Venus: A Lush Tropical Paradise?

Venus is bathed in more intense sunlight than Earth because it is closer to the Sun. If Venus had no atmosphere, and if its surface had an albedo like that of Mercury or the Moon, heating by the Sun would bring its surface temperature to around 45°C (113°F)—comparable

> **Venus was once thought to be steamy and tropical, and Mars was thought to have vegetation that varied with the seasons**

to that found in the hottest regions on Earth. What was unclear until the 1960s was how much effect Venus's atmosphere has on the planet's surface temperature.

We saw in Section 9-1 that gases such as water vapor (H_2O) and carbon dioxide (CO_2) in the Earth's atmosphere trap some of the infrared radiation that is emitted from our planet's surface. This elevates the temperatures of both the atmosphere and its surface, a phenomenon called the greenhouse effect. In 1932, Walter S. Adams and Theodore Dunham Jr. at the Mount Wilson Observatory found absorption lines of CO_2 in the spectrum of sunlight reflected from Venus, indicating that carbon dioxide is also present in the Venusian atmosphere. (Section 7-3 describes how astronomers use spectroscopy to determine the chemical compositions of atmospheres.) Thus, Venus's surface, like the Earth's, should be warmed by the greenhouse effect.

However, Venus's perpetual cloud cover reflects back into space a substantial amount of the solar energy reaching the planet. This effect by itself acts to cool Venus's surface, just as clouds in our atmosphere can make overcast days cooler than sunny days. Some scientists suspected that this cooling kept the greenhouse effect in check, leaving Venus with surface temperatures below the boiling point of water. Venus might then have warm oceans and possibly even life in the form of tropical vegetation. But if the greenhouse effect were strong enough, surface temperatures could be so high that any liquid water would boil away. Then Venus would have a dry, desertlike surface with no oceans, lakes, or rivers.

Close-Up Observations of Venus: Revealing a Broiled Planet

This scientific controversy was resolved in 1962 when the unmanned U.S. spacecraft *Mariner 2* (Figure 11-15) made the first close flyby of Venus. As we learned in Sections 5-3 and 5-4, a dense object (like a planet's surface) emits radiation whose intensity and spectrum depend on the temperature of the object. *Mariner 2* carried instruments that measured radiation coming from the planet at two microwave wavelengths, 1.35 cm and 1.9 cm. Venus's atmosphere is transparent to both these wavelengths, so *Mariner 2* saw radiation that had been emitted by the

Figure 11-15

The *Mariner 2* Spacecraft *Mariner 2* was the first of all spacecraft from Earth to make a successful flyby of another planet. After coming within 34,773 km of Venus on December 14, 1962, it went into a permanent orbit around the Sun. (GSFC/NASA)

WEB LINK 11.13

planet's surface and then passed through the atmosphere like visible light through glass.

From the amount of radiation that *Mariner 2* detected at these two wavelengths, astronomers concluded that the surface temperature of Venus was more than 400°C—well above the boiling point of water and higher than even daytime temperatures on Mercury. Thus, there cannot be any liquid water on the planet's surface. Water vapor absorbs microwaves at 1.35 cm, so if there were substantial amounts of water vapor in Venus's atmosphere, it would have blocked this wavelength from reaching the detectors on board *Mariner 2*. In fact, the spacecraft did detect strong 1.35-cm emissions from Venus. Thus, the atmosphere, too, must be all but devoid of water.

This picture of Venus as a dry, hellishly hot world was confirmed in 1970 by the Soviet spacecraft *Venera 7*, which survived a descent through the Venusian atmosphere and managed to transmit data for a few seconds directly from the planet's surface. *Venera 7* and other successful Soviet landers during the early 1970s finally determined the surface temperature to be a nearly constant 460°C (860°F). Thus, Venus has an extraordinarily strong greenhouse effect, and the planet's surface is no one's idea of a tropical paradise.

Speculations About Mars: A World With Plant Life and Canals?

As we saw in Section 11-2, Mars's axis of rotation is tilted by almost the same angle as the Earth's axis, and so Mars goes through seasons much like those on Earth. Observations with Earth-based telescopes show that the colors of Mars (see Figure 11-4c) vary with the seasons: During spring and summer in a Martian hemisphere, the polar cap in that hemisphere shrinks and the dark markings (which sometimes look greenish) become very distinct. Half a Martian year later, with the approach of fall and winter, the dark markings fade and the polar cap grows. Could these seasonal variations mean that there is vegetation on Mars that changes seasonally, just as on the Earth? And if there is plant life on Mars, might there not also be intelligent beings?

Such speculations were fueled by the Italian astronomer Giovanni Schiaparelli, who observed Mars during the favorable opposition of 1877. He reported that during moments of "good seeing," when the air is exceptionally calm and clear, he was able to see 40 straight-line features crisscrossing the Martian surface (Figure 11-16). He called these dark linear features *canali*, an Italian word for "channels," which was soon mistranslated into English as "canals." The alleged discovery of canals implied that there were intelligent creatures on Mars capable of substantial engineering feats. This speculation helped motivate the American millionaire Percival Lowell to finance a major new observatory near Flagstaff, Arizona, primarily to study Mars. By the end of

the nineteenth century, Lowell had reported observations of 160 Martian canals.

Although many astronomers observed Mars, not all saw the canals. In 1894, the American astronomer Edward Barnard, working at Lick Observatory in California, complained that "to save my soul I can't believe in the canals as Schiaparelli draws them." But objections by Barnard and others did little to sway the proponents of the canals.

As the nineteenth century drew to a close, speculation about Mars grew more and more fanciful. Perhaps the reddish-brown color of the planet meant that Mars was a desert world, and perhaps the Martian canals were an enormous planetwide irrigation network. From these ideas, it was a small leap to envision Mars as a dying world with canals carrying scarce water from melting polar caps to farmlands near the equator. The terrible plight of the Martian race formed the basis of inventive science fiction by Edgar Rice Burroughs, Ray Bradbury, and many others. It also led to the notion that the Martians might be a warlike race who schemed to invade the Earth for its abundant resources. (In Roman mythology, Mars was the god of war.) Stories of alien invasions, from H. G. Wells's 1898 novel *The War of the Worlds* down to the present day, all owe their existence to the *canali* of Schiaparelli.

Observations of Mars: Revealing a Windblown, Cratered World

However, during the first decades of the twentieth century the Greek astronomer Eugène Antoniadi observed Mars with a state-of-the-art telescope and found that the "canals" were actually unconnected dark spots. There was no evidence whatsoever of linear canals. Why, then, were Schiaparelli, Lowell, and others so certain that the canals were there? One reason is that their observations were made from the Earth's surface, at the bottom of an often turbulent atmosphere that blurs images seen through a telescope. A second reason is that the human eye and brain can readily be deceived. The brain can connect together two dark streaks on the Martian surface to give the perception of a single "canal." In this way did astronomers such as Schiaparelli and Lowell delude themselves.

The other speculation, that Mars was covered with vegetation, was conclusively ruled out several decades later. In the 1960s, three American spacecraft flew past Mars and sent back the first close-up pictures of the planet's surface. (Figure 11-17 shows even better images from a subsequent mission to Mars.) These images showed that the dark surface markings are just different-colored terrain.

Why did many Earth-based observers report that dark areas on Mars had the greenish hue of vegetation? The probable explanation is a quirk of human color vision. When a neutral gray area is viewed next to a bright red or brown area, the eye perceives the gray to have a blue-green color.

The seasonal variations in color can be explained by winds in the Martian atmosphere that blow in different directions in different seasons. These seasonal winds blow fine dust across the Martian surface, producing planetwide dust storms that can be seen from Earth (Figure 11-18). The motion of dust covers some parts of the Martian terrain and exposes others. Thus, Earth observers see large areas of the planet vary from dark to light with

Schiaparelli's imaginary "canals"

Figure 11-16

The Mirage of the Martian Canals Giovanni Schiaparelli studied the Martian surface using a 20-cm (8-in.) telescope, the same size used by many amateur astronomers today. He recorded his observations in drawings like this one, which shows a network of linear features that Percival Lowell and others interpreted as irrigation canals. Later observations with larger telescopes revealed that the "canals" were mere illusions. (Michael Hoskin, ed., *The Cambridge Illustrated History of Astronomy*, Cambridge University Press, 1997, p. 286. Illustration by G. V. Schiaparelli. Courtesy Institute of Astronomy, University of Cambridge, UK)

(a) Mars from the Hubble Space Telescope

(b) Closeup of Sinus Sabeus region

Figure 11-17   R I **V** U X G

Martian Craters (a) This image, made during the favorable opposition of 2003, shows cratered regions on the side of Mars opposite to that in the image that opens this chapter. Arabia Terra is dotted with numerous flat-bottom craters. Syrtis Major was first identified by Christiaan Huygens in 1659. A single titanic impact carved out Hellas Planitia, which is 5 times

the size of Texas. **(b)** This mosaic of images from the *Viking Orbiter 1* and *2* spacecraft shows an extensively cratered region located south of the Martian equator. (a: NASA; J. Bell, Cornell University; and M. Wolff, SSI; b: USGS)

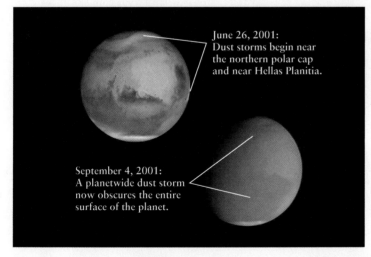

Figure 11-18   R I **V** U X G

A Martian Dust Storm Dust storms on Mars can expand to cover the entire planet. These Hubble Space Telescope images show a particularly intense storm that took place in 2001, when it was spring in the Martian northern hemisphere and autumn in the southern hemisphere. (NASA; J. Bell, Cornell University; M. Wolff, SSI; and the Hubble Heritage Team, STScI/AURA)

the passage of the seasons, mimicking the color changes that would be expected from vegetation.

The most remarkable discovery of the missions to Mars in the 1960s was that the planet's surface is pockmarked with craters, as Figure 11-17*b* and Figure 7-10*c* show. (Although many of the Martian craters are quite large, they had escaped detection from Earth because the thin Martian atmosphere obscures the planet's surface somewhat. That is why even Hubble Space Telescope images like Figure 11-17*a* do not show Martian craters distinctly, and why it took close-up images made by spacecraft near Mars to reveal these craters.) During that same decade of the 1960s scientists were learning that most of the craters on the Moon were formed by interplanetary debris that struck the lunar surface more than 3 billion years ago. The Martian craters, too, are the result of long-ago impacts from space. Since so many craters survive to the present day, at least part of the Martian surface must be extremely ancient (see Section 7-6).

The episode of the Martian canals and Martian vegetation was not the first time that a number of scientists came to very wrong conclusions. Like any human activity, science will always involve human error and human frailties. But scientific investigation is fundamentally self-correcting, because new observations and experiments can always call into question the scientific theories of the past.

11-5 Both Venus and Mars have volcanoes and craters—but no Earthlike plate tectonics

From a distance, Venus and Mars appear radically different: Venus is nearly the size of Earth and has a thick atmosphere, while Mars is much smaller and has only a thin atmosphere. Yet spacecraft observations reveal that these two worlds have many surface features in common, including volcanoes and impact craters. By comparing the surfaces of these two worlds to the Earth we can gain insight into the similarities and differences between the three largest terrestrial planets.

> **Venus has no plate tectonics because its crust is too thin, while on Mars the crust is too thick**

Observing Venus and Mars from Orbit

To make a detailed study of a planet's surface, a spacecraft that simply flies by the planet will not suffice. Instead, it is necessary to place a spacecraft in orbit around the planet. Since the 1970s a number of spacecraft have been placed in Venus orbit and Mars orbit.

In order to map the surface of Venus through the perpetual cloud layer, several of the Venus orbiters carried radar devices. A beam of microwave radiation from the orbiter easily penetrates Venus's clouds and reflects off the planet's surface; a receiver on the orbiter then detects the reflected beam. Different types of terrain reflect microwaves more or less efficiently, so astronomers are able to construct a map of the Venusian surface by analyzing the reflected radiation.

 Magellan, the most recent spacecraft to orbit Venus, also carried a radar altimeter that bounced microwaves directly off the ground directly below the spacecraft. By measuring the time delay of the reflected waves, astronomers can determine the height and depth of Venus's terrain.

By contrast, a Mars orbiter can use more conventional telescopes to view the Martian surface through that planet's thin atmosphere. Instead of using a radar altimeter to map surface elevations, the *Mars Global Surveyor* spacecraft (which entered Mars's orbit in 1997) used a laser beam for the same purpose.

The Topography of Venus and Mars

Figure 11-19 and **Figure 11-20** show topographic maps of Venus and Mars derived using radar and laser altimeters respectively. (Note that the scale of these two maps is different: The diameter of Mars is only 56% as large as the diameter of Mars.) The topographies of both worlds differ in important ways from that of our Earth. Our planet has two broad classes of terrain: About 71% of the surface is oceanic crust and about 27% is continental crust that rises above the ocean floors by about 4–6 km on average. On Venus, by contrast, about 60% of the terrain lies within

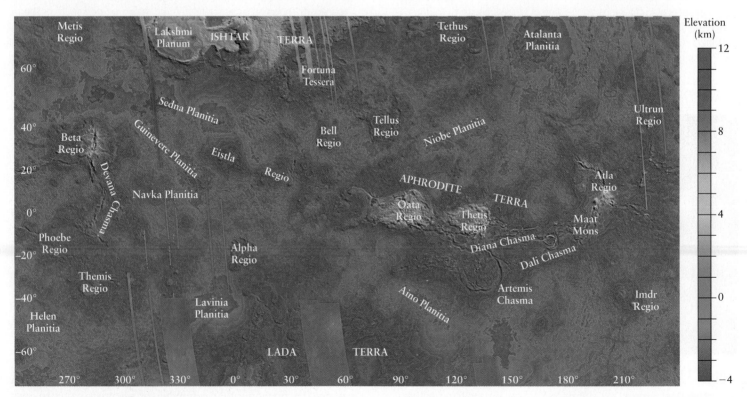

Figure 11-19   R I V U X G

A Topographic Map of Venus Radar altimeter measurements by *Magellan* were used to produce this topographic map of Venus. Color indicates elevations above (positive numbers) or below (negative numbers) the planet's average radius. (The blue areas are *not* oceans!)

Gray areas were not mapped by *Magellan.* Flat plains of volcanic origin cover most of the planet's surface, with only a few continentlike highlands. (Peter Ford, MIT; NASA/JPL)

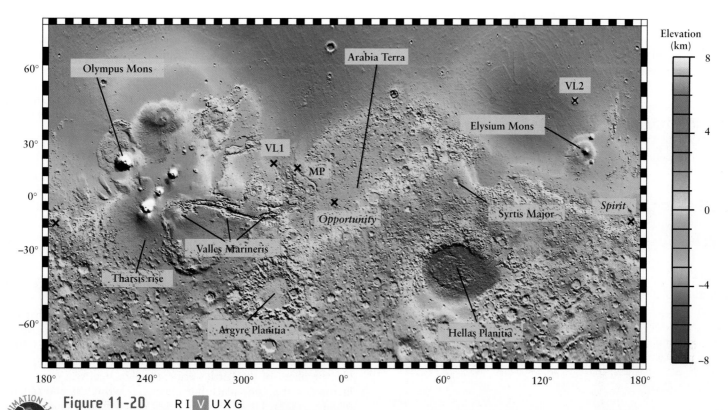

ANIMATION 11.7

Figure 11-20   R I V U X G

A Topographic Map of Mars This map was generated from measurements made by the laser altimeter on board the *Mars Global Surveyor* spacecraft. As in Figure 11-19, color indicates elevations above or below the planet's average radius. Most of the southern hemisphere lies several kilometers above the northern hemisphere, with the exception of the immense impact feature called Hellas Planitia. The landing sites for *Viking Landers 1* and *2* (VL1 and VL2), *Mars Pathfinder* (MP), and the *Mars Exploration Rovers* (*Spirit* and *Opportunity*) are each marked with an X. (MOLA Science Team, NASA/GSFC)

500 m of the average elevation, with only a few localized highlands (shown in yellow and red in Figure 11-19). Mars is different from both Earth and Venus: Rather than having continents scattered among low-lying ocean floors, all of the high terrain on Mars (shown in red and orange in Figure 11-20) is in the southern hemisphere. Hence, planetary scientists refer to Mars as having **northern lowlands** and **southern highlands.** The implication is that the Martian crust is about 5 km thicker in the southern hemisphere than in the northern hemisphere, a situation called the **crustal dichotomy.** In this sense Mars resembles the Moon, which has a thicker crust on the far side than on the side that faces the Earth (see Section 10-1).

Tectonics on Venus: A Light, Flaky Crust

Before radar maps like Figure 11-19 were available, scientists wondered whether Venus had plate tectonics like those that have remolded the face of the Earth. Venus is only slightly smaller than the Earth, and should have retained enough heat to sustain a molten interior and the convection currents that drive tectonic activity on Earth (see Section 9-3, especially Figure 9-16). If this were the case, then the same tectonic effects might also have shaped the surface of Venus. As we saw in Section 9-3, the Earth's hard outer shell, or lithosphere, is broken into about a dozen large plates that slowly shuffle across the globe. Long mountain ranges, like the Mid-Atlantic Ridge (see Figure 9-13), are created where fresh magma wells up from the Earth's interior to push the plates apart.

Radar images from *Magellan* show *no* evidence of Earthlike tectonics on Venus. On Earth, long chains of volcanic mountains (like the Cascades in North America or the Andes in South America) form along plate boundaries where subduction is taking place. Mountainous features on Venus, by contrast, do not appear in chains. There are also no structures like the Earth's Mid-Atlantic Ridge, which suggests that there is no seafloor spreading on Venus. With no subduction or seafloor spreading, there have been only limited horizontal displacement of Venus's lithosphere. Thus, like the Moon (see Section 10-1) and Mercury (see Section 11-3), Venus is a one-plate planet.

Unlike the Moon, however, Venus has had local, small-scale deformations and reshaping of the surface. One piece of evidence for this is that roughly a fifth of Venus's surface is covered by folded and faulted ridges. Further evidence comes from close-up *Magellan* images that show that Venus has about a thousand craters larger than a few kilometers in diameter, many more than have been found on Earth but only a small fraction of the number on the Moon or Mercury. We saw in Section 7-6 that the number of impact craters is a clue to the age of a planet's surface. Such craters formed at a rapid rate during the early history of the solar system, when considerable interplanetary debris still orbited

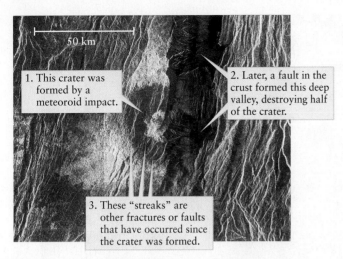

Figure 11-21   R I V U X G

A Partially Obliterated Crater on Venus This *Magellan* image shows how the right half of an old impact crater 37 km (20 mi) in diameter was erased by a fault in the crust. This crater lies in Beta Regio (see the left side of Figure 11-19). Most features on Venus are named for women in history and legend; this crater commemorates Emily Greene Balch, an American economist and sociologist who won the 1946 Nobel Peace Prize. (NASA/JPL)

the Sun, and have formed at a much slower rate since then. Consequently, the more craters a planet has, the older its surface. The number of craters on Venus indicates that the Venusian surface is roughly 500 million years old. This is about twice the age of the Earth's surface but much younger than the surfaces of the Moon or Mercury, each of which is billions of years old. No doubt Venus was more heavily cratered in its youth, but localized activity in its crust has erased the older craters (Figure 11-21).

Surprisingly, Venus's craters are uniformly scattered across the planet's surface. We would expect that older regions on the surface—which have been exposed to bombardment for a longer time—would be more heavily cratered, while younger regions would be relatively free of craters. For example, the ancient highlands on the Moon are much more heavily cratered than the younger maria (see Section 10-4). Because such variations are not found on Venus, scientists conclude that the entire surface of the planet has essentially the *same* age. This is very different from the Earth, where geological formations of widely different ages can be found.

One model that can explain these features suggests that the convection currents in Venus's interior are actually more vigorous than inside the Earth, but that the Venusian crust is much thinner than the continental crust on Earth. Rather than sliding around like the plates of the Earth's crust, the thin Venusian crust stays in roughly the same place but undergoes wrinkling and flaking (Figure 11-22). Hence, this model is called **flake tectonics.** The Earth, too, may have displayed flake tectonics billions of years ago when its interior was hotter.

Figure 11-22

Plate Tectonics Versus Flake Tectonics This illustration shows the difference between plate tectonics on Earth and the model of flake tectonics on Venus. (Adapted from T. Grotzinger, T. H. Jordan, F. Press, and R. Siever, *Understanding Earth,* 5th ed., W. H. Freeman, 2007)

Although Venus almost certainly has molten material in its interior, it has no planetwide magnetic field. As we discussed in Section 7-7, this may be because Venus rotates too slowly to generate the kind of internal motions that would produce a magnetic field. With no magnetic field, Venus has no magnetosphere. The planet is nonetheless shielded from the solar wind by ions in its upper atmosphere.

Tectonics on Mars: A Thick, Rigid Crust

Like Venus, Mars lacks the global network of ridges and subduction zones that plate tectonics has produced on the Earth. Hence, the entire crust of Mars makes up a single tectonic plate, as is the case on the Moon, Mercury, and Venus. The explanation is quite different from that for Venus, however: Because Mars is a much smaller world than Earth or Venus, the outer layers of the red planet have cooled more extensively. Thus, Mars lacks plate tectonics because its crust is too thick for one part of the crust to be subducted beneath another. We see that for a terrestrial planet to have plate tectonics, the crust must not be too thin (like Venus) or too thick (like Mars), but just right (like Earth).

This idea has been verified by carefully monitoring the motion of a spacecraft orbiting Mars. If there is a concentration of mass (such as a thicker crust) at one location on the planet, gravitational attraction will make the spacecraft speed up as it ap-

proaches the concentration and slow down as it moves away. A team of scientists headed by David Smith of NASA's Goddard Space Flight Center and Maria Zuber of MIT analyzed the orbit of *Mars Global Surveyor* in just this way. They found that unlike the Earth's crust, which varies in thickness from 5 to 35 km, the Martian crust is about 40 km thick under the northern lowlands but about 70 km thick under the southern highlands. Both regions of the Martian crust are too thick to undergo subduction, making plate tectonics impossible.

Although there is no plate motion on Mars, there are features on the planet's surface that indicate there has been substantial motion of material in the Martian mantle. The Tharsis rise, visible on the left-hand side of Figure 11-20, is a dome-shaped bulge that has been lifted 5 to 6 km above the planet's average elevation. Apparently, a massive plume of magma once welled upward from a hot spot under this region. East of the Tharsis rise, a vast chasm runs roughly parallel to the Martian equator (**Figure 11-23**). If this canyon were located on Earth, it would stretch from Los Angeles to New York. In honor of the *Mariner 9* spacecraft that first revealed its presence in 1971, this chasm has been named Valles Marineris.

Valles Marineris has heavily fractured terrain at its western end near the Tharsis rise. At its eastern end, by contrast, it is dominated by ancient cratered terrain. Many geologists suspect that Valles Marineris was caused by the same upwelling of material

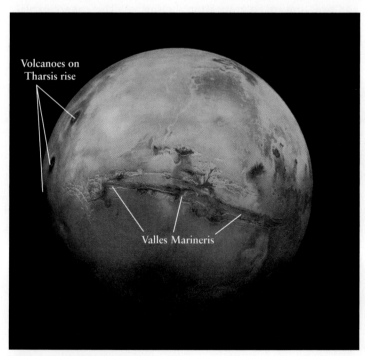

(a) Mars and Valles Marineris

(b) The central region of Valles Marineris

Figure 11-23   R I [V] U X G

Valles Marineris (a) This mosaic of *Viking Orbiter* images shows the huge rift valley of Valles Marineris, which extends from west to east for more than 4000 km (2500 mi) and is 600 km (400 mi) wide at its center. Its deepest part is 8 km (5 mi) beneath the surrounding plateau. At its western end is the

Tharsis rise. Compare this image with Figure 11-20. **(b)** This perspective image from the *Mars Express* spacecraft shows what you would see from a point high above the central part of Valles Marineris. (a: USGS/NASA; b: ESA/DLR/FU Berlin, G. Neukum)

that formed the Tharsis rise. As the Martian surface bulged upward at Tharsis, there would have been tremendous stresses on the crust, which would have caused extensive fracturing. Thus, Valles Marineris may be a **rift valley,** a feature created when a planet's crust breaks apart along a fault line. Rift valleys are found on Earth; two examples are the Red Sea (see Figure 9-17) and the Rhine River valley in Europe. Other, smaller rifts in the Martian crust are found all around the Tharsis rise. All of these features are very old, however, and it is thought that there has been little geologic activity on Mars for billions of years.

Additional evidence that may point to ancient geologic activity on Mars is the crustal dichotomy between the northern lowlands and the southern highlands. The southern highlands are much more heavily cratered than either Earth or Venus, though less so than the Moon, while the northern lowlands are remarkably smooth and free of craters. (You can see this difference in Figure 11-20.) Since most cratering occurred early in the history of the solar system, this implies that the northern lowlands are relatively young, while the more heavily cratered southern highlands are relatively old. One proposed explanation is that older craters in the northern lowlands were erased by tectonic activity that took place long ago when the Martian crust had not yet cooled to its present thickness. Other hypotheses are that older craters could have been erased by a giant asteroid impact or by lava flows.

A final piece of evidence about ancient Mars is the existence of locally magnetized areas in the old southern highlands. These areas presumably solidified during an era in which the interior of Mars was much hotter, and moving, molten material in the interior developed a magnetic field. Today, by contrast, Mars has no planetwide magnetic field. There is probably still molten material in the core, but the Martian core is thought to contain substantial amounts of sulfur in addition to iron. While temperatures in the core should be high enough to melt sulfur and sulfur compounds, the electrical properties of these substances are such that their motions do not generate the electric currents needed to produce a magnetic field.

Volcanoes on Venus and Mars

Radar images of Venus and visible-light images of Mars show that both planets have a number of large volcanoes (**Figure 11-24**). *Magellan* observed more than 1600 major volcanoes and volcanic features on Venus, two of which are shown in Figure 11-24a. Both of these volcanoes have gently sloping sides. A volcano with this characteristic is called a **shield volcano,** because in profile it resembles an ancient Greek warrior's shield lying on the ground. Martian volcanoes are less numerous than those on Venus, but they are also shield volcanoes; the largest of these, Olympus Mons, is the largest volcano in the solar system (Figure 11-24b). Olympus Mons rises 24 km (15 mi) above the surrounding plains. By comparison, the highest volcano on Earth, Mauna Loa in the Hawaiian Islands, has a summit only 8 km (5 mi) above the ocean floor.

Most volcanoes on Earth are found near the boundaries of tectonic plates, where subducted material becomes molten magma and rises upward to erupt from the surface. This cannot explain the volcanoes of Venus and Mars, since there are no subduction

(a) Volcanoes and lava flows on Venus   R I V U X G

(b) Cloud-topped volcanoes on Mars   R I V U X G

Figure 11-24

Volcanoes on Venus and Mars (a) The false color in this computer-generated perspective view suggests the actual color of sunlight that penetrates Venus's thick clouds. The brighter color of the extensive lava flows indicates that they reflect radio waves more strongly. To emphasize the gently sloping volcanoes, the vertical scale has been exaggerated 10 times. **(b)** The volcanoes of Mars also have gently sloping sides. In this view looking down from Mars orbit you can see bluish clouds topping the summits of the volcanoes. These clouds, made of water ice crystals, form on most Martian afternoons. (a: NASA, JPL Multimission Image Processing Laboratory; b: NASA/JPL/Malin Space Science Systems)

Figure 11-25

Hot-Spot Volcanoes on Earth A hot spot under the Pacific plate has remained essentially stationary for 70 million years while the plate has moved some 6000 km to the northwest. The upwelling magma has thus produced a long chain of volcanoes. The Hawaiian Islands are the newest of these; the oldest, the Emperor Seamount Chain, have eroded so much that they no longer protrude above the ocean surface.

zones on those planets. Instead, Venusian and Martian volcanoes probably formed by **hot-spot volcanism.** In this process, magma wells upward from a hot spot in a planet's mantle, elevating the overlying surface and producing a shield volcano.

On Earth, hot-spot volcanism is the origin of the Hawaiian Islands. These islands are part of a long chain of shield volcanoes that formed in the middle of the Pacific tectonic plate as that plate moved over a long-lived hot spot (**Figure 11-25**). On Venus and Mars, by contrast, the absence of plate tectonics means that the crust remains stationary over a hot spot. On Mars, a single hot spot under Olympus Mons probably pumped magma upward through the same vent for millions of years, producing one giant volcano rather than a long chain of smaller ones. The Tharsis rise and its volcanoes (see Figure 11-20, Figure 11-23a, and Figure 11-24b) may have formed from the same hot spot as gave rise to Olympus Mons; a different hot spot on the opposite side of Mars produced a smaller bulge centered on the volcano Elysium Mons, shown near the right-hand side in Figure 11-20. The same process of hot-spot volcanism presumably gave rise to large shield volcanoes on Venus like those shown in Figure 11-24a.

Volcanic Activity on Venus

About 80% of the surface of Venus is composed of flat plains of volcanic origin. In other words, essentially the entire planet is covered with lava! This observation shows the tremendous importance of volcanic activity in Venusian geology.

To verify the volcanic nature of the Venusian surface, it is necessary to visit the surface and examine rock samples. **Figure 11-26** is a panoramic view taken in 1981 by the Soviet spacecraft *Venera 13*, one of 10 unmanned spacecraft that the Soviet Union landed successfully on the surface of Venus. Russian scientists hypothesize that this region was covered with a thin layer of lava that fractured upon cooling to create the rounded, interlocking shapes seen in the photograph. This hypothesis agrees with information obtained from chemical analyses of surface material made by the spacecraft's instruments. These analyses indicate that the surface composition is similar to lava rocks called basalt, which are common on the Earth (see Figure 9-19a) and in the maria of the Moon (see Figure 10-13). The results from *Venera 13* and other landers are consistent with the picture of Venus as a world whose surface and atmosphere have been shaped by volcanic activity.

Most of the volcanoes on Venus are probably inactive at present, just as is the case with most volcanoes on Earth. But *Magellan* found evidence of recent volcanic activity, some of which may be continuing today. The key to estimating the amount of recent volcanic activity is that the radar reflectivity of volcanic materials depends on whether the material is relatively fresh or relatively old. By mapping these reflectivity variations on Venus, *Magellan* found many areas with young lava flows. Some of the youngest material found by *Magellan* caps the volcano Maat Mons (see Figure 11-24a). Geologists estimate that the topmost material is no more than 10 million years old and could be much younger. The presence of such young lava flows suggests that Venus, like the Earth, has some present-day volcanic activity.

Another piece of evidence for ongoing volcanic activity on Venus comes from the planet's atmosphere. An erupting volcano on Earth, like the one shown in Figure 9-15, ejects substantial amounts of sulfur dioxide, sulfuric acid, and other sulfur compounds into the air. Many of these substances are highly reactive and short-lived, forming sulfate compounds that become part of the planet's surface rocks. For these substances to be relatively abundant in a planet's atmosphere, they must be constantly replenished by new eruptions. Sulfur compounds make up about 0.015% of the Venusian atmosphere, compared to less than 0.0001% of Earth's atmosphere. This suggests that ongoing volcanic eruptions on Venus are ejecting sulfur compounds into the atmosphere to sustain the high sulfur content.

Volcanic Activity on Mars

Spectroscopic observations from Mars's orbit confirm that the planet's rocks and sands are made almost entirely of the three minerals feldspar, pyroxene, and olivine. These are the components of basalt, or solidified lava. Thus, Mars, like Venus, had a volcanic past.

Unlike lava flows on Venus, however, most of the lava flows on Mars have impact craters on them. The interpretation is that most Martian lava flows are very old and that most of the volcanoes on the red planet are no longer active. This is what we would expect from a small planet whose crust has cooled and solidified to a greater depth than on Earth, making

(a) Image from *Venera 13*

(b) Color-corrected image

Figure 11-26   R I $\boxed{\text{V}}$ U X G

A Venusian Landscape (a) This wide-angle color photograph from *Venera 13* shows the rocky surface of Venus. The thick atmosphere absorbs the blue component of sunlight, giving the image an orange tint: The stripes on the spacecraft's color calibration bar that appear yellow are actually white in color. **(b)** This color-corrected view shows that the rocks are actually gray in color. The rocky plates covering the ground may be fractured segments of a thin layer of lava. (Courtesy of C. M. Pieters and the Russian Academy of Sciences)

it difficult for magma to travel from the Martian mantle to the surface.

However, a few Martian lava flows are crater-free, which suggests that they are only a few million years old. If volcanoes erupted on Mars within the past few million years, are they erupting now? Scientists have used infrared telescopes to search for telltale hot spots on the Martian surface, but have yet to discover any. Perhaps volcanism on Mars is rare but not yet wholly extinct.

Spacecraft that have landed on Mars have taught us a great deal about the geology and history of the red planet. Before we explore the Martian surface in detail, however, it is useful to examine the unique atmospheres of both Mars and Venus.

11-6 The dense atmosphere of Venus and the thin Martian atmosphere are dramatically different but have similar chemical compositions

We have seen that the three large terrestrial planets—Earth, Venus, and Mars—display very different styles of geology. As we will discover, these differences help to explain why each of these three worlds has a distinctively different atmosphere.

To explore a planet's atmosphere it is necessary to send a spacecraft to make measurements as it descends through that atmosphere. Both Soviet and American spacecraft have done this

for Venus (see Figure 11-26), and five U.S. spacecraft have successfully landed on Mars. The results of these missions are summarized in **Figure 11-27**, which shows how pressure and temperature vary with altitude in the atmospheres of Earth, Venus, and Mars.

> On Venus, sulfuric acid rain evaporates before reaching the ground; on Mars, it snows frozen carbon dioxide

The Venusian Atmosphere: Dense, Hot, and Corrosive

Figure 11-27*b* depicts just how dense the Venusian atmosphere is. At the surface, the pressure is 90 atmospheres—that is, 90 times greater than the average air pressure at sea level on Earth. This is about the same as the water pressure at a depth of about 1 kilometer below the surface of the Earth's oceans. The density of the atmosphere at the surface of Venus is likewise high, more than 50 times greater than the sea-level density of our atmosphere. The atmosphere is so massive that once heated by the Sun, it retains its heat throughout the long Venusian night. As a result, temperatures on the day and night sides of Venus are almost identical.

The composition of atmosphere explains why Venus's surface temperatures are so high: 96.5% of the molecules in Venus's atmosphere are carbon dioxide, with nitrogen (N_2) making up most of the remaining 3.5% (see Table 9-4 for a comparison of the atmospheres of Venus, Earth, and Mars). Because Venus's atmosphere is so thick, and because most of the atmosphere is the

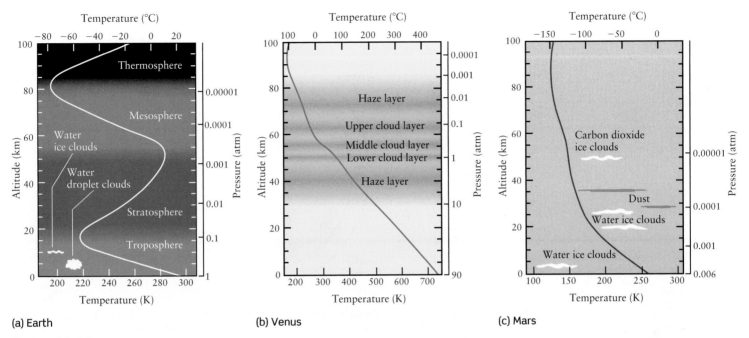

(a) Earth   (b) Venus   (c) Mars

Figure 11-27

Atmospheres of the Terrestrial Planets In each graph the curve shows how temperature varies with altitude from 0 to 100 km above the planet's surface. The scale on the right-hand side of each graph shows how pressure varies with altitude. **(a)** Clouds in the Earth's atmosphere are seldom found above 12 km (40,000 ft). **(b)** Venus's perpetual cloud layers lie at much higher altitudes. The atmosphere is so dense that the pressure 50 km above the Venusian surface is 1 atm, the same as at sea level on Earth. **(c)** By contrast, the Martian atmosphere is so thin that the surface pressure is the same as the pressure at an altitude of 35 km on Earth. Wispy clouds can be found at extreme altitudes.

greenhouse gas CO_2, the greenhouse effect (Section 9-1) has run wild. On Earth the greenhouse gases H_2O and CO_2 together make up only about 1% of our relatively sparse atmosphere, and the greenhouse effect has elevated the surface temperature by about 33°C (59°F). But on Venus, the dense shroud of CO_2 traps infrared radiation from the surface so effectively that it has raised the surface temperature by more than 400°C (720°F).

Soviet and American spacecraft also discovered that Venus's clouds are primarily confined to three high-altitude layers. An upper cloud layer lies at altitudes between 68 and 58 km, a denser and more opaque cloud layer from 58 to 52 km, and an even more dense and opaque layer between 52 and 48 km. Above and below the clouds are 20-km-thick layers of haze. Below the lowest haze layer, the atmosphere is remarkably clear all the way down to the surface of Venus.

We saw in Section 11-5 that sulfur, which is not found in any appreciable amount in our atmosphere, plays an important role in the Venusian atmosphere. Sulfur combines with other elements to form gases such as sulfur dioxide (SO_2) and hydrogen sulfide (H_2S), along with sulfuric acid (H_2SO_4), the same acid used in automobile batteries. While the Earth's clouds are composed of water droplets, Venusian clouds contain almost no water. Instead, they are composed of droplets of concentrated, corrosive sulfuric acid. Thanks to the high temperatures on Venus, these droplets never rain down on the planet's surface; they simply evaporate at high altitude. (You can see a similar effect on Earth. On a hot day in the desert of the American Southwest, streamers of rain called *virga* appear out of the bottoms of clouds but evaporate before reaching the ground.)

Convection in the Venusian Atmosphere

We saw in Section 9-6 that the Earth's atmosphere is in continuous motion: Hot gases rise and cooler gases sink, forming convection cells in the atmosphere. The same is true on Venus, but with some important differences. On Venus, gases in the equatorial regions are warmed by the Sun, then rise upward and travel in the upper cloud layer toward the cooler polar regions. At the polar latitudes, the cooled gases sink to the lower cloud layer, in which they are transported back toward the equator. The result is two huge convection cells, one in the northern hemisphere and another in the southern hemisphere (**Figure 11-28a**). These cells are almost entirely contained within the main cloud layers shown in Figure 11-27b. The circulation is so effective at transporting heat around Venus's atmosphere that there is almost no temperature difference between the planet's equator and its poles. By contrast, the Earth's atmosphere has a more complicated convection pattern (see Figure 11-28b and Figure 9-24) because the Earth's rotation—which is far more rapid than Venus's—distorts the convection cells.

In addition to the north-south motions due to convection, high-altitude winds with speeds of 350 km/h (220 mi/h) blow from east to west, the same direction as Venus's retrograde rotation. Thus, the upper atmosphere rotates around the planet once every 4 days. The fast-moving winds stretch out the convection cells and produce V-shaped, chevronlike patterns in the clouds (**Figure 11-29**). In a similar way, the Earth's rotation stretches out the convection cells in our atmosphere to produce the complicated circulation patterns shown in Figure 9-24.

Figure 11-28

Atmospheric Circulation on Venus and Earth
Solar heating causes convection in the atmospheres of both Venus and Earth, with warm air rising at the equator and cold air descending at the poles. **(a)** Because Venus rotates very slowly, it has little effect on the circulation. **(b)** Earth's rapid rotation distorts the atmospheric circulation into a more complex pattern (compare Figure 9-24).

(a) Atmospheric circulation on Venus:
two convection cells

(b) Atmospheric circulation on Earth:
six convection cells

On the Earth, friction between the atmosphere and the ground causes wind speeds at the surface to be much less than at high altitude. The same is true on Venus: The greatest wind speed measured by spacecraft on Venus's surface is only about 5 km/h (3 mi/h). Thus, only slight breezes disturb the crushing pressures and infernal temperatures found on Venus's dry, lifeless surface.

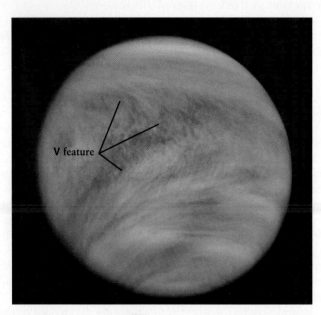

Figure 11-29   R I V U X G

Venus's Cloud Patterns This false-color image of Venus was made at ultraviolet wavelengths, at which the planet's atmospheric markings stand out best. The dark V feature is produced by the rapid motion of clouds around the planet's equator. The clouds move in the same retrograde direction as the rotation of the planet itself but at a much greater speed. (GSFC/NASA)

The Martian Atmosphere: Not a Drop to Drink

As Figure 11-27c shows, the atmosphere of Mars is very cold and thin compared to the atmosphere of either Earth or Venus: The surface pressure is a mere 0.006 atmosphere. Yet its chemical composition is very close to that of Venus: The Martian atmosphere is 95.3% carbon dioxide (versus 96.5% for Venus) and 2.7% nitrogen (versus 3.5% for Venus). The predominance of carbon dioxide means that Mars, like Earth and Venus, is warmed by the greenhouse effect. However, the Martian atmosphere is so thin that the greenhouse effect is very weak and warms the Martian surface by only 5°C (versus 33°C on Earth). While the perpetual cloud cover on Venus maintains a steady surface temperature, the thin atmosphere on Mars provides very little thermal insulation. While daytime highs on Mars can be as high as 20°C (68°F), at night the temperature can plummet to −140°C (−220°F).

Water vapor makes up about 0.03% of the Martian atmosphere (versus about 1% for the Earth), and this vapor can form clouds. Figure 11-24b shows clouds that form near the tops of Martian volcanoes when air moves up the volcanic slopes. The rising air cools until the water vapor condenses into ice crystals. The same process occurs on Earth, but the clouds that form in this way are made of droplets of liquid water rather than ice crystals. Why is there a difference?

The explanation is that liquid water cannot exist anywhere on the Martian surface or in the Martian atmosphere. Water is liquid over only a limited temperature range: If the temperature is too low, water becomes ice, and if the temperature is too high, it becomes water vapor. What determines this temperature range is the atmospheric pressure above a body of water or around a water drop. If the pressure is very low, molecules easily escape from the liquid's surface, causing the water to vaporize. Thus, at low pressures, water more easily becomes water vapor. The average surface temperature on Mars is only about 250 K (−23°C, or

−10°F), and the average pressure is only 0.006 atmosphere. With this combination of temperature and pressure, water can exist as a solid (ice) and as a gas (water vapor) but not as a liquid. (You can see this same situation inside a freezer, where water vapor swirls around over ice cubes.) Hence, it never rains on Mars and there are no bodies of standing water anywhere on the planet.

In order to keep water on Mars in the liquid state, it would be necessary to increase both the temperature (to keep water from freezing) and the pressure (to keep the liquid water from evaporating). Later in this chapter we will see evidence that higher temperatures and pressures did exist on Mars billions of years ago, and that there was once liquid water on the surface.

The Martian Poles: Freezing the Atmosphere

Figure 11-27c shows that at very high altitudes above the Martian surface, atmospheric carbon dioxide can freeze into crystals and form clouds. (Frozen carbon dioxide can be made on Earth and is called "dry ice.") Such frozen carbon dioxide is also found in larger quantities in ice caps at the north and south poles of Mars, one of which you can see in Figure 11-4c. Although no spacecraft has successfully landed on the ice caps, their composition can be measured from orbit by comparing the spectrum sunlight reflected from the ice caps with the known reflection spectrum of frozen carbon dioxide. (Figure 7-4 shows another application of this idea.)

The size of the ice caps varies substantially with the seasons (**Figure 11-30**). When it is winter at one of the Martian poles, temperatures there are so low that atmospheric carbon dioxide freezes and the ice cap grows. (Think of it: It gets so cold at the Martian poles that the atmosphere freezes!) With the coming of spring, much of the carbon dioxide returns to the gaseous state and the ice cap shrinks. With the arrival of summer, however, the rate of shrinkage slows abruptly and a **residual ice cap** remains

solid throughout the summer. Scientists had long suspected that the north and south residual polar caps contain a large quantity of frozen water that is less easily evaporated. The European *Mars Express* spacecraft confirmed this idea in 2004. It used its infrared cameras to peer through the carbon dioxide layer at the south polar ice cap and detect the characteristic spectrum of water ice. Because we do not know the thickness of the layer of water ice, the amount of frozen water stored in the Martian polar caps is still unknown.

The wintertime freezing of atmospheric carbon dioxide was measured by the *Viking Lander 1* and *Viking Lander 2* spacecraft, which made the first successful landings on Mars in 1976. Both of these spacecraft touched down in the northern hemisphere of Mars when it was spring in that hemisphere and autumn in the southern hemisphere. (You can see the landing sites in Figure 11-20.) **Figure 11-31a** shows the springtime view from *Viking Lander 1*. Initially, both spacecraft measured pressures around 0.008 atmosphere. After only a few weeks on the Martian surface, however, atmospheric pressure at both landing sites was dropping steadily. Mars seemed to be rapidly losing its atmosphere, and some scientists joked that soon all the air would be gone!

The actual explanation was that winter was coming to the southern hemisphere, causing carbon dioxide to freeze in that hemisphere's atmosphere and fall as dry-ice "snow." The formation of this "snow" removed gas from the Martian atmosphere, thereby lowering the atmospheric pressure across the planet. Several months later, when spring came to the southern hemisphere, the dry-ice snow evaporated rapidly and the atmospheric pressure returned to its prewinter levels. The pressure dropped again in the southern hemisphere's summer, because it is then winter in the northern hemisphere, and dry-ice snow condenses at northern latitudes (Figure 11-31b). Therefore, both temperature and atmospheric pressure change significantly with the Martian seasons.

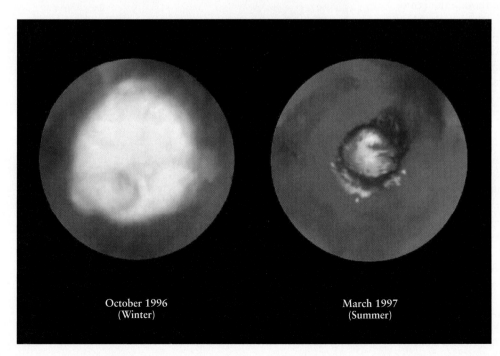

October 1996
(Winter)

March 1997
(Summer)

Figure 11-30   R I V̄ U X G

Changing Seasons at the Martian North Pole During the Martian winter, the temperature drops so low that carbon dioxide freezes out of the Martian atmosphere. A thin coating of carbon dioxide frost covers a broad region around Mars's north pole. During summer in the northern hemisphere, the ice cap is much smaller because carbon dioxide goes back into the atmosphere. The shrinking of the ice cap exposes a ring of dark sand dunes around the Martian north pole. (S. Lee/ J. Bell/M. Wolff/Space Science Institute/NASA)

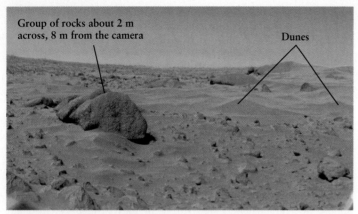

Group of rocks about 2 m
across, 8 m from the camera

Dunes

(a) A view from *Viking Lander 1*

(b) A wintertime view from *Viking Lander 2*

Figure 11-31   R I V U X G

Seasons on the Martian Surface (a) This view from *Viking Lander 1* shows rocks that resemble volcanic rocks on Earth (see Figure 9-19a). They are thought to be part of an ancient lava flow that was broken apart by crater-forming asteroid impacts. Fine-grained debris has formed sand dunes. **(b)** This picture was taken during midwinter at the *Viking Lander 2* site. Freezing carbon dioxide adheres to water-ice crystals and dust grains in the atmosphere, causing them to fall to the ground and coat the surface with frost. This frost lasted for about a hundred days. (a: Dr. Edwin Bell II/NSSDC/GSFC/NASA; b: NASA/JPL)

Martian Dust

Figure 11-31a shows that the landscape around the *Viking Lander 1* is covered with a fine-grained dust. The dust particles have piled up in places to form drifts that resemble sand dunes on Earth. The dust particles are much smaller than ordinary household dust on Earth: They are only about 1 μm (10^{-6} m) across, about the size of the particles in cigarette smoke. The rocks at the *Viking Lander 2* site are also covered with dust, though it does not pile up in dunes. More recent missions to Mars have also found dust around their landing sites.

Because Martian dust is so fine, it can be carried aloft and spread over the surface by Martian winds even though the atmosphere is so tenuous. (A wind of 100 km/h, or 60 mi/h, on Mars is only about as strong as an Earth breeze of 10 km/h or 6 mi/h.) Windblown dust can sometimes be seen from Earth as a yellow haze that obscures Martian surface features. Figure 11-18 shows a dust storm that covered the entire planet, triggered by the flow of carbon dioxide evaporating from the north polar ice cap with the coming of northern spring. On Earth, stormy weather means cloudy skies, wind, and falling rain; on Mars, where no raindrop falls, stormy weather means wind and dust.

Dust also plays a role in the ordinary daily weather on Mars. Each afternoon, parcels of warm air rise from the heated surface and form whirlwinds called **dust devils**. (A similar phenomenon with the same name occurs in dry or desert terrain on Earth.) Martian dust devils can reach an altitude of 6 km (20,000 ft) and help spread dust particles across the planet's surface. All three landers measured changes in air pressure as dust devils swept past. Often several of these would pass in a single afternoon. Dust devils are large enough to be seen by spacecraft orbiting Mars (**Figure 11-32**). As the sun sets and afternoon turns to evening, the dust devils subside until the following day.

Airborne dust also affects the color of the Martian sky, shown in Figure 11-31. The Earth's largely dust-free sky is blue because molecules in our atmosphere scatter blue sunlight more effectively than other colors (see Box 5-4). On Mars, however, dust particles in the air do a good job of scattering all visible wavelengths, which by itself would make the sky appear white. But, in addition, the dust particles contain a mineral called magnetite, which absorbs blue light. The result is that the Martian sky has a yellowish color reminiscent of butterscotch.

Dust devil

Dark streak

Shadow

500 m

Figure 11-32   R I V U X G

A Martian Dust Devil This *Mars Global Surveyor* image shows a dust devil as seen from almost directly above. This tower of swirling air and dust casts a long shadow in the afternoon sun. The dust devil had been moving from right to left before the picture was taken, leaving a dark, curlicue-shaped trail in its wake. (NASA/JPL/Malin Space Science Systems)

11-7 The atmospheres of Venus and Mars were very different billions of years ago

Why do Earth, Venus, and Mars have such dramatically different atmospheres? As we will see, the original atmospheres of all three worlds were essentially the same. However, each atmosphere evolved in a unique way determined by the planet's size and distance from the Sun. The *Cosmic Connections* figure summarizes the processes that led to the present-day atmospheres of the three large terrestrial planets.

> The early atmospheres of Earth, Venus, and Mars were predominantly water vapor and carbon dioxide

The Origin of Atmospheres on Earth, Venus, and Mars

The original atmospheres of Earth, Venus, and Mars all derived from gases that were emitted, or *outgassed*, from volcanoes (**Figure 11-33**). The gases released by present-day Earth volcanoes are predominantly water vapor (H_2O), carbon dioxide (CO_2), sulfur dioxide (SO_2), and nitrogen (N_2). These gases should therefore have been important parts of the original atmospheres of all three planets. Much of the water on all three planets is thought to have come from impacts from comets, icy bodies from the outer solar system (see Section 7-5 and Section 8-5).

In fact, water probably once dominated the atmospheres of all three planets. Venus and the Earth are similar in size and mass, so Venusian volcanoes may well have outgassed as much water vapor as on the Earth, and both planets would have had about the same number of comets strike their surfaces. Studies of how stars evolve suggest that the early Sun was only about 70% as luminous as it is now, so the temperature in Venus's early atmosphere must have been quite a bit lower. Water vapor may actually have been able to liquefy and form oceans on Venus. Mars is smaller than either Earth or Venus, but its large volcanoes could also have emitted substantial amounts of water vapor. With a thicker atmosphere and hence greater pressure, liquid water could also have existed on Mars.

The Evolution of Earth's Atmosphere

On the present-day Earth most of the water is in the oceans. Nitrogen, which is not very reactive, is still in the atmosphere. By contrast, carbon dioxide dissolves in water, which can fall as rain; as a result, rain removed most of the H_2O and CO_2 out of our planet's atmosphere long ago. The dissolved CO_2 reacted with rocks in rivers and streams, and the residue was ultimately deposited on the ocean floors, where it turned into carbonate rocks, such as limestone. Consequently, most of the Earth's carbon dioxide is tied up in the Earth's crust, and only about 1 in every 3000 molecules in our atmosphere is a CO_2 molecule. If the Earth became as hot as Venus, much of its CO_2 would be boiled out of the oceans and baked out of the crust, and our planet would soon develop a thick, oppressive carbon dioxide atmosphere much like that of Venus.

The small amount of carbon dioxide in our atmosphere is sustained in part by plate tectonics. Tectonic activity causes carbonate rocks to cycle through volcanoes, where they are heated and forced to liberate their trapped CO_2. This liberated gas then rejoins the atmosphere. Without volcanoes, rainfall would remove all of the CO_2 from our atmosphere in only a few thousand years. This would dramatically reduce the greenhouse effect, our planet's surface temperature would drop precipitously, and the oceans would freeze. Plate tectonics is thus essential for maintaining our planet's livable temperatures.

The oxygen (O_2) in our atmosphere is the result of photosynthesis by plant life, which absorbs CO_2 and releases O_2. The net result is an atmosphere that is predominantly N_2 and O_2, with enough pressure and a high enough temperature (thanks to the greenhouse effect) for water to remain a liquid—and hence for life, which requires liquid water, to exist.

The Evolution of the Venusian Atmosphere: A Runaway Greenhouse Effect

When Venus was young and had liquid water, the amount of atmospheric CO_2 was kept in check much as on Earth today: CO_2 was released by volcanoes, then dissolved in the oceans and bound up in carbonate rocks. Enough of the liquid water would have vaporized to create a thick cover of water vapor clouds. Since water vapor is a greenhouse gas, this humid atmosphere—perhaps denser than the Earth's present-day atmosphere, but far less dense than the atmosphere that surrounds Venus today—

Figure 11-33   R I V U X G

A Volcanic Eruption on Earth The eruption of Mount St. Helens in Washington state on May 18, 1980 released a plume of ash and gas 40 km (25 mi) high. Eruptions of this kind on Earth, Venus, and Mars probably gave rise to those planets' original atmospheres. (USGS)

COSMIC CONNECTIONS

Earth, Venus, and Mars formed with similar original atmospheres. However, these atmospheres changed dramatically over time due to factors such as planetary size and distance from the Sun.

Evolution of Terrestrial Atmospheres

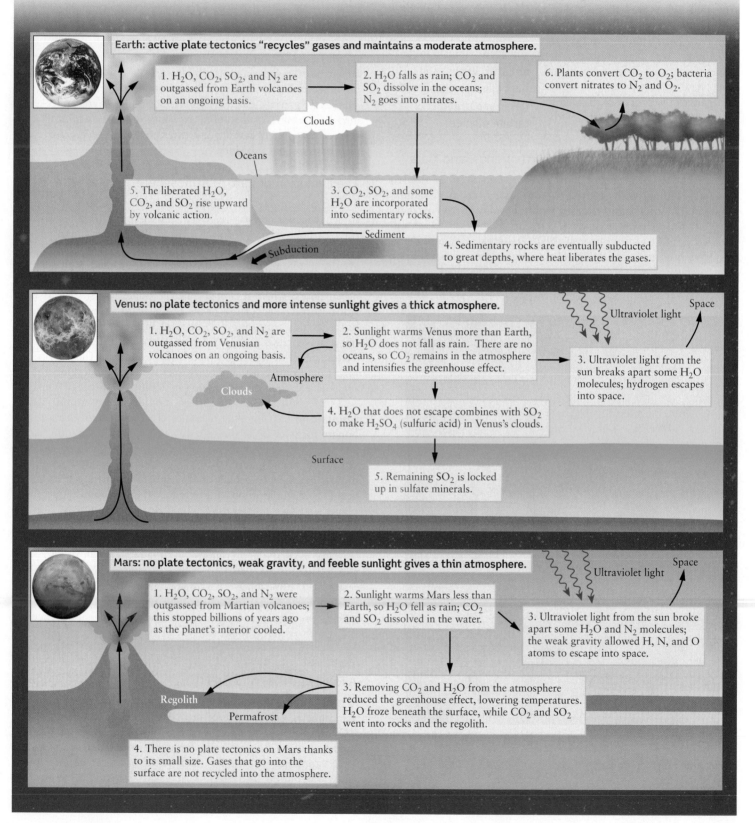

Earth: active plate tectonics "recycles" gases and maintains a moderate atmosphere.

1. H_2O, CO_2, SO_2, and N_2 are outgassed from Earth volcanoes on an ongoing basis.

2. H_2O falls as rain; CO_2 and SO_2 dissolve in the oceans; N_2 goes into nitrates.

6. Plants convert CO_2 to O_2; bacteria convert nitrates to N_2 and O_2.

Clouds

Oceans

5. The liberated H_2O, CO_2, and SO_2 rise upward by volcanic action.

3. CO_2, SO_2, and some H_2O are incorporated into sedimentary rocks.

Sediment

Subduction

4. Sedimentary rocks are eventually subducted to great depths, where heat liberates the gases.

Venus: no plate tectonics and more intense sunlight gives a thick atmosphere.

1. H_2O, CO_2, SO_2, and N_2 are outgassed from Venusian volcanoes on an ongoing basis.

2. Sunlight warms Venus more than Earth, so H_2O does not fall as rain. There are no oceans, so CO_2 remains in the atmosphere and intensifies the greenhouse effect.

Ultraviolet light

Space

3. Ultraviolet light from the sun breaks apart some H_2O molecules; hydrogen escapes into space.

Atmosphere

Clouds

4. H_2O that does not escape combines with SO_2 to make H_2SO_4 (sulfuric acid) in Venus's clouds.

Surface

5. Remaining SO_2 is locked up in sulfate minerals.

Mars: no plate tectonics, weak gravity, and feeble sunlight gives a thin atmosphere.

1. H_2O, CO_2, SO_2, and N_2 were outgassed from Martian volcanoes; this stopped billions of years ago as the planet's interior cooled.

2. Sunlight warms Mars less than Earth, so H_2O fell as rain; CO_2 and SO_2 dissolved in the water.

Ultraviolet light

Space

3. Ultraviolet light from the sun broke apart some H_2O and N_2 molecules; the weak gravity allowed H, N, and O atoms to escape into space.

Regolith

Permafrost

3. Removing CO_2 and H_2O from the atmosphere reduced the greenhouse effect, lowering temperatures. H_2O froze beneath the surface, while CO_2 and SO_2 went into rocks and the regolith.

4. There is no plate tectonics on Mars thanks to its small size. Gases that go into the surface are not recycled into the atmosphere.

would have efficiently trapped heat from the Sun. At first, this would have had little effect on the oceans of Venus. Although the temperature would have climbed above 100°C, the boiling point of water at sea level on the Earth, the added atmospheric pressure from water vapor would have kept the water in Venus's oceans in the liquid state.

This hot and humid state of affairs may have persisted for several hundred million years. But as the Sun's energy output slowly increased over time, the temperature at the surface would eventually have risen above 374°C (647 K, or 705°F). Above this temperature, no matter what the atmospheric pressure, Venus's oceans would have begun to evaporate. The added water vapor in the atmosphere would have increased the greenhouse effect, making the temperature even higher and causing the oceans to evaporate faster. That would have added even more water vapor to the atmosphere, further intensifying the greenhouse effect and making the temperature climb higher still. This is an example of a **runaway greenhouse effect,** in which an increase in temperature causes a further increase in temperature, and so on.

ANALOGY A runaway greenhouse effect is like a house in which the thermostat has accidentally been connected backward. If the temperature in such a house gets above the value set on the thermostat, the heater comes on and makes the house even hotter.

Once Venus's oceans disappeared, so did the mechanism for removing carbon dioxide from the atmosphere. With no oceans to dissolve it, outgassed CO_2 began to accumulate in the atmosphere, making the greenhouse effect "run away" even faster. Temperatures eventually became high enough to "bake out" any CO_2 that was trapped in carbonate rocks. This liberated carbon dioxide formed the thick atmosphere of present-day Venus.

Sulfur dioxide (SO_2) is also a greenhouse gas released by volcanoes. Although present in far smaller amounts than carbon dioxide, it would have contributed to Venus's rising temperatures. Like CO_2, it dissolves in water, a process that helps moderate the amount of sulfur dioxide in our atmosphere. But when the oceans evaporated on Venus, the amount of atmospheric SO_2 would have increased.

The temperature in Venus's atmosphere would not have increased indefinitely. In time, solar ultraviolet radiation striking molecules of water vapor—the dominant cause of the runaway greenhouse effect—would have broken them into hydrogen and oxygen atoms. The lightweight hydrogen atoms would have then escaped into space (see Box 7-2 for a discussion of why lightweight atoms can escape a planet more easily than heavy atoms). The remaining atoms of oxygen, which is one of the most chemically active elements, would have readily combined with other substances in Venus's atmosphere. Eventually, almost all of the water vapor would have been irretrievably lost from Venus's atmosphere. With all the water vapor gone, and essentially all the carbon dioxide removed from surface rocks, the greenhouse effect would no longer have "run away," and the rising temperature would have leveled off.

Today, the infrared-absorbing properties of CO_2 have stabilized Venus's surface temperature at its present value of 460°C. Only minuscule amounts of water vapor—about 30 parts per million, or 0.003%—remain in the atmosphere. As on Earth, volcanic

outgassing still adds small amounts of water vapor to the atmosphere, along with carbon dioxide and sulfur dioxide. But on Venus, these water molecules either combine with sulfur dioxide to form sulfuric acid clouds or break apart due to solar ultraviolet radiation. The great irony is that this state of affairs is the direct result of an earlier Venusian atmosphere that was predominantly water vapor!

The Evolution of the Martian Atmosphere: A Runaway Icehouse Effect

Mars probably had a thicker atmosphere 4 billion years ago. Thanks to Mars's greater distance from the Sun and hence less intense sunlight, temperatures would have been lower than on the young Earth, and any water in the atmosphere would more easily have fallen as rain or snow. This would have washed much of the planet's carbon dioxide from its atmosphere, perhaps creating carbonate minerals in which the CO_2 is today chemically bound. Measurements from Mars orbit show only small amounts of carbonate materials on the surface, suggesting that the amount of atmospheric CO_2 that rained out was small. Hence, even the original Martian atmosphere was relatively thin, though thicker than the present-day atmosphere.

Because Mars is so small, it cooled early in its history and volcanic activity came to an end. Thus, any solid carbonates were not recycled through volcanoes as they are on Earth. The depletion of carbon dioxide from the Martian atmosphere into the surface would therefore have been permanent.

As the amount of atmospheric CO_2 declined, the greenhouse effect on Mars would have weakened and temperatures begun to fall. This temperature decrease would have caused more water vapor to condense into rain or snow and fall to the surface, taking even more CO_2 with them and further weakening the greenhouse effect. Thus, a decrease in temperature would have caused a further decrease in temperature—a phenomenon sometimes called a **runaway icehouse effect.** (This is the reverse of the runaway greenhouse effect that has taken place on Venus.) Ultimately, both water vapor and most of the carbon dioxide would have been removed from the Martian atmosphere. With only a very thin CO_2 atmosphere remaining, surface temperatures on Mars eventually stabilized at their present frigid values.

As both water vapor and carbon dioxide became depleted, ultraviolet light from the Sun could penetrate the thinning Martian atmosphere to strip it of nitrogen. Nitrogen molecules (N_2) normally do not have enough thermal energy to escape from Mars, but they can acquire that energy from ultraviolet photons, which break the molecules in two. Ultraviolet photons can also split carbon dioxide and water molecules, giving their atoms enough energy to escape (Figure 13-16b). Indeed, in 1971 the Soviet *Mars 2* spacecraft found a stream of oxygen and hydrogen atoms (from the breakup of water molecules) escaping into space. Oxygen atoms that did not escape into space could have combined with iron-bearing minerals in the surface, forming rustlike compounds. Such compounds have a characteristic reddish-brown color and may be responsible for the overall color of the planet.

Was the early Martian atmosphere thick enough and warm enough for water to remain as a liquid on the planet's surface? As we will see in the next section, scientists have found evidence on Mars that the answer is a qualified yes.

11-8 Rovers have found evidence of ancient Martian water

Just as ancient fossils on Earth provide evidence of how life has evolved on our planet, ancient rocks and geological formations help us understand how our planet's surface and atmosphere have evolved. (For example, the oldest rocks on Earth have a chemical structure which shows that they formed in a time when there was little oxygen in the atmosphere and hence before the appearance of photosynthetic planet life.) Planetary scientists who want to learn about the history of Venus and Mars are therefore very interested in looking at ancient rocks on those worlds.

> **Parts of Mars are far drier than a bone, while others had standing water for extended periods**

On Venus we cannot look very far into the past, since the planet's volcanic activity and flake tectonics has erased features more than about 500 million years old (see Section 11-5). On Mars, by contrast, we expect to find rocks that are billions of years old: The thin atmosphere causes very little erosion, there has been little recent volcanic activity to cover ancient rocks, and there is no subduction to drag old surface features back into the planet's interior. (The heavy cratering of the southern highlands bears testament to the great age of much of the Martian surface.) In a quest to find these ancient rocks and learn about the early history of Mars, scientists have sent spacecraft to land on the red planet and explore it at close range.

Landing on Mars

As of this writing (2006) five spacecraft have successfully landed on the Martian surface. *Viking Lander 1* and *Viking Lander 2*, which we first discussed in Section 11-6, touched down in 1976. They had no capability to move around on the surface, and so each *Viking Lander* could only study objects within reach of the scoop at the end of its mechanical arm (Figure 11-34a). Figure 11-31 shows the view from the two *Viking Lander* sites.

Three more recent unmanned explorers, *Mars Pathfinder* (landed 1997) and two *Mars Exploration Rovers* (landed 2004), were equipped with electrically driven wheels that allow travel over the surface (Figure 11-34b). Unlike the *Viking Landers*, these rovers did not have a heavy and expensive rocket engine to lower it gently to the Martian surface. Instead, the spacecraft was surrounded by airbags like those used in automobiles. Had there been anyone to watch *Mars Pathfinder* or one of the *Mars Exploration Rovers* land, they would have seen an oversized beach ball hit the surface at 50 km/h (30 mi/h), then bounce for a kilometer before finally rolling to a stop. Unharmed by its wild ride, the spacecraft then deflated its airbags and began to study the geology of Mars.

Figure 11-20 shows the landing sites of the *Mars Pathfinder* rover (named *Sojourner*) and the two *Mars Exploration Rovers* (named *Spirit* and *Opportunity*). Following directions from Earth, *Sojourner* spent three months exploring the rocks around its landing site. The *Mars Exploration Rovers* were also designed for a three-month lifetime, but both were still functioning more than 2½ years after landing. Over that time *Spirit* and *Opportunity* had each traveled several kilometers over the Martian surface.

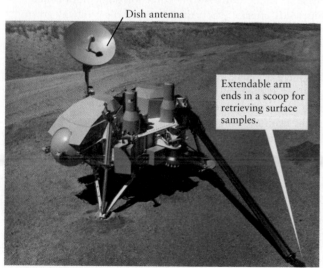

Dish antenna

Extendable arm ends in a scoop for retrieving surface samples.

(a) A *Viking Lander*

Cameras *Mars Exploration Rover* *Mars Pathfinder* rover

(b) Two generations of rovers

Figure 11-34   R I V U X G

Martian Explorers (a) Each *Viking Lander* is about 2 m (6 ft) tall from the base of its footpads to the top of its dish antenna, used for sending data to an orbiter, which in turn relayed the information back to Earth. (This photograph shows a full-size replica under test on Earth.) **(b)** The technicians in this photograph pose with one of the *Mars Exploration Rovers* and with a replica of its predecessor, the *Mars Pathfinder* rover. (a: NASA; b: NASA/JPL)

In addition to cameras, all five landers carried a device called an X-ray spectrometer to measure the chemical composition of rock samples. *Spirit* and *Opportunity* also carried a tool designed to grind holes into rocks, thus revealing parts of the rock's interior that have not been weathered by exposure to the Martian atmosphere.

The Martian Regolith

The *Viking Landers* were the first to reveal the nature of the Martian regolith, or surface material. The scoop on the extendable arm (Figure 11-34a) was used to dig into the regolith and retrieve samples for analysis. Bits of the regolith clung to a magnet mounted on the scoop, indicating that the regolith contains iron. The X-ray spectrometers showed that samples collected at both *Viking Lander* sites were rich in iron, silicon, and sulfur.

The *Viking Landers* each carried a miniature chemistry laboratory for analyzing the regolith. Curiously, when a small amount of water was added to a sample of regolith, the sample began to fizz and release oxygen. This experiment indicated that the regolith contains unstable chemicals called peroxides and superoxides, which break down in the presence of water to release oxygen gas.

The unstable chemistry of the Martian regolith probably comes from solar ultraviolet radiation that beats down on the planet's surface. (Unlike Earth, Mars has no ozone layer to screen out ultraviolet light.) Ultraviolet photons easily break apart molecules of carbon dioxide (CO_2) and water vapor (H_2O) by knocking off oxygen atoms, which then become loosely attached to chemicals in the regolith. Ultraviolet photons also produce highly reactive ozone (O_3) and hydrogen peroxide (H_2O_2), which also become incorporated in the regolith.

Among the most important scientific goals of the *Viking Landers* was to search for evidence of life on Mars. Although it was clear that Mars has neither civilizations nor vast fields of plants, it still seemed possible that simple microorganisms could have evolved on Mars during its early brief epoch of Earthlike climate. If so, perhaps some species had survived to the present day. To test this idea, the *Viking Landers* looked for biologically significant chemical reactions in samples of the Martian regolith, but they failed to detect any reactions of this type. Perhaps life never existed on Mars, or perhaps it existed once but was destroyed by the presence of peroxides and superoxides in the regolith.

On Earth, hydrogen peroxide is commonly used as an antiseptic. When poured on a wound, it fizzes and froths as it reacts with organic material and destroys germs. Thus, the *Viking Landers* may have failed to detect any biochemistry on Mars because the regolith has literally been sterilized! Hence, we cannot yet tell whether life ever existed on Mars, or whether it might yet exist elsewhere on or beneath the planet's surface. (In Chapter 28 we discuss the *Viking Lander* experiments in detail, as well as the controversial proposal that fossil Martian microorganisms may have traveled to Earth aboard meteorites.)

Exploring Martian Rocks

Images made from Mars orbit show a number of features that suggest water flowed there in the past (Figure 11-35). To investigate these, *Mars Pathfinder* landed in an area of the northern lowlands called Ares Valles, which appears to be an ancient flood plain. Figure 11-35b shows a portion of this plain. Flood waters can carry rocks of all kinds great distances from their original locations, so scientists expected that this landing site (Figure 11-36) would show great geologic diversity. They were not disappointed.

(a)

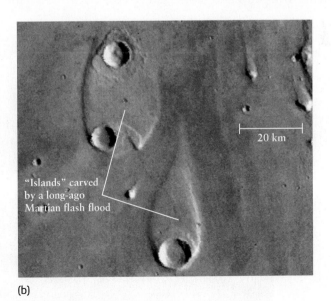

20 km

"Islands" carved by a long-ago Martian flash flood

(b)

Figure 11-35   R I V̄ U X G

Sign of Ancient Martian Water (a) This *Viking Orbiter* image shows a network of dried riverbeds extending across the cratered southern highlands. Liquid water would evaporate in the sparse, present-day Martian atmosphere, so these must date from an earlier era when the atmosphere was thicker. (b) These teardrop-shaped islands rise above the floor of Ares Valles. They were carved out by a torrent of water that flowed from the bottom of the image toward the top. Similar flood-carved islands are found on Earth in eastern Washington state. (a: Michael Carr, USGS; b: NASA/USGS)

Mars Pathfinder rover   Hill, 1 km (0.6 mi) away

Figure 11-36   R I ⊻ U X G

Roving the Martian Surface This view from the stationary part of *Mars Pathfinder* shows the rover, called *Sojourner*, deploying its X-ray spectrometer against a rock named Moe. Scientists gave other rocks whimsical names such as Jiminy Cricket and Pooh Bear. (NASA/JPL)

Many of the rocks appeared to be igneous rock produced by volcanic action. Other rocks, which resemble the impact breccias found in the ancient highlands of the Moon (see Figure 10-15), appear to have come from areas that were struck by crater-forming meteoroids. Still others have a layered structure like sedimentary rocks on Earth (see Figure 9-19b). Such rocks form

gradually at the bottom of bodies of water, suggesting that liquid water was stable in at least some regions of Mars for a substantial period of time.

Searching for Martian Water

The two *Mars Exploration Rovers* also landed in sites that were suspected to have been under water in the distant past. *Spirit* touched down in Gusev Crater, a large impact basin. A valley opens into this crater, resembling a river that runs into a lake. The destination for *Opportunity* was the flat plain of Sinus Meridianii, where orbiting spectrometers had spotted evidence of hematite—a type of iron oxide that, on Earth, tends to form in wet locations.

Contrary to expectations, most of the rocks encountered by *Spirit* show *no* evidence that they have been exposed to water. The volcanic rocks found in and around Gusev Crater (**Figure 11-37a**) are composed of the minerals olivine and pyroxene, which break down if they are exposed to liquid water. In a few locations *Spirit* found deposits of sulfur salts, which can be formed by the action of water (Figure 11-37b). However, olivine and pyroxene were still found among the sulfur salts. While it remains a possibility that Gusev Crater was once a river-fed lake, the results from *Spirit* show that for the past billion years this region has been drier than any desert on Earth.

The results from *Opportunity* were entirely different. This rover discovered outcrops of sedimentary rocks on the plains of Sinus Meridianii (**Figure 11-38**). These rocks are heavily laden with sulfur compounds, strongly suggesting that they were produced by the evaporation of sulfur-rich water. *Opportunity* also found millimeter-sized spheres of hematite that, like analogous spheres found on Earth, probably formed from iron-rich water circulating through sediment. These observations suggest Sinus

(a) Volcanic rocks

(b) Sulfur salts

Figure 11-37   R I ⊻ U X G

Dry Mars At the *Spirit* landing site, both **(a)** the dusty volcanic rocks that dominate the site and **(b)** these light-colored sulfur salts contain minerals that break apart upon contact with liquid water. Hence, this area must

have been dry for billions of years. (a: NASA/JPL-Caltech/Cornell; b: NASA/JPL/Cornell)

Figure 11-38   R I V U X G

Wet Mars Sedimentary rocks are formed when minerals collect in standing water. Hence, the site where *Opportunity* found this extensive outcrop of light-colored sedimentary rocks must have been under water for an extended period. This panoramic view spans about 120 degrees from side to side. (NASA/JPL-Caltech/Cornell)

Meridianii, which is among the flattest places known in the solar system, was once a lake bed. It is not known how long ago this area was under water, but the density of craters found there suggests that the surface formed 3 to 4 billion years ago.

When combined with observations from Mars orbit, the *Spirit* and *Opportunity* results show that ancient Mars was a very diverse planet. Most of the planet was probably very dry, like Gusev Crater, but there were isolated areas like Sinus Meridianii that had liquid water for some period of time. It remains a challenge to planetary scientists to explain why liquid water appeared on Mars only in certain places and at certain times.

Where Is the Water Now?

Whatever liquid water once coursed over parts of the Martian surface is presumably now frozen, either at the polar ice caps or beneath the planet's surface. To check this idea, scientists used a device on board the *Mars Odyssey* orbiter to measure the result of cosmic rays hitting Mars.

Cosmic rays are fast-moving subatomic particles that enter our solar system from interstellar space. When these particles strike and penetrate into the Martian soil, they collide with the nuclei of atoms in the soil and knock out neutrons. (Recall from Section 5-7 that atomic nuclei contain both positively charged protons and electrically neutral neutrons.) If the ejected neutrons originate within a meter or so of the surface, they can penetrate through the regolith and escape into space. However, if there are water molecules (H_2O) in the surface, the nuclei of the hydrogen atoms in these molecules will absorb the neutrons and prevent them from escaping. An instrument on board *Mars Odyssey* detects neutrons coming from Mars; where the instrument finds a *deficiency* of neutrons is where the soil contains hydrogen and, presumably, water.

The results from the *Mars Odyssey* neutron measurements show that there is abundant water beneath the surface at both Martian poles (Figure 11-39). But they also show a surprising amount of subsurface water in two regions near the equator, Arabia Terra (see Figure 11-17*a* and Figure 11-20) and another region on the opposite side of the planet. One possible explanation is that about a million years ago, gravitational forces from other planets caused a temporary change in the tilt of Mars's rotation axis. The ice caps were then exposed to more direct sunlight, causing some of the water to evaporate and eventually refreeze elsewhere on the planet.

Percent abundance of water (by mass)

2   4   6   8   10   12   14   16   18

High concentration of subsurface water in the polar regions

High concentration of subsurface water in some locations near the equator

Figure 11-39

Water Beneath the Martian Surface This map of the abundance of water (probably in the form of ice) just below the surface of Mars is based on data collected from orbit by the *Mars Odyssey* spacecraft. These measurements can only reveal the presence of water or water ice to a depth of about a meter; there may be much more water at greater depths. (LANL)

There are other surprising features on Mars related to the presence of water. A number of *Mars Global Surveyor* images show gullies apparently carved by water flowing down the walls of pits or craters (Figure 11-40). These could have formed when water trapped underground—where the pressure is greater and water can remain a liquid—seeped onto the surface. The gullies appear to be geologically young, so it is possible that even today some liquid water survives below the Martian surface. Alternatively, these gullies may have formed during periods when the Martian climate was colder than usual and the slopes of craters were covered with a layer of dust and snow. Melting snow on the underside of this layer could have carved out gullies, which were later exposed when the climate warmed and the snow evaporated.

The total amount of water on Mars is not known. But by examining flood channels at various locations around Mars, Michael Carr of the U.S. Geological Survey has estimated that there is enough water to cover the planet to a depth of 500 meters (1500 ft). (By comparison, Earth has enough water to cover our planet to a depth of 2700 meters, or 8900 ft.) Thus, while the total amount of water on Mars is less than on Earth, there may be enough to have once formed lakes or seas—one of which may have covered the plains of Sinus Meridianii and produced the sedimentary rocks shown in Figure 11-38. Future missions to Mars will delve more deeply into the past climate and geology of the red planet.

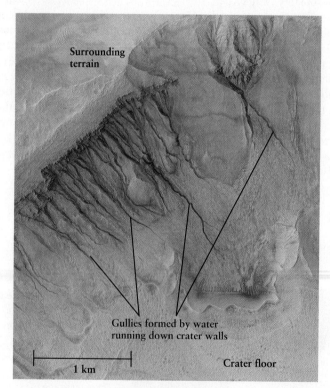

Figure 11-40   R I V U X G

Martian Gullies When the *Mars Global Surveyor* spacecraft looked straight down into this crater in the Martian southern hemisphere, it saw a series of gullies along the crater wall. These gullies may have formed by subsurface water seeping out to the surface, or by the melting of snow that fell on crater walls. (NASA/JPL/Malin Space Science Systems)

11-9 The two Martian moons resemble asteroids

Two moons move around Mars in orbits close to the planet's surface. These satellites are so tiny that they were not discovered until the favorable opposition of 1877. While Schiaparelli was seeing canals, the American astronomer Asaph Hall spotted the two moons, which orbit almost directly above its equator. He named them Phobos ("fear") and Deimos ("panic"), after the mythical horses that drew the chariot of the Greek god of war.

The Orbits of Phobos and Deimos

Phobos is the inner and larger of the Martian moons. It circles Mars in just 7 hours and 39 minutes at an average distance of only 9378 km (5828 mi) from the center of Mars. (Recall that our Moon orbits at an average distance of 384,400 km from the Earth's center.) This orbital period is much shorter than the Martian day; hence, Phobos rises in the *west* and gallops across the sky in only 5½ hours, as viewed by an observer near the Martian equator. During this time, Phobos appears several times brighter in the Martian sky than Venus does from the Earth.

> As seen from Mars, Deimos rises in the east but Phobos rises in the west

Deimos, which is farther from Mars and somewhat smaller than Phobos, appears about as bright from Mars as Venus does from the Earth. Deimos orbits at an average distance of 23,460 km (14,580 mi) from the center of Mars, slightly beyond the right distance for a synchronous orbit—an orbit in which a satellite seems to hover above a single location on the planet's equator. As seen from the Martian surface, Deimos rises in the east and takes about 3 full days to creep slowly from one horizon to the other. (Figure 11-41*a*).

Images from spacecraft in Mars's orbit reveal Phobos and Deimos to be jagged and heavily cratered (Figure 11-41*b*). Both moons are somewhat elongated from a spherical shape: Phobos is approximately 28 by 23 by 20 km in size, while Deimos is slightly smaller, roughly 16 by 12 by 10 km. Observations from the *Viking Orbiters* showed that Phobos and Deimos are both in synchronous rotation. The tidal forces that Mars exerts on these elongated moons cause them to always keep the same sides toward Mars, just as the Earth's tidal forces ensure that the same side of the Moon always faces us (see Section 4-8).

We saw in Section 10-5 that the Moon raises a tidal bulge on the Earth. Because the Earth rotates on its axis more rapidly than the Moon orbits the Earth, this bulge is dragged ahead of a line connecting the Earth and Moon (see Figure 10-17). The gravitational force of this bulge pulls the Moon into an ever larger orbit. In the same way, and for the same reason, Deimos is slowly moving away from Mars. But for fast-moving Phobos, the situation is reversed: Mars rotates on its axis more slowly than Phobos moves around Mars. This slow rotation holds back the tidal bulge that Phobos raises in the solid body of Mars, and this bulge thus is *behind* a line connecting the centers of Mars and Phobos. (Imagine that the bulge in Figure 10-17 is tilted clockwise rather than counterclockwise.) This bulge exerts a gravitational force on

(a) As seen from Mars

(b) As seen from Mars orbit

Figure 11-41   R I [V] U X G

The Moons of Mars (a) The *Spirit* rover recorded seven images of the Martian night sky at 15-minute intervals to produce this composite view of the motions of Deimos and Phobos. **(b)** These *Viking Orbiter* images show Deimos and Phobos to the same scale. What appears to be a fold at the top of the image of Deimos is actually a side view of an impact crater 10 km across, comparable to the Moon's 22-km average diameter. An only slightly more powerful impact would have broken Deimos apart. (a: NASA/JPL/Cornell/Texas A&M; b: JPL/NASA)

Phobos that pulls it back, slowing the moon's motion and making it slowly spiral inward toward Mars. This inward spiral will cease in about 40 million years when Phobos smashes into the Martian surface.

The Nature of the Martian Moons

The origin of the Martian moons is unknown. One idea is that they are asteroids that Mars captured from the nearby asteroid belt (see Section 7-5). Like Phobos and Deimos, many asteroids reflect very little sunlight (typically less than 10%) because of a high carbon content. Phobos and Deimos also have very low densities (1900 kg/m^3 and 1760 kg/m^3, respectively), which are also characteristic of carbon-rich asteroids. Perhaps two of these asteroids wandered close enough to Mars to become permanently trapped by the planet's gravitational field. Alternatively, Phobos and Deimos may have formed in orbit around Mars out of debris left over from the formation of the planets. Some of this debris might have come from the asteroid belt, giving these two moons an asteroidlike character. Future measurements of the composition of the Martian moons—perhaps by landing spacecraft on their surfaces—may shed light on their origin.

Future Exploration of Mars

The exploration of Mars and its two moons has really just begun. Recent discoveries from orbiters and landers have only whetted our appetite to know more about Mars. Fortunately, every 25 months the Earth and Mars are ideally placed for sending a spacecraft from one planet to another with the minimum thrust. For the next several years, one or more spacecraft will be launched toward Mars at each of these opportunities. One of these spacecraft may land on Mars, collect samples with an advanced rover, and return those samples to Earth. Detailed analy-

sis of Martian samples in laboratories on Earth could answer many open questions about the red planet, its formation, and its evolution.

It is ironic that at the end of the nineteenth century, the English novelist H. G. Wells envisioned a fleet of Martian spaceships coming to Earth on a mission of conquest. In the twenty-first century, it is we on Earth who are sending robotic spacecraft on a peaceful scientific "invasion" of Mars.

Key Words

1-to-1 spin-orbit coupling, p. 261
3-to-2 spin-orbit coupling, p. 262
cosmic rays, p. 289
crustal dichotomy, p. 273
dust devil, p. 282
favorable opposition, p. 258
flake tectonics, p. 274
greatest eastern elongation, p. 256
greatest western elongation, p. 256

hot-spot volcanism, p. 277
northern lowlands, p. 273
prograde rotation, p. 263
residual polar cap, p. 281
retrograde rotation, p. 263
rift valley, p. 276
runaway greenhouse effect, p. 285
runaway icehouse effect, p. 285
scarp, p. 265
shield volcano, p. 276
southern highlands, p. 273

Key Ideas

Motions of Mercury, Venus, and Mars in the Earth's Sky: Mercury and Venus can be seen in the morning or evening sky only, while it is possible to see Mars at any time of night depending on its position in its orbit.

• At their greatest eastern and western elongations, Mercury is only 28° from the Sun and Venus is only 47° from the Sun.

• The best Earth-based views of Mars are obtained at favorable oppositions, when Mars is simultaneously at opposition and near perihelion.

Rotation of Mercury, Venus, and Mars: Poor telescopic views of Mercury's surface led to the mistaken impression that the planet always keeps the same face toward the Sun (1-to-1 spin-orbit coupling).

• Radio and radar observations revealed that Mercury in fact has 3-to-2 spin-orbit coupling: The planet rotates on its axis three times every two orbits.

• Venus rotates slowly in a retrograde direction. Its rotation period is longer than its orbital period.

• Mars rotates at almost the same rate as the Earth, and its rotation axis is tilted by almost the same angle as the Earth's axis.

Mercury's Surface, Interior, and Magnetic Field: The Mercurian surface is pocked with craters, but there are extensive smooth plains between these craters.

• Long cliffs called scarps meander across the surface of Mercury. These probably formed as the planet's crust cooled, solidified, and shrank.

• Mercury has an iron core with a diameter equal to about $\frac{3}{4}$ of the planet's diameter. By contrast, the diameter of the Earth's core is only slightly more than $\frac{1}{2}$ of Earth's diameter.

• Mercury has a weak magnetic field, which indicates that at least part of the iron core is liquid.

Comparing Venus and Mars: Most of the surface of Venus is at about the same elevation, with just a few elevated regions. On Mars, the southern highlands rise several kilometers above the northern lowlands.

• Venus has a thick atmosphere and a volcanically active surface. Mars has a very thin atmosphere and little or no current volcanism.

• There is no evidence of plate tectonics on either Venus or Mars. On Venus, there is vigorous convection in the planet's interior, but the crust is too thin to move around in plates; instead, it wrinkles and flakes. On Mars, the planet's smaller size means the crust has cooled and become too thick to undergo subduction.

• Volcanoes on both Venus and Mars were produced by hot spots in the planet's interior.

• The entire Venusian surface is about 500 million years old and has relatively few craters. By contrast, most of the Martian surface is cratered and is probably billions of years old. The southern highlands on Mars are the most heavily cratered and hence the oldest part off the planet's surface.

The Atmospheres of Venus and Mars: Both planetary atmospheres are over 95% carbon dioxide, with a few percent of nitrogen.

• The pressure at the surface of Venus is about 90 atmospheres. The greenhouse effect is very strong, which raises the surface temperature to 460°C. The pressure at the surface of Mars is only 0.006 atmosphere, and the greenhouse effect is very weak.

• The permanent high-altitude clouds on Venus are made primarily of sulfuric acid. By contrast, the few clouds in the Martian atmosphere are composed of water ice and carbon dioxide ice.

• The circulation of the Venusian atmosphere is dominated by two huge convection currents in the cloud layers, one in the northern hemisphere and one in the southern hemisphere. The upper cloud layers of the Venusian atmosphere move rapidly around the planet in a retrograde direction, with a period of only about 4 Earth days.

• Weather on Mars is dominated by the north and south flow of carbon dioxide from pole to pole with the changing seasons. This can trigger planetwide dust storms.

Evolution of Atmospheres: Earth, Venus and Mars all began with relatively thick atmospheres of carbon dioxide, water vapor, and sulfur dioxide.

• On Earth, most of the carbon dioxide went into carbonate rocks and most of the water into the oceans. Ongoing plate tectonics recycles atmospheric gases through the crust.

• On Venus, more intense sunlight and the absence of plate tectonics led to a thick carbon dioxide atmosphere and a runaway greenhouse effect.

• On Mars, a runaway icehouse effect resulted from weaker sunlight and the absence of plate tectonics.

Water on Mars: Liquid water cannot exist on present-day Mars because the atmosphere is too thin and cold. But there is evidence for frozen water at the polar ice caps and beneath the surface of the regolith.

• Geological evidence from unmanned rovers shows that much of the Martian surface has been dry for billions of years, but some regions had substantial amounts of water.

The Moons of Mars: Mars has two small, football-shaped satellites that move in orbits close to the surface of the planet. They may be captured asteroids or may have formed in orbit around Mars out of solar system debris.

Questions

Review Questions

1. Why is it impossible to see Mercury or Venus in the sky at midnight?

2. Why are naked-eye observations of Mercury best made at dusk or dawn, while telescopic observations are best made during the day?

3. Why can't you see any surface features on Mercury when it is closest to the Earth?

4. Why is it best to view Mars near opposition? Why are some oppositions better than others?

5. Explain why Mars can best be viewed from Earth while it is undergoing retrograde motion.

6. Explain why the photograph in Figure 11-4b must have been made during the daytime.

7. In his 1964 science fiction story "The Coldest Place," author Larry Niven described the "dark side" of Mercury as the coldest place in the solar system. What assumption did he

make about the rotation of Mercury? Did this assumption turn out to be correct?

8. What is 3-to-2 spin-orbit coupling? How is the rotation period of an object exhibiting 3-to-2 spin-orbit coupling related to its orbital period? What aspects of Mercury's orbit cause it to exhibit 3-to-2 spin-orbit coupling? What telescopic observations proved this?

9. Why was it so difficult to determine the rate and direction of Venus's rotation? How were these finally determined? What is one proposed explanation for the slow, retrograde rotation of Venus?

10. Explain why Mercury does not have a substantial atmosphere.

11. What kind of surface features are found on Mercury? How do they compare to surface features on the Moon? Why are they probably much older than most surface features on the Earth?

12. How do we know that the scarps on Mercury are younger than the lava flows? How can you tell that the scarp in Figure 11-10 is younger than the vertically distorted crater at the center of the figure?

13. If Mercury is the closest planet to the Sun and has such a high average surface temperature, how is it possible that ice might exist on its surface?

14. Why do astronomers think that Mercury has a very large iron core?

15. Why is it surprising that Mercury has a global magnetic field? Why does the 58.646-day rotation period of Mercury imply that the planet can have only a weak magnetic field?

16. How is Mercury's magnetosphere similar to that of the Earth? How is it different? Why do you suppose Mercury does not have Van Allen belts?

17. Why was it difficult to determine Venus's surface temperature from Earth? How was this finally determined?

18. The *Mariner 2* spacecraft did not enter Venus's atmosphere, but it was nonetheless able to determine that the atmosphere is very dry. How was this done?

19. Why did Earth observers report that they had seen straight-line features ("canals") on Mars? How did the seasonal winds trick them into thinking that Mars had vegetation?

20. Do Venus and Mars have continents like those on Earth?

21. What is the Martian crustal dichotomy? What is the evidence that the southern highlands are older than the northern lowlands?

22. What is the evidence that the surface of Venus is only about 500 million years old?

23. What is flake tectonics? Why does Venus exhibit flake tectonics rather than plate tectonics?

24. What geologic features (or lack thereof) on Mars have convinced scientists that plate tectonics did not significantly shape the Martian surface?

25. What geologic processes are thought to have created Valles Marineris?

26. Compare the volcanoes of Venus, the Earth, and Mars. Cite evidence that hot-spot volcanism is or was active on all three worlds.

27. Describe the evidence that there has been recent volcanic activity (a) on Venus and (b) on Mars.

28. Suppose all of Venus's volcanic activity suddenly stopped. (a) How would this affect Venus's clouds? (b) How would this affect the overall Venusian environment?

29. Why are the patterns of convection in the Venusian atmosphere so different from those in our atmosphere?

30. Why is it impossible for liquid water to exist on Mars today? If liquid water existed on Mars in the past, what must have been different then?

31. Why does the atmospheric pressure on Mars vary with the seasons? What is the relationship between this pressure variation and Martian dust storms?

32. Why is it reasonable to assume that the primordial atmospheres of the Earth, Venus, and Mars were roughly the same?

33. Carbon dioxide accounts for about 95% of the present-day atmospheres of both Mars and Venus. Why, then, is there a strong greenhouse effect on Venus but only a weak greenhouse effect on Mars?

34. (a) What is a runaway greenhouse effect? (b) What is a runaway icehouse effect?

35. What is a dust devil? Why would you feel much less breeze from a Martian dust devil than from a dust devil on Earth?

36. (a) Why is Mars red? (b) Why is the Martian sky the color of butterscotch?

37. Were the *Viking Landers* able to determine whether life currently exists on Mars or whether it once existed there? Why or why not?

38. (a) The *Spirit* rover found the minerals olivine and pyroxene at its landing site on Mars. Explain how this shows that there has been no liquid water at that site for billions of years. (b) What evidence did the *Opportunity* rover find at its landing site to suggest that liquid water had once been present there?

39. How was the *Mars Odyssey* spacecraft able to detect water beneath the Martian surface without landing on the planet?

40. A full moon on Earth is bright enough to cast shadows. As seen from the Martian surface, would you expect a full Phobos or full Deimos to cast shadows? Why or why not?

Advanced Questions

Problem-solving tips and tools

We discussed the relationship between angular distance and linear distance in Box 1-1 and the idea of angular resolution in Section 6-3. You may need to refresh your memory about Kepler's third law, described in Section 4-4. You may also need to review the form of Kepler's third law that explicitly includes mass (see Section 4-7 and Box 4-4). Box 4-1 describes the relationship between a planet's sidereal orbital period and its synodic period (the time from one inferior conjunction to the next). You should recall that Wien's law (Section 5-4) relates the temperature of a blackbody to λ_{max}, its wavelength of maximum emission. Section 5-9 and Box 5-6 explain the Doppler effect and how to do calculations using it. The linear speed of a point on a planet's equator is the planet's circumference divided by its rotation period; Recall that the circumference of a circle of radius r is $2\pi r$. The volume of a sphere of radius r is $4\pi r^3/3$. The speed of light is given in Appendix 7.

41. Figure 11-1 shows Mercury with a greatest eastern elongation of 18° and a greatest western elongation of 28°. On November 25, 2006, Mercury was at a greatest western elongation of 20°. Was Mercury at perihelion, aphelion, or some other point on its orbit? Explain.

42. Venus takes 440 days to move from greatest western elongation to greatest eastern elongation, but it needs only 144 days to go from greatest eastern elongation to greatest western elongation. With the aid of a diagram like Figure 11-1, explain why.

43. Venus's sidereal rotation period is 243.01 days and its orbital period is 224.70 days. Use these data to prove that a solar day on Venus lasts 116.8 days. (*Hint:* Develop a formula relating Venus's solar day to its sidereal rotation period and orbital period similar to the first formula in Box 4-1.)

44. In Section 11-2 we described the relationship between the length of Venus's synodic period and the length of an apparent solar day on Venus. Using this and a diagram, explain why at each inferior conjunction the same side of Venus is turned toward the Earth.

45. This time-lapse photograph was taken on May 7, 2003, during a *solar transit* of Mercury. Over a period of 5 hours and 19 minutes, Mercury appeared to move across the face of the Sun. Such solar transits of Mercury occur 13 or 14 times each century; they do *not* happen each time that Mercury is at inferior conjunction. Explain why not. (*Hint:* For a solar transit to occur, the Sun, Mercury, and the Earth must be in a nearly perfect alignment. Does the orbit of Mercury lie in the plane of the ecliptic?)

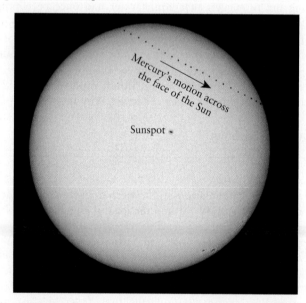

Mercury's motion across the face of the Sun

Sunspot

R I V U X G

(Dominique Dierick)

46. Find the largest angular size that Mercury can have as seen from the Earth. In order for Mercury to have this apparent size, at what point in its orbit must it be?

47. (a) Suppose you have a telescope with an angular resolution of 1 arcsec. What is the size (in kilometers) of the smallest feature you could have seen on the Martian surface during the opposition of 2005, when Mars was 0.464 AU from Earth? (b) Suppose you had access to the Hubble Space Telescope, which has an angular resolution of 0.1 arcsec. What is the size (in kilometers) of the smallest feature you could have seen on Mars with the HST during the 2005 opposition?

48. For a planet to appear to the naked eye as a disk rather than as a point of light, its angular size would have to be 1 arcmin, or 60 arcsec. (This is the same as the angular separation between lines in the bottom row of an optometrist's eye chart.) (a) How close would you have to be to Mars in order to see it as a disk with the naked eye? Does Mars ever get this close to Earth? (b) Would the Earth ever be visible as a disk to an astronaut on Mars? Would she be able to see the Earth and the Moon separately, or would they always appear as a single object? Explain.

49. Using Figure 11-3, explain why oppositions of Mars are most favorable when they occur in August.

50. Imagine that you are part of the scientific team monitoring a spacecraft that has landed on Mars. At 5:00 P.M. in your control room on Earth, the spacecraft reports that the Sun is highest in the sky as seen from its location on Mars. When the Sun is next at its highest point as seen from the spacecraft, what time will it be in the control room on Earth?

51. (a) Mercury has a 58.646-day rotation period. What is the speed at which a point on the planet's equator moves due to this rotation? (*Hint:* Remember that speed is distance divided by time. What distance does a point on Mercury's equator travel as the planet makes one rotation?) (b) Use your answer to (a) to answer the following: As a result of rotation, what difference in wavelength is observed for a radio wave of wavelength 12.5 cm (such as is actually used in radar studies of Mercury) emitted from either the approaching or receding edge of the planet?

52. The orbital period of *Mariner 10* is twice that of Mercury. Use this fact to calculate the length of the semimajor axis of the spacecraft's orbit.

53. Much of the area shown in Figure 11-10 is pockmarked with small impact craters. By contrast, there are very few small craters within the large crater at the upper right of this figure. Explain why this suggests that the interior of this crater was flooded with lava while the area outside the crater was not.

54. Consider the idea that Mercury has a solid iron-bearing mantle that is permanently magnetized like a giant bar magnet. Using the fact that iron demagnetizes at temperatures above 770°C, present an argument against this explanation of Mercury's magnetic field.

55. (a) At what wavelength does Venus's surface emit the most radiation? (b) Do astronomers have telescopes that can detect this radiation? (c) Why can't we use such telescopes to view the planet's surface?

56. The *Mariner 2* spacecraft detected more microwave radiation when its instruments looked at the center of Venus's disk than when it looked at the edge, or limb, of the planet. (This effect is called *limb darkening.*) Explain how these observations show that the microwaves are emitted by the planet's surface rather than its atmosphere.

57. In the classic Ray Bradbury science-fiction story "All Summer in a Day," human colonists on Venus are subjected to con-

tinuous rainfall except for one day every few years when the clouds part and the Sun comes out for an hour or so. Discuss how our understanding of Venus's atmosphere has evolved since this story was first published in 1954.

58. Suppose that Venus had no atmosphere at all. How would the albedo of Venus then compare with that of Mercury or the Moon? Explain your answer.

59. A hypothetical planet has an atmosphere that is opaque to visible light but transparent to infrared radiation. How would this affect the planet's surface temperature? Contrast and compare this hypothetical planet's atmosphere with the greenhouse effect in Venus's atmosphere.

60. The Earth's northern hemisphere is 39% land and 61% water, while its southern hemisphere is only 19% land and 81% water. Thus, the southern hemisphere could also be called the "water hemisphere." The Moon also has two distinct hemispheres, the near side (which has a number of maria) and the far side (which has almost none). How are these hemispheric differences on Earth and on the Moon similar to the Martian crustal dichotomy? How are they different?

61. For a group of properly attired astronauts equipped with oxygen tanks, a climb to the summit of Olympus Mons would actually be a relatively easy (albeit long) hike rather than a true mountain climb. Give two reasons why.

62. On Mars, the difference in elevation between the highest point (the summit of Olympus Mons) and the lowest point (the bottom of the Hellas Planitia basin) is 30 km. On Earth, the corresponding elevation difference (from the peak of Mount Everest to the bottom of the deepest ocean) is only 20 km. Discuss why the maximum elevation difference is so much greater on Mars.

63. The *Mars Global Surveyor* (MGS) spacecraft is in a nearly circular orbit around Mars with an orbital period of 117 minutes. (a) Using the data in Table 11-3, find the radius of the orbit. (b) What is the average altitude of MGS above the Martian surface? (c) The orbit of MGS passes over the north and south poles of Mars. Explain how this makes it possible for the spacecraft to observe the entire surface of the planet.

64. The elevations in Figure 11-20 were measured using an instrument called MOLA (Mars Orbiter Laser Altimeter) on board *Mars Global Surveyor*. MOLA fires a laser beam downward, then measures how long it takes for the beam to return to the spacecraft after reflecting off the surface. Suppose MOLA measures this round-trip time for the laser beam to reflect off the summit of Olympus Mons, as well as the round-trip time to reflect off the bottom of Hellas Planitia (see Question 61). Which round-trip time is longer? How much longer is it? Do you need to know the distance from *Mars Global Surveyor* to the surface to answer this question?

65. (a) The Grand Canyon in Arizona was formed over 15 to 20 million years by the flowing waters of the Colorado River, as well as by rain and wind. Contrast this formation scenario to that of Valles Marineris on Mars. (b) Valles Marineris is sometimes called "the Grand Canyon of Mars." Is this an appropriate description? Why or why not?

66. Water has a density of 1000 kg/m^3, so a column of water n meters tall and 1 meter square at its base has a mass of $n \times$ 1000 kg. On either the Earth or Venus, which have nearly the same surface gravity, a mass of 1 kg weighs about 9.8

newtons (2.2 lb). Calculate how deep you would have to descend into the Earth's oceans for the pressure to equal the atmospheric pressure on Venus's surface, 90 atm or 9×10^6 newtons per square meter. Give your answer in meters.

67. Marine organisms produce sulfur-bearing compounds, some of which escape from the oceans into the Earth's atmosphere. (These compounds are largely responsible for the characteristic smell of the sea.) Even more sulfurous gases are injected into our atmosphere by the burning of sulfur-rich fossil fuels, such as coal, in electric power plants. Both of these processes add more sulfur compounds to the atmosphere than do volcanic eruptions. On lifeless Venus, by contrast, volcanoes are the only source for sulfurous atmospheric gases. Why, then, are sulfur compounds so much rarer in our atmosphere than in the Venusian atmosphere?

68. The classic 1950 science-fiction movie *Rocketship X-M* shows astronauts on the Martian surface with oxygen masks for breathing but wearing ordinary clothing. Would this be a sensible choice of apparel for a walk on Mars? Why or why not?

69. Suppose that the only information you had about Mars was the images of the surface in Figure 11-31. Describe at least two ways that you could tell from these images that Mars has an atmosphere.

70. Although the *Viking Lander 1* and *Viking Lander 2* landing sites are 6500 km apart and have different geologic histories, the chemical compositions of the dust at both sites are nearly identical. (a) What does this suggest about the ability of the Martian winds to transport dust particles? (b) Would you expect that larger particles such as pebbles would also have identical chemical compositions at the two *Viking Lander* sites? Why or why not?

71. Why do you suppose that Phobos and Deimos are not round like our Moon?

72. The orbit of Phobos has a semimajor axis of 9378 km. Use this information and the orbital period given in the text to calculate the mass of Mars. How does your answer compare with the mass of Mars given in Table 11-3?

73. Calculate the angular sizes of Phobos and Deimos as they pass overhead, as seen by an observer standing on the Martian equator. How do these sizes compare with that of the Moon seen from the Earth's surface? Would Phobos and Deimos appear as impressive in the Martian sky as the Moon does in our sky?

74. You are to put a spacecraft into a synchronous circular orbit around the Martian equator, so that its orbital period is equal to the planet's rotation period. Such a spacecraft would always be over the same part of the Martian surface. (a) Find the radius of the orbit and the altitude of the spacecraft above the Martian surface. (b) Suppose Mars had a third moon that was in a synchronous orbit. Would tidal forces make this moon tend to move toward Mars, away from Mars, or neither? Explain.

Discussion Questions

75. Before about 350 B.C., the ancient Greeks did not realize that Mercury seen in the morning sky (which they called Apollo) and seen in the evening sky (which they called Hermes) were

actually the same planet. Discuss why you think it took some time to realize this.

76. If you were planning a new mission to Mercury, what features and observations would be of particular interest to you?

77. What evidence do we have that the surface features on Mercury were not formed during recent geological history?

78. Describe the apparent motion of the Sun during a "day" on Venus relative to (a) the horizon and (b) the background stars. (Assume that you can see through the cloud cover.)

79. If you could examine rock samples from the surface of Venus, would you expect them to be the same as rock samples from Earth? Would you expect to find igneous, sedimentary, and metamorphic rocks like those found on Earth (see Section 9-3)? Explain your answers.

80. In 1978 the *Pioneer Venus Orbiter* spacecraft arrived at Venus. It carried an ultraviolet spectrometer to measure the chemical composition of the Venusian atmosphere. This instrument recorded unexpectedly high levels of sulfur dioxide and sulfuric acid, which steadily declined over the next several years. Discuss how this observation suggests that volcanic eruptions occurred on Venus not long before *Pioneer Venus Orbiter* arrived there.

81. The total cost of the *Mars Global Surveyor* mission was about $154 million. (To put this number into perspective, in 2000 the U.S. Mint spent about $40 million to advertise its new $1 coin, which failed to be accepted by the public. Several recent Hollywood movies have had larger budgets than *Mars Global Surveyor*.) Does this expenditure seem reasonable to you? Why or why not?

Web/eBook Questions

82. **Elongations of Mercury.** Access the animation "Elongations of Mercury" in Chapter 11 of the *Universe* Web site or eBook. (a) View the animation and notice the dates of the greatest eastern and greatest western elongations. Which time interval is greater: from a greatest eastern elongation to a greatest western elongation, or vice versa? (b) Based on what you observe in the animation, draw a diagram to explain your answer to the question in (a).

83. Just as Mercury can pass in front of the Sun as seen from Earth (see Question 45), so can Venus. Transits of Venus are quite rare. The dates of the only transits in the twenty-first century are June 8, 2004, and June 6, 2012; the next ones will occur in 2117 and 2125, A number of European astronomers traveled to Asia and the Pacific islands to observe the transits of Venus in 1761 and 1769. Search the World Wide Web for information about these expeditions. Why were these events of such interest to astronomers? How definitive were the results of these observations?

84. **Surface Temperature of Venus.** Access the Active Integrated Media Module "Wien's Law" in Chapter 5 of the *Universe* Web site or eBook. (a) Using the Wien's Law calculator, determine Venus's approximate temperature if it emits blackbody radiation with a peak wavelength of 3866 nm. (b) By trial and error, find the wavelength of maximum emission for a surface temperature of

733 K (for present-day Venus) and a surface temperature of 833 K (as it might be in the event of a global catastrophe that released more greenhouse gases into Venus's atmosphere). In what part of the electromagnetic spectrum do these wavelengths lie?

85. Search the World Wide Web for information about possible manned missions to Mars. How long might such a mission take? How expensive would such a project be? What would be the advantages of a manned mission compared to an unmanned one?

86. **Conjunctions of Mars.** Access and view the animation "The Orbits of Earth and Mars" in Chapter 11 of the *Universe* Web site or eBook. (a) The animation highlights three dates when Mars is in opposition, so that the Earth lies directly between Mars and the Sun. By using the "Stop" and "Play" buttons in the animation, find two times during the animation when Mars is in *conjunction*, so that the Sun lies directly between Mars and the Earth (see Figure 4-6). For each conjunction, make a drawing showing the positions of the Sun, the Earth, and Mars, and record the month and year when the conjunction occurs. (*Hint:* See Figure 11-3.) (b) When Mars is in conjunction, at approximately what time of day does it rise as seen from Earth? At what time of day does it set? Is Mars suitably placed for telescopic observation when in conjunction?

Activities

Observing Projects

> **Observing tips and tools**
>
> Mercury and Venus are visible in the morning sky when it is at or near greatest western elongation and in the evening sky when at or near greatest eastern elongation. These are also the times when Venus can be seen at the highest altitude above the horizon during the hours of darkness. Mars is most easily seen around an opposition. At other times Mars may be visible only in the early morning hours before sunrise or in the early evening just after sunset. Consult such magazines as *Sky & Telescope* and *Astronomy* or their Web sites for more detailed information about when and where to look for Mercury, Venus, and Mars during a given month. You can also use the *Starry Night Enthusiast™* program on the CD-ROM that certain printed copies of accompanies this textbook.

87. Refer to the *Universe* Web site or eBook for a link to a Web site that calculates the dates of upcoming greatest elongations of Mercury. Consult such magazines as *Sky & Telescope* and *Astronomy*, or the Web sites for these magazines, to determine if any of these greatest elongations is going to be a favorable one. If so, make plans to be one of those rare individuals who has actually seen the innermost planet of the solar system. Set

aside several evenings (or mornings) around the date of the favorable elongation to reduce the chances of being "clouded out." Select an observing site that has a clear, unobstructed view of the horizon where the Sun sets (or rises). If possible, make arrangements to have a telescope at your disposal. Search for the planet on the dates you have selected, and make a drawing of its appearance through your telescope.

88. *This observing project should be performed only under the direct supervision of an astronomer who knows how to point a telescope safely at Mercury.* Make arrangements to view Mercury during broad daylight. This is best done by visiting an observatory where the coordinates (right ascension and declination) of Mercury's position can be used to point the telescope. **DO NOT LOOK AT THE SUN! Looking directly at the Sun can cause blindness.**

89. Refer to the *Universe* Web site or eBook for a link to a Web site that calculates the dates of greatest elongations of Venus. If Venus is near a greatest elongation, view the planet through a telescope. Make a sketch of the planet's appearance. From your sketch, can you determine if Venus is closer to us or farther from us than the Sun?

90. Using a small telescope, observe Venus once a week for a month and make a sketch of the planet's appearance on each occasion. From your sketches, can you determine whether Venus is approaching us or moving away from us?

91. *This observing project should be performed only under the direct supervision of an astronomer who knows how to point a telescope safely at Venus.* Make arrangements to view Venus during broad daylight. This is best done by visiting an observatory where the coordinates (right ascension and declination) of Venus's position can be used to point the telescope. **DO NOT LOOK AT THE SUN! Looking directly at the Sun can cause blindness.**

92. If Mars is suitably placed for observation, arrange to view the planet through a telescope. Draw a picture of what you see. What magnifying power seems to give you the best image? Can you distinguish any surface features? Can you see a polar cap or dark markings? If not, can you offer an explanation for Mars's bland appearance?

93. Use the *Starry Night Enthusiast*™ program to observe solar transits of Mercury (see Question 45) when the planet passes in front of the Sun as viewed from Earth. Display the entire celestial sphere (select **Guides > Atlas** in the **Favourites** menu). Open the **Find** pane and click the menu button for Mercury and select **Centre** from the menu. Using the controls at the right-hand end of the Toolbar, zoom in until the field of view is about 1° × 1°. (a) In the toolbar, stop time flow and set the date and time to November 8, 2006, at 12:00:00 A.M. Set the **Time Flow Rate** to **1 hour.** Step backward or forward through time using the single-step buttons (the leftmost and the rightmost buttons) and record the times at which the solar transit begins and ends. You can zoom in to a field of view of about 11 × 8 arc minutes centered upon Mercury and change the **Time Flow Rate** to **1 minute** and use the **Time Flow** controls in the toolbar to move back and forth in time to increase the accuracy of your measurement. What is the total duration of

the solar transit? (b) Set the date and time to May 9, 2016, at 12:00:00 A.M. Again, step backward or forward through time, record the times when the solar transit begins and ends, and find the total duration of the solar transit. (c) The maximum duration of a solar transit of Mercury is 9 hours. Explain why the transits you observed in (a) and (b) last a substantially shorter time.

94. Use the *Starry Night Enthusiast*™ program to examine Mercury. Select **Favourites > Solar System > Mercury** from the menu. Select **View > Feet** from the menu to remove the image of the astronaut's spacesuit from the view. Select **Options > Solar System > Planets-Moons. . .** from the menu to allow you to examine the complete surface of the planet. In the Planets-Moons Options dialog box, slide the control next to the label **Show dark side** all the way to the right (**Brighter**) end of the scale. Click **OK** to close the dialog box. Use the **Zoom** controls in the toolbar to zoom in and out on the view. Rotate the image of Mercury by placing the mouse cursor over the image of the planet, holding down the mouse button, and moving the mouse. (On a two-button mouse, hold down the left mouse button.) As you explore the surface of the planet Mercury, estimate the diameter of the largest craters by measuring their size on the screen with a ruler and comparing to the diameter of Mercury (see Table 11-1). (Note that the rosette patterns surrounding both planetary poles are an artifact of the technique used to produce these images).

95. Use the *Starry Night Enthusiast*™ program to observe solar transits of Venus (see Question 83). Display the entire celestial sphere (select **Guides > Atlas** in the **Favourites** menu). Open the **Find** pane and click the menu button for Venus (the downward-pointing blue arrow to the left of Venus) in the list. Select **Centre** from the menu. Use the zoom controls in the toolbar to adjust the field of view to about 1° × 1°. (a) In the toolbar, **Stop time flow** and then set the **Time** and **Date** to June 8, 2004, at 12:00:00 A.M. Set the **Time Flow Rate** to **1 hour.** Step backward or forward through time using the single-step buttons (the leftmost and the rightmost buttons) and record the times at which the solar transit begins and ends, changing the Time Flow Rate to 1 minute and the field of view to increase the accuracy of your measurement, as necessary. What is the total duration of the solar transit? (b) The ecliptic appears in *Starry Night Enthusiast*™ as a green line. During the transit, is Venus precisely on the ecliptic? If not, about how far off is it? (*Hint:* The Sun has an angular diameter of about 30 arcmin.) (c) Repeat parts (a) and (b) for the solar transit of Venus on June 6, 2012.

96. Use the *Starry Night Enthusiast*™ program to observe the apparent motion of Venus on the celestial sphere. Display the entire celestial sphere (select **Guides > Atlas** in the **Favourites** menu). Open the **Find** pane and click the menu button in the list to the left of the label for the Sun. Select **Centre** from the menu that appears. Using the controls at the right-hand end of the toolbar, zoom out until the field of view is 100°. **Stop** Time Flow and in the toolbar, set the date and time to January 1, 2007, at 12:00:00 A.M. and the **Time Flow Rate** to **1 day.**

(a) Use the **Run Time Forward** and **Stop time** buttons to find the first date after January 1, 2007, when Venus is as far to the right of the Sun as possible, and the first date after January 1, 2007, when Venus is as far to the left of the Sun as possible. What is your interpretation of these two dates and how would you label them? (b) Set the date to December 1, 2007, and start the animation by clicking on the **Run Time Forward** button. Based on your observations, explain why Venus has neither a greatest western elongation nor a greatest eastern elongation during 2008.

97. Use the *Starry Night Enthusiast*™ program to compare the orbits of Venus and the Earth. Select **Options > Viewing Location. . .** from the menu. In the Viewing Location dialog box, set the **View from** to a **position moving with** the **Sun** and choose the option **Above orbital plane.** Then click the **Set Location** button to close the dialog. Click on and hold the **Increase current elevation** button in the Viewing Location section of the toolbar until the distance shown in the Viewing Location display pane is approximately **1.5 au** from the Sun. Open the **Find** pane and click both of the checkboxes on either side of the listing for the Earth and for Venus. This labels the two planets and draws their orbits in the view. Close the Find pane. **Stop** time flow and set the time and the date to 0:00:00 UT on August 19, 2007 AD. You can zoom in and zoom out on these two planets and their orbits using the buttons in the **Zoom** section of the toolbar. You can also rotate the solar system by holding down the **Shift** key and then holding down the mouse button and moving the mouse. (On a two-button mouse, hold down the left mouse button.) Are the orbits of Venus and of the Earth in the same plane? At the time shown in the image, is Venus nearest to inferior conjunction, superior conjunction, greatest eastern elongation, or greatest western elongation as seen from Earth? Explain your answers. Rotate your view to look down upon the orbits from above the pole of the Sun. Are the orbits of Earth and Venus circular?

98. Use the *Starry Night Enthusiast*™ program to observe the appearance of Mars. Select **Favourites > Guides > Atlas** from the menu. Select **View > Celestial Grid** from the menu to turn this option off. Open the **Find** pane and click the menu button for **Mars** and choose **Centre** from the menu. Close the Find pane and then use the Zoom controls in the toolbar to set a field of view of approximately 58″ × 40″. (a) Set the **Time Flow Rate** to **1 hour** and then run **Time Forward**. Describe what you see. (b) **Stop** time flow. Change the **Time Flow Rate** to **1 lunar month. Run Time Forward** again. Describe what you see. Using a diagram like Figure 4-6, explain the changes in the apparent size of the planet. (c) **Stop** time flow and **zoom out** to a field of view of approximately 2′ × 1′. Change the time and date in the toolbar to 12:00:00 A.M. on August 28, 2003, to see Mars during a very favourable opposition. You will see Mars and its two moons, Phobos and Deimos. In the toolbar, set the **Time Flow Rate** to **1 minute**. Record the date and time in the display, and note the position of Phobos (the inner moon). Click the **Run Time Forward** and single time step button (the rightmost time control button) to advance time until Phobos returns to approximately the same

position relative to Mars. Record the date and time in the display. From your observations, what is the orbital period of Phobos? How does your result compare with the orbital period given in Appendix 3? (d) Repeat part (c) for Deimos (the outer moon).

99. Use the *Starry Night Enthusiast*™ program to compare the orbits of Mars and the Earth. **Select Options > Viewing Location. . .** from the menu. In the Viewing Location dialog box, set the **View from** to a **position moving with** the **Sun** and choose the option **Above orbital plane** and click the **Set Location** button to close the dialog. Click on and hold the **Increase current elevation** button in the Viewing Location section of the toolbar until the distance shown in the Viewing Location display pane is approximately **3.5 au** from the Sun. Open the **Find** pane and click the two checkboxes on either side of the listings for the **Earth** and for **Mars** to label these two planets and draw their orbits in the view. Close the Find pane. You can zoom in and zoom out using the buttons in the **Zoom** section of the toolbar. You can also rotate the solar system by holding down the **Shift** key and then holding down the mouse button and moving the mouse. (On a two-button mouse, hold down the left mouse button.) Stop time flow and set the time and date to 0:00:00 UT on October 16, 2005 AD. (a) Is Mars a few weeks before an opposition or a few weeks after an opposition at this time on this date? (*Hint:* Arrowheads on the orbits show you which way each planet moves around its orbit.) (b) Are the orbits of Mars and of the Earth in the same plane? Use your answer to explain the shape of the apparent path of Mars on the celestial sphere in 2011–2012, shown in Figure 4-2.

Collaborative Exercises

100. Figures 11-6*b* and 11-6*c* show a planet in synchronous rotation and Mercury with a 3-to-2 spin-orbit coupling, respectively. Stand up and demonstrate how planets move in each of these rotations by "orbiting" around a stationary classmate who represents our Sun. How would Mercury's motion be different if it had a 4-to-2 spin-orbit coupling instead?

101. In the nineteenth century, French mathematician and astronomer U. J. J. Le Verrier led a failed search for a hypothetical planet named Vulcan (after the mythical blacksmith of the gods) orbiting closer to our Sun than Mercury. If Vulcan had an orbit with the same eccentricity as Mercury's orbit but only one-half the size, what would have been its maximum eastern and western elongations?

102. Figure 11-7 shows prograde and retrograde rotation. Stand up and demonstrate how Venus rotates as it orbits our Sun, using a classmate as our stationary Sun. How is this different from how Earth rotates in its orbit?

103. The image of Mars that opens this chapter is from the Hubble Space Telescope. Draw a circle on your paper roughly 5 cm in diameter and, taking turns, have each person in your group sketch a different region of Mars. How is your collaborative sketch different than Schiaparelli's drawing shown in Figure 11-16?

Robotic Geologists and the Search for Habitable Environments on Mars

by John Grotzinger

Scientists discover that which exists; engineers create that which never was.

Theodor von Kármán, founder of NASA's Jet Propulsion Lab

Humans are explorers by nature. The great quests of human history have presented grand challenges that by necessity have inspired creative design of ships of exploration. In the search to understand the history of water on Mars, and its potential role in enabling the evolution of life on the most Earthlike terrestrial planet, human imagination has been no less bridled. Since humans cannot yet travel to other planets, we have developed robots that can survive interplanetary travel and the harsh surface conditions on other planets. These robotic explorers are designed to act as geologists, which was the case for the 2004 *Mars Exploration Rover* mission.

In June 2003, two golf cart–sized rovers destined to land on the Martian surface were launched from Cape Canaveral, Florida, and began their 300-million-kilometer journey to the Red Planet. This mission succeeded beyond anyone's expectations, making 2004, 2005, and 2006 three of the greatest years in the history of space exploration. The *Mars Exploration Rovers—Spirit* and *Opportunity*—were designed to survive 3 months under the hostile Martian surface conditions and drive no farther than 300 m. Remarkably, at the time this article was written, each rover has operated for more than two and a half years, and driven a combined distance of more than 15 km! The rovers have had to survive nighttime temperatures below –150 °C, dust devils that could have tipped them over, global dust storms that diminished their solar power, and drives along rocky slopes of almost 30° and through piles of treacherous windblown sand and dust. The rovers have also discovered a treasure trove of geologic wonders. These discoveries include compelling evidence for water on the ancient Martian surface—a necessary condition for life, at least as it exists on Earth, and can be contemplated for Mars.

In particular, the rover *Opportunity* has discovered evidence from the Meridiani Planum area that water may have pooled in shallow lakes, known as playas, and infiltrated pores in sediments and soils to cause aqueous alteration. So we now know that water was present in the shallow subsurface environment and was at least intermittently present at the surface as shallow lakes and streams. But the evidence for water alone is not enough to demonstrate that life may have existed. One must also ask if the water would have had physical attributes that would have allowed life to develop.

Mars today is a forbidding place. Temperature and atmospheric pressure are so low that liquid water is not stable on the present-day Martian surface. The surface is also chemically harsh and subject to strong radiation. It is doubtful that organisms thrive today at the Martian surface. *Opportunity*'s discoveries indicate that surface environments may have, unfortunately, been biologically challenging for *most* of Mars's history. The salty and transient streams and lakes that covered Meridiani 3 to 4 billion years ago indicate that while chemical weathering and erosion provided many of the elements required for life, ambient environments were arid, acidic, and oxidizing. Terrestrial ecology suggests that microorganisms could survive many aspects of the inferred Meridiani environment, but habitability would depend critically on the time scale of sufficiently high water activity to support cell biology—a time scale that is currently unknown.

Whether Meridiani is broadly representative of the Martian surface 3 to 4 billion years ago is unknown, but remote sensing from orbiting spacecraft suggests that it could be. The sulfate minerals that are so widespread at Meridiani seem to be widespread in others areas of the planet, particularly the equatorial lowlands. However, there appear to be hints that other combinations of minerals—some dominated by clays—may be associated with the oldest rocks on Mars. From an astrobiological perspective this is encouraging because it allows for less acidic conditions. All considered, the formerly

John Grotzinger is a field geologist interested in the evolution of Earth's surficial environments and biosphere. He received a B.S. in geoscience from Hobart College in 1979, an M.S. in geology from University of Montana in 1981, and a Ph.D. in geology from Virginia Tech in 1985. He currently works as a geologist on the *Mars Exploration Rover* team. This mission is the first to conduct ground-based exploration of the bedrock geology of another planet, resulting in the discovery of sedimentary rocks formed in aqueous depositional environments.

(continued on the next page)

Robotic Geologists and the Search for Habitable Environments on Mars *(continued)*

water-drenched rocks at Meridiani are biologically permissive, but may record the close of a biologically favorable window in the history of Mars, not its beginning.

For future missions, such as the *Mars Science Laboratory* rover, due to launch in 2009, the most promising places to look for evidence of Mars surface life are probably sedimentary basins that preserve a record of Mars's earliest history, when water was most abundant and persistent, and both oxi-dation and acidity were least developed. *Mars Science Laboratory* will come with a science payload that is designed to search for large organic molecules that would provide unambiguous tests for past life. If present, their discovery would surely rank as one of the greatest benchmarks in the history of scientific exploration and would no doubt significantly influence the way that we think about the origin and evolution of life in the universe as a whole.

12

Jupiter and Saturn: Lords of the Planets

 R I V U X G

Jupiter (left) and Saturn (right) to scale. (NASA/JPL/Space Science Institute)

Among the most remarkable sights in our solar system are the colorful, turbulent atmosphere of Jupiter and the ethereal beauty of Saturn's rings. Both of these giant worlds dwarf our own planet: More than 1200 Earths would fit inside Jupiter's immense bulk, and more than 700 Earths inside Saturn. Unlike the Earth, both Jupiter and Saturn are composed primarily of the lightweight elements hydrogen and helium, in abundances very similar to those in the Sun.

As these images show, the rapid rotations of both Jupiter and Saturn stretch their weather systems into colorful bands that extend completely around each planet. Both planets also have intense magnetic fields that are generated in their interiors, where hydrogen is so highly compressed that it becomes a metal.

Saturn is not merely a miniature version of Jupiter, however. Its muted colors and more flattened shape are clues that Saturn's atmosphere and interior have important differences from those of Jupiter. The most striking difference is Saturn's elaborate system of rings, composed of countless numbers of icy fragments orbiting in the plane of the planet's equator. The rings display a complex and elegant structure, which is shaped by subtle gravitational influences from Saturn's retinue of moons. Jupiter, too, has rings, but they are made of dark, dustlike particles that reflect little light. These systems of rings, along with the distinctive properties of the planets themselves, make Jupiter and Saturn highlights of our tour of the solar system.

12-1 Jupiter and Saturn are the most massive planets in the solar system

Jupiter and Saturn are respectively the largest and second largest of the planets, and the largest objects in the solar system other

Learning Goals

By reading the sections of this chapter, you will learn

12-1 What observations from Earth reveal about Jupiter and Saturn

12-2 How Jupiter and Saturn rotate differently from terrestrial planets like Earth

12-3 The nature of the immense storms seen in the clouds of Jupiter and Saturn

12-4 How the internal heat of Jupiter and Saturn drives activity in their atmospheres

12-5 What the *Galileo* space probe revealed about Jupiter's atmosphere

12-6 How the shapes of Jupiter and Saturn indicate the sizes of their rocky cores

12-7 How Jupiter and Saturn's intense magnetic fields are produced by an exotic form of hydrogen

12-8 The overall structure and appearance of Saturn's ring system

12-9 What kinds of particles form the rings of Jupiter and Saturn

12-10 How spacecraft observations revealed the intricate structure of Saturn's rings

12-11 How Saturn's rings are affected by the presence of several small satellites

Table 12-1   Jupiter Data

Average distance from Sun:	5.203 AU $= 7.783 \times 10^8$ km
Maximum distance from Sun:	5.455 AU $= 8.160 \times 10^8$ km
Minimum distance from Sun:	4.950 AU $= 7.406 \times 10^8$ km
Eccentricity of orbit:	0.048
Average orbital speed:	13.1 km/s
Orbital period:	11.86 years
Rotation period:	$9^h \, 50^m \, 28^s$ (equatorial) $9^h \, 55^m \, 29^s$ (internal)
Inclination of equator to orbit:	3.12°
Inclination of orbit to ecliptic:	1.30°
Diameter:	142,984 km = 11.209 Earth diameters (equatorial) 133,708 km = 10.482 Earth diameters (polar)
Mass:	1.899×10^{27} kg = 317.8 Earth masses
Average density:	1326 kg/m^3
Escape speed:	60.2 km/s
Surface gravity (Earth = 1):	2.36
Albedo:	0.44
Average temperature at cloudtops:	$-108°C = -162°F = 165$ K
Atmospheric composition (by number of molecules):	86.2% hydrogen (H_2), 13.6% helium (He), 0.2% methane (CH_4), ammonia (NH_3), water vapor (H_2O), and other gases

WEB LINK 12.2

R I V UX G

(NASA/JPL)

than the Sun itself. Astronomers have known for centuries about the huge diameters and immense masses of these two planets. Given the distance to each planet and their angular sizes, they used the small-angle formula (Box 1-1) to calculate that Jupiter is about 11 times larger in diameter than Earth, while Saturn's diameter is about 9 times larger than Earth's (see Figure 7-2). By observing the orbits of Jupiter's four large moons and Saturn's large moon Titan (see Table 7-2) and applying Newton's form of Kepler's third law (see Section 4-7 and Box 4-4), astronomers also determined that Jupiter and Saturn are, respectively, 318 times and 95 times more massive than the Earth. In fact, Jupiter has 2½

times the combined mass of all the other planets, satellites, asteroids, meteoroids, and comets in the solar system. A visitor from interstellar space might well describe our solar system as the Sun, Jupiter, and some debris! Table 12-1 and Table 12-2 list some basic data about Jupiter and Saturn.

Observing Jupiter and Saturn

As for any planet whose orbit lies outside the Earth's orbit, the best time to observe Jupiter or Saturn is when they are at opposition. At opposition, Jupiter can appear nearly 3 times brighter than Sirius, the brightest star in the sky. Only the Moon and

Table 12-2 Saturn Data

Average distance from Sun:	$9.572 \text{ AU} = 1.432 \times 10^9 \text{ km}$
Maximum distance from Sun:	$10.081 \text{ AU} = 1.508 \times 10^9 \text{ km}$
Minimum distance from Sun:	$9.063 \text{ AU} = 1.356 \times 10^9 \text{ km}$
Eccentricity of orbit:	0.053
Average orbital speed:	9.64 km/s
Orbital period:	29.37 years
Rotation period:	$10^h 13^m 59^s$ (equatorial)
	$10^h 39^m 25^s$ (internal)
Inclination of equator to orbit:	26.73°
Inclination of orbit to ecliptic:	2.48°
Diameter:	120,536 km = 9.449 Earth diameters (equatorial)
	108,728 km = 8.523 Earth diameters (polar)
Mass:	$5.685 \times 10^{26} \text{ kg} = 95.16$ Earth masses
Average density:	687 kg/m^3
Escape speed:	35.5 km/s
Surface gravity (Earth = 1):	0.92
Albedo:	0.46
Average temperature at cloudtops:	$-180°C = -292°F = 93 \text{ K}$
Atmospheric composition (by number of molecules):	96.3% hydrogen (H_2), 3.3% helium (He), 0.4% methane (CH_4), ammonia (NH_3), water vapor (H_2O), and other gases

R I **V** U X G

(STScI/Hubble Heritage Team)

Venus can outshine Jupiter at opposition. Saturn's brightness at opposition is only about 1/7 that of Jupiter, but it still outshines all of the stars except Sirius and Canopus.

Through a telescope, Jupiter at opposition presents a disk nearly 50 arcsec in diameter, approximately twice the angular diameter of Mars under the most favorable conditions (**Figure 12-1a**). Because Jupiter takes almost a dozen Earth years to orbit the Sun, it appears to meander slowly across the 12 constellations of the zodiac at the rate of approximately one constellation per year. Successive oppositions occur at intervals of about 13 months.

Saturn's orbital period is even longer, more than 29 Earth years, so it moves even more slowly across the zodiac. If Saturn did not move at all, successive oppositions would occur exactly one Earth year apart; thanks to its slow motion, Saturn's oppositions occur at intervals of about one year and two weeks. At opposition Saturn appears as a disk about 20 arcsec in diameter, with a dramatic difference from Jupiter: Saturn is surrounded by a magnificent system of rings (Figure 12-1b). Later in this chapter we will examine these rings in detail.

Even a small telescope reveals colored bands on Jupiter and rings around Saturn

(a)

(b)

Figure 12-1   R I [V] U X G

Jupiter and Saturn as Viewed from Earth (a) The disk of Jupiter at opposition appears about $2\frac{1}{2}$ times larger than **(b)** the disk of Saturn at opposition. Both planets display dark and light bands, though these are fainter on Saturn. (a: Courtesy of Stephen Larson; b: NASA)

Dark Belts and Light Zones

As seen through an Earth-based telescope, both Jupiter and Saturn display colorful bands that extend around each planet parallel to its equator (Figure 12-1). More detailed close-up views of Jupiter from a spacecraft (**Figure 12-2***a*) show alternating dark and light bands parallel to Jupiter's equator in subtle tones of red, orange, brown, and yellow. The dark, reddish bands are called **belts,** and the light-colored bands are called **zones.** The image of

Saturn in Figure 12-2*b* shows similar features, but with colors that are much less pronounced. (We will see in Section 12-4 how differences between the atmospheres of Jupiter and Saturn can explain Saturn's washed-out appearance.)

In addition to these conspicuous stripes, a huge, red-orange oval called the **Great Red Spot** is often visible in Jupiter's southern hemisphere. This remarkable feature was first seen by the English scientist Robert Hooke in 1664 but may be much older. It appears to be an extraordinarily long-lived storm in the planet's

(a) Jupiter

(b) Saturn

Figure 12-2   R I [V] U X G

Jupiter and Saturn as Viewed from Space
(a) This view of Jupiter is a composite of four images made by the *Cassini* spacecraft as it flew past Jupiter in 2000.

(b) Images from the *Voyager 2* spacecraft as it flew past Saturn in 1981 were combined to show the planet in approximately natural color. (a: NASA/JPL/University of Arizona; b: USGS/NASA/JPL) dynamic atmosphere. Saturn has no long-lived storm systems of this kind. However, many careful observers have reported smaller spots and blemishes in the atmospheres of both Jupiter and Saturn that last for only a few weeks or months.

12-2 Unlike the terrestrial planets, Jupiter and Saturn exhibit differential rotation

Observations of features like the Great Red Spot and smaller storms allow astronomers to determine how rapidly Jupiter and Saturn rotate. At its equator, Jupiter completes a full rotation in only 9 hours, 50 minutes, and 28 seconds, making it not only the largest and most massive planet in the solar system but also the one with the fastest rotation. However, Jupiter rotates in a strikingly different way from the Earth, the Moon, Mercury, Venus, or Mars.

Differential Rotation

If Jupiter were a solid body like a terrestrial planet (or, for that matter, a billiard ball), all parts of Jupiter's surface would rotate through one complete circle in this same amount of time (**Figure 12-3a**). But by watching features in Jupiter's cloud cover, Gian Domenico Cassini discovered in 1690 that the polar regions of the planet rotate a little more slowly than do the equatorial regions. (You may recall this Italian astronomer from Section 11-2 as the gifted observer who first determined Mars's rate of rotation.) Near the poles, the rotation period of Jupiter's atmosphere is about 9 hours, 55 minutes, and 41 seconds. Saturn, too, has a

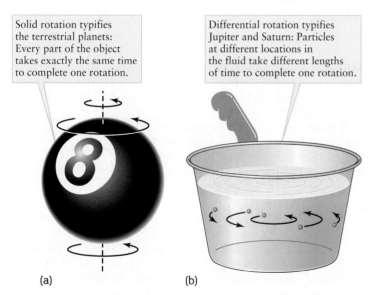

Solid rotation typifies the terrestrial planets: Every part of the object takes exactly the same time to complete one rotation.

Differential rotation typifies Jupiter and Saturn: Particles at different locations in the fluid take different lengths of time to complete one rotation.

(a)   (b)

Figure 12-3

Solid Rotation versus Differential Rotation (a) All parts of a solid object rotate together, but **(b)** a rotating fluid displays differential rotation. To see this, put some grains of sand, bread crumbs, or other small particles in a pot of water. Stir the water with a spoon to start it rotating, then take out the spoon. The particles near the center of the pot take less time to make a complete rotation than those away from the center.

longer rotation period near its poles (10 hours, 39 minutes, and 24 seconds) than at its equator (10 hours, 13 minutes, and 59 seconds).

ANALOGY You can see this kind of rotation, called **differential rotation,** in the kitchen. As you stir the water in a pot, different parts of the liquid take different amounts of time to make one "rotation" around the center of the pot (Figure 12-3b). Differential rotation shows that neither Jupiter nor Saturn can be solid throughout their volumes: They must be at least partially fluid, like water in a pot.

The Compositions of Jupiter and Saturn

If Jupiter and Saturn have partially fluid interiors, they cannot be made of the rocky materials that constitute the terrestrial planets. An important clue to the compositions of Jupiter and Saturn are their average densities, which are only 1326 kg/m^3 for Jupiter and 687 kg/m^3 for Saturn. (By comparison, the Earth's average density is 5515 kg/m^3.) To explain these low average densities, Rupert Wildt of the University of Göttingen in Germany suggested in the 1930s that Jupiter and Saturn are composed mostly of hydrogen and helium atoms—the two lightest elements in the universe—held together by their mutual gravitational attraction to form a planet. Wildt was motivated in part by his observations of prominent absorption lines of methane and ammonia in Jupiter's spectrum. (We saw in Section 7-3 how spectroscopy plays an important role in understanding the planets.) A molecule of methane (CH_4) contains four hydrogen atoms, and a molecule of ammonia (NH_3) contains three. The presence of these hydrogen-rich molecules was strong, but indirect, evidence of abundant hydrogen in Jupiter's atmosphere.

Direct evidence for hydrogen and helium in Jupiter's atmosphere, however, was slow in coming. The problem was that neither gas produces prominent spectral lines in the visible sunlight reflected from the planet. To show the presence of these elements conclusively, astronomers had to look for spectral lines in the ultraviolet part of the spectrum. These lines are very difficult to measure from the Earth, because almost no ultraviolet light penetrates our atmosphere (see Figure 6-25). Astronomers first detected the weak spectral lines of hydrogen molecules in Jupiter's spectrum in 1960. The presence of helium on Jupiter and Saturn was finally confirmed in the 1970s and 1980s, when spacecraft first flew past these planets and measured their hydrogen spectra in detail. (Collisions between helium and hydrogen atoms cause small but measurable changes in the hydrogen spectrum, which is what the spacecraft instruments detected.)

> **Spacecraft observations were needed to determine the compositions of Jupiter and Saturn**

Today we know that the chemical composition of Jupiter's atmosphere is 86.2% hydrogen molecules (H_2), 13.6% helium atoms, and 0.2% methane, ammonia, water vapor, and other gases. The percentages in terms of *mass* are somewhat different because a helium atom is twice as massive as a hydrogen molecule. Hence, by mass, Jupiter's atmosphere is approximately 75% hydrogen, 24% helium, and 1% other substances, quite similar to that of the Sun. We will see evidence in Section 12-6 that Jupiter has a large rocky core made of heavier elements. It is estimated that the breakdown by mass of the planet as a whole (atmosphere plus interior) is approximately 71% hydrogen, 24% helium, and 5% all heavier elements.

Saturn's Missing Helium

Like Jupiter, Saturn is thought to have a large rocky core. But unlike Jupiter, data from Earth-based telescopes and spacecraft show that the atmosphere of Saturn has a serious helium deficiency: Its chemical composition is 96.3% hydrogen molecules, 3.3% helium, and 0.4% other substances (by mass, 92% hydrogen, 6% helium, and 2% other substances). This is a puzzle because Jupiter and Saturn are thought to have formed in similar ways from the gases of the solar nebula (see Section 8-4), and so both planets (and the Sun) should have essentially the same abundances of hydrogen and helium. So where did Saturn's helium go?

The explanation may be simply that Saturn is smaller than Jupiter, and as a result Saturn probably cooled more rapidly. (We saw in Section 7-6 why a small world cools down faster than a large one.) This cooling would have triggered a process analogous to the way rain develops here on Earth. When the air is cool enough, humidity in the Earth's atmosphere condenses into raindrops that fall to the ground. On Saturn, however, it is droplets of liquid helium that condense within the planet's cold, hydrogen-rich outer layers. In this scenario, helium is deficient in Saturn's upper atmosphere simply because it has fallen farther down into the planet. By contrast, Jupiter's helium has not rained out because its upper atmosphere is warmer and the helium does not form droplets.

ANALOGY An analogy to helium "rainfall" within Saturn is what happens when you try to sweeten tea by adding sugar. If the tea is cold, the sugar does not dissolve well and tends to sink to the bottom of the glass even if you stir the tea with a spoon. But if the tea is hot, the sugar dissolves with only a little stirring. In the same way, it is thought that the descending helium droplets once again dissolve in hydrogen once they reach the warmer depths of Saturn's interior.

In this scenario, Jupiter and Saturn both have about the same overall chemical composition. But Saturn's smaller mass, less than a third that of Jupiter, means that there is less gravitational force tending to compress its hydrogen and helium. This explains why Saturn's density is only about half that of Jupiter, and is in fact the lowest of any planet in the solar system.

CAUTION! Because Jupiter and Saturn are almost entirely hydrogen and helium, it would be impossible to land a spacecraft on either planet. An astronaut foolish enough to try would notice the hydrogen and helium around the spacecraft becoming denser, the temperature rising, and the pressure increasing as the spacecraft descended. But the hydrogen and helium would never solidify into a surface on which the spacecraft could touch down. Long before reaching the planet's rocky core, the pressure of the hydrogen and helium would reach such unimaginably high levels that any spacecraft, even one made of the strongest known materials, would be crushed.

12-3 Spacecraft images show remarkable activity in the clouds of Jupiter and Saturn

Most of our detailed understanding of Jupiter and Saturn comes from a series of robotic spacecraft that have examined these remarkable planets at close range. They found striking evidence of stable, large-scale weather patterns in both planets' atmospheres, as well as evidence of dynamic changes on smaller scales.

Spacecraft to Jupiter and Saturn

 The first several spacecraft to visit Jupiter and Saturn each made a single flyby of the planet. *Pioneer 10* flew past Jupiter in December 1973; it was followed a year later by the nearly identical *Pioneer 11*, which went on to make the first-ever flyby of Saturn in 1979. Also in 1979, another pair of spacecraft, *Voyager 1* and *Voyager 2*, sailed past Jupiter. These spacecraft sent back spectacular close-up color pictures of Jupiter's dynamic atmosphere. Both *Voyagers* subsequently flew past Saturn.

The first spacecraft to go into Jupiter orbit was *Galileo*, which carried out an extensive program of observations from 1995 to 2003. The *Cassini* spacecraft went into orbit around Saturn in 2004. A cooperative project of NASA, the European Space Agency, and the Italian Space Agency, this spacecraft is named for astronomer Gian Domenico Cassini. On the way to its destination it viewed Jupiter at close range, recording detailed images such as Figure 12-2*a* and the image of Jupiter that opens this chapter.

Observing Jupiter's Dynamic Atmosphere

While the general pattern of Jupiter's atmosphere stayed the same during the four years between the *Pioneer* and *Voyager* flybys, there were some remarkable changes in the area surrounding the centuries-old Great Red Spot. During the *Pioneer* flybys, the Great Red Spot was embedded in a broad white zone that dominated the planet's southern hemisphere (**Figure 12-4***a*). By the time of the *Voyager* missions, a dark belt had broadened and encroached on the Great Red Spot from the north (Figure 12-4*b*). In the same way that colliding weather systems in our atmosphere can produce strong winds and turbulent air, the interaction in Jupiter's atmosphere between the belt and the Great Red Spot embroiled the entire region in turbulence (**Figure 12-5**). By 1995, the Great Red Spot was once again centered within a white zone (Figure 12-4*c*); when *Cassini* flew past Jupiter in 2000, the dark belt to the north of the Great Red Spot embroiled the entire region in turbulence.

> **Immense storms on Jupiter and Saturn can last for months or years**

Over the past three centuries, Earth-based observers have reported many long-term variations in the Great Red Spot's size and color. At its largest, it measured 40,000 by 14,000 km—so large that three Earths could fit side by side across it. At other times (as in 1976 and 1977), the spot almost faded from view. During the *Voyager* flybys of 1979, the Great Red Spot was comparable in size to the Earth (see Figure 12-5).

Clouds at different levels in Jupiter's atmosphere reflect different wavelengths of infrared light. Using this effect, astronomers

The Great Red Spot lies within a white zone.

(a) *Pioneer 11*, December 1974

A dark belt has encroached on the Great Red Spot.

(b) *Voyager 2*, July 1979

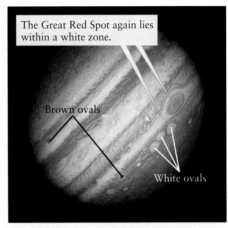

The Great Red Spot again lies within a white zone.

Brown ovals

White ovals

(c) HST, February 1995

Figure 12-4   R I V U X G

Jupiter's Changing Appearance These images from **(a)**, **(b)** spacecraft and **(c)** the Hubble Space Telescope show major changes in the planet's upper atmosphere over a 20-year period. (a and b: NASA/JPL; c: Reta Beebe and Amy Simon, New Mexico State University, and NASA)

used an infrared telescope on board the *Galileo* spacecraft to help clarify the vertical structure of the Great Red Spot. Most of the Great Red Spot is made of clouds at relatively high altitudes, surrounded by a collar of very low-level clouds about 50 km (30 miles, or 160,000 feet) below the high clouds at the center of the

Winds on the north side of the Great Red Spot flow westward.

Turbulence downwind of the Great Red Spot

Winds within the Great Red Spot circulate counterclockwise.

Winds on the south side of the Great Red Spot flow eastward.

20,000 km   Earth's diameter

Figure 12-5   R I V U X G

Circulation around the Great Red Spot This 1979 image from *Voyager 2* shows atmospheric turbulence and the direction of winds around the Great Red Spot (compare Figure 12-4b, taken at approximately the same time). Winds within the Great Red Spot itself make it spin counterclockwise, completing a full revolution in about six days. (NASA/JPL; adapted from A. P. Ingersoll)

spot. This same kind of structure is seen in high-pressure areas in the Earth's atmosphere, although on a very much smaller scale.

Cloud motions in and around the Great Red Spot reveal that the spot rotates counterclockwise with a period of about six days. Furthermore, winds to the north of the spot blow to the west, and winds south of the spot move toward the east. The circulation around the Great Red Spot is thus like a wheel spinning between two oppositely moving surfaces (see Figure 12-5). Weather patterns on the Earth tend to change character and eventually dissipate when they move between plains and mountains or between land and sea. But because no solid surface or ocean underlies Jupiter's clouds, no such changes can occur for the Great Red Spot—which may explain how this wind pattern has survived for at least three centuries.

Other persistent features in Jupiter's atmosphere are the **white ovals.** Several white ovals are visible in Figure 12-4b. As in the Great Red Spot, wind flow in white ovals is counterclockwise. White ovals are also apparently long-lived; Earth-based observers have reported seeing them in the same location since 1938.

Most of the white ovals are observed in Jupiter's southern hemisphere, whereas **brown ovals** are more common in Jupiter's northern hemisphere. Brown ovals appear dark in a visible-light image like Figure 12-4b, but they appear bright in an infrared image. For this reason, brown ovals are understood to be holes in Jupiter's cloud cover. They permit us to see into the depths of the Jovian atmosphere, where the temperature is higher and the atmosphere emits infrared light more strongly. White ovals, by contrast, have relatively low temperatures. They are areas with cold, high-altitude clouds that block our view of the lower levels of the atmosphere.

Between 1998 and 2000 three white ovals merged in Jupiter's southern hemisphere. The merger of these ovals, each of which had been observed for 60 years, led to the formation of a massive storm about half the size of the Great Red Spot (**Figure 12-6**). Time will tell whether this new feature is as long-lived as its larger cousin.

Figure 12-6   R I V U X G

Jupiter's "Red Spot Jr." This 2006 image of Jupiter and a new storm that formed between 1998 and 2000 was made using adaptive optics on the Gemini North telescope in Hawaii (see Section 6-3). Because Jupiter displays differential rotation (see Section 12-2), the two storms travel around Jupiter at different rates; they are not always side-by-side as shown here. (Gemini Observatory ALTAIR Adaptive Optics image/Chris Go)

Great Red Spot   New storm

Regularities in Jupiter's Atmosphere

The *Voyager*, *Galileo*, and *Cassini* images might suggest a state of incomprehensible turmoil, but, surprisingly, there is also great regularity in the Jovian atmosphere. The views in **Figure 12-7** (assembled from a large number of *Cassini* images) show how Jupiter would look to someone located directly over either the planet's north pole or its south pole. Note the regular spacing of such cloud features as ripples, plumes, colored wisps, and white ovals in the southern hemisphere. These regularities are probably the result of stable, large-scale weather patterns that encircle Jupiter.

Storms on Saturn

While Saturn has belts and zones like Jupiter, it has no storm systems as long-lived as Jupiter's Great Red Spot. But about every 30 years—roughly the orbital period of Saturn—Earth-based observers have reported seeing storms in Saturn's clouds that last for

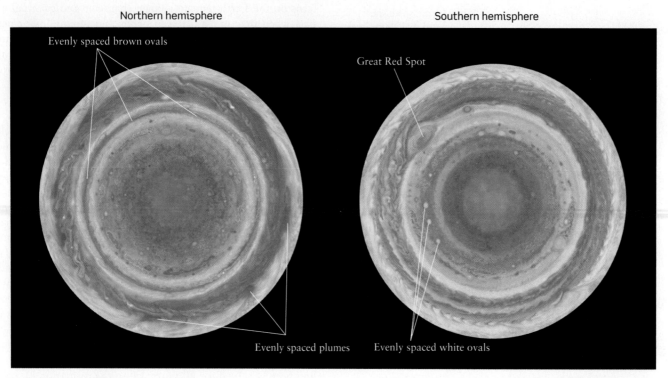

Northern hemisphere   Southern hemisphere

Evenly spaced brown ovals   Great Red Spot

Evenly spaced plumes   Evenly spaced white ovals

Figure 12-7   R I V U X G

Jupiter's Northern and Southern Hemispheres *Cassini* images were combined and computer processed to construct these views that look straight down onto Jupiter's north and south poles. Various cloud features are evenly spaced in longitude. (NASA/JPL/Space Science Institute)

Figure 12-8   R I V U X G

A New Storm on Saturn This infrared image from the *Cassini* spacecraft shows a storm the size of the Earth that appeared in Saturn's southern hemisphere in 2004. *Cassini* has observed several storms in this same region of Saturn. (NASA/JPL/Space Science Institute)

several days or even months. **Figure 12-8** shows one such storm that appeared in Saturn's southern hemisphere in 2004.

Saturn's storms are thought to form when warm atmospheric gases rise upward and cool, causing gaseous ammonia to solidify into crystals and form white clouds. On Earth, similar rapid lifting of air occurs within thunderstorms and can cause water droplets to solidify into hailstones. Earth thunderstorms produce electrical discharges (lightning) that generate radio waves, which you hear as "static" on an AM radio. Radio receivers on board *Cassini* have recorded similar "static" emitted by storms on Saturn, which strongly reinforces the idea that these storms are like thunderstorms on Earth.

12-4 The internal heat of Jupiter and Saturn has a major effect on the planets' atmospheres

Weather patterns on Earth are the result of the motions of air masses. Winter storms in the midwestern United States occur when cold, moist air moves southward from Canada, and tropical storms in the southwest United States are caused by the northeastward motion of hot, moist air from Hawaii. All such motions, as well as the overall global circulation of our atmosphere, are powered by the Sun. Sunlight is absorbed by the Earth's surface,

and the heated surface in turn warms the atmosphere and stirs it into motion (see Sections 9-1 and 9-5). The energy of sunlight also powers the motions of the atmospheres of Venus and Mars (see Section 11-6). On both Jupiter and Saturn, however, atmospheric motions are powered both by solar energy and by the internal energy of the planet.

Jupiter and Saturn: Radiating Energy into Space

In the late 1960s, astronomers using Earth-based telescopes made the remarkable discovery that Jupiter emits more energy in the form of infrared radiation than it absorbs from sunlight—in fact, about twice as much. (By comparison, the internal heat that the Earth radiates into space is only about 0.005% of what it absorbs from the Sun.) The excess heat presently escaping from Jupiter is thought to be energy left over from the formation of the planet 4.56 billion years ago. As gases from the solar nebula fell into the protoplanet, vast amounts of gravitational energy were converted into thermal energy that heated the planet's interior. Jupiter has been slowly radiating this energy into space ever since. Because Jupiter is so large, it has retained substantial thermal energy even after billions of years, and so still has plenty of energy to emit as infrared radiation.

Saturn, too, radiates into space more energy than it receives as sunlight. Because Saturn is smaller than Jupiter, it should have begun with less internal heat trapped inside, and it should have radiated that heat away more rapidly. Hence, we would expect Saturn to be radiating very little energy today. In fact, when we take Saturn's smaller mass into account, we find that Saturn releases about 25% *more* energy from its interior on a per-kilogram basis than does Jupiter. The explanation for this seeming paradox may be that helium is raining out of Saturn's upper atmosphere, as we described in Section 12-2. The helium condenses into droplets at cold upper altitudes, but generates heat at lower altitudes due to friction between the falling droplets and the surrounding gases. This heat eventually escapes from the planet's surface as infrared radiation. This "raining out" of helium from Saturn's upper layers is calculated to have begun 2 billion years ago. The amount of thermal energy released by this process adequately accounts for the extra heat radiated by Saturn since that time.

Heat naturally flows from a hot place to a colder place, never the other way around. Hence, in order for heat to flow upward through the atmospheres of Jupiter and Saturn and radiate out into space, the temperature must be warmer deep inside the atmosphere than it is at the cloudtops. Infrared measurements have confirmed that the temperature within the atmospheres of both planets does indeed increase with increasing depth (**Figure 12-9**).

ANALOGY You can understand the statement that "heat naturally flows from a hot place to a colder place" by visualizing an ice cube placed on a hot frying pan. Experience tells you that the hot frying pan cools down a little while the ice warms up and melts. If heat were to flow the other way, the ice would get colder and the frying pan would get hotter—which never happens in the real world.

A Convection Paradox

When a fluid is warm at the bottom and cool at the top, like water being warmed in a pot, one effective way for energy to be

Figure 12-9

The Upper Atmospheres of Jupiter and Saturn The black curves in these graphs show temperature versus altitude in each atmosphere, as well as the probable arrangements of the cloud layers. Zero altitude in each atmosphere is chosen to be the point where the pressure is 100 millibars, or one-tenth of Earth's atmospheric pressure. Beneath both planets' cloud layers, the atmosphere is composed almost entirely of hydrogen and helium. (Adapted from A. P. Ingersoll)

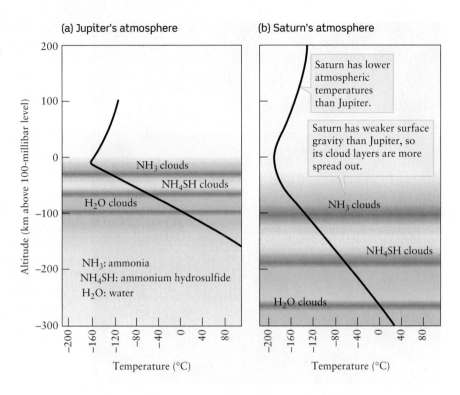

distributed through the fluid is by the up-and-down motion called *convection.* (We introduced the idea of convection in Section 9-1; see, in particular, Figure 9-5.) For many years it was thought that Jupiter's light-colored zones are regions where warm gas from low levels is rising and cooling, forming high-altitude clouds, while the dark-colored belts are regions where cool high-altitude gas is descending and being heated. However, this picture has been called into question by images made by *Cassini* during its 2000–2001 flyby of Jupiter, which showed details as small as 58 km (36 mi) across. The *Cassini* images revealed numerous white cloud cells welling upward in Jupiter's atmosphere, almost all of which appeared in the dark belts.

One interpretation of these observations is that air is actually rising in the *belts* and sinking in the *zones.* Another suggestion is that air is rising in the zones and producing clouds, but these are hidden from our view by a high-altitude layer of opaque white ammonia crystals. In this model the belts are regions in which air is descending except in the localized regions where the clouds seen by *Cassini* are found.

One thing that is well known about Jupiter's atmosphere is that the belts are regions in which we can see into the atmosphere's lower levels. You can see this in Figure 12-10, which compares visible-light and infrared images of Jupiter. The cold Jovian clouds emit radiation primarily at infrared wavelengths, with an intensity that depends on the cloud temperature (see Section 5-4) and hence on the depth of the cloud within the atmosphere (see Figure 12-9a). The infrared image in Figure 12-10b shows stronger emission from the belts, where we are looking at lower-lying, warmer levels of Jupiter's atmosphere.

The Fastest Winds in the Solar System

In addition to the vertical motion of gases within the belts and zones, the very rapid rotation of both Jupiter and Saturn creates a global pattern of eastward and

westward **zonal winds.** Wind speeds on Jupiter can exceed 500 km/h (300 mi/h). The Earth's atmospheric circulation has a similar pattern of eastward and westward flow (see Figure 9-24), but with slower wind speeds. The faster winds on Jupiter presumably result from the planet's more rapid rotation as well as the substantial flow of heat from the planet's interior.

The zonal winds on Jupiter are generally strongest at the boundaries between belts and zones. Within a given belt or zone, the wind reverses direction between the northern and southern boundaries. As Figure 12-5 shows, such reversals in wind direction are associated with circulating storms such as the Great Red Spot.

Saturn rotates at about the same rate as Jupiter, but its interior releases less heat and it receives only a quarter as much energy in sunlight as does Jupiter. With less energy available to power the motions of Saturn's atmosphere, we would expect its zonal winds to be slower than Jupiter's. Surprisingly, Saturn's winds are *faster!* In 1980 the *Voyager* spacecraft measured wind speeds near Saturn's equator that approach 1800 km/h (1100 mi/h), approximately two-thirds the speed of sound in Saturn's atmosphere and faster than winds on any other planet. Since then the equatorial winds seem to have abated somewhat: Between 1996 and 2004, astronomers using the Hubble Space Telescope showed that the wind speed had slowed to about 1000 km/h, and in 2005 *Cassini* measured wind speeds of about 1350 km/h at the equator. (These changes may be due to seasonal variations or changes in Saturn's clouds.) Why Saturn's winds should be even faster than Jupiter's is not yet completely understood.

Jupiter and Saturn's Cloud Layers

From spectroscopic observations and calculations of atmospheric temperature and pressure, scientists conclude that both Jupiter and Saturn have three main cloud layers of differing chemical

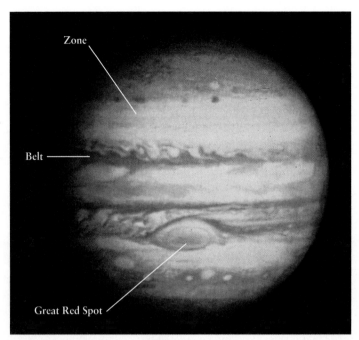

(a) Visible-light image   R I **V** U X G

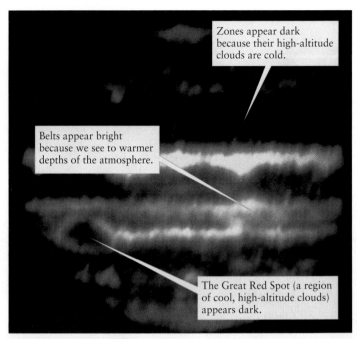

(b) Infrared image   R **I** V U X G

Figure 12-10

Jupiter's Warm Belts and Cool Zones (a) This visible-light image of Jupiter was made by the *Voyager 1* spacecraft. **(b)** Bright and dark areas in this Earth-based infrared image, taken at the same time as the image in (a), correspond to high and low temperatures, respectively. (NASA/JPL)

composition. The uppermost cloud layer is composed of crystals of frozen ammonia. Deeper in the atmosphere, ammonia (NH_3) and hydrogen sulfide (H_2S)—a compound of hydrogen and sulfur—combine to produce ammonium hydrosulfide (NH_4SH) crystals. At even greater depths, the clouds are composed of crystals of frozen water.

Jupiter's strong surface gravity compresses these cloud layers into a region just 75 km deep in the planet's upper atmosphere (Figure 12-9a). But Saturn has a smaller mass and hence weaker surface gravity, so the atmosphere is less compressed and the same three cloud layers are spread out over a range of nearly 300 km (Figure 12-9b). The colors of Saturn's clouds are less dramatic than Jupiter's (see Figure 12-2) because deeper cloud layers are partially obscured by the hazy atmosphere above them.

The colors of clouds on Jupiter and Saturn depend on the temperatures of the clouds and, therefore, on the depth of the clouds within the atmosphere. Brown clouds are the warmest and are thus the deepest layers that we can see. Whitish clouds form the next layer up, followed by red clouds in the highest layer. The whitish zones on each planet are therefore somewhat higher than the brownish belts, while the red clouds in Jupiter's Great Red Spot are among the highest found anywhere on that planet.

Despite all the data from various spacecraft, we still do not know what gives the clouds of Jupiter and Saturn their colors. The crystals of NH_3, NH_4SH, and water ice in the three main cloud layers are all white. Thus, other chemicals must cause the browns, reds, and oranges. Certain molecules closely related to ammonium hydrosulfide, which can form into long chains that have a yellow-brown color, may play a role. Compounds of sulfur or phosphorus, which can assume many different colors, de-

pending on their temperature, might also be involved. The Sun's ultraviolet radiation may induce the chemical reactions that produce these colorful compounds.

12-5 A space probe has explored Jupiter's deep atmosphere

By observing Jupiter's and Saturn's cloud layers from a distance, we have been able to learn a great deal about the atmospheres beneath the clouds. But there is no substitute for making measurements on site, and for many years scientists planned to send a spacecraft to explore deep into Jupiter's atmosphere. While still 81.52 million kilometers (50.66 million miles) from Jupiter, the *Galileo* spacecraft released the *Galileo Probe*, a cone-shaped body about the size of an office desk. While *Galileo* itself later fired its rockets to place itself into an orbit around Jupiter, the *Galileo Probe* continued on a course that led it on December 7, 1995, to a point in Jupiter's clouds just north of the planet's equator.

Exploring Jupiter's Atmosphere

A heat shield protected the *Galileo Probe* as air friction slowed its descent speed from 171,000 km/h (106,000 mi/h) to 40 km/h (25 mi/h) in just 3 minutes. The spacecraft then deployed a parachute and floated down through the atmosphere (**Figure 12-11**). For the next hour, the probe observed its surroundings and radioed its findings back to the main *Galileo* spacecraft, which in turn radioed them to Earth. The mission ceased at a point some 200 km (120 mi) below Jupiter's upper cloud layer, where the

Figure 12-11

The *Galileo Probe* Enters Jupiter's Atmosphere This artist's impression shows the *Galileo Probe* descending under a parachute through Jupiter's clouds. The jettisoned heat shield (shown to the right of the probe) protected the probe during its initial high-speed entry into the atmosphere. The *Galileo Probe* returned data for 58 minutes before it was crushed and melted by the pressure and temperature of the atmosphere. (NASA)

tremendous pressure (24 atmospheres) and high temperature (152°C = 305°F) finally overwhelmed the probe's electronics.

Although the *Galileo Probe* did not have a camera, it did carry a variety of instruments that made several new discoveries. A radio-emissions detector found evidence for lightning discharges that, while less frequent than on Earth, are individually much stronger than lightning bolts in our atmosphere. Other measurements showed that Jupiter's winds, which are brisk in the atmosphere above the clouds, are even stronger beneath the clouds: The *Galileo Probe* measured a nearly constant wind speed of 650 km/h (400 mi/h) throughout its descent. This shows convincingly that the energy source for the winds is Jupiter's interior heat. If the winds were driven primarily by solar heating, as is the case on Earth, the wind speed would have decreased with increasing depth.

A central purpose of the *Galileo Probe* mission was to test the three-layer cloud model depicted in Figure 12-9a by making direct measurements of Jupiter's clouds. But the probe saw only traces of the NH_3 and NH_4SH cloud layers and found no sign at all of the low-lying water clouds. One explanation of this surprising result is that the probe by sheer chance entered a **hot spot,** an unusually warm and cloud-free part of Jupiter's atmosphere. Indeed, observations from the Earth showed strong infrared emission from the probe entry site, as would be expected where a

break in the cloud cover allows a view of Jupiter's warm interior (see Figure 12-10b).

The *Galileo Probe* also made measurements of the chemical composition of Jupiter's atmosphere. These revealed that the relative proportions of hydrogen and helium are almost exactly the same as in the Sun, in line with the accepted picture of how Jupiter formed from the solar nebula (see Section 8-4). But a number of heavy elements, including carbon, nitrogen, and sulfur, were found to be significantly more abundant on Jupiter than in the Sun. This was an expected result. Over the past several billion years, Jupiter's strong gravity should have pulled in many pieces of interplanetary debris, thus building up a small but appreciable abundance of heavy elements.

Water is a major constituent of small bodies in the outer solar system and thus should likewise be present inside Jupiter in noticeable quantities. Surprisingly, the amount of water actually detected by the *Galileo Probe* was less than half the expected amount. One proposed explanation for this disparity is that the hot spot where the probe entered the atmosphere was also unusually dry. If this explanation is correct, Jupiter's hot spots are analogous to hot, dry deserts on Earth. An alternative idea is that more water than expected has rained out of Jupiter's upper atmosphere and coalesced deep within the planet. For now, the enigma of Jupiter's missing water remains unresolved.

Noble Gases and the Origin of Jupiter

Another surprising and potentially important result from the *Galileo Probe* concerns three elements, argon (Ar), krypton (Kr), and xenon (Xe). These elements are three of the so-called **noble gases,** which do not combine with other atoms to form molecules. (The noble gases, which also include helium, appear in the rightmost column of the periodic table in Box 5-5.) All three of these elements appear in tiny amounts in the Earth's atmosphere and in the Sun, and so must also have been present in the solar nebula. But the *Galileo Probe* found that argon, krypton, and xenon are about 3 times as abundant in Jupiter's atmosphere as in the Sun's.

If these elements were incorporated into Jupiter directly from the gases of the solar nebula, they should be equally as abundant in Jupiter as in the Sun (which formed from the gas at the center of the solar nebula; see Section 8-3). So, the excess amounts of argon, krypton, and xenon must have entered Jupiter in the form of solid planetesimals, just as did carbon, nitrogen, and sulfur. But at Jupiter's distance from the Sun, the presumed temperature of the solar nebula was too high to permit argon, krypton, or xenon to solidify. How, then, did the excess amounts of these noble gases become part of Jupiter?

Several hypotheses have been proposed to explain these results. One idea is that icy bodies, including solid argon, krypton, and xenon, formed in an interstellar cloud even before the cloud collapsed to form the solar nebula. A second notion is that the solar nebula may actually have been colder than previously thought. If either of these ideas proves to be correct, it would mean that Jovian planets can form closer to their stars than had heretofore been thought. This would help to explain why astronomers find massive, presumably Jovian planets orbiting close to stars other than the Sun (see Section 8-5).

Measuring trace amounts of rare gases on Jupiter provides clues about the origins of the planets

A third hypothesis proposes that our models of the solar nebula are correct, but that Jupiter originally formed much farther from the Sun. Out in the cold, remote reaches of the solar nebula, argon, krypton, and xenon could have solidified into icy particles and fallen into the young Jupiter, thus enriching it in those elements. Interactions between Jupiter and the material of the solar nebula could then have forced the planet to spiral inward, eventually reaching its present distance from the Sun. As we saw in Section 8-5, a model of this kind has been proposed to explain why extrasolar planets orbit so close to their stars. More research will be needed to decide which of these hypotheses is closest to the truth.

The story of the *Galileo Probe* exemplifies two of the most remarkable aspects of scientific research. First, when new scientific observations are made, the results often pose many new questions at the same time that they provide answers to some old ones. And second, it sometimes happens that investigating what might seem to be trivial details—such as the abundances of trace elements in Jupiter's atmosphere—can reveal clues about truly grand issues, such as how the planets formed.

12-6 The oblateness of Jupiter and Saturn reveals their rocky cores

The *Galileo Probe* penetrated only a few hundred kilometers into Jupiter's atmosphere before being crushed. To learn about Jupiter's structure at greater depths as well as that of Saturn, astronomers must use indirect clues. The shapes of Jupiter and Saturn are important clues, since they indicate the size of the rocky cores at the centers of the planets.

Oblateness and Rotation

Even a casual glance through a small telescope shows that Jupiter and Saturn are not spherical but slightly flattened or **oblate**. (You can see this in Figures 12-1 and 12-2.) The diameter across Jupiter's equator (142,980 km) is 6.5% larger than its diameter from pole to pole (133,700 km). Thus, Jupiter is said to have an **oblateness** of 6.5%, or 0.065. Saturn has an even larger oblateness of 9.8%, or 0.098, making it the most oblate of all the planets. By comparison, the oblateness of the Earth is just 0.34%, or 0.0034.

If Jupiter and Saturn did not rotate, both would be perfect spheres. A massive, nonrotating object naturally settles into a spherical shape so that every atom on its surface experiences the same gravitational pull aimed directly at the object's center. Because Jupiter and Saturn do rotate, however, the body of each planet tends to fly outward and away from the axis of rotation.

ANALOGY You can demonstrate this effect for yourself. Stand with your arms hanging limp at your sides, then spin yourself around. Your arms will naturally tend to fly outward, away from the vertical axis of rotation of your body. Likewise, if you drive your car around a circular road, you feel thrown toward the outside of the circle; the car as a whole is rotating around the center of the circle, and you tend to move away from the rotation axis. (We discussed the physical principles behind this in Box 4-3.) For Jupiter and Saturn, this same effect deforms the planets into their nonspherical, oblate shapes.

Modeling Jupiter and Saturn's Interiors

The oblateness of a planet depends both on its rotation rate and on how the planet's mass is distributed over its volume. The challenge to planetary scientists is to create a model of a planet's mass distribution that gives the right value for the oblateness. One such model for Jupiter suggests that 2.6% of its mass is concentrated in a dense, rocky core (**Figure 12-12a**). Although this core has about 8 times the mass of the Earth, the crushing weight of the remaining bulk of Jupiter compresses it to a diameter of just 11,000 km, slightly smaller than the Earth's diameter of 12,800 km. The pressure at the center of the core is estimated to be about 70 million atmospheres, and the central temperature is probably about 22,000 K. By contrast, the temperature at Jupiter's cloudtops is only 165 K.

If Jupiter formed by accretion of gases onto a rocky "seed," the present-day core is presumably that very seed. However, additional amounts of rocky material were presumably added later

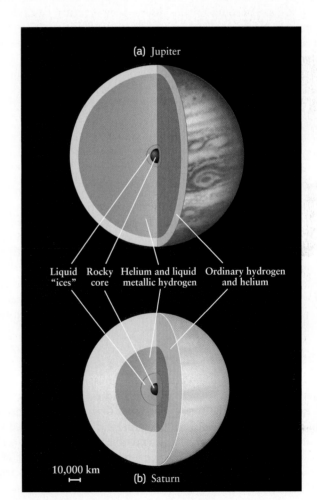

(a) Jupiter

Liquid "ices"   Rocky core   Helium and liquid metallic hydrogen   Ordinary hydrogen and helium

10,000 km

(b) Saturn

Figure 12-12

The Internal Structures of Jupiter and Saturn These diagrams of the interiors of Jupiter and Saturn are drawn to the same scale. Each planet's rocky core is surrounded by an outer core of liquid "ices," a layer of helium and liquid metallic hydrogen, and a layer of helium and ordinary molecular hydrogen (H_2). Saturn's rocky core contains a larger fraction of the planet's mass than does Jupiter's, while Saturn has a smaller volume of liquid metallic hydrogen than does Jupiter.

INTERACTIVE EXERCISE 12.1

by meteoritic material that fell into the planet and sank to the center. (If Jupiter formed directly from the gases of the solar nebula, this additional material makes up the present-day core.) Material from icy planetesimals, too, would have sunk deep within Jupiter and added to the core. In the model we have been describing, these "ices"—principally water (H_2O), methane (CH_4), ammonia (NH_3), and other molecules made by chemical reactions among these substances—form a layer some 3000 km thick that surrounds the rocky core. (Because these substances are less dense than rock, they "float" on top of the rocky core.) Despite the high pressures in this layer, the temperature is so high that the "ices" are probably in a liquid state.

Saturn rotates at about the same rate as Jupiter but has less mass, and therefore less gravity to pull its material inward. Hence, Saturn's rotation should cause material at its equator to bulge outward more than on Jupiter, giving Saturn a greater oblateness. In fact, if Saturn and Jupiter had the same internal structure, we would actually expect Saturn to be even more oblate than it really is. We can account for this discrepancy if Saturn has a different mass distribution than Jupiter. Detailed calculations suggest that about 10% of Saturn's mass is contained in its rocky core, compared to 2.6% for Jupiter's rocky core (see Figure 12-12b). As for Jupiter, the rocky core of Saturn is probably surrounded by an outer core of liquid "ices."

12-7 Metallic hydrogen inside Jupiter and Saturn endows the planets with strong magnetic fields

The oblateness of Jupiter and Saturn gives information about the size of the planets' rocky cores. It does not, however, provide much insight into the planets' bulk between their cores and their colorful cloudtops. Most of our knowledge of these parts of the interiors of Jupiter and Saturn comes from observations with radio telescopes. By using these telescopes to detect radio waves from Jupiter and Saturn, astronomers have found evidence of electric currents flowing within both planets' hydrogen-rich interiors.

Radio Emissions from Jupiter

Astronomers began discovering radio emissions coming from Jupiter in the 1950s. A small portion of these emissions is **thermal radiation,** with the sort of spectrum that would be expected from an opaque object due to its temperature. (We discussed the electromagnetic radiation from heated objects in Sections 5-3 and 5-4.) However, most of the radio energy emitted from Jupiter is **nonthermal radiation.** This has a very different sort of spectrum from radiation from a heated object and is found in two distinct wavelength ranges.

At wavelengths of a few meters, Jupiter emits sporadic bursts of radio waves. This **decametric (10-meter) radiation** is probably caused by electrical discharges associated with powerful electric currents in Jupiter's ionosphere. These discharges result from electromagnetic interactions between Jupiter and its large satellite Io. At shorter wavelengths of a few tenths of a meter, Jupiter emits a nearly constant stream of electromagnetic radiation. The prop-

erties of this **decimetric** (tenth-meter) **radiation** show that it is emitted by electrons moving through a strong magnetic field at speeds comparable to the speed of light. The magnetic field deflects the electrons into spiral-shaped trajectories, and the spiraling electrons radiate energy like miniature radio antennas. Energy produced in this fashion is called **synchrotron radiation.** Astronomers now realize that synchrotron radiation is very important in a wide range of situations throughout the universe, including radio emissions from entire galaxies.

Jupiter's Magnetic Field

If some of Jupiter's radiation is indeed synchrotron radiation, then Jupiter must have a magnetic field. Measurements by the *Pioneer* and *Voyager* spacecraft verified this and indicated that the field at Jupiter's equator is 14 times stronger than at the Earth's equator. Like the Earth's magnetic field, Jupiter's field is thought to be generated by motions of an electrically conducting fluid in the planet's interior. But wholly unlike Earth, the moving fluid within Jupiter is an exotic form of hydrogen called **liquid metallic hydrogen.**

As we discussed in Section 7-7, hydrogen becomes a liquid metal only when the pressure exceeds about 1.4 million atmospheres. This occurs at depths more than about 7000 km below Jupiter's cloudtops. Thus, the internal structure of Jupiter consists of four distinct regions, as shown in Figure 12-12a: a rocky core, a roughly 3000-km thick layer of liquid "ices," a layer of helium and liquid metallic hydrogen about 56,000 km thick, and a layer of helium and ordinary hydrogen about 7000 km thick. The colorful cloud patterns that we can see through telescopes are located in the outermost 100 km of the outer layer.

Figure 12-12a shows that much of the enormous bulk of Jupiter is composed of liquid metallic hydrogen. Jupiter's rapid rotation sets the liquid metallic hydrogen into motion, giving rise to a powerful dynamo 20,000 times stronger than Earth's and generating an intense magnetic field.

Saturn's Magnetic Field

Saturn also has a substantial magnetic field, but it pales by Jupiter standards: Saturn's internal dynamo is only 3% as strong as Jupiter's. (To be fair, it is still 600 times stronger than Earth's dynamo.) This difference between the planets suggests that even though Saturn rotates nearly as rapidly as Jupiter, there must be far less liquid metallic hydrogen within Saturn to be stirred up by the rotation and thereby generate a magnetic field. This makes sense: Compared to Jupiter, Saturn has less mass, less gravity, and less internal pressure to compress ordinary hydrogen into liquid metallic hydrogen. As a consequence, as Figure 10-14b shows, liquid metallic hydrogen probably makes up only a relatively small fraction of Saturn's bulk.

Jupiter's Magnetosphere

Jupiter's strong magnetic field surrounds the planet with a magnetosphere so large that it envelops the orbits of many of its moons. The *Pioneer* and *Voyager* spacecraft, which carried instruments to detect charged particles and magnetic fields, revealed the awesome dimensions of the Jovian magnetosphere. The shock wave that surrounds Jupiter's magnetosphere is nearly 30 million

Figure 12-13

Jupiter's Magnetosphere The shock wave marks where the supersonic solar wind is abruptly slowed to subsonic speeds. Most solar wind particles are deflected around Jupiter in the turbulent region (shown in purple) between the shock wave and magnetopause (where the pressure within the magnetosphere balances the pressure of the solar wind). Particles trapped inside Jupiter's magnetosphere are spread out into a vast current sheet by the planet's rapid rotation.

kilometers across. If you could see Jupiter's magnetosphere from the Earth, it would cover an area in the sky 16 times larger than the Moon. **Figure 12-13** shows the structure of Jupiter's magnetosphere. (Compare to Earth's much smaller magnetosphere, shown in Figure 9-19.)

The outer boundary of Jupiter's magnetosphere (on the left in Figure 12-13) lies at distances ranging from 3 to 7 million kilometers (2 to 4 million miles) above the planet. The "downstream" side of the magnetosphere (on the right in Figure 12-13) extends for more than a billion kilometers (650 million miles), even beyond the orbit of Saturn.

> **Both Jupiter and Saturn have magnetospheres that are millions of kilometers across**

The inner regions of the Earth's magnetosphere are dominated by two huge Van Allen belts (recall Figure 9-20), which are filled with charged particles. In a similar manner, Jupiter entraps immense quantities of charged particles in belts of its own. An astronaut venturing into these belts would quickly receive a lethal dose of radiation. In addition, Jupiter's rapid rotation spews the charged particles out into a huge **current sheet,** which lies close to the plane of Jupiter's magnetic equator (see Figure 12-13). Jupiter's magnetic axis is inclined 11° from the planet's axis of rotation, and the orientation of Jupiter's magnetic field is the reverse of the Earth's magnetic field—a compass would point toward the south on Jupiter.

Figure 12-14 shows radio emissions from charged particles in the densest regions of Jupiter's magnetosphere. These emissions vary slightly with a period of 9 hours, 55 minutes, and 30 seconds, as Jupiter's rotation changes the angle at which we view its magnetic field. Because the magnetic field is anchored deep within the planet, this variation reveals Jupiter's **internal rotation period.** This period, indicative of the rotation of the bulk of the planet's mass, is slightly slower than the atmospheric rotation at the equator (period 9 hours, 50 minutes, 28 seconds) but about the same as the atmospheric rotation at the poles (period 9 hours, 55 minutes, 41 seconds). The faster motion of the atmosphere at the equator relative to the interior is driven by Jupiter's internal heat.

The *Voyager* spacecraft discovered that the inner regions of Jupiter's magnetosphere contain a hot, gaslike mixture of charged particles called a **plasma.** A plasma is formed when a gas is heated

to such extremely high temperatures that electrons are torn off the atoms of the gas. As a result, a plasma is a mixture of positively charged ions and negatively charged electrons. The plasma that envelops Jupiter consists primarily of electrons and protons, with some ions of helium, sulfur, and oxygen. A major source of these particles is Jupiter's volcanically active moon Io (see Table 7-2), which lies deep within the magnetosphere. Io ejects a ton of material into Jupiter's magnetosphere each second from its surface and volcanic plumes. These charged particles are caught up in Jupiter's rapidly rotating magnetic field and are accelerated to high speeds.

The fast-moving particles in Jupiter's plasma exert a substantial pressure that holds off the solar wind. The *Voyager* data suggest that the pressure balance between the solar wind and the hot plasma inside the Jovian magnetosphere is precarious. A gust in the solar wind can blow away some of the plasma, at which point the magnetosphere deflates rapidly to as little as one-half its

Strongest radio emission (yellow) from electrons trapped in the current sheet.

Figure 12-14   R I V U X G

A Radio View of Jupiter Instruments on board the *Cassini* spacecraft recorded this false-color map of decimetric emission from Jupiter at a wavelength of 2.2 cm. The emission comes from a region elongated parallel to the planet's magnetic equator, which is tilted relative to its geographic equator. A visible-light image of Jupiter is superimposed on the radio image. (NASA/JPL) original size. However, additional electrons and ions accelerated by Jupiter's rotating magnetic field soon replenish the plasma, and the magnetosphere expands again.

Saturn's Magnetosphere

Saturn's magnetosphere is similar to Jupiter's, but is only about 10 to 20% as large and contains far fewer charged particles than Jupiter's. There are two reasons for this deficiency. First, Saturn lacks a highly volcanic moon like Io to inject particles into the magnetosphere. Second, and more important, many charged particles are absorbed by the icy particles that make up Saturn's rings (see Section 12-10). The charged particles that do exist in Saturn's magnetosphere are concentrated in radiation belts similar to the Van Allen belts in the Earth's magnetosphere.

Cassini measurements show, however, that the moons of Saturn do influence the magnetosphere. The inner magnetosphere is dominated by water and atomic oxygen, material that is thought to come from the planet's icy moons (which we will discuss in Chapter 13), as well as from Saturn's rings (which we will learn about later in this chapter).

 As for Jupiter, the motion of Saturn's magnetic field allows us to track the planet's internal rotation. *Voyager* measurements in 1980 showed Saturn is similar to Jupiter in that its internal rotation period (10 hours, 39 minutes, and 25 seconds) is slower than the rotation period of the visible surface at the equator (10 hours, 13 minutes, and 59 seconds). Surprisingly, when *Cassini* repeated these measurements in 2003–2004, it found that the magnetosphere's rotation had slowed down: The new rotation period is 10 hours, 45 minutes, and 45 seconds, more than 6 minutes longer than in 1980. It seems unlikely that the rotation of the planet's interior could have slowed so much in such a short time. Instead, the magnetic field inside Saturn may have slipped and reattached itself at higher latitudes, where the differentially rotating planet spins at a slower rate. The reason for this slippage is not known.

Aurorae on Jupiter and Saturn

On Earth, particles trapped in our planet's magnetic field stream onto the north and south magnetic poles, creating the aurora (see Section 9-4). The same effect takes place on both Jupiter and Saturn. **Figure 12-15** shows images from the Hubble Space Telescope that reveal glowing auroral rings centered on the north and south magnetic poles of Jupiter and of Saturn. As on Earth, the emission comes from a region high in each planet's atmosphere, hundreds of kilometers above the cloudtops.

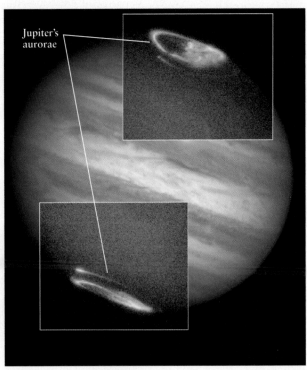

(a) Jupiter R I V U X G

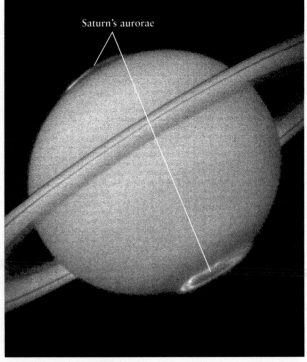

(b) Saturn R I V U X G

Figure 12-15

Aurorae on Jupiter and Saturn On both Jupiter and Saturn, charged particles from the magnetosphere are funneled onto the planet's magnetic poles. When the particles collide with and excite molecules in the upper atmosphere, the molecules emit ultraviolet light.

(The ultraviolet images in (a) are superimposed on a visible-light image of Jupiter.) (a: J. Clarke, University of Michigan, and NASA; b: J. Trauger, JPL, and NASA)

12-8 Earth-based observations reveal three broad rings encircling Saturn

Compared to Saturn, Jupiter is unquestionably larger, more massive, more colorful, and more dynamic in both its atmosphere and magnetosphere. But as seen through even a small telescope, Saturn stands out as perhaps the most beautiful of all the worlds of the solar system, thanks to its majestic rings (**Figure 12-16**). As we will see in Section 12-9, Jupiter also has rings, but they are so faint that they were not discovered until spacecraft could observe Jupiter at close range.

Discovering Saturn's Rings

In 1610, Galileo became the first person to see Saturn through a telescope. He saw few details, but he did notice two puzzling lumps protruding from opposite edges of the planet's disk. Curiously, these lumps disappeared in 1612, only to reappear in 1613. Other observers saw similar appearances and disappearances over the next several decades.

In 1655, the Dutch astronomer Christiaan Huygens began to observe Saturn with a better telescope than was available to any of his predecessors. (We discussed in Section 11-2 how Huygens used this same telescope to observe Mars.) On the basis of his observations, Huygens suggested that Saturn was surrounded by a thin, flattened ring. At times this ring was edge-on as viewed from the Earth, making it almost impossible to see. At other times Earth observers viewed Saturn from an angle either above or below the plane of the ring, and the ring was visible, as in Figure 12-1. (The lumps that Galileo saw were the parts of the ring to either side of Saturn, blurred by the poor resolution of his rather small telescope.) Astronomers confirmed this brilliant deduction over the next several years as they watched the ring's appearance change just as Huygens had predicted.

As the quality of telescopes improved, astronomers realized that Saturn's "ring" is actually a *system* of rings, as Figure 12-16 shows. In 1675, the Italian astronomer Cassini discovered a dark division in the ring. The **Cassini division** is an apparent gap about 4500 kilometers wide that separates the outer **A ring** from the brighter **B ring** closer to the planet. In the mid-1800s, astronomers managed to identify the faint **C ring**, or *crepe* ring, that lies just inside the B ring.

Saturn and its rings are best seen when the planet is at or near opposition. A modest telescope gives a good view of the A and B rings, but a large telescope and excellent observing conditions are needed to see the C ring.

How the Rings Appear from Earth

Earth-based views of the Saturnian ring system change as Saturn slowly orbits the Sun. Huygens was the first to understand that this change occurs because the rings lie in the plane of Saturn's equator, and this plane is tilted 27° from the plane of Saturn's orbit. As Saturn orbits the Sun, its rotation axis and the plane of its equator maintain the same orientation in space. (The same is true for the Earth; see Figure 2-12.) Hence, over the course of a Saturnian year, an Earth-based observer views the rings from various angles (**Figure 12-17**). At certain times Saturn's north pole is tilted toward the Earth and the observer looks "down" on the "top side" of the rings. Half a Saturnian year later, Saturn's south pole is tilted toward us and the "underside" of the rings is exposed to our Earth-based view.

> As seen from Earth, Saturn's rings change their orientation as the planet orbits the Sun

When our line of sight to Saturn is in the plane of the rings, the rings are viewed edge-on and seem to disappear entirely (see the images at the upper left and lower right corners of Figure 12-17).

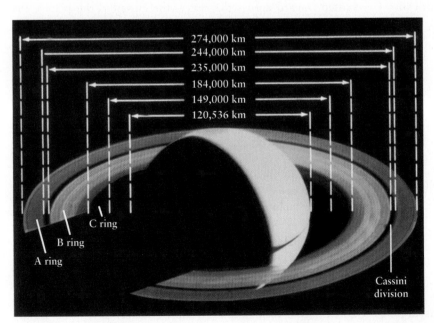

Figure 12-16   R I **V** U X G

Saturn's System of Rings This *Voyager 1* image shows many details of Saturn's rings. Saturn's equatorial diameter is labeled, as are the diameters of the inner and outer edges of the rings. The C ring is so faint that it is almost invisible in this view. Saturn is visible through the rings (look near the bottom of the image), which shows that the rings are not solid. (NASA/JPL)

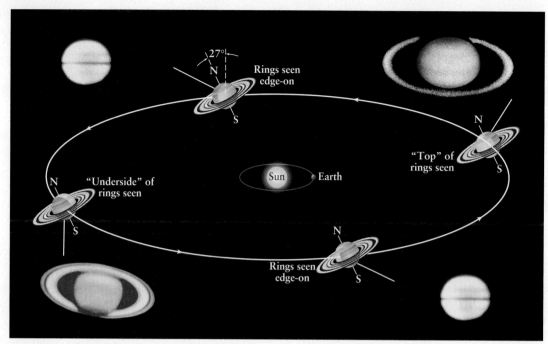

Figure 12-17 R I **V** U X G

The Changing Appearance of Saturn's Rings Saturn's rings are aligned with its equator, which is tilted 27° from Saturn's orbital plane. As Saturn moves around its orbit, the rings maintain the same orientation in space, so observers on Earth see the rings at various angles. The accompanying Earth-based photographs show Saturn at various points in its orbit. Note that the rings seem to disappear entirely when viewed edge-on, which occurs about every 15 years. (Lowell Observatory)

This disappearance indicates that the rings are very thin. In fact, they are thought to be only a few tens of meters thick. In proportion to their diameter, Saturn's rings are thousands of times thinner than the sheets of paper used to print this book.

The last edge-on presentation of Saturn's rings of the 20th century was in 1995–1996; the first two of the 21st century are in 2008–2009 and in 2025. From 1996 to 2008, we Earth observers see the "underside" of the rings (as in the image at the lower left corner of Figure 12-17); from 2009 until 2025 we see the "top" of the rings.

12-9 Saturn's rings are composed of numerous icy fragments, while Jupiter's rings are made of small rocky particles

Astronomers have long known that Saturn's rings could not possibly be solid sheets of matter. In 1857, the Scottish physicist James Clerk Maxwell (whom we last encountered in Section 5-2) proved theoretically that if the rings were solid, differences in Saturn's gravitational pull on different parts of the rings would cause the rings to tear apart. He concluded that Saturn's rings are made of "an indefinite number of unconnected particles."

Ring Particles

In 1895, James Keeler at the Allegheny Observatory in Pittsburgh became the first to confirm by observation that the rings are not rigid. He did this by photographing the spectrum of sunlight reflected from Saturn's rings. As the rings orbit Saturn, the spectral lines from the side approaching us are blueshifted by the Doppler effect (recall Section 5-9, especially Figure 5-26). At the same time, spectral lines from the receding side of the rings are redshifted. Keeler noted that the size of the wavelength shift increased inward across the rings—the closer to the planet, the greater the shift. This variation in Doppler shift proved that the inner portions of Saturn's rings are moving around the planet more rapidly than the outer portions. Indeed, the orbital speeds across the rings are in complete agreement with Kepler's third law: The square of the orbital period about Saturn at any place in the rings is proportional to the cube of the distance from Saturn's center (see Section 4-7 and Box 4-4). This is exactly what would be expected if the rings consisted of numerous tiny "moonlets," or **ring particles,** each individually circling Saturn.

Saturn's rings are quite bright; they reflect 80% of the sunlight that falls on them. (By comparison, Saturn itself reflects 46% of incoming sunlight.) Astronomers therefore long suspected that the ring particles are made of ice and ice-coated rock. This hunch was confirmed in the 1970s, when the American astronomers Gerard P. Kuiper and Carl Pilcher identified absorption features of frozen water in the rings' near-infrared spectrum. The *Voyager* and *Cassini* spacecraft have made even more detailed infrared measurements that indicate the temperature of the rings ranges from –180°C (–290°F) in the sunshine to less than –200°C (–330°F) in Saturn's shadow. Water ice is in no danger of melting or evaporating at these temperatures.

To determine the sizes of the particles that make up Saturn's rings, astronomers analyzed the radio signals received from a spacecraft as it passes behind the rings. How easily radio waves can travel through the rings depends on the relationship between the wavelength and the particle size. The results show that most of the particles range in size from pebble-sized fragments about 1 cm in diameter to chunks about 5 m across, the size of large boulders. Most abundant are snowball-sized particles about 10 cm in diameter.

The Roche Limit

All of Saturn's material may be ancient debris that failed to accrete (fall together) into satellites. The total amount of material in the rings is quite small. If Saturn's entire ring system were compressed together to make a satellite, it would be no more than 100 km (60 mi) in diameter. But, in fact, the ring particles are so close to Saturn that they will never be able to form moons.

To see why, imagine a collection of small particles orbiting a planet. Gravitational attraction between neighboring particles tends to pull the particles together. However, because the various particles are at differing distances from the parent planet, they also experience different amounts of gravitational pull from the planet. This difference in gravitational pull is a **tidal force** that tends to keep the particles separated. (We discussed tidal forces in detail in Section 4-8. You may want to review that section, and in particular Figure 4-24.)

> **Tidal forces prevent the material of Saturn's rings from coalescing into a moon**

The closer a pair of particles is to the planet, the greater the tidal force that tries to pull the pair apart. At a certain distance from the planet's center, called the **Roche limit,** the disruptive tidal force is just as strong as the gravitational force between the particles. (The concept of this limit was developed in the mid-1800s by the French mathematician Edouard Roche.) Inside the Roche limit, the tidal force overwhelms the gravitational pull between neighboring particles, and these particles cannot accrete to form a larger body. Instead, they tend to spread out into a ring around the planet. (The *Cosmic Connections* figure on the next page depicts this process.) Indeed, most of Saturn's system of rings visible in Figure 12-16 lies within the planet's Roche limit.

All large planetary satellites are found outside their planet's Roche limit. If any large satellite were to come inside its planet's Roche limit, the planet's tidal forces would cause the satellite to break up into fragments. We will see in Chapter 14 that such a catastrophic tidal disruption may be the eventual fate of Neptune's large satellite, Triton.

CAUTION! It may seem that it would be impossible for any object to hold together inside a planet's Roche limit. But the ring particles inside Saturn's Roche limit survive and do not break apart. The reason is that the Roche limit applies only to objects held together by the *gravitational* attraction of each part of the object for the other parts. By contrast, the forces that hold a rock or a ball of ice together are *chemical* bonds between the object's atoms and molecules. These chemical forces are much stronger than the disruptive tidal force of a nearby planet, so the

VIDEO 12.7 **Figure 12-18**   R I V U X G

Jupiter's Main Ring This *Galileo* image shows Jupiter's main ring almost edge-on. The ring's outer radius, about 129,000 km from the center of Jupiter, is close to the orbit of Adrastea, the second closest (after Metis) of Jupiter's moons. Not shown is an even larger and even more tenuous pair of "gossamer rings," one of which extends out to 181,000 km and the other out to 222,000 km. (NASA/JPL; Cornell University)

rock or ball of ice does not break apart. In the same way, people walking around on the Earth's surface (which is inside the Earth's Roche limit) are in no danger of coming apart, because we are held together by comparatively strong chemical forces rather than gravity.

Jupiter's Ring and the Roche Limit

When the *Voyager 1* spacecraft flew past Jupiter in 1979, it trained its cameras not just on the planet but also on the space around the planet's equator. It discovered that Jupiter, too, has a system of rings that lies within its Roche limit (Figure 12-18). These rings differ from Saturn's in two important ways. First, Jupiter's rings are composed of tiny particles of rock with an average size of only about 1 μm (= 0.001 mm = 10^{-6} m) and that reflect less than 5% of the sunlight that falls on them. Second, there is very little material in the rings of Jupiter, less than 1/100,000 (10^{-5}) the amount of material in Saturn's rings. As a result, Jupiter's rings are extremely faint, which explains why their presence was first revealed by a spacecraft rather than an Earth-based telescope. The ring particles are thought to originate from meteorite impacts on Jupiter's four small, inner satellites, two of which are visible in Figure 12-18.

We will see in Chapter 14 that the rings of Uranus and Neptune are also made of many individual particles orbiting inside each planet's Roche limit. Like the rings of Jupiter, these rings are quite dim and difficult to see from Earth.

12-10 Saturn's rings consist of thousands of narrow, closely spaced ringlets

The photograph in Figure 12-1*b*, which was made in 1974, indicates what astronomers understood about the structure of Saturn's rings in the mid-1970s. Each of the A, B, and C rings appeared to be rather uniform, with little or no evidence of any internal structure. Within a few years, however, close-up observations from spacecraft revealed the true complexity of the rings, as well as their chemical composition.

COSMIC CONNECTIONS

If a small moon wanders too close to a planet, tidal forces tear the moon into smaller particles. These particles form a ring around the planet. The critical distance from the planet at which this happens is called the Roche limit. This helps us understand why the rings of Jupiter and Saturn are made of small particles, as are the rings of Uranus and Neptune (discussed in Chapter 14).

If a satellite is outside the planet's Roche limit, the tidal forces exerted by the planet may deform the satellite but will not tear it apart.

If the satellite crosses the planet's Roche limit, the tidal forces overwhelm the gravitational forces that hold the satellite together and the satellite fragments.

Fragments that are closer to the planet orbit faster (in accordance with Kepler's third law). This spreads the fragments around the planet...

...and the result is a ring.

Jupiter, Saturn, Uranus, and Neptune all have systems of rings that lie mostly within the Roche limit. This diagram shows each of the four ring systems scaled to the radius of the planet.

A planet's Roche limit is about 2.4 times the radius of the planet, provided the material orbiting the planet has the same density as the planet itself. For denser material the Roche limit is closer to the planet.

Spacecraft Views of Saturn's Rings

Pioneer 11, *Voyager 1*, and *Voyager 2* recorded images of Saturn's rings during their flybys in 1979, 1980, and 1981 respectively. *Pioneer 11* had only a relatively limited capability to make images, but cameras on board the two *Voyager* spacecraft sent back a number of pictures showing the detailed structure of Saturn's rings. Since entering Saturn orbit in 2004, the *Cassini* spacecraft has provided scientists with even higher-resolution images of the rings.

Figure 12-19 shows a *Voyager 1* image. Some of the features seen in this image were expected, including the **Encke gap**, a 270-km-wide division in the outer A ring observed by the German astronomer Johann Franz Encke in 1838. But to the amazement of scientists, images like the one in Figure 12-19 revealed that the broad A, B, and C rings are not uniform at all but instead consist of hundreds upon hundreds of closely spaced bands or **ringlets**. These ringlets are arrangements of ring particles that have evolved from the combined gravitational forces of neighboring particles, of Saturn's moons, and of the planet itself.

Pioneer 11 first detected the narrow **F ring**, which you can see in Figure 12-19. It is only about 100 km wide and lies 4000 km beyond the outer edge of the A ring. Close-up views from the *Voyagers* and *Cassini* show that the F ring is made up of several narrow ringlets. Small moons that orbit close to the F ring exert gravitational forces on the ring particles, thus warping and deforming these ringlets (**Figure 12-20**).

Exploring Saturn's Ring Particles

Spacecraft have done more than show Saturn's rings in greater detail; they have also viewed the rings from perspectives not possible from Earth. Through Earth-based telescopes, we can see only

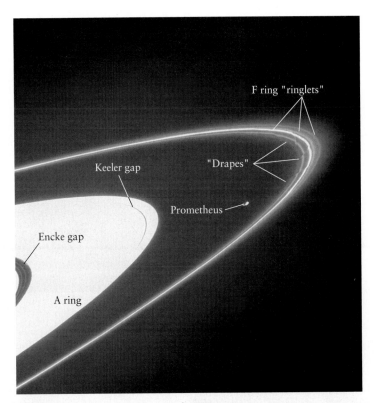

VIDEO 12.8 **Figure 12-20**   R I **V** U X G

Details of Saturn's F Ring This *Cassini* image shows Saturn's F ring (compare Figure 12-19). The bright central ringlet is about 50 km wide. As the small satellite Prometheus orbits just inside the F ring, its gravitational pull causes disturbances called "drapes" in the ringlets. The narrowness of the F ring is the result of gravitational forces exerted by Prometheus and another small satellite called Pandora (see Section 12-11). This image also shows two divisions in the A ring, the Encke gap and the Keeler gap. (NASA/JPL/Space Science Institute)

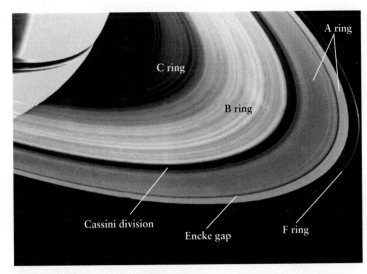

Figure 12-19   R I **V** U X G

Details of Saturn's Rings The colors in this *Voyager 1* image have been enhanced to emphasize small differences between different portions of the rings. The broad Cassini division is clearly visible, as is the narrow Encke gap in the outer A ring. The very thin F ring lies just beyond the outer edge of the A ring. (NASA/JPL)

the sunlit side of the rings. From this perspective, the B ring appears very bright, the A ring moderately bright, the C ring dim, and the Cassini division dark (see Figure 12-19). The fraction of sunlight reflected back toward the Sun (the *albedo*) is directly related to the concentration and size of the particles in the ring. The B ring is bright because it has a high concentration of relatively large, icy, reflective particles, whereas the darker Cassini division has a lower concentration of such particles.

The *Voyagers* and *Cassini* have expanded our knowledge of the ring particles by imaging the shaded side of the rings, which cannot be seen from Earth. **Figure 12-21** shows such an image. Because the spacecraft was looking back toward the Sun, this picture shows sunlight that has passed through the rings. The B ring looks darkest in this image because little sunlight gets through its dense collection of particles. If the Cassini division were completely empty, it would look black, because we would then see through it to the blackness of empty space. But the Cassini division looks *bright* in Figure 12-21. This shows that the Cassini division is not empty but contains a relatively small

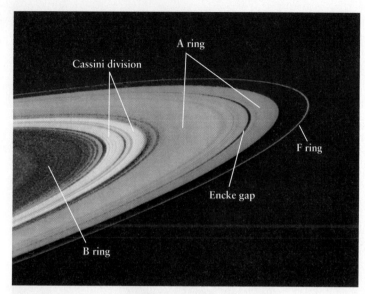

Figure 12-21   R I **V** U X G

The View from the Far Side of Saturn's Rings This false-color view of the side of the rings away from the Sun was taken by *Voyager 1*. The Cassini division appears white, not black like the empty gap between the A and F rings. Hence, the Cassini division cannot be empty, but must contain a number of relatively small particles that scatter sunlight like dust motes. (NASA/JPL)

number of particles. (In the same way, dust particles in the air in your room become visible when a shaft of sunlight passes through them.)

The process by which light bounces off particles in its path is called **light scattering.** The proportion of light scattered in various directions depends upon both the size of the particles and the wavelength of the light. (See Box 5-4 for examples of light scattering on Earth.) Because visible light has a much smaller wavelength than radio waves, light scattering allows scientists to measure the sizes of particles too small to detect using the radio technique described in Section 12-10. As an example, by measuring the amount of light scattered from the rings at various wave-

lengths and from different angles as the two *Voyager* spacecraft sped past Saturn, scientists showed that the F ring contains a substantial number of tiny particles about 1 μm (= 10^{-6} m) in diameter. This is about the same size as the particles found in smoke.

There are also subtle differences in color from one ring to the next, as the computer-enhanced image in Figure 12-22 shows. Although the main chemical constituent of the ring particles is frozen water, trace amounts of other chemicals—perhaps coating the surfaces of the ice particles—are probably responsible for the different colors. These trace chemicals have not yet been identified.

The color variations in Figure 12-22 suggest that the icy particles do not migrate substantially from one ringlet to another. Had such migration taken place, the color differences would have been smeared out over time. The color differences may also indicate that different sorts of material were added to the rings at different times. In this scenario, the rings did not all form at the same time as Saturn but

> **Different colors in the rings reveal variations in chemical composition**

were added to over an extended period. New ring material could have come from small satellites that shattered after being hit by a stray asteroid or comet.

Discovering New Ring Systems

In addition to revealing new details about the A, B, C, and F rings, the *Voyager* cameras also discovered three new ring systems: the D, E, and G rings. The drawing in Figure 12-23 shows the layout of all of Saturn's known rings along with the orbits of some of Saturn's satellites. The **D ring** is Saturn's innermost ring system. It consists of a series of extremely faint ringlets located between the inner edge of the C ring and the Saturnian cloudtops. The **E ring** and the **G ring** both lie far from the planet, well beyond the narrow F ring. Both of these outer ring systems are extremely faint, fuzzy, and tenuous. Each lacks the ringlet structure so prominent in the main ring systems. The E ring encloses the orbit of Enceladus, one of Saturn's icy satellites. Some scientists suspect that water geysers on Enceladus are the source of ice particles in the E ring, much as Io's volcanoes supply material to Jupiter's magnetosphere (see Section 12-8).

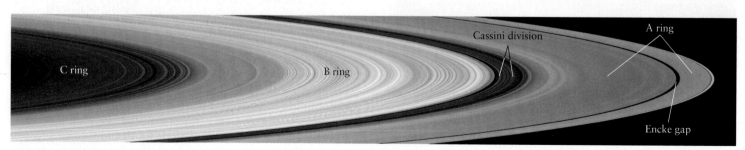

Figure 12-22   R I **V** U X G

Color Variations in Saturn's Rings This mosaic of *Cassini* images shows the rings in natural color. The color variations are indicative of slight differences in chemical composition among particles in different parts of the rings. (NASA/JPL/Space Science Institute)

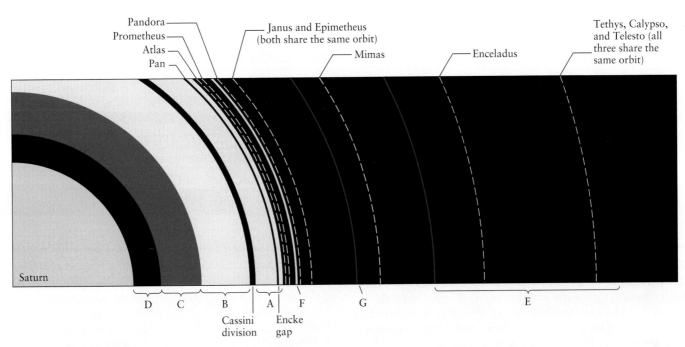

Figure 12-23

The Arrangement of Saturn's Rings This scale drawing shows the location of Saturn's rings along with the orbits of the 11 inner satellites. Only the A, B, and C rings can readily be seen from the Earth. The remaining rings are very faint, and their existence was only confirmed by spacecraft flybys.

12-11 Saturn's inner satellites affect the appearance and structure of its rings

Why do Saturn's rings have such a complex structure? The answer is that, like any object in the universe, the particles that make up the rings are affected by the force of gravity. Saturn's gravitational pull keeps the ring particles in orbit around the planet. But the satellites of Saturn also exert gravitational forces on the rings. These forces can shape the orbits of the ring particles and help give rise to the rings' structure.

Resonance and Shepherd Satellites

Astronomers have long known that one of Saturn's moons, Mimas, has a gravitational effect on its ring system. Mimas is a moderate-sized satellite that orbits Saturn every 22.6 hours. According to Kepler's third law, ring particles in the Cassini division should orbit Saturn approximately every 11.3 hours. Consequently, a given set of ring particles in the Cassini division will find itself between Saturn and Mimas on every second orbit (that is, every 22.6 hours). In other words, particles in the Cassini division are in a 2-to-1 resonance with Mimas. Because of these repeated alignments, the combined gravitational forces of Saturn and Mimas cause small particles to deviate from their original orbits, sweeping them out of the Cassini division. This helps explain why the Cassini division is relatively empty (although not completely so, as Figure 12-21 shows).

While Mimas tends to spread ring particles apart, other satellites exert gravitational forces that herd particles together. The *Voyager* spacecraft discovered two such satellites orbiting on ei-

ther side of the narrow F ring (Figure 12-24). Pandora, the outer of the two satellites, moves around Saturn at a slightly slower speed than the particles in the ring. As ring particles pass Pandora, they experience a tiny gravitational tug that slows them

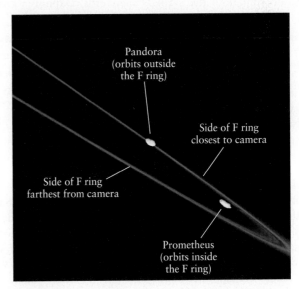

Figure 12-24   R I V U X G

Saturn's F Ring and Its Two Shepherds This *Cassini* image shows the tiny satellites Prometheus (102 km across) and Pandora (84 km across). (Compare with Figure 12-20.) These orbit Saturn on either side of its F ring, shepherding the ring particles into a narrow band. The satellites lap each other every 25 days. (NASA/JPL/Space Science Institute)

Figure 12-25   R I ☑V U X G

Daphnis and the Keeler Gap This *Cassini* spacecraft shows Saturn's tiny moon Daphnis orbiting within the Keeler gap (see also Figure 12-20). Although it is just 7 km across, Daphnis exerts enough gravitational force to keep the 42-km wide gap open and induce wavelike disturbances in the two edges of the gap. (NASA/JPL/Space Science Institute)

Keeler gap

Daphnis

As Daphnis orbits, its gravity induces scalloping along the edges of the gap

Outer edge of the A ring

down slightly. Hence, these particles lose a little energy and fall into orbits a bit closer to Saturn. At the same time, Prometheus, the inner satellite, orbits the planet somewhat faster than the F ring particles. The gravitational pull of Prometheus makes the ring particles speed up a little, giving them some extra energy and nudging them into slightly higher orbits. The competition between the pulls exerted by these two satellites focuses the F ring particles into a well-defined, narrow band.

Because of their confining influence, Prometheus and Pandora are called **shepherd satellites.** We will see in Chapter 14 that Uranus is encircled by narrow rings that are likewise kept confined by shepherd satellites.

An effect similar to shepherding explains two divisions in the A ring, the Encke gap and the even narrower Keeler gap (Figure 12-20). In 2005, *Cassini* images revealed a miniature moon, Daphnis, which actually orbits inside the Keeler gap (Figure 12-25). In a fashion similar to the F ring's shepherds, Daphnis's gravitational forces cause ring particles outside its orbit to move farther out and cause those inside its orbit to move farther inward toward Saturn. In this way, Daphnis helps to maintain a gap in the rings that is 42 km (26 mi) wide, many times wider than Daphnis's 7-km (4-mi) diameter. In a similar way, a 20-km wide moon called Pan orbits within and maintains the 270-km wide Encke gap. Both of these moons cause wavelike disturbances in the ring material on either side of the gap through which they travel (see Figure 12-25).

> **Gravitational effects carve gaps in the rings and focus ring particles into narrow ringlets**

In addition to their dynamic ring systems, Jupiter and Saturn together have an exotic collection of moons. These include Saturn's moon Titan, with a thick atmosphere rich in hydrocarbons; Jupiter's satellite Io, the most volcanic world in the solar system; and Saturn's satellite Enceladus, which has exotic geysers that spew a mixture of water vapor and ice. We will examine these alien worlds in Chapter 13.

Key Words

Key Ideas

Composition and Structure: Jupiter and Saturn are both much larger than Earth. Each is composed of 71% hydrogen, 24% helium, and 5% all other elements by mass. Both planets have a higher percentage of heavy elements than does the Sun.

• Jupiter probably has a rocky core several times more massive than the Earth. The core is surrounded by a layer of liquid "ices" (water, ammonia, methane, and associated compounds). On top of this is a layer of helium and liquid metallic hydrogen and an outermost layer composed primarily of ordinary hydrogen and helium. All of Jupiter's visible features are near the top of this outermost layer.

• Saturn's internal structure is similar to that of Jupiter, but its core makes up a larger fraction of its volume and its liquid metallic hydrogen mantle is shallower than that of Jupiter.

• Jupiter and Saturn both rotate so rapidly that the planets are noticeably flattened. The rotation of both planets' interiors is revealed by variations in the radio emissions from the planets.

• Both Jupiter and Saturn emit more energy than they receive from the Sun. Presumably both planets are still cooling.

Atmospheres: The visible "surfaces" of Jupiter and Saturn are actually the tops of their clouds. The rapid rotation of the planets twists the clouds into dark belts and light zones that run parallel to the equator. Strong zonal winds run along the belts and zones.

• The outer layers of both planets' atmospheres show differential rotation: The equatorial regions rotate slightly faster than the polar regions. For both Jupiter and Saturn, the polar rotation rate is nearly the same as the internal rotation rate.

• The colored ovals visible in the Jovian atmosphere represent gigantic storms. Some, such as the Great Red Spot, are quite stable and persist for many years. Storms in Saturn's atmosphere seem to be shorter-lived.

• There are presumed to be three cloud layers in the atmospheres of Jupiter and Saturn. The reasons for the distinctive colors of these different layers are not yet known. The cloud layers in Saturn's atmosphere are spread out over a greater range of altitude than those of Jupiter, giving Saturn a more washed-out appearance.

• Saturn's atmosphere contains less helium than Jupiter's atmosphere. This lower abundance may be the result of helium raining downward into the planet. Helium "rainfall" may also account for Saturn's surprisingly strong heat output.

Magnetic Fields and Magnetospheres: Jupiter has a strong magnetic field created by currents in the metallic hydrogen layer. Its huge magnetosphere contains a vast current sheet of electrically charged particles. Charged particles in the densest portions of Jupiter's magnetosphere emit synchrotron radiation at radio wavelengths.

• The Jovian magnetosphere encloses a low-density plasma of charged particles. The magnetosphere exists in a delicate balance between pressures from the plasma and from the solar wind. When this balance is disturbed, the size of the magnetosphere fluctuates drastically.

• Saturn's magnetic field and magnetosphere are much less extensive than Jupiter's.

Rings: Saturn is circled by a system of thin, broad rings lying in the plane of the planet's equator. This system is tilted away from the plane of Saturn's orbit, which causes the rings to be seen at various angles by an Earth-based observer over the course of a Saturnian year.

Structure of the Rings: Saturn has three major, broad rings (A, B, and C) that can be seen from Earth. Other, fainter rings were found by the *Voyager* spacecraft.

• The principal rings of Saturn are composed of numerous particles of ice and ice-coated rock ranging in size from a few micrometers to about 10 m. Most of the rings exist inside the Roche limit of Saturn, where disruptive tidal forces are stronger than the gravitational forces attracting the ring particles to each other.

• Each of Saturn's major rings is composed of a great many narrow ringlets. The faint F ring, which is just outside the A ring, is kept narrow by the gravitational pull of shepherd satellites.

• Jupiter's faint rings are composed of a relatively small amount of small, dark, rocky particles that reflect very little light.

Questions

Review Questions

1. Mars passes closer to the Earth than Jupiter does, but with an Earth-based telescope it is easier to see details on Jupiter than on Mars. Why is this?

2. Saturn is the most distant of the planets visible without a telescope. Is there any way we could infer this from naked-eye observations? Explain. (*Hint:* Think about how Saturn's position on the celestial sphere must change over the course of weeks or months.)

3. As seen from Earth, does Jupiter or Saturn undergo retrograde motion more frequently? Explain your answer.

4. In what ways are the motions of Jupiter's atmosphere like the motion of water stirred in a pot (see Figure 12-3b)? In what ways are they different?

5. Is the chemical composition of Jupiter as a whole the same as that of its atmosphere? Explain any differences.

6. Astronomers can detect the presence of hydrogen in stars by looking for the characteristic absorption lines of hydrogen in the star's visible spectrum (Figure 5-21). They can also detect hydrogen in glowing gas clouds by looking for hydrogen's characteristic emission lines (Figure 5-18). Explain why neither of these techniques helped Earth-based astronomers to detect hydrogen in Jupiter's atmosphere.

7. On a warm, humid day, water vapor remains in the atmosphere. But if the temperature drops suddenly, the water vapor forms droplets, clouds appear, and it begins to rain. Relate this observation to why there is relatively little helium in Saturn's atmosphere compared to the atmosphere of Jupiter.

8. What would happen if you tried to land a spacecraft on the surface of Jupiter?

9. What are the belts and zones in the atmospheres of Jupiter and Saturn? Is the Great Red Spot more like a belt or a zone? Explain your answer.

10. Give one possible explanation why weather systems on Jupiter are longer-lived than weather systems on Earth.

11. What are white ovals and brown ovals? What can we infer about them from infrared observations?

12. Compare and contrast the source of energy for motions in the Earth's atmosphere with the energy source for motions in the atmospheres of Jupiter and Saturn.

13. Both Jupiter and Saturn emit more energy than they receive from the Sun in the form of sunlight. Compare the internal energy sources of the two planets that produce this emission.

14. What observations from the *Cassini* spacecraft contradict the accepted picture of convection in Jupiter's atmosphere?

15. Compare the atmospheres of Jupiter and Saturn. Why does Saturn's atmosphere look "washed out" in comparison to that of Jupiter?

16. Which data from the *Galileo Probe* were in agreement with astronomers' predictions? Which data were surprising?

17. Fewer than one in every 10^5 atoms in Jupiter's atmosphere is an argon atom, and fewer than one in 10^8 is an atom of krypton or xenon. If these atoms are so rare, why are scientists concerned about them? How do the abundances of these elements in Jupiter's atmosphere compare to the abundances in the Sun? What hypotheses have been offered to explain these observations?

18. Why is Jupiter oblate? What do astronomers learn from the value of Jupiter's oblateness?

19. What is liquid metallic hydrogen? What is its significance for Jupiter?

20. Describe the internal structures of Jupiter and Saturn, and compare them with the internal structure of the Earth.

21. Explain why Saturn is more oblate than Jupiter, even though Saturn rotates more slowly.

22. If Jupiter does not have any observable solid surface and its atmosphere rotates differentially, how are astronomers able to determine the planet's internal rotation rate?

23. Compare and contrast Jupiter's magnetosphere with the magnetosphere of a terrestrial planet like Earth. Why is the size of the Jovian magnetosphere highly variable, while that of the Earth's magnetosphere is not?

24. What is a plasma? Where are plasmas found in the vicinity of Jupiter?

25. Why is Saturn's magnetosphere less extensive than Jupiter's?

26. What observations of Saturn's rings proved that they are not solid?

27. If Saturn's rings are not solid, why do they look solid when viewed through a telescope?

28. Although the *Voyager* and *Cassini* spacecraft did not collect any samples of Saturn's ring particles, measurements from these spacecraft allowed scientists to determine the sizes of the particles. Explain how this was done.

29. The Space Shuttle and other spacecraft orbit the Earth well within the Earth's Roche limit. Explain why these spacecraft are not torn apart by tidal forces.

30. How do Jupiter's rings differ from those of Saturn?

31. Describe the structure of Saturn's rings. What evidence is there that ring particles do not migrate significantly between ringlets?

32. During the planning stages for the *Pioneer 11* mission, when relatively little was known about Saturn's rings, it was proposed to have the spacecraft fly through the Cassini division. Why would this have been a bad idea?

33. What is the relationship between Saturn's satellite Mimas and the Cassini division?

34. Why is the term "shepherd satellite" appropriate for the objects so named? Explain how a shepherd satellite operates.

Advanced Questions

The question preceded by an asterisk (*) involves topics discussed in Box 7-2.

Problem-solving tips and tools

Box 4-4 describes how to use Newton's form of Kepler's third law. Newton's universal law of gravitation, discussed in Section 4-7, is the basic equation from which you can calculate a planet's surface gravity. Box 7-2 discusses escape speed, and Sections 5-3 and 5-4 discuss the properties of thermal radiation (including the Stefan-Boltzmann law, which relates the temperature of a body to the amount of thermal radiation that it emits).

---

35. The angular diameter of Jupiter at opposition varies little from one opposition to the next. By contrast, the angular diameter of Mars at opposition is quite variable. Explain why there is a difference between these two planets.

36. Jupiter was at opposition on June 5, 2007. On that date Jupiter appeared to be in the constellation Ophiuchus. Ap-proximately when will Jupiter next be at opposition in this same region of the celestial sphere? Explain your answer.

37. Jupiter's equatorial diameter and the rotation period at Jupiter's equator are both given in Table 12-1. Use these data to calculate the speed at which an object at the cloudtops along Jupiter's equator moves around the center of the planet.

38. Using orbital data for a Jovian satellite of your choice (see Appendix 3), calculate the mass of Jupiter. How does your answer compare with the mass quoted in Table 12-1?

39. It has been claimed that Saturn would float if one had a large enough bathtub. Using the mass and size of Saturn given in Table 12-2, confirm that the planet's average density is about 690 kg/m^3, and comment on this somewhat fanciful claim.

40. Roughly speaking, Jupiter's composition (by mass) is three-quarters hydrogen and one-quarter helium. The mass of a single hydrogen atom is given in Appendix 7; the mass of a single helium atom is about 4 times greater. Use these numbers to calculate how many hydrogen atoms and how many helium atoms there are in Jupiter.

41. Use the information given in Section 12-3 to estimate the wind velocities in the Great Red Spot, which rotates with a period of about 6 days.

42. An astronaut floating above Saturn's cloudtops would see a blue sky, even though Saturn's atmosphere has a very different chemical composition than Earth's. Explain why. (*Hint:* See Box 5-4.)

43. If Jupiter emitted just as much energy per second (as infrared radiation) as it receives from the Sun, the average temperature of the planet's cloudtops would be about 107 K. Given that Jupiter actually emits twice this much energy per second, calculate what the average temperature must actually be.

44. (a) From Figure 12-9a, by how much does the temperature increase as you descend from the 100-millibar level in Jupiter's atmosphere to an altitude 100 km below that level? (b) Use Figure 12-9b to answer the same question for Saturn's atmosphere. (c) In the Earth's troposphere (see Section 9-5), the air temperature increases by 6.4°C for each kilometer that you descend. In which planet's atmosphere—Earth, Jupiter, or Saturn—does the temperature increase most rapidly with decreasing altitude?

*45. Consider a hypothetical future spacecraft that would float, suspended from a balloon, for extended periods in Jupiter's upper atmosphere. If we want this spacecraft to return to Earth after completing its mission, calculate the speed at which the spacecraft's rocket motor would have to accelerate it in order to escape Jupiter's gravitational pull. Compare with the escape speed from the Earth, equal to 11.2 km/s.

46. The *Galileo Probe* had a mass of 339 kg. On Earth, its weight (the gravitational force exerted on it by Earth) was 3320 newtons, or 747 lb. What was the gravitational force that Jupiter exerted on the *Galileo Probe* when it entered Jupiter's clouds?

47. From the information given in Section 12-6, calculate the average density of Jupiter's rocky core. How does this compare with the average density of the Earth? With the average density of the Earth's solid inner core? (See Table 9-1 and Table 9-2 for data about the Earth.)

48. In the outermost part of Jupiter's outer layer (shown in yellow in the upper part of Figure 12-12), hydrogen is principally in the form of molecules (H_2). Deep within the liquid metallic hydrogen layer (shown in orange in Figure 12-12), hydrogen is in the form of single atoms. Recent laboratory experiments suggest that there is a gradual transition between these two states, and that the transition layer overlaps the boundary between the ordinary hydrogen and liquid metallic hydrogen layers. Use this information to redraw the upper part of Figure 12-12 and to label the following regions in Jupiter's interior: (i) ordinary (nonmetallic) hydrogen molecules; (ii) nonmetallic hydrogen with a mixture of atoms and molecules; (iii) liquid metallic hydrogen with a mixture of atoms and molecules; (iv) liquid metallic hydrogen atoms.

49. When Saturn is at different points in its orbit, we see different aspects of its rings because the planet has a 27° tilt. If the tilt angle were different, would it be possible to see the upper and lower sides of the rings at all points in Saturn's orbit? If so, what would the tilt angle have to be? Explain your answers.

50. As seen from Earth, the intervals between successive edge-on presentations of Saturn's rings alternate between about 13 years, 9 months, and about 15 years, 9 months. Why do you think these two intervals are not equal?

51. (a) Use Newton's form of Kepler's third law to calculate the orbital periods of particles at the outer edge of Saturn's A ring and at the inner edge of the B ring. (b) Saturn's rings orbit in the same direction as Saturn's rotation. If you were floating along with the cloudtops at Saturn's equator, would the outer edge of the A ring and the inner edge of the B ring appear to move in the same or opposite directions? Explain.

 R I **V** U X G
(NASA/JPL)

52. The above *Voyager 2* close-up image of Saturn's rings shows a number of dark, straight features called *spokes*. As these features orbit around Saturn, they tend to retain their shape like the rigid spokes on a rotating bicycle wheel. (The black dots were added by the *Voyager* camera system to help scientists calibrate the electronic image.) The spokes rotate at the same rate as Saturn's magnetic field and are thought to be clouds of tiny, electrically charged particles kept in orbit by magnetic forces. Explain why the spokes could *not* maintain their shape if they were kept in orbit by gravitational forces alone.

53. The Cassini division involves a 2-to-1 resonance with Mimas. Does the location of the Encke gap—133,500 km from Saturn's center—correspond to a resonance with one of the other satellites? If so, which one? (See Appendix 3.)

Discussion Questions

54. Describe some of the semipermanent features in Jupiter's atmosphere. Compare and contrast these long-lived features with some of the transient phenomena seen in Jupiter's clouds.

55. Suppose you were asked to design a mission to Jupiter involving an unmanned airplanelike vehicle that would spend many days (months?) flying through the Jovian clouds. What observations, measurements, and analyses should this aircraft be prepared to make? What dangers might the aircraft encounter, and what design problems would you have to overcome?

56. What sort of experiment or space mission would you design in order to establish definitively whether Jupiter has a rocky core?

57. The classic science-fiction films *2001: A Space Odyssey* and *2010: The Year We Make Contact* both involve manned spacecraft in orbit around Jupiter. What kinds of observations could humans make on such a mission that cannot be made by robotic spacecraft? What would be the risks associated with such a mission? Do you think that a manned Jupiter mission would be as worthwhile as a manned mission to Mars? Explain your answers.

58. Suppose that Saturn were somehow moved to an orbit around the Sun with semimajor axis 1 AU, the same as the Earth's. Discuss what long-term effects this would have on the planet and its rings.

Web/eBook Questions

59. On Jupiter, the noble gases argon, krypton, and xenon provide important clues about Jupiter's past. On Earth, xenon is used in electronic strobe lamps because it emits a very white light when excited by an electric current. Argon, by contrast, is one of the gases used to fill ordinary incandescent lightbulbs. Search the World Wide Web for information about why argon is used in this way. Why do premium, long-life lightbulbs use krypton rather than argon?

60. Search the World Wide Web, especially the Web sites for NASA's Jet Propulsion Laboratory and the European Space Agency, for information about the current status of the *Cassini* mission. What recent discoveries has *Cassini* made? What ideas are being considered for the *Cassini* extended mission, to begin in 2008?

61. The two *Voyager* spacecraft were launched from Earth along a trajectory that took them directly to Jupiter. The force of Jupiter's gravity then gave the two spacecraft a "kick" that helped push them onward to Saturn. The much larger *Cassini* spacecraft, by contrast, was first launched on a trajectory that took it past Venus. Search the World Wide Web, especially the Web sites for NASA's Jet Propulsion Laboratory and the European Space Agency, for information about the trajectory that *Cassini* took through the solar system to reach

Saturn. Explain why this trajectory was so different from that of the *Voyagers*.

 62. The Rotation Rate of Saturn. Access and view the video "Saturn from the Hubble Space Telescope" in Chapter 12 of the *Universe* Web site or eBook. The total amount of time that actually elapses in this video is 42.6 hours. Using this information, identify and follow an atmospheric feature and determine the rotation period of Saturn. How does your answer compare with the value given in Table 12-2?

Activities

Observing Projects

> Observing tips and tools
>
>   Like Mars, Jupiter and Saturn are most easily seen around opposition. (The dates of Jupiter's and Saturn's oppositions are available on the World Wide Web.) Each planet is visible in the night sky for several months before or after its opposition. At other times, Jupiter and Saturn may be visible in either the predawn morning sky or the early evening sky. Consult such magazines as *Sky & Telescope* and *Astronomy* or their Web sites for more detailed information about when and where to look for Jupiter during a given month. You can also use the *Starry Night Enthusiast™* program on the CD-ROM that accompanies selected copies of this textbook. A relatively small telescope with an objective diameter of 15 to 20 cm (6 to 8 inches), used with a medium-power eyepiece to give a magnification of 25X or so, should enable you to see some of the dark belts on Jupiter and the rings of Saturn.

63. If Jupiter is visible in the night sky, make arrangements to view the planet through a telescope. What magnifying power seems to give you the best view? Draw a picture of what you see. Can you see any belts and zones? How many? Can you see the Great Red Spot?

64. Make arrangements to view Jupiter's Great Red Spot through a telescope. Consult the *Sky & Telescope* Web site, which lists the times when the center of the Great Red Spot passes across Jupiter's central meridian. The Great Red Spot is well placed for viewing for 50 minutes before and after this happens. You will need a refractor with an objective of at least 15 cm (6 in.) diameter or a reflector with an objective of at least 20 cm (8 in.) diameter. Using a pale blue or green filter can increase the color contrast and make the spot more visible. For other useful hints, see the article "Tracking Jupiter's Great Red Spot" by Alan MacRobert (*Sky & Telescope*, September 1997).

65. View Saturn through a small telescope. Make a sketch of what you see. Estimate the angle at which the rings are tilted to your line of sight. Can you see the Cassini division? Can you see any belts or zones in Saturn's clouds? Is there a faint, starlike object near Saturn that might be Saturn's satellite Titan? What observations could you perform to test whether the starlike object is a Saturnian satellite?

 66. Use the *Starry Night Enthusiast™* program to observe the appearance of Jupiter. Select **Favourites > Guides > Atlas** from the menu. Open the **Find** pane and click the menu button for **Jupiter** and select **Magnify** from the drop-down menu. **Stop** time flow and, in the toolbar, set the date and time to **March 4, 2004, at 12:00:00 A.M.** to see Jupiter at opposition. Set the **Time Flow Rate** to **1 minute.** Then **Run Time Forward.** Use the Time Flow controls to determine the rotation period of the planet. (*Hint:* You may want to track the motion of an easily recognizable feature, such as the Great Red Spot.) How does your answer compare with the rotation period given in Table 12-1?

67. Use the *Starry Night Enthusiast™* program to observe the appearance of Saturn. Select **Favourites > Guides > Atlas** from the menu. Open the **Find** pane and click the menu button for **Saturn** and select **Magnify** from the drop-down menu. **Stop** time flow. Set the **Time Flow Rate** to 1 lunar month. Then **Run Time Forward.** Describe how Saturn's appearance changes over time. Explain what causes these changes.

68. Use the *Starry Night Enthusiast™* program to examine Jupiter and Saturn. (**a**) Select **Favourites > Solar System > Jupiter** from the menu. Select **View > Feet** to remove the image of the spacesuit. You can zoom in and zoom out on the view and you can also rotate Jupiter by putting the mouse cursor over the image, holding down the mouse button, and moving the mouse. (On a two-button mouse, hold down the left mouse button.) How many white ovals and brown ovals can you count in Jupiter's southern hemisphere (the hemisphere in which the Great Red Spot is located)? How many white ovals and brown ovals can you count in the northern hemisphere? What general rule can you state about the abundance of these storms in the two hemispheres? (**b**) Select **Favourites > Solar System > Saturn** from the menu. Select **View > Feet** to remove the image of the spacesuit. You can zoom in and zoom out on the view and you can also rotate Saturn by placing the mouse cursor over the image, holding down the mouse button, and moving the mouse. (On a two-button mouse, hold down the left mouse button.) Rotate Saturn so that you are looking straight down on the plane of the rings. Draw a copy of what you see and label the different rings and divisions. (Compare Figure 12-23.)

Collaborative Exercises

69. Using a ruler with millimeter markings on five various images of Jupiter in the text, Figures 12-1a, 12-2a, and 12-4, determine the ratio of the longest width of the Great Red Spot to the full diameter of Jupiter. Each group member should measure a different image and all values should be averaged.

70. The text provides different years that spacecraft have flown by Jupiter and Saturn. List these dates and create a time line by listing one important event that was occurring on Earth during each of those years.

71. If the largest circle you can draw on a piece of paper represents the largest diameter of Saturn's rings, about how large would Saturn be if scaled appropriately? Which item in a group member's backpack is closest to this size?

13

Jupiter and Saturn's Satellites of Fire and Ice

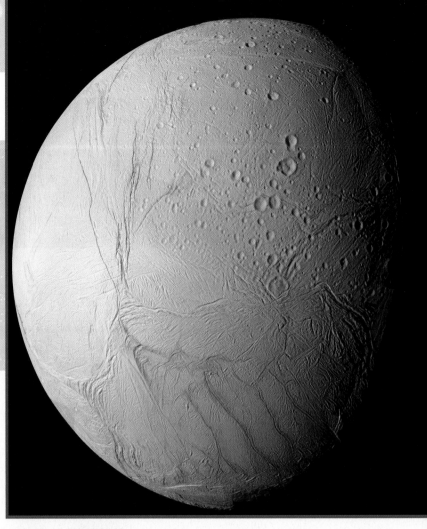

A color-enhanced view of Saturn's satellite Enceladus. (NASA/JPL/Space Science Institute)

WEB LINK 13.1

The delicately colored world shown here is neither the creation of an abstract artist nor the product of a science fiction writer's imagination. It is Enceladus, a satellite of Saturn whose icy surface has a split personality: One hemisphere is ancient and covered with impact craters, while the other hemisphere is nearly crater-free and crisscrossed by stress faults and fractures. The latter hemisphere is actually the site of active volcanoes that eject icy particles rather than lava.

Enceladus is not the only satellite of the giant planets to display exotic geologic activity. Io, a large satellite of Jupiter discovered by Galileo in 1610, is the most geologically active world in the solar system. Its surface is covered with colorful sulfur compounds deposited by dozens of active volcanoes. Europa, another of Jupiter's large satellites, has no volcanoes but does have a weirdly cracked surface of ice. Beneath Europa's ice there may be a worldwide ocean, within which there could possibly be single-celled organisms.

No less exotic is Saturn's large satellite Titan, which is enveloped by a thick, nitrogen-rich atmosphere from which liquid hydrocarbons fall as rain. In this chapter we will explore these and other intriguing aspects of Jupiter and Saturn's extensive collection of satellites.

13-1 Jupiter's Galilean satellites are easily seen with Earth-based telescopes

When Galileo trained his telescope on Jupiter in January 1610, he became the first person to observe satellites orbiting another planet (see Section 4-5). He called them the Medicean Stars, in

Learning Goals

By reading the sections of this chapter, you will learn

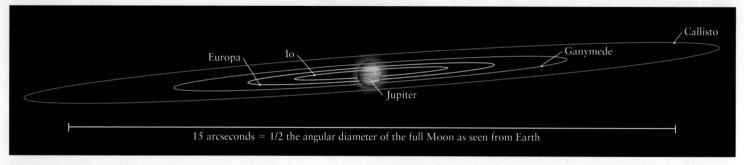

15 arcseconds = 1/2 the angular diameter of the full Moon as seen from Earth

Figure 13-1

The Orbits of the Galilean Satellites This illustration shows the orbits of the four Galilean satellites as seen from Earth. All four orbits lie in nearly the same plane as Jupiter's equator. The apparent angular size and orientation of the orbits depends on the relative positions of Earth and Jupiter.

order to curry the favor of a wealthy Florentine patron of the arts and sciences. We now call them the **Galilean satellites**. Individually, they are named Io, Europa, Ganymede, and Callisto, after four mythical lovers and companions of the god whom the Greeks called Zeus and the Romans called Jupiter.

When viewed through an Earth-based telescope, the Galilean satellites look like pinpoints of light (see Figure 4-16). All four satellites orbit Jupiter in nearly the same plane as the planet's equator (Figure 13-1). From Earth, we always see that plane nearly edge-on, so the satellites appear to move back and forth relative to Jupiter (see Figure 4-17). The orbital periods are fairly short, ranging from 1.8 (Earth) days for Io to 16.7 days for Callisto, so we can easily see the satellites move from one night to the next and even during a single night. These motions led Galileo to realize that he was seeing objects orbiting around Jupiter, in much the same way that Copernicus said that planets move around the Sun (see Section 4-2).

Each of the four Galilean satellites is bright enough to be visible to the naked eye. Why, then, did Galileo need a telescope to discover them? The reason is that as seen from the Earth, the angular separation between Jupiter and the satellites is quite small—never more than 10 arcminutes for Callisto and even less for the other three Galilean satellites. To the naked eye, these satellites are lost in the overwhelming glare of Jupiter. But a small telescope or even binoculars increases the apparent angular separation by enough to make the Galilean satellites visible.

Synchronous Rotation

The brightness of each satellite varies slightly as it orbits Jupiter. This is because the satellites also spin on their axes as they orbit Jupiter, and dark and light areas on their surfaces are alternately exposed to and hidden from our view. Remarkably, each Galilean satellite goes through one complete cycle of brightness during one orbital period. This tells us that each satellite is in *synchronous rotation*, so that it rotates exactly once on its axis during each trip around its orbit (see Figure 3-4*b*). As we saw in Section 4-8, our Moon's synchronous rotation is the result of gravitational forces exerted on the Moon by Earth. Likewise, Jupiter's gravitational forces keep the Galilean satellites in synchronous rotation.

Synchronous rotation means that the rotation period and orbital period are in a 1-to-1 ratio for each Galilean satellite. Remarkably, there is also a simple ratio of the different orbital periods of the three inner Galilean satellites, Io, Europa, and Ganymede. During the 7.155 days that Ganymede takes to complete one orbit around Jupiter, Europa makes two orbits and Io makes four orbits. Thus, the orbital periods of Io, Europa, and Ganymede are in the ratio 1:2:4, which you can verify from the data in Table 13-1.

This special relationship among the satellites' orbits is maintained by the gravitational forces that they exert on one another. Those forces can be quite strong, because the three inner Galilean satellites pass relatively close to one another; at their closest approach, Io and Europa are separated by only two-thirds the distance from the Earth to the Moon. Such a close approach occurs once for every two of Io's orbits, so Europa's gravitational pull acts on Io in a rhythmic way. Indeed, Io, Europa, and Ganymede all exert rhythmic gravitational tugs on one another. Just as a drummer's rhythm keeps musicians on the same beat, this gravitational rhythm maintains the simple ratio of orbital periods among the satellites.

> **Gravitational interactions between Io, Europa, and Ganymede induce a rhythmic relationship among their orbits**

By contrast, Callisto orbits at a relatively large distance from the other three large satellites. Hence, the gravitational forces on Callisto from the other satellites are relatively weak, and there is no simple relationship between Callisto's orbital period and the period of Io, Europa, or Ganymede.

Transits, Eclipses, and Occultations

Because the orbital planes of the Galilean satellites are nearly edge-on to our line of sight, we see these satellites undergoing transits, eclipses, and occultations. In a *transit*, a satellite passes between us and Jupiter, and we see the satellite's shadow as a black dot against the planet's colorful cloudtops. (Figure 12-2*a* shows Europa's shadow on Jupiter.) In an *eclipse*, one of the satellites disappears and then reappears as it passes into and out of Jupiter's enormous shadow. In an **occultation**, a satellite passes

Table 13-1   Jupiter's Galilean Satellites Compared with the Moon, Mercury, and Mars

	Average distance from Jupiter (km)	Orbital period (days)	Diameter (km)	Mass (kg)	Mass (Moon = 1)	Average density (kg/m³)	Albedo
Io	421,600	1.769	3642	8.932×10^{22}	1.22	3529	0.63
Europa	670,900	3.551	3120	4.791×10^{22}	0.65	3018	0.64
Ganymede	1,070,000	7.155	5268	1.482×10^{23}	2.02	1936	0.43
Callisto	1,883,000	16.689	4800	1.077×10^{23}	1.47	1851	0.17
Moon	—	—	3476	7.349×10^{22}	1.00	3344	0.11
Mercury	—	—	4880	3.302×10^{23}	4.49	5430	0.12
Mars	—	—	6794	6.419×10^{23}	8.73	3934	0.15

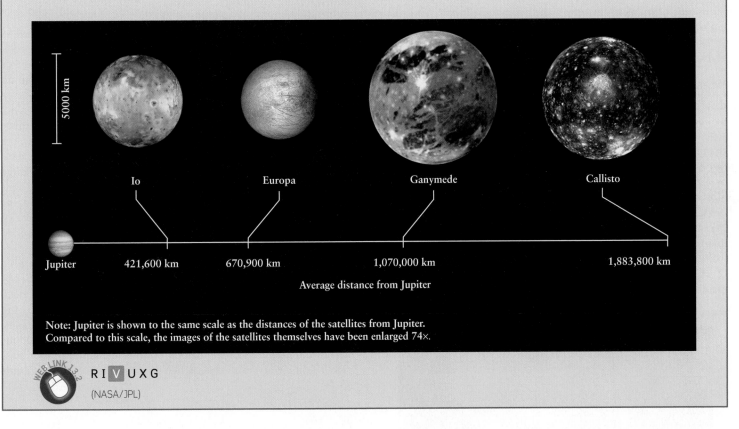

5000 km

Io   Europa   Ganymede   Callisto

Jupiter   421,600 km   670,900 km   1,070,000 km   1,883,800 km

Average distance from Jupiter

Note: Jupiter is shown to the same scale as the distances of the satellites from Jupiter. Compared to this scale, the images of the satellites themselves have been enlarged 74×.

WEB LINK 13.2   R I V UX G

(NASA/JPL)

completely behind Jupiter, and the satellite is temporarily blocked from our Earth-based view. (The word *occultation* comes from a Latin verb meaning "to cover.")

Before spacecraft first ventured to Jupiter, astronomers used eclipses to estimate the diameters of the Galilean satellites. When a satellite emerges from Jupiter's shadow, it does not blink on instantly. Instead, there is a brief interval during which the satellite gets progressively brighter as more of its surface is exposed to sunlight. The satellite's diameter is calculated from the duration of this interval and the satellite's orbital speed (which is known from Kepler's laws).

Occultations can also be used to determine the diameter of a satellite by measuring the time it takes to disappear behind Jupiter or to reappear from behind the planet. In addition, the Galilean satellites occasionally occult one another, because all four orbit Jupiter in the plane of the planet's equator. Every six years, when the Earth passes through this equatorial plane, there is a period of a few months during which Earth observers can see mutual occultations of the satellites. Once again, timing the occultations enables astronomers to calculate the diameters of the satellites.

As Table 13-1 shows, the smallest Galilean satellite (Europa) is slightly smaller than our Moon; the largest satellite (Ganymede) is larger than Mercury and more than three-quarters the size of Mars. Thus, the Galilean satellites truly are worlds in their own right.

13-2 Data from spacecraft reveal the unique properties of the Galilean satellites

Even the finest images made with Earth-based telescopes have revealed relatively few details about the Galilean satellites. Almost everything we know about these satellites has come from observations made at close range by spacecraft.

The first close-range observations of the Galilean satellites were made by the *Pioneer 10* and *Pioneer 11* spacecraft as they flew past Jupiter in 1973 and 1974. The images from these missions were of relatively low resolution, however, and much more information has come from three subsequent missions. *Voyager 1* and *Voyager 2* flew past Jupiter in 1979 and recorded tens of thousands of images, one of which is shown in Figure 13-2. The *Galileo* spacecraft, which made 35 orbits around Jupiter between 1995 and 2003, carried out an even more extensive investigation of Jupiter's satellites.

Measuring Satellite Densities

One key goal of these missions was to determine the densities of the Galilean satellites to very high accuracy. Given the density of a planet or satellite, astronomers can draw conclusions about its chemical composition and internal structure. To find the densities, scientists first determined the satellite masses. They did this by measuring how the gravity of Io, Europa, Ganymede, and Callisto deflected the trajectories of the *Voyager* and *Galileo* spacecraft. Given the masses and diameters of the satellites, they then calculated each satellite's average density (mass divided by volume). Table 13-1 lists our current knowledge about the sizes and masses of the Galilean satellites.

Of the four Galilean satellites, Europa proves to be the least massive: Its mass is only two-thirds that of our Moon. Ganymede

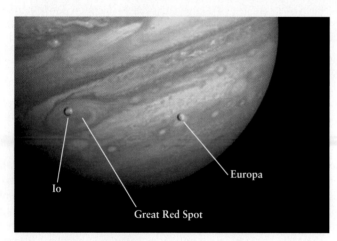

Figure 13-2   R I [V] U X G

Io and Europa from *Voyager 1* *Voyager 1* recorded this image of Jupiter and the Galilean satellites Io and Europa in February 1979, when it was 20 million kilometers (12.4 million miles) from the giant planet. The satellites are actually smaller than some of Jupiter's cloud features, such as the Great Red Spot (see Section 14-3). *Voyager 1* and *Voyager 2* also recorded many images of the Galilean satellites at closer range. (NASA/JPL)

is by far the most massive of the four, with more than double the mass of our Moon. In fact, Ganymede is the most massive satellite anywhere in the solar system. In second place is Callisto, with about $1\frac{1}{2}$ times the Moon's mass.

However, Ganymede and Callisto are *not* merely larger versions of our Moon. The reason is that both Ganymede and Callisto have very low average densities of less than 2000 kg/m³. By comparison, typical rocks in the Earth's crust have densities around 3000 kg/m³, and the average density of our Moon is 3344 kg/m³. The low densities of Ganymede and Callisto mean that these satellites cannot be composed primarily of rock. Instead, they are probably roughly equal parts of rock and water ice. Water ice is substantially less dense than rock, but it becomes as rigid as rock under high pressure (such as is found in the interiors of the Galilean satellites). Furthermore, water molecules are relatively common in the solar system. Enough water would have been available in the early solar system to make up a substantial portion of a large satellite such as Ganymede or Callisto.

> **Three of the Galilean satellites are made of a mixture of ice and rock; only Io is ice-free**

Water ice cannot be a major constituent of the two inner Galilean satellites, however. The innermost satellite, Io, has the highest average density, 3529 kg/m³, slightly greater than the density of our Moon. The next satellite out, Europa, has an average density of 3018 kg/m³. Both of these values are close to the densities of typical rocks in the Earth's crust. Hence, it is reasonable to suppose that both Io and Europa are made primarily of rocky material.

The most definitive evidence for water ice on Jupiter's satellites has come from spectroscopic observations. The spectra of sunlight reflected from Europa, Ganymede, and Callisto show absorption at the infrared wavelengths characteristic of water ice molecules. You can see this in Figure 7-4, which compares the spectrum of Europa to the spectrum of ice. Europa's density shows that it is composed mostly of rock, so its ice must be limited to the satellite's outer regions. Among the Galilean satellites, only Io shows no trace of water ice on its surface or in its interior.

13-3 The Galilean satellites formed like a solar system in miniature

As Table 13-1 shows, the more distant a Galilean satellite is from Jupiter, the greater the proportion of ice within the satellite and the lower its average density. But why should this be? An important clue is that the average densities of the planets show a similar trend: Moving outward from the Sun, the average density of the planets steadily decreases, from more than 5000 kg/m³ for Mercury to less than 1000 kg/m³ for Saturn (recall Table 7-1). The best explanation for this similarity is that the Galilean satellites formed around Jupiter in much the same way that the planets formed around the Sun, although on a much smaller scale.

> **The differences in composition among the Galilean satellites mimic those among planets orbiting the Sun**

We saw in Section 8-4 that dust grains in the solar nebula played an important role in planet formation. In the inner parts of the nebula, close to the glowing protosun, only dense, rocky grains were able to survive. These accumulated over time into the dense, rocky inner planets. But in the cold outer reaches of the nebula, dust grains were able to retain icy coatings of low-density material like water and ammonia. Hence, when the Jovian planets formed in the outer solar nebula and incorporated these ice-coated grains, the planets ended up with both rock and ices in their cores as well as substantial amounts of ammonia and water throughout their interiors and atmospheres (see Section 14-1 and Section 14-6).

As Jupiter coalesced, the gas accumulating around it formed a rotating "Jovian nebula." The central part of this nebula became the huge envelope of hydrogen and helium that makes up most of Jupiter's bulk. But in the outer parts of this nebula, dust grains could have accreted to form small solid bodies. These grew to become the Galilean satellites.

Although the "Jovian nebula" was far from the protosun, not all of it was cold. Jupiter, like the protosun, must have emitted substantial amounts of radiation due to Kelvin-Helmholtz contraction, which we discussed in Section 8-3. (We saw in Section 12-4 that even today, Jupiter emits more energy due to its internal heat than it absorbs from sunlight.) Hence, temperatures very close to Jupiter must have been substantially higher than at locations farther from the planet. Calculations show that only rocky material would condense at the orbital distances of Io and Europa, but frozen water could be retained and incorporated into satellites at the distances of Ganymede and Callisto. In this way Jupiter ended up with two distinct classes of Galilean satellites (Figure 13-3).

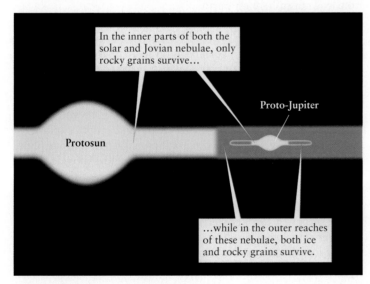

Figure 13-3

Formation of the Galilean Satellites Heat from the protosun made it impossible for icy grains to survive within the innermost 2.5 to 4 AU of the solar nebula. In the same way, Jupiter's heat evaporated any icy grains that were too close to the center of the "Jovian nebula." Hence, the two inner Galilean satellites were formed primarily from rock, while the outer two incorporated both rock and ice.

The analogy between the solar nebula and the "Jovian nebula" is not exact. Unlike the Sun, Jupiter is a "failed star." Its internal temperatures and pressures never became high enough to ignite nuclear reactions that convert hydrogen into helium. (Jupiter's mass would have had to be about 80 times larger for these reactions to have begun.) Furthermore, the icy worlds that formed in the outer region of the "Jovian nebula"—namely, Ganymede and Callisto—were of relatively small mass. Hence, they were unable to attract gas from the nebula and become Jovian planets in their own right. Nonetheless, it is remarkable how the main processes of "planet" formation seem to have occurred twice in our solar system: once in the solar nebula around the Sun, and once in microcosm in the gas cloud around Jupiter.

The diverse densities of the Galilean satellites suggest that these four worlds may be equally diverse in their geology. This turns out to be entirely correct. Because each of the Galilean satellites is unique, we devote the next several sections to each satellite in turn.

13-4 Io is covered with colorful sulfur compounds ejected from active volcanoes

Before the two *Voyager* spacecraft flew past Jupiter, the nature of Io was largely unknown. On the basis of Io's size and density, however, it was expected that Io would be a world much like our own Moon—geologically dead, with little internal heat available to power tectonic or volcanic activity (see Section 7-6). Hence, it was thought that Io's surface would be extensively cratered, because there would have been little geologic activity to erase those craters over the satellite's history.

Io proved to be utterly different from these naive predictions. On March 5, 1979, *Voyager 1* came within 21,000 kilometers (13,000 miles) of Io and began sending back a series of bizarre and unexpected pictures of the satellite (Figure 13-4). These images showed that Io has *no* impact craters at all! Instead, the surface is pockmarked by irregularly shaped pits and is blotched with color.

The scientists were seeing confirmation of a prediction published in the journal *Science* just three days before *Voyager 1* flew past Io. In this article, Stanton Peale of the University of California, Santa Barbara, and Patrick Cassen and Ray Reynolds of NASA's Ames Research Center reported their conclusion that Io's interior must be kept hot by Jupiter's tidal forces, and that this should lead to intense volcanic activity.

Tidal Heating on Io

As we saw in Section 4-8, tidal forces are differences in the gravitational pull on different parts of a planet or satellite. These forces tend to deform the shape of the planet or satellite (see Figure 4-25). Io, for example, is somewhat distorted from a spherical shape by tidal forces from Jupiter. If Io were in a perfectly circular orbit so that it traveled around Jupiter at a steady speed, Io's rotation would always be in lockstep with its orbital motion. In this case, Io would be in perfect synchronous rotation, with its long axis always pointed directly toward Jupiter. But as Io moves

Figure 13-4   R I **V** U X G

Io These mosaics of Io's two hemispheres were built up from individual *Voyager* images. The extraordinary colors are probably caused by deposits of sulfur and sulfur compounds ejected from Io's numerous volcanoes. The labels show several of these volcanoes, which are named for sun gods and fire gods of different cultures. (NASA/JPL)

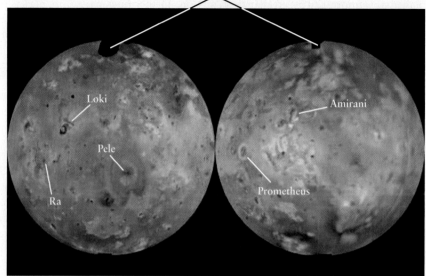

Areas not observed by the *Voyager* spacecraft

around its orbit, Europa and Ganymede exert gravitational tugs on it in a regular, rhythmic fashion, thanks to the 1:2:4 ratio among the orbital periods of these three satellites. These tugs distort Io's orbit into an ellipse, and so its speed varies as it moves around its orbit. Consequently, Io's long axis "nods" back and forth by about $\frac{1}{2}°$ as seen from Jupiter. Due to this "nodding," the tidal stresses that Jupiter exerts on Io vary rhythmically as Io moves around its orbit. These varying tidal stresses alternately squeeze and flex Io. Just as a ball of clay or bread dough gets warm as you knead it between your fingers, this squeezing and flexing causes **tidal heating** of Io's interior. (The Earth's Moon also "nods" as it orbits our planet, but experiences almost no tidal heating because the Earth's tidal forces are very weak compared to those of massive Jupiter.)

> The theory of tidal forces helped predict Io's volcanic nature

Tidal heating adds energy to Io at a rate of about 10^{14} watts, equivalent to 24 tons of TNT exploding every second. As this energy makes its way to the satellite's surface, it provides about 2.5 watts of power to each square meter of Io's surface. By comparison, the average global heat flow through the Earth's crust is 0.06 watts per square meter. Only in volcanically active areas on Earth do we find heat flows that even come close to Io's average. Thus, Peale, Cassen, and Reynolds predicted "widespread and recurrent surface volcanism" on Io.

Io's Active Volcanoes

No one expected that *Voyager 1* would obtain images of erupting volcanoes on Io. After all, a spacecraft making a single trip past the Earth would be highly unlikely to catch a large volcano actually erupting. But, in fact, *Voyager 1* images of Io revealed *eight* different giant plumes of gas from volcanic eruptions. (Figure 13-5a shows two of these.) More recently, the *Galileo* spacecraft, which came as close as 200 km to Io's surface during its eight years in orbit around Jupiter, returned detailed images of

several such plumes (Figure 13-5b). These observations resoundingly confirm the Peale-Cassen-Reynolds prediction: Io is by far the most volcanic world in the solar system.

Io's volcanic plumes rise to astonishing heights of 70 to 280 km above Io's surface. To reach such altitudes, the material must emerge from volcanic vents with speeds between 300 and 1000 m/s (1100 to 3600 km/h, or 700 to 2200 mi/h). Even the most violent terrestrial volcanoes have eruption speeds of only around 100 m/s (360 km/h, or 220 mi/h). Scientists thus began to suspect that Io's volcanic activity must be fundamentally different from volcanism here on Earth.

The Nature of Io's Volcanic Eruptions

An important clue about volcanism on Io came from the infrared spectrometers aboard *Voyager 1*, which detected abundant sulfur and sulfur dioxide in Io's volcanic plumes. This led to the idea that the plumes are actually more like geysers than volcanic eruptions. In a geyser on Earth, water seeps down to volcanically heated rocks, where it changes to steam and erupts explosively through a vent. Planetary geologists Susan Kieffer, Eugene Shoemaker, and Bradford Smith suggested that sulfur dioxide rather than water could be the principal propulsive agent driving volcanic plumes on Io. Sulfur dioxide is a solid at the frigid temperatures found on most of Io's surface, but it should be molten at depths of only a few kilometers. Just as the explosive conversion of water into steam produces a geyser on the Earth, the conversion of liquid sulfur dioxide into a high-pressure gas could result in eruption velocities of up to 1000 m/s.

Io's dramatic coloration (see Figure 13-4) is probably due to sulfur and sulfur dioxide, which are ejected in volcanic plumes and later fall back to the surface. Sulfur is normally bright yellow, which explains the dominant color of Io's surface. But if sulfur is heated and suddenly cooled, as would happen if it were ejected from a volcanic vent and allowed to fall to the surface, it can assume a range of colors from orange

(a) *Voyager 1*, March 1979

(b) *Galileo*, November 1997

Figure 13-5   R I V U X G

Volcanic Plumes on Io **(a)** Io's volcanic eruptions were first discovered on this *Voyager 1* image. The plume at upper left, which extends 260 km (150 mi) above Io's surface, came from the volcano Pele (see Figure 13-4). **(b)** This *Galileo* image shows plumes from the volcanoes Prometheus (see Figure 13-4) and Zamama. The plumes rise to about 100 km (60 mi) above Io's surface and are about 250 km (150 mi) wide. (a: NASA; b: Planetary Image Research Laboratory/University of Arizona/JPL/NASA)

and red to black. Indeed, these colors are commonly found around active volcanic vents (Figure 13-6). Whitish surface deposits (examine Figure 13-4 and Figure 13-6*a*), by contrast, are probably due to sulfur dioxide (SO₂). Volcanic vents on Earth commonly discharge SO₂ in the form of an acrid gas. But on Io, when hot SO₂ gas is released by an eruption into the cold vacuum of space, it crystallizes into white snowflakes. This sulfur dioxide "snow" then rains back onto Io's surface.

The plumes are not the whole story of volcanism on Io, however. The sources of the plumes appear in *Voyager* images as black spots 10 to 50 km in diameter, which are actually volcanic vents (see Figure 13-4). Many of these black spots, which cover 5% of Io's surface, are surrounded by dark lava flows. Figure 13-7 is a *Galileo* image of one of Io's most active volcanic regions, where in 1999 lava was found spouting to altitudes of thousands of meters along a fissure 40 km (25 mi) in length. The glow from this lava fountain was so intense that it was visible with Earth-

based telescopes. Thus, Io has two distinct styles of volcanic activity: unique geyserlike plumes and lava flows that are more dramatic versions of volcanic flows on Earth.

While sulfur and sulfur compounds are the key constituents of Io's volcanic plumes, the lava flows must be made of something else. Infrared measurements from the *Galileo* spacecraft (Figure 13-8) show that lava flows on Io are at temperatures of 1700 to 2000 K (about 1450 to 1750°C, or 2600 to 3150°F). At these temperatures sulfur could not remain molten but would evaporate almost instantly. Io's lavas are also unlikely to have the same chemical composition as typical lavas on Earth, which have temperatures of only 1300 to 1450 K. Instead, lavas on Io are probably **ultramafic lavas.** These are enriched in magnesium and iron, which give the lava a higher melting temperature.

Solidified ultramafic lavas are found on Earth, but primarily in lava beds that formed billions of years ago when the Earth's interior was much hotter than today. The presence of molten

(a)

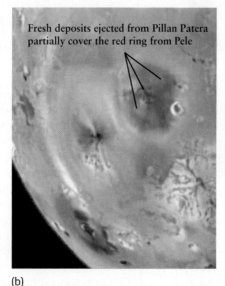

(b)

Figure 13-6   R I V U X G

Rapid Changes on Io **(a)** This *Galileo* image shows a portion of Io's southern hemisphere. Sulfur ejected from the active volcanic vent Pele produced a red ring that resembles paint sprayed from a can. **(b)** A few months later, a volcanic plume erupted from a vent called Pillan Patera, spraying dark material over an area some 400 km (250 mi) in diameter (about the size of Arizona). Other than Earth, Io is the only world in the solar system that shows such noticeable changes over short time spans. (NASA/JPL)

Figure 13-7 R I V U X G

Io's Lava Flows and a Curtain of Fire This mosaic of *Galileo* spacecraft images shows a nested chain of volcanic calderas (pits) with black lava flows. The red streak at upper left is a fissure in the surface about 40 km (25 mi) long, through which lava is erupting upward in a vertical sheet or curtain. The erupting lava reaches the amazing height of 1500 m (5000 ft). (University of Arizona/JPL/NASA)

ultramafic lava on Io suggests that its interior, too, is substantially hotter than that of the present-day Earth. One model proposes that Io has a 100-km (60-mi) thick crust floating atop a worldwide ocean of liquid magma 800 km (500 mi) deep. (As we saw in Section 9-2, the mantle that lies underneath the Earth's crust is

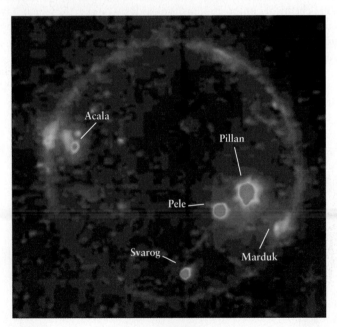

WEBLINK 13.6

Figure 13-8 R I V U X G

Io's Glowing Volcanoes This false-color *Galileo* image (made while Io was in the darkness of Jupiter's shadow) shows the infrared and visible glow from several volcanic regions, including Pele and Pillan (see Figure 13-6). These regions are substantially hotter than typical lava flows on Earth. (Planetary Image Research Laboratory/University of Arizona/JPL/NASA)

a plastic solid rather than a true liquid.) The global extent of this "magma ocean" would explain why volcanic activity is found at all points on Io's surface, rather than being concentrated in pockets as on Earth (see Section 9-3). It could also explain the origin of Io's mountains, which reach heights up to 10 km (30,000 ft). In this model, these mountains are blocks of Io's crust that have tilted before sinking into the depths of the magma ocean.

Io's worldwide volcanic activity is remarkably persistent. When *Voyager 2* flew through the Jovian system in July 1979, four months after *Voyager 1*, almost all of the volcanic plumes seen by *Voyager 1* were still active. And when *Galileo* first viewed Io in 1996, about half of the volcanoes seen by the *Voyager* spacecraft were still ejecting material, while other, new volcanoes had become active.

Io may have as many as 300 active volcanoes, each of which ejects an estimated 10,000 tons of material per second. Altogether, volcanism on Io may eject as much as 10^{13} tons of matter each year. This is sufficient to cover the satellite's entire surface to a depth of 1 meter in a century, or to cover an area of 1000 square kilometers in a few weeks (see Figure 13-6). Thanks to this continual "repaving" of the surface, there are probably no long-lived features on Io, and any impact craters are quickly obliterated.

13-5 Jupiter's magnetic field makes electric currents flow through Io

Most of the material ejected from Io's volcanoes falls back onto the satellite's surface. But some material goes on a remarkable journey into space, by virtue of Io's location deep within Jupiter's magnetosphere.

The Io Torus

The Jovian magnetosphere contains charged particles, some of which collide with Io and its volcanic plumes. The impact of these collisions knocks ions out of the plumes and off the surface, and these ions become part of Jupiter's magnetosphere. Indeed, material from Io is the main source for Jupiter's magnetosphere as a whole. Some of the ejected material forms the **Io torus**, a huge doughnut-shaped ring of ionized gas, or plasma (see Section 12-7), that circles Jupiter at the distance of Io's orbit. The plasma's glow can be detected from Earth (**Figure 13-9a**).

Jupiter's magnetic field has other remarkable effects on Io. As Jupiter rotates, its magnetic field rotates with it, and this field sweeps over Io at high speed. This generates a voltage of 400,000 volts across the satellite, which causes 5 million amperes of electric current to flow through Io.

> **Jupiter and Io together act like an immense electric generator**

ANALOGY Whenever you go shopping with a credit card, you use the same physical principle that generates an electric current within Io. The credit card's number is imprinted as a magnetic code in the dark stripe on the back of the card. The salesperson who swipes your card through the card reader is actually moving the card's magnetic field past a coil of wire within the reader. This generates within the coil an electric current that carries the same coded information about your credit card number. That

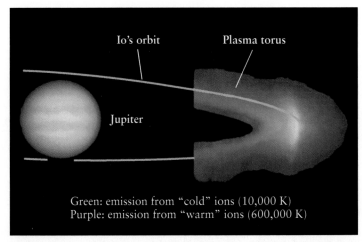

(a)

Green: emission from "cold" ions (10,000 K)
Purple: emission from "warm" ions (600,000 K)

(b)

Figure 13-9   R I V U X G

The Io Torus **(a)** An Earth-based telescope recorded this false-color infrared image of Io's plasma torus. Because of Jupiter's glare, only the outer edge of the torus could be photographed; an artist added the visible-light picture of Jupiter and the yellow line indicating the rest of the torus. **(b)** A flux tube (a region of concentrated magnetic field)

carries plasma between Jupiter and Io, forming an immense electric circuit. (a: Courtesy of J. Trauger; b: Alfred T. Kamajian and Torrence V. Johnson, "The Galileo Mission to Jupiter and Its Moons," *Scientific American,* February 2000, p. 44)

information is transmitted to the credit card company, which (it is hoped) approves your purchase. An electric generator at a power plant works in the same way. A coil of wire is moved through a strong magnetic field, which makes current flow in the coil. This current is delivered to transmission lines and eventually to your home.

In fact, a current flows not only through Io but also through the sea of charged particles in Jupiter's magnetosphere and through Jupiter's atmosphere. This forms a gigantic, oval-shaped electric circuit that connects Io and Jupiter in the same way that wires connect the battery and the lightbulb in a flashlight (Figure 13-9b). Jupiter's aurora, which we discussed in Section 12-7, is particularly strong at the locations where this current strikes the upper atmosphere of Jupiter.

Part of the electric current that flows between Io and Jupiter is made up of electrons that spiral around Jupiter's magnetic field lines. As they spiral, the electrons act like miniature antennas and emit radio waves. This is the source of Jupiter's bursts of decametric radio radiation, which we discussed in Section 12-7.

Probing Io's Interior

The electric current within and near Io also creates a weak magnetic field, which was first detected by *Voyager 1*. Remarkably, measurements made by *Galileo* during its close approaches to Io suggest that the magnetic field near Io may be too strong—comparable to that of Mercury, although weaker than that of Earth—to be generated by Io's electric current alone. This has led scientists to speculate that Io generates its own magnetic field through the motions of molten material in its interior, just as the Earth does (see Section 7-7 and Section 9-4). If these speculations are correct, Io is the smallest world in the solar system to generate its own magnetic field.

If Io has undergone enough tidal heating to melt its interior, chemical differentiation must have taken place and the satellite should have a dense core. (We described chemical differentiation in Box 7-1 and Section 8-4.) To test for this, scientists measured how Io's gravity deflected the trajectory of *Galileo* as the spacecraft flew past. From these measurements, they could determine not only Io's mass but also the satellite's oblateness (how much it deviates from a spherical shape because of its rotation). The amount of oblateness indicates the size of the core; the greater the fraction of the satellite's mass contained in its core, the less oblate the satellite will be. The *Galileo* observations suggest that Io has a dense core composed of iron and iron sulfide (FeS), with a radius of about 900 km (about half the satellite's overall radius). Surrounding the core is a mantle of partially molten rock, on top of which is Io's visible crust.

13-6 Europa is covered with a smooth layer of ice that may cover a worldwide ocean

Europa, the second of the Galilean satellites, is the smoothest body in the solar system. There are no mountains and no surface features greater than a few hundred meters high (Figure 13-10). There are almost no craters, indicating a young surface that has been reprocessed by geologic activity. The dominant surface feature is a worldwide network of stripes and cracks (Figure 13-11). Like Io, Europa is an exception to the general rule that a small world should be cratered and geologically dead (see Section 7-6).

Europa's Surface: Icy but Active

Neither *Voyager 1* nor *Voyager 2* flew very near Europa, so most of our knowledge of this satellite comes from the close passes

Figure 13-10   R I V U X G

Europa Dark lines crisscross Europa's smooth, icy surface in this false-color composite of visible and infrared images from *Galileo*. These are fractures in Europa's crust that can be as much as 20 to 40 km (12 to 25 mi) wide. Only a few impact craters are visible on Europa, which indicates that this satellite has a very young surface on which all older craters have been erased. (NASA/JPL/University of Arizona)

Europa's mass. Hence, only a small fraction of the mass, about 10 to 15%, is water ice. Because the surface is icy, we can conclude that the rocky material is found within Europa's interior.

Europa is too small to have retained much of the internal heat that it had when it first formed. But there must be internal heat nonetheless to power the geologic processes that erase craters and bring fresh water to Europa's surface. What keeps Europa's interior warm? The most likely answer, just as for Io, is heating by Jupiter's tidal forces. The rhythmic gravitational tugs exerted by Io and Ganymede on Europa deform its orbit into an ellipse; the varying speed of Europa as it goes around its orbit causes Europa to "nod" back and forth, causing variations in tidal stresses that make Europa flex. But because it is farther from Jupiter, tidal effects on Europa are only about one-fourth as strong as those on Io, which may explain why no ongoing volcanic activity has yet been seen on Europa.

Some features on Europa's surface, such as the fracture patterns shown in Figure 13-11, may be the direct result of the crust being stretched and compressed by tidal flexing. Figure 13-12 shows other features, such as networks of ridges and a young, very smooth circular area, that were probably caused by the internal heat that tidal flexing generates. The rich variety of terrain depicted in Figure 13-12, with stress ridges going in every direction, shows that Europa has a complex geologic history.

Among the unique structures found on Europa's surface are **ice rafts**. The area shown in Figure 13-13a was apparently subjected to folding, producing the same kind of linear features as those in Figure 13-12. But a later tectonic disturbance broke the surface into small chunks of crust a few kilometers across, which then "rafted" into new positions. A similar sort of rafting hap-

made by *Galileo*. (Both Figure 13-10 and Figure 13-11 are *Galileo* images.) But even before these spacecraft visited Jupiter, spectroscopic observations from Earth indicated that Europa's surface is almost pure frozen water (see Figure 7-4). This was confirmed by instruments on board *Galileo*, which showed that Europa's infrared spectrum is a close match to that of a thin layer of fine-grained water ice frost on top of a surface of pure water ice. (The brown areas in Figure 13-10 show where the icy surface contains deposits of rocky material from meteoritic impacts, from Europa's interior, or from a combination of these sources.)

> **Geology on Europa is based on solid ice and liquid water, not rock and molten magma**

The purity of Europa's ice suggests that water is somehow brought upward from the moon's interior to the surface, where it solidifies to make a fresh, smooth layer of ice. Indeed, some *Galileo* images show what appear to be lava flows on Europa's surface, although the "lava" in this case is mostly ice. This idea helps to explain why Europa has very few craters (any old ones have simply been covered up) and why its surface is so smooth. Europa's surface may thus represent a water-and-ice version of plate tectonics.

Although Europa's surface is almost pure water ice, keep in mind that Europa is *not* merely a giant ice ball. The satellite's density shows that rocky material makes up about 85 to 90% of

Smooth ice plains with a network of fractures

Linear ridges (dark colors are caused by minerals in the ice)

500 km

Figure 13-11   R I V U X G

Europa's Fractured Crust False colors in this *Galileo* composite image emphasize the difference between the linear ridges and the surrounding plains. The smooth ice plains (shown in blue) are the basic terrain found on Io. (NASA/JPL)

Ridges produced by folding and faulting of the surface.

5 km

Smooth area formed by fluid erupting onto the surface.

Rugged patch created by a subsurface disturbance.

Figure 13-12   R I V U X G

Europa in Close-up This high-resolution *Galileo* image shows a network of overlapping ridges, part of which has been erased to leave a smooth area and part of which has been jumbled into a rugged patch of terrain. Europa's interior must be warm enough to power this complex geologic activity. (NASA/JPL)

pens in the Earth's Arctic Ocean every spring, when the winter's accumulation of surface ice breaks up into drifting ice floes (Figure 13-13*b*). The existence of such structures on Europa strongly suggests that there is a subsurface layer of liquid water or soft ice over which the ice rafts can slide with little resistance.

An Underground Ocean?

The ridges, faults, ice rafts, and other features strongly suggest that Europa has substantial amounts of internal heat. This heat could prevent water from freezing beneath Europa's surface, creating a worldwide ocean beneath the crust. If this picture is correct, geologic processes on Europa may be an exotic version of those on Earth, with the roles of solid rock and molten magma being played by solid ice and liquid water.

A future spacecraft may use radar to penetrate through Europa's icy crust to search for definitive proof of liquid water beneath the surface. But magnetic field measurements made by the *Galileo* spacecraft have already provided some key evidence favoring this picture. Unlike the Earth or Jupiter, Europa does not seem to create a steady magnetic field of its own. But as Europa moves through Jupiter's intense magnetic field, electric currents are induced within the satellite's interior, just as they are within Io (see Section 13-5), and these currents generate a weak but measurable field. (The strength and direction of this induced field vary as Europa moves through different parts of the Jovian magnetosphere, which would not be the case if Europa generated its field by itself.)

To explain these observations, there must be an electrically conducting fluid beneath Europa's crust—a perfect description of an underground ocean of water with dissolved minerals. (Pure water is a very poor conductor of electricity, so other substances must be present to make the water conducting.) If some of this water should penetrate upward to Europa's surface through cracks in the crust, it would vaporize and spread the dissolved minerals across the terrain. This could explain the reddish-brown colors in Figure 13-13*a*.

Ice rafts (fractured segments of Europa's icy crust)

Linear features show how rafts once fit together.

10 km

(a) Ice rafts on Europa

Ice floes

Akpatok Island

10 km

(b) Ice floes on Earth

Figure 13-13   R I V U X G

Moving Ice on Europa and Earth (a) Some time after a series of ridges formed in this region of Europa's surface, the icy crust broke into "rafts" that were moved around by an underlying liquid or plastic layer. The colors in this *Galileo* image may be due to

minerals that were released from beneath the surface after the crust broke apart. **(b)** Europa's ice rafts are analogous to ice floes created when pack ice breaks up, as in this spacecraft view of part of the Canadian arctic. (a: NASA/JPL; b: USGS and NASA)

By combining measurements of Europa's induced magnetic field, gravitational pull, and oblateness, scientists conclude that Europa's outermost 100 to 200 km are ice and water. (It is not clear how much of this is liquid and how much solid.) Within this outer shell is a rocky mantle surrounding a metallic core some 600 km (400 mi) in radius.

Remarkably, Europa also has an extremely thin atmosphere of oxygen. (By Earth standards, this atmosphere would qualify as a near-vacuum.) We saw in Section 9-5 that oxygen in the Earth's atmosphere is produced by plants through photosynthesis. But Europa's oxygen atmosphere is probably the result of ions from Jupiter's magnetosphere striking the satellite's icy surface. These collisions break apart water molecules, liberating atoms of hydrogen (which escape into space) and oxygen.

The existence of a warm, subsurface ocean on Europa, if proved, would make Europa the only world in the solar system other than the Earth on which there is liquid water. This would have dramatic implications. On Earth, water and warmth are essentials for the existence of life. Perhaps single-celled organisms have evolved in the water beneath Europa's crust, where they would use dissolved minerals and organic compounds as food sources. In light of this possibility, NASA has taken steps to prevent biological contamination of Europa. At the end of the *Galileo* mission in 2003, the spacecraft (which may have carried traces of organisms from Earth) was sent to burn up in Jupiter's atmosphere, rather than remaining in orbit where it might someday crash into Europa. An appropriately sterilized spacecraft may one day visit Europa and search for evidence of life within this exotic satellite.

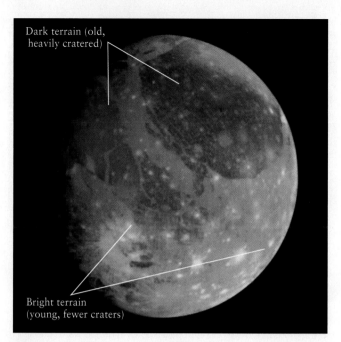

Dark terrain (old, heavily cratered)

Bright terrain (young, fewer craters)

Figure 13-14   R I ⊻ U X G

Ganymede Two distinct types of terrain—one dark, heavily cratered, and hence old, the other bright, less cratered, and hence younger—are visible in this *Galileo* image of the hemisphere of Ganymede that always faces away from Jupiter. Craters in general appear bright, suggesting that the impacts that formed the craters excavated the surface to reveal ice underneath. (NASA/JPL)

13-7 Liquid water may also lie beneath the cratered surfaces of Ganymede and Callisto

Unlike Io and Europa, the two outer Galilean satellites have cratered surfaces. In this respect, Ganymede and Callisto bear a superficial resemblance to our own Moon (see the illustration with Table 13-1). But unlike the Moon's craters, the craters on both Ganymede and Callisto are made primarily of ice rather than rock, and Ganymede has a number of surface features that indicate a geologically active past.

Both Ganymede and Callisto show evidence of past geologic activity

The Two Faces of Ganymede's Surface

Ganymede is the largest satellite in our solar system and is even larger than the planet Mercury. Like our Moon, Ganymede has two distinct kinds of terrain, called **dark terrain** and **bright terrain** (Figure 13-14). On the Moon, the dark maria are younger than the light-colored lunar highlands (see Section 10-1). On Ganymede, by contrast, the dark terrain is older, as indicated by its higher density of craters. The bright terrain is much less cratered and therefore younger. Because ice is more reflective than rock, even Ganymede's dark terrain is substantially brighter than the lunar surface.

Voyager and *Galileo* images show noticeable differences between young and old impact craters on Ganymede. The youngest

craters are surrounded by bright rays of freshly exposed ice that has been excavated by the impact (see the lower left of Figure 13-14). Older craters are darker, perhaps because of chemical processes triggered by exposure to sunlight.

The close-up *Galileo* image in Figure 13-15 shows the contrast between dark and bright terrain. The area on the left-hand side of this image is strewn with large and small craters, reinforcing the idea that the dark terrain is billions of years old. The dark terrain is also marked by long, deep furrows. These are deformations of Ganymede's crust that have partially erased some of the oldest craters. Since other craters lie on top of the furrows, the stresses that created the furrows must have occurred long ago. Since geologic activity is a result of a planet or satellite's internal heat, this shows that Ganymede must have had a warm interior in the distant past.

Linear features are also seen in the bright terrain on the right in Figure 13-15. The *Voyager* missions discovered that the bright terrain is marked by long grooves, some of which extend for hundreds of kilometers and are as much as a kilometer deep. Images from *Galileo* show that some adjacent grooves run in different directions, indicating that the bright terrain has been subjected to a variety of tectonic stresses. *Galileo* also found a number of small craters overlying the grooves, some of which are visible in Figure 13-15. The density of craters shows that the bright terrain, while younger than the dark terrain, is at least a billion years old.

After the *Voyager* missions, it was thought that the bright terrain represented fresh ice that had flooded through cracks in

Figure 13-15   R I V U X G

Evidence for Geologic Activity on Ganymede Stresses in Ganymede's crust have produced linear features that extend for hundreds of kilometers across both the dark and bright terrain. These features must be quite ancient, since numerous impact craters have formed on top of them. (NASA)

Ganymede's crust, analogous to the maria on our Moon (see Section 10-1). However, high-resolution *Galileo* images do not show the kinds of surface features that would be expected if this terrain was volcanic in origin. Yet there are several low-lying areas within the bright terrain that appear to have been flooded by a watery fluid that subsequently froze. These low-lying areas may be *rift valleys* where the crust has been pulled apart, like Valles Marineris on Mars (see Section 11-5). In order to flood the floors of these valleys, there must have been liquid water beneath Ganymede's surface when the valleys formed. This is additional evidence that Ganymede must have had a warm interior in the past, and that this internal warmth persisted at least until the era when the bright terrain formed a billion years ago. This is rather surprising, because Ganymede is too small to have retained much of the heat left over from its formation or from radioactive elements, and it orbits too far from Jupiter to have any substantial tidal heating.

Ganymede's Interior and Bizarre Magnetic Field

An equally great surprise from *Galileo* observations was the discovery that Ganymede has its own magnetic field, and that this field is twice as strong as that of Mercury. This field is sufficiently strong to trap charged particles, giving Ganymede its own "mini-magnetosphere" within Jupiter's much larger magnetosphere. The presence of a magnetic field shows that electrically conducting material must be in motion within Ganymede to act as a dynamo, which means that the satellite must have substantial internal heat even today. A warm interior causes differentiation, and *Galileo* measurements indeed showed that Ganymede is highly differentiated. It has a metallic core some 500 km in radius, surrounded by a rocky mantle and by an outer shell of ice some 800 km thick.

An even more surprising result is that Ganymede's magnetic field varies somewhat as it orbits Jupiter, and that these variations are correlated with Ganymede's location within Jupiter's mag-

netic field. One explanation for this variation is that Ganymede has a layer of salty, electrically conducting liquid water some 170 km beneath its surface. As the satellite moves through Jupiter's magnetic field, electric currents are induced in this layer just as happens within Europa (see Section 13-6), and these currents generate a weak field that adds to the stronger field produced by Ganymede's dynamo.

A possible explanation for Ganymede's internal heat is that gravitational forces from the other Galilean satellites may have affected its orbit. While Ganymede's present-day orbit around Jupiter is quite circular, calculations show that the orbit could have been more elliptical in the past. Just as for Io and Europa, such an elliptical orbit would have given rise to substantial tidal heating of Ganymede's interior. It may have retained a substantial amount of that heat down to the present day.

Callisto's Perplexing Surface

WEB LINK 13-1   In marked contrast to Io, Europa, and Ganymede is Callisto, Jupiter's outermost Galilean satellite. Images from *Voyager* showed that Callisto has numerous impact craters scattered over an icy crust (**Figure 13-16**). Callisto's ice is not as reflective as that on Europa or Ganymede, however; the surface appears to be covered with some sort of dark mineral deposit.

CAUTION! Figure 13-16, as well as the images that accompany Table 13-1, shows that Callisto is the darkest of the Galilean satellites. Nonetheless, its surface is actually more than twice as reflective as that of our own Moon. Callisto's surface may be made of "dirty" ice, but even dirty ice reflects more light than the gray rock found on the Moon. If you could somehow replace our own Moon with Callisto, a full moon as seen from Earth would be 4 times brighter than it is now (thanks also in part to Callisto's larger size).

Figure 13-16 R I [V] U X G

Callisto Numerous craters pockmark Callisto's icy surface. The inset shows a portion of a huge impact basin called Valhalla. Most of the very smallest craters in this close-up view have been completely obliterated, and the terrain between the craters is blanketed by dark, dusty material. (Left: Arizona State University/JPL/NASA; right: JPL/NASA)

The *Voyager* images led scientists to conclude that Callisto is a dead world, with an ancient surface that has never been reshaped by geologic processes. There is no simple relationship between Callisto's orbital period and those of the other Galilean satellites, so it would seem that Callisto never experienced any tidal heating and hence was unable to power any geologic processes at its surface. But once the *Galileo* spacecraft went into orbit around Jupiter and began returning high-resolution images, scientists realized that Callisto's nature is not so simple. In fact, Callisto proves to be the most perplexing of the Galilean satellites.

One curious discovery from *Galileo* is that while Callisto has numerous large craters, there are very few with diameters less than 1 km. Such small craters are abundant on Ganymede, which has probably had the same impact history as Callisto. This implies that most of Callisto's small craters have been eroded away. How this could have happened is not known. Another surprising result is that Callisto's surface is covered by a blanket of dark, dusty material (see the inset in Figure 13-16). Where this material came from, and how it came to be distributed across the satellite's surface, is not understood.

Callisto's Interior: Hot or Cold?

Callisto's most unexpected feature is that, like Europa and Ganymede, it appears to have an induced magnetic field that varies as the satellite moves through different parts of Jupiter's magnetosphere. Callisto must have a layer of electrically conducting material in order to generate such a field. One model suggests that this material is in the form of a subsurface ocean like Europa's but only about 10 km (6 mi) deep. However, without tidal heating, Callisto's interior is probably so cold that a layer of liquid water would quickly freeze. One proposed explanation is that Callisto's ocean is a mixture of water and a form of "antifreeze." In cold climates on Earth, antifreeze is added to the coolant in automobile radiators to lower the liquid's freezing temperature and prevent it from solidifying. Ammonia in Callisto's proposed subsurface ocean could play the same role.

The presence of a liquid layer beneath Callisto's surface is difficult to understand. Even with ammonia acting as an antifreeze, this layer cannot be too cold or it would solidify. This argues for a relatively *warm* interior. But *Galileo* measurements suggest that Callisto's interior is not highly differentiated. This

Figure 13-17

Interiors of the Galilean Satellites These cross-sectional diagrams show the probable internal structures of the four Galilean satellites of Jupiter, based on information from the *Galileo* mission. (NASA/JPL)

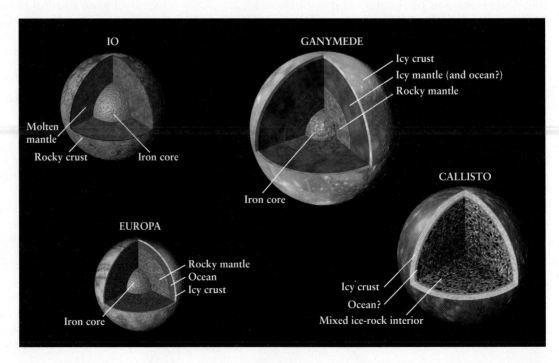

implies that the interior is *cold*, and may never have been heated enough to undergo complete chemical differentiation. The full story of Callisto's interior, it would seem, has yet to be told.

Although Ganymede and Callisto have many differences, one feature they share is an extremely thin and tenuous atmosphere. The molecules that make up Ganymede's atmosphere are oxygen (O_2) and ozone (O_3). These are thought to be released from gas bubbles trapped in Ganymede's surface ice. On Callisto, by contrast, the atmosphere is composed of carbon dioxide (CO_2). This gas may have evaporated from Callisto's poles, where temperatures are low enough to form CO_2 ice.

The diagrams in Figure 13-17 summarize the probable interior structures of the Galilean satellites. They demonstrate the remarkable variety of these terrestrial worlds.

13-8 Titan has a thick atmosphere and streams carved by liquid hydrocarbons

Unlike Jupiter, Saturn has only one large satellite that is comparable in size to our own Moon. This satellite, Titan, was discovered by Christiaan Huygens in 1665. Titan's diameter of 5150 km makes it second in size among satellites to Ganymede (Section 13.7). Like Ganymede, Titan has a low density that suggests it is made of a mixture and ice and rock. (See Table 7-2 for a comparison between Titan, the Moon, and the Galilean satellites.) But unlike Ganymede or any other satellite in the solar system, Titan has a thick atmosphere with a unique chemical composition.

Titan's Atmosphere

By the early 1900s, scientists had begun to suspect that Titan (Figure 13-18) might have an atmosphere, because it is cool enough and massive enough to retain heavy gases. (Box 7-2 explains the criteria for a planet or satellite to retain an atmosphere.) These suspicions were confirmed in 1944, when Gerard Kuiper discovered the spectral lines of methane in sunlight reflected from Titan. (We discussed Titan's spectrum and its interpretation in Section 7-3.)

Because of its atmosphere, Titan was a primary target for the *Voyager* missions. To the chagrin of mission scientists and engineers, however, *Voyager 1* spent hour after precious hour of its limited, preprogrammed observation time sending back featureless images like the one at the center of Figure 13-18. Titan's atmosphere is so thick—about 200 km (120 mi) deep, or about 10 times the depth of the Earth's troposphere—that it blocked *Voyager*'s view of the surface. The haze surrounding Titan is so dense that little sunlight penetrates to the ground; noon on Titan is only about 1/1000 as bright as noon on Earth, though still 350 times brighter than night on Earth under a full moon.

By examining how the radio signal from *Voyager 1* was affected by passing through Titan's atmosphere, scientists inferred that the atmospheric pressure at Titan's surface is 50% greater than the atmospheric pressure at sea level on the Earth. Titan's surface gravity is weaker than that of Earth, so to produce this pressure Titan must have considerably more gas in its atmosphere than Earth. Indeed, calculations show that about 10 times more gas (by mass) lies above a square meter of Titan's surface than above a square meter of Earth.

Voyager data show that more than 95% of Titan's atmosphere is nitrogen. Most of this nitrogen probably came from ammonia (NH_3), a compound of nitrogen and hydrogen that is quite common in the outer solar system. Ammonia is easily broken into hydrogen and nitrogen atoms by the Sun's ultraviolet radiation. Titan's gravity is too weak to retain hydrogen atoms (see Box 7-2), so these atoms escape into space and leave the more massive nitrogen atoms behind.

The remainder of Titan's atmosphere is predominantly methane (CH_4), which is the principal component of the "natural gas" used on Earth as fuel. The interaction of methane with ultraviolet light from the Sun produces a variety of other carbon-hydrogen compounds, or **hydrocarbons.** For example, the *Voyager* infrared spectrometers detected small amounts of ethane

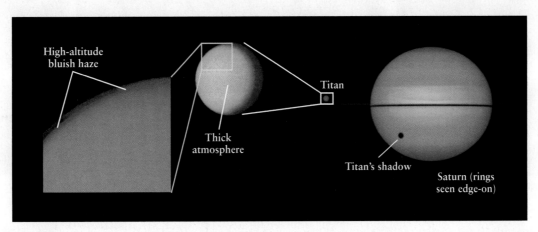

Figure 13-18   R I V U X G

Titan The Hubble Space Telescope image at far right shows Titan and the shadow it casts on Saturn's clouds. The *Voyager* image in the middle shows how Titan's atmosphere gives it a nearly featureless appearance. At far left is a color-enhanced *Voyager* image that shows a haze extending well above the satellite's visible edge, or limb. (Left: JPL/NASA; center: NASA; right: Erich Karkoschka, LPL/STScI/NASA)

(C_2H_6), acetylene (C_2H_2), ethylene (C_2H_4), and propane (C_3H_8) in Titan's atmosphere.

On Earth, water can be a gas (like water vapor in the atmosphere), a liquid (as in the oceans), or a solid (as in the polar ice caps). On Titan, the atmospheric pressure is so high and the surface temperature is so low—about 95 K (–178°C, or –288°F)—that any water is frozen solid. But conditions on Titan are just right for methane and ethane to exist as gases, liquids, or solids. This raises the tantalizing possibility that Titan could have methane rain and lakes of liquid ethane.

> **A faint mist of methane rain falls continually on Titan**

Nitrogen in Titan's atmosphere can combine with airborne hydrocarbons to produce other compounds, such as hydrogen cyanide (HCN). Hydrogen cyanide, along with other molecules, can join together in long, repeating molecular chains to form substances called **polymers**. Droplets of some polymers remain suspended in the atmosphere, forming an **aerosol**. (Common aerosols on the Earth include fog, mist, and paint sprayed from a can.) The polymers in Titan's airborne aerosol are thought to be responsible for the reddish-brown color seen in Figure 13-18.

Exploring Titan with *Cassini* and *Huygens*

The designers of the *Cassini* mission to Saturn made the exploration of Titan a central goal. During its primary mission in Saturn orbit from 2004 to 2008, the *Cassini* spacecraft will make 44 close flybys of Titan. Titan's atmosphere is more transparent to infrared wavelengths than to visible light, so *Cassini* has used its infrared telescope to obtain the first detailed images of the satellite's surface (Figure 13-19*a*). Most of the surface is light-colored, but a swath of dark terrain extends around most of Titan's equator.

Cassini has used its on-board radar to map the dark terrain. (Like infrared light, radio waves can pass through Titan's atmosphere.) These images reveal a series of long, parallel lines of sand dunes about 1 to 2 km apart. These are aligned west to east, in the same direction that winds blow at Titan's equator, and so presumably formed by wind action. Unlike Earth sand, which is made of small particles of silicate rock, sand on Titan is probably small particles of water ice combined with polymers that have fallen from Titan's atmosphere. These polymers give the dark terrain its color.

The infrared telescope aboard *Cassini* also detected a surface feature that appears to be a volcano. The presence of volcanism on Titan can explain why there is so much methane in the satellite's atmosphere. Methane molecules are broken apart by ultraviolet photons from the Sun, and the hydrogen atoms escape into space. This process would destroy all of Titan's methane within 10 million years unless a fresh supply is added to the atmosphere by volcanic activity. Volcanic activity can also explain the presence of isolated very bright patches on Titan's surface (see Figure 13-19*a*). A volcanic eruption would outgas not just methane but

(a) R I V U X G

1. Liquid flowed in a number of small streams...

2. ...which merged into a river...

...3. which emptied into a large, dark-colored outflow channel.

4. Dark hydrocarbon polymers from Titan's atmosphere fell onto the surface, then was washed by methane rains into the streams, river, and outflow channel. Hence these surface features have a dark color.

(b) R I V U X G

15 cm

(c) R I V U X G

Figure 13-19

Beneath Titan's Clouds (a) This mosaic of Titan was constructed from *Cassini* images made at visible and infrared wavelengths during a close flyby in December 2005. **(b)** The *Huygens* lander recorded this view of features on Titan carved by flowing liquid. It is too cold on Titan for water to be a liquid, so these streams and river must have carried liquid methane or ethane. **(c)** This nearly true-color *Huygens* image shows the view from Titan's surface. The 15-cm (6-in.) wide "rock" is about 85 cm (33 in.) from the camera. Unlike Earth rocks, Titan "rocks" are chunks of water ice. (NASA/JPL/ESA/University of Arizona) also water vapor and carbon dioxide. These gases would quickly freeze onto the surface, making a patch of highly reflective ice.

The most remarkable images of Titan's surface have come from *Huygens*, a small lander that *Cassini* carried on its journey from Earth and released before entering Saturn orbit. Named for the discoverer of Titan, *Huygens* entered Titan's atmosphere on January 14, 2005. The lander took $2\frac{1}{2}$ hours to descend to the surface under a parachute, during which time it made detailed images of the terrain (see Figure 13-19*b*). After touchdown, *Huygens* continued to return data for another 70 minutes (see Figure 13-19*c*) before its batteries succumbed to Titan's low temperatures.

Huygens images such as Figure 13-19*b* show that liquids have indeed flowed on Titan like water on Earth. Yet none of the images made by the lander during its descent or after touchdown showed any evidence of standing liquid. There was nonetheless evidence of recent methane rainfall at the *Huygens* landing site. A lander instrument designed to measure atmospheric composition was equipped with a heater for its gas inlet. During the three minutes it took to warm this inlet, the measured methane concentration jumped by 30% and then remained steady. The explanation is that the ground was soaked with liquid methane that had fallen as rain and which evaporated as it was warmed by the heater. *Huygens* measurements indicate that the average annual methane rainfall on Titan is about 5 cm (2 in.), about the same as the amount of water rain that falls each year at Death Valley in California. The rivers and outflow channels that *Huygens* observed during its descent may date from an earlier period in Titan's history when methane was more abundant and rainfall was more intense.

Before the *Cassini* mission scientists expected to find lakes of liquid ethane across Titan's surface. As of this writing (2006) the only location where *Cassini* appears to have located such lakes is near Titan's north pole (**Figure 13-20**). It is currently winter in the northern hemisphere, so temperatures are lowest there. Scientists also expected to find extensive clouds of ethane in Titan's atmosphere; these, too, have only been found near the north pole. Over the course of a Titan year (which, like that of its parent planet Saturn, lasts 29.37 years) the supply of liquid ethane may migrate with the seasons from pole to pole, much as frozen carbon dioxide migrates seasonally on Mars (see Section 11-6).

Impact craters are also conspicuously absent on Titan: Only a handful of craters have been observed by *Cassini*. The majority of craters may have been eroded by wind or rainfall, or covered by material ejected from volcanoes.

Titan's Geologic History

We have seen that Titan must have been volcanically active to provide its atmosphere with methane. But what has powered the volcanic activity? Titan is too small to have retained the internal heat from its formation. Furthermore, although Titan's rotation is synchronous like that of Io, its orbit around Saturn is so circular that it cannot be heated by tidal flexing in the way that Io is (see Section 13-4).

One explanation has been proposed by planetary scientists Gabriel Tobie, Jonathan Lunine, and Christophe Sotin. In their model, Titan developed a solid rocky core, a mantle of liquid water, and an icy crust soon after it formed some 4.5 billion years ago. The internal heat left over from formation facilitated volcanic activity that outgassed Titan's original complement of atmospheric methane molecules. Some of this methane was reabsorbed into the interior; the rest was destroyed by ultraviolet light from the Sun within the first billion years. From 3.5 billion to 2 billion years ago, Titan's atmosphere was essentially free of methane.

Over time, the rocky core increased in temperature due to the decay of radioactive elements like uranium. (These elements are concentrated in Titan's core to a greater extent than on Earth, which has rocky substances throughout its volume.) About 2 billion years ago, the core became warm enough to partially melt, and convection began in the melted portion. This allowed heat to flow from the core into the mantle, triggering a second episode of methane outgassing. This, too, eventually came to an end, and methane once again disappeared from Titan's atmosphere.

The third and final release of methane began about 500 million years ago, as the last of Titan's internal heat from radioactivity caused convection to take place in the satellite's ice crust. The amount of methane released during the current episode is enough to sustain the gas in the atmosphere, but not enough to provide extensive lakes of methane or ethane.

Within the next several hundred million years, methane outgassing on Titan will cease for good. In time, the hydrocarbon haze will disappear, the skies will clear, and the surface of Titan will be revealed to whatever outside observers may gaze upon it in that distant era.

Dark areas: Possible lakes of methane or ethane

100 km

Figure 13-20   R I V U X G

Hydrocarbon Seas on Titan? *Cassini* used its onboard radar to record this view of a region near Titan's north pole. The dark regions show where the surface does not reflect radio waves at all, which would be the case if these regions were lakes filled with liquid methane or liquid ethane. In support of this idea, many of the dark regions appear to have liquid-carved inflow or outflow channels. (NASA/JPL)

13-9 Jupiter has dozens of small satellites that have different origins

Besides the Galilean satellites, Jupiter has at least 59 other small satellites. Each is named for a mythological character associated with Jupiter. Appendix 3 gives details of some of the satellites' orbits. It is helpful to sort Jupiter's 63 known satellites (as of 2006) into three groups according to their sizes and orbits.

Four planet-sized satellites, Io, Europa, Ganymede, and Callisto, can be thought of as four of the terrestrial worlds of the solar system. All are comparable in size to the planet Mercury (see Table 13-1). The semimajor axes of their orbits range from 5.9 to 26 times the radius of Jupiter. (By comparison, the distance from the Earth to the Moon is about 60 times the Earth's radius.)

Four small inner satellites, Metis, Adrastea, Amalthea, and Thebe, have orbital semimajor axes between 1.8 and 3.1 times the radius of Jupiter. As **Figure 13-21** shows, these satellites have irregular shapes, much like a potato (or an asteroid). The largest of the inner satellites, Amalthea, measures about 270 by 150 km (170 by 95 mi)—about 10 times larger than the moons of Mars, but only about one-tenth the size of the Galilean satellites. Discovered in 1892, Amalthea has a distinct reddish color due to sulfur that Jupiter's magnetosphere removed from Io's volcanic plumes, then deposited onto Amalthea's surface. The three other inner satellites are even smaller than Amalthea. They were first discovered in images from *Voyager 1* and *Voyager 2.*

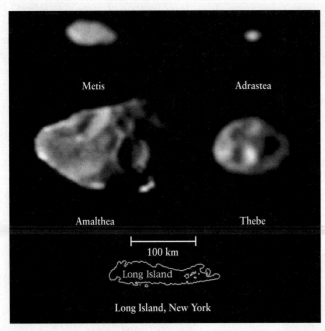

Figure 13-21   R I **V** U X G

Jupiter's Four Inner Satellites First observed in 1892, Amalthea was the last satellite of any planet to be discovered without the aid of photography. Metis, Adrastea, and Thebe, by contrast, are so small that they were not discovered until 1979. These images from the *Galileo* spacecraft are the first to show these three satellites as anything more than points of light. (NASA/JPL; Cornell University)

Fifty-five small outer satellites have diameters from 184 km for the largest to about 1 km for the smallest. All orbit far beyond Callisto, with semimajor axes that range from about 100 to 360 times the radius of Jupiter. (Twenty-three of these satellites were discovered in 2003 alone, and even more may be discovered as telescope technology continues to evolve.)

The Galilean satellites and the four inner satellites all orbit Jupiter in the plane of the planet's equator. (The same is true of Jupiter's rings, discussed in Section 12-9.) Furthermore, these are all **prograde orbits,** which means that these objects orbit Jupiter in the same direction as Jupiter's rotation. This is what we would expect if these satellites formed from the same primordial, rotating cloud as Jupiter itself. (In a similar way, the planets all orbit the Sun in nearly the same plane and in the same direction as the Sun's rotation. As we saw in Section 8-3, this reinforces the idea that the Sun and planets formed from the same solar nebula.)

> Only a few of Jupiter's satellites orbit the planet in the same direction as Jupiter rotates

In contrast, the 55 outer satellites all circle Jupiter along orbits that are inclined at steep angles to the planet's equatorial plane. These satellites are thought not to have formed along with Jupiter; rather, they are probably wayward asteroids captured by Jupiter's powerful gravitational field. One piece of evidence for this is that 48 of the outer satellites are in **retrograde orbits:** They orbit Jupiter in a direction opposite the planet's rotation. Calculations show that it is easier for Jupiter to capture a satellite into a retrograde orbit than a prograde one.

13-10 The icy surfaces of Saturn's six moderate-sized moons provide clues to their histories

Saturn, too, has an extensive collection of satellites: 56 were known as of 2006. As we did for Jupiter's satellites in Section 13-9, we can subdivide the satellites of Saturn into three categories based on their sizes and their orbits. (Appendix 3 has information about the orbits of all of Saturn's satellites.) We will see that while Saturn does not have a set of large moons like the Galilean satellites of Jupiter, it does have a remarkably diverse collection of moderate-sized satellites.

Saturn's Retinue of Moons

One planet-sized satellite, Titan, is actually one of the terrestrial worlds of our solar system. With its diameter of 5150 km, Titan is intermediate in size between Mercury and Mars. Its prograde orbit has a semimajor axis of just over 20 times the radius of Saturn and lies in nearly the same plane as Saturn's equator and rings.

Six moderate-sized satellites were all discovered before 1800. These six satellites have properties and diameters that lead us to group them in pairs: Mimas and Enceladus (400 to 500 km in diameter), Tethys and Dione (roughly 1000 km in diameter), and Rhea and Iapetus (about 1500 km in diameter). The semimajor axes of their orbits range from 3.1 to 59 times the radius of

Saturn. Like Titan, all of their orbits are prograde and close to the plane of Saturn's equator.

Forty-nine small satellites have sizes from 3 to 266 km. Some are in prograde orbits as small as 2.2 times the radius of Saturn, and lie within or close to the rings (see Figure 12-25). These may be jagged fragments of ice and rock produced by impacts and collisions. The shepherd satellites that shape Saturn's F ring (see Figure 12-24) may be such fragments. Others are in retrograde orbits with semimajor axes as large as 380 times Saturn's radius, and may be captured asteroids.

Six Strange Satellites

The most curious and mysterious of these worlds may be Saturn's six moderate-sized satellites (**Figure 13-22**). These six satellites have very low average densities of less than 1400 kg/m³. This implies that unlike the satellites of the inner solar system or the Galilean satellites of Jupiter, they are composed primarily of ices (mostly frozen water and ammonia) with very little rock. The motions of these mid-sized satellites are not surprising: All six orbit the planet in the plane of Saturn's equator and in the direction of the planet's rotation, and all six exhibit synchronous rotation. What is surprising is the curious variety of surface features that the *Voyager* and *Cassini* spacecraft discovered on these satellites.

Mimas, the innermost of the six middle-sized moons, has the heavily cratered surface we would expect on such a small world. Its most remarkable feature is a crater so large (Figure 13-22*a*) that the impact that formed it must have come close to shattering Mimas into fragments.

In contrast to Mimas's heavily cratered surface, its neighbor and near twin, Enceladus, has extensive crater-free regions (Figure 13-22*b*). Some form of geologic activity has resurfaced these areas within the past 100 million years, and may still be under way. Geologic activity may also explain why Enceladus has an amazingly high albedo of 0.95, making it the most

> Although just 500 km across, Enceladus has powerful volcanoes at its south pole

reflective large object in our solar system. Some process has freed Enceladus's icy surface of rock and dust, whose presence on the other icy satellites is the likeliest explanation for their lower albedos. Most remarkably, *Cassini* discovered that Enceladus is

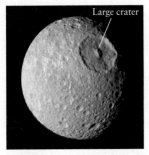

(a) Mimas
(diameter 392 km)

(b) Enceladus
(diameter 500 km)

(c) Tethys
(diameter 1060 km)

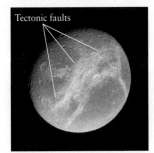

(d) Dione
(diameter 1120 km)

(e) Rhea
(diameter 1530 km)

(f) Iapetus
(diameter 1460 km)

Figure 13-22   R I [V] U X G

Saturn's Six Mid-Sized Satellites **(a)–(f)** These images from *Cassini* show the variety of surface features found on Saturn's six moderate-sized satellites. The satellites depicted in parts (a)–(f) are *not* shown to scale. **(g)** These images show the relative sizes of Saturn's moderate-sized satellites. (NASA/JPL/Space Science Institute)

(g) Satellites to scale

(a)

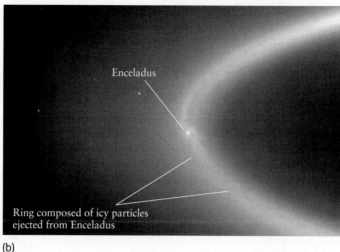

(b)

Figure 13-23   R I **V** U X G

Volcanoes on Enceladus (a) This *Cassini* image shows plumes of icy particles being ejected from ice volcanoes near the south pole of Enceladus. The faint plumes are best seen when they are backlit by the Sun, so Enceladus appears as a crescent. **(b)** A wide-angle backlit view from *Cassini* shows how particles ejected from Enceladus become part of Saturn's rings. This structure is actually part of the E ring, within which Enceladus orbits (see Figure 12-21). (NASA/JPL/Space Science Institute)

volcanically active, though with ice taking the place of lava. **Figure 13-23a** shows several distinct plumes ejecting icy particles from near the satellite's south pole, the same region where the surface is laced with bluish stress cracks (see Figure 13-22b). Some of the particles are ejected with such force that they escape Enceladus altogether and go into orbit around Saturn, forming a tenuous ring (Figure 13-23b).

What could be powering the geologic activity on Enceladus? One guess is tidal heating, which provides the internal heat for the inner Galilean satellites of Jupiter (see Section 13-4). The satellite Dione orbits Saturn in 65.7 hours, almost exactly twice Enceladus's orbital period of 32.9 hours, and the 1:2 ratio of orbital periods should set Enceladus up to be caught in a tidal tug-of-war between Saturn and Dione. Calculations suggest that the amount of tidal heating produced in this way is modest, but should be enough to melt the ices of Enceladus at least part of the time. Curiously, there is also a 1:2 ratio between the orbital periods of Mimas and Saturn's moon Tethys, but the cratered surface of Mimas shows no sign that this moon was ever tidally heated.

Tethys and Dione are the next-largest pair of Saturn's six moderate-sized satellites. Most of Tethys, like Mimas, is heavily cratered. But Tethys also has an immense rift valley called Ithaca Chasma that extends for 2000 km, or about $3/4$ of the way around Tethys (Figure 13-22c). This rift may have formed when the water inside Tethys cooled and solidified, or it may be the result of a shock wave that traveled through Tethys after an immense asteroid impact. (A large impact crater lies on the side of Tethys opposite Ithaca Chasma.) Dione, slightly larger than Tethys, has a strange surface dichotomy: The surfaces of its leading hemisphere (that is, the hemisphere facing toward its direction of orbital motion) and trailing hemisphere are quite different. Dione's leading hemisphere is heavily cratered, but its trailing hemisphere is cov-

ered by a network of bright stress cracks caused by some sort of tectonic activity (Figure 13-22d).

Because Tethys and Dione have about double the diameter of Enceladus, the energy to power their geologic activity might have come from within. These larger satellites retain heat more effectively, including heat from the decay of radioactive materials in each satellite's interior. Such materials are scarce within the icy satellites of Saturn, but only relatively little heat is needed to power ice volcanism.

Rhea and Iapetus are the largest and most distant of Saturn's moderate-sized moons. Rhea is heavily cratered, including a bright, fresh-looking crater on its leading hemisphere (Figure 13-22e). This may be the result of a more recent impact than those that formed most of Rhea's craters.

Iapetus has been known to be unusual ever since its discovery in 1671: Unlike the other satellites, its brightness varies greatly as it orbits Saturn. Images from *Cassini* show extreme differences between the leading and trailing hemispheres of Iapetus (Figure 13-22f). Most of the leading hemisphere is as black as asphalt (albedo = 0.05), but the trailing hemisphere is highly reflective (albedo = 0.50), like the surfaces of the other moderate-sized moons.

What could account for the two-sided nature of Iapetus? One possibility is that the dark material covering Iapetus's leading hemisphere may have come from Phoebe, a smaller satellite that orbits about $3\frac{1}{2}$ times farther from Saturn than Iapetus. Phoebe moves in a retrograde direction along an orbit tilted well away from the plane of Saturn's equator. This motion leads scientists to suspect that Phoebe is a captured asteroid, and its dark surface does indeed resemble the surfaces of a particular class of carbon-rich asteroids. Bits and pieces of this dark charcoal-like material drifting toward Saturn may have been swept up onto the leading hemisphere of Iapetus. However, a sticking point with this scenario is that the dark material on Iapetus is reddish in color, while Phoebe's surface is not. An alternative explanation is that a large portion of the leading hemisphere of Iapetus was coated with methane ice, which darkened when exposed to solar radiation.

Key Words

aerosol, p. 344
bright terrain (Ganymede), p. 340
dark terrain (Ganymede), p. 340
Galilean satellites, p. 330
hydrocarbon, p. 343
ice rafts (Europa), p. 338

Io torus, p. 336
occultation, p. 330
polymer, p. 344
prograde orbit, p. 346
retrograde orbit, p. 346
tidal heating, p. 334
ultramafic lava, p. 335

Key Ideas

Nature of the Galilean Satellites: The four Galilean satellites orbit Jupiter in the plane of its equator. All are in synchronous rotation.

• The orbital periods of the three innermost Galilean satellites, Io, Europa, and Ganymede, are in the ratio 1:2:4.

• The two innermost Galilean satellites, Io and Europa, have roughly the same size and density as our Moon. They are composed principally of rocky material. The two outermost Galilean satellites, Ganymede and Callisto, are roughly the size of Mercury. Lower in density than either the Moon or Mercury, they are made of roughly equal parts ice and rock.

• The Galilean satellites probably formed in a similar fashion to our solar system but on a smaller scale.

Io: Io is covered with a colorful layer of sulfur compounds deposited by frequent explosive eruptions from volcanic vents. These eruptions resemble terrestrial geysers.

• The energy to heat Io's interior and produce the satellite's volcanic activity comes from tidal forces that flex the satellite. This tidal flexing is aided by the 1:2:4 ratio of orbital periods among the inner three Galilean satellites.

• The Io torus is a ring of electrically charged particles circling Jupiter at the distance of Io's orbit. Interactions between this ring and Jupiter's magnetic field produce strong radio emissions. Io may also have a magnetic field of its own.

Europa: While composed primarily of rock, Europa is covered with a smooth layer of water ice.

• The surface has hardly any craters, indicating a geologically active history. Other indications are a worldwide network of long cracks and ice rafts that indicate a subsurface layer of liquid water or soft ice. As for Io, tidal heating is responsible for Europa's internal heat.

• An ocean may lie beneath Europa's frozen surface. Minerals dissolved in this ocean may explain Europa's induced magnetic field.

Ganymede: Two types of terrain are found on the icy surface of Ganymede: areas of dark, ancient, heavily cratered surface and regions of heavily grooved, lighter-colored, younger terrain.

• Ganymede is highly differentiated, and probably has a metallic core. It has a surprisingly strong magnetic field and a magnetosphere of its own.

• While there is at present little tidal heating of Ganymede, it may have been heated in this fashion in the past. An induced magnetic field suggests that it, too, has a layer of liquid water beneath the surface.

Callisto: Callisto has a heavily cratered crust of water ice. The surface shows little sign of geologic activity, because there was never any significant tidal heating of Callisto. However, some unknown processes have erased the smallest craters and blanketed the surface with a dark, dusty substance.

• Magnetic field data seem to suggest that Callisto has a shallow subsurface ocean.

Titan: The largest Saturnian satellite, Titan, is a terrestrial world with a dense nitrogen atmosphere. A variety of hydrocarbons are produced there by the interaction of sunlight with methane. These compounds form an aerosol layer in Titan's atmosphere and fall as a gentle rain on the surface.

• Titan's surface shows that liquid hydrocarbons have flowed over its surface, forming streams, rivers, and outflow channels. Very little of this liquid appears to be present on Titan's surface today.

Other Satellites: As of 2006, Jupiter has a total of 63 known satellites and Saturn has a total of 56.

• In addition to the Galilean satellites, Jupiter has four small inner satellites that lie inside Io's orbit. Like the Galilean satellites, these orbit in the plane of Jupiter's equator. The remaining satellites are small and move in much larger orbits that are noticeably inclined to the plane of Jupiter's equator. Many of these orbit in the direction opposite to Jupiter's rotation.

• In addition to Titan, six moderate-sized moons circle Saturn in regular orbits: Mimas, Enceladus, Tethys, Dione, Rhea, and Iapetus. They are probably composed largely of ice, but their surface features and histories vary significantly. The other, smaller moons include shepherd satellites that control the shapes of Saturn's rings and captured asteroids in large retrograde orbits.

Questions

Review Questions

1. If you observed the Galilean satellites through a telescope for a single night, could you notice their motions around Jupiter?
2. Why can't the Galilean satellites be seen with the naked eye?
3. During the time it takes Ganymede to complete one orbit, how many orbits do Io and Europa complete?
4. In what ways does the system of Galilean satellites resemble our solar system? In what ways is it different?
5. No spacecraft from Earth has ever landed on any of the Galilean satellites. How, then, can we know anything about the chemical compositions of these satellites?
6. In what ways did the formation of the Galilean satellites mimic the formation of the planets? In what ways were the two formation processes different?

7. In the classic science-fiction film *2010: The Year We Make Contact,* an alien intelligence causes Jupiter to contract so much that nuclear reactions begin at its center. As a result, Jupiter becomes a star like the Sun. Is this possible in principle? Explain your answer.

8. All of the Galilean satellites orbit Jupiter in the same direction. Furthermore, the planes of their orbits all lie within 0.5° of Jupiter's equatorial plane. Explain why these observations are consistent with the idea that the Galilean satellites formed from a "Jovian nebula."

9. What is the source of energy that powers Io's volcanoes? How is it related to the orbits of Io and the other Galilean satellites?

10. Io has no impact craters on its surface, while our Moon is covered with craters. What is the explanation for this difference?

11. Despite all the gases released from its interior by volcanic activity, Io does not possess a thick atmosphere. Explain why not.

12. Long before the *Voyager* flybys, Earth-based astronomers reported that Io appeared brighter than usual for the few hours after it emerged from Jupiter's shadow. From what we know about the material ejected from Io's volcanoes, suggest an explanation for this brief brightening of Io.

13. How do lavas on Io differ from typical lavas found on Earth? What does this difference tell us about Io's interior?

14. What accounts for the different colors found on Io's surface?

15. What is the Io torus? What is its source?

16. What is the origin of the electric current that flows through Io?

17. How was the *Galileo* spacecraft used to determine the internal structure of Io and the other Galilean satellites?

18. What surface features on Europa provide evidence for geologic activity?

19. What is the evidence for an ocean of liquid water beneath Europa's icy surface? What is the evidence that substances other than water are dissolved in this ocean?

20. What aspects of Europa lead scientists to speculate that life may exist there? What precautions have been taken to prevent contaminating Europa with Earth organisms?

21. Why is ice an important constituent of Ganymede and Callisto, but not of the Earth's Moon?

22. How do scientists know that the dark terrain on Ganymede is younger than the bright terrain?

23. In what ways is Ganymede like our own Moon? In what ways is it different? What are the reasons for the differences?

24. Why were scientists surprised to learn that Ganymede has a magnetic field? What does this field tell us about Ganymede's history?

25. What leads scientists to suspect that there is liquid water beneath Ganymede's surface?

26. Why are numerous impact craters found on Ganymede and Callisto but not on Io or Europa?

27. Describe the surprising aspects of Callisto's surface and interior that were revealed by the *Galileo* spacecraft. Why did these come as a surprise?

28. Compare and contrast the surface features of the four Galilean satellites. Discuss their relative geological activity and the evolution of these four satellites.

29. The larger the orbit of a Galilean satellite, the less geologic activity that satellite has. Explain why.

30. Explain how the 1:2:4 ratio of the orbital periods of Io, Europa, and Ganymede is related to the geologic activity on these satellites.

31. Describe Titan's atmosphere. What effect has the Sun's ultraviolet radiation had on Titan's atmosphere?

32. In what ways do hydrocarbons on Titan behave like water on Earth?

33. Why does the presence of methane in Titan's atmosphere imply that Titan has had recent volcanic activity?

34. What is the evidence that there were liquid hydrocarbons on Titan in the past? That there are liquid hydrocarbons now?

35. How would you account for the existence of the satellites of Jupiter other than the Galilean ones?

36. Which of Saturn's moderate-sized satellites show evidence of geologic activity? What might be the energy source for this activity?

37. Explain why debris from Phoebe would be expected to pile up only on the leading hemisphere of Iapetus. (*Hint:* How do the orbits of these two satellites compare? How does the orbital motion of debris falling slowly inward toward Saturn compare with the orbital motion of Iapetus?)

38. Saturn's equator is tilted by 27° from the ecliptic, while Jupiter's equator is tilted by only 3°. Use these data to explain why we see fewer transits, eclipses, and occultations of Saturn's satellites than of the Galilean satellites.

Advanced Questions

Questions preceded by an asterisk (*) involve topics discussed in Box 1-1 or Box 7-2.

> ### Problem-solving tips and tools
>
> Newton's form of Kepler's third law (Box 4-4) relates the masses of two objects in orbit around each other to the period and size of the orbit. The small-angle formula is discussed in Box 1-1. The best seeing conditions on the Earth give a limiting angular resolution of ¼ arcsec. Because the orbits of the Galilean satellites are almost perfect circles, you can easily calculate the orbital speeds of these satellites from the data listed in Table 13-1. Data about Jupiter itself are given in Table 12-1 and data about all of the satellites of Jupiter and Saturn can be found in Appendix 3. For a discussion of escape speed and how planets retain their atmospheres, see Box 7-2.

39. Using the orbital data in Table 13-1, demonstrate that the Galilean satellites obey Kepler's third law.

*40. What is the size of the smallest feature you should be able to see on a Galilean satellite through a large telescope under conditions of excellent seeing when Jupiter is near opposition? How does this compare with the best *Galileo* images, which have resolutions around 25 meters?

41. Using the diameter of Io (3642 km) as a scale, estimate the height to which the plume of Pele rises above the surface of Io in Figure 13-5a. (You will need to make measurements on this figure using a ruler.) Compare your answer to the value given in the figure caption.

42. Explain why the volcanic plumes in Figure 13-5*b* have a bluish color. (*Hint:* See Box 5-4.)

43. Jupiter, its magnetic field, and the charged particles that are trapped in the magnetosphere all rotate together once every 10 hours. Io takes 1.77 days to complete one orbit. Using a diagram, explain why particles from Jupiter's magnetosphere hit Io primarily from behind (that is, on the side of Io that trails as it orbits the planet).

44. Assuming material is ejected from Io into Jupiter's magnetosphere at the rate of 1 ton per second (1000 kg/s), how long will it be before Io loses 10% of its mass? How does your answer compare with the age of the solar system?

45. Volcanic eruptions from Io add about 1 ton of material to Jupiter's magnetosphere every second. Based on the information given in Section 13-4, what fraction of the material ejected from Io's volcanoes goes into Jupiter's magnetosphere?

46. How long does it take for Ganymede to entirely enter or entirely leave Jupiter's shadow? Assume that the shadow has a sharp edge.

47. Figure 13-15 shows that Ganymede's dark terrain has both more craters and larger craters than the bright terrain. Explain what this tells you about the sizes of meteoroids present in the solar system in the distant past and in the more recent past.

48. Put a few cubes of ice in a glass, then fill the glass with warm water to make the ice cubes crack. The fracture lines in each cube are clearly visible because they reflect light. Use this observation to suggest why Ganymede's icy bright terrain, which may have been badly fractured by the stresses that produced the grooves (Figure 13-15), is 25% more reflective than the dark terrain.

*49. Find the escape speed on Titan. What is the limiting molecular weight of gases that could be retained by Titan's gravity? (*Hint:* Use the ideas presented in Box 7-2 and assume an average atmospheric temperature of 95 K.)

50. Many of the gases in the atmosphere of Titan, such as methane, ethane, and acetylene, are highly flammable. Why, then, doesn't Titan's atmosphere catch fire? (*Hint:* What gas in our atmosphere is needed to make wood, coal, or gasoline burn?)

51. Suppose the Earth's Moon were removed and replaced in its orbit by Titan. What changes would you expect to occur in Titan's atmosphere? Would solar eclipses be more or less common as seen from Earth? Explain your answers.

52. At infrared wavelengths, the Hubble Space Telescope can see details on Titan's surface as small as 580 km (360 mi) across. Determine the angular resolution of the Hubble Space Telescope using infrared light. If visible light is used, is the angular resolution better, worse, or the same? Explain your answer.

53. (a) To an observer on Enceladus, what is the time interval between successive oppositions of Dione? Explain your answer. (b) As seen from Enceladus, what is the angular diameter of Dione at opposition? How does this compare to the angular diameter of the Moon as seen from Earth (about ½°)?

Discussion Questions

54. If you could replace our Moon with Io, and if Io could maintain its present amount of volcanic activity, what changes would this cause in our nighttime sky? Do you think that Io

could in fact remain volcanically active in this case? Why or why not?

55. Speculate on the possibility that Europa, Ganymede, or Callisto might harbor some sort of life. Explain your reasoning.

56. Comment on the suggestion that Titan may harbor life-forms.

57. Jupiter's satellite Io and Saturn's satellite Enceladus are both geologically active, and both are in 2-to-1 resonances with other satellites. However, the amount of geologic activity of Enceladus is far less than on Io. Discuss some possible reasons for this difference.

58. Imagine that you are in charge of planning a successor to the *Cassini* spacecraft to further explore the Saturnian system. In your opinion, which satellites in the system should be examined more closely? What data should be collected? What kinds of questions should the new mission attempt to answer?

Web/eBook Questions

59. Various spacecraft missions have been proposed to explore Europa in greater detail. Search the World Wide Web for information about these. How would these missions test for the presence of an ocean beneath Europa's surface?

60. Twenty-one new satellites of Jupiter were discovered in 2003. Search the World Wide Web for information about how these satellites were discovered. How was it determined that these satellites are actually in orbit around Jupiter?

61. **The Surface of Ganymede.** Access and view the video "Jupiter's Moon Ganymede" in Chapter 13 of the *Universe* Web site or eBook. Describe the different surface features that you see, and explain how each type of feature was formed.

Activities

Observing Projects

Observing tips and tools

You can easily find the apparent positions of the Galilean satellites for any date and time using the *Starry Night Enthusiast*™ software on the CD-ROM that accompanies certain printed copies of this textbook. For even more detailed information about satellite positions, consult the "Satellites of Jupiter" section in the *Astronomical Almanac* for the current year. If your goal is to view Saturn's satellites, consult the section entitled "Satellites of Saturn" in the *Astronomical Almanac* for the current year. This includes a diagram showing the orbits of Mimas, Enceladus, Tethys, Dione, Rhea, Titan, and Hyperion. Plan your observing session by looking up the dates and times of the most recent greatest eastern elongations of the various satellites. You will have to convert from universal time (UT), also known as Greenwich Mean Time, to your local time zone. Then, using the tick marks along the orbits in the diagram, estimate the positions of the satellites relative to Saturn at the time you will be at the telescope. Another useful resource is the "Celestial Calendar" section of *Sky & Telescope*. During months when Saturn is visible in the night sky, this section of the magazine includes a chart of Saturn's satellites.

62. Observe Jupiter through a pair of binoculars. Can you see all four Galilean satellites? Make a drawing of what you observe. If you look again after an hour or two, can you see any changes?

63. Observe Jupiter through a small telescope on three or four consecutive nights. Make a drawing each night showing the positions of the Galilean satellites relative to Jupiter. Record the time and date of each observation. Consult the sources listed above in the "Observing tips and tools" to see if you can identify the satellites by name.

64. If you have access to a moderately large telescope, make arrangements to observe several of Saturn's satellites. At the telescope, you should have no trouble identifying Titan. Tethys, Dione, and Rhea are about one-sixth as bright as Titan and should be the next easiest satellites to find. Can you confidently identify any of the other satellites?

65. Use the *Starry Night Enthusiast*™ program to observe the Galilean satellites of Jupiter. Open the **Favourites pane** and click on **Guides > Atlas** to display the entire celestial sphere. Open the **Find** pane and double-click the entry for Jupiter to center this planet in the view. Using the controls at the right-hand end of the toolbar, zoom in to a field of view of approximately **13′ × 9′**. Stop time flow and in the toolbar, set the date and time to March 4, 2004, at 12:00:00 A.M to see Jupiter at opposition. Set the **Time Flow Rate** to 20 minutes by clicking on the number in the Time Flow Rate box and setting the value with the keyboard. Then click on the Run Time Forward button (a triangle that points to the right). You will see the four Galilean satellites orbiting Jupiter. (If these moons appear to move too quickly, adjust the Time Flow Rate to 10 minutes.) (a) Are all four satellites ever on the same side of Jupiter? (b) Observe the satellites passing in front of and behind Jupiter and look for their shadows upon the planet. (Zoom in as needed). Explain how your observations tell you that all four satellites orbit Jupiter in the same direction.

66. Use the *Starry Night Enthusiast*™ program to view Saturn from its satellite Enceladus. First click the **Home** button in the toolbar. Select **Options > Viewing Location . . .** from the menu. In the dialog window that appears at the top of the Viewing Location, set the drop-down box next to the "**View from:**" label to read "**the surface of**" and set the second drop-down box to read "**Enceladus**" (listed under Saturn). In the list of locations that appears under the **List** tab on Enceladus, scroll to "**Ahmad.**" Click on this entry to highlight it and then click the **Set Location** button. To see Saturn from this location on Enceladus, center on Saturn by double-clicking the entry for this planet in the **Find** pane. (a) Stop the time flow and set the Time Flow Rate to **1 minute** in the toolbar, then click on the **Run Time Forward** button (a triangle that points to the

right). How do the stars appear to move as seen from this location on Enceladus? How does Saturn appear to move? What do these observations tell you about the relationship between the orbital period and rotation period of Enceladus? (b) Set the Time Flow Rate to **1 hour.** By stepping forward through time using the rightmost single-step button, determine how much time elapses from when Saturn appears full from this location to when it next appears full. Explain why this is the same as the orbital period of Enceladus, and compare this to the value of the orbital period given in Appendix 3.

67. Use the *Starry Night Enthusiast*™ program to examine the satellites of Saturn. Open the **Favourites** pane and select **Solar System > Saturn**. Remove the astronaut's feet from this view by clicking on **View > Feet**. In this view you can rotate Saturn by placing the mouse cursor over the image of the planet, holding down the mouse button, and moving the mouse (on a two-button mouse, hold down the left mouse button). Use this technique to rotate Saturn so that you are viewing the rings edge-on. Then use the elevation controls in the toolbar (the buttons to the left of the Home button) to move closer to and further away from Saturn. This allows you to identify the satellites of Saturn, since they will appear to move whereas the distant stars will remain stationary as you move with respect to the planet. Alternately, you can reduce the confusion in identifying these moons by clicking on **View > Stars > Stars** to remove the stars from the view. You should be able to see at least eight satellites. Which satellites are these? (Move the mouse to center the cursor on a satellite and its name will appear. If you cannot see all eight, try moving further from Saturn (increase the viewing location elevation with the elevation button showing the Up arrowhead). (b) The plane of Saturn's rings is the same as the plane of Saturn's equator. Which satellites appear to be the farthest from this plane?

Collaborative Exercises

68. Imagine that scientists are proposing to send a robotic lander to Jupiter's satellite Callisto. Create a 100-word written proposal describing a robotic lander mission to another of the Galilean satellites, explaining why your group found it to be the most interesting and why the government should allocate the money for your alternative project. In your proposal, be sure to demonstrate your knowledge of Callisto and the other satellite.

69. From the data and the accompanying images for Table 13-1, "The Galilean Satellites Compared to the Moon, Mercury, and Mars," use someone's shoe to represent the 150,000 km diameter of Jupiter and determine about how many "shoes" away would each of the Galilean satellites be from Jupiter.

14

Uranus, Neptune, Pluto, and the Kuiper Belt: Remote Worlds

R I V U X G

Earth, Uranus, Neptune, Pluto, and Eris to scale. (Alan Stern, Southwest Research Institute; Marc Buie, Lowell Observatory; NASA; and ESA)

Beyond Saturn, in the cold, dark recesses of the solar system, orbit three planets that have long been shrouded in mystery. These planets—Uranus, Neptune, and Pluto (all shown here to the same scale as the Earth)—are so distant, so dimly lit by the Sun, and so slow in their motion against the stars that they were unknown to ancient astronomers and were discovered only after the invention of the telescope. Even then, little was known about Uranus and Neptune until *Voyager 2* flew past these planets during the 1980s.

Surrounded by a system of small moons and thin, dark rings, Uranus is tipped on its side so that its axis of rotation lies nearly in its orbital plane. This remarkable orientation suggests that Uranus may have been the victim of a staggering impact by a mas-

sive planetesimal. Neptune is, at first glance, a denser, more massive version of Uranus, but it is a far more active world. It has an internal energy source that Uranus seemingly lacks, as well as atmospheric bands and storm activity resembling those on Jupiter.

Neptune also has dark rings, small, icy moons, and an intriguing large satellite, Triton, which is nearly devoid of impact craters and has geysers that squirt nitrogen-rich vapors. Triton's retrograde orbit suggests that this strange world may have been gravitationally captured by Neptune.

Close cousins of Triton are Pluto and thousands of other trans-Neptunian objects, among the most remote members of the solar system. These objects harbor many mysteries, in part because they have not yet been visited by a spacecraft.

Learning Goals

By reading the sections of this chapter, you will learn

14-1 How Uranus and Neptune were discovered

14-2 The unusual properties of the orbit and atmosphere of Uranus

14-3 What gives Neptune's clouds and atmosphere their distinctive appearance

14-4 The internal structures of Uranus and Neptune

14-5 The unique orientations of the magnetic fields of Uranus and Neptune

14-6 Why the rings of Uranus and Neptune are hard to see

14-7 What could have powered geologic activity on Uranus's moderate-sized moons

14-8 Why Neptune's satellite Triton is destined to be torn apart

14-9 How Pluto came to be discovered

14-10 What shapes the orbits of Pluto and the thousands of other objects that orbit beyond Neptune

Table 14-1	Uranus Data
Average distance from Sun:	19.194 AU = 2.871 × 10⁹ km
Maximum distance from Sun:	20.017 AU = 2.995 × 10⁹ km
Minimum distance from Sun:	18.371 AU = 2.748 × 10⁹ km
Eccentricity of orbit:	0.0429
Average orbital speed:	6.83 km/s
Orbital period:	84.099 years
Rotation period (internal):	17.24 hours
Inclination of equator to orbit:	97.86°
Inclination of orbit to ecliptic:	0.77°
Diameter:	51,118 km = 4.007 Earth diameters (equatorial)
Mass:	8.682 × 10²⁵ kg = 14.53 Earth masses
Average density:	1318 kg/m³
Escape speed:	21.3 km/s
Surface gravity (Earth = 1):	0.90
Albedo:	0.56
Average temperature at cloudtops:	−218°C = −360°F = 55 K
Atmospheric composition (by number of molecules):	82.5% hydrogen (H₂), 15.2% helium (He), 2.3% methane (CH₄)
Average distance from Sun:	30.066 AU = 4.498 × 10⁹ km
Maximum distance from Sun:	30.367 AU = 4.543 × 10⁹ km

R I V U X G

(NASA/JPL)

14-1 Uranus was discovered by chance, but Neptune's existence was predicted by applying Newtonian mechanics

Before the eighteenth century, only six planets were known to orbit the Sun: Mercury, Venus, Earth, Mars, Jupiter, and Saturn. That state of affairs changed thanks to William Herschel, a German-born musician who moved to England in 1757 and became fascinated by astronomy.

Discovering the "Georgian Star"

Using a telescope that he built himself, Herschel was systematically surveying the sky on March 13, 1781, when he noticed a faint, fuzzy object that he first thought to be a distant comet. By the end of 1781, however, his observations had revealed that the object's orbit was relatively circular and was larger than Saturn's orbit. Comets, by contrast, can normally be seen only when they follow elliptical orbits that bring them much closer to the Sun.

Herschel had discovered the seventh planet from the Sun. In doing so, he had doubled the radius of the known solar system from 9.5 AU (the semimajor axis of Saturn's orbit) to 19.2 AU (the distance from the Sun to Uranus). Herschel originally named his discovery Georgium Sidus (Latin for "Georgian star") in honor of the reigning monarch, George III. The name Uranus—in Greek mythology, the personification of Heaven—came into currency only some decades later.

Although Herschel received the credit for discovering Uranus, he was by no means the first person to have seen it. At opposition, Uranus is just barely bright enough to be seen with the naked eye under good observing conditions, so it was probably seen by the ancients. Many other astronomers with telescopes had

Table 14-2	Neptune Data
Average distance from Sun:	30.066 AU = 4.498×10^9 km
Maximum distance from Sun:	30.367 AU = 4.543×10^9 km
Minimum distance from Sun:	29.765 AU = 4.453×10^9 km
Eccentricity of orbit:	0.010
Average orbital speed:	5.5 km/s
Orbital period:	164.86 years
Rotation period (internal):	16.11 hours
Inclination of equator to orbit:	29.56°
Inclination of orbit to ecliptic:	1.77°
Diameter:	49,528 km = 3.883 Earth diameters (equatorial)
Mass:	1.024×10^{26} kg = 17.15 Earth masses
Average density:	1638 kg/m³
Escape speed:	23.5 km/s
Surface gravity (Earth = 1):	1.1
Albedo:	0.51
Average temperature at cloudtops:	−218°C = −360°F = 55 K
Atmospheric composition (by number of molecules):	79% hydrogen (H_2), 18% helium (He), 3% methane (CH_4)
Average distance from Sun:	30.066 AU = 4.498×10^9 km
Maximum distance from Sun:	30.367 AU = 4.543×10^9 km

WEB LINK 14.3   R I V U X G
(NASA/JPL)

sighted this planet before Herschel; it is plotted on at least 20 star charts drawn between 1690 and 1781. But all these other observers mistook Uranus for a dim star. Herschel was the first to track its motion relative to the stars and recognize it as a planet. This was no small task, because Uranus moves very slowly on the celestial sphere, just over 4° in the space of a year (compared to about 12° for Saturn and about 35° for Jupiter).

The Discovery of Neptune

It was by carefully tracking Uranus's slow motions that astronomers were led to discover Neptune. By the beginning of the nineteenth century, it had become painfully clear to astronomers that they could not accurately predict the orbit of Uranus using Newtonian mechanics. By 1830 the discrepancy between the planet's predicted and observed positions had become large

enough (2 arcmin) that some scientists suspected that Newton's law of gravitation might not be accurate at great distances from the Sun.

By the mid-1840s, two scientists working independently—the French astronomer Urbain Jean Joseph Le Verrier and the English mathematician John Couch Adams—were exploring an earlier and sounder suggestion. Perhaps the gravitational pull of an as yet undiscovered planet was causing Uranus to deviate slightly from its predicted orbit. Calculations by both scientists concluded that Uranus had indeed caught up with and had passed a more distant planet. Uranus had accelerated slightly as it approached the unknown planet, then decelerated slightly as it receded from the planet.

Inspired by Le Verrier's results, astronomers at Cambridge University Observatory undertook a six-week search for the proposed new planet in the summer of 1846. They were unsuccessful, in part because they lacked accurate star maps of the part of the sky being searched. The Cambridge astronomers also did not receive supporting information from Adams in a timely manner.

Meanwhile, Le Verrier wrote to Johann Gottfried Galle at the Berlin Observatory with detailed predictions of where to search for the new planet. Galle received the letter on September 23, 1846. That very night, aided by more complete star maps and after just a half-hour of searching, Galle and Heinrich d'Arrest located an uncharted star with the expected brightness in the predicted location. Subsequent observations confirmed that this new "star" showed a planetlike motion with respect to other stars.

Le Verrier proposed that the planet be called Neptune. After years of debate between English and French astronomers, the credit for its discovery came to be divided between Adams and Le Verrier. Neptune is thus the only planet whose existence was revealed by calculation rather than chance discovery.

At opposition, Neptune can be bright enough to be visible through small telescopes. The first person thus to have seen it was probably Galileo. His drawings from January 1613, when he was using his telescope to observe Jupiter and its four large satellites, show a "star" less than 1 arcmin from Neptune's location. Galileo even noted in his observation log that on one night this star seemed to have moved in relation to the other stars. But Galileo would have been hard pressed to identify Neptune as a planet, because its motion against the background stars is so slow (just over 2° per year).

Observing Uranus and Neptune

Through a large, modern telescope, both Uranus and Neptune are dim, uninspiring sights. Each planet appears as a hazy, featureless disk with a faint greenish-blue tinge. Although Uranus and Neptune are both about 4 times larger in diameter than Earth, they are so distant that their angular diameters as seen from the Earth are tiny—no more than 4 arcsec for Uranus and just over 2 arcsec for Neptune. To an Earth-based observer, Uranus is roughly the size of a golf ball seen at a distance of 1 kilometer.

> **Uranus and Neptune are in such large orbits that they move very slowly on the celestial sphere**

From 2007 through 2011, Uranus and Neptune are less than 40° apart in the sky in the adjacent constellations of Pisces, Aquarius, and Capricornus (the Sea Goat). During these years, the two planets are at opposition in either August or September, which are thus the best months to view them with a telescope. Table 14-1 and Table 14-2 give basic data about Uranus and Neptune, respectively.

14-2 Uranus is nearly featureless and has an unusually tilted axis of rotation

Scientists had hoped that *Voyager 2* would reveal cloud patterns in Uranus's atmosphere when it flew past the planet in January 1986. But even images recorded at close range showed Uranus to be remarkably featureless (Figure 14-1). Faint cloud markings became visible in images of Uranus only after extreme computer enhancement (Figure 14-2).

South pole

Figure 14-1   R I **V** U X G

Uranus from *Voyager 2* This image looks nearly straight down onto Uranus's south pole, which was pointed almost directly at the Sun when *Voyager 2* flew past in 1986. None of the *Voyager 2* images of Uranus show any pronounced cloud patterns. The color is due to methane in the planet's atmosphere, which absorbs red light but reflects green and blue. (JPL/NASA)

Uranus's Atmosphere

Voyager 2 data confirmed that the Uranian atmosphere is dominated by hydrogen (82.5%) and helium (15.2%), similar to the atmospheres of Jupiter and Saturn. Uranus differs, however, in that 2.3% of its atmosphere is methane (CH_4), which is 5 to 10 times the percentage found on Jupiter and Saturn. In fact, Uranus has a higher percentage of all heavy elements—including carbon atoms, which are found in molecules of methane—than Jupiter and Saturn. (In Section 14-4 we will investigate how this could have come about.)

Methane preferentially absorbs the longer wavelengths of visible light, so sunlight reflected from Uranus's upper atmosphere is depleted of its reds and yellows. This gives the planet its distinct greenish-blue appearance. As on Saturn's moon Titan (see Section 13-8), ultraviolet light from the Sun turns some of the methane gas into a hydrocarbon haze, making it difficult to see the lower levels of the atmosphere.

Ammonia (NH_3), which constitutes 0.01 to 0.03% of the atmospheres of Jupiter and Saturn, is almost completely absent from the Uranian atmosphere. The reason is that Uranus is colder than Jupiter or Saturn: The temperature in its upper atmosphere is only 55 K ($-218°C = -360°F$). Ammonia freezes at these very low temperatures, so any ammonia has long since precipitated out of the atmosphere and into the planet's interior. For the same reason, Uranus's atmosphere is also lacking in water. Hence, the substances that make up the clouds on Jupiter and Saturn—ammonia, ammonium hydrosulfide (NH_4SH), and water—are not available on Uranus. This helps to explain the bland, uniform appearance of the planet shown in Figure 14-1.

The few clouds found on Uranus are made primarily of methane, which condenses into droplets only if the pressure is suf-

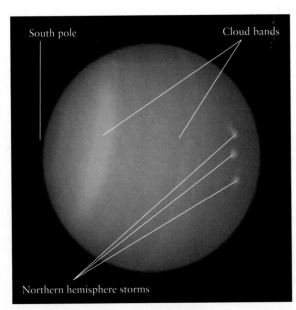

WEB LINK 14.5

Figure 14-2   R I **V** **U** x G

Uranus from the Hubble Space Telescope Images made at ultraviolet, visible, and infrared wavelengths were combined and enhanced to give this false-color view of cloud features on Uranus. These images were captured in August 2004 (18 years after the image in Figure 14-1), during springtime in Uranus's northern hemisphere. (NASA and Erich Karkoschka, University of Arizona)

ficiently high. Hence, methane clouds form only at lower levels within the atmosphere, where they are difficult to see.

An Oddly Tilted Planet

The rotation period of Uranus's atmosphere is about 16 hours. Like Jupiter and Saturn, Uranus rotates differentially, so this period depends on the latitude. This can be measured by tracking the motions of clouds. To determine the rotation period for the underlying body of the planet, scientists looked to Uranus's magnetic field, which is presumably anchored in the planet's interior, or at least in the deeper and denser layers of its atmosphere. Data from *Voyager 2* indicate that Uranus's internal period of rotation is 17.24 hours.

Voyager 2 also confirmed that Uranus's rotation axis is tilted in a unique and bizarre way. Herschel found the first evidence of this in 1787, when he discovered two moons orbiting Uranus in a plane that is almost perpendicular to the plane of the planet's orbit around the Sun. Because the large moons of Jupiter and Saturn were known to orbit in the same plane as their planet's equator and in the same direction as their planet's rotation, it was thought that the same must be true for the moons of Uranus. Thus, Uranus's equator must be almost perpendicular to the plane of its orbit, and its rotation axis must lie very nearly in that plane (**Figure 14-3**).

Careful measurement shows that Uranus's axis of rotation is tilted by 98°, as compared to 23½° for Earth (compare Figure 14-3 with Figure 2-12). A tilt angle greater than 90° means that

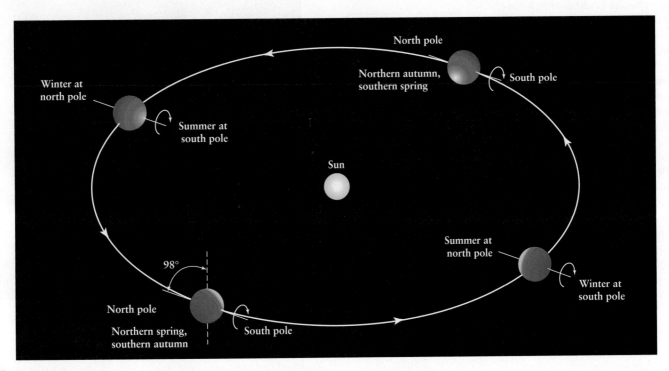

ANIMATION 14.1

Figure 14-3

Exaggerated Seasons on Uranus For most planets, the rotation axis is roughly perpendicular to the plane of the planet's orbit around the Sun. But for Uranus the rotation axis is tilted by 98° from the perpendicular. This causes severely exaggerated seasons.

For example, during midsummer at Uranus's south pole, the Sun appears nearly overhead for many Earth years, while the planet's northern regions are in continuous darkness. Half an orbit later, the seasons are reversed.

Uranus exhibits retrograde (backward) rotation like that of Venus, shown in Figure 11-7b. Astronomers suspect that Uranus might have acquired its large tilt angle billions of years ago, when another massive body collided with Uranus while the planet was still forming.

The radical tilt of its axis means that as Uranus moves along its 84-year orbit, its north and south poles alternately point toward or away from the Sun. This produces highly exaggerated seasonal changes. For example, when *Voyager 2* flew by in 1986, Uranus's south pole was pointed toward the Sun. Most of the planet's southern hemisphere was bathed in continuous sunlight, while most of the northern hemisphere was subjected to a continuous, frigid winter night. But over the following quarter of a Uranian year later (21 of our years), sunlight has gradually been returning to the northern hemisphere, triggering immense storms there. Figure 14-2 shows some of these storms, which are much more visible at infrared wavelengths than with visible light.

Atmospheric Motions on Uranus

By following the motions of clouds and storm systems on Uranus, scientists find that the planet's winds flow to the east—that is, in the same direction as the planet's rotation—at northern and southern latitudes, but to the west near the equator. This is quite unlike the situation on Jupiter and Saturn, where the zonal winds alternate direction many times between the north and south poles (see Section 12-4). The fastest Uranian winds (about 700 km/h, or 440 mi/h) are found at the equator.

We saw in Section 12-4 that the internal heat of Jupiter and Saturn plays a major role in driving atmospheric activity on those worlds. Uranus is different: It appears to have little or no internal source of thermal energy. Measurements to date show that unlike Jupiter or Saturn, Uranus radiates into space only as much energy as it receives from the Sun. With only feeble sunlight to provide energy to its atmosphere, Uranus lacks the dramatic wind and cloud patterns found on Jupiter and Saturn. It is not known why Uranus should be different from Jupiter and Saturn.

> **Unlike the other Jovian planets, Uranus's interior heat has no effect on its atmosphere's motions**

Although Uranus's equatorial region was receiving little sunlight at the time of the *Voyager 2* flyby, the atmospheric temperature there (about 55 K = −218°C = −359°F) was not too different from that at the sunlit pole. Heat must therefore be efficiently transported from the poles to the equator. This north-south heat transport, which is perpendicular to the wind flow, may have mixed and homogenized the atmosphere to make Uranus nearly featureless.

14-3 Neptune is a cold, bluish world with Jupiterlike atmospheric features

 At first glance, Neptune appears to be the twin of Uranus. (See the image that opens this chapter, and compare Tables 14-1 and 14-2.) But these two planets are by no means identical. While Neptune and Uranus have almost the same

diameter, Neptune is 18% more massive. Neptune's axis of rotation also has a more modest 29½° tilt. When *Voyager 2* flew past Neptune in August 1989, it revealed that the planet has a more active and dynamic atmosphere than Uranus. This activity suggests that Neptune, unlike Uranus, has a powerful source of energy in its interior.

Neptune's Atmosphere

The *Voyager 2* data showed that Neptune has essentially the same atmospheric composition as Uranus: 79% hydrogen, 18% helium, 3% methane, and almost no ammonia or water vapor. As for Uranus, the presence of methane gives Neptune a characteristic bluish-green color. The temperature in the upper atmosphere is also the same as on Uranus, about 55 K. That this should be so, even though Neptune is much farther from the Sun, is further evidence that Neptune has a strong internal heat source.

Unlike Uranus, however, Neptune has clearly visible cloud patterns in its atmosphere. At the time that *Voyager 2* flew past Neptune, the most prominent feature in the planet's atmosphere was a giant storm called the Great Dark Spot (Figure 14-4). The Great Dark Spot had a number of similarities to Jupiter's Great Red Spot (see Sections 12-1 and 12-3). The storms

> **Unlike its close relative Uranus, Neptune has truly immense storms**

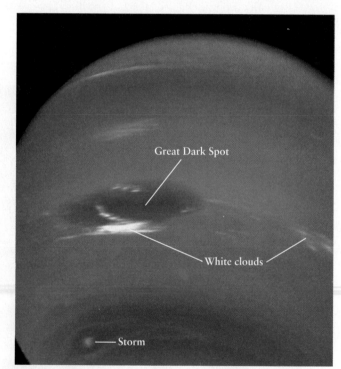

Great Dark Spot

White clouds

Storm

WEB LINK 14.6   **VIDEO 14.2**   **Figure 14-4**   R I V̲ U X G

Neptune from *Voyager 2* When this picture of Neptune's southern hemisphere was taken in 1989, the Great Dark Spot measured about 12,000 by 8000 km, comparable in size to the Earth. (A smaller storm appears at the lower left.) The white clouds are thought to be composed of crystals of methane ice. The color contrast in this image has been exaggerated to emphasize the differences between dark and light areas. (NASA/JPL)

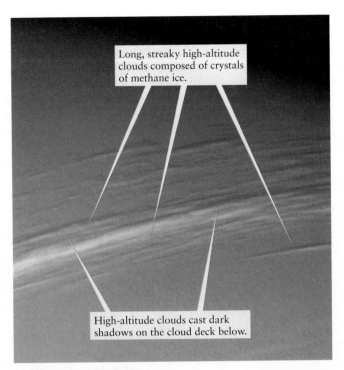

Figure 14-5   R I V U X G

Cirrus Clouds over Neptune *Voyager 2* recorded this image of clouds near Neptune's terminator (the border between day and night on the planet). Like wispy, high-altitude cirrus clouds in the Earth's atmosphere, these clouds are thought to be made of ice crystals. The difference is that Neptune's cirrus clouds are probably methane ice, not water ice as on Earth. (NASA/JPL)

on both planets were comparable in size to the Earth's diameter, both appeared at about the same latitude in the southern hemisphere, and the winds in both storms circulated in a counterclockwise direction (see Figure 12-5). But Neptune's Great Dark Spot appears not to have been as long-lived as the Great Red Spot on Jupiter. When the Hubble Space Telescope first viewed Neptune in 1994, the Great Dark Spot had disappeared. Another dark storm appeared in 1995 in Neptune's northern hemisphere.

Voyager 2 also saw a few conspicuous whitish clouds on Neptune. These clouds are thought to be produced when winds carry methane gas into the cool, upper atmosphere, where it condenses into crystals of methane ice. *Voyager 2* images show these high-altitude clouds casting shadows onto lower levels of Neptune's atmosphere (**Figure 14-5**). Images from the Hubble Space Telescope also show the presence of high-altitude clouds (**Figure 14-6**).

Neptune's Internal Heat

Thanks to its greater distance from the Sun, Neptune receives less than half as much energy from the Sun as Uranus. But with less solar energy available to power atmospheric motions, why are there high-altitude clouds and huge, dark storms on Neptune but not on Uranus? At least part of the answer is probably that Neptune is still slowly contracting, thus converting gravitational energy into thermal energy that heats the planet's core. (The same is true for Jupiter and Saturn, as we saw in Section 12-4.) The ev-

idence for this is that Neptune, like Jupiter and Saturn but unlike Uranus, emits more energy than it receives from the Sun. The combination of a warm interior and a cold outer atmosphere can cause convection in Neptune's atmosphere, producing the up-and-down motion of gases that generates clouds and storms. Neptune also resembles Jupiter in having faint belts and zones parallel to the planet's equator (see Figure 14-6).

Like those on Uranus, most of Neptune's clouds are probably made of droplets of liquid methane. Because these droplets form fairly deep within the atmosphere, the clouds are more difficult to see than the ones on Jupiter. Hence, Neptune's belts and zones are less pronounced than Jupiter's, although more so than those on Uranus (thanks to the extra cloud-building energy from Neptune's interior). As described above, Neptune's high-altitude clouds (see Figure 14-5) are probably made of *frozen* methane.

Although Neptune displays more evidence of up-and-down motion in its atmosphere than Uranus, the global pattern of east and west winds is almost identical on the two planets. This is rather strange. The two planets are heated very differently by the Sun, thanks to their different distances from the Sun and the different tilts of their axes of rotation, so we might have expected the wind patterns on Uranus and Neptune also to be different. Perhaps the explanation of these wind patterns will involve understanding how heat is transported not only within the atmospheres of Uranus and Neptune but within their interiors as well.

Figure 14-6   R I V U X G

Neptune's Banded Structure This enhanced-color Hubble Space Telescope image shows Neptune's belts and zones. White areas denote high-altitude clouds; the very highest clouds (near the top of the image) are shown in yellow-red. The green belt near the south pole is a region where the atmosphere absorbs blue light, perhaps indicating a different chemical composition there. (Lawrence Sromovsky, University of Wisconsin-Madison; STScI/NASA)

14-4 Uranus and Neptune contain a higher proportion of heavy elements than Jupiter and Saturn

At first glance, Uranus and Neptune might seem to be simply smaller and less massive versions of Jupiter or Saturn (see Table 7-1). But like many first impressions, this one is misleading because it fails to take into account what lies beneath the surface. We saw in Section 14-2 that the interiors of both Jupiter and Saturn are composed primarily of hydrogen and helium, in nearly the same abundance as the Sun. But Uranus and Neptune must have a different composition, because their average densities are too high.

Uranus and Neptune: Curiously Dense

If Uranus and Neptune had solar abundances of the elements, the smaller masses of these planets would produce less gravitational compression, and Uranus and Neptune would both have lower average densities than Jupiter or Saturn. In fact, however, Uranus and Neptune have average densities (1320 kg/m^3 and 1640 kg/m^3, respectively) that are comparable to or greater than those of Jupiter (1330 kg/m^3) or Saturn (690 kg/m^3). Therefore, we can conclude that Uranus and Neptune contain greater proportions of the heavier elements than do Jupiter or Saturn.

This picture is not what we might expect. According to our discussion in Section 8-4 of how the solar system formed, hydrogen and helium should be relatively more abundant as distance from the vaporizing heat of the Sun increases. But Uranus and Neptune contain a greater percentage of heavy elements, and, therefore, a *smaller* percentage of hydrogen and helium, than Jupiter and Saturn.

A related problem is explaining the masses of Uranus and Neptune. At the locations of Uranus (19 AU from the Sun) and Neptune (30 AU from the Sun) the solar nebula was probably so sparse that these planets would have taken tens of hundreds of millions of years to grow to their present sizes around a core of icy planetesimals. But protoplanetary disks observed around other stars (see Section 8-3) do not appear to survive for that long before they dissipate. In other words, the problem is not how to explain why Uranus and Neptune are smaller than Jupiter and Saturn; it is how to explain why Uranus and Neptune should exist at all!

Models of How Uranus and Neptune Formed

In one model for the origin of Uranus and Neptune, the planets formed in denser regions of the solar nebula between 4 and 10 AU from the Sun. In this region they could have grown rapidly to their present sizes. Had the planets remained at these locations, they could eventually have accumulated enough hydrogen and helium to become as large as Jupiter or Saturn. But before that could happen, gravitational interactions with Jupiter and Saturn would have pushed Uranus and Neptune outward into their present large orbits. Because the solar nebula was very sparse at those greater distances, Uranus and Neptune stopped growing and remained at the sizes we see today.

> Explaining the origins of Uranus and Neptune poses a challenge to planetary scientists

An alternative idea is that Uranus and Neptune formed in their present locations, not around a core of planetesimals but directly from the gases of the solar nebula. (We introduced this idea in Section 8-4.) Once a ball of gas formed, icy particles within the ball would have settled to its center, forming a solid core. This

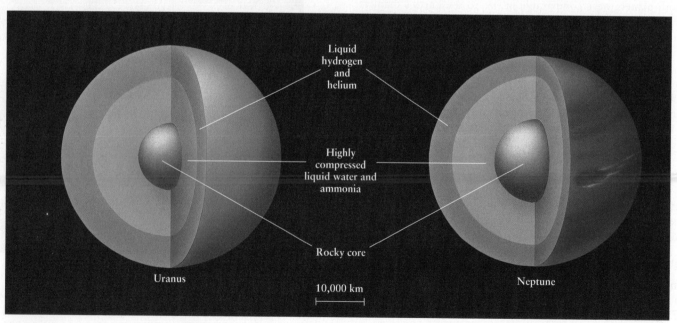

Figure 14-7

The Internal Structures of Uranus and Neptune In the model shown here, both Uranus and Neptune have a rocky core, resembling a terrestrial planet; a mantle of liquid water with ammonia dissolved in it; and an outer layer of liquid molecular hydrogen and liquid helium. The atmospheres are very thin shells on top of the hydrogen-helium layer. Uranus and Neptune have about the same diameter, but since Neptune is more massive it may have a somewhat larger core.

process could have formed Uranus and Neptune in only a few hundred years. In this model, Uranus and Neptune would actually have ended up larger than they are today, with substantially more hydrogen and helium. But if there happened to be a hot star relatively near the solar nebula, it would have emitted substantial amounts of ultraviolet radiation (see Section 5-3 and Figure 5-11). Ultraviolet photons have high energy (see Section 5-5), and a shower of them could have stripped Uranus and Neptune of many of their lightweight hydrogen and helium atoms, leaving shrunken planets with a larger percentage of heavy elements. (In this picture, Jupiter and Saturn were protected from the star's ultraviolet radiation because they were shrouded by the thicker clouds of material in their part of the solar nebula.) More research will be needed to better determine the origins of Uranus and Neptune.

Figure 14-7 shows one model for the present-day internal structures of Uranus and Neptune. (Compare this to the internal structures of Jupiter and Saturn, shown in Figure 12-12.) In this model each planet has a rocky core roughly the size of Earth, although more massive. Each planet's core is surrounded by a mantle of liquid water and ammonia. (This means that the mantle is chemically similar to household window cleaning fluid.) Around the mantle is a layer of liquid molecular hydrogen and liquid helium, with a small percentage of liquid methane. This layer is relatively shallow compared to those on Jupiter and Saturn, and the pressure is not high enough to convert the liquid hydrogen into liquid metallic hydrogen.

14-5 The magnetic fields of both Uranus and Neptune are oriented at unusual angles

Astronomers were quite surprised by the data sent back from *Voyager 2*'s magnetometer as the spacecraft sped past Uranus and Neptune. These data, as well as radio emissions from charged particles in their magnetospheres, showed that the magnetic fields of both Uranus and Neptune are tilted at steep angles to their axes of rotation. Uranus's **magnetic axis**, the line connecting its north and south magnetic poles, is inclined by 59° from its axis of rotation; Neptune's magnetic axis is tilted by 47°. By contrast, the magnetic axes of Earth, Jupiter, and Saturn are all nearly aligned with their rotation axes; the angle between their magnetic and rotation axes is 12° or less (Figure 14-8). Scientists were also surprised to find that the magnetic fields of Uranus and Neptune are offset from the centers of the planets.

CAUTION! The drawing of the Earth's magnetic field at the far left of Figure 14-8 may seem to be mislabeled, because it shows the *south* pole of a magnet at the Earth's *north* pole. But, in fact, this is correct, as you can understand by thinking about how magnets work. On a magnet that is free to swivel, like the magnetized needle in a compass, the "north pole" is called that because it tends to point north on Earth. Likewise, a compass needle's south pole points toward the south on Earth. Furthermore, opposite magnetic poles attract. If you take two magnets and put them next to each other, they try to align themselves so that one magnet's north pole is next to the other magnet's south pole. Now, if you think of a compass needle as one magnet and the entire Earth as the other magnet, it makes sense that the compass needle's north pole is being drawn toward a magnetic *south* pole—which happens to be located near the Earth's geographic north pole. The Earth's magnetic pole nearest its geographic North Pole is called the "magnetic north pole." Note that the magnets drawn inside Jupiter, Saturn, and Neptune are oriented opposite to the Earth's. On any of these planets, the north pole of a compass needle would point south, not north!

Explaining Misaligned Magnetism

Why are the magnetic axes and axes of rotation of Uranus and Neptune so badly misaligned? And why are the magnetic fields offset from the centers of the planets? One possibility is that their

Figure 14-8

The Magnetic Fields of Five Planets This illustration shows how the magnetic fields of the Earth, Jupiter, Saturn, Uranus, and Neptune are tilted relative to their rotation axes (shown in red). For both Uranus and Neptune, the magnetic fields are offset from the planet's center and steeply inclined to the rotation axis.

magnetic fields might be undergoing a reversal; geological data show that the Earth's magnetic field has switched north to south and back again many times in the past. Another possibility is that the misalignments resulted from catastrophic collisions with planet-sized bodies. As we discussed in Section 14-2, the tilt of Uranus's rotation axis and its system of moons suggest that such collisions occurred long ago. As we will see in Section 14-7, Neptune may have gravitationally captured its largest moon, Triton, but no one knows if that incident was responsible for offsetting Neptune's magnetic axis.

Because neither Uranus nor Neptune is massive enough to compress hydrogen to a metallic state, their magnetic fields cannot be generated in the same way as those of Jupiter and Saturn. Instead, under the high pressures found in the watery mantles of Uranus and Neptune, dissolved molecules such as ammonia lose one or more electrons and become electrically charged (that is, they become *ionized*; see Section 5-8). Water is a good conductor of electricity when it has such electrically charged molecules dissolved in it, and electric currents in this fluid are probably the source of the magnetic fields of Uranus and Neptune.

> **Unlike any other planets, Uranus and Neptune have magnetic fields caused by ionized ammonia**

The *Cosmic Connections* figure summarizes the key properties of Uranus and Neptune and how they compare with those of Jupiter and Saturn.

14-6 Uranus and Neptune each has a system of thin, dark rings

On March 10, 1977, Uranus was scheduled to move in front of a faint star, as seen from the Indian Ocean. A team of astronomers headed by James L. Elliot of Cornell University observed this event, called an **occultation,** from a NASA airplane equipped with a telescope. They hoped that by measuring how long the star was hidden, they could accurately measure Uranus's size. In addition, by carefully measuring how the starlight faded when Uranus passed in front of the star, they planned to deduce important properties of Uranus's upper atmosphere.

The Surprising Rings of Uranus

To everyone's surprise, the background star briefly blinked on and off several times just before the star passed behind Uranus and again immediately after (Figure 14-9). The astronomers

Figure 14-9

How the Rings of Uranus Were Discovered As seen from Earth, Uranus occasionally appears to move in front of a distant star. Such an event is called an occultation. The star's light is completely blocked when it is behind the planet. But before and after the occultation by Uranus, the starlight dims temporarily as the rings pass in front of the star.

COSMIC CONNECTIONS

The Outer Planets: A Comparison

Uranus and Neptune are not simply smaller versions of Jupiter and Saturn. This table summarizes the key differences among the four Jovian planets.

	Interior	Surface	Rings	Atmosphere	Magnetic Field
Jupiter	Terrestrial core, liquid metallic hydrogen shell, liquid hydrogen mantle	No solid surface, atmosphere gradually thickens to liquid state, belt and zone structure, hurricanelike features	Yes	Primarily H, He	19,000× Earth's total field; at its cloud layer, 14× stronger than Earth's surface field
Saturn	Similar to Jupiter, with bigger terrestrial core and less metallic hydrogen	No solid surface, less distinct belt and zone structure than Jupiter	Yes	Primarily H, He	570× Earth's total field; at its cloud layer, $^2/_3×$ Earth's surface field
Uranus	Terrestrial core, liquid water shell, liquid hydrogen and helium mantle	No solid surface, weak belt and zone system, hurricanelike features, color from methane absorption of red, orange, yellow	Yes	Primarily H, He, some CH_4	50× Earth's total field; at its cloud layer, 0.7× Earth's surface field
Neptune	Similar to Uranus	Like Uranus	Yes	Primarily H, He, some CH_4	35× Earth's total field; at its cloud layer, 0.4× Earth's surface field

For detailed comparisons between planets, see Appendices 1 and 2.
* To see the orientations of these magnetic fields relative to the rotation axes of the planets, see Figure 14-8.

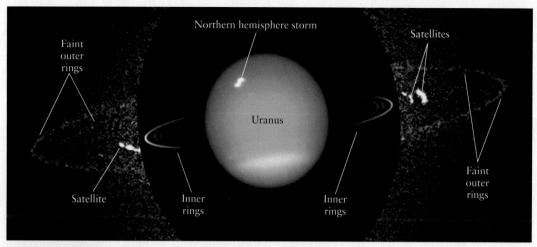

Figure 14-10 R I **V** U X G

Uranus's Rings from the Hubble Space Telescope Several Hubble Space Telescope images were assembled into this view of Uranus's rings. Relatively short exposure times reveal the planet and the inner rings, while much longer exposure times were needed to reveal the outer rings.

During these long exposures several of Uranus's satellites moved noticeably, leaving bright trails on the image. (NASA; ESA; and M. Showalter, SETI Institute)

concluded that Uranus must be surrounded by a series of nine narrow rings. In addition to these nine rings, *Voyager 2* discovered two others that lie even closer to the cloudtops of Uranus. Two other extremely faint rings, much larger in diameter than any of the others, were found in 2005 using the Hubble Space Telescope (Figure 14-10).

> **Like the planet itself, the rings of Uranus were discovered by accident**

Unlike Saturn's rings, the rings of Uranus are dark and narrow: Most are less than 10 km wide. Typical particles in Saturn's rings are chunks of reflective ice the size of snowballs (a few centimeters across), but typical particles in Uranus's rings are 0.1 to 10 meters in size and are no more reflective than lumps of coal. It is not surprising that these narrow, dark rings escaped detection for so long. Figure 14-11 is a *Voyager 2* image of the rings from the side of the planet away from the Sun, where light scattering from exceptionally small ring particles makes them more visible. (We discussed light scattering in Section 5-6 and Box 5-4.) Figure 14-10 shows the Sun-facing side of the rings.

The Uranian rings are so dark because sunlight at Uranus is only one-quarter as intense as at Saturn. As a result, Uranus's ring particles are so cold that they can retain methane ice. Scientists speculate that eons of impacts by electrons trapped in the magnetospheres of the planets have converted this methane ice into dark carbon compounds. This **radiation darkening** can account for the low reflectivity of the rings.

Uranus's major rings are located less than 2 Uranian radii from the planet's center, well within the planet's Roche limit (see Section 12-9). Some sort of mechanism, possibly one involving shepherd satellites, efficiently confines particles to their narrow orbits. (In Section 12-11 we discussed how shepherd satellites help keep planetary rings narrow.) *Voyager 2* searched for shepherd satellites but found only two. The others may be so small and dark that they have simply escaped detection.

The two faint rings discovered in 2005 (see Figure 14-10) are different: They lie well *outside* Uranus's Roche limit. The outer of these two rings owes its existence to a miniature satellite called Mab (named for a diminutive fairy in English folklore) that orbits within this ring. Meteorites colliding with Mab knock dust off the satellite's surface. Since Mab is so small (just 36 km in diameter) and hence has little gravity, the ejected dust escapes from

Figure 14-11 R I **V** U X G

The Shaded Side of Uranus's Rings This view from *Voyager 2*, taken when the spacecraft was in Uranus's shadow, looks back toward the Sun. Numerous fine-grained dust particles gleam in the spaces between the main rings. This dust is probably debris from collisions between larger particles in the main rings. The short horizontal streaks are star images blurred by the spacecraft's motion during the exposure. (NASA/JPL)

In 2002 and 2003, a team led by Imke de Pater of the University of California, Berkeley, used the 10-meter Keck telescope in Hawaii to observe in detail the rings of Neptune. Remarkably, they found that all of the rings had become fainter since the *Voyager 2* flyby in 1989. Apparently, the rings are losing particles faster then new ones are being added. If the rings continue to decay at the same rate, one of them may vanish completely within a century. More research into the nature of Neptune's rings is needed to explain this curious and rapid decay.

14-7 Some of Uranus's satellites show evidence of past tidal heating

Before the *Voyager 2* flyby of Uranus, five moderate-sized satellites—Titania, Oberon, Ariel, Umbriel, and Miranda—were known to orbit the planet (Figure 14-13). Most are named after sprites and spirits in Shakespeare's plays. They range in diameter from about 1600 km (1000 mi) for Titania and Oberon to less than 500 km (300 mi) for Miranda. All these moons have average densities around 1500 kg/m³, which is consistent with a mixture of ice and rock.

Voyager 2 discovered 11 other small Uranian satellites, most of which have diameters of less than 100 km (60 mi); 11 more were found using ground-based telescopes and the Hubble Space Telescope between 1997 and 2003. Only Jupiter and Saturn have more known satellites than Uranus. Figure 14-14 shows several of the smaller satellites. Appendix 3 summarizes information about all 27 known moons of Uranus.

Uranus's Unusual Moderate-Sized Satellites

Umbriel and Oberon both appear to be geologically dead worlds, with surfaces dominated by impact craters. By contrast, Ariel's surface appears to have been cracked at some time in the past, allowing some sort of ice lava to flood low-lying areas. A similar process appears to have taken place on Titania. This geologic activity may be due to a combination of the satellites' internal heat and tidal heating like that which powers volcanism on Jupiter's satellite Io (see Section 13-4). Io's tidal heating is only possible because of the 1:2:4 ratio of the orbital periods of Io, Europa, and Ganymede. While there are no such simple ratios between the

Figure 14-12   R I V U X G

Neptune's Rings The two main rings of Neptune and a faint, inner ring can easily be seen in this composite of two *Voyager 2* images. There is also a faint sheet of particles whose outer edge is located between the two main rings and extends inward toward the planet. The overexposed image of Neptune itself has been blocked out. (NASA)

the satellite and goes into orbit around Uranus, forming a ring. The inner of the two faint rings may be formed in the same way; however, no small satellite has yet been found within this ring.

The Decaying Rings of Neptune

Like Uranus, Neptune is surrounded by a system of thin, dark rings that were first detected in stellar occultations. Figure 14-12 is a *Voyager 2* image of Neptune's rings. The ring particles vary in size from a few micrometers (1 μm = 10^{-6} m) to about 10 meters. As for Uranus's rings, the particles that make up Neptune's rings reflect very little light because they have undergone radiation darkening.

Miranda   Ariel   Umbriel   Titania   Oberon

Figure 14-13   R I V U X G

Uranus's Principal Satellites This "family portrait" (a montage of five *Voyager 2* images) shows Uranus's five largest moons to the same scale and correctly displays their respective reflectivities. (The darkest satellite, Umbriel, is actually more reflective than the Earth's Moon.) All five satellites have grayish surfaces, with only slight variations in color. (NASA/JPL)

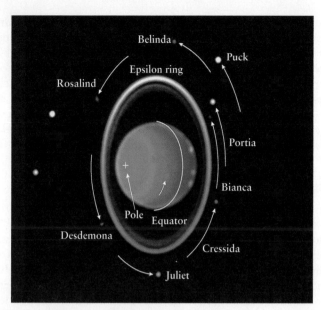

Figure 14-14   R [I] V U X G

Uranus's Rings and Small Satellites This false-color infrared image from the Hubble Space Telescope shows eight of Uranus's satellites, all of which were discovered by *Voyager 2* when it flew past Uranus in 1986. They all lie within 86,000 km of the planet's center (only about one-fifth of the distance from the Earth to our Moon) and all are less than 160 km (100 mi) in diameter. The arcs show how far each satellite moves around its orbit in 90 minutes. (Erich Karkoschka, University of Arizona; and NASA)

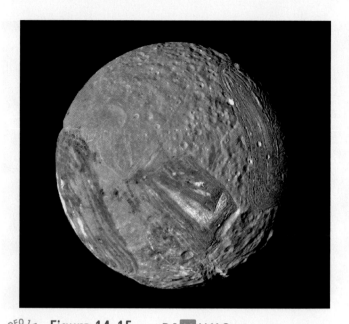

Figure 14-15   R I [V] U X G

Miranda This composite of *Voyager 2* images shows that part of Miranda's surface is ancient and heavily cratered, while other parts are dominated by parallel networks of valleys and ridges. At the very bottom of the image—where a "bite" seems to have been taken out of Miranda—is a range of enormous cliffs that jut upward to an elevation of 20 km, twice as high as Mount Everest. (NASA/JPL)

present-day orbital periods of Uranus's satellites, there may have been in the past. If so, tidal heating could have helped reshape the surfaces of some of the satellites.

Unique among Uranus's satellites is Miranda, which has a landscape unlike that of any other world in the solar system. Much of the surface is heavily cratered, as we would expect for a satellite only 470 km in diameter, but several regions have unusual and dramatic topography (Figure 14-15). Detailed analysis of Miranda's geology suggests that this satellite's orbital period was once in a whole-number ratio with that of more massive Umbriel or Ariel. The resulting tidal heating melted Miranda's interior, causing dense rocks in some locations on the surface to settle toward the satellite's center as blocks of less dense ice were forced upward toward the surface, thus creating Miranda's resurfaced terrain. Tidal heating must have ceased before this process could run its full course, which explains why some ancient, heavily cratered regions remain on Miranda's surface.

> **Uranus's satellite Miranda has cliffs twice as high as Mount Everest**

Uranus's Perplexing Small Satellites

While Uranus's small satellites are unlikely to show the kind of geology found in Miranda, some of them move in curious orbits. Eight of the nine small satellites in large orbits beyond Oberon are in retrograde orbits: They go around Uranus oppositely to the direction in which Uranus rotates. Many of the small outer satellites of Jupiter and Saturn have retrograde orbits and are proba-

bly captured asteroids (see Section 13-9 and Section 13-10); the same is probably true of the outer satellites of Uranus.

The 13 satellites that orbit closest to Uranus (inside the orbit of Miranda) are all in prograde orbits, so they travel around Uranus in the same direction as the planet rotates. However, when Mark Showalter and Jack Lissauer compared 11 of these satellites' orbits in 2005 with their orbits in 1994, they found surprisingly large differences. (The other two inner satellites were only discovered in 2003, so it is not known whether their orbits underwent similar changes.) These satellites can pass within a few thousand to a few hundred kilometers from each other, so they can exert strong gravitational forces on each other. Over time these forces can modify their orbits. The magnitude of the orbital changes is such that the inner satellites will begin colliding with each other within a few million years. If these inner satellites are in such unstable orbits, is it possible that they have been in orbit around Uranus since it formed more than 4.5 billion years ago? Or did Uranus somehow acquire these satellites in the relatively recent past? More research will be needed to answer these questions.

14-8 Neptune's satellite Triton is an icy world with a young surface and a tenuous atmosphere

Neptune has 13 known satellites, listed in Appendix 3. They are named for mythological beings related to bodies of water. (Neptune itself is named for the Roman god of the sea.) Most of these worlds are small, icy bodies, probably similar to the smaller satellites of Uranus. The one striking exception is Triton, Neptune's largest satellite (see Table 7-2). In many ways Triton is quite unlike any other world in the solar system.

Triton's Backward Orbit and Young Surface

Like many of the small satellites of Jupiter, Saturn, Uranus, and Neptune, Triton is in a retrograde orbit, so that it moves around Neptune oppositely to the direction of Neptune's rotation. Furthermore, the plane of Triton's orbit is inclined by 23° from the plane of Neptune's equator. It is difficult to imagine how a satellite might form out of the same cloud of material as a planet and end up orbiting in a direction opposite the planet's rotation and in such a tilted plane. Hence, Triton probably formed elsewhere in the solar system, collided long ago with a now-vanished satellite of Neptune, and was captured by Neptune's gravity. With a diameter of 2706 km, a bit smaller than our Moon but much larger than any other satellite in a retrograde orbit, Triton is certainly the largest captured satellite.

Figure 14-16 shows the icy, reflective surface of Triton as imaged by *Voyager 2*. There is a conspicuous absence of large craters, which immediately tells us that Triton has a young surface on which the scars of ancient impacts have largely been erased by tectonic activity. There are areas that resemble frozen lakes and may be the calderas of extinct ice volcanoes. Some of Triton's surface features resemble the long cracks seen on Europa (Section 13-6) and Ganymede (Section 13-7). Still other features are unique to Triton. For example, in the upper portion of Figure 14-16 you can see dimpled, wrinkled terrain that resembles the skin of a cantaloupe.

Triton's tectonically active history is probably related to its having been captured into orbit around Neptune. After its capture, Triton most likely started off in a highly elliptical orbit, but today the satellite's orbit is quite circular. The satellite's original elliptical orbit would have been made circular by tidal forces exerted on Triton by Neptune's gravity. These tidal forces would also have stretched and flexed Triton, causing enough tidal heating to melt much of the interior. The resulting volcanic activity (with lavas made of ice rather than molten rock) would have obliterated Triton's original surface features, including craters.

There may still be warmth in Triton's interior today. *Voyager 2* observed plumes of dark material being ejected from the surface to a height of 8 km (5 mi). These plumes may have been generated from a hot spot far below Triton's surface, similar to geysers on the Earth. Alternatively, the energy source for the plumes may be sunlight that warms the surface, producing subsurface pockets of gas and creating fissures in the icy surface through which the gas can escape.

Triton's surface temperature is only 38 K (−235°C = −391°F), the lowest of any world yet visited by spacecraft. This temperature is low enough to solidify nitrogen, and indeed the spectrum of sunlight reflected from Triton's surface shows absorption lines due to nitrogen ice as well as methane ice. But Triton is also warm enough to allow some nitrogen to evaporate from the surface, like the steam that rises from ice cubes when you first take them out of the freezer. *Voyager 2* confirmed that Triton has a very thin nitrogen atmosphere with a surface pressure of only 1.6×10^{-5} atmosphere, about the same as at an altitude of 100 km above the Earth's surface. Despite its thinness, Triton's atmosphere has noticeable effects. *Voyager 2* saw areas on Triton's surface where dark material has been blown downwind by a steady breeze (see Figure 14-16). It also observed dark material ejected from the geyserlike plumes being carried as far as 150 km by high-altitude winds.

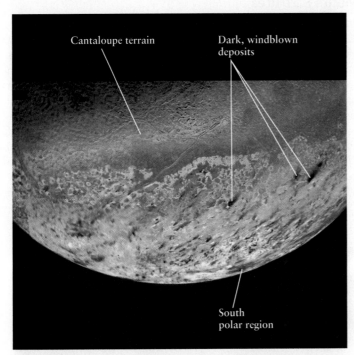

Cantaloupe terrain

Dark, windblown deposits

South polar region

Figure 14-16   R I **V** U X G

Triton Several high-resolution *Voyager 2* images were combined to create this mosaic. The pinkish material surrounding Triton's south polar region is probably nitrogen frost. Some of this presumably evaporates when summer comes to the south pole; the northward flow of the evaporated gas may cause the dark surface markings. Farther north is a brown area of "cantaloupe terrain" that resembles the skin of a melon. (NASA/JPL)

Tidal Forces and Triton's Doom

Just as tidal forces presumably played a large role in Triton's history, they also determine its future. Triton raises a tidal bulge on Neptune, just as our Moon distorts the Earth (recall Figure 10-17). In the case of the Earth-Moon system, the gravitational pull of the Earth's tidal bulge causes the Moon to spiral away from the Earth. But because Triton's orbit is retrograde, the tidal bulge on Neptune exerts a force on Triton that makes the satellite slow down rather than speed up. (In Figure 10-17, imagine that the moon is orbiting toward the bottom of the figure rather than toward the top.) This is causing Triton to spiral gradually in toward Neptune. In approximately 100 million years, Triton will be inside Neptune's Roche limit, and the satellite will eventually be torn to pieces by tidal forces. When this happens, the planet will develop

> **Within 100 million years Triton will be torn apart by Neptune's tidal forces**

a spectacular ring system—overshadowing by far its present-day set of narrow rings—as rock fragments gradually spread out along Triton's former orbit.

Prior to *Voyager 2*, only one other satellite was known to orbit Neptune. Nereid, which was first sighted in 1949, is in a prograde orbit. Hence, it orbits Neptune in the direction opposite to Triton. Nereid also has the most eccentric orbit of any satellite in the solar system; its distance from Neptune varies from 1.4 million to 9.7 million kilometers. One possible explanation is that when Triton was captured by Neptune's gravity, the interplay of gravitational forces exerted on Nereid by both Neptune and Triton moved Nereid from a relatively circular orbit (like those of Neptune's other, smaller moons) into its present elliptical one.

14-9 Pluto is smaller than any planet

Speculations about worlds beyond Neptune date back to the late 1800s, when a few astronomers suggested that Neptune's orbit was being perturbed by an unknown object. Encouraged by the fame of Adams and Le Verrier, several people set out to discover "Planet X." Two prosperous Boston gentlemen, William Pickering and Percival Lowell, were prominent in this effort. (Lowell also enthusiastically promoted the idea of canals on Mars; see Section 11-4.) Modern calculations show that there are, in fact, no unaccounted perturbations of Neptune's orbit. It is thus not surprising that no orbiting body was found at the positions predicted by Pickering, Lowell, and others. Yet the search continued.

The Discovery of Pluto

Before he died in 1916, Lowell urged that a special camera with a wide field of view be constructed to help search for Planet X. After many delays, the camera was finished in 1929 and installed at the Lowell Observatory in Flagstaff, Arizona, where a young astronomer, Clyde W. Tombaugh, had joined the staff to carry on the project. On February 18, 1930, Tombaugh finally discovered the long-sought object. It was disappointingly faint—a thousand times dimmer than the dimmest object visible with the naked eye and 250 times dimmer than Neptune at opposition—and presented no discernible disk. The new world was named for Pluto, the mythological god of the underworld; fittingly, this name has Percival Lowell's initials as its first two letters. The discovery was publicly announced on March 13, 1930, the 149th anniversary of the discovery of Uranus. The two photographs in **Figure 14-17** show one day's motion of Pluto.

Pluto's orbit about the Sun is more elliptical (eccentricity 0.2501) and more steeply inclined to the plane of the ecliptic (17.15°) than the orbit of any of the planets. In fact, Pluto's orbit is so eccentric that it is sometimes closer to the Sun than Neptune. This was the case from 1979 until 1999; indeed, when Pluto was at perihelion in 1989, it was more than 10^8 km closer to the Sun than was Neptune. Happily, the orbits of Neptune and Pluto are such that the two worlds will never collide.

Pluto is so far away that it subtends only a very small angle of 0.15 arcsec. Hence, it is extraordinarily difficult to resolve any of its surface features. But by observing Pluto with the Hubble Space Telescope over the course of a solar day on Pluto (6.3872 Earth days) and using computer image processing, Alan Stern and Marc Buie generated the maps of Pluto's surface shown in **Figure 14-18**. These maps show bright polar ice caps as well as regions of different reflectivity near the planet's equator. Observations of Pluto's rotation confirm that the planet's rotation axis is tipped by more than 90°, so that Pluto has retrograde rotation like Uranus.

Pluto's Satellites

In 1978, while examining some photographs of Pluto, James W. Christy of the U.S. Naval Observatory noticed that the image of the planet on a photographic plate appeared slightly elongated, as though Pluto had a lump on one side. He promptly examined a number of other photographs of Pluto, and found a series that showed the lump moving clockwise around Pluto with a period of about 6 days.

Christy concluded that the lump was actually a satellite of Pluto. He proposed that the newly discovered moon be christened Charon (pronounced KAR-en), after the mythical boatman who ferried souls across the River Styx to Hades, the domain ruled by Pluto. (Christy also chose the name because of its similarity to Charlene, his wife's name.) The average distance between Charon

Day 1: Pluto appears between these two stars.

Day 2: Pluto has moved relative to the stars.

Figure 14-17   R I **V** U X G

Pluto's Motion across the Sky Pluto was discovered in 1930 by searching for a dim, starlike object that moves slowly in relation to the background stars. These two photographs were taken one day apart. Even when its apparent motion on the celestial sphere is fastest, Pluto moves only about 1 arcmin per day relative to the stars. (Lick Observatory)

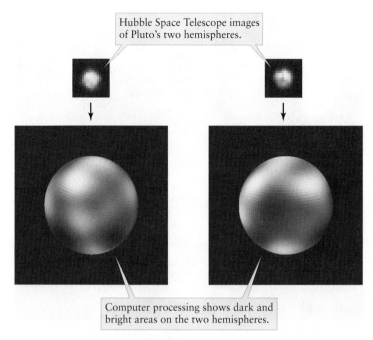

Hubble Space Telescope images of Pluto's two hemispheres.

Computer processing shows dark and bright areas on the two hemispheres.

Figure 14-18   R I V U X G

Pluto's Surface The bright regions at the top and bottom of each of Pluto's hemispheres may be polar ice caps. The bright regions nearer the equator may be impact basins where more reflective subsurface ice has been exposed. (Alan Stern, Southwest Research Institute; Marc Buie, Lowell Observatory; NASA; and ESA)

and Charon added together. The brightness variations during eclipses suggest that Charon has a bright southern polar cap.

The average densities of Pluto and Charon, at about 2000 kg/m^3, are essentially the same as that of Triton (2070 kg/m^3). All three worlds are therefore probably composed of a mixture of rock and ice, as we might expect for small bodies that formed in the cold outer reaches of the solar nebula.

Pluto's spectrum shows absorption lines of various solid ices that cover the planet's surface, including nitrogen (N_2), methane (CH_4), and carbon monoxide (CO). Stellar occultation measurements have shown that Pluto, like Triton, has a very thin atmosphere. Exposed to a daytime temperature of around 40 K, N_2 and CO ices turn to gas more easily than frozen CH_4. For this reason, most of Pluto's tenuous atmosphere probably consists of N_2 and CO. Charon's weaker gravity has apparently allowed most of the N_2, CH_4, and CO to escape into space; only water ice is found on its surface.

The Origin of Pluto's Satellites

Pluto and Charon are remarkably like each other in mass, size, and density. Throughout the rest of the solar system, planets are many times larger and more massive than any of their satellites.

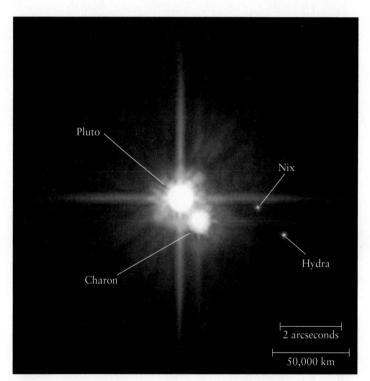

Pluto

Nix

Hydra

Charon

2 arcseconds

50,000 km

and Pluto is a scant 19,640 km, less than 5% of the distance between the Earth and our Moon. The best pictures of Pluto and Charon have been obtained using the Hubble Space Telescope (HST). In 2005 a team of astronomers using HST discovered two additional, smaller satellites that orbit Pluto at two to three times the distance of Charon (**Figure 14-19**).

Observations show that Charon's orbital period of 6.3872 days is the same as the rotational period of Pluto *and* the rotational period of Charon. In other words, both Pluto and Charon rotate synchronously with their orbital motion, and so both always keep the same face toward each other. As seen from the Charon-facing side of Pluto, the satellite remains perpetually suspended above the horizon. Likewise, Pluto neither rises nor sets as seen from Charon.

> **Pluto and its satellite Charon both rotate in lockstep, always keeping the same faces toward each other**

Soon after the discovery of Charon, astronomers witnessed an alignment that occurs only once every 124 years. From 1985 through 1990, Charon's orbital plane appeared nearly edge-on as seen from the Earth, allowing astronomers to view mutual eclipses of Pluto and its moon. As the bodies passed in front of each other, their combined brightness diminished in ways that revealed their sizes and surface characteristics. Data from the eclipses combined with subsequent measurements give Pluto's diameter as about 2274 km and Charon's as about 1190 km. For comparison, our Moon's diameter (3476 km) nearly equals the diameters of Pluto

VIDEO 14.7   VIDEO 14.8   **Figure 14-19**   R I V U X G

Pluto and Its Satellites This false-color Hubble Space Telescope shows Pluto (discovered 1930), its large satellite Charon (discovered 1978), and the two small satellites Nix and Hydra discovered in 2005. Nix is named for the mythological goddess of the night who was the mother of Charon; Hydra in mythology was a nine-headed, poisonous serpent who guarded the entrance to Hades. (NASA; ESA; H. Weaver, JHU/APL; A. Stern, SwRI; and the HST Pluto Companion Search Team)

The exceptional similarities between Pluto and Charon suggest to some astronomers that this binary system probably formed when Pluto collided with a similar body. Perhaps chunks of matter were stripped from the second body, leaving behind a mass, now called Charon, that was captured into orbit by Pluto's gravity. This same collision probably also left behind the small satellites Nix and Hydra shown in Figure 14-19.

In support of this model is the observation that all three satellites have almost the same color and reflection spectrum, suggesting a common origin. Pluto, by contrast, has a redder color than its satellites, which is consistent with the idea that Pluto and the satellites formed in different ways.

For this scenario to be feasible, there must have been many Plutolike objects in the outer regions of the solar system. Astronomers estimate that in order for a collision or close encounter between two of them to have occurred at least once since the solar system formed some 4.56 billion years ago, there must have been at least a thousand Plutos. As we will see, astronomers have begun to discover this population of Plutolike objects in the dark recesses of the solar system beyond Neptune.

14-10 Pluto is just one of thousands of icy objects that orbit the Sun beyond Neptune

For many years astronomers attempted to find other worlds beyond Neptune using the same technique used to discover Pluto (see Figure 14-17). The first to succeed were David Jewitt and Jane Luu, who in 1992 found an object with a semimajor axis of 42 AU and a diameter estimated to be only 240 km. This object, named 1992 QB1, has a reddish color like Pluto, possibly because frozen methane has been degraded by eons of radiation exposure. As of this writing (2006), more than 1100 more **trans-Neptunian objects**—icy worlds whose orbits have semimajor axes larger than that of Neptune—have been discovered. Most of these are relatively small like 1992 QB1, but a number are larger than Charon and one is larger than Pluto itself (**Figure 14-20**). Thus, Pluto is by no means unique: We can best regard it as a particularly large (but not the largest) trans-Neptunian object rather than a planet.

With improvements in the sensitivity of telescopes, new objects beyond Neptune are being discovered at a rapid pace; 67 new trans-Neptunian objects were found in 2005 alone. Based on these observations, it is thought that there could be 35,000 or more objects beyond Neptune with diameters greater than 100 km. One of the larger objects could have collided with Pluto in the past, giving rise to Pluto's retinue of satellites (see Section 14-9). At least 29 trans-Neptunian objects have satellites, which suggests that such collisions have taken place many times.

The Kuiper Belt and Beyond

Most of the trans-Neptunian objects lie within the **Kuiper belt,** which extends from about 30 to 50 AU from the Sun and is relatively close to the ecliptic. When the solar system was young, a large number of icy planetesimals formed in the region beyond Jupiter. Over time, the gravitational forces of the massive Jovian planets pushed most of these planetesimals beyond Neptune's orbit, concentrating them into a

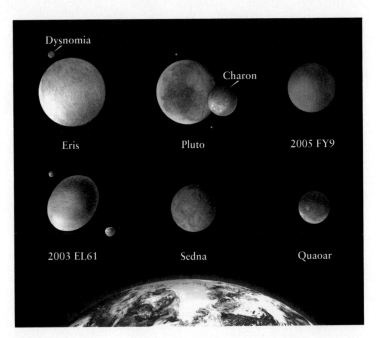

Figure 14-20

The Largest Trans-Neptunian Objects This artist's impression depicts the Earth and the largest objects known beyond Neptune (as of 2006) to the same scale. The largest of these, Eris, has a diameter of about 2900 km versus 2274 km for Pluto. Note the differences in color among these objects and that three of them (Eris, Pluto, and 2003 EL61) have satellites of their own. Table 7-4 in Section 7-5 lists the properties of these objects. (NASA; ESA; and A. Field, STScI)

belt centered on the plane of the ecliptic. Most of the trans-Neptunian objects within the belt are in orbits that are only slightly inclined to the ecliptic; these are thought to have formed beyond Neptune and to be in roughly their original orbits. Other objects such as 2005 FY9 and 2003 EL61 (see Figure 14-20 and Table 7-4) are in orbits that are inclined by about 30° from the ecliptic. These are thought to have been pushed into their steeply inclined orbits by gravitational interactions with Neptune.

The processes that gave rise to the Kuiper belt in our solar system also appear to have taken place around other stars. Figure 14-21 shows a disk of material surrounding the young star HD 139664. This disk has dimensions comparable to our Kuiper belt. A number of other young stars have been found with disks of this same type.

There are relatively few members of the Kuiper belt between the orbits of Neptune and Pluto. Remarkably, there are about 100 objects that have nearly the same semimajor axis as Pluto. These so-called **plutinos,** which include Pluto itself, have the property that they complete nearly two orbits around the Sun in the same time that Neptune completes three orbits. The plutinos thus experience rhythmic gravitational pulls from Neptune, and these pulls keep them

Neptune's gravity shapes the orbits of many icy worlds, including Pluto

in their orbits. (In Section 13-1 we saw a similar relationship among the orbital periods of Jupiter's satellites Io, Europa, and Ganymede, though the ratio of their orbital periods is 1:2:4 rather

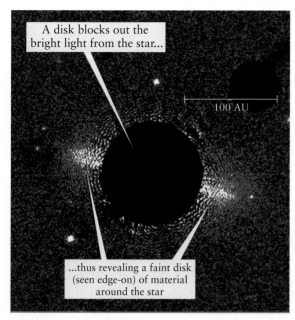

Figure 14-21 R I V U X G

A "Kuiper Belt" Around Another Star The star HD 139664 is only slightly more massive than the Sun but is thought to be just 300 million years old. This Hubble Space Telescope image shows a ring of material around HD 139664 that is similar in size to the Kuiper belt in our solar system. HD 139664 is 57 light-years (17 parsecs) from Earth in the constellation Lupus (the Wolf). (NASA; ESA; and P. Kalas/University of California, Berkeley)

than the 2:3 ratio for Neptune and the plutinos.) Most Kuiper belt objects orbit at distances beyond the plutinos but within about 50 AU from the Sun, at which distance an object completes one orbit for every two orbits of Neptune. At this distance the rhythmic gravitational forces of Neptune appear to pull small objects inward, giving the Kuiper belt a relatively sharp outer edge.

Two examples of trans-Neptunian objects that are not members of the Kuiper belt are shown in Figure 14-20. Eris, the largest known trans-Neptunian object, has a semimajor axis of more than 67 AU and an orbital eccentricity of 0.442. This orbit takes Eris from inside the orbit of Pluto to far beyond the Kuiper belt. An even more extreme case is Sedna: Its orbital semimajor axis is 489 AU, giving it an orbital period of more than 10,000 years, and the orbital eccentricity has the remarkably high value of 0.844. It is not well understood how Sedna could have been moved into such an immense and elongated orbit.

Excitement about the worlds beyond Neptune has motivated the development of a spacecraft called *New Horizons*. Launched in 2006, *New Horizons* swung by Jupiter in 2007 and will make the first-ever flyby of Pluto and Charon in 2015. It will then be aimed toward one or two other targets in the Kuiper belt. The choice of targets will not be made until a year before the Pluto-Charon flyby, giving astronomers extra time to discover new members of the Kuiper belt that might

be suitable destinations. The high-resolution images and other data to be returned by *New Horizons* promise to revolutionize our understanding of these most remote members of the solar system.

Key Words

Kuiper belt, p. 370
magnetic axis, p. 361
occultation, p. 362

plutino, p. 370
radiation darkening, p. 364
trans-Neptunian object, p. 370

Key Ideas

Discovery of the Outer Planets: Uranus was discovered by chance, while Neptune was discovered at a location predicted by applying Newtonian mechanics. Pluto was discovered after a long search.

Atmospheres of Uranus and Neptune: Both Uranus and Neptune have atmospheres composed primarily of hydrogen, helium, and a few percent methane.

• Methane absorbs red light, giving Uranus and Neptune their greenish-blue color.

• No white ammonia clouds are seen on Uranus or Neptune. Presumably the low temperatures have caused almost all the ammonia to precipitate into the interiors of the planets. All of these planets' clouds are composed of methane.

• Much more cloud activity is seen on Neptune than on Uranus. This is because Uranus lacks a substantial internal heat source.

Interiors and Magnetic Fields of Uranus and Neptune: Both Uranus and Neptune may have a rocky core surrounded by a mantle of water and ammonia. Electric currents in these mantles may generate the magnetic fields of the planets.

• The magnetic axes of both Uranus and Neptune are steeply inclined from their axes of rotation. The magnetic and rotational axes of all the other planets are more nearly parallel. The magnetic fields of Uranus and Neptune are also offset from the centers of the planets.

Uranus's Unusual Rotation: Uranus's axis of rotation lies nearly in the plane of its orbit, producing greatly exaggerated seasonal changes on the planet.

• This unusual orientation may be the result of a collision with a planetlike object early in the history of our solar system. Such a collision could have knocked Uranus on its side.

Ring Systems of Uranus and Neptune: Uranus and Neptune are both surrounded by systems of thin, dark rings. The low reflectivity of the ring particles may be due to radiation-darkened methane ice.

Satellites of Uranus and Neptune: Uranus has five satellites similar to the moderate-sized moons of Saturn, plus at least 22 more small satellites. Neptune has 13 satellites, one of which (Triton) is comparable in size to our Moon or the Galilean satellites of Jupiter.

• Triton has a young, icy surface indicative of tectonic activity. The energy for this activity may have been provided by tidal heating that occurred when Triton was captured by Neptune's gravity into a retrograde orbit.

• Triton has a tenuous nitrogen atmosphere.

Worlds Beyond Neptune: Pluto and its moon, Charon, move together in a highly elliptical orbit steeply inclined to the plane of the ecliptic.

• More than a thousand icy worlds have been discovered beyond Neptune. Pluto and Charon are part of this population.

• Most trans-Neptunian objects lie in a band called the Kuiper belt that extends from 30 to 50 AU from the Sun. Neptune's gravity shapes the orbits of objects within the Kuiper belt.

Questions

Review Questions

1. Could astronomers in antiquity have seen Uranus? If so, why was it not recognized as a planet?

2. Why do you suppose that the discovery of Neptune is rated as one of the great triumphs of science, whereas the discoveries of Uranus and Pluto are not?

3. Why is it so difficult to see features in the atmosphere of Uranus?

4. (a) Draw a figure like Figure 14-3, and indicate on it where Uranus was in 1986 and 2004. Explain your reasoning. (*Hint:* See Figure 14-1 and Figure 14-2.) (b) In approximately what year will the Sun next be highest in the sky as seen from Uranus's south pole? Explain your reasoning.

5. Why do you suppose the tilt of Uranus's rotation axis was deduced from the orbits of its satellites and not by observing the rotation of the planet itself?

6. Describe the seasons on Uranus. In what ways are the Uranian seasons different from those on Earth?

7. Explain the statement "Methane is to Uranus's atmosphere as water is to Earth's atmosphere."

8. A number of storms in the Uranian atmosphere can be seen in Figure 14-2, but none are visible in Figure 14-1. How can you account for the difference?

9. Why are Uranus and Neptune distinctly blue-green in color, while Jupiter or Saturn are not?

10. How does the energy source for Uranus's atmospheric motions differ from those from Jupiter, Saturn, and Neptune?

11. Why are fewer white clouds seen on Uranus and Neptune than on Jupiter and Saturn?

12. Why do Uranus and Neptune have higher densities than Jupiter and Saturn?

13. Discuss some competing explanations of why Uranus and Neptune are substantially smaller than Jupiter and Saturn.

14. How do the orientations of Uranus's and Neptune's magnetic axes differ from those of other planets?

15. Briefly describe the evidence supporting the idea that Uranus was struck by a large planetlike object several billion years ago.

16. Compare the rings that surround Jupiter, Saturn, Uranus, and Neptune. Briefly discuss their similarities and differences.

17. The 1977 occultation that led to the discovery of Uranus's rings was visible from the Indian Ocean. Explain why it could not be seen from other parts of the Earth's night side.

18. As *Voyager 2* flew past Uranus, it produced images only of the southern hemispheres of the planet's satellites. Why do you suppose this was?

19. Why do astronomers think that the energy needed to resurface parts of Miranda came from tidal heating rather than the satellite's own internal heat?

20. Using the data in Appendix 3, explain why Uranus's satellites Caliban and Sycorax (both discovered in 1997) were probably captured from space rather than having formed at the same time as the planet itself.

21. Briefly describe the evidence supporting the idea that Triton was captured by Neptune.

22. If you were floating in a balloon in Neptune's upper atmosphere, in what part of the sky would you see Triton rise? Explain your reasoning.

23. Why is it reasonable to suppose that Neptune will someday be surrounded by a broad system of rings, perhaps similar to those that surround Saturn?

24. How can astronomers distinguish a faint solar system object like Pluto from background stars within the same field of view?

25. What is the evidence that the three moons of Pluto have a common origin?

26. How do the presence of Pluto's moons suggest that there must be other worlds beyond Neptune?

27. Why are there a large number of objects with the same semimajor axis as Pluto?

28. Are there any trans-Neptunian objects that are not members of the Kuiper belt? Are there any members of the Kuiper belt that are not trans-Neptunian objects? Explain.

Advanced Questions

Questions preceded by an asterisk (*) involve topics discussed in Box 1-1.

Problem-solving tips and tools

See Box 1-1 for the small-angle formula. Recall that the volume of a sphere of radius r is $4\pi r^3/3$. Section 4-7 discusses the gravitational force between two objects and Newton's form of Kepler's third law. You will find discussions of the original form of Kepler's third law in Section 4-4, tidal forces in Section 4-8, Wien's law for blackbody radiation in Section 5-4, and the transparency of the Earth's atmosphere to various wavelengths of light in Section 6-7.

---

29. For which configuration of the Sun, Uranus, and Neptune is the gravitational force of Neptune on Uranus at a maximum? For this configuration, calculate the gravitational force exerted by the Sun on Uranus and by Neptune on Uranus. Then

calculate the fraction by which the sunward gravitational pull on Uranus is reduced by Neptune at that configuration. Based on your calculations, do you expect that Neptune has a relatively large or relatively small effect on Uranus's orbit?

30. At certain points in its orbit, a stellar occultation by Uranus would *not* reveal the existence of the rings. What points are those? How often does this circumstance arise? Explain using a diagram.

31. According to one model for the internal structure of Uranus, the rocky core and the surrounding shell of water and methane ices together make up 80% of the planet's mass. This interior region extends from the center of Uranus to about 70% of the planet's radius. (a) Find the average density of this interior region. (b) How does your answer to (a) compare with the average density of Uranus as a whole? Is this what you would expect? Why?

32. Uranus's epsilon (ε) ring has a radius of 51,150 km. (a) How long does it take a particle in the ε ring to make one complete orbit of Uranus? (b) If you were riding on one of the particles in the ε ring and watching a cloud near Uranus's equator, would the cloud appear to move eastward or westward as Uranus rotates? Explain your answer.

33. Suppose you were standing on Pluto. Describe the motions of Charon relative to the Sun, the stars, and your own horizon. Would you ever be able to see a total eclipse of the Sun? (*Hint:* You will need to calculate the angles subtended by Charon and by the Sun as seen by an observer on Pluto.) In what circumstances would you *never* see Charon?

34. It is thought that Pluto's tenuous atmosphere may become even thinner as the planet moves toward aphelion (which it will reach in 2113), then regain its present density as it again moves toward perihelion. Why should this be?

35. The brightness of sunlight is inversely proportional to the square of the distance from the Sun. For example, at a distance of 4 AU from the Sun, sunlight is only $(^1/_4)^2 = ^1/_{16} = 0.0625$ as bright as at 1 AU. Compared with the brightness of sunlight on the Earth, what is its brightness (a) on Pluto at perihelion (29.649 AU from the Sun) and (b) on Pluto at aphelion (49.425 AU from the Sun)? (c) How much brighter is it on Pluto at perihelion compared with aphelion? (Even this brightness is quite low. Noon on Pluto is about as dim as it is on the Earth a half hour after sunset on a moonless night.)

*36. If Earth-based telescopes can resolve angles down to 0.25 arcsec, how large could a trans-Neptunian object be at Pluto's average distance from the Sun and still not present a resolvable disk?

37. The observations of Pluto shown in Figure 14-19 were made using blue and ultraviolet light. What advantages does this have over observations made with red or infrared light?

*38. Calculate the maximum angular separation between Pluto and Charon as seen from Earth. Assume that Pluto is at perihelion (29.649 AU from the Sun) and that Pluto is at opposition as seen from Earth.

39. Presumably Pluto and Charon raise tidal bulges on each other. Explain why the average distance between Pluto and Charon is probably constant, rather than increasing like the Earth-Moon distance or decreasing like the Neptune-Triton distance. Include a diagram like Figure 10-17 as part of your answer.

40. (a) Find the semimajor axis of the orbit of an object whose period is 3/2 of the orbital period of Neptune. How does your result compare to the semimajor axis of Pluto's orbit? (b) A number of Kuiper belt objects called plutinos have been discovered with the same orbital period and hence the same semimajor axis as Pluto. Explain how these objects can avoid colliding with Pluto.

41. Find the semimajor axis of the orbit of an object whose period is twice the orbital period of Neptune. How does your result compare to the outer limit of the Kuiper belt?

42. Suppose you wanted to search for trans-Neptunian objects. Why might it be advantageous to do your observations at infrared rather than visible wavelengths? (*Hint:* At visible wavelengths, the light we see from planets is reflected sunlight. At what wavelengths would you expect distant planets to *emit* their own light most strongly? Use Wien's law to calculate the wavelength range best suited for your search.) Could such observations be done at an observatory on the Earth's surface? Explain.

43. The *New Horizons* spacecraft will swing by Jupiter to get a boost from that planet's gravity, enabling it to reach Pluto relatively quickly. To see what would happen if this technique were not used, consider a spacecraft trajectory that is an elliptical orbit around the Sun. The perihelion of this orbit is at 1 AU from the Sun (at the Earth) and the aphelion is at 30 AU (at Pluto's position). Calculate how long it would take a spacecraft in this orbit to make the one-way trip from Earth to Pluto. Based on the information in Section 14-10, how much time is saved by making a swing by Jupiter instead?

Discussion Questions

44. Discuss the evidence presented by the outer planets that suggests that catastrophic impacts of planetlike objects occurred during the early history of our solar system.

45. Some scientists are discussing the possibility of placing spacecraft in orbit about Uranus and Neptune. What kinds of data should be collected, and what questions would you like to see answered by these missions?

46. If Triton had been formed along with Neptune rather than having been captured, would you expect it to be in a prograde or retrograde orbit? Would you expect the satellite to show signs of tectonic activity? Explain your answers.

47. Would you expect the surfaces of Pluto and Charon to be heavily cratered? Explain why or why not.

48. Imagine that you are in charge of planning the *New Horizons* flyby of Pluto and Charon. In your opinion, what data should be collected and what kinds of questions should the mission attempt to answer?

49. In 2006 the International Astronomical Union changed Pluto's designation from *planet* to *dwarf planet*. One criterion that Pluto failed to meet was that a planet must have "cleared the neighborhood" around its orbit. In what sense has Pluto not done so? In what sense have the eight planets (Mercury

through Uranus) cleared their neighborhoods? Do you agree with this criterion?

Web/eBook Questions

50. **Miranda.** Access and view the video "Uranus's Moon Miranda" in Chapter 14 of the *Universe* Web site or eBook. Discuss some of the challenges that would be involved in launching a spacecraft from Earth to land on the surface of Miranda.

51. Charon was discovered by an astronomer at the U.S. Naval Observatory. Why do you suppose the U.S. Navy carries out work in astronomy? Search the World Wide Web for the answer.

52. Search the World Wide Web for a list of trans-Neptunian objects. What are the largest and smallest objects of this sort that have so far been found, and how large are they? Have any objects larger than Eris been found?

53. **Separation of Pluto and Charon.** Pluto is located about 4.5 billion km from Earth and has a maximum observable separation from Charon of about 0.9 arcseconds. Access the AIMM (Active Integrated Media Module) called "Small-Angle Toolbox" in Chapter 1 of the *Universe* Web site or eBook. Use this AIMM and the above data to determine the distance between Pluto and Charon. How does your answer compare with the value given in the text?

Activities

Observing Projects

> **Observing tips and tools**
>
> During the period 2007–2011, Uranus will be at opposition in September, Neptune in August, and Pluto in June. You can find Uranus and Neptune with binoculars if you know where to look (a good star chart is essential), but Pluto is so dim that it can be a challenge to spot even with a 25-cm (10-inch) telescope. Each year, star charts that enable you to find these planets are printed in the issue of *Sky & Telescope* for the month in which each planet is first visible in the nighttime sky. You can also locate the outer planets using the *Starry Night Enthusiast™* program on the CD-ROM that accompanies certain printed copies of this textbook.

54. Make arrangements to view Uranus through a telescope. The planet is best seen at or near opposition. Use a star chart at the telescope to find the planet. Are you certain that you have found Uranus? Can you see a disk? What is its color?

55. If you have access to a large telescope, make arrangements to view Neptune. Like Uranus, Neptune is best seen at or near opposition and can most easily be found using a star chart. Can you see a disk? What is its color?

56. If you have access to a large telescope (at least 25 cm in diameter), make arrangements to view Pluto. Using the star chart from *Sky & Telescope* referred to above, view the part

of the sky where Pluto is expected to be seen and make a careful sketch of all of the stars that you see. Repeat this process on a later night. Can you identify the "star" that has moved?

57. Use the *Starry Night Enthusiast™* program to observe the five large satellites of Uranus. Open the **Favourites pane** and click on **Guides > Atlas** to display the entire celestial sphere. Open the **Find** pane and double-click the entry for Uranus to center this planet in the view. (Clicking once on the **Space** bar will speed up this centering). You can reduce the confusion in this view by removing the background stars by clicking on **View > Stars > Stars** and by ensuring that the celestial grid is removed by clicking on **View > Celestial Grid.** Using the controls at the right-hand end of the toolbar, zoom in to a field of view of about $2' \times 1'$. In the toolbar, set the year to 1986 and the **Time Flow Rate** to 1 hour. Then click on the **Run Time Forward** button, the right-pointing triangle on the toolbar. You can scroll on and off the labels for the moons by clicking on **Labels > Planets-Moons.** (a) Describe how the satellites move, and relate your observations to Kepler's third law (see Sections 4-4 and 4-7). (b) Set the year to 2007 and again click on the **Run Time Forward** button. How do the orbits look different than in (a)? Explain any differences.

58. Use the *Starry Night Enthusiast™* program to examine the satellites of Uranus. (a) Select **Solar System > Uranus** from the **Favourites** menu. Remove the image of the astronaut's spacesuit by clicking on **View > Feet** in the menu and remove the background stars by selecting **View > Stars > Stars** from the menu. Use the Elevation buttons in the **Viewing Location** section of the toolbar to change the distance from the planet to about 0.004 AU. You should now be able to see at least five satellites of the planet Uranus. Which satellites are these? Select **Label > Planets-Moons** from the menu to confirm your identification of these satellites. (b) You can rotate the image of the planet and its moons by holding down the **Shift** key while clicking the mouse button and moving the mouse. Use this technique to rotate Uranus until you are looking at the plane of the satellites' orbits edge-on. Do all of the satellites appear to lie in the same plane? (To display the orbits of each of the moons, open the **Find** pane, expand the layer for Uranus, and click in the right-hand box next to each moon.) How do you imagine that this plane relates to the plane of Uranus's equator? Why do you suspect that this is so?

59. Use the *Starry Night Enthusiast™* program to observe Neptune and its satellites Triton and Nereid. Display the entire celestial sphere (select **Guides > Atlas** in the **Favourites** menu). Open the **Find** pane and double-click on the entry for Neptune to center this planet in the view. In the Find Pane, expand the layer for Neptune and then click the box to the left of the entries for Triton and Nereid to label them in the view. To reduce confusion, remove the background stars and the coordinate grid by selecting **Stars > Stars** and **> Celestial Grid** under the View menu. Use the controls at the right-hand end of the toolbar to zoom in to a field of view of about $9' \times 6'$. In the toolbar, set the Time Flow Rate to 1 hour and click on the **Run**

Time Forward button (a triangle that points to the right). (a) The satellite you see orbiting close to Neptune is Triton. In which direction (clockwise or counterclockwise) does it orbit Neptune? (b) Nereid is in a rather elongated orbit that takes it far from Neptune. You can use another viewpoint to show the relative positions and orientations of the orbits of these moons more clearly. In the **Options** menu, open the **Viewing Location** pane and, in the **View from** boxes, select **position hovering over** and **Neptune**. On the map that appears, you will note that you will be hovering over Neptune's equator. To choose this location, click on **Set Location**. Increase your elevation to **0.05 a.u.** above Neptune by clicking and holding the **up** arrow in the **Viewing Location**. Again, you can remove the background stars by clicking on **View > Stars > Stars**. Open the **Find** pane and expand the list of moons under Neptune. You can now label **Triton** and **Nereid** and display their orbits by clicking in the boxes to the left and the right of their names respectively. Set the Time Flow Rate to 1 day and click on the **Run Time Forward** button. In which direction (clockwise or counterclockwise) does Nereid orbit Neptune? When it is closest to Neptune, does Nereid approach closer to Uranus than does Triton? You can view these orbits from different perspectives by holding down the Shift key while moving the mouse over Neptune to tilt this planetary system to different angles. (Note: You are at a point directly above a moon's orbit when its orbital track appears uniformly bright.) You can reduce the elevation above Neptune and set the Time Flow Rate to 10 minutes to show the planet's rotation and see the different clouds and dark spots on its surface. What is the planet's rotation period?

60. Use the *Starry Night Enthusiast*™ program to observe Pluto and Charon. First select **Options > Viewing Location** from the menu. At the top of the Viewing Location dialog box, select **position hovering over** and **Pluto** in the drop boxes. Then click on the center of the map of Pluto that appears in the dialog window and click the **Set Location** button. Use the elevation buttons in the toolbar to increase the distance from the surface of Pluto to about **35,000 km**. Use the **Location scroller** (hold down the Shift key while holding down the mouse button and moving the mouse, to rotate the view around Pluto. In the toolbar, set the **Time Flow Rate** to 1 hour, then click on the **Run Time Forward** button (a triangle that points to the right). (a) Estimate Charon's orbital period. (b) By following a spot on Pluto's surface, estimate Pluto's rotation period. How does it compare to your answer in part (a)? (c) Select **Options > Viewing Location** from the menu and set the dropdown boxes at the top of the Viewing Location dialog window to read **position moving with** and **Pluto**. Then select the **Above orbital plane** option and click the **Set Location** button. Open the **Find** pane and double-click the entry for the **Sun** to center the Sun in the view. Set the Time Flow Rate to 1 year, and click on the **Run Time Forward** button to see the apparent motion of the Sun as seen from Pluto. Observe the motion for several centuries of simulated time. Does the Sun always appear to move at the same speed? Use the properties of Pluto's orbit to explain why or why not.

Collaborative Exercise

61. Sir William Herschel, a British astronomer, discovered Uranus in 1781 and named it Georgium Sidus, after the reigning monarch, George III. What name might Uranus have been given in 1781 if an astronomer in your country had discovered it? Why? What if it had been discovered in your country in 1881? In 1991?

15

Vagabonds of the Solar System

R I V UXG

Comet Hale-Bopp, the great comet of 1997. (Courtesy of Johnny Horne)

On the Tuesday night after Easter in the year 1066, a strange new star appeared in the European sky. Seemingly trailing fire, this "star" hung in the night sky for weeks. Some, including the English king Harold, may have taken it as an ill omen; others, such as Harold's rival William, Duke of Normandy, may have regarded it as a sign of good fortune. Perhaps they were both right, because by the end of the year Harold was dead and William was installed on the English throne.

Neither man ever knew that he was actually seeing the trail of a city-sized ball of ice and dust, part of which evaporated as it rounded the Sun to produce a long tail. They were seeing a comet—remarkably, the same Comet Halley that last appeared in 1986 and will next be seen in 2061. Although we now understand that comets are natural phenomena rather than supernatural omens, they still have the power to awe and inspire us, as did Comet Hale-Bopp (shown here) in 1997.

While comets are made of ice and dust, asteroids are rocky bits of the solar nebula that never formed into a full-sized planet. Some asteroids break into fragments, as do some comets, and some of these fragments fall to the Earth as meteorites. On extraordinarily rare occasions, an entire asteroid strikes Earth. Such an asteroid impact may have led to the extinction of the dinosaurs some 65 million years ago. Thus, these minor members of our solar system can have major consequences for our planet.

15-1 A search for a planet between Mars and Jupiter led to the discovery of asteroids

After William Herschel's discovery of Uranus in 1781 (which we described in Section 14-1), many astronomers began to wonder if there were other, as yet undiscovered, planets. If these planets were too dim to be seen by the naked eye, they might still be visible through telescopes. These planets would presumably be found close to the plane of the ecliptic, because all the other planets orbit the Sun in or near that plane (see Section 7-1). But how far from the Sun might these additional planets be found?

The Hunt for the "Missing Planet"

 Astronomers in the late eighteenth century had a simple rule of thumb, called the *Titius-Bode Law*, relating the sizes of planetary orbits: From one planet to

Learning Goals

By reading the sections of this chapter, you will learn

15-1 What led to the discovery of the asteroids

15-2 Why the asteroids never formed into a planet

15-3 What asteroids look like

15-4 How an asteroid led to the demise of the dinosaurs

15-5 What meteorites tell us about the nature of asteroids

15-6 What meteorites may reveal about the origin of the solar system

15-7 What comets are and why they have tails

15-8 How comets originate from the outer solar system

15-9 The connection between comets and meteor showers

the next, the semimajor axis of the orbit increases by a factor between approximately 1.4 and 2. For example, the semimajor axis of Mercury's orbit is 0.39 AU. Venus's orbit has a semimajor axis of 0.72 AU, which is larger by a factor of (0.72)/(0.39) = 1.85. (Unlike Newton's laws, the Titius-Bode "Law" is not a fundamental law of nature. It is probably just a reflection of how our solar system happened to form.)

The glaring exception is Jupiter (semimajor axis 5.20 AU), which is more than *3 times* farther from the Sun than Mars (semimajor axis 1.52 AU). Table 7-1 depicts this large space between the orbits of Mars and Jupiter. If the rule of thumb is correct, astronomers reasoned, there should be a "missing planet" with an orbit about 1.4 to 2 times larger than that of Mars. This planet should therefore have a semimajor axis between 2 and 3 AU.

Six German astronomers, who jokingly called themselves the "Celestial Police," organized an international group to begin a careful search for this missing planet. Before their search got under way, however, surprising news reached them from Giuseppe Piazzi, a Sicilian astronomer. Piazzi had been carefully mapping faint stars in the constellation of Taurus when on January 1, 1801, the first night of the nineteenth century, he noticed a dim, previously uncharted star. This star's position shifted slightly over the next several nights. Suspecting that he might have found the "missing planet," Piazzi excitedly wrote to Johann Bode, the director of the Berlin Observatory and a member of the Celestial Police.

Unfortunately, Piazzi's letter did not reach Bode until late March. By that time, the Earth had moved around its orbit so that Piazzi's object appeared too near the Sun to be visible in the night sky. With no way of knowing where to look after it emerged from the Sun's glare, astronomers feared that Piazzi's object might have been lost.

Upon hearing of this dilemma, the brilliant young German mathematician Karl Friedrich Gauss took up the challenge. He developed a general method of computing an object's orbit from only three separate measurements of its position on the celestial sphere. (With slight modifications, this same method is used by astronomers today.) In November 1801, Gauss predicted that Piazzi's object would be found in the predawn sky in the constellation of Virgo. And indeed Piazzi's object was sighted again on December 31, 1801, only a short distance from the position Gauss had calculated. Piazzi named the object Ceres (pronounced SEE-reez), after the patron goddess of Sicily in Roman mythology.

Ceres orbits the Sun once every 4.6 years at an average distance of 2.77 AU, just where astronomers had expected to find the missing planet. But Ceres is very small; its equatorial diameter is a mere 974.6 km (Figure 15-1). (The pole-to-pole diameter is even smaller, just 909.4 km.) Hence, Ceres reflects only a little sunlight, which is why it cannot be seen with the naked eye even at opposition; it can be seen with binoculars, but it looks like just another faint star. Because of its small size, it is known today as a **minor planet,** or **asteroid.**

On March 28, 1802, Heinrich Olbers discovered another faint, starlike object that moved against the background stars. He called it Pallas, after the Greek goddess of wisdom. Like Ceres, Pallas orbits the Sun every 4.6 years at an average distance of 2.77 AU, but its orbit is more steeply inclined from the plane of

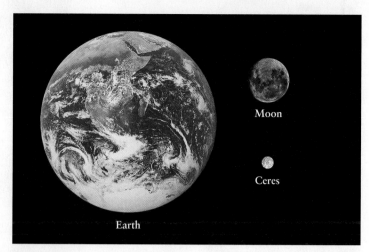

Figure 15-1   R I V U X G

Ceres Compared with the Earth and Moon A drawing of Ceres, the largest of the asteroids and the first one discovered, is shown here to the same scale as the Earth and the Moon. Ceres is too small to be considered a planet, and it cannot be regarded as a moon because it does not orbit any other body. To denote their status, asteroids are also called minor planets. (NASA)

the ecliptic and is somewhat more eccentric. Pallas is even smaller and dimmer than Ceres, with an estimated diameter of only 522 km. Obviously, Pallas was also not the missing planet.

Did the missing planet even exist? Some astronomers speculated that perhaps there had once been such a full-size planet, but it had somehow broken apart or exploded to produce a population of asteroids orbiting between Mars and Jupiter. The search was on to discover this population. Two more such asteroids were discovered in the next few years, Juno in 1804 and Vesta in 1807. Several hundred more were discovered beginning in the mid-1800s, by which time telescopic equipment and techniques had improved.

An Abundance of Asteroids

The next major breakthrough came in 1891, when the German astronomer Max Wolf began using photographic techniques to search for asteroids. Before this, asteroids had to be painstakingly discovered by scrutinizing the skies for faint, uncharted, starlike objects whose positions move slightly from one night to the next. With photography, an astronomer simply aims a camera-equipped telescope at the stars and takes a long exposure. If an asteroid happens to be in the field of view, its orbital motion during the long exposure leaves a distinctive trail on the photographic plate (Figure 15-2). Using this technique, Wolf alone discovered 228 asteroids.

Today, more than 300,000 asteroids have been discovered, of which 100,000 have been studied well enough so that their orbits are known. About 5000 more are discovered every month, some by amateur astronomers. After the discoverer reports a new find to the Minor Planet Center of the Smithsonian Astrophysical Observatory, the new asteroid is given

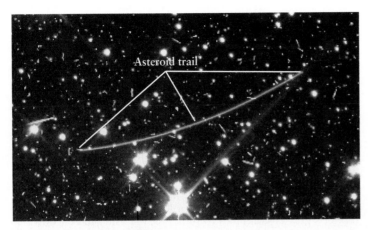

Figure 15-2   R I V̄ U X G

The Trail of an Asteroid Telescopes used to photograph the stars are motorized to follow the apparent motion of the celestial sphere. Because asteroids orbit the Sun, their positions change with respect to the stars, and they leave blurred trails on time exposure images of the stars. This bluish asteroid trail was recorded quite by accident by the Hubble Space Telescope. (R. Evans and K. Stapelfeldt, Jet Propulsion Laboratory; and NASA)

a provisional designation. For example, asteroid 1980 JE was the fifth ("E") to be discovered during the second half of May (the tenth half-month of the year, "J") 1980. The same scheme for provisional designations is also used for small objects found in the outer solar system beyond Neptune, or trans-Neptunian objects (see Section 14-10).

If the asteroid is located again on at least four succeeding oppositions—a process that can take decades—the asteroid is assigned an official sequential number (Ceres is 1, Pallas is 2, and so forth), and the discoverer is given the privilege of suggesting a name for the asteroid. The suggested name must then be approved by the International Astronomical Union. For example, the asteroid 1980 JE was officially named 3834 Zappafrank (after the American musician Frank Zappa) in 1994.

Ceres—or, in modern nomenclature, 1 Ceres—is unquestionably the largest asteroid; it has about 25% of the mass of all the asteroids combined. In recognition of its size, 1 Ceres is also called a **dwarf planet,** a designation that it shares with the large trans-Neptunian objects Eris and Pluto (see Section 14-10).

Besides 1 Ceres, only two asteroids—2 Pallas, and 4 Vesta—have diameters greater than 300 km. Thirty other asteroids have diameters between 200 and 300 km, and 200 more are bigger than 100 km across. The vast majority of asteroids are less than 1 km across.

CAUTION! Be careful not to confuse asteroids with the trans-Neptunian objects that we discussed in Section 14-10. Asteroids are found in the inner solar system, are made primarily of rocky materials, and are generally quite small. By contrast, trans-Neptunian objects are found primarily in the Kuiper belt beyond Neptune, are a mixture of ices and rock, and range in size from a few kilometers across to 2900 km in diameter—more than 3 times the diameter of 1 Ceres. The combined mass of all trans-Neptunian objects is not known, but is probably thousands of times greater than the combined mass of all asteroids. Asteroids make up only a tiny fraction of the total mass of the solar system!

The Asteroid Belt

Like Ceres, Pallas, Vesta, and Juno, most asteroids orbit the Sun at distances between 2 and 3.5 AU. This region of our solar system between the orbits of Mars and Jupiter is called the **asteroid belt** (Figure 15-3).

CAUTION! You may wonder how spacecraft such as *Voyager 1, Voyager 2, Galileo,* and *Cassini* were able to cross the asteroid belt to reach Jupiter or Saturn. Aren't asteroids hazards to space navigation, as they are sometimes depicted in science-fiction movies? Wouldn't these spacecraft have been likely to run into an asteroid? Happily, the answer to both questions is no! While it is true that there are more than 10^5 asteroids, they are spread over a belt with a total area (as seen from above the plane of the ecliptic) of about 10^{17} square

> **Science-fiction movies to the contrary, asteroids are not hazards to space navigation**

kilometers—about 10^8 times greater than the Earth's entire surface area. Hence, the average distance between asteroids in the ecliptic plane is about 10^6 kilometers, about twice the distance between the Earth and Moon. Furthermore, many asteroids have orbits that are tilted out of the ecliptic. The *Galileo* spacecraft did pass within a few thousand kilometers of two asteroids, but these passes were intentional and required that the spacecraft be carefully aimed.

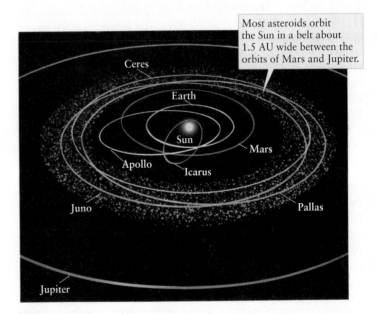

Most asteroids orbit the Sun in a belt about 1.5 AU wide between the orbits of Mars and Jupiter.

Figure 15-3

The Asteroid Belt Most asteroids, including the large asteroids 1 Ceres, 2 Pallas, and 3 Juno, have roughly circular orbits that lie within the asteroid belt. By contrast, some asteroids, such as 1862 Apollo and 1566 Icarus, have eccentric orbits that carry them inside the orbit of Earth.

What of the nineteenth-century notion of a missing planet? This idea has long since been discarded, because the combined matter of all the asteroids (including an estimate for those not yet officially known) would produce an object barely 1500 km in diameter. This is considerably smaller than anything that could be considered a missing planet. We now understand that asteroids are actually debris left over from the formation of the solar system.

15-2 Jupiter's gravity helped shape the asteroid belt

Where did the asteroid belt come from? Why did asteroids, rather than a planet, form in the solar nebula (see Section 8-5) in the region between Mars and Jupiter? Important insights into these questions come from supercomputer simulations of the formation of the terrestrial planets, like the one depicted in Figure 8-12 but more elaborate.

Jupiter and the Formation of the Asteroid Belt

A typical simulation starts off with a billion (10^9) or so planetesimals, each with a mass of 10^{15} kg or more, so that their combined mass equals that of the terrestrial planets. The computer then follows these planetesimals as they collide and accrete to form the planets. By selecting different speeds and positions for the planetesimals at the start of the simulation, a scientist can study alternative scenarios of the development of the inner solar system.

If the effects of Jupiter's gravity are not included in a simulation, an Earth-sized planet usually forms in the asteroid belt, giving us five terrestrial planets instead of four. But when Jupiter's gravity is added, this fifth terrestrial planet is less likely to form. Jupiter's strong pull "clears out" the asteroid belt by disrupting the orbits of planetesimals in the belt, ejecting most of them from the solar system altogether. As a result, the asteroid belt becomes depleted of planetesimals before a planet has a chance to form. The few planetesimals that remain in the simulation—only a small fraction of those that originally orbited between Mars and Jupiter—become the asteroids that we see today.

The asteroids were never part of a larger planet

Jupiter's gravitational influence, however, probably cannot explain why asteroid orbits have the wide variety of semimajor axes, eccentricities, and inclinations to the ecliptic shown in Figure 15-3. Even with the influence of Jupiter, collisions between planetesimals in the solar nebula would have left the asteroids in roughly circular orbits close to the ecliptic plane, like the orbits of the planets. Therefore, something else must have "stirred up" the asteroids to put them into their current state. Recent analyses suggest that one or more Mars-sized objects—that is, terrestrial planets intermediate in size between the Earth and the Moon— did in fact form within the asteroid belt. Planetesimals that passed within close range of these Mars-sized objects would have experienced strong gravitational forces, and these forces could have deflected the planetesimals into the eccentric or inclined orbits that many asteroids follow today.

Mars-sized objects would also have helped clear out the asteroid belt by deflecting planetesimals into orbits that crossed Jupiter's orbit. Once a planetesimal wandered close to Jupiter, it could be ejected from the solar system by the giant planet's gravitational force.

In this scheme, the Mars-sized objects disappeared when gravitational forces from Jupiter either accelerated them out of the solar system entirely or caused them to slow down and fall into the Sun. A Mars-sized object colliding with the Earth is thought to have produced the Moon (see Figure 10-18). Perhaps this object was an "escapee" from the asteroid belt.

Kirkwood Gaps

Jupiter's gravity continues to influence the asteroid belt down to the present day. The American astronomer Daniel Kirkwood found the first evidence for this in 1867, when he discovered gaps in the asteroid belt. These features, today called **Kirkwood gaps,** can best be seen in a histogram like **Figure 15-4**, which shows asteroid orbital periods. The gaps in this histogram show regions where there are relatively few asteroids. Curiously, these gaps occur for asteroid orbits whose periods are simple fractions (such as 1/3, 2/5, 3/7, and 1/2) of Jupiter's orbital period.

To understand why the Kirkwood gaps exist, imagine an asteroid within the belt that circles the Sun once every 5.93 years, exactly half of Jupiter's orbital period. On every second trip around the Sun, the asteroid finds itself lined up between Jupiter

Figure 15-4

The Kirkwood Gaps This graph displays the numbers of asteroids at various distances from the Sun. Very few asteroids are found in orbits whose orbital periods are equal to the simple fractions 1/3, 2/5, 3/7, and 1/2 of Jupiter's orbital period. The Trojan asteroids (at far right) follow the same orbit as Jupiter and thus have the same orbital period. We describe these in Section 15-4.

and the Sun, always at the same location and with the same orientation. Because of these repeated alignments, called a *2-to-1 resonance,* Jupiter's gravity deflects the asteroid from its original 5.93-year orbit, ultimately ejecting it from the asteroid belt. Similar resonances help clear out the Cassini division in Saturn's rings (see Section 12-11), drive the tidal heating of Jupiter's moon Io (see Section 13-4), and give an outer edge to the Kuiper belt beyond Neptune (see Section 14-10). According to Kepler's third law, a period of 5.93 years corresponds to a semimajor axis of 3.28 AU. Because of Jupiter's gravity, almost no asteroids orbit the Sun at this average distance.

Another Kirkwood gap corresponds to an orbital period of one-third Jupiter's period, or 3.95 years (a *3-to-1 resonance*). Additional gaps exist for other simple ratios between the periods of asteroids and Jupiter. Any Mars-sized bodies that existed in the asteroid belt were probably ejected after wandering into one of the Kirkwood gaps.

15-3 Astronomers use a variety of techniques to study asteroids

Although the asteroid belt is mostly empty space, collisions between asteroids should occur from time to time. Asteroids move in orbits with a variety of different eccentricities and inclinations (see Figure 15-3), and some of these orbits intersect each other. A collision between asteroids must be an awe-inspiring event. Typical collision speeds are estimated to be from 1 to 5 km/s (3600 to 18,000 km/h, or 2000 to 11,000 mi/h), which is more than sufficient to shatter rock. Recent observations show that these titanic impacts play an important role in determining the nature of asteroids.

> **Many asteroids have been battered by collisions with other asteroids**

Observing Asteroids with Telescopes, Radar, and Spacecraft

State-of-the-art optical telescopes have good enough resolution to reveal surface details on large asteroids such as 1 Ceres, 2 Pallas, and 4 Vesta (Figure 15-5). These asteroids resemble miniature terrestrial planets. They have enough mass, and hence enough gravity, to pull themselves into roughly spherical shapes. These large asteroids are likely to have extensively cratered surfaces due to collisions with other asteroids in the asteroid belt.

Optical telescopes reveal much less detail about the smaller asteroids. But by studying how the brightness of an asteroid changes as it rotates, astronomers can infer the asteroid's shape. An elongated asteroid appears dimmest when its long axis points toward the Sun so that it presents less of its surface to reflect sunlight. By contrast, a rotating spherical asteroid will have a more constant brightness. Such observations show that most smaller asteroids are not spherical at all. Instead, they retain the odd shapes produced by previous collisions with other asteroids.

An important recent innovation in studying asteroids has been the use of radar. An intense radio beam is sent toward an asteroid, and the reflected signal is detected and analyzed to determine

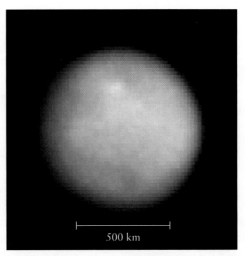

Figure 15-5   R I V U X G

A Hubble Space Telescope View of a Large Asteroid With an equatorial diameter of 974.6 km (605.6 mi), 1 Ceres is the largest of all the asteroids. Its very nearly spherical shape suggests that it was once molten throughout its volume, in which case it probably underwent chemical differentiation like the terrestrial planets (see Box 7-1 and Section 8-5). (NASA; ESA; J. Parker, SwRI; P. Thomas, Cornell U., L. McFadden, U. of Maryland; and M. Mutchler and Z. Levay, STScI)

the asteroid's shape (Figure 15-6). The amount of the radio signal that is reflected also gives clues about the texture of the surface.

The most powerful technique for studying asteroids is to send a spacecraft to make observations at close range. The Jupiter-bound *Galileo* spacecraft gave us our first close-up view of an asteroid

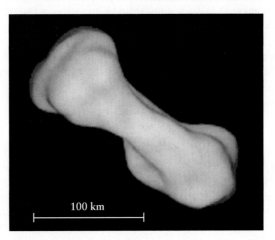

Figure 15-6   R I V U X G

A Radar View of a Medium-Sized Asteroid This radar image reveals the curious dog-bone shape of the asteroid 216 Kleopatra. The 300-m Arecibo radio telescope in Puerto Rico was used both to send radio waves toward Kleopatra and to detect the waves that were reflected back. Scientists then created this image by computer analysis of the reflected waves. (Arecibo Observatory, JPL/NASA)

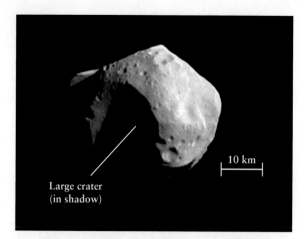

Figure 15-7 R I V̄ U X G

A Spacecraft View of a Small Asteroid The asteroid 253 Mathilde has an albedo of only 0.04, making it half as reflective as a charcoal briquette. It has an irregular shape, with dimensions 66 × 48 × 46 km. The partially shadowed crater indicated by the label is about 20 km across and 10 km deep. The *NEAR Shoemaker* spacecraft imaged Mathilde at close range during its close flyby in June 1997. (Johns Hopkins University, Applied Physics Laboratory)

in 1991, when it flew past 951 Gaspra. Two years later *Galileo* made a close pass by 243 Ida. The *NEAR Shoemaker* spacecraft made a close approach to 253 Mathilde in 1997 (**Figure 15-7**). It later went into orbit around 433 Eros (see Figure 7-7) and touched down on Eros in 2001. It was the first spacecraft to orbit and land on an asteroid. A number of other spacecraft have since made close flybys of asteroids.

One of the most ambitious asteroid missions to date is *Hayabusa* ("Falcon"), a project of the Japan Aerospace Exploration Agency. In 2005 it touched down briefly on the surface of 25143 Itokawa, where it was to have gathered small samples of the asteroid's surface. Scientists will not know whether this attempt was successful until *Hayabusa* returns to Earth in 2010. In 2007 the *Dawn* spacecraft was scheduled to be launched on a mission that would place it in orbit around 4 Vesta in 2011 and in orbit around 1 Ceres in 2015.

Collisions and the Nature of Asteroids

What have spacecraft and Earth-based radar told us about the asteroids? To appreciate the answer, we should first consider what scientists expected to find with these new tools. A few years ago, the consensus among planetary scientists was that the smaller asteroids were solid, rocky objects. The surfaces were presumed to be cratered by impacts, but only relatively small craters were expected. The reason is that an impact forceful enough to make a very large crater would probably fracture a rocky asteroid into two or more smaller asteroids.

The truth proves to be rather different. It is now suspected that most asteroids larger than about a kilometer are not entirely solid, but are actually composed of several pieces. An extreme example is Mathilde (see Figure 15-7). This asteroid has such a low density (1300 kg/m³) that it cannot be made of solid rock, which typically has densities of 2500 kg/m³ or higher. Instead, it is prob-

ably a "rubble pile" of small fragments that fit together loosely. Billions of years of impacts have totally shattered this asteroid. But the impacts were sufficiently gentle that the resulting fragments drifted away relatively slowly, only to be pulled back onto the asteroid by gravitational attraction.

Another surprising discovery is that some asteroids have extremely large craters, comparable in size to the asteroid itself. An example is the immense crater on Mathilde, shown in Figure 15-7. The object that formed this crater probably collided with Mathilde at a speed of 1 to 5 km/s (3600 to 18,000 km/h, or 2000 to 11,000 mi/h), which is more than sufficient to shatter rock. Why didn't this collision break the asteroid completely apart, sending the pieces flying in all directions? The explanation may be that the asteroid had already been broken into a loose collection of fragments before this major impact took place. Such a fragmented asteroid can more easily absorb the energy of a collision than can a solid, rocky one.

ANALOGY If you fire a rifle bullet at a wine glass, the glass will shatter into tiny pieces. But if you fire the same bullet into a bag full of sand (which has much the same chemical composition as glass), the main damage will be a hole in the bag. In a similar way, an asteroid that is a collection of small pieces can survive a major impact more easily than can a solid asteroid.

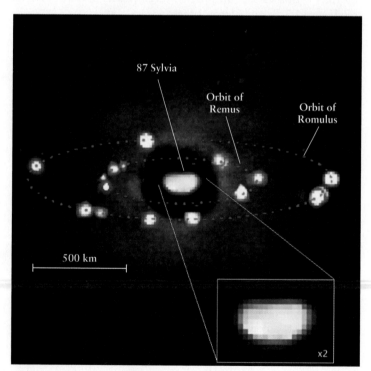

Figure 15-8 R I V U X G

An Asteroid with Two Satellites Although it is only about 280 km across, the potato-shaped asteroid 87 Sylvia has two satellites of its own. The outer satellite, Romulus, is 18 km across; the inner satellite, Remus, is just 7 km across. This image is actually a composite of 9 observations made with the adaptive optics system at the Yepun telescope of the European Southern Observatory (see Figure 6-17). (F. Marchis, U. of California; P. Descamps, D. Hestroffer, and J. Berthier, Observatoire de Paris; and ESO)

In some cases a collision that produces a rubble-pile asteroid can eject a piece that goes into orbit around the asteroid. This may explain why a handful of asteroids are known to have satellites. There is even one known case of an asteroid with two satellites, like a miniature planetary system (Figure 15-8).

If a collision is sufficiently energetic, it may shatter an asteroid permanently. Evidence for this was first pointed out in 1918 by the Japanese astronomer Kiyotsugu Hirayama. He drew attention to families of asteroids that share nearly identical orbits. Each of these groupings, now called **Hirayama families,** presumably resulted from a parent asteroid that was broken into fragments by a high-speed, high-energy collision with another asteroid.

15-4 Asteroids are found outside the asteroid belt—and have struck the Earth

While the majority of asteroids lie within the asteroid belt, there are some notable exceptions. A few actually share the same orbit as Jupiter. Other asteroids venture well inside the orbit of Mars, and can pose a serious threat to life on Earth.

The Lagrange Points and Trojan Asteroids

We saw in Section 15-3 that Jupiter's gravitational pull by itself depletes certain orbits in the asteroid belt, forming the Kirkwood gaps. But the gravitational forces of the Sun and Jupiter work to-gether to *capture* asteroids at two locations outside the asteroid belt called the **stable Lagrange points.** One of these points is one-sixth of the way around Jupiter's orbit ahead of the planet, while the other point is the same distance behind the planet (Figure 15-9a). The French mathematician Joseph Louis Lagrange predicted the existence of these points in 1772; asteroids were first discovered there in 1906.

The asteroids trapped at Jupiter's Lagrange points are called **Trojan asteroids,** named individually after heroes of the Trojan War. More than 1600 Trojan asteroids have been catalogued so far (see Figure 15-9b). (Four objects have also been found at the Lagrange points of Neptune. These are icy trans-Neptunian objects rather than rocky asteroids.)

Near-Earth Objects and Impacts from Space

Some asteroids have highly elliptical orbits that bring them into the inner regions of the solar system. Asteroids that cross Mars's orbit, or whose orbits lie completely within that of Mars, are called **near-Earth objects** or **NEOs.** These are shown in red in Figure 15-9b.

Occasionally, a near-Earth object may pass relatively close to the Earth. On October 30, 1937, Hermes passed the Earth at a distance of 900,000 km (560,000 mi), only a little more than twice the distance to the Moon. One of the closest calls to date occurred on December 9, 1994, when an asteroid called 1994 XM1 passed within a mere 105,000 km (60,000 mi) of Earth.

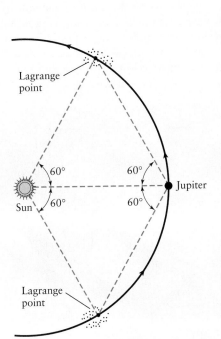

(a) A clump of asteroids is found at each of Jupiter's Lagrange points

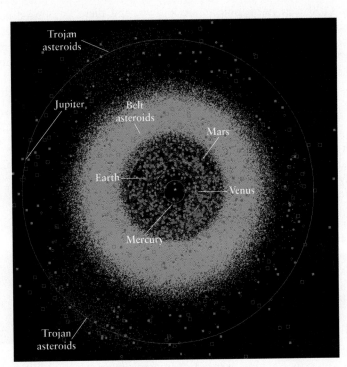

(b) A map of all asteroids within Jupiter's orbit

Figure 15-9

Asteroids Outside the Belt (a) The combined gravitational forces of Jupiter and the Sun tend to trap asteroids near the two stable Lagrange points along Jupiter's orbit. Note that the Sun, Jupiter, and either one of these points lie at the vertices of an equilateral triangle. **(b)** This plot shows the actual positions of all known asteroids at Jupiter's orbit or closer. Green dots denote belt asteroids, deep blue squares denote Trojan asteroids, and red circles denote asteroids that come within 1.3 AU of the Sun. Comets are filled and unfilled light-blue squares. (b: Minor Planet Center)

Figure 15-10   R I V U X G

WEB LINK 15.9

The Barringer Crater This 1.2-km-(³⁄₄-mi-) wide crater was formed by a meteoroid impact about 50,000 years ago. Fewer than 200 other impact craters between 3 and 150 km across have been identified on the Earth. None of these is more than 500 million years old, because the reshaping of the Earth's surface by erosion and by tectonic processes tends to erase craters over time. (Meteor Crater Enterprises)

Because 1994 XM1 is only about 10 meters across (about the size of a house), even if it had been aimed directly at the Earth it would probably have broken up in the atmosphere before reaching the surface.

More than 4100 near-Earth objects are known, and there may be thousands more that we have not yet detected. Fortunately, space is a big place and the Earth is a rather small target, so asteroids strike the Earth only very rarely.

However, collisions between asteroids produce numerous smaller chunks of rock, and many of these do eventually rain down on the terrestrial planets. Fortunately for us, the majority of these fragments, usually called **meteoroids,** are quite small. But on rare occasions a large fragment collides with our planet. When such a collision takes place, the result is an impact crater whose diameter depends on the mass and speed of the impinging object.

A relatively young, pristine terrestrial impact crater is the Barringer Crater near Winslow, Arizona (**Figure 15-10**). This crater was formed 50,000 years ago when an iron-rich object approximately 50 m across struck the ground at a speed of more than 11 km/s (40,000 km/h, or 25,000 mi/h). The resulting blast was equivalent to the detonation of a 20-megaton hydrogen bomb and left a crater 1.2 km (³⁄₄ mi) wide and 200 m (650 ft) deep at its center. (Figure 7-10*b* shows a much larger but much older crater in Canada.)

Iridium and the Demise of the Dinosaurs

Iron is an important constituent of asteroids and meteoroids. Another element, iridium, is common in iron-rich minerals but rare in other types of rocks. Measurements of iridium in the Earth's crust can thus tell us the rate at which meteoritic material has been deposited on the Earth over the ages. The geologist Walter Alvarez and his physicist father, Luis Alvarez, from the University of California, Berkeley, made such measurements in the late 1970s.

Working at a site of exposed marine limestone in the Apennine Mountains in Italy, the Alvarez team discovered an exceptionally high abundance of iridium in a dark-colored layer of clay between limestone strata (**Figure 15-11**). Since this discovery in 1979, a comparable layer of iridium-rich material has been uncovered at numerous sites around the world. In every case, geological dating reveals that this apparently worldwide layer of iridium-rich clay was deposited about 65 million years ago.

> An asteroid impact probably ended the age of dinosaurs 65 million years ago

Paleontologists were quick to realize the significance of this date, because it was 65 million years ago that all the dinosaurs rather suddenly became extinct. In fact, more than 75% of all the species on the Earth disappeared within a relatively brief span of time.

The Alvarez discovery suggests a startling explanation for this dramatic mass extinction: Perhaps an asteroid hit the Earth at that time. An asteroid 10 km in diameter slamming into our planet would have produced a fiery plume that ignited tremendous wildfires, killing untold numbers of animals and consuming much of the Earth's vegetation. The dust thrown up from the impact and soot from wildfires hung in the atmosphere, blackening the sky for months after the impact. The dust eventually drifted to the ground, forming a worldwide layer of iridium-rich clay. (Evidence for this picture is that a layer of soot 65 million years old is also found at various sites around the world. From the thickness of this layer, scientists calculate that the wildfires produced nearly 70 *billion* tons of soot.) Temperatures would have plummeted due to the blocking out of sunlight by dust and soot, only to increase later due to global warming brought on by greenhouse gases (see Section 9-1 and Section 9-7) released by the wildfires. The *Cosmic Connections* figure depicts the sequence of events that may have followed the asteroid impact.

10 cm

Figure 15-11   R I V U X G

Iridium-Rich Clay This photograph of strata in Italy's Apennine Mountains shows a dark-colored layer of iridium-rich clay. This is sandwiched between older layers of white limestone (lower right) and younger layers of grayish limestone (upper left). The iridium-rich layer is thought to be the result of an asteroid impact 65 million years ago. The coin is the size of a U.S. quarter. (Courtesy of W. Alvarez)

COSMIC CONNECTIONS — A Killer Asteroid

There is compelling evidence that 65 million years ago the Earth was struck by an asteroid about 10 km in diameter. This impact triggered a biological cataclysm that wiped out 75 percent of all plant and animal species on our planet. (From "The Day the World Burned" by D. A. Kring and D. D. Durda, *Scientific American*, December 2003; art by Chris Butler)

The Day Before

Late Cretaceous swamps and rivers in North America had a mix of coniferous, broad-leaved evergreen, and deciduous trees. They formed canopied forests and open woodlands with understories of ferns, aquatic plants and flowering shrubs.

Impact

The Chicxulub Impact occurred in a shallow sea and immediately lofted rocky, molten and vaporous debris into the atmosphere. The bulk of the debris rained down on nearby continental regions, but much of it rose all the way into space.

40 Minutes

The vapor-rich plume of material expanded to envelop Earth. As material in that plume fell back to the ground, it streaked through the atmosphere like trillions of meteors, heating it in some places by hundreds of degrees.

One Week

After fires had ravaged the landscape, only a few stark trunks and skeletons remained. Soot from the fires and dust from the impact slowly settled to the ground. Sunlight was dramatically, if not totally, attenuated for months.

One Year

The postimpact environment was less diverse. Ferns and algae were the first to recover. Plant species in swamps and swamp margins generally survived better than species in other types of ecosystems. Conifers fared particularly badly.

50 Years

Shrubs took advantage of the vacant landscape and began to cover it. Species pollinated by the wind did better than those that relied on insects. Trees began to grow, but it took years for forest canopies to rebuild. The recovery time is uncertain.

Tiny rodentlike creatures capable of ferreting out seeds and nuts were among the animals that managed to survive this catastrophe, setting the stage for the rise of mammals in the eons that followed. As the dominant mammals on the planet, we may owe our very existence to an ancient asteroid.

In 1992 a team of geologists suggested that this asteroid crashed into a site in Mexico. They based this conclusion on glassy debris and violently shocked grains of rock ejected from the 180-km-diameter Chicxulub Crater on the Yucatán Peninsula. Using radioactive dating techniques (see Section 8-3 and Box 8-1), the scientists found that the asteroid struck 64.98 million years ago, in remarkable agreement with the age of the iridium-rich layer of clay.

Some geologists and paleontologists are not convinced that an asteroid impact led to the extinction of the dinosaurs. But many scientists agree that this hypothesis fits the available evidence better than any other explanation that has been offered so far.

Tunguska and the Threat of Asteroid Impact

On June 30, 1908, a spectacular explosion occurred over the Tunguska region of Siberia that released about 10^{15} joules of energy, equivalent to a nuclear detonation of several hundred kilotons. The blast knocked a man off his porch some 60 km away and could be heard more than 1000 km away. Millions of tons of dust were injected into the atmosphere, darkening the sky as far away as California.

Preoccupied with political upheaval and World War I, Russia did not send a scientific expedition to the site until 1927. Researchers found that trees had been seared and felled radially outward in an area about 50 km in diameter (Figure 15-12). There was no clear evidence of a crater. In fact, the trees at "ground zero" were left standing upright, although they were completely stripped of branches and leaves.

Several teams of modern astronomers have argued that the Tunguska explosion was caused by an asteroid traveling at supersonic speed. They arrived at this conclusion after assessing the effects of various impactor sizes, speeds, and compositions. Even data from above-ground nuclear detonations of the 1940s and 1950s were worked into the calculations. The Tunguska event is well-matched by an asteroid about 80 m (260 ft) in diameter entering the Earth's atmosphere at 22 km/s (80,000 km/h = 50,000 mph).

Large objects, such as that which formed the Chicxulub Crater and may have led to the demise of the dinosaurs, occasionally strike the Earth with the destructive force of as much as a million megatons of TNT. An asteroid 1 or 2 kilometers in diameter, striking the Earth at 30 km/s, would destroy an area the size of California. Such an impact would throw more than 10^{13} kg of microscopic dust particles into the atmosphere, which would decrease the amount of sunlight reaching the Earth's surface by enough to threaten the health of the world's agriculture. The loss of a year's crops could lead to the demise of a quarter of the world's population and

Figure 15-12   R I $\boxed{\text{V}}$ U X G

Aftermath of the Tunguska Event In 1908 a stony asteroid with a mass of about 10^8 kilograms entered the Earth's atmosphere over the Tunguska region of Siberia. Its passage through the atmosphere made a blazing trail in the sky some 800 km long. The asteroid apparently exploded before reaching the surface, blowing down trees for hundreds of kilometers around "ground zero." (Courtesy of Sovfoto)

would place our civilization—although perhaps not the survival of our species—in grave jeopardy.

The good news is that such large objects strike the Earth *very* infrequently. An asteroid 1 km in diameter might be expected to hit the Earth only once every 300,000 years. Statistically, the probability that an average person will die because of such an event is about 1 in 20,000, or only about 1% as great as the probability of dying in an automobile accident. While the threat of asteroid or comet impact exists, a catastrophic impact is very unlikely to occur in our lifetimes.

15-5 Meteorites are classified as stones, stony irons, or irons, depending on their composition

Asteroids can only be studied at close range using spacecraft. But from time to time pieces of asteroids fall to Earth where they can be examined by scientists. These asteroid fragments are called *meteorites*.

Meteoroids, Meteors, and Meteorites

A *meteoroid*, like an asteroid, is a chunk of rock in space. There is no official dividing line between meteoroids and asteroids, but the term *asteroid* is generally applied only to objects larger than 50 m across.

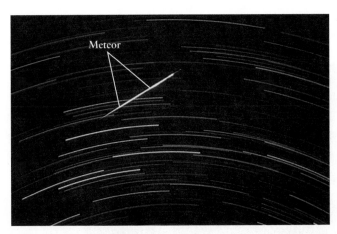

Figure 15-13   R I V U X G

A Meteor A meteor is produced when a meteoroid—a piece of interplanetary rock or dust—strikes the Earth's atmosphere at high speed (typically 5 to 30 km/s). The heat of air friction makes the meteoroid glow as it disintegrates. Most meteoroids burn up completely at altitudes of 100 km (60 mi) or so. This long exposure (notice the star trails) shows an exceptionally bright meteor called a fireball. (Courtesy of R. A. Oriti)

A **meteor** is the brief flash of light (sometimes called a *shooting star*) that is visible at night when a meteoroid enters the Earth's atmosphere (**Figure 15-13**). Frictional heat is generated as the meteoroid plunges through the atmosphere, leaving a fiery trail across the night sky.

WEB LINK 15.2   If a piece of rock survives its fiery descent through the atmosphere and reaches the ground, it is called a **meteorite**. As incredible as it may seem, an estimated total of 300 tons of extraterrestrial matter falls on the Earth each day. Most of this is dust rather than meteorites, however, and is hardly noticeable. It is fortunate that large meteorite falls are rather rare events, because these fast-moving rock fragments can cause substantial damage when they strike the ground (**Figure 15-14**). Because they are so rare, meteorites are prized by scientists and collectors alike. Almost all meteorites are asteroid fragments and thus provide information about the chemical composition of asteroids. A very few meteorites have been identified as pieces of the Moon or Mars that were blasted off the surfaces of those worlds by asteroid impacts and eventually landed on the Earth.

> Studying meteorites reveals the properties of the asteroids from which they came

People have been finding specimens of meteorites for thousands of years, and descriptions of them appear in ancient Chinese, Greek, and Roman literature. Many civilizations have regarded meteorites as objects of veneration. The sacred Black Stone, enshrined at the Ka'aba in the Great Mosque at Mecca, may be a relic of a meteorite impact.

The extraterrestrial origin of meteorites was hotly debated until as late as the eighteenth century. Upon hearing a lecture by two Yale professors, President Thomas Jefferson is said to have remarked, "I could more easily believe that two Yankee professors could lie than that stones could fall from Heaven." Although several falling meteorites had been widely witnessed and specimens from them had been collected, many scientists were reluctant to accept the idea that rocks could fall to the Earth from outer space. Conclusive evidence for the extraterrestrial origin of meteorites came on April 26, 1803, when fragments pelted the French town of L'Aigle. This event was analyzed by the noted physicist Jean-Baptiste Biot, and his conclusions helped finally to convince scientists that meteorites were indeed extraterrestrial.

Types of Meteorites

Meteorites are classified into three broad categories: stones, stony irons, and irons. As their name suggests, **stony meteorites,** or **stones,** look like ordinary rocks at first glance, but they are sometimes covered with a **fusion crust** (**Figure 15-15a**). This crust is produced by the melting of the meteorite's outer layers during its fiery descent through the atmosphere. When a stony meteorite is cut in two and polished, tiny flecks of iron can sometimes be found in the rock (Figure 15-15b).

Although stony meteorites account for about 95% of all the meteoritic material that falls to the Earth, they are the most difficult meteorite specimens to find. If they lie undiscovered and become exposed to the weather for a few years, they look almost indistinguishable from common terrestrial rocks. (The asteroid that caused the Tunguska explosion, described in Section 15-4, was probably of stony composition—which would account for the lack of obvious meteoritic debris on the ground.) Meteorites with high iron content can be found much more easily, because they can be located with a metal detector. Consequently, iron and stony iron meteorites dominate most museum collections.

Figure 15-14   R I V U X G

A Meteorite "Fender-Bender" On the evening of October 9, 1992, Michelle Knapp of Peekskill, New York, heard a noise from outside that sounded like a car crash. She discovered that the trunk of her car had been smashed by a 12-kilogram (27-pound) meteorite, which was lying beside the car and was still warm to the touch. Meteorite damage is not covered by automobile insurance, but Ms. Knapp was offered several tens of thousands of dollars for the meteorite (and the car). (R. A. Langheinrich Meteorites, Ilion, N.Y.)

Many stony meteorites are coated with dark fusion crusts...

(a)

...but when cut and polished they reveal tiny specks of iron in the rock.

(b)

Figure 15-15   R I V U X G

Stony Meteorites Of all meteorites that fall on the Earth, 95% are stones. **(a)** Many freshly discovered specimens, like the one shown here, are coated with dark fusion crusts. This particular stone fell in Texas.

(b) Some stony meteorites are found to contain tiny specks of iron mixed in the rock. This specimen was discovered in California. (From the collection of R. A. Oriti)

Stony iron meteorites consist of roughly equal amounts of rock and iron. Figure 15-16, for example, shows the mineral olivine suspended in a matrix of iron. Only about 1% of the meteorites that fall to the Earth are stony irons.

Iron meteorites (Figure 15-17), or **irons**, account for about 4% of the material that falls on the Earth. Iron meteorites usually contain no stone, but many contain from 10% to 20% nickel. Before humans began extracting iron from ores around 2000 B.C., the only source of the metal was from iron meteorites. Because these are so rare, iron was regarded as a precious metal like gold or silver.

Figure 15-16   R I V U X G

A Stony Iron Meteorite Stony irons account for about 1% of all the meteorites that fall to the Earth. This particular specimen, a variety of stony iron called a pallasite, fell in Chile. (Chip Clark)

In 1808, Count Alois von Widmanstätten, director of the Imperial Porcelain Works in Vienna, discovered a conclusive test for the most common type of iron meteorite. About 75% of all iron meteorites have a unique crystalline structure. These **Widmanstätten patterns** (pronounced VIT-mahn-shtetten) become visible when the meteorite is cut, polished, and briefly dipped into a dilute solution of acid (Figure 15-17b). Nickel-iron crystals of this type can form only if the molten metal cools slowly over many millions of years. Therefore, Widmanstätten patterns are never found in counterfeit meteorites, or "meteorwrongs."

Meteorites and the Early History of Asteroids

Widmanstätten patterns suggest that some asteroids were originally at least partly molten inside and remained partly molten long after they were formed. As soon as an asteroid accreted from planetesimals some 4.56 billion years ago, rapid decay of short-lived radioactive isotopes could have heated the asteroid's interior to temperatures above the melting point of rock. If the asteroid was large enough—at least 200 to 400 km in diameter—to insulate the interior and prevent cooling, the interior would have remained molten over the next few million years and chemical differentiation would have occurred (see Box 7-1 and Section 8-5). Iron and nickel would have sunk toward the asteroid's center, forcing less dense rock upward toward the asteroid's surface.

Like a terrestrial planet, such a **differentiated asteroid** would have a distinct core and crust. After a differentiated asteroid cooled and its core solidified, collisions with other asteroids would have fragmented the parent body into meteoroids. Iron meteorites are therefore specimens from a differentiated asteroid's core, while stony irons come from regions between a differentiated asteroid's core and its crust.

Unlike irons and stony irons, stony meteorites may or may not come from differentiated asteroids. Smaller asteroids would

Figure 15-17   R I V̄ U X G

Iron Meteorites (a) The surface of a typical iron is covered with small depressions caused by ablation (removal by melting) during its high-speed descent through the atmosphere. Note the one-cent coin for scale. (b) When cut, polished, and etched with a weak acid solution, most iron meteorites exhibit interlocking crystals in designs called Widmanstätten patterns. Both of these iron meteorites were found in Australia. (From the collection of R. A. Oriti)

Iron meteorites are composed of nickel-iron minerals and are characterized by a surface covered with depressions…

…and when cut and polished, by interlocking crystals in a Widmanstätten pattern.

(a)   (b)

not have retained their internal heat long enough for chemical differentiation to occur. Some stony meteorites are fragments of such **undifferentiated asteroids** (also called *primitive* asteroids), while others are relics of the crust of differentiated asteroids.

15-6 Some meteorites retain traces of the early solar system

A class of rare stony meteorites called **carbonaceous chondrites** shows no evidence of ever having been melted as part of asteroids. As suggested by their name, carbonaceous chondrites contain substantial amounts of carbon and carbon compounds, including complex organic molecules and as much as 20% water bound into the minerals. These compounds would have been broken down and the water driven out if these meteorites had been subjected to heating and melting. Carbonaceous chondrites may therefore be samples of the original material from which our solar system was created. The asteroid 253 Mathilde, shown in Figure 15-7, has a very dark gray color and the same sort of spectrum as a carbonaceous chondrite. It, too, is likely composed of material that predates the formation of the solar system.

Amino acids, the building blocks of proteins upon which terrestrial life is based, are among the organic compounds occasionally found inside carbonaceous chondrites. Interstellar organic material has thus probably been falling on our planet since its formation. Some scientists suspect that carbonaceous chondrites may have played a role in the origin of life on Earth.

Shortly after midnight on February 8, 1969, the night sky around Chihuahua, Mexico, was illuminated by a brilliant blue-white light moving across the heavens. As the light crossed the sky, it disintegrated in a spectacular, noisy explosion that dropped thousands of rocks and pebbles over terrified onlookers. Within hours, teams of scientists were on their way to collect specimens of a carbonaceous chondrite, collectively named the Allende meteorite after the locality (**Figure 15-18**).

Specimens were scattered in an elongated ellipse approximately 50 km long by 10 km wide. Most fragments were coated

with a fusion crust, but minerals immediately beneath the crust showed no signs of damage. Surface material is peeled away as it becomes heated during flight through the atmosphere; this process forms the fusion crust. Because the heat has little time to penetrate the meteorite's interior, compounds there are left intact.

One of the most striking discoveries to come from the Allende meteorite was evidence suggesting that a nearby star exploded into a **supernova** about 4.6 billion years ago. In a supernova, a massive star reaches the end of its life cycle and blows itself apart in a cataclysm that hurls matter outward at tremendous speeds (see Figure 1-8). During this detonation, violent collisions between nuclei produce a host of radioactive elements, including ^{26}Al, a radioactive isotope of aluminum.

Figure 15-18   R I V̄ U X G

A Piece of the Allende Meteorite This carbonaceous chondrite fell near Chihuahua, Mexico, in February 1969. The meteorite's dark color is due to its high abundance of carbon. Radioactive age-dating indicates that this meteorite is 4.56 billion years old, suggesting that this meteorite is a specimen of primitive planetary material that predates the formation of the planets. The ruler is 15 cm (6 in.) long. (Courtesy of J. A. Wood)

Researchers found clear evidence for the former presence of ^{26}Al in the Allende meteorite. Chemical analyses revealed a high abundance of a stable isotope of magnesium (^{26}Mg), which is produced by the radioactive decay of ^{26}Al. Some astronomers interpret this as evidence for a supernova in our vicinity at about the time the Sun was born. Indeed, by compressing interstellar gas and dust, the supernova's shock wave may have triggered the birth of our solar system.

15-7 A comet is a chunk of ice and dust that partially vaporizes as it passes near the sun

Just as heat from the protosun produced two classes of planets, the terrestrial and the Jovian, two main types of small bodies formed in the solar system. Near the Sun, interplanetary debris consists of the rocky objects called asteroids or meteoroids. Far from the Sun, where temperatures in the early solar system were low enough to permit ices of water, methane, ammonia, and carbon dioxide to form, interplanetary debris took the form of loose collections of ices and small rocky particles. These objects are like icy versions of "rubble pile" asteroids such as 253 Mathilde, which we described in Section 15-3. The Harvard astronomer Fred Whipple, who devised this model in 1950, called them "icy conglomerates" or "dirty snowballs." Some of these "snowballs" are trans-Neptunian objects that orbit forever in the outer regions of the solar system. Some of them, however, are in noncircular orbits that take them inward toward the Sun, where the ice begins to evaporate. These objects are known as **comets.**

The Structure of a Comet

Kuiper belt objects and asteroids travel around the Sun along roughly circular orbits that lie close to the plane of the ecliptic. In sharp contrast, comets travel around the Sun along highly elliptical orbits inclined at random angles to the ecliptic. Comets are just a few kilometers across, so when they are far from the Sun they are difficult to spot with even a large telescope. As a comet approaches the Sun, however, solar heat begins to vaporize the comet's ices, liberating gases as well as dust particles. The liberated gases begin to glow, producing a fuzzy, luminous ball called a **coma** that is typically 1 million km in diameter. Some of these luminous gases stream outward into a long, flowing **tail.** This tail, which can be more than 100 million kilometers in length—comparable to the distance from the Earth to the Sun—is one of the most awesome sights that can be seen in the night sky (see **Figure 15-19** and the image that opens this chapter).

Many comets are discovered when they are still far from the Sun, long before their tails grow to full size. A number of these discoveries are made by amateur astronomers, many of whom use only binoculars or small telescopes to aid in their search.

Figure 15-20 depicts the structure of a comet. The solid part of the comet, from which the coma and tail emanate, is called the **nucleus.** It is a mixture of ice

Figure 15-19   R I [V] U X G

Comet Hyakutake Using binoculars, Japanese amateur astronomer Yuji Hyakutake first noticed this comet on the morning of January 30, 1996. Two months later, Comet Hyakutake came within 0.1 AU (15 million km, or 9 million mi) of Earth and became one of the brightest comets of the twentieth century. This image was made in April 1996, when the comet's tail extended more than 30° across the sky (more than 60 times the diameter of the full moon). The stars appear as short streaks because the camera followed the comet's motion during the exposure. (Courtesy of Gary Goodman)

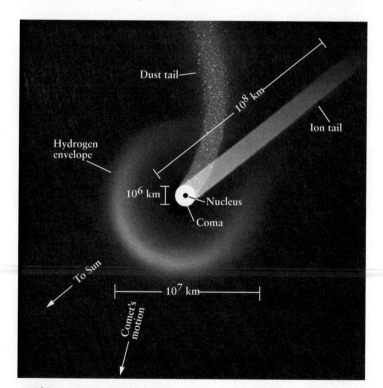

Figure 15-20

The Structure of a Comet The solid part of a comet is the nucleus. It is typically about 10 km in diameter. The coma typically measures 1 million kilometers in diameter, and the hydrogen envelope usually spans 10 million kilometers. A comet's tail can be as long as 1 AU.

Figure 15-21   R I V U X G

Comet Nuclei **(a)** This image from the European Space Agency spacecraft *Giotto* shows the dark, potato-shaped nucleus of Comet Halley in March 1986. Sunlight from the left illuminates half of the nucleus, which is about 15 km long and 8 km wide. Two bright jets of dust extend from the nucleus toward the Sun. **(b)** The *Deep Impact* spacecraft recorded this image of the nucleus of Comet Tempel 1 in July 2005. This nucleus, which measures 7.6 by 4.9 km, has a variety of different terrain on its surface. (a: Max-Planck-Institut für Aeronomie; b: NASA/JPL/U. of Maryland)

(a) Comet Halley

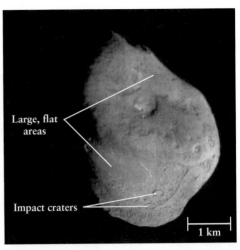

(b) Comet Tempel 1

and dust that typically measures a few kilometers across. Before 1986, no one had seen the nucleus of a comet. It is small, dim, and buried in the glare of the coma. In that year, the first close-up pictures of a comet's nucleus were obtained by a spacecraft that flew past Comet Halley (Figure 15-21*a*). Comet Halley's potato-shaped nucleus is darker than coal, reflecting only about 4% of the light that strikes it. This dark color is probably caused by a layer of carbon-rich compounds and dust that is left behind as the comet's ice evaporates.

Several 15-km-long jets of dust particles were seen emanating from bright areas on Halley's nucleus. These seem to be active only when exposed to the Sun. The bright areas, which probably cover about 10% of the surface of the nucleus, are presumably places where the dark layer covering the nucleus is particularly thin. When exposed to the Sun, these areas evaporate rapidly, producing the jets. When the nucleus's rotation brings the bright areas into darkness, away from the Sun, the jets shut off.

A few other comet nuclei have been imaged at close range at spacecraft. The best images to date have come from the *Deep Impact* spacecraft, which flew to within 500 km of Comet Tempel 1 in 2005 (Figure 15-21*b*). As it approached, *Deep Impact* launched a 372-kg copper projectile that slammed into the surface of the nucleus at 37,000 km/h (23,000 mi/h). The collision excavated about 11,000 tons of material from the nucleus. The character of the ejected material confirmed that the nucleus of Tempel 1 is a "rubble pile" held together not by chemical forces between its component parts (like a chocolate chip cookie) but by the gravitational forces between those parts.

The *Deep Impact* mission also measured the mass and hence the density of the comet nucleus. The very low density of 600 kg/m^3 (60% of the density of water) shows that the nucleus cannot be solid. It is best regarded as a porous jumble of rock, fine dust, and ice.

An even more ambitious mission to explore a comet is the European Space Agency's *Rosetta*. This spacecraft was lunched in 2004 and is scheduled to fly past at least one asteroid on its way to intercept Comet 67P/Churyumov-Gerasimenko in 2014. If all goes well, *Rosetta* will go into orbit around the comet nucleus and release a small probe that will actually land on the nucleus.

Comet Envelopes and Comet Tails

A part of a comet that is not visible to the human eye is the **hydrogen envelope,** a huge sphere of tenuous gas surrounding the nucleus. This hydrogen comes from water molecules (H_2O) that escape from the comet's evaporating ice and then break apart when they absorb ultraviolet photons from the Sun. The hydrogen atoms also absorb solar ultraviolet photons, which excite the atoms from the ground state into an excited state (see Section 5-8). When the atoms return to the ground state, they emit ultraviolet photons.

Unfortunately for astronomers, these emissions from a comet's hydrogen envelope cannot be detected by Earth-based telescopes because our atmosphere is largely opaque to ultraviolet light (see Section 6-7). Instead, cameras above the Earth's atmosphere must be used. Figure 15-22 shows two views of the same comet, one as it appeared in visible light to Earth-based observers and one as photographed by an ultraviolet camera aboard a rocket. The ultraviolet image shows the enormous extent of the hydrogen envelope, which can span 10 million kilometers.

As the diagram in Figure 15-20 suggests, comet tails always point away from the Sun. This is true regardless of the direction of the comet's motion (Figure 15-23). The implication that something from the Sun was "blowing" the comet's gases radially outward led Ludwig Biermann to predict the existence of a *solar wind*, a stream of particles rushing away from the Sun (see Section 8-5). A decade later, in 1962, Biermann's prediction was confirmed when the solar wind was detected for the first time by instruments on the *Mariner 2* spacecraft (see Figure 11-15).

> A comet's tail does not stream behind it; instead, it points generally away from the Sun

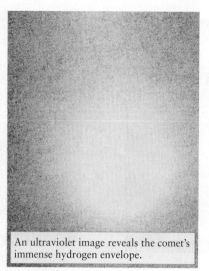

A visible-light image shows the comet's tail.

An ultraviolet image reveals the comet's immense hydrogen envelope.

R I V U X G   R I V U X G

Figure 15-22

A Comet and Its Hydrogen Envelope These visible-light and ultraviolet images of Comet Kohoutek, which made an appearance in 1974, are reproduced to the same scale. The hydrogen envelope is only visible in the ultraviolet image. (Johns Hopkins University, Naval Research Laboratory)

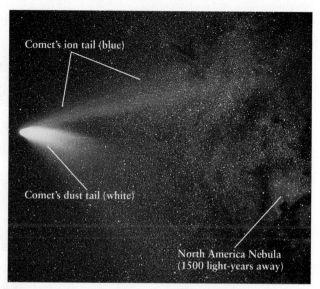

Comet's ion tail (blue)

Comet's dust tail (white)

North America Nebula (1500 light-years away)

Figure 15-24   R I V U X G

The Two Tails of Comet Hale-Bopp A comet's dust tail is white because it reflects sunlight, while the molecules in the ion tail emit their own light with a characteristic blue color. When this picture was taken on March 8, 1997, the ion tail extended more than 10° across the sky. The red object to the right is the North America Nebula, a star-forming region some 1500 light-years beyond the solar system. (Courtesy of Tony and Daphne Hallas Astrophotos)

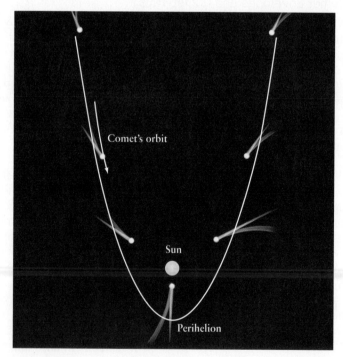

Comet's orbit

Sun

Perihelion

Figure 15-23

The Orbit and Tail of a Comet The solar wind and the pressure of sunlight blow a comet's dust particles and ionized atoms away from the Sun. Consequently, a comet's tail points generally away from the Sun. In particular, the tail does *not* always stream behind the nucleus. At the upper right of this figure, the comet is moving upward along its orbit and is literally chasing its own tail.

In fact, the Sun usually produces *two* comet tails—an **ion tail** and a **dust tail** (**Figure 15-24**). Ionized atoms and molecules—that is, atoms and molecules missing one or more electrons—are swept directly away from the Sun by the solar wind to form the relatively straight ion tail. The distinct blue color of the ion tail is caused by emissions from carbon-bearing molecules such as CN and C_2. The dust tail is formed when photons of light strike dust particles freed from the evaporating nucleus. (Figure 8-6 shows a dust particle of this type.) Light exerts a pressure on any object that absorbs or reflects it. This pressure, called **radiation pressure,** is quite weak but is strong enough to make fine-grained dust particles in a comet's coma drift away from the comet, thus producing a dust tail. The solar wind has less of an effect on dust particles than on (much smaller) ions, so the dust tail ends up being curved rather than straight. On rare occasions a comet is oriented in such a way that its dust tail appears to stick out in front of the comet (**Figure 15-25**).

After the comet passes perihelion, it recedes from the Sun back into the cold regions of the outer solar system. The ices stop vaporizing, the coma and tail disappear, and the comet goes back into an inert state—until the next time its orbit takes it toward the inner solar system. An object in an elliptical orbit spends most of its time far from the Sun, so it is only during a relatively brief period before and after perihelion that a comet can have a prominent tail. An example is Comet Halley, which orbits the Sun along a highly elliptical path that stretches from just inside the Earth's orbit to slightly beyond the orbit of Neptune (**Figure 15-26**). The comet's tail is visible to the naked eye or with binoc-

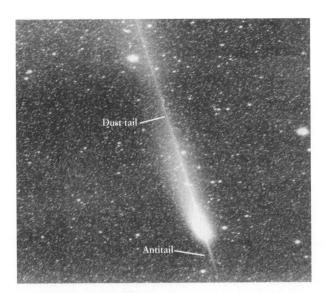

Figure 15-25   R I [V] U X G

The Antitail of Comet Hale-Bopp In January 1998, nine months after passing perihelion, Comet Hale-Bopp was oriented in such a way that the end of its arched dust tail looked like a spike sticking out of the comet's head. This spike is called an "antitail" because it appears to protrude in front of the comet. (European Southern Observatory)

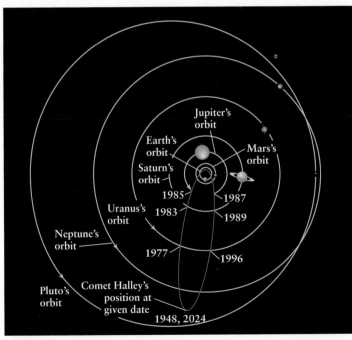

Figure 15-26

Comet Halley's Eccentric Orbit Like most comets, Comet Halley has an elongated orbit and spends most of its time far from the Sun. Figure 4-23 shows the comet's tail, which appears only when the comet is close to perihelion. Comet Halley has been observed at intervals of about 76 years—the period of its orbit—since 88 B.C. (see Section 4-7).

ulars only during a few months around perihelion, which last occurred in 1986 and will happen again in 2061.

Other comets have orbits that are larger and more elongated, with even longer periods. The orbit of Comet Hyakutake (see Figure 15-19) takes it to a distance of about 2000 AU from the Sun, 50 times the size of Pluto's orbit, with an orbital period of around 30,000 years!

15-8 Comets originate either from the Kuiper belt or from a vast cloud in near-interstellar space

Both comets and Kuiper belt objects are mixtures of ice and rock that must have formed in the outer solar system. (Had they formed in the warm inner solar system, they would not have included ices.) But most Kuiper belt objects have relatively circular orbits that lie close to the plane of the ecliptic, while comets have very elongated orbits that are often steeply inclined to the ecliptic. What gives comets their distinctive orbits? As we will see, the answer to this question depends on whether the comet's orbital period is relatively short or relatively long.

Jupiter-Family Comets and the Kuiper Belt

As we saw in Section 14-10, the gravitational effects of Neptune shape the orbits of objects in the **Kuiper belt.** Most of these objects are found between the orbit of Pluto (where an object makes two orbits for every three orbits of Neptune) and about 50 AU from the Sun (where an object makes one orbit for every two orbits of Neptune).

However, collisions between Kuiper belt objects can break off relatively small chunks. Gravitational perturbations from Neptune can occasionally launch one of these chunks into a highly elliptical orbit that takes it close to the Sun. When one of these chunks approaches the Sun, it develops a visible tail and appears as a comet. These are called **Jupiter-family comets** because their orbits tend to be influenced strongly by Jupiter's gravitational pull. Jupiter-family comets orbit the Sun in fewer than 20 years. One example is Comet Tempel 1 (see Figure 15-21*b*), which has an orbital period of 5.5 years.

There may be tens of thousands of objects in the Kuiper belt that could eventually be perturbed into an elongated orbit and become comets. It is thought that a few of these refugees from the Kuiper belt have been captured by the gravitational pull of Saturn, becoming that planet's small outer satellites (see Section 13-10).

Intermediate-Period Comets, Long-Period Comets, and the Oort Cloud

The majority of comets are **intermediate-period comets,** with orbital periods between 20 and 200 years, and **long-period comets,** which take more than 200 years to complete one orbit of the Sun. Comet Halley, shown in Figure 15-21*a*, has an orbital period of 76 years and so is an intermediate-period comet; Comet Hyakutake (see Figure 15-19) is a long-period comet that takes more than 70,000 years to complete one orbit.

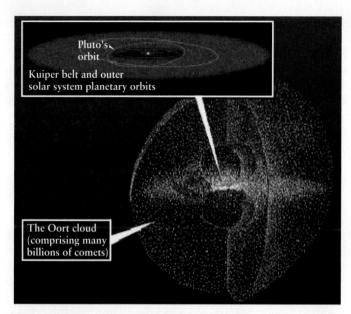

Figure 15-27

The Kuiper Belt and the Oort Cloud Objects in the Kuiper belt orbit close to the plane of the ecliptic. The Oort cloud, by contrast, is a spherical distribution of icy objects whose orbits are randomly oriented relative to the ecliptic. (NASA and A. Field/Space Telescope Science Institute)

Long-period comets travel along extremely elongated orbits and consequently spend most of their time at distances of roughly 10^4 to 10^5 AU from the Sun, or about one-fifth of the way to the nearest star. These orbits extend far beyond the Kuiper belt. Furthermore, intermediate-period and long-period comets have orbital planes that are often steeply inclined to the plane of the ecliptic. (Comet Halley's orbital plane is inclined by 162° to the ecliptic, which means that it orbits the Sun in the opposite direction to the planets.)

The best explanation for these observations is that intermediate-period and long-period comets come from a reservoir that extends from the Kuiper belt to some 50,000 AU from the Sun. This reservoir, first hypothesized by Dutch astronomer Jan Oort in 1950 and now called the **Oort cloud**, does not lie in or near the ecliptic; rather, it is a spherical distribution centered on the Sun (Figure 15-27). This explains why many intermediate-period and long-period comets have steeply inclined orbits.

Because astronomers discover long-period comets at the rate of about one per month, it is reasonable to suppose that there is an enormous population of comets in the Oort cloud. Estimates of the number of "dirty snowballs" in the Oort cloud range as high as 5 trillion (5×10^{12}). Only such a large reservoir of comet nuclei would explain why we see so many long-period comets, even though each one takes several million years to travel once around its orbit. Because the Oort cloud is so distant, it has not yet been possible to detect objects in the Oort cloud directly.

The Oort cloud was probably created 4.56 billion years ago from numerous icy planetesimals that orbited the Sun in the vicin-ity of the newly formed Jovian planets. During near-collisions with the giant planets, many of these chunks of ice and dust were catapulted by gravity into highly elliptical orbits, in much the same way that *Pioneer* and *Voyager* spacecraft were flung far from the Sun during their flybys of the Jovian planets. Gravitational perturbations from nearby stars could have tilted the planes of the orbits in all directions, giving the Oort cloud its spherical shape.

Changing a Comet's Orbit

The distinction between long-period, intermediate-period, and Jupiter-family comets can be blurred by the effects of gravitational perturbations. During a return trip toward the Sun, an encounter with a Jovian planet may force a comet into a much larger orbit. Alternatively, such an encounter can move a long-period comet into a smaller orbit (Figure 15-28). Several comets have been perturbed in this way into orbits that always remain within the inner solar system. One such is Comet Tempel 1 (see Figure 15-21b), whose orbit extends from the orbit of Mars out to the orbit of Jupiter.

Our description of cometary origins suggests that different comets are of different ages. Computer simulations suggest that many Jupiter-family comets were produced no more than a few hundred million years ago by collisions between larger icy objects in the Kuiper belt. Comets emanating from the Oort cloud may

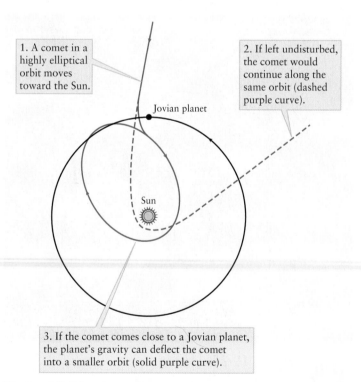

1. A comet in a highly elliptical orbit moves toward the Sun.

2. If left undisturbed, the comet would continue along the same orbit (dashed purple curve).

Jovian planet

Sun

3. If the comet comes close to a Jovian planet, the planet's gravity can deflect the comet into a smaller orbit (solid purple curve).

Figure 15-28

Transforming a Comet's Orbit The gravitational force of a planet can deflect a comet from a highly elliptical orbit into a less elliptical one. In some cases a comet ends up with an orbit that keeps it within the inner solar system.

Figure 15-29   R I V̄ U X G

The Fragmentation of Comet LINEAR The Lincoln Near Earth Asteroid Research (LINEAR) project discovered this comet in 1999. As it passed the perihelion of its orbit in July 2000, Comet LINEAR broke into more than a dozen small fragments. These continue to orbit the Sun along the same trajectory that the comet followed prior to its breakup. This image was made using the Very Large Telescope (see Figure 6-18). (European Southern Observatory)

date from the very early solar system, but may have originated at somewhat different times at different distances from the center of the solar nebula before being catapulted into the Oort cloud. This variety of ages is one reason why comets are of particular interest to planetary scientists: Comets carry information about the history of the cold, outer regions of the solar system, but are brought into the inner solar system where we can examine them at closer range using telescopes or spacecraft.

Comet Breakup and Meteor Showers

Because they evaporate away part of their mass each time they pass near the Sun, comets cannot last forever. A typical comet may lose about 0.5% to 1% of its ice each time it passes near the Sun. Hence, the ice completely vaporizes after about 100 or 200 perihelion passages, leaving only a swarm of dust and pebbles. Astronomers have observed some comet nuclei in the process of fragmenting (**Figure 15-29**).

> **Comets eventually break apart, and their fragments give rise to meteor showers**

One remarkable comet, called Shoemaker-Levy 9, broke apart for a different reason. As the comet swung by Jupiter in July 1992, the giant planet's tidal forces tore the nucleus into more than 20 fragments. Jupiter's gravity then deflected the trajectories of these fragments so that they plummeted into the planet's atmosphere.

As a comet's nucleus evaporates, residual dust and rock fragments form a **meteoritic swarm,** a loose collection of debris that continues to circle the Sun along the comet's orbit (**Figure 15-30**). If the Earth's orbit happens to pass through this swarm, a **meteor shower** is seen as the dust particles strike the Earth's upper atmosphere.

(a)

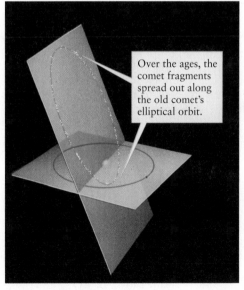

(b)

Figure 15-30

Meteoritic Swarms Rock fragments and dust from "burned out" comets continue to circle the Sun. **(a)** The most spectacular meteor showers occur when the Earth passes through compact swarms of debris from recently extinct comets. **(b)** Meteor showers associated with older comets are more predictable, because the Earth passes through the evenly distributed swarm on each trip around the Sun.

Table 15-1 Prominent Yearly Meteor Showers

Shower name	Date of maximum intensity*	Typical hourly rate	Average speed (km/s)	Radiant constellation
Quadrantids	January 3	40	40	Boötes
Lyrids	April 22	15	50	Lyra
Eta Aquarids	May 4	20	64	Aquarius
Delta Aquarids	July 30	20	40	Aquarius
Perseids	August 12	50	60	Perseus
Orionids	October 21	20	66	Orion
Taurids	November 4	15	30	Taurus
Leonids	November 16	15	70	Leo
Geminids	December 13	50	35	Gemini
Ursids	December 22	15	35	Ursa Minor

*The date of maximum intensity is the best time to observe a particular shower, although good displays can often be seen a day or two before or after the maximum. The typical hourly rate is given for an observer under optimum viewing conditions. The average speed refers to how fast the meteoroids are moving when they strike the atmosphere.

Nearly a dozen meteor showers can be seen each year (Table 15-1). Note that the **radiants** for these showers—that is, the places among the stars from which the meteors appear to come—are not confined to the constellations of the zodiac. This means that the swarms do not all orbit in the plane of the ecliptic. The reason is that the orbits of their parent comets were inclined at random angles to the plane of the ecliptic (see Figure 15-30).

Key Words

amino acids, p. 389
asteroid, p. 378
asteroid belt, p. 379
carbonaceous chondrite, p. 389
coma (of a comet), p. 390
comet, p. 390
differentiated asteroid, p. 388
dust tail, p. 392
dwarf planet, p. 379
fusion crust, p. 387
Hirayama family, p. 383
hydrogen envelope, p. 391
intermediate-period comet, p. 393
iron meteorite (iron), p. 388
ion tail, p. 392
Jupiter-family comet, p. 393
Kirkwood gaps, p. 380
Kuiper belt, p. 393
long-period comet, p. 393
meteor, p. 387
meteor shower, p. 395

meteorite, p. 387
meteoritic swarm, p. 395
meteoroid, p. 384
minor planet, p. 378
near-Earth object (NEO), p. 383
nucleus (of a comet), p. 390
Oort cloud, p. 394
radiant (of a meteor shower), p. 396
radiation pressure, p. 392
stable Lagrange points, p. 383
stony iron meteorite, p. 388
stony meteorite (stone), p. 387
supernova, p. 389
tail (of a comet), p. 390
Trojan asteroid, p. 383
undifferentiated asteroid, p. 389
Widmanstätten patterns, p. 388

Key Ideas

Discovery of the Asteroids: Astronomers first discovered the asteroids while searching for a "missing planet."

• Thousands of asteroids with diameters ranging from a few kilometers up to 1000 kilometers orbit within the asteroid belt between the orbits of Mars and Jupiter.

Origin of the Asteroids: The asteroids are the relics of planetesimals that failed to accrete into a full-sized planet, thanks to the effects of Jupiter and other Mars-sized objects.

• Even today, gravitational perturbations by Jupiter deplete certain orbits within the asteroid belt. The resulting gaps, called Kirkwood gaps, occur at simple fractions of Jupiter's orbital period.

• Jupiter's gravity also captures asteroids in two locations, called Lagrangian points, along Jupiter's orbit.

Asteroid Collisions: Asteroids undergo collisions with each other, causing them to break up into smaller fragments.

• Some asteroids, called near-Earth objects, move in elliptical orbits that cross the orbits of Mars and Earth. If such an asteroid strikes the Earth, it forms an impact crater whose diameter depends on both the mass and the speed of the asteroid.

• An asteroid struck the Earth 65 million years ago, possibly causing the extinction of the dinosaurs and many other species.

Meteoroids, Meteors, and Meteorites: Small rocks in space are called meteoroids. If a meteoroid enters the Earth's atmosphere, it produces a fiery trail called a meteor. If part of the object survives the fall, the fragment that reaches the Earth's surface is called a meteorite.

• Meteorites are grouped into three major classes, according to composition: iron, stony iron, and stony meteorites. Irons and stony irons are fragments of the core of an asteroid that was large enough and hot enough to have undergone chemical differentiation, just like a terrestrial planet. Some stony meteorites come from the crust of such differentiated meteorites, while others are fragments of small asteroids that never underwent differentiation.

• Rare stony meteorites called carbonaceous chondrites may be relatively unmodified material from the solar nebula. These meteorites often contain organic material and may have played a role in the origin of life on Earth.

• Analysis of isotopes in certain meteorites suggests that a nearby supernova may have triggered the formation of the solar system 4.56 billion years ago.

Comets: A comet is a chunk of ice with imbedded rock fragments that generally moves in a highly elliptical orbit about the Sun.

• As a comet approaches the Sun, its icy nucleus develops a luminous coma, surrounded by a vast hydrogen envelope. An ion tail and a dust tail extend from the comet, pushed away from the Sun by the solar wind and radiation pressure, respectively.

Origin and Fate of Comets: Comets are thought to originate from two regions, the Kuiper belt and the Oort cloud.

• The Kuiper belt lies in the plane of the ecliptic at distances between 30 and 50 AU from the Sun. It is thought to contain many tens of thousands of comet nuclei. Many Jupiter-family comets probably come from the Kuiper belt, and hundreds of larger objects have been observed in the Kuiper belt.

• The Oort cloud contains billions of comet nuclei in a spherical distribution that extends out to 50,000 AU from the Sun. Intermediate-period and long-period comets are thought to originate in the Oort cloud. As yet no objects in the Oort cloud have been detected directly.

• Fragments of "burned out" comets produce meteoritic swarms. A meteor shower is seen when the Earth passes through a meteoritic swarm.

Questions

Review Questions

1. What led astronomers to suspect that there were members of the solar system that orbit between Mars and Jupiter?
2. How did the first asteroid come to be discovered? What role did theoretical calculations play in confirming the discovery? How did this discovery differ from what astronomers had expected to find?
3. How do modern astronomers discover new asteroids?
4. What are the differences between asteroids and trans-Neptunian objects?
5. Describe the asteroid belt. Does it lie completely within the plane of the ecliptic? What are its inner and outer radii?

6. If Jupiter was not present in our solar system, would the asteroid belt exist? Why or why not?
7. What are Kirkwood gaps? What causes them?
8. Compare the explanation of the Kirkwood gaps in the asteroid belt to the way in which Saturn's moons help produce divisions in that planet's rings (see Section 12-11).
9. How is it possible to tell that some asteroids are nonspherical even though we do not have images of those asteroids?
10. What is the evidence that some asteroids are made of a loose conglomeration of smaller pieces?
11. The asteroid 243 Ida, which was viewed by the *Galileo* spacecraft, is a member of a Hirayama family. Discuss what this tells us about the history of this asteroid. Where might you look to find other members of the same family?
12. What are the Trojan asteroids, and where are they located? What holds them in this location?
13. What are near-Earth objects? What is the evidence that the Earth has been struck by these objects?
14. What is the evidence that an asteroid impact could have contributed to the demise of the dinosaurs?
15. What is the difference between a meteoroid, a meteor, and a meteorite?
16. Is there anywhere on the Earth where you might find large numbers of stony meteorites that are not significantly weathered? If so, where? If not, why not?
17. Scientists can tell that certain meteorites came from the interior of an asteroid rather than from its outer layers. Explain how this is done.
18. Why are some asteroids differentiated while others are not?
19. Suppose you found a rock you suspect might be a meteorite. Describe some of the things you could do to determine whether it was a meteorite or a "meteorwrong."
20. What is the evidence that carbonaceous chondrites are essentially unaltered relics of the early solar system? What do they suggest about how the solar system may have formed?
21. With the aid of a drawing, describe the structure of a comet.
22. Why is the phrase "dirty snowball" an appropriate characterization of a comet's nucleus?
23. What did scientists learn about the structure of comet nuclei from the *Deep Impact* mission?
24. Why do the ion tail and dust tail of a comet point in different directions?
25. Why do comets have prominent tails for only a short time during each orbit?
26. Why is it that Jupiter and Saturn can be seen in the night sky every year, while seeing specific comets such as Halley and Hyakutake is a once-in-a-lifetime event?
27. What is the relationship between the Kuiper belt and comets?
28. What is the Oort cloud? How might it be related to planetesimals left over from the formation of the solar system?
29. Why are comets more likely to break apart at perihelion than at aphelion?
30. Why do astronomers think that meteorites come from asteroids, while meteor showers are related to comets?
31. Why are asteroids, meteorites, and comets all of special interest to astronomers who want to understand the early history and subsequent evolution of the solar system?

Advanced Questions

Questions preceded by an asterisk (*) involve topics discussed in Box 1-1 or 7-2.

Problem-solving tips and tools

The small-angle formula is described in Box 1-1. We discussed retrograde motion in Section 4-1 and described its causes in Section 4-2. You will need to use Kepler's third law, described in Section 4-4 and Box 4-2, in some of the problems below. Box 7-2 discusses the concept of escape speed. A spherical object of radius r intercepts an amount of sunlight proportional to its cross-sectional area, equal to πr^2. The volume of a sphere of radius r is $4\pi r^3/3$.

---

32. When Olbers discovered Pallas in March 1802, the asteroid was moving from east to west relative to the stars. At what time of night was Pallas highest in the sky over Olbers's observatory? Explain your reasoning.

33. Consider the Kirkwood gap whose orbital period is two-fifths of Jupiter's 11.86-year period. Calculate the distance from the Sun to this gap. Does your answer agree with Figure 15-4?

*34. When the image in Figure 15-5 was made, the asteroid 1 Ceres was 1.63 AU, or 2.44×10^8 km, from the Hubble Space Telescope. (a) What was the angular size of the asteroid as a whole? (b) You can see individual pixels in the image shown in Figure 15-5. Using a ruler and the scale bar in the figure, determine how many kilometers on the surface of 1 Ceres are contained in the width of one pixel. (c) What is the angular width of each pixel in arcseconds?

35. Suppose that a binary asteroid (two asteroids orbiting each other) is observed in which one member is 16 times brighter than the other. Suppose that both members have the same albedo and that the larger of the two is 120 km in diameter. What is the diameter of the other member?

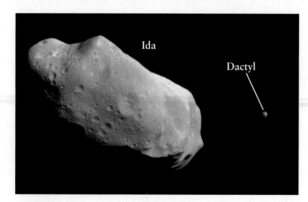

RI V UXG

(JPL/NASA)

36. The accompanying image from the *Galileo* spacecraft shows the asteroid 243 Ida, which has dimensions $56 \times 24 \times 21$ km. *Galileo* discovered a tiny moon called Dactyl, just $1.6 \times 1.4 \times 1.2$ km in size, which orbits Ida at a distance of

about 100 km. (In Greek mythology, the Dactyli were beings who lived on the slopes of Mount Ida.) Describe a scenario that could explain how Ida came to have a moon.

*37. Assume that Ida's tiny moon Dactyl (see Question 36) has a density of 2500 kg/m³. (a) Calculate the mass of Dactyl in kilograms. For simplicity, assume that Dactyl is a sphere 1.4 km in diameter. (b) Calculate the escape speed from the surface of Dactyl. If you were an astronaut standing on Dactyl's surface, could you throw a baseball straight up so that it would never come down? Professional baseball pitchers can throw at speeds around 40 m/s (140 km/h, or 90 mi/h); your throwing speed is probably a bit less.

38. Imagine that you are an astronaut standing on the surface of a Trojan asteroid. How will you see the phase of Jupiter change with the passage of time? How will you see Jupiter move relative to the distant stars? Explain your answers.

39. Use the percentages of stones, irons, and stony iron meteorites that fall to Earth to estimate what fraction of their parent asteroids' interior volume consisted of an iron core. Assume that the percentages of stones and irons that fall to the Earth indicate the fractions of a parent asteroid's interior volume occupied by rock and iron, respectively. How valid do you think this assumption is?

40. On March 8, 1997, Comet Hale-Bopp was 1.39 AU from the Earth and 1.00 AU from the Sun. Use this information and that given in the caption to Figure 15-24 to estimate the length of the comet's ion tail on that date. Give your answer in kilometers and astronomical units.

41. *Sun-grazing comets* come so close to the Sun that their perihelion distances are essentially zero. Find the orbital periods of Sun-grazing comets whose aphelion distances are (a) 100 AU, (b) 1000 AU, (c) 10,000 AU, and (d) 100,000 AU. Assuming that these comets can survive only a hundred perihelion passages, calculate their lifetimes. (*Hint:* Remember that the semimajor axis of an orbit is one-half the length of the orbit's long axis.)

42. Comets are generally brighter a few weeks after passing perihelion than a few weeks before passing perihelion. Explain why might this be. (*Hint:* Water, including water ice, does an excellent job of retaining heat.)

43. The hydrogen envelopes of comets are especially bright at an ultraviolet wavelength of 122 nm. Use Figure 5-24 to explain why.

44. A *very* crude model of a typical comet nucleus is a cube of ice (density 1000 kg/m³) 10 km on a side. (a) What is the mass of this nucleus? (b) Suppose 1% of the mass of the nucleus evaporates away to form the comet's tail. Suppose further that the tail is 100 million (10^8) km long and 1 million (10^6) km wide. Estimate the average density of the tail (in kg/m³). For comparison, the density of the air you breathe is about 1.2 kg/m³. (c) In 1910 the Earth actually passed through the tail of Comet Halley. At the time there was some concern among the general public that this could have deleterious effects on human health. Was this concern justified? Why or why not?

45. For many years it was thought that the Tunguska event was caused by a comet striking the Earth. This idea was rejected because a small comet would have broken up too high in the

atmosphere to cause significant damage on the ground. Explain why, using your knowledge of a comet's structure.

Discussion Questions

46. Discuss the idea that 1 Ceres should be regarded as the smallest dwarf planet rather than the largest asteroid. What are the advantages of this scheme? What are the disadvantages?

47. From the abundance of craters on the Moon and Mercury, we know that numerous asteroids and meteoroids struck the inner planets early in the history of our solar system. Is it reasonable to suppose that numerous comets also pelted the planets 3.5 to 4.5 billion years ago? Speculate about the effects of such a cometary bombardment, especially with regard to the evolution of the primordial atmospheres on the terrestrial planets.

48. In the 1998 movie *Armageddon,* an asteroid "the size of Texas" is on a collision course with Earth. The asteroid is first discovered by astronomers just 18 days prior to impact. To avert disaster, a team of astronauts blasts the asteroid into two pieces just 4 hours before impact. Discuss the plausibility of this scenario. (*Hint:* On average, the state of Texas extends for about 750 km from north to south and from east to west. How does this compare with the size of the largest known asteroids?)

49. Suppose astronomers discover that a near-Earth object the size of 1994 XM1 is on a collision course with Earth. Describe what humanity could do within the framework of present technology to counter such a catastrophe.

Web/eBook Questions

50. A NASA spacecraft called *Dawn* is intended to go into orbit around two asteroids, 1 Ceres and 4 Vesta. Search the World Wide Web for information about this mission. Why were these two particular asteroids selected for study? What types of observations will the spacecraft make?

51. Search the World Wide Web to find out why some scientists disagree with the idea that a tremendous impact led to the demise of the dinosaurs. (They do not dispute that the impact took place, only what its consequences were.) What are their arguments? From what you learn, what is your opinion?

52. Several scientific research programs are dedicated to the search for near-Earth objects (NEOs), especially those that might someday strike our planet. Search the World Wide Web for information about at least one of these programs. How does this program search for NEOs? How many NEOs has this program discovered? Will any of these pose a threat in the near future?

53. **Estimating the Speed of a Comet.** Access and view the video "Two Comets and an Active Sun" in Chapter 15 of the *Universe* Web site or eBook. (a) Why don't the comets reappear after passing the Sun? (b) The white circle shows the size of the Sun, which has the diameter 1.39×10^6 km. Using this to set the scale, step through the video and measure how the position of one of the comets changes. Use your measurements and the times displayed in the video to estimate the comet's speed in km/h and km/s. (Assume that the comet moved in the same plane

as that shown in the video.) As part of your answer, explain the technique and calculations you used. (c) How does your answer in (b) compare with the orbital speed of Mercury, the innermost and fastest-moving planet (see Table 11-1)? Why is there a difference? (d) If the comet's motion was not in the same plane as that shown in the video, was its actual speed more or less than your estimate in (b)? Explain.

Activities

Observing Projects

> **Observing tips and tools**
>
> *Meteors:* Informative details concerning upcoming meteor showers appear on the Web sites for *Sky & Telescope* and *Astronomy* magazines.
>
> *Comets:* Because astronomers discover dozens of comets each year, there is usually a comet visible somewhere in the sky. Unfortunately, they are often quite dim, and you will need to have access to a moderately large telescope (at least 35 cm, or 14 in.). You can get up-to-date information from the Minor Planet Center Web site. Also, if there is an especially bright comet in the sky, useful information about it might be found at the Web sites for *Sky & Telescope* and *Astronomy* magazines.
>
> *Asteroids:* At opposition, some of the largest asteroids are bright enough to be seen through a modest telescope. Check the Minor Planet Center Web site to see if any bright asteroids are near opposition. If so, check the current issue as well as the most recent January issue of *Sky & Telescope* magazine for a star chart showing the asteroid's path among the constellations. You will need such a chart to distinguish the asteroid from background stars. Also, you can locate Ceres, Pallas, Juno, and Vesta using the *Starry Night Enthusiast*™ program on the CD-ROM that accompanies certain printed copies of this textbook.

54. Make arrangements to view a meteor shower. Table 15-1 lists the dates of major meteor showers. Choose a shower that will occur near the time of a new moon. Set your alarm clock for the early morning hours (1 to 3 A.M.). Get comfortable in a reclining chair or lie on your back so that you can view a large portion of the sky. Record how long you observe, how many meteors you see, and what location in the sky they seem to come from. How well does your observed hourly rate agree with published estimates, such as those in Table 15-1? Is the radiant of the meteor shower apparent from your observations?

55. If a comet is visible with a telescope at your disposal, make arrangements to view it. Can you distinguish the comet from background stars? Can you see its coma? Can you see a tail?

56. Make arrangements to view an asteroid. Observe the asteroid on at least two occasions, separated by a few days. On each night, draw a star chart of the objects in your telescope's field of view. Has the position of one starlike object shifted between observing sessions? Does the position of the

moving object agree with the path plotted on published star charts? Do you feel confident that you have in fact seen the asteroid?

57. Use the *Starry Night Enthusiast*™ program to observe two asteroids. Display the entire celestial sphere by selecting **Guides > Atlas** in the **Favourites** menu. Center on **Ceres** by typing its name in the **Search All Databases** box in the **Find** pane and label it by clicking in the box to the left of its name as it appears in the Find list. In the toolbar, set the Time Flow Rate to 1 day. Then click the Run Time Forward button. (**a**) Watch Ceres for at least two years of simulated time. Describe how Ceres moves. How can you tell that Ceres orbits the Sun in the same direction as the planets? Does Ceres remain as close to the ecliptic (shown as a solid green line) as the planets do? (If you do not see a green line representing the ecliptic, make sure that the item **The Ecliptic** is checked in the View menu.) (**b**) Click on the **Stop** button and use the **Find** pane to locate and center the view on Pallas. Again click on the Run Time Forward button, and watch how Pallas moves for at least two years of simulated time. How does the motion of Pallas differ from that of Ceres? How are the two motions similar? Which asteroid's orbit is more steeply inclined to the plane of the Earth's orbit? How can you tell?

58. Use the *Starry Night Enthusiast*™ program to study the motion of a comet. First set up the field of view so that you are observing the inner solar system from a distance (select **Solar System > Inner Solar system** in the **Favourites** menu). In the toolbar, click on the **Stop** button to halt the animation, and then set the date to January 1, 1995, and the time step to 1 day. Select **View > Solar System > Asteroids** in the menu to remove the asteroids from the view. Open the **Find** pane and center on Comet Hyakutake by typing "*Hyakutake*" in the **Search All Databases** box and then pressing the **Enter** key. Use the Zoom controls to decrease the field of view to about 25° × 17°. Then click on the **Run Time Forward** button. (**a**) Watch the motion of Comet Hyakutake for at least two years of simulated time. Describe what you see. Is the comet's orbit in about the same plane as the orbits of the inner planets, or is it steeply inclined to that plane? (You can tilt the plane of the solar system by holding down the **Shift** key while clicking on

and moving the mouse to investigate this off-ecliptic motion.) How does the comet's speed vary as it moves along its orbit? During which part of the orbit is the tail visible? In what direction does the tail point? (**b**) Click on the **Stop** button to halt the animation, and set up the field of view so that you are observing from the center of a transparent Earth by selecting **Guides > Atlas** in the **Favourites** menu. Set the date to **January 1, 1995,** and the **Time Flow Rate** to **1 day,** and again center on Comet Hyakutake. Use the controls at the right-hand end of the toolbar to zoom out as far as possible. Then click on the **Run Time Forward** button and watch the comet's motion for at least two years of simulated time. Describe the motion, and explain why it is more complicated than the motion you observed in part (**a**). (**c**) Stop the animation, set the date to today's date, set the **Time Flow Rate** to **1 month** ("lunar m."), and restart the animation. Comet Hyakutake is currently moving almost directly away from the Sun and so, as seen from the Sun, its position on the celestial sphere should not change. Is this what you see in *Starry Night Enthusiast*™? Explain any differences. (*Hint:* You are observing from the Earth, not the Sun.)

59. Use the *Starry Night Enthusiast*™ program to examine Comet Halley as it would have been seen during its last visit to the Sun and the inner solar system. Display the entire celestial sphere as seen from the center of a transparent Earth by clicking on **Guides > Atlas** in the **Favourites** pane. **Stop** time flow and set the date to **February 28, 1986,** and the time to midnight, **12:00:00 A.M.** Center on Halley's Comet by using **Edit > Find** and entering the name *Halley* in the **Search** box. Set the **Time Flow Rate** to **6 hours.** From this view at the Earth's center with daylight turned off, you will note that the Sun is to the left of the view, on the ecliptic plane, as expected. (**a**) Based on the direction of the comet's tail, can you tell in which direction the comet is moving at this instant in time? Step time forward by a few steps to check your prediction. (**b**) Did Halley's Comet move in the direction that you predicted in (**a**) above? Run time backward to about January 7, 1986, and then run time forward and backward again several times to observe Halley's comet and particularly the direction of its tail during this encounter with the Sun. (**c**) In what direction does the tail of the comet point? Explain why.

Pluto and the Kuiper Belt

by Scott Sheppard

The "what is a planet" controversy began with the discovery of Ceres and several other objects between Mars and Jupiter in the early 1800s. At first, Ceres and then Pallas, Juno, and Vesta were considered planets, even though they were orders of magnitude less massive than the other known planets and all had similar orbits. The definition of a planet at this time was simple. Any object in orbit about the Sun that did not show cometary effects was a planet.

By the mid 1800s, the discovery of more and more objects between Mars and Jupiter saw the words "asteroid" and "minor planet" start to be used in order to signify these objects as an ensemble of bodies. That is, astronomers felt a major planet should be a rare and unique thing. The major planet club thus was left with eight members: Mercury, Venus, Earth, Mars, Jupiter, Saturn, Uranus, and Neptune.

The planet controversy arose once again in 1930 with the discovery of Pluto, where it was quickly regarded as the ninth planet. During the ensuing years, new observations downgraded Pluto's size from being near that of Earth to being smaller than our Moon. Its orbit also proved to be peculiar because it was highly inclined and eccentric, having it even crossing Neptune's orbit. Its uniqueness kept Pluto in the planet club with relatively few detractors until 1992, when the first Kuiper belt object was discovered. As it became apparent that hundreds of thousands of Kuiper belt objects existed on stable orbits just beyond Neptune, Pluto's planet status was becoming less and less certain. With the discovery of Eris (2003 UB313) in 2003, an object larger than Pluto, Pluto's last unique claim of being the largest Kuiper belt object was lost. Thus, the planet controversy had finally reached a point where a major planet's minimum size must be precisely defined.

No matter how you count them, there are no longer nine major planets in our solar system. The same situation of the main belt asteroids in the mid 1800's is now apparent with Pluto, Eris, and the rest of the Kuiper belt objects. Although the largest Kuiper belt objects are significantly bigger than the largest main belt asteroids, they are still an ensemble of bodies with a continuous size distribution and similar formation and evolution histories. The reason Pluto's status as a planet is a controversy now is simply because it took 62 years between the discovery of Pluto and the next Kuiper belt object in 1992. Within that time span Pluto became known as the ninth planet through several generations and was ingrained in society as such.

Classification is not that important to scientists but it is the understanding of how the object formed and its history. Even so, any classification should have a scientific base and not be based on tradition, as seems to be the case for Pluto in the recent past. Eris as well as Pluto and the rest of the Kuiper belt objects had very similar histories and formation scenarios and thus they should be considered as an ensemble of objects just like the main belt asteroids between Mars and Jupiter. As technology advances, our knowledge and thus understanding advances and so should our classification.

Several minimum-size planet definitions were debated recently. The first is that anything larger than a certain minimum size is a planet. The chosen size would likely be arbitrary and may be a nice round number like 1000 km or the size of Pluto. This stems from the traditional belief that Pluto has been considered a planet in the past and should continue as such. This would give us ten current planets including Eris. A second possible definition is that anything that is spherical should be a planet. This definition is based on physics in which the gravitational force of an object is large enough to overcome any other forces in determining the overall shape of the object. Determining if an object meets this criteria or not is complicated. This definition would also significantly change the status quo by increasing the number of known planets by several factors overnight. A third definition is that any object with a unique orbit and gravitationally dominates its local orbital environment should be a planet. This definition is the most scientifically based, in that objects with similar formation and evolutionary histories would be grouped together. Using this definition would give us eight known major planets, dropping Pluto from the list.

A final proposal as to what is a planet takes into account that the current known major planets themselves are remarkably different and can be split into individual categories themselves.

Scott S. Sheppard is an astronomer at the Carnegie Institution of Washington in the Department of Terrestrial Magnetism located in Washington, D.C. His main research currently involves discovering small bodies in our solar system to understand their dynamical and physical properties that help constrain solar system and planet formation. He has discovered several small moons around all the giant planets, been a pioneer in the discovery of Neptune Trojans, as well as found tens of Kuiper Belt objects.

(continued on the next page)

Pluto and the Kuiper Belt *(continued)*

Mercury, Venus, Earth and Mars are terrestrial planets whose compositions are dominated by rock. Jupiter and Saturn are gas giant planets dominated by their hydrogen and helium envelopes. Uranus and Neptune are ice giant planets dominated by gases other than hydrogen and helium. The trans-Neptunian objects or "ice dwarf planets," as some are calling them, are probably composed of large amounts of volatiles such as methane ice and water ice.

The International Astronomical Union (IAU) recently choose to use a combination of the second and third definitions above to define a planet. That is, a planet is an object that is both spherical and has a unique orbit in which it is gravitationally dominant. An object that only satisfies definition two above and not three is now being called a dwarf planet, of which Ceres, Pluto and Eris qualify, as well as several more objects in the Kuiper belt. In reality, it's the public that must accept this definition, and we will only know if that is the case a few generations from now.

Defining the word "planet" is like defining what an ocean is. At first it appears to be a simple term, but is very hard to precisely define. Further planet discoveries will be made and there are sure to be borderline cases. This is just the way nature is; it does not have little bins to nicely classify objects, but there is usually a continuous array of objects. The important thing to take away from all of this is to understand this array of objects by using our rapidly expanding scientific knowledge and understanding our place in the solar system.

16

Our Star,
the Sun

R I **V U** X G

A composite view of the Sun (at ultraviolet wavelengths) and an upheaval in the Sun's outer atmosphere, or corona (at visible wavelengths). (*SOHO/LASCO/EIT/ESA/NASA*)

The Sun is by far the brightest object in the sky. By earthly standards, the temperature of its glowing surface is remarkably high, about 5800 K. Yet there are regions of the Sun that reach far higher temperatures of tens of thousands or even millions of kelvins. Gases at such temperatures emit ultraviolet light, which makes them prominent in the accompanying image from an ultraviolet telescope in space. Some of the hottest and most energetic regions on the Sun spawn immense disturbances. These can propel solar material across space to reach the Earth and other planets.

In recent decades, we have learned that the Sun shines because at its core hundreds of millions of tons of hydrogen are converted to helium every second. We have confirmed this picture by detecting the by-products of this transmutation—strange, ethereal particles called neutrinos—streaming outward from the Sun into space. We have discovered that the Sun has a surprisingly violent atmosphere, with a host of features such as sunspots whose numbers rise and fall on a predictable 11-year cycle. By studying the Sun's vibrations, we have begun to probe beneath its surface into hitherto unexplored realms. And we have just begun to investigate how changes in the Sun's activity can affect the Earth's environment as well as our technological society.

16-1 The Sun's energy is generated by thermonuclear reactions in its core

The Sun is the largest member of the solar system. It has almost a thousand times more mass than all the planets, moons, asteroids, comets, and meteoroids put together. But the Sun is also a star. In fact, it is a remarkably typical star, with a mass, size, surface temperature, and chemical composition that are roughly midway between the extremes exhibited by the myriad other stars in the heavens. Table 16-1 lists essential data about the Sun.

Learning Goals

By reading the sections of this chapter, you will learn

16-1 The source of the Sun's heat and light

16-2 How scientists model the Sun's internal structure

16-3 How the Sun's vibrations reveal what lies beneath its glowing surface

16-4 How scientists are able to probe the Sun's energy-generating core

16-5 Why the gaseous Sun appears to have a sharp outer edge

16-6 Why the upper regions of the solar atmosphere have an emission spectrum

16-7 The relationship between the Sun's corona and the solar wind

16-8 The nature of sunspots

16-9 The connection between sunspots and the Sun's magnetic field

16-10 How magnetic reconnection can power immense solar eruptions

403

Table 16-1	Sun Data
Distance from the Earth:	Mean: 1 AU = 149,598,000 km
	Maximum: 152,000,000 km
	Minimum: 147,000,000 km
Light travel time to the Earth:	8.32 min
Mean angular diameter:	32 arcmin
Radius:	696,000 km = 109 Earth radii
Mass:	1.9891×10^{30} kg = 3.33×10^5 Earth masses
Composition (by mass):	74% hydrogen, 25% helium, 1% other elements
Composition (by number of atoms):	92.1% hydrogen, 7.8% helium, 0.1% other elements
Mean density:	1410 kg/m³
Mean temperatures:	Surface: 5800 K; Center: 1.55×10^7 K
Luminosity:	3.90×10^{26} W
Distance from center of Galaxy:	8000 pc = 26,000 ly
Orbital period around center of Galaxy:	220 million years
Orbital speed around center of Galaxy:	220 km/s

WEB LINK 16.1

R I V UX G
(NOAO)

Solar Energy

For most people, what matters most about the Sun is the energy that it radiates into space. Without the Sun's warming rays, our atmosphere and oceans would freeze into an icy layer coating a desperately cold planet, and life on Earth would be impossible. To understand why we are here, we must understand the nature of the Sun.

Why is the Sun such an important source of energy? An important part of the answer is that the Sun has a far higher surface temperature than any of the planets or moons. The Sun's spectrum is close to that of an idealized blackbody with a temperature of 5800 K (see Sections 5-3 and 5-4, especially Figure 5-12). Thanks to this high temperature, each square meter of the Sun's surface emits a tremendous amount of radiation, principally at visible wavelengths. Indeed, the Sun is the only object in the solar system that emits substantial amounts of visible light. The light that we see from the Moon and planets is actually sunlight that struck those worlds and was reflected toward Earth.

The Sun's size also helps us explain its tremendous energy output. Because the Sun is so large, the total number of square meters of radiating surface—that is, its surface area—is immense. Hence, the total amount of energy emitted by the Sun each second, called its **luminosity**, is very large indeed: about 3.9×10^{26} watts, or 3.9×10^{26} joules of energy emitted per second. (We discussed the relation among the Sun's surface temperature, radius, and luminosity in Box 5-2.) Astronomers denote the Sun's luminosity by the symbol $L_\odot$. (A circle with a dot in the center is the astronomical symbol for the Sun and was also used by ancient astrologers.)

The Source of the Sun's Energy: Early Ideas

These ideas lead us to a more fundamental question: What keeps the Sun's visible surface so hot? Or, put another way, what is the fundamental source of the tremendous energies that the Sun radiates into space? For centuries, this was one of the greatest mysteries in science. The mystery deepened in the nineteenth century, when geologists and biologists found convincing evidence that life had existed on Earth for at least several hundred million years. (We now know that the Earth is 4.56 billion years old and that life has existed on it for most of its history.) Since life as we know

Figure 16-1   R I V U X G

The Sun The Sun's visible surface has a temperature of about 5800 K. At this temperature, all solids and liquids vaporize to form gases. It was only in the twentieth century that scientists discovered what has kept the Sun so hot for billions of years: the thermonuclear fusion of hydrogen nuclei in the Sun's core. (Jeremy Woodhouse/PhotoDisc)

it depends crucially on sunlight, the Sun must be as old. This posed a severe problem for physicists. What source of energy could have kept the Sun shining for so long (Figure 16-1)?

One attempt to explain solar energy was made in the mid-1800s by the English physicist Lord Kelvin (for whom the temperature scale is named) and the German scientist Hermann von Helmholtz. They argued that the tremendous weight of the Sun's outer layers should cause the Sun to contract gradually, compressing its interior gases. Whenever a gas is compressed, its temperature rises. (You can demonstrate this with a bicycle pump: As you pump air into a tire, the temperature of the air increases and the pump becomes warm to the touch.) Kelvin and Helmholtz thus suggested that gravitational contraction could cause the Sun's gases to become hot enough to radiate energy out into space.

This process, called *Kelvin-Helmholtz contraction,* actually does occur during the earliest stages of the birth of a star like the Sun (see Section 8-4). But Kelvin-Helmholtz contraction cannot be the major source of the Sun's energy today. If it were, the Sun would have had to be much larger in the relatively recent past. Helmholtz's own calculations showed that the Sun could have started its initial collapse from the solar nebula no more than about 25 million years ago. But the geological and fossil record shows that the Earth is far older than that, and so the Sun must be as well. Hence, this model of a Sun that shines because it shrinks cannot be correct.

On Earth, a common way to produce heat and light is by burning fuel, such as a log in a fireplace or coal in a power plant. Is it possible that a similar process explains the energy released by the Sun? The answer is no, because this process could not continue for a long enough time to explain the age of the Earth. The chemical reactions involved in burning release roughly 10^{-19} joule of energy per atom. Therefore, the number of atoms that would have to undergo chemical reactions each second to gener-

ate the Sun's luminosity of 3.9×10^{26} joules per second is approximately

$$\frac{3.9 \times 10^{26} \text{ joules per second}}{10^{-19} \text{ joule per atom}} = 3.9 \times 10^{45} \text{ atoms per second}$$

From its mass and chemical composition, we know that the Sun contains about 10^{57} atoms. Thus, the length of time that would be required to consume the entire Sun by burning is

$$\frac{10^{57} \text{ atoms}}{3.9 \times 10^{45} \text{ atoms per second}} = 3 \times 10^{11} \text{ seconds}$$

There are about 3×10^7 seconds in a year. Hence, in this model, the Sun would burn itself out in a mere 10,000 (10^4) years! This is far shorter than the known age of the Earth, so chemical reactions also cannot explain how the Sun shines.

The Source of the Sun's Energy: Discovering Thermonuclear Fusion

The source of the Sun's luminosity could be explained if there were a process that was like burning but released much more energy per atom. Then the rate at which atoms would have to be consumed would be far less, and the lifetime of the Sun could be long enough to be consistent with the known age of the Earth. Albert Einstein discovered the key to such a process in 1905. According to his *special theory of relativity,* a quantity m of mass can in principle be converted into an amount of energy E according to a now-famous equation:

Einstein's mass-energy equation

$$E = mc^2$$

m = quantity of mass, in kg

c = speed of light = 3×10^8 m/s

E = amount of energy into which the mass can be converted, in joules

The speed of light c is a large number, so c^2 is huge. Therefore, a small amount of matter can release an awesome amount of energy.

Inspired by Einstein's ideas, astronomers began to wonder if the Sun's energy output might come from the conversion of matter into energy. The Sun's low density of 1410 kg/m^3 indicates that it must be made of the very lightest atoms, primarily hydrogen and helium. In the 1920s, the British astronomer Arthur Eddington showed that temperatures near the center of the Sun must be so high that atoms become completely ionized. Hence, at the Sun's center we expect to find hydrogen nuclei and electrons flying around independent of each other.

Another British astronomer, Robert Atkinson, suggested that under these conditions hydrogen nuclei could fuse together to produce helium nuclei in a *nuclear reaction* that transforms a tiny amount of mass

Ideas from relativity and nuclear physics led to an understanding of how the Sun shines

BOX 16-1

Converting Mass into Energy

The *Cosmic Connections* figure shows the steps involved in the thermonuclear fusion of hydrogen at the Sun's center. In these steps, four protons are converted into a single nucleus of ^{4}He, an isotope of helium with two protons and two neutrons. (As we saw in Box 5-5, different isotopes of the same element have the same number of protons but different numbers of neutrons.) The reaction depicted in Figure 16-2a also produces a neutral, nearly massless particle called the *neutrino*. Neutrinos respond hardly at all to ordinary matter, so they travel almost unimpeded through the Sun's massive bulk. Hence, the energy that neutrinos carry is quickly lost into space. This loss is not great, however, because the neutrinos carry relatively little energy. (See Section 16-4 for more about these curious particles.)

Most of the energy released by thermonuclear fusion appears in the form of gamma-ray photons. The energy of these photons remains trapped within the Sun for a long time, thus maintaining the Sun's intense internal heat. Some gamma-ray photons are produced by the reaction shown as Step 2 in the *Cosmic Connections* figure. Others appear when an electron in the Sun's interior annihilates a positively charged electron, or **positron**, which is a by-product of the reaction shown in Step 1 in the *Cosmic Connections* figure. An electron and a positron are respectively matter and antimatter, and they convert entirely into energy when they meet. (You may have thought that "antimatter" was pure science fiction. In fact, tremendous amounts of antimatter are being created and annihilated in the Sun as you read these words.)

We can summarize the thermonuclear fusion of hydrogen as follows:

$$4\ ^1\text{H} \rightarrow\ ^4\text{He} + \text{neutrinos} + \text{gamma-ray photons}$$

To calculate how much energy is released in this process, we use Einstein's mass-energy formula: The energy released is equal to the amount of mass consumed multiplied by c^2, where c is the speed of light. To see how much mass is consumed, we compare the combined mass of four hydrogen atoms (the ingredients) to the mass of one helium atom (the product):

4 hydrogen atoms =	6.693×10^{-27} kg
−1 helium atom =	6.645×10^{-27} kg
Mass lost =	0.048×10^{-27} kg

Thus, a small fraction (0.7%) of the mass of the hydrogen going into the nuclear reaction does not show up in the mass of the helium. This lost mass is converted into an amount of energy $E = mc^2$:

$$\begin{aligned} E = mc^2 &= (0.048 \times 10^{-27}\ \text{kg})(3 \times 10^8\ \text{m/s})^2 \\ &= 4.3 \times 10^{-12}\ \text{joule} \end{aligned}$$

This is the amount of energy released by the formation of a single helium atom. It would light a 10-watt lightbulb for almost one-half of a trillionth of a second.

EXAMPLE: How much energy is released when 1 kg of hydrogen is converted to helium?

Situation: We are given the initial mass of hydrogen. We know that a fraction of the mass is lost when the hydrogen undergoes fusion to make helium; our goal is to find the quantity of energy into which this lost mass is transformed.

Tools: We use the equation $E = mc^2$ and the result that 0.7% of the mass is lost when hydrogen is converted into hydrogen.

Answer: When 1 kilogram of hydrogen is converted to helium, the amount of mass lost is 0.7% of 1 kg, or 0.007 kg. (This means that 0.993 kg of helium is produced.) Using Einstein's equation, we find that this missing 0.007 kg of matter is transformed into an amount of energy equal to

$$E = mc^2 = (0.007\ \text{kg})(3 \times 10^8\ \text{m/s})^2 = 6.3 \times 10^{14}\ \text{joules}$$

Review: The energy released by the fusion of 1 kilogram of hydrogen is the same as that released by burning *20,000 metric tons* (2×10^7 kg) of coal! Hydrogen fusion is a *much* more efficient energy source than ordinary burning.

The Sun's luminosity is 3.9×10^{26} joules per second. To generate this much power, hydrogen must be consumed at a rate of

$$\frac{3.9 \times 10^{26}\ \text{joules per second}}{6.3 \times 10^{14}\ \text{joules per kilogram}}$$
$$= 6 \times 10^{11}\ \text{kilograms per second}$$

That is, the Sun converts 600 million metric tons of hydrogen into helium every second.

into a large amount of energy. Experiments in the laboratory using individual nuclei show that such reactions can indeed take place. The process of converting hydrogen into helium is called **hydrogen fusion.** (It is also sometimes called *hydrogen burning,* even though nothing is actually burned in the conventional sense. Ordinary burning involves chemical reactions that rearrange the

outer electrons of atoms but have no effect on the atoms' nuclei.) Hydrogen fusion provides the devastating energy released in a hydrogen bomb (see Figure 1-4).

The fusing together of nuclei is also called **thermonuclear fusion,** because it can take place only at extremely high temperatures. The reason is that all nuclei have a positive electric charge

COSMIC CONNECTIONS

The most common form of hydrogen fusion in the Sun involves three steps, each of which releases energy.

The Proton-Proton Chain

Hydrogen fusion in the Sun usually takes place in a sequence of steps called the proton-proton chain. Each of these steps releases energy that heats the Sun and gives it its luminosity.

STEP 1

(a) Two protons (hydrogen nuclei, ^{1}H) collide.

(b) One of the protons changes into a neutron (shown in blue). The proton and neutron form a hydrogen isotope (^{2}H).

(c) One byproduct of converting a proton to a neutron is a neutral, nearly massless neutrino (ν). This escapes from the Sun.

(d) The other byproduct of converting a proton to a neutron is a positively charged electron, or positron (e^+). This encounters an ordinary electron (e^-), annihilating both particles and converting them into gamma-ray photons (γ). The energy of these photons goes into sustaining the Sun's internal heat.

STEP 2

(a) The ^{2}H nucleus produced in Step 1 collides with a third proton (^{1}H).

(b) The result of the collision is a helium isotope (^{3}He) with two protons and one neutron.

(c) This nuclear reaction releases another gamma-ray photon (γ). Its energy also goes into sustaining the internal heat of the Sun.

STEP 3

(a) The ^{3}He nucleus produced in Step 2 collides with another ^{3}He nucleus produced from three other protons.

(b) Two protons and two neutrons from the two ^{3}He nuclei rearrange themselves into a different helium isotope (^{4}He).

(c) The two remaining protons are released. The energy of their motion contributes to the Sun's internal heat.

(d) Six ^{1}H nuclei went into producing the two ^{3}He nuclei, which combine to make one ^{4}He nucleus. Since two of the original ^{1}H nuclei are returned to their original state, we can summarize the three steps as:

$$4\ ^1\text{H} \longrightarrow\ ^4\text{He} + \text{energy}$$

Hydrogen fusion also takes place in all of the stars visible to the naked eye. (Fusion follows a different sequence of steps in the most massive stars, but the net result is the same.)

(Courtesy of Wally Pacholka)

and so tend to repel one another. But in the extreme heat and pressure at the Sun's center, hydrogen nuclei (protons) are moving so fast that they can overcome their electric repulsion and actually touch one another. When that happens, thermonuclear fusion can take place.

ANALOGY You can think of protons as tiny electrically charged spheres that are coated with a very powerful glue. If the spheres are not touching, the repulsion between their charges pushes them apart. But if the spheres are forced into contact, the strength of the glue "fuses" them together.

CAUTION! Be careful not to confuse thermonuclear fusion with the similar-sounding process of *nuclear fission*. In nuclear fusion, energy is released by joining together nuclei of lightweight atoms such as hydrogen. In nuclear fission, by contrast, the nuclei of very massive atoms such as uranium or plutonium release energy by fragmenting into smaller nuclei. Nuclear power plants produce energy using fission, not fusion. (Generating power using fusion has been a goal of researchers for decades, but no one has yet devised a commercially viable way to do this.)

Converting Hydrogen to Helium

We learned in Section 5-8 that the nucleus of a hydrogen atom (H) consists of a single proton. The nucleus of a helium atom (He) consists of two protons and two neutrons. In the nuclear process that Atkinson described, four hydrogen nuclei combine to form one helium nucleus, with a concurrent release of energy:

$$4\ H \rightarrow He + energy$$

In several separate reactions, two of the four protons are changed into neutrons, and eventually combine with the remaining protons to produce a helium nucleus. This sequence of reactions is called the **proton-proton chain.** The *Cosmic Connections* figure depicts the proton-proton chain in detail.

Each time this process takes place, a small fraction (0.7%) of the combined mass of the hydrogen nuclei does not show up in the mass of the helium nucleus. This "lost" mass is converted into energy. **Box 16-1** describes how to use Einstein's mass-energy equation to calculate the amount of energy released.

CAUTION! You may have heard the idea that mass is always conserved (that is, it is neither created nor destroyed), or that energy is always conserved in a reaction. Einstein's ideas show that neither of these statements is quite correct, because mass can be converted into energy and vice versa. A more accurate statement is that the total amount of mass *plus* energy is conserved. Hence, the destruction of mass in the Sun does not violate any laws of nature.

For every four hydrogen nuclei converted into a helium nucleus, 4.3×10^{-12} joule of energy is released. This may seem like only a tiny amount of energy, but it is about 10^7 times larger than the amount of energy released in a typical chemical reaction, such as occurs in ordinary burning. Thus, thermonuclear fusion can explain how the Sun could have been shining for billions of years.

To produce the Sun's luminosity of 3.9×10^{26} joules per second, 6×10^{11} kg (600 million metric tons) of hydrogen must be converted into helium each second. This rate is prodigious, but there is a literally astronomical amount of hydrogen in the Sun. In particular, the Sun's core contains enough hydrogen to have been giving off energy at the present rate for as long as the solar system has existed, about 4.56 billion years, and to continue doing so for more than 6 billion years into the future.

The proton-proton chain is also the energy source for many of the stars in the sky. In stars with central temperatures that are much hotter than that of the Sun, however, hydrogen fusion proceeds according to a different set of nuclear reactions, called the **CNO cycle,** in which carbon, nitrogen, and oxygen nuclei absorb protons to produce helium nuclei. Still other thermonuclear reactions, such as helium fusion, carbon fusion, and oxygen fusion, occur late in the lives of many stars.

16-2 A theoretical model of the Sun shows how energy gets from its center to its surface

While thermonuclear fusion is the source of the Sun's energy, this process cannot take place everywhere within the Sun. As we have seen, extremely high temperatures—in excess of 10^7 K—are required for atomic nuclei to fuse together to form larger nuclei. The temperature of the Sun's visible surface, about 5800 K, is too low for these reactions to occur there. Hence, thermonuclear fusion can be taking place only within the Sun's interior. But precisely where does it take place? And how does the energy produced by fusion make its way to the surface, where it is emitted into space in the form of photons?

To answer these questions, we must understand conditions in the Sun's interior. Ideally, we would send an exploratory spacecraft to probe deep into the Sun; in practice, the Sun's intense heat would vaporize even the sturdiest spacecraft. Instead, astronomers use the laws of physics to construct a theoretical model of the Sun. (We discussed the use of models in science in Section 1-1.) Let's see what ingredients go into building a model of this kind.

Hydrostatic Equilibrium

Note first that the Sun is not undergoing any dramatic changes. The Sun is not exploding or collapsing, nor is it significantly heating or cooling. The Sun is thus said to be in both *hydrostatic equilibrium* and *thermal equilibrium*.

To understand what is meant by **hydrostatic equilibrium,** imagine a slab of material in the solar interior (**Figure 16-2a**). In equilibrium, the slab on average will move neither up nor down. (In fact, there are upward and downward motions of material inside the Sun, but these motions average out.) Equilibrium is maintained by a balance among three forces that act on this slab:

1. The downward pressure of the layers of solar material above the slab.

2. The upward pressure of the hot gases beneath the slab.

3. The slab's weight—that is, the downward gravitational pull it feels from the rest of the Sun.

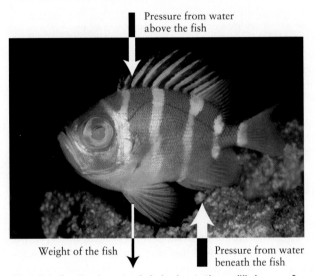

(a) Material inside the sun is in hydrostatic equilibrium, so forces balance

(b) A fish floating in water is in hydrostatic equilibrium, so forces balance

Figure 16-2

Hydrostatic Equilibrium **(a)** Material in the Sun's interior tends to move neither up nor down. The upward forces on a slab of solar material (due to pressure of gases below the slab) must balance the downward forces (due to the slab's weight and the pressure of gases above the slab). Hence, the pressure must increase with increasing depth. **(b)** The same principle applies to a fish floating in water. In equilibrium, the forces balance and the fish neither rises nor sinks. (Ken Usami/PhotoDisc)

The pressure from below must balance both the slab's weight and the pressure from above. Hence, the pressure below the slab must be greater than that above the slab. In other words, pressure has to increase with increasing depth. For the same reason, pressure increases as you dive deeper into the ocean (Figure 16-2*b*) or as you move toward lower altitudes in our atmosphere.

Hydrostatic equilibrium also tells us about the density of the slab. If the slab is too dense, its weight will be too great and it will sink; if the density is too low, the slab will rise. To prevent this, the density of solar material must have a certain value at each depth within the solar interior. (The same principle applies to objects that float beneath the surface of the ocean. Scuba divers wear weight belts to increase their average density so that they will neither rise nor sink but will stay submerged at the same level.)

Thermal Equilibrium

Another consideration is that the Sun's interior is so hot that it is completely gaseous. Gases compress and become more dense when you apply greater pressure to them, so density must increase along with pressure as you go to greater depths within the Sun. Furthermore, when you compress a gas, its temperature tends to rise, so the temperature must also increase as you move toward the Sun's center.

While the temperature in the solar interior is different at different depths, the temperature at each depth remains constant in time. This principle is called **thermal equilibrium.** For the Sun to be in thermal equilibrium, all the energy generated by thermonuclear reactions in the Sun's core must be transported to the Sun's glowing surface, where it can be radiated into space. If too much energy flowed from the core to the surface to be radiated away, the Sun's interior would cool down; alternatively, the Sun's interior would heat up if too little energy flowed to the surface.

Transporting Energy Outward from the Sun's Core

But exactly how is energy transported from the Sun's center to its surface? There are three methods of energy transport: *conduction, convection,* and *radiative diffusion.* Only the last two are important inside the Sun.

If you heat one end of a metal bar with a blowtorch, energy flows to the other end of the bar so that it too becomes warm. The efficiency of this method of energy transport, called **conduction,** varies significantly from one substance to another. For example, copper is a good conductor of heat, but wood is not (which is why copper pots often have wooden handles). Conduction is not an efficient means of energy transport in substances with low average densities, including the gases inside stars like the Sun.

Inside stars like our Sun, energy moves from center to surface by two other means: convection and radiative diffusion. **Convection** is the circulation of fluids—gases or liquids—between hot and cool regions. Hot gases rise toward a star's surface, while cool gases sink back down toward the star's center. This physical movement of gases transports heat energy outward in a star, just as the physical movement of water boiling in a pot transports energy from the bottom of the pot (where the heat is applied) to the cooler water at the surface.

In **radiative diffusion,** photons created in the thermonuclear inferno at a star's center diffuse outward toward the star's surface. Individual photons are absorbed and reemitted by atoms and electrons inside the star. The overall result is an outward migration from the hot core, where photons are constantly created, toward the cooler surface, where they escape into space.

Modeling the Sun

To construct a model of a star like the Sun, astrophysicists express the ideas of hydrostatic equilibrium, thermal equilibrium, and energy transport as a set of equations. To ensure that the model applies to the particular star under study, they also make

Table 16-2 A Theoretical Model of the Sun

Distance from the Sun's center (solar radii)	Fraction of luminosity	Fraction of mass	Temperature ($\times 10^6$ K)	Density (kg/m^3)	Pressure relative to pressure at center
0.0	0.00	0.00	15.5	160,000	1.00
0.1	0.42	0.07	13.0	90,000	0.46
0.2	0.94	0.35	9.5	40,000	0.15
0.3	1.00	0.64	6.7	13,000	0.04
0.4	1.00	0.85	4.8	4,000	0.007
0.5	1.00	0.94	3.4	1,000	0.001
0.6	1.00	0.98	2.2	400	0.0003
0.7	1.00	0.99	1.2	80	4×10^{-5}
0.8	1.00	1.00	0.7	20	5×10^{-6}
0.9	1.00	1.00	0.3	2	3×10^{-7}
1.0	1.00	1.00	0.006	0.00030	4×10^{-13}

Note: The distance from the Sun's center is expressed as a fraction of the Sun's radius ($R_\odot$). Thus, 0.0 is at the center of the Sun and 1.0 is at the surface. The fraction of luminosity is that portion of the Sun's total luminosity produced within each distance from the center; this is equal to 1.00 for distances of 0.25 $R_\odot$ or more, which means that all of the Sun's nuclear reactions occur within 0.25 solar radius from the Sun's center. The fraction of mass is that portion of the Sun's total mass lying within each distance from the Sun's center. The pressure is expressed as a fraction of the pressure at the center of the Sun.

use of astronomical observations of the star's surface. (For example, to construct a model of the Sun, they use the data that the Sun's surface temperature is 5800 K, its luminosity is 3.9×10^{26} W, and the gas pressure and density at the surface are almost zero.) The astrophysicists then use a computer to solve their set of equations and calculate conditions layer by layer in toward

the star's center. The result is a model of how temperature, pressure, and density increase with increasing depth below the star's surface.

Table 16-2 and Figure 16-3 show a theoretical model of the Sun that was calculated in just this way. Different models of the Sun use slightly different assumptions, but all models give essen-

Figure 16-3

A Theoretical Model of the Sun's Interior These graphs depict what percentage of the Sun's total luminosity is produced within each distance from the center (upper left), what percentage of the total mass lies within each distance from the center (lower left), the temperature at each distance (upper right), and the density at each distance (lower right). (See Table 16-2 for a numerical version of this model.)

---

tially the same results as those shown here. From such computer models we have learned that at the Sun's center the density is 160,000 kg/m³ (14 times the density of lead!), the temperature is 1.55×10^7 K, and the pressure is 3.4×10^{11} atm. (One atmosphere, or 1 atm, is the average atmospheric pressure at sea level on Earth.)

Table 16-2 and Figure 16-3 show that the solar luminosity rises to 100% at about one-quarter of the way from the Sun's center to its surface. In other words, the Sun's energy production occurs within a volume that extends out only to 0.25 $R_\odot$. (The symbol $R_\odot$ denotes the solar radius, or radius of the Sun as a whole, equal to 696,000 km.) Outside 0.25 $R_\odot$, the density and temperature are too low for thermonuclear reactions to take place. Also note that 94% of the total mass of the Sun is found within the inner 0.5 $R_\odot$. Hence, the outer 0.5 $R_\odot$ contains only a relatively small amount of material.

How energy flows from the Sun's center toward its surface depends on how easily photons move through the gas. If the solar gases are comparatively transparent, photons can travel moderate distances before being scattered or absorbed, and energy is thus transported by radiative diffusion. If the gases are comparatively opaque, photons cannot get through the gas easily and heat builds up. Convection then becomes the most efficient means of energy transport. The gases start to churn, with hot gas moving upward and cooler gas sinking downward.

From the center of the Sun out to about 0.71 $R_\odot$, energy is transported by radiative diffusion. Hence, this region is called the **radiative zone.** Beyond about 0.71 $R_\odot$, the temperature is low enough (a mere 2×10^6 K or so) for electrons and hydrogen nuclei to join into hydrogen atoms. These atoms are very effective at absorbing photons, much more so than free electrons or nuclei,

and this absorption chokes off the outward flow of photons. Therefore, beyond about 0.71 $R_\odot$, radiative diffusion is not an effective way to transport energy. Instead, convection dominates the energy flow in this outer region, which is why it is called the **convective zone.** Figure 16-4 shows these aspects of the Sun's internal structure.

Although energy travels through the radiative zone in the form of photons, the photons have a difficult time of it. Table 16-2 shows that the material in this zone is extremely dense, so photons from the Sun's core take a long time to diffuse through the radiative zone. As a result, it takes approximately 170,000 years for energy created at the Sun's center to travel 696,000 km to the solar surface and finally escape as sunlight. The energy flows outward at an average rate of 50 centimeters per hour, or about 20 times slower than a snail's pace.

Once the energy escapes from the Sun, it travels much faster—at the speed of light. Thus, solar energy that reaches you today took only 8 minutes to travel the 150 million kilometers from the Sun's surface to the Earth. But this energy was actually produced by thermonuclear reactions that took place in the Sun's core hundreds of thousands of years ago.

> **The sunlight that reaches Earth today results from thermonuclear reactions that took place about 170,000 years ago**

16-3 Astronomers probe the solar interior using the Sun's own vibrations

 We have described how astrophysicists construct models of the Sun. But since we cannot see into the Sun's opaque interior, how can we check these models to see if they are accurate? What is needed is a technique for probing the Sun's interior. A very powerful technique of just this kind involves measuring vibrations of the Sun as a whole. This field of solar research is called **helioseismology.**

Vibrations are a useful tool for examining the hidden interiors of all kinds of objects. Food shoppers test whether melons are ripe by tapping on them and listening to the vibrations. Geologists can determine the structure of the Earth's interior by using seismographs to record vibrations during earthquakes.

Although there are no true "sunquakes," the Sun does vibrate at a variety of frequencies, somewhat like a ringing bell. These vibrations were first noticed in 1960 by Robert Leighton of the California Institute of Technology, who made high-precision Doppler shift observations of the solar surface. These measurements revealed that parts of the Sun's surface move up and down about 10 meters every 5 minutes. Since the mid-1970s, several astronomers have reported slower vibrations, having periods ranging from 20 to 160 minutes. The detection of extremely slow vibrations has inspired astronomers to organize networks of telescopes around and in orbit above the Earth to monitor the Sun's vibrations on a continuous basis.

The vibrations of the Sun's surface can be compared with sound waves. If you could somehow survive within the Sun's outermost layers, you would first notice a deafening roar, somewhat like a jet engine, produced by

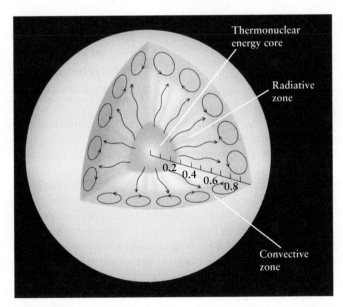

Figure 16-4

The Sun's Internal Structure Thermonuclear reactions occur in the Sun's core, which extends out to a distance of 0.25 $R_\odot$ from the center. Energy is transported outward, via radiative diffusion, to a distance of about 0.71 $R_\odot$. In the outer layers between 0.71 $R_\odot$ and 1.00 $R_\odot$, energy flows outward by convection.

turbulence in the Sun's gases. But superimposed on this noise would be a variety of nearly pure tones. You would need greatly enhanced hearing to detect these tones, however; the strongest has a frequency of just 0.003 hertz, 13 octaves below the lowest frequency audible to humans. (Recall from Section 5-2 that one hertz is one oscillation per second.)

In 1970, Roger Ulrich, at UCLA, pointed out that sound waves moving upward from the solar interior would be reflected back inward after reaching the solar surface. However, as a reflected sound wave descends back into the Sun, the increasing density and pressure bend the wave severely, turning it around and aiming it back toward the solar surface. In other words, sound waves bounce back and forth between the solar surface and layers deep within the Sun. These sound waves can reinforce each other if their wavelength is the right size, just as sound waves of a particular wavelength resonate inside an organ pipe.

The Sun oscillates in millions of ways as a result of waves resonating in its interior. **Figure 16-5** is a computer-generated illustration of one such mode of vibration. Helioseismologists can deduce information about the solar interior from measurements of these oscillations. For example, they have been able to set limits on the amount of helium in the Sun's core and con-

> The Sun's oscillations resemble those inside an organ pipe, but are much more complex

vective zone and to determine the thickness of the transition region between the radiative zone and convective zone. They have also found that the convective zone is thicker than previously thought.

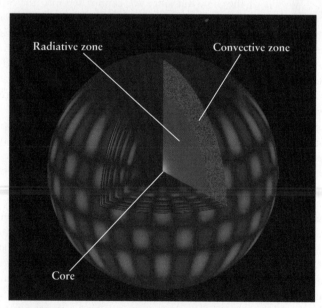

Figure 16-5

A Sound Wave Resonating in the Sun This computer-generated image shows one of the millions of ways in which the Sun's interior vibrates. The regions that are moving outward are colored blue, those moving inward, red. As the cutaway shows, these oscillations are thought to extend into the Sun's radiative zone (compare Figure 16-4). (National Solar Observatory)

16-4 Neutrinos reveal information about the Sun's core—and have surprises of their own

We have seen circumstantial evidence that thermonuclear fusion is the source of the Sun's power. To be certain, however, we need more definitive evidence. How can we show that thermonuclear fusion really is taking place in the Sun's core?

What Sunlight Cannot Tell Us

Although the light energy that we receive from the Sun originates in the core, it provides few clues about conditions there. The problem is that this energy has changed form repeatedly during its passage from the core: It appeared first as photons diffusing through the radiative zone, then as heat transported through the outer layers by convection, and then again as photons emitted from the Sun's glowing surface. As a result of these transformations, much of the information that the Sun's radiated energy once carried about conditions in the core has been lost.

ANALOGY If you make a photocopy of a photocopy of a photocopy of an original document, the final result may be so blurred as to be unreadable. In an analogous way, because solar energy is transformed many times while en route to Earth, the story it could tell us about the Sun's core is hopelessly blurred.

Helioseismology also cannot reveal what is happening in the core. The vibrations that astronomers see on the Sun's surface do not penetrate that far into the interior.

Solar Neutrinos

Happily, there is a way for scientists to learn about conditions in the Sun's core and to get direct evidence that thermonuclear fusion really does happen there. The trick is to detect the subatomic by-products of thermonuclear fusion reactions.

> Scientists use the most ethereal of subatomic particles to learn about the Sun, and vice versa

As part of the process of hydrogen fusion, protons change into neutrons and release **neutrinos** (see the *Cosmic Connections* figure in Section 16-1 as well as Box 16-1). Like photons, neutrinos are particles that have no electric charge. Unlike photons, however, neutrinos interact only very weakly with matter. Even the vast bulk of the Sun offers little impediment to their passage, so neutrinos must be streaming out of the core and into space. Indeed, the conversion of hydrogen into helium at the Sun's center produces 10^{38} neutrinos each second. Every second, about 10^{14} neutrinos from the Sun—that is, **solar neutrinos**—must pass through each square meter of the Earth.

If it were possible to detect these solar neutrinos, we would have direct evidence that thermonuclear reactions really do take place in the Sun's core. Beginning in the 1960s, scientists began to build neutrino detectors for precisely this purpose.

The challenge is that neutrinos are exceedingly difficult to detect. Just as neutrinos pass unimpeded through the Sun, they also pass through the Earth almost as if it were not there. We stress the word "almost," because neutrinos can and do interact with matter, albeit infrequently.

On rare occasions a neutrino will strike a neutron and convert it into a proton. This effect was the basis of the original solar neutrino detector, designed and built by Raymond Davis of the Brookhaven National Laboratory in the 1960s. This device used 100,000 gallons of perchloroethylene (C_2Cl_4), a fluid used in dry cleaning, in a huge tank buried deep underground. Most of the solar neutrinos that entered Davis's tank passed right through it with no effect whatsoever. But occasionally a neutrino struck the nucleus of one of the chlorine atoms (^{37}Cl) in the cleaning fluid and converted one of its neutrons into a proton, creating a radioactive atom of argon (^{37}Ar).

The rate at which argon is produced is related to the neutrino flux—that is, the number of neutrinos from the Sun arriving at the Earth per square meter per second. By counting the number of newly created argon atoms, Davis was able to determine the neutrino flux from the Sun. (Other subatomic particles besides neutrinos can also induce reactions that create radioactive atoms. By placing the experiment deep underground, however, the body of the Earth absorbs essentially all such particles—with the exception of neutrinos.)

The Solar Neutrino Problem

Davis and his collaborators found that solar neutrinos created one radioactive argon atom in the tank every three days. But this rate corresponded to only one-third of the neutrino flux predicted from standard models of the Sun. This troubling discrepancy between theory and observation, called the **solar neutrino problem,** motivated scientists around the world to conduct further experiments to measure solar neutrinos.

One key question was whether the neutrinos that Davis had detected had really come from the Sun. (The Davis experiment had no way to determine the direction from which neutrinos had entered the tank of cleaning fluid.) This was resolved by an experiment in Japan called Kamiokande, which was designed by the physicist Masatoshi Koshiba. A large underground tank containing 3000 tons of water was surrounded by 1100 light detectors. From time to time, a high-energy solar neutrino struck an electron in one of the water molecules, dislodging it and sending it flying like a pin hit by a bowling ball. The recoiling electron produced a flash of light, which was sensed by the detectors. By analyzing the flashes, scientists could tell the direction from which the neutrinos were coming and confirmed that they emanated from the Sun. These results in the late 1980s gave direct evidence that thermonuclear fusion is indeed occurring in the Sun's core. (Davis and Koshiba both received the 2002 Nobel Prize in Physics for their pioneering research on solar neutrinos.)

Like Davis's experiment, however, Koshiba and his colleagues at Kamiokande detected only a fraction of the expected flux of neutrinos. Where, then, were the missing solar neutrinos?

One proposed solution had to do with the energy of the detected neutrinos. The vast majority of neutrinos from the Sun are created during the first step in the proton-proton chain, in which two protons combine to form a heavy isotope of hydrogen (see the *Cosmic Connections* figure in Section 16-1). But these neutrinos have too little energy to convert chlorine into argon. Both Davis's and Koshiba's experiments responded only to high-energy neutrinos produced by reactions that occur only part of the time near the end of the proton-proton chain. (The *Cosmic Connections* figure in Section 16-1 does not show these reactions.) Could it be that the discrepancy between theory and observation would go away if the flux of low-energy neutrinos could be measured?

To test this idea, two teams of physicists constructed neutrino detectors that used several tons of gallium (a liquid metal) rather than cleaning fluid. Low-energy neutrinos convert gallium (^{71}Ga) into a radioactive isotope of germanium (^{71}Ge). By chemically separating the germanium from the gallium and counting the radioactive atoms, the physicists were able to measure the flux of low-energy solar neutrinos. These experiments—GALLEX in Italy and SAGE (Soviet-American Gallium Experiment) in Russia—detected only 50% to 60% of the expected neutrino flux. Hence, the solar neutrino problem was a discrepancy between theory and observation for neutrinos of all energies.

Another proposed solution to the neutrino problem was that the Sun's core is cooler than predicted by solar models. If the Sun's central temperature were only 10% less than the current estimate, fewer neutrinos would be produced and the neutrino flux would agree with experiments. However, a lower central temperature would cause other obvious features, such as the Sun's size and surface temperature, to be different from what we observe.

Finding the Missing Neutrinos

Only very recently has the solution to the neutrino problem been found. The answer lies not in how neutrinos are produced, but rather in what happens to them between the Sun's core and detectors on the Earth. Physicists have found that there are actually three types of neutrinos. Only one of these types is produced in the Sun, and it is only this type that can be detected by the experiments we have described. But if some of the solar neutrinos change in flight into a different type of neutrino, the detectors in these experiments would record only a fraction of the total neutrino flux. This effect is called **neutrino oscillation.** In June 1998, scientists at the Super-Kamiokande neutrino observatory (a larger and more sensitive device than Kamiokande) revealed evidence that neutrino oscillation does indeed take place.

The best confirmation of this idea has come from the Sudbury Neutrino Observatory (SNO) in Canada. Like Kamiokande and Super-Kamiokande, SNO uses a large tank of water placed deep underground (**Figure 16-6**). But unlike those earlier experiments, SNO can detect all of the three types of neutrinos. It does this by using *heavy water*. In ordinary, or "light," water, each hydrogen atom in the H_2O molecule has a solitary proton as its nucleus. In heavy water, by contrast, each of the hydrogen nuclei has a nucleus made up of a proton and a neutron. (This is the isotope ^2H shown in the *Cosmic Connections* figure in Section 16-1.) If a high-energy solar neutrino of any type passes through SNO's tank of heavy water, it can knock the neutron out of one of the ^2H nuclei. The ejected neutron can then be captured by

WEB LINK 16.7

Figure 16-6 R I **V** U X G

The Sudbury Neutrino Observatory Under Construction The transparent acrylic sphere holds 1000 tons of heavy water. Any of the three types of solar neutrino produces a flash of light when it interacts with the heavy water. The flash is sensed by 9600 light detectors surrounding the tank. (The detectors were not all installed when this photograph was taken.) (Photo courtesy of SNO)

Hollow transparent sphere to hold heavy water

Frame for light detectors

Workers

another nucleus, and this capture releases energy that manifests itself as a tiny burst of light. As in Kamiokande, detectors around SNO's water tank record these light flashes.

Scientists using SNO have found that the combined flux of all three types of neutrinos coming from the Sun is *equal* to the theoretical prediction. Together with the results from earlier neutrino experiments, this strongly suggests that the Sun is indeed producing neutrinos at the predicted rate as a by-product of thermonuclear reactions. But before these neutrinos can reach the Earth, about two-thirds of them undergo an oscillation and change their type. Thus, there is really no solar neutrino problem—scientists merely needed the right kind of detectors to observe all the neutrinos, including the ones that transformed in flight.

The story of the solar neutrino problem illustrates how two different branches of science—in this case, studies of the solar interior and investigations of subatomic particles—can sometimes interact, to the mutual benefit of both. While there is still much we do not understand about the Sun and about neutrinos, a new generation of neutrino detectors in Japan, Canada, and elsewhere promises to further our knowledge of these exotic realms of astronomy and physics.

16-5 The photosphere is the lowest of three main layers in the Sun's atmosphere

Although the Sun's core is hidden from our direct view, we can easily see sunlight coming from the high-temperature gases that make up the Sun's atmosphere. These outermost layers of the Sun prove to be the sites of truly dramatic activity, much of which has a direct impact on our planet. By studying these layers, we gain further insight into the character of the Sun as a whole.

Observing the Photosphere

A visible-light photograph like **Figure 16-7** makes it appear that the Sun has a definite surface. This is actually an illusion; the Sun is gaseous throughout its volume because of its high internal temperature, and the gases simply become less and less dense as you move farther away from the Sun's center.

Figure 16-7 R I **V** U X G

The Photosphere The photosphere is the layer in the solar atmosphere from which the Sun's visible light is emitted. Note that the Sun appears darker around its limb, or edge; here we are seeing the upper photosphere, which is relatively cool and thus glows less brightly. (The dark sunspots, which we discuss in Section 16-8, are also relatively cool regions.) (Celestron International)

Why, then, does the Sun appear to have a sharp, well-defined surface? The reason is that essentially all of the Sun's visible light emanates from a single, thin layer of gas called the **photosphere** ("sphere of light"). Just as you can see only a certain distance through the Earth's atmosphere before objects vanish in the haze, we can see only about 400 km into the photosphere. This distance is so small compared with the Sun's radius of 696,000 km that the photosphere appears to be a definite surface.

The photosphere is actually the lowest of the three layers that together constitute the solar atmosphere. Above it are the *chromosphere* and the *corona*, both of which are transparent to visible light. We can see them only using special techniques, which we discuss later in this chapter. Everything below the photosphere is called the *solar interior*.

The Lesson of Limb Darkening

The photosphere is heated from below by energy streaming outward from the solar interior. Hence, temperature should decrease as you go upward in the photosphere, just as in the solar interior (see Table 16-2 and Figure 16-3). We know this is the case because the photosphere appears darker around the edge, or *limb*, of the Sun than it does toward the center of the solar disk, an effect called **limb darkening** (examine Figure 16-7). This happens because when we look near the Sun's limb, we do not see as deeply into the photosphere as we do when we look near the center of the disk (**Figure 16-8**). The high-altitude gas we observe at the limb is not as hot and thus does not glow as brightly as the deeper, hotter gas seen near the disk center.

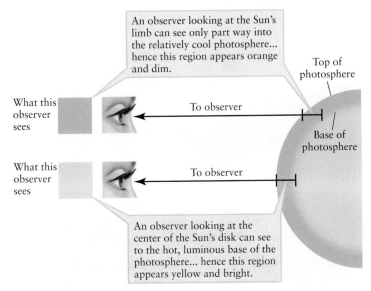

An observer looking at the Sun's limb can see only part way into the relatively cool photosphere... hence this region appears orange and dim.

Top of photosphere

What this observer sees

To observer

Base of photosphere

What this observer sees

To observer

An observer looking at the center of the Sun's disk can see to the hot, luminous base of the photosphere... hence this region appears yellow and bright.

Figure 16-8

The Origin of Limb Darkening Light from the Sun's limb and light from the center of its disk both travel about the same straight-line distance through the photosphere to reach us. Because of the Sun's curved shape, light from the limb comes from a greater height within the photosphere, where the temperature is lower and the gases glow less brightly. Hence, the limb appears darker and more orange.

The spectrum of the Sun's photosphere confirms how its temperature varies with altitude. As we saw in Section 16-1, the photosphere shines like a nearly perfect blackbody with an average temperature of about 5800 K. However, superimposed on this spectrum are many dark absorption lines (see Figure 5-12). As discussed in Section 5-6, we see an *absorption line spectrum* of this sort whenever we view a hot, glowing object through a relatively cool gas. In this case, the hot object is the lower part of the photosphere; the cooler gas is in the upper part of the photosphere, where the temperature declines to about 4400 K. All the absorption lines in the Sun's spectrum are produced in this relatively cool layer, as atoms selectively absorb photons of various wavelengths streaming outward from the hotter layers below.

CAUTION! You may find it hard to think of 4400 K as "cool." But keep in mind that the ratio of 4400 K to 5800 K, the temperature in the lower photosphere, is the same as the ratio of the temperature on a Siberian winter night to that of a typical day in Hawaii.

Granules and Supergranules in the Photosphere

We can learn still more about the photosphere by examining it with a telescope—but only when using special dark filters to prevent eye damage. *Looking directly at the Sun without the correct filter, whether with the naked eye or with a telescope, can cause permanent blindness!* Under good observing conditions, astronomers using such filter-equipped telescopes can often see a blotchy pattern in the photosphere, called **granulation** (**Figure 16-9**). Each light-colored **granule** measures about 1000 km (600 mi) across—equal in size to the areas of Texas and Oklahoma combined—and is surrounded by a darkish boundary. The difference in brightness between the center and the edge of a granule corresponds to a temperature drop of about 300 K.

Granulation is caused by convection of the gas in the photosphere. The inset in Figure 16-9 shows how gas from lower levels rises upward in granules, cools off, spills over the edges of the granules, and then plunges back down into the Sun. This can occur only if the gas is heated from below, like a pot of water being heated on a stove (see Section 16-2). Along with limb darkening and the Sun's absorption line spectrum, granulation shows that the upper part of the photosphere must be cooler than the lower part.

> Like water boiling on a stove, the photosphere bubbles with convection cells

Time-lapse photography reveals more of the photosphere's dynamic activity. Granules form, disappear, and reform in cycles lasting only a few minutes. At any one time, about 4 million granules cover the solar surface.

Superimposed on the pattern of granulation are even larger convection cells called **supergranules** (**Figure 16-10**). As in granules, gases rise upward in the middle of a supergranule, move horizontally outward toward its edge, and descend back into the Sun. The difference is that a typical supergranule is about 35,000 km in diameter, large enough to enclose several hundred granules. This large-scale convection moves at only about 0.4 km/s (1400 km/h, or 900 mi/h), about one-tenth the speed of gases churning in a granule. A given supergranule lasts about a day.

Figure 16-9 R I V U X G

Solar Granulation High-resolution photographs of the Sun's surface reveal a blotchy pattern called granulation. Granules are convection cells about 1000 km (600 mi) wide in the Sun's photosphere. **Inset:** Rising hot gas produces bright granules.

Cooler gas sinks downward along the boundaries between granules; this gas glows less brightly, giving the boundaries their dark appearance. This convective motion transports heat from the Sun's interior outward to the solar atmosphere. (MSFC/NASA; inset: Goran Scharmer, Lund Observatory)

ANALOGY Similar patterns of large-scale and small-scale convection can be found in the Earth's atmosphere. On the large scale, air rises gradually at a low-pressure area, then sinks gradually at a high-pressure area, which might be hundreds of kilometers away. This is analogous to the flow in a supergranule. Thunderstorms in our atmosphere are small but intense convection cells within which air moves rapidly up and down. Like granules, they last only a relatively short time before they dissipate.

The Photosphere: Hot, Thin, and Opaque

Although the photosphere is a very active place, it actually contains relatively little material. Careful examination of the spectrum shows that it has a density of only about 10^{-4} kg/m^3,

Blue: areas of rising gas
Red: areas of sinking gas

Figure 16-10 R I V U X G

Supergranules and Large-Scale Convection Supergranules display relatively little contrast between their center and edges, so they are hard to observe in ordinary images. But they can be seen in a false-color Doppler image like this one. Light from gas that is approaching us (that is, rising) is shifted toward shorter wavelengths, while light from receding gas (that is, descending) is shifted toward longer wavelengths (see Section 5-9). (David Hathaway, MSFC/NASA) roughly 0.01% the density of the Earth's atmosphere at sea level. The photosphere is made primarily of hydrogen and helium, the most abundant elements in the solar system (see Figure 8-3).

Despite being such a thin gas, the photosphere is surprisingly opaque to visible light. If it were not so opaque, we could see into the Sun's interior to a depth of hundreds of thousands of kilometers, instead of the mere 400 km that we can see down into the photosphere. What makes the photosphere so opaque is that its hydrogen atoms sometimes acquire an extra electron, becoming **negative hydrogen ions**. The extra electron is only loosely attached and can be dislodged if it absorbs a photon of any visible wavelength. Hence, negative hydrogen ions are very efficient light absorbers, and there are enough of these light-absorbing ions in the photosphere to make it quite opaque. Because it is so opaque, the photosphere's spectrum is close to that of an ideal blackbody.

16-6 Spikes of rising gas extend through the Sun's chromosphere

An ordinary visible-light image such as Figure 16-7 gives the impression that the Sun ends at the top of the photosphere. But during a total solar eclipse, the Moon blocks the photosphere from our view, revealing a glowing, pinkish layer of gas above the photosphere (**Figure 16-11**). This is the tenuous **chromosphere**

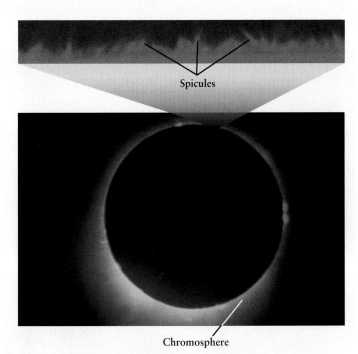

Spicules

Chromosphere

Figure 16-11   R I V U X G

The Chromosphere During a total solar eclipse, the Sun's glowing chromosphere can be seen around the edge of the Moon. It appears pinkish because its hot gases emit light at only certain discrete wavelengths, principally the H$_\alpha$ emission of hydrogen at a red wavelength of 656.3 nm. The expanded area above shows spicules, jets of chromospheric gas that surge upward into the Sun's outer atmosphere. (NOAO)

("sphere of color"), the second of the three major levels in the Sun's atmosphere. The chromosphere is only about one ten-thousandth (10^{-4}) as dense as the photosphere, or about 10^{-8} as dense as our own atmosphere. No wonder it is normally invisible!

Comparing the Chromosphere and Photosphere

Unlike the photosphere, which has an absorption line spectrum, the chromosphere has a spectrum dominated by emission lines. An emission line spectrum is produced by the atoms of a hot, thin gas (see Section 5-6 and Section 5-8). As their electrons fall from higher to lower energy levels, the atoms emit photons.

One of the strongest emission lines in the chromosphere's spectrum is the H$_\alpha$ line at 656.3 nm, which is emitted by a hydrogen atom when its single electron falls from the $n = 3$ level to the $n = 2$ level (recall Figure 5-21b). This wavelength is in the red part of the spectrum, which gives the chromosphere its characteristic pinkish color. The spectrum also contains emission lines of singly ionized calcium, as well as lines due to ionized helium and ionized metals. In fact, helium was originally discovered in the chromospheric spectrum in 1868, almost 30 years before helium gas was first isolated on Earth.

Analysis of the chromospheric spectrum shows that temperature *increases* with increasing height in the chromosphere. This is just the opposite of the situation in the photosphere, where temperature decreases with increasing height. The temperature is about 4400 K at the top of the photosphere; 2000 km higher, at the top of the chromosphere, the temperature is nearly 25,000 K. This is very surprising, since temperatures should decrease as you move away from the Sun's interior. In Section 16-10 we will see how solar scientists explain this seeming paradox.

 The photospheric spectrum is dominated by absorption lines at certain wavelengths, while the spectrum of the chromosphere has emission lines at these same wavelengths. In other words, the photosphere appears dark at the wavelengths at which the chromosphere emits most strongly, such as the H$_\alpha$ wavelength of 656.3 nm. By viewing the Sun through a special filter that is transparent to light only at the wavelength of H$_\alpha$, astronomers can screen out light from the photosphere and make the chromosphere visible. (The same technique can be used with other wavelengths at which the chromosphere emits strongly, including nonvisible wavelengths.) This makes it possible to see the chromosphere at any time, not just during a solar eclipse.

Spicules

The top photograph in Figure 16-11 is a high-resolution image of the Sun's chromosphere taken through an H$_\alpha$ filter. This image shows numerous vertical spikes, which are actually jets of rising gas called **spicules**. A typical spicule lasts just 15 minutes or so: It rises at the rate of about 20 km/s (72,000 km/h, or 45,000 mi/h), can reach a height of several thousand kilometers, and then collapses and fades away (**Figure 16-12**). Approximately 300,000 spicules exist at any one time, covering about 1% of the Sun's surface.

> **Jets of gas thousands of kilometers in height rise through the chromosphere**

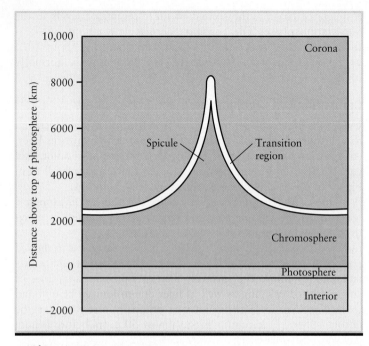

Figure 16-12

The Solar Atmosphere This schematic diagram shows the three layers of the solar atmosphere. The lowest, the photosphere, is about 400 km thick. The chromosphere extends about 2000 km higher, with spicules jutting up to nearly 10,000 km above the photosphere. Above a transition region is the Sun's outermost layer, the corona, which we discuss in Section 16-7. It extends many millions of kilometers out into space. (Adapted from J. A. Eddy)

Spicules are generally located directly above the edges of supergranules (see Figure 16-10). This is a surprising result, because chromospheric gases are rising in a spicule while photospheric gases are *descending* at the edge of a supergranule. What, then, is pulling gases upward to form spicules? As we will see in Section 16-10, the answer proves to be the Sun's intense magnetic field. But before we delve into how this happens, let us complete our tour of the solar atmosphere by exploring its outermost, least dense, most dynamic, and most bizarre layer—the region called the corona.

16-7 The corona ejects mass into space to form the solar wind

The **corona**, or outermost region of the Sun's atmosphere, begins at the top of the chromosphere. It extends out to a distance of several million kilometers. Despite its tremendous extent, the corona is only about one-millionth (10^{-6}) as bright as the photosphere—no brighter than the full moon. Hence, the corona can be viewed only when the light from the photosphere is blocked out, either by use of a specially designed telescope or during a total eclipse.

Figure 16-13 is an exceptionally detailed photograph of the Sun's corona taken during a solar eclipse. It shows that the corona is not merely a spherical shell of gas surrounding the Sun.

Rather, numerous streamers extend in different directions far above the solar surface. The shapes of these streamers vary on time scales of days or weeks. (For another view of the corona during a solar eclipse, see Figure 3-10*b*.)

Comparing the Corona, Chromosphere, and Photosphere

Like the chromosphere that lies below it, the corona has an emission line spectrum characteristic of a hot, thin gas. When the spectrum of the corona was first measured in the nineteenth century, astronomers found a number of emission lines at wavelengths that had never been seen in the laboratory. Their explanation was that the corona contained elements that had not yet been detected on the Earth. However, laboratory experiments in the 1930s revealed that these unusual emission lines were in fact caused by the same atoms found elsewhere in the universe—but in highly ionized states. For example, a prominent green line at 530.3 nm is caused by highly ionized iron atoms, each of which has been stripped of 13 of its 26 electrons. In order to strip that many electrons from atoms, temperatures in the corona must reach 2 million kelvins (2×10^6 K) or even higher—far greater than the temperatures in the chromosphere. Figure 16-14 shows how temperature in both the chromosphere and corona varies with altitude.

CAUTION! The corona is actually not very "hot"—that is, it contains very little thermal energy. The reason is that the corona is nearly a vacuum. In the corona there are only about 10^{11} atoms

Figure 16-13   R I V U X G

The Solar Corona This striking photograph of the corona was taken during the total solar eclipse of July 11, 1991. Numerous streamers extend for millions of kilometers above the solar surface. The unearthly light of the corona is one of the most extraordinary aspects of experiencing a solar eclipse. (Courtesy of R. Christen and M. Christen, Astro-Physics, Inc.)

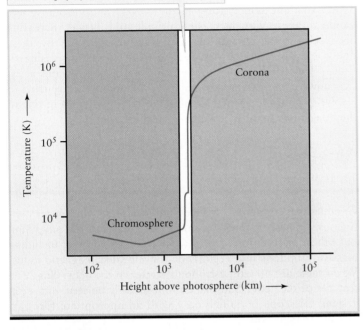

In this narrow transition region between the chromosphere and corona, the temperature rises abruptly by about a factor of 100.

Figure 16-14

Temperatures in the Sun's Upper Atmosphere This graph shows how temperature varies with altitude in the Sun's chromosphere and corona and in the narrow transition region between them. In order to show a large range of values, both the vertical and horizontal scales are nonlinear. (Adapted from A. Gabriel)

per cubic meter, compared with about 10^{23} atoms per cubic meter in the Sun's photosphere and about 10^{25} atoms per cubic meter in the air that we breathe. Because of the corona's high temperature, the atoms there are moving at very high speeds. But because there are so few atoms in the corona, the total amount of energy in these moving atoms (a measure of how "hot" the gas is) is rather low. If you flew a spaceship into the corona, you would have to worry about becoming overheated by the intense light coming from the photosphere, but you would notice hardly any heating from the corona's ultra-thin gas.

ANALOGY The situation in the corona is similar to that inside a conventional oven that is being used for baking. Both the walls of the oven and the air inside the oven are at the same high temperature, but the air contains very few atoms and thus carries little energy. If you put your hand in the oven momentarily, the lion's share of the heat you feel is radiation from the oven walls.

The low density of the corona explains why it is so dim compared with the photosphere. In general, the higher the temperature of a gas, the brighter it glows. But because there are so few atoms in the corona, the net amount of light that it emits is very feeble compared with the light from the much cooler, but also much denser, photosphere.

The Solar Wind and Coronal Holes

The Earth's gravity keeps our atmosphere from escaping into space. In the same way, the Sun's powerful gravitational attraction keeps most of the gases of the photosphere, chromosphere, and corona from escaping. But the corona's high temperature means that its atoms and ions are moving at very high speeds, around a million kilometers per hour. As a result, some of the coronal gas can and does escape. This outflow of gas, which we first encountered in Section 8-5, is called the **solar wind.**

Each second the Sun ejects about a million tons (10^9 kg) of material into the solar wind. But the Sun is so massive that, even over its entire lifetime, it will eject only a few tenths of a percent of its total mass. The solar wind is composed almost entirely of electrons and nuclei of hydrogen and helium. About 0.1% of the solar wind is made up of ions of more massive atoms, such as silicon, sulfur, calcium, chromium, nickel, iron, and argon. The aurorae seen at far northern or southern latitudes on Earth are produced when electrons and ions from the solar wind enter our upper atmosphere.

Special telescopes enable astronomers to see the origin of the solar wind. To appreciate what sort of telescopes are needed, note that because the temperature of the coronal gas is so high, ions in the corona are moving very fast (see Box 7-2). When ions collide, the energy of the impact is so great that the ion's electrons are boosted to very high energy levels. As the electrons fall back to lower levels, they emit high-energy photons in the ultraviolet and X-ray portions of the spectrum—wavelengths at which the photosphere and chromosphere are relatively dim. Hence, telescopes sensitive to these short wavelengths are ideal for studying the corona and the flow of the solar wind.

WEB LINK 16.9 The Earth's atmosphere is opaque to ultraviolet light and X rays, so telescopes for these wavelengths must be placed on board spacecraft (see Section 6-7, especially Figure 6-25). **Figure 16-15** shows an ultraviolet view of the corona from the *SOHO* spacecraft (*Solar and Heliospheric Observatory*), a joint project of the European Space Agency (ESA) and NASA.

Figure 16-15 reveals that the corona is not uniform in temperature or density. The densest, highest-temperature regions appear bright, while the thinner, lower-temperature regions are dark. Note the large dark area, called a **coronal hole** because it is almost devoid of luminous gas. Particles streaming away from the Sun can most easily flow outward through these particularly thin regions.

> Unlike the lower levels of the Sun's atmosphere, the corona has immense holes that shift and reshape

Therefore, it is thought that coronal holes are the main corridors through which particles of the solar wind escape from the Sun.

WEB LINK 16.10 Evidence in favor of this picture has come from the *Ulysses* spacecraft, another joint ESA/NASA mission. In 1994 and 1995, *Ulysses* became the first spacecraft to fly over the Sun's north and south poles, where there are apparently permanent coronal holes. The spacecraft indeed measured a stronger solar wind emanating from these holes.

The temperatures in the corona and the chromosphere are not at all what we would expect. Just as you feel warm if you stand close to a campfire but cold if you move away, we would expect that the temperature in the corona and chromosphere would *decrease* with increasing altitude and, hence, increasing distance from the warmth of the Sun's photosphere. Why, then, does the temperature in these regions *increase* with increasing altitude? This has been one of the major unsolved mysteries in astronomy for the past half-century. As astronomers have tried to resolve this dilemma, they have found important clues in one of the Sun's most familiar features—sunspots.

16-8 Sunspots are low-temperature regions in the photosphere

Granules, supergranules, spicules, and the solar wind occur continuously. These features are said to be aspects of the *quiet* Sun. But other, more dramatic features appear periodically, including massive eruptions and regions of concentrated magnetic fields. When these are present, astronomers refer to the *active* Sun. The features of the active Sun that can most easily be seen with even a small telescope (although only with an appropriate filter attached) are sunspots.

Observing Sunspots

Sunspots are irregularly shaped dark regions in the photosphere. Sometimes sunspots appear in isolation (**Figure 16-16a**), but frequently they are found in sunspot groups (Figure 16-16b; see also Figure 16-7). Although sunspots vary greatly in size, typical ones

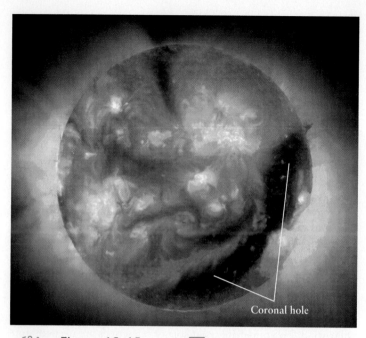

Figure 16-15   R I V U X G

The Ultraviolet Corona The *SOHO* spacecraft recorded this false-color ultraviolet view of the solar corona. The dark feature running across the Sun's disk from the top is a coronal hole, a region where the coronal gases are thinner than elsewhere. Such holes are often the source of strong gusts in the solar wind. (*SOHO*/EIT/ESA/NASA)

(a)

(b)

Figure 16-16   R I V U X G

Sunspots (a) This high-resolution photograph of the photosphere shows a mature sunspot. The dark center of the spot is called the umbra. It is bordered by the penumbra, which is less dark and has a featherlike appearance. **(b)** In this view of a typical sunspot group, several sunspots are close enough to overlap. In both images you can see granulation in the surrounding, undisturbed photosphere. (NOAO) measure a few tens of thousands of kilometers across—comparable to the diameter of the Earth. Sunspots are not permanent features of the photosphere but last between a few hours and a few months.

Each sunspot has a dark central core, called the *umbra*, and a brighter border called the *penumbra*. We used these same terms in Section 3-4 to refer to different parts of the Earth's or the Moon's shadow. But a sunspot is not a shadow; it is a region in the photosphere where the temperature is relatively low, which makes it appear darker than its surroundings. If the surrounding photosphere is blocked from view, a sunspot's umbra appears red and the penumbra appears orange. As we saw in Section 5-4, Wien's law relates the color of a blackbody (which depends on the wavelength at which it emits the most light) to the blackbody's temperature. The colors of a sunspot indicate that the temperature of the umbra is typically 4300 K and that of the penumbra is typically 5000 K. While high by earthly standards, these temperatures are quite a bit lower than the average photospheric temperature of 5800 K.

The Stefan-Boltzmann law (see Section 5-4) tells us that the energy flux from a blackbody is proportional to the fourth power of its temperature. This law lets us compare the amounts of light energy emitted by a square meter of a sunspot's umbra and by a square meter of undisturbed photosphere. The ratio is:

$$\frac{\text{flux from umbra}}{\text{flux from photosphere}} = \left(\frac{4300 \text{ K}}{5800 \text{ K}}\right)^4 = 0.30$$

That is, the umbra emits only 30% as much light as an equally large patch of undisturbed photosphere. This is why sunspots appear so dark.

November 9
November 12
November 14
November 15
November 17
November 19

Figure 16-17   R I V U X G

Tracking the Sun's Rotation with Sunspots This series of photographs taken in 1999 shows the rotation of the Sun. By observing the same group of sunspots from one day to the next, Galileo found that the Sun rotates once in about four weeks. (The equatorial regions of the Sun actually rotate somewhat faster than the polar regions.) Notice how the sunspot group shown here changed its shape. (The Carnegie Observatories)

(a)

(b) Near sunspot maximum

(c) Near sunspot minimum

Figure 16-18   R I V U X G

The Sunspot Cycle (a) The number of sunspots on the Sun varies with a period of about 11 years. The most recent sunspot maximum occurred in 2000. **(b)** This photograph, taken near sunspot maximum in 1989, shows a number of sunspots and large sunspot groups. The sunspot group visible near the bottom of the Sun's disk has about the same diameter as the planet Jupiter. **(c)** Near sunspot minimum, as in this 1986 photograph, essentially no sunspots are visible. (NOAO)

Figure 16-19

Variations in the Average Latitude of Sunspots The dots in this graph (sometimes called a "butterfly diagram") record how far north or south of the Sun's equator sunspots were observed. At the beginning of each sunspot cycle, most sunspots are found near latitudes 30° north or south. As the cycle goes on, sunspots typically form closer to the equator. (NASA Marshall Space Flight Center)

Sunspots and the Sun's Rotation

Occasionally, a sunspot group is large enough to be seen without a telescope. Chinese astronomers recorded such sightings 2000 years ago, and huge sunspot groups visible to the naked eye (with an appropriate filter) were seen in 1989 and 2003. But it was not until Galileo introduced the telescope into astronomy (see Section 4-5) that anyone was able to examine sunspots in detail.

Galileo discovered that he could determine the Sun's rotation rate by tracking sunspots as they moved across the solar disk (Figure 16-17). He found that the Sun rotates once in about four weeks. A typical sunspot group lasts about two months, so a specific one can be followed for two solar rotations.

Further observations by the British astronomer Richard Carrington in 1859 demonstrated that the Sun does not rotate as a rigid body. Instead, the equatorial regions rotate more rapidly than the polar regions. This phenomenon is known as **differential rotation.** Thus, while a sunspot near the solar equator takes only 25 days to go once around the Sun, a sunspot at 30° north or south of the equator takes 27½ days. The rotation period at 75° north or south is about 33 days, while near the poles it may be as long as 35 days.

The Sunspot Cycle

The average number of sunspots on the Sun is not constant, but varies in a predictable **sunspot cycle** (Figure 16-18a). This phenomenon was first reported by the German astronomer Heinrich Schwabe in 1843 after many years of observing. As Figure 16-18a shows, the average number of sunspots varies with a period of about 11 years. A period of exceptionally many sunspots is a **sunspot maximum** (Figure 16-18b), as occurred in 1979, 1989, and 2000. Conversely, the Sun is almost devoid of sunspots at a **sunspot minimum** (Figure 16-18c), as occurred in 1976, 1986, and 1996 and is projected to occur in 2007.

> The number of sunspots increases and decreases on an 11-year cycle

The locations of sunspots also vary with the same 11-year sunspot cycle. At the beginning of a cycle, just after a sunspot minimum, sunspots first appear at latitudes around 30° north and south of the solar equator (Figure 16-19). Over the succeeding years, the sunspots occur closer and closer to the equator.

16-9 Sunspots are produced by a 22-year cycle in the Sun's magnetic field

Why should the number of sunspots vary with an 11-year cycle? Why should their average latitude vary over the course of a cycle? And why should sunspots exist at all? The first step toward answering these questions came in 1908, when the American astronomer George Ellery Hale discovered that sunspots are associated with intense magnetic fields on the Sun.

Probing Solar Magnetism

When Hale focused a spectroscope on sunlight coming from a sunspot, he found that many spectral lines appear to be split into several closely spaced lines (Figure 16-20). This "splitting" of spectral lines is called the **Zeeman effect,** after the Dutch physicist Pieter Zeeman, who first observed it in his laboratory in 1896. Zeeman showed that a spectral line splits when the atoms are subjected to an intense magnetic field. The more intense the magnetic field, the wider the separation of the split lines.

Hale's discovery showed that sunspots are places where the hot gases of the photosphere are bathed in a concentrated magnetic field. Many of the atoms of the Sun's atmosphere are ionized due to the high temperature. The solar atmosphere is thus a special type of gas called a **plasma,** in which electrically charged ions and electrons can move freely. Like any moving, electrically charged objects, they can be deflected by magnetic fields. Figure 16-21 shows how a magnetic field in the laboratory bends a beam of fast-moving electrons into a curved trajectory. Similarly, the paths of moving ions and electrons in the photosphere are deflected by the Sun's magnetic field. In particular, magnetic forces act on the hot plasma that rises from the Sun's interior due to convection. Where the magnetic field is particularly strong, these

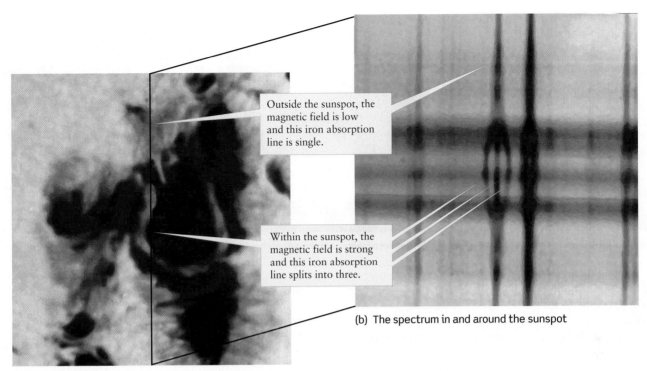

Outside the sunspot, the magnetic field is low and this iron absorption line is single.

Within the sunspot, the magnetic field is strong and this iron absorption line splits into three.

(b) The spectrum in and around the sunspot

(a) A sunspot

Figure 16-20   R I V U X G

Sunspots Have Strong Magnetic Fields (a) A black line in this image of a sunspot shows where the slit of a spectrograph was aimed. **(b)** This is a portion of the resulting spectrum, including a dark absorption line caused by iron atoms in the photosphere. The splitting of this line by the sunspot's magnetic field can be used to calculate the field strength. Typical sunspot magnetic fields are over 5000 times stronger than the Earth's field at its north and south poles. (NOAO)

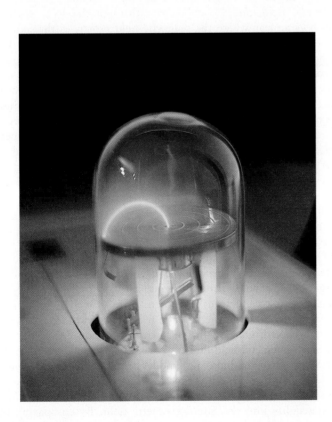

forces push the hot plasma away. The result is a localized region where the gas is relatively cool and thus glows less brightly—in other words, a sunspot.

To get a fuller picture of the Sun's magnetic fields, astronomers take images of the Sun at two wavelengths, one just less than and one just greater than the wavelength of a magnetically split spectral line. From the difference between these two images, they can construct a picture called a **magnetogram,** which displays the magnetic fields in the solar atmosphere. Figure 16-22*a* is an ordinary white-light photograph of the Sun taken at the same time as the magnetogram in Figure 16-22*b*. In the magnetogram, dark blue indicates areas of the photosphere with one magnetic polarity (north), and yellow indicates areas with the opposite (south) magnetic polarity. This image shows that many sunspot groups have roughly comparable areas covered by north

Figure 16-21   R I V U X G

Magnetic Fields Deflect Moving, Electrically Charged Objects In this laboratory experiment, a beam of negatively charged electrons (shown by a blue arc) is aimed straight upward from the center of the apparatus. The entire apparatus is inside a large magnet, and the magnetic field deflects the beam into a curved path. (Courtesy of Central Scientific Company)

(a) Visible-light image

(b) Magnetogram

(c) Magnetogram of a sunspot group

Figure 16-22   R I [V] U X G

Mapping the Sun's Magnetic Field (a) This visible-light image and **(b)** this false-color magnetogram were recorded at the same time. Dark blue and yellow areas in the magnetogram have north and south magnetic polarity, respectively; blue-green regions have weak magnetic fields. The highly magnetized regions in (b) correlate with the sunspots in (a). **(c)** The two ends of this large sunspot group have opposite magnetic polarities (colored blue and yellow), like the ends of a giant bar magnet. (NOAO)

and south magnetic polarities (see also Figure 16-22*c*). Thus, a sunspot group resembles a giant bar magnet, with a north magnetic pole at one end and a south magnetic pole at the other.

If different sunspot groups were unrelated to one another, their magnetic poles would be randomly oriented, like a bunch of compass needles all pointing in random directions. As Hale discovered, however, there is a striking regularity in the magnetization of sunspot groups. As a given sunspot group moves with the Sun's rotation, the sunspots in front are called the "preceding members" of the group. The spots that follow behind are referred to as the "following members." Hale compared the sunspot groups in the two solar hemispheres, north or south of the Sun's equator. He found that the preceding members in one solar hemisphere all have the same magnetic polarity, while the preceding members in the other hemisphere have the opposite polarity. Furthermore, in the hemisphere where the Sun has its north magnetic pole, the preceding members of all sunspot groups have north magnetic polarity. In the opposite hemisphere, where the Sun has its south magnetic pole, the preceding members all have south magnetic polarity.

Along with his colleague Seth B. Nicholson, Hale also discovered that the Sun's polarity pattern completely reverses itself every 11 years—the same interval as the time from one solar maximum to the next. The hemisphere that has preceding north magnetic poles during one 11-year sunspot cycle will have preceding south magnetic poles during the next 11-year cycle, and vice versa. The north and south magnetic poles of the Sun itself also reverse every 11 years. Thus, the Sun's magnetic pattern repeats itself only after two sunspot cycles, which is why astronomers speak of a **22-year solar cycle**.

The Magnetic-Dynamo Model

In 1960, the American astronomer Horace Babcock proposed a description that seems to account for many features of this 22-year solar cycle. Babcock's scenario, called a **magnetic-dynamo model**, makes use of two basic properties of the Sun's photosphere—differential rotation and convection. Differential rotation causes the magnetic field in the photosphere to become wrapped around the Sun (**Figure 16-23**). As a result, the magnetic field becomes concentrated at certain latitudes on either side of the solar equator. Convection in the photosphere creates tangles in the concentrated magnetic field, and "kinks" erupt through the solar surface. Sunspots appear where the magnetic field protrudes through the photosphere. The theory suggests that sunspots should appear first at northern and southern latitudes and later form nearer to the equator. This is just what is observed (see Figure 16-19). Note also that as shown on the far right in Figure 16-23, the preceding member of a sunspot group has the same polarity (N or S) as the Sun's magnetic pole in that hemisphere. This is just as Hale observed.

Differential rotation eventually undoes the twisted magnetic field. The preceding members of sunspot groups move toward the Sun's equator, while the following members migrate toward the poles. Because the preceding members from the two hemispheres have opposite magnetic polarities, their magnetic fields cancel each other out when they meet at the equator. The following members in each hemisphere have the opposite polarity to the Sun's pole in that hemisphere; hence, when they converge on the pole, the following members first cancel out and then

> The Sun's differential rotation makes the magnetic field twist like a rubber band

reverse the Sun's overall magnetic field. The fields are now completely relaxed. Once again, differential rotation begins to twist the Sun's magnetic field, but now with all magnetic polarities reversed. In this way, Babcock's model helps to explain the change in field direction every 11 years.

Recent discoveries in helioseismology (Section 16-3) offer new insights into the Sun's magnetic field. By comparing the

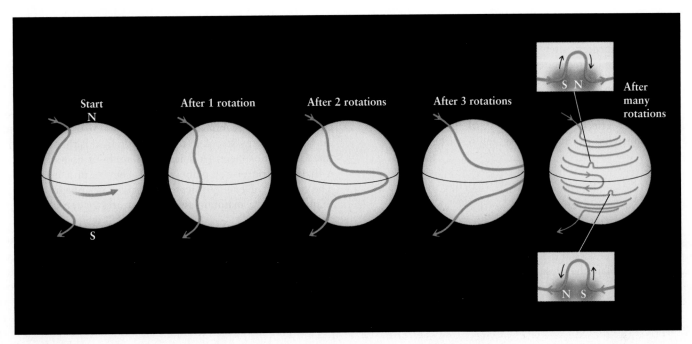

Figure 16-23

Babcock's Magnetic Dynamo Model Magnetic field lines tend to move along with the plasma in the Sun's outer layers. Because the Sun rotates faster at the equator than near the poles, a field line that starts off running from the Sun's north magnetic pole (N) to its south magnetic pole

(S) ends up wrapped around the Sun like twine wrapped around a ball. The insets on the far right show how sunspot groups appear where the concentrated magnetic field rises through the photosphere.

speeds of sound waves that travel with and against the Sun's rotation, helioseismologists have been able to determine the Sun's rotation rate at different depths and latitudes. As shown in **Figure 16-24**, the Sun's surface pattern of differential rotation persists through the convective zone. Farther in, within the radiative zone, the Sun seems to rotate like a rigid object with a period of 27 days at all latitudes. Astronomers suspect that the Sun's magnetic field originates in a relatively thin layer where the radiative and convective zones meet and slide past each other due to their different rotation rates.

One dilemma about sunspots is that compressed magnetic fields tend to push themselves apart, which means that sunspots should dissipate rather quickly. Yet observations show that sunspots can persist for many weeks. The resolution of this paradox may have been found using helioseismology (see Section 16-3). Analysis of the vibrations of the Sun around sunspots shows that beneath the surface of the photosphere, the gases surrounding each sunspot are circulating at high speed—rather like a hurricane as large as the Earth. The circulation of charged gases around the magnetic field holds the fields in place, thus stabilizing the sunspot.

Much about sunspots and solar activity remains mysterious. There are perplexing irregularities in the solar cycle. For example, the overall reversal of the Sun's magnetic field is often piecemeal and haphazard. One pole may reverse polarity long before the other. For several weeks the Sun's surface may have two north magnetic poles and no south magnetic pole at all.

What is more, there seem to be times when all traces of sunspots and the sunspot cycle vanish for many years. For

25 Days　　　　　　　　35 Days

Figure 16-24

Rotation of the Solar Interior This cutaway picture of the Sun shows how the solar rotation period (shown by different colors) varies with depth and latitude. The surface and the convective zone have differential rotation (a short period at the equator and longer periods near the poles). Deeper within the Sun, the radiative zone seems to rotate like a rigid sphere. (Courtesy of K. Libbrecht, Big Bear Solar Observatory) example, virtually no sunspots were seen from 1645 through 1715. Curiously, during these same years Europe experienced record low temperatures, often referred to as the Little Ice Age, whereas the western United States was subjected to severe drought. By contrast, there was apparently a period of increased sunspot activity during the eleventh and twelfth centuries, during which the Earth was warmer than it is today. Thus, variations in solar activity appear to affect climates on the Earth. The origin of this Sun-Earth connection is a topic of ongoing research.

16-10 The Sun's magnetic field heats the corona, produces flares, and causes massive eruptions

Astronomers now understand that the Sun's magnetic field does more than just explain the presence of sunspots. It is also responsible for the existence of spicules, as well as a host of other dramatic phenomena in the chromosphere and corona.

Magnetic Reconnection

In a plasma, magnetic field lines and the material of the plasma tend to move together. This means that as convection pushes material toward the edge of a supergranule, it pushes magnetic field lines as well. The result is that vertical magnetic field lines pile up around a supergranule. Plasma that "sticks" to these magnetic field lines thus ends up lifted upward, forming a spicule (see Figures 16-12 and 16-13).

The tendency of plasma to follow the Sun's magnetic field can also explain why the temperature of the chromosphere and corona is so high. Spacecraft observations show magnetic field arches extending tens of thousands of kilometers into the corona, with streamers of electrically charged particles moving along each arch (Figure 16-25a). If the magnetic fields of two arches come into proximity, their magnetic fields can rearrange in a process called **magnetic reconnection** (Figure 16-25b). The tremendous amount of energy stored in the magnetic field is then released into the solar atmosphere. (A single arch contains as much energy as a hydroelectric power plant would generate in a million years.) The amount of energy released in this way appears to be more than enough to maintain the temperatures of the chromosphere and corona.

ANALOGY The idea that a magnetic field can heat gases has applications on Earth as well as on the Sun. In an automobile engine's ignition system an electric current is set up in a coil of wire, which produces a magnetic field. When the current is shut off, the magnetic field collapses and its energy is directed to a spark plug in one of the engine's cylinders. The released energy heats the mixture of air and gasoline around the plug, causing the mixture to ignite. This drives the piston in that cylinder and makes the automobile go.

 Magnetic heating can also explain why the parts of the corona that lie on top of sunspots are often the most prominent in ultraviolet images. (Some examples are the

(a)

(b)

Figure 16-25   R I V U X G

Magnetic Arches and Magnetic Reconnection (a) This false-color ultraviolet image from the *TRACE* spacecraft (*Transition Region and Coronal Explorer*) shows magnetic field loops suspended high above the solar surface. The loops are made visible by the glowing gases trapped within them. **(b)** When the magnetic fields in these loops change their arrangement, a tremendous amount of energy is released and solar material can be ejected upward. (a: Stanford-Lockheed Institute for Space Research; *TRACE;* and NASA)

Violent, eruptive events on the Sun, called **solar flares,** occur in complex sunspot groups. Within only a few minutes, temperatures in a compact region may soar to 5×10^6 K and vast quantities of particles and radiation—including as much material as is in the prominence shown in Figure 16-27—are blasted out into space. These eruptions can also cause disturbances that spread outward in the solar atmosphere, like the ripples that appear when you drop a rock into a pond.

The most energetic flares carry as much as 10^{30} joules of energy, equivalent to 10^{14} one-megaton nuclear weapons being exploded at once! However, the energy of a solar flare does not come from thermonuclear fusion in the solar atmosphere; instead, it appears to be released from the intense magnetic field around a sunspot group.

> **A solar flare can have as much energy as 100 trillion nuclear bombs**

As energetic as solar flares are, they are dwarfed by **coronal mass ejections.** One such event is shown in the image that opens this chapter; **Figure 16-28a** shows another. In a coronal mass ejection, more than 10^{12} kilograms (a billion tons) of high-temperature coronal gas is blasted into space at speeds of hundreds of kilometers per second. A typical coronal mass ejection lasts a few hours. These explosive events seem to be related to large-scale alterations in the Sun's magnetic field, like the magnetic

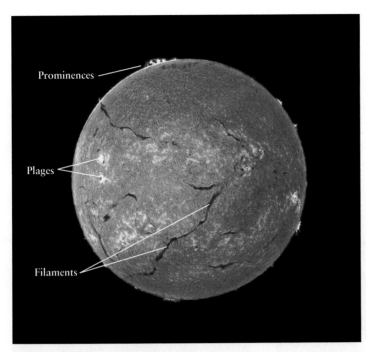

Figure 16-26   R I [V] U X G

The Active Sun Seen through an H$_\alpha$ Filter This image was made using a red filter that only passes light at a wavelength of 656 nm. The spectrum of the photosphere has an absorption line at this wavelength and so appears dark. Hence, this filter reveals the photosphere and corona. Prominences, plages, and filaments are associated with strong magnetic fields. (NASA)

bright regions in Figure 16-15.) The intense magnetic field of the sunspots helps trap and compress hot coronal gas, giving it such a high temperature that it emits copious amounts of high-energy ultraviolet photons and even more energetic X-ray photons.

Prominences, Flares, ands Coronal Mass Ejections

Spicules and coronal heating occur even when the Sun is quiet. But magnetic fields can also explain many aspects of the active Sun in addition to sunspots. **Figure 16-26** is an image of the chromosphere made with an H$_\alpha$ filter during a sunspot maximum. The bright areas are called **plages** (from the French word for "beach"). These are bright, hot regions in the chromosphere that tend to form just before the appearance of new sunspots. They are probably created by magnetic fields that push upward from the Sun's interior, compressing and heating a portion of the chromosphere. The dark streaks, called **filaments,** are relatively cool and dense parts of the chromosphere that have been pulled along with magnetic field lines as they arch to high altitudes.

When seen from the side, so that they are viewed against the dark background of space, filaments appear as bright, arching columns of gas called **prominences** (**Figure 16-27**). They can extend for tens of thousands of kilometers above the photosphere. Some prominences last for only a few hours, while others persist for many months. The most energetic prominences break free of the magnetic fields that confined them and burst into space.

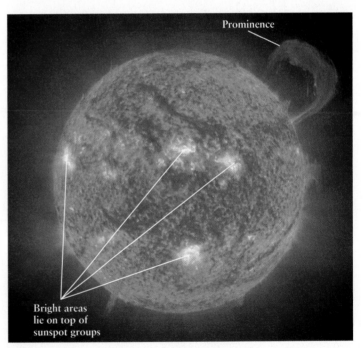

Figure 16-27   R I V [U] X G

A Solar Prominence A huge prominence arches above the solar surface in this ultraviolet image from the *SOHO* spacecraft. The image was recorded using light at a wavelength of 30.4 nm, emitted by singly ionized helium atoms at a temperature of about 60,000 K. By comparison, the material within the arches in Figure 16-25 reaches temperatures in excess of 2×10^6 K. (*SOHO*/EIT/ESA/NASA)

Figure 16-28   R I V [U] X G

A Coronal Mass Ejection **(a)** *SOHO* recorded this coronal mass ejection in an X-ray image. (The image of the Sun itself was made at ultraviolet wavelengths.) **(b)** Within two to four days the fastest-moving ejected material reaches a distance of 1 AU from the Sun. Most particles are deflected by the Earth's magnetosphere, but some are able to reach the Earth. (The ejection shown in (a) was not aimed toward the Earth and did not affect us.) (*SOHO/EIT/LASCO/ESA/NASA*)

(a) A coronal mass ejection

Material ejected from the corona

Ejected material encounters the Earth's magnetosphere

Earth

(b) Two to four days later

reconnection shown in Figure 16-25*b*. Coronal mass ejections occur every few months; smaller eruptions may occur almost daily.

If a solar flare or coronal mass ejection happens to be aimed toward Earth, a stream of high-energy electrons and nuclei reaches us a few days later (Figure 16-28*b*). When this plasma arrives, it can interfere with satellites, pose a health hazard to astronauts in orbit, and disrupt electrical and communications equipment on the Earth's surface. Telescopes on Earth and on board spacecraft now monitor the Sun continuously to provide warnings of dangerous levels of solar particles.

The numbers of plages, filaments, solar flares, and coronal mass ejections all vary with the same 11-year cycle as sunspots. But unlike sunspots, coronal mass ejections never completely cease, even when the Sun is at its quietest. Astronomers are devoting substantial effort to understanding these and other aspects of our dynamic Sun.

Key Words

The term preceded by an asterisk (*) is discussed in Box 16-1.

22-year solar cycle, p. 424
chromosphere, p. 417
CNO cycle, p. 408
conduction, p. 409
convection, p. 409
convective zone, p. 411
corona, p. 418
coronal hole, p. 419
coronal mass ejection, p. 427

differential rotation, p. 422
filament, p. 427
granulation, p. 415
granule, p. 415
helioseismology, p. 411
hydrogen fusion, p. 406
hydrostatic equilibrium, p. 408
limb darkening, p. 415

luminosity (of the Sun), p. 404
magnetic-dynamo model, p. 424
magnetogram, p. 423
magnetic reconnection, p. 426
negative hydrogen ion, p. 417
neutrino, p. 412
neutrino oscillation, p. 413
photosphere, p. 415
plage, p. 427
plasma, p. 422
*positron, p. 406
prominence, p. 427
proton-proton chain, p. 408
radiative diffusion, p. 409

radiative zone, p. 411
solar flare, p. 427
solar neutrino, p. 412
solar neutrino problem, p. 413
solar wind, p. 419
spicule, p. 417
sunspot, p. 420
sunspot cycle, p. 422
sunspot maximum, p. 422
sunspot minimum, p. 422
supergranule, p. 415
thermal equilibrium, p. 409
thermonuclear fusion, p. 406
Zeeman effect, p. 422

Key Ideas

Hydrogen Fusion in the Sun's Core: The Sun's energy is produced by hydrogen fusion, a sequence of thermonuclear reactions in which four hydrogen nuclei combine to produce a single helium nucleus.

• The energy released in a nuclear reaction corresponds to a slight reduction of mass according to Einstein's equation $E = mc^2$.

• Thermonuclear fusion occurs only at very high temperatures; for example, hydrogen fusion occurs only at temperatures in excess of about 10^7 K. In the Sun, fusion occurs only in the dense, hot core.

Models of the Sun's Interior: A theoretical description of a star's interior can be calculated using the laws of physics.

• The standard model of the Sun suggests that hydrogen fusion takes place in a core extending from the Sun's center to about 0.25 solar radius.

• The core is surrounded by a radiative zone extending to about 0.71 solar radius. In this zone, energy travels outward through radiative diffusion.

• The radiative zone is surrounded by a rather opaque convective zone of gas at relatively low temperature and pressure. In this zone, energy travels outward primarily through convection.

Solar Neutrinos and Helioseismology: Conditions in the solar interior can be inferred from measurements of solar neutrinos and of solar vibrations.

• Neutrinos emitted in thermonuclear reactions in the Sun's core have been detected, but in smaller numbers than expected. Recent neutrino experiments explain why this is so.

• Helioseismology is the study of how the Sun vibrates. These vibrations have been used to infer pressures, densities, chemical compositions, and rotation rates within the Sun.

 The Sun's Atmosphere: The Sun's atmosphere has three main layers: the photosphere, the chromosphere, and the corona. Everything below the solar atmosphere is called the solar interior.

• The visible surface of the Sun, the photosphere, is the lowest layer in the solar atmosphere. Its spectrum is similar to that of a blackbody at a temperature of 5800 K. Convection in the photosphere produces granules.

• Above the photosphere is a layer of less dense but higher-temperature gases called the chromosphere. Spicules extend upward from the photosphere into the chromosphere along the boundaries of supergranules.

• The outermost layer of the solar atmosphere, the corona, is made of very high-temperature gases at extremely low density. Activity in the corona includes coronal mass ejections and coronal holes. The solar corona blends into the solar wind at great distances from the Sun.

The Active Sun: The Sun's surface features vary in an 11-year cycle. This is related to a 22-year cycle in which the surface magnetic field increases, decreases, and then increases again with the opposite polarity.

• Sunspots are relatively cool regions produced by local concentrations of the Sun's magnetic field. The average number of sunspots increases and decreases in a regular cycle of approximately 11 years, with reversed magnetic polarities from one 11-year cycle to the next. Two such cycles make up the 22-year solar cycle.

• The magnetic-dynamo model suggests that many features of the solar cycle are due to changes in the Sun's magnetic field. These changes are caused by convection and the Sun's differential rotation.

• A solar flare is a brief eruption of hot, ionized gases from a sunspot group. A coronal mass ejection is a much larger eruption that involves immense amounts of gas from the corona.

Questions

Review Questions

1. What is meant by the luminosity of the Sun?
2. What is Kelvin-Helmholtz contraction? Why is it ruled out as a source of the present-day Sun's energy?
3. Why is it impossible for the burning of substances like coal to be the source of the Sun's energy?
4. What is hydrogen fusion? Why is hydrogen fusion fundamentally unlike the burning of a log in a fireplace?
5. If the electric force between protons were somehow made stronger, what effect would this have on the temperature required for thermonuclear fusion to take place?
6. Why do thermonuclear reactions occur only in the Sun's core, not in its outer regions?
7. Describe how the net result of the reactions shown in the *Cosmic Connections* figure (Section 16-1) is the conversion of four protons into a single helium nucleus. What other particles are produced in this process? How many of each particle are produced?
8. Give an everyday example of hydrostatic equilibrium. Give an example of thermal equilibrium. Explain how these equilibrium conditions apply to each example.
9. If thermonuclear fusion in the Sun were suddenly to stop, what would eventually happen to the overall radius of the Sun? Justify your answer using the ideas of hydrostatic equilibrium and thermal equilibrium.
10. Give some everyday examples of conduction, convection, and radiative diffusion.
11. Describe the Sun's interior. Include references to the main physical processes that occur at various depths within the Sun.
12. Suppose thermonuclear fusion in the Sun stopped abruptly. Would the intensity of sunlight decrease just as abruptly? Why or why not?
13. Explain how studying the oscillations of the Sun's surface can give important, detailed information about physical conditions deep within the Sun.
14. What is a neutrino? Why is it useful to study neutrinos coming from the Sun? What do they tell us that cannot be learned from other avenues of research?
15. Unlike all other types of telescopes, neutrino detectors are placed deep underground. Why?
16. What was the solar neutrino problem? What solution to this problem was suggested by the results from the Sudbury Neutrino Observatory?
17. Describe the dangers in attempting to observe the Sun. How have astronomers learned to circumvent these observational problems?
18. Briefly describe the three layers that make up the Sun's atmosphere. In what ways do they differ from each other?
19. What is solar granulation? Describe how convection gives rise to granules.
20. High-resolution spectroscopy of the photosphere reveals that absorption lines are blueshifted in the spectrum of the central, bright regions of granules but are redshifted in the spectrum of the dark boundaries between granules. Explain how these observations show that granulation is due to convection.

21. What is the difference between granules and supergranules?

22. What are spicules? Where are they found? How can you observe them? What causes them?

23. How do astronomers know that the temperature of the corona is so high?

24. How do astronomers know when the next sunspot maximum and minimum will occur?

25. Why do astronomers say that the solar cycle is really 22 years long, even though the number of sunspots varies over an 11-year period?

26. Explain how the magnetic-dynamo model accounts for the solar cycle.

27. Describe one explanation for why the corona has a higher temperature than the chromosphere.

28. Why should solar flares and coronal mass ejections be a concern for businesses that use telecommunication satellites?

Advanced Questions

Questions preceded by an asterisk (*) involve the topic discussed in Box 16-1.

> Problem-solving tips and tools
>
> You may have to review Wien's law and the Stefan-Boltzmann law, which are the subjects of Section 5-4. Section 5-5 discusses the properties of photons. As we described in Box 5-2, you can simplify calculations by taking ratios, such as the ratio of the flux from a sunspot to the flux from the undisturbed photosphere. When you do this, all the cumbersome constants cancel out. Figure 5-7 shows the various parts of the electromagnetic spectrum. We introduced the Doppler effect in Section 5-9 and Box 5-6. For information about the planets, see Table 7-1.

29. Calculate how much energy would be released if each of the following masses were converted *entirely* into their equivalent energy: (a) a carbon atom with a mass of 2×10^{-26} kg, (b) 1 kilogram, and (c) a planet as massive as the Earth (6×10^{24} kg).

30. Use the luminosity of the Sun (given in Table 16-1) and the answers to the previous question to calculate how long the Sun must shine in order to release an amount of energy equal to that produced by the complete mass-to-energy conversion of (a) a carbon atom, (b) 1 kilogram, and (c) the Earth.

31. Assuming that the current rate of hydrogen fusion in the Sun remains constant, what fraction of the Sun's mass will be converted into helium over the next 5 billion years? How will this affect the overall chemical composition of the Sun?

32. (a) Estimate how many kilograms of hydrogen the Sun has consumed over the past 4.56 billion years, and estimate the amount of mass that the Sun has lost as a result. Assume that the Sun's luminosity has remained constant during that time. (b) In fact, the Sun's luminosity when it first formed was only about 70% of its present value. With this in mind, explain whether your answers to part (a) are an overestimate or an underestimate.

*33. To convert one kilogram of hydrogen (^{1}H) into helium (^{4}He) as described in Box 16-1, you must start with 1.5 kg of hy-

drogen. Explain why, and explain what happens to the other 0.5 kg. (*Hint:* How many ^{1}H nuclei are used to make the two ^{3}He nuclei shown in the *Cosmic Connections* figure in Section 16-1? How many of these ^{1}H nuclei end up being incorporated into the ^{4}He nucleus shown in the figure?)

*34. (a) A positron has the same mass as an electron (see Appendix 7). Calculate the amount of energy released by the annihilation of an electron and positron. (b) The products of this annihilation are two photons, each of equal energy. Calculate the wavelength of each photon, and confirm from Figure 5-7 that this wavelength is the gamma-ray range.

35. Sirius is the brightest star in the night sky. It has a luminosity of 23.5 $L_\odot$, that is, it is 23.5 times as luminous as the Sun and burns hydrogen at a rate 23.5 times greater than the Sun. How many kilograms of hydrogen does Sirius convert into helium each second?

36. (Refer to the preceding question.) Sirius has 2.3 times the mass of the Sun. Do you expect that the lifetime of Sirius will be longer, shorter, or the same length as that of the Sun? Explain your reasoning.

37. (a) If the Sun were not in a state of hydrostatic equilibrium, would its diameter remain the same? Explain your reasoning. (b) If the Sun were not in a state of thermal equilibrium, would its luminosity remain the same? What about its surface temperature? Explain your reasoning.

38. Using the mass and size of the Sun given in Table 16-1, verify that the average density of the Sun is 1410 kg/m^3. Compare your answer with the average densities of the Jovian planets.

39. Use the data in Table 16-2 to calculate the average density of material within 0.1 $R_\odot$ of the center of the Sun. (You will need to use the mass and radius of the Sun as a whole, given in Table 16-1.) Explain why your answer is not the same as the density at 0.1 $R_\odot$ given in Table 16-2.

40. In a typical solar oscillation, the Sun's surface moves up or down at a maximum speed of 0.1 m/s. An astronomer sets out to measure this speed by detecting the Doppler shift of an absorption line of iron with wavelength 557.6099 nm. What is the maximum wavelength shift that she will observe?

41. Explain why the results from the Sudbury Neutrino Observatory (SNO) only provide an answer to the solar neutrino problem for relatively high-energy neutrinos. (*Hint:* Can SNO detect solar neutrinos of all energies?)

42. The amount of energy required to dislodge the extra electron from a negative hydrogen ion is 1.2×10^{-19} J. (a) The extra electron can be dislodged if the ion absorbs a photon of sufficiently short wavelength. (Recall from Section 5-5 that the higher the energy of a photon, the shorter its wavelength.) Find the longest wavelength (in nm) that can accomplish this. (b) In what part of the electromagnetic spectrum does this wavelength lie? (c) Would a photon of visible light be able to dislodge the extra electron? Explain. (d) Explain why the photosphere, which contains negative hydrogen ions, is quite opaque to visible light but is less opaque to light with wavelengths longer than the value you calculated in (a).

43. Astronomers often use an H_α filter to view the chromosphere. Explain why this can also be accomplished with filters that are transparent only to the wavelengths of the H and K lines

of ionized calcium. (*Hint:* The H and K lines are dark lines in the spectrum of the photosphere.)

44. Calculate the wavelengths at which the photosphere, chromosphere, and corona emit the most radiation. Explain how the results of your calculations suggest the best way to observe these regions of the solar atmosphere. (*Hint:* Treat each part of the atmosphere as a perfect blackbody. Assume average temperatures of 50,000 K and 1.5×10^6 K for the chromosphere and corona, respectively.)

45. On November 15, 1999, the planet Mercury passed in front of the Sun as seen from Earth. The TRACE spacecraft made these time-lapse images of this event using ultraviolet light (top) and visible light (bottom). (Mercury moved from left to right in these images. The time between successive views of Mercury is 6 to 9 minutes.) Explain why the Sun appears somewhat larger in the ultraviolet image than in the visible-light image.

R I V **U** X G

R I **V** U X G

(K. Schrijver, Stanford-Lockheed Institute for Space Research, TRACE, and NASA)

46. The moving images on a television set that uses a tube (as opposed to a LCD or plasma flat-screen TV) are made by a fast-moving electron beam that sweeps over the back of the screen. Explain why placing a strong magnet next to the screen distorts the picture. (*Caution:* Don't try this with your television! You can do permanent damage to your set.)

47. Find the wavelength of maximum emission of the umbra of a sunspot and the wavelength of maximum emission of a sunspot's penumbra. In what part of the electromagnetic spectrum do these wavelengths lie?

48. (**a**) Find the ratio of the energy flux from a patch of a sunspot's penumbra to the energy flux from an equally large patch of undisturbed photosphere. Which patch is brighter? (**b**) Find the ratio of the energy flux from a patch of a sunspot's penumbra to the energy flux from an equally large patch of umbra. Again, which patch is brighter?

49. Suppose that you want to determine the Sun's rotation rate by observing its sunspots. Is it necessary to take the Earth's orbital motion into account? Why or why not?

50. (**a**) Using a ruler to make measurements of Figure 16-25, determine how far the arches in that figure extend above the Sun's surface. The diameter of the Earth is 12,756 km. (**b**) In Figure 16-25 (an ultraviolet image) the photosphere appears dark compared to the arches. Explain why.

51. The amount of visible light emitted by the Sun varies only a little over the 11-year sunspot cycle. But the amount of X rays emitted by the Sun can be 10 times greater at solar maximum than at solar minimum. Explain why these two types of radiation should be so different in their variability.

Discussion Questions

52. Discuss the extent to which cultures around the world have worshiped the Sun as a deity throughout history. Why do you suppose there has been such widespread veneration?

53. In the movie *Star Trek IV: The Voyage Home,* the starship *Enterprise* flies on a trajectory that passes close to the Sun's surface. What features should a real spaceship have to survive such a flight? Why?

54. Discuss some of the difficulties in correlating solar activity with changes in the Earth's climate.

55. Describe some of the advantages and disadvantages of observing the Sun (**a**) from space and (**b**) from the Earth's south pole. What kinds of phenomena and issues might solar astronomers want to explore from these locations?

Web/eBook Questions

56. Search the World Wide Web for information about features in the solar atmosphere called *sigmoids*. What are they? What causes them? How might they provide a way to predict coronal mass ejections?

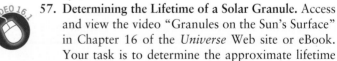

57. **Determining the Lifetime of a Solar Granule.** Access and view the video "Granules on the Sun's Surface" in Chapter 16 of the *Universe* Web site or eBook. Your task is to determine the approximate lifetime of a solar granule on the photosphere. Select an area, then slowly and rhythmically repeat "start, stop, start, stop" until you can consistently predict the appearance and disappearance of granules. While keeping your rhythm, move to a different area of the video and continue monitoring the appearance and disappearance of granules. When you are confident you have the timing right, move your eyes (or use a partner) to look at the clock shown in the video. Determine the length of time between the appearance and disappearance of the granules and record your answer.

Activities

Observing Projects

Observing tips and tools

At the risk of repeating ourselves, we remind you to *never look directly at the Sun, because it can easily cause permanent blindness.* You can view the Sun safely without a telescope just by using two pieces of white cardboard. First, use a pin to poke a small hole in one piece of cardboard; this will be your "lens," and the other piece of cardboard will be your "viewing screen." Hold the "lens" piece of cardboard so that it is face-on to the Sun and sunlight can pass through the hole. With your other hand, hold the "viewing screen" so that the sunlight from the "lens" falls on it. Adjust the distance between the two pieces of cardboard so that you see a sharp image of the Sun on the "viewing screen." This image is perfectly safe to view. It is actually possible to see sunspots with this low-tech apparatus.

For a better view, use a telescope with a solar filter that fits on the front of the telescope. A standard solar filter is a piece of glass coated with a thin layer of metal to give it a mirrorlike appearance. This coating reflects almost all the sunlight that falls on it, so that only a tiny, safe amount of sunlight enters the telescope. An H_α filter, which looks like a red piece of glass, keeps the light at a safe level by admitting only a very narrow range of wavelengths. (Filters that fit on the back of the telescope are *not* recommended. The telescope focuses concentrated sunlight on such a filter, heating it and making it susceptible to cracking—and if the filter cracks when you are looking through it, your eye will be ruined instantly and permanently.)

To use a telescope with a solar filter, first aim the telescope away from the Sun, then put on the filter. Keep the lens cap on the telescope's secondary wide-angle "finder scope" (if it has one), because the heat of sunlight can fry the finder scope's optics. Next, aim the telescope toward the Sun, using the telescope's shadow to judge when you are pointed in the right direction. You can then safely look through the telescope's eyepiece. When you are done, make sure you point the telescope away from the Sun before removing the filter and storing the telescope.

Note that the amount of solar activity that you can see (sunspots, filaments, flares, prominences, and so on) will depend on where the Sun is in its 11-year sunspot cycle.

---

58. Use a telescope with a solar filter to observe the surface of the Sun. Do you see any sunspots? Sketch their appearance. Can you distinguish between the umbrae and penumbrae of the sunspots? Can you see limb darkening? Can you see any granulation?

59. If you have access to an H_α filter attached to a telescope especially designed for viewing the Sun safely, use this instrument to examine the solar surface. How does the appearance of the Sun differ from that in white light? What do sunspots look like in H_α? Can you see any prominences? Can you see any filaments? Are the filaments in the H_α image near any sunspots seen in white light? (Note that the amount of activity that you see will be much greater at some times during the solar cycle than at others.)

60. Use the *Starry Night Enthusiast*™ program to measure the Sun's rotation. Display the entire celestial sphere by selecting **Guides > Atlas** in the **Favourites** menu and center on the Sun by double-clicking on **Sun** in the **Find** pane. Using the controls at the right-hand end of the toolbar, zoom in until you can see details on the Sun's surface clearly. In the toolbar, set the **Time Flow Rate** to 1 day. Using the time forward and backward buttons in the toolbar, step through enough time to determine the rotation period of the Sun. Which part of the actual Sun's surface rotates at the rate shown in *Starry Night Enthusiast*™? (Note: The program does not show the Sun's differential rotation.)

61. Use the *Starry Night Enthusiast*™ program to examine the Sun. Open the **Favourites** pane and double-click on **Solar System > Inner Solar System** to display the inner planets surrounding the Sun. Stop **Time Advance** and use the **down** arrow in the toolbar under **Viewing Location** to approach to within about 0.015 AU of the Sun. You can rotate the Sun by placing the mouse cursor over the image and, while holding down the **Shift** key, hold down the mouse button while moving the mouse. (On a two-button mouse, hold down the left mouse button.) (a) The Sun's equator lies close to the plane of the ecliptic. Where do most of the sunspots visible on the image of the Sun lie relative to the solar equator? Check Figure 16-18 for more realistic images of sunspots and Figure 16-19 for the latitude distribution of sunspots. (b) Based on your observations in (a), does the image in *Starry Night Enthusiast*™ show the Sun near the beginning, middle, or end of the 11-year sunspot cycle? Explain your reasoning. You can see current solar images from both ground and space-based solar telescopes by opening the LiveSky pane if you have an Internet connection on your computer.

Collaborative Exercises

62. Figure 16-19 shows variations in the average latitude of sunspots. Estimate the average latitude of sunspots in the year you were born and estimate the average latitude on your twenty-first birthday. Make rough sketches of the Sun during those years to illustrate your answers.

63. Create a diagram showing a sketch of how limb darkening on the Sun would look different if the Sun had either a thicker or thinner photosphere. Be sure to include a caption explaining your diagram.

64. Solar granules, shown in Figure 16-9, are about 1000 km across. What city is about that distance away from where you are right now? What city is that distance from the birthplace of each group member?

65. Magnetic arches in the corona are shown in Figure 16-25a. How many Earths high are these arches, and how many Earths could fit inside one arch?

28

The Search for Extraterrestrial Life

WEB LINK 28.1

R I V U X G

An "alien" life-form on Earth: pink, eyeless worms in an underwater mound of yellow solid methane. (Dr. Charles Fisher, Eberly College of Science, Pennsylvania State University)

One of the most compelling questions in science is also one of the simplest: Are we alone? That is, does life exist beyond the Earth? As yet, we have no definitive answer to this question. None of our spacecraft has found life elsewhere in the solar system, and radio telescopes have yet to detect signals of intelligent origin coming from space. Reports of aliens visiting our planet and abducting humans make compelling science fiction, but none of these reports has ever been verified.

Yet there are reasons to suspect that life might indeed exist beyond the Earth. One is that biologists find living organisms in some of the most "unearthly" environments on our planet. An example (shown at the top of this page) is at the bottom of the Gulf of Mexico, where the crushing pressure and low temperature cause methane—normally a gas—to form solid, yellowish mounds. Amazingly, these mounds teem with colonies of pink, eyeless, alien-looking worms the size of your thumb. If life can flourish here, might it not also flourish in the equally hostile conditions found on other worlds?

In this chapter we will look for places in our solar system where life may once have originated, and where it may exist today. We will see how scientists estimate the chances of finding life beyond our solar system, and how they search for signals from other intelligent species. And we will learn how a new generation of telescopes may make it possible to detect the presence of even single-celled organisms on worlds many light-years away.

28-1 The chemical building blocks of life are found in space

WEB LINK 28.2

Suppose you were the first visitor to a new and alien planet. How would you recognize which of the strange objects around you were living, and which were inanimate? Questions such as these are central to **astrobiology,** the study of life in the universe. Most astrobiologists suspect that if we find living organisms on other worlds, they will be "life as we know it"—that is, their biochemistry will be based on the unique properties of the carbon atom, as is the case for all terrestrial life.

Organic Molecules in the Universe

Why carbon? The reason is that carbon has the most versatile chemistry of any element. Carbon atoms can form chemical bonds to create especially long and complex molecules (**Figure 28-1**). These carbon-based compounds, called **organic molecules,** include all the molecules of which living organisms are made. (Silicon has some chemical similarities to carbon, and it can also

Learning Goals

By reading the sections of this chapter, you will learn

28-1 How comets and meteorites could have aided the origin of life on Earth

28-2 On which other worlds of our solar system life might have evolved

28-3 About the controversy over proposed fossil life from Mars

28-4 How astronomers estimate the number of other civilizations in the Galaxy

28-5 Why radio telescopes may be a useful tool for contacting alien civilizations

28-6 How astronomers hope to detect Earthlike planets orbiting other stars

$$Z - X - X - X - X - Y$$

(a) Linear molecule

$$\underset{H}{\overset{O}{\underset{\|}{C}}} - \underset{\underset{H}{|}}{\overset{\overset{OH}{|}}{C}} - \underset{\underset{OH}{|}}{\overset{\overset{H}{|}}{C}} - \underset{\underset{H}{|}}{\overset{\overset{OH}{|}}{C}} - \underset{\underset{H}{|}}{\overset{\overset{OH}{|}}{C}} - CH_2OH$$

(b) Glucose

Figure 28-1

Complex Molecules and Carbon (a) Atoms that can bond to only two other atoms, like the atoms denoted X shown here, can form a chain of atoms called a linear molecule. The chain stops where we introduce an atom, such as those labeled Y and Z, that can bond to only one other atom. **(b)** A carbon atom (denoted C) can bond with up to *four* other atoms. Hence, carbon atoms can form more complex, nonlinear molecules like glucose. All organic molecules that are found in living organisms have backbones of carbon atoms.

form complex molecules. But as **Figure 28-2** shows, complex silicon molecules do not have the right properties to make up complex systems such as living organisms.)

Organic molecules can be linked together to form elaborate structures, such as chains, lattices, and fibers. Some of these structures are capable of complex, self-regulating chemical reactions. Furthermore, the primary constituents of organic molecules—carbon, hydrogen, nitrogen, oxygen, sulfur, and phosphorus—are among the most abundant elements in the universe. The versatility and abundance of carbon suggest that extraterrestrial life is also likely to be based on organic chemistry.

If life is based on organic molecules, then these molecules must initially be present on a planet in order for life to arise from nonliving matter. We now understand that many organic molecules originate from nonbiological processes in interstellar space. One such molecule is carbon monoxide (CO), which is made when a carbon atom and an oxygen atom collide and bond together. Carbon monoxide is found in abundance within giant interstellar clouds that lie along the spiral arms of our Milky Way Galaxy (see Figure 1-7) as well as in other galaxies (see Figure 1-9). Carbon atoms have also combined with other elements to produce an impressive array of interstellar organic molecules, including ethyl alcohol (CH_3CH_2OH), formaldehyde (H_2CO), methyl cyanoacetylene (CH_3C_3N), and acetaldehyde (CH_3CHO). Radio astronomers have detected these by looking for the telltale

(a)

(b)

Figure 28-2 R I U X G

Why Silicon is Unsuitable for Making Living Organisms Like carbon, silicon atoms can bond with up to four other atoms. However, the resulting compounds are either too soft or too hard, or too reactive or too inert, to be suitable for use in living organisms. **(a)** Silicone has a backbone of silicon and oxygen atoms. The molecules form a gel or liquid rather than a solid, and are too inert to undergo the rapid chemical changes required of molecules in organisms. **(b)** Molecules can also be made with a silicon-carbon-oxygen backbone, but the results (like this quartz crystal) are too rigid for use in organisms. (a: Richard Megna/Fundamental Photographs, New York; b: © Mark A. Schneider/Visuals Unlimited)

(a)

(b)

Figure 28-3   R I V U X G

A Carbonaceous Chondrite (a) Carbonaceous chondrites are primitive meteorites that date back to the very beginning of the solar system. This sample is a piece of the Allende meteorite, a large carbonaceous chondrite that fell in Mexico in 1969. (b) Chemical analyses of newly fallen specimens disclose that they are rich in organic molecules (arrow), many of which are the chemical building blocks of life. (a: From the collection of Ronald A. Oriti; b: Harvard-Smithsonian Center for Astrophysics)

microwave emission lines of carbon-based chemicals in interstellar clouds.

The planets of our solar system formed out of interstellar material (see Section 8-5), and some of the organic molecules in that material must have ended up on the planets' surfaces. Evidence for this comes from meteorites called **carbonaceous chondrites,** like the one shown in Figure 28-3a. These are ancient meteorites that date from the formation of the solar system and that are often found to contain a variety of carbon-based molecules (Figure 28-3b). The spectra of comets (see Section 7-5)—which are also among the oldest objects in the solar system—show that they, too, contain an assortment of organic compounds.

Comets and meteoroids were much more numerous in the early solar system than they are today, and they were correspondingly more likely to collide with a planet. These collisions would have seeded the planets with organic compounds from the very beginning of our solar system's history. Similar processes are thought to take place in other planetary systems, which are thought to form in basically the same way as did our own (see Figure 8-13 and the image that opens Chapter 8).

The Miller-Urey Experiment

Comets and meteorites would not have been the only sources of organic material on the young planets of our solar system. In 1952, the American chemists Stanley Miller and Harold Urey demonstrated that under conditions that are thought to have prevailed on the primitive Earth, simple chemicals can combine to form the chemical building blocks of life. In a closed container, they prepared a sample of "atmosphere": a mixture of hydrogen (H_2), ammonia (NH_3), methane (CH_4), and water vapor (H_2O), the most common molecules in the solar system. Miller and Urey then exposed this mixture of gases to an electric arc (to simulate atmospheric lightning) for a week. At the end of this period, the inside of the container had become coated with a reddish-brown substance rich in amino acids and other compounds essential to life.

Since Miller and Urey's original experiment, most scientists have come to the conclusion that Earth's primordial atmosphere was composed of carbon dioxide (CO_2), nitrogen (N_2), and water vapor outgassed from volcanoes, along with some hydrogen. Modern versions of the Miller-Urey experiment (Figure 28-4) using these common gases have also succeeded in synthesizing a wide variety of organic compounds. The combination of comets and meteorites falling from space and chemical synthesis in the atmosphere could have made the chemical building blocks of life available in substantial quantities on the young Earth.

> Organic molecules do not have to come from living organisms—they can also be synthesized in nature from simple chemicals

CAUTION! It is important to emphasize that scientists have *not* created life in a test tube. While organic molecules may have been available on the ancient Earth, biologists have yet to figure out how these molecules gathered themselves into cells and developed systems for self-replication. Nevertheless, because so many chemical components of life are so easily synthesized under conditions that simulate the primordial Earth, it seems reasonable to suppose that life could have originated as the result of chemical processes. Furthermore, because the molecules that combine to form these compounds are rather common, it seems equally reasonable that life could have originated in the same way on other planets.

Organic building blocks are commonplace throughout the universe, but this does not guarantee that life is equally commonplace. If a planet's environment is hostile, life may never get started or may quickly be extinguished. But we now have evidence

Figure 28-4

An Updated Miller-Urey Experiment Modern versions of this classic experiment prove that numerous organic compounds important to life can be synthesized from gases that were present in the Earth's primordial atmosphere. This experiment supports the hypothesis that life on the Earth arose as a result of ordinary chemical reactions.

Labels in figure:
- Electrodes
- Primitive atmosphere (H_2O, CO_2, N_2, H_2)
- Condenser (maintained at a low temperature)
- Boiling water
- Organic molecules accumulate here

that Jupiter-sized planets orbit other stars (see Section 8-6 and Geoff Marcy's essay "Alien Planets" following Chapter 8) and that additional planetary systems are forming around young stars (see Section 8-4, especially Figure 8-8). It seems probable that there are Earthlike planets orbiting other stars, and that conditions on some of these worlds may be suitable for life as we know it.

28-2 Europa and Mars have the potential for life to have evolved

If life evolved on Earth from nonliving organic molecules, might the same process have taken place elsewhere in our solar system? Scientists have carefully scrutinized the planets and satellites in an attempt to answer this question, and most of the answers have been disappointing.

The Importance of Liquid Water

One major problem is that liquid water is essential for the survival of life as we know it. The water need not be pleasant by human standards—terrestrial organisms have been found in water that is boiling hot, fiercely acidic, or ice cold (see the image that opens this chapter)—but it must be liquid. In order for water on a planet's surface to remain liquid, the temperature cannot be too hot or too cold. Furthermore, there must be a relatively thick atmosphere to provide enough pressure to keep liquid water from evaporating. Of all the worlds of the present-day solar system, only Earth has the right conditions for water to remain liquid on its surface.

WEB LINK 28.4 However, there is now compelling evidence that Europa, one of the large satellites of Jupiter (see Table 7-2 and Figure 7-4), has an ocean of water *beneath* its icy surface. As it orbits Jupiter, Europa is caught in a tug-of-war between Jupiter's gravitational influence and those of the other large satellites. This flexes the interior of Europa, and this flexing generates enough heat to keep subsurface water from freezing. Chunks of ice on the surface can float around on this underground ocean, rearranging themselves into a pattern that reveals the liquid water beneath.

No one knows whether life exists in Europa's ocean. But interest in this exotic little world is great, and scientists have proposed several missions to explore Europa in more detail.

Searching for Life on Mars

The next best possibility for the existence of life is Mars. The present-day Martian atmosphere is so thin that water can exist only as ice or as a vapor. However, images made from Martian orbit show dried-up streambeds, flash flood channels, and sediment deposits. These features are evidence that the Martian atmosphere was once thicker and that water once coursed over the planet's surface. Could life have evolved on Mars during its "wet" period? If so, could life—even in the form of microorganisms—have survived as the Martian atmosphere thinned and the surface water either froze or evaporated?

Some areas on Mars were covered with liquid water for extended periods

In 1976, two spacecraft landed on different parts of Mars in search of answers to these questions. *Viking Lander 1* and *Viking Lander 2* each carried a scoop at the end of a mechanical arm to retrieve surface samples (**Figure 28-5**). These samples were deposited into a compact on-board biological laboratory that carried out three different tests for Martian microorganisms.

1. The *gas-exchange experiment* was designed to detect any processes that might be broadly considered as respiration. A surface sample was placed in a sealed container along with a controlled amount of gas and nutrients. The gases in the container were then monitored to see if their chemical composition changed.

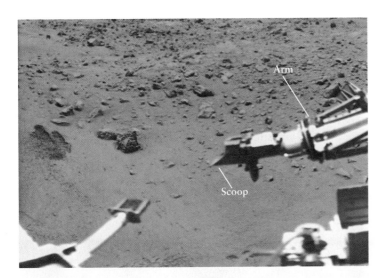

Figure 28-5   R I V U X G

Digging in the Martian Surface This view from the *Viking Lander 1* spacecraft shows the mechanical arm with its small scoop against the backdrop of the Martian terrain. The scoop was able to dig about 30 cm (12 in.) beneath the surface. (NASA)

2. The *labeled-release experiment* was designed to detect metabolic processes. A sample was moistened with nutrients containing radioactive carbon atoms. If any organisms in the sample consumed the nutrients, their waste products should include gases containing the telltale radioactive carbon.

3. The *pyrolytic-release experiment* was designed to detect photosynthesis, the biological process by which terrestrial plants use solar energy to help synthesize organic compounds from carbon dioxide. In the *Viking* experiments, a surface sample was placed in a container along with radioactive carbon dioxide and exposed to artificial sunlight. If plantlike photosynthesis occurred, microorganisms in the sample would take in some of the radioactive carbon from the gas.

The first data returned from these experiments caused great excitement, for in almost every case, rapid and extensive changes were detected inside the sealed containers. Further analysis of the data, however, led to the conclusion that these changes were due solely to nonbiological chemical processes. It appears that the Martian surface is rich in unstable chemicals that react with water to release oxygen gas. Because the present-day surface of Mars is bone-dry, these chemicals had nothing to react with until they were placed inside the moist interior of the *Viking Lander* laboratory.

At best, the results from the *Viking Lander* biological experiments were inconclusive. Perhaps life never existed on Mars at all. Or perhaps it did originate there, but failed to survive the thinning of the Martian atmosphere, the unstable chemistry of the planet's surface, and exposure to ultraviolet radiation from the Sun. (Unlike Earth, Mars has no ozone layer to block ultraviolet rays.) Another possibility is that Martian microorganisms have survived only in certain locations that the *Viking Landers* did not sample, such as isolated spots on the surface or deep beneath the

ground. And yet another option is that there is life on Mars, but the experimental apparatus on board the *Viking Lander* spacecraft was not sophisticated enough to detect it.

An entirely different set of biological experiments were designed for the British spacecraft *Beagle 2*, which landed on Mars in December 2003. (The spacecraft's name commemorated HMS *Beagle,* the survey ship from which Charles Darwin made many of the observations that led to the theory of evolution.)

Unlike the *Viking Lander* experiments, the apparatus on board *Beagle 2* was designed to test for the presence of either living or dead microorganisms. To do this, the spacecraft was to bore into the interiors of rocks to gather pristine, undisturbed samples, then heat these samples in the presence of oxygen gas. All carbon compounds decompose and form carbon dioxide (CO_2) when treated in this way, but biologically important molecules signal their presence by decomposing at a lower temperature. As a further test for the chemicals of life, *Beagle 2* was to check to see how many of the CO_2 molecules contain the isotope ^{12}C (which appears preferentially in biological molecules) and how many contain ^{13}C (which does not). (See Box 5-5 for a description of isotopes.) A positive result to these experiments would indicate that Martian rocks contain microorganisms that either survive to the present day or that died out at some point in the past.

Another *Beagle 2* experiment was to search for traces of methane in the Martian atmosphere. Microorganisms on Earth can gain energy by converting carbon dioxide to methane, and presumably Martian microorganisms could do the same. Left to itself, methane rapidly decomposes in the Martian atmosphere. Hence, if any methane is found on Mars, it must necessarily have been freshly formed—which would strongly suggest that life exists on Mars today.

Unfortunately, scientists on Earth were unable to establish contact with *Beagle 2* after landing, and so no results were returned from these experiments. But scientists hope to send a replacement lander to Mars using a set of experiments like those on board *Beagle 2*.

New Evidence for Martian Water

While the promise of *Beagle 2* will have to be fulfilled by a future mission, scientists searching for evidence of Martian life have been encouraged by other spacecraft that have provided new evidence of water on Mars. The European Space Agency's *Mars Express* spacecraft, which went into orbit around Mars in December 2003, used its infrared cameras to examine the ice cap at the Martian south pole (**Figure 28-6**). These cameras allowed scientists to see through the ice cap's surface layer of frozen carbon dioxide and reveal an underlying layer with the characteristic spectrum of water ice. (Figure 7-4 shows a similar spectrum obtained from Europa.) In January 2004, NASA successfully landed two robotic rovers named *Spirit* and *Opportunity* at two very different sites on opposite sides of Mars. While the terrain around the *Spirit* landing site appears to have been dry for billions of years, *Opportunity* landed in an area that appears to have been under water for extended periods (**Figure 28-7a**). Measurements made by

(a) Visible-light image   (b) Infrared image of carbon dioxide ice   (c) Infrared image of water ice

R I V U X G   R I V U X G   R I V U X G

Figure 28-6

Water at the Martian South Pole (a) This visible-light image shows the south polar cap of Mars, but does not indicate its chemical composition. But by using a camera tuned to different wavelengths of infrared light, the *Mars Express* spacecraft was able to identify the distinctive reflections of **(b)** an upper layer of carbon dioxide ice and **(c)** a deeper layer of water ice. Other observations have shown that there is also water ice at the Martian north pole. (ESA-OMEGA)

Opportunity confirm that some of the very dark surface material at its landing site contains an iron-rich mineral called gray hematite (Figure 28-7b). On Earth, deposits of gray hematite are commonly found at the bottoms of lakes or mineral hot springs. The presence of gray hematite at the *Opportunity* site reinforces the argument that Mars once had liquid water on its surface, and helps hold open the possibility that living organisms could have evolved on Mars.

The Martian Civilization That Never Was

In 1976, while the *Viking Landers* were carrying out their biological experiments on the Martian surface, the companion *Viking Orbiter* spacecraft photographed some surface features that at first glance seemed to have been crafted by *intelligent* life on Mars. The *Viking Orbiter 1* image in **Figure 28-8a** shows what appears to be a humanlike face, perhaps the product of an advanced and artistic civilization. However, when the more

(a)

(b)

Figure 28-7   R I V U X G

Evidence of Ancient Martian Water (a) The Mars rover *Opportunity* photographed these sedimentary layers in a region called Meridiani Planum. Some of the layers are made of dust deposited by the Martian winds, but others were laid down by minerals that precipitated out of standing water. **(b)** In this false-color image from *Opportunity* of Martian sand dunes, the bluish color shows the presence of millimeter-sized spheres of gray hematite. Such spheres naturally form in water-soaked deposits. (NASA/JPL/Cornell)

(a)

(b)

(c)

Figure 28-8   R I ⟨V⟩ U X G

A "Face" on Mars? **(a)** This 1976 image from *Viking Orbiter 1* shows a Martian surface feature that resembles a human face. Some suggested that this feature might have been made by intelligent beings. **(b)** This 1998 *Mars Global Surveyor* (MGS) image, made under different lighting conditions with a far superior camera, reveals the "face" to be just an eroded hill. **(c)** This MGS image shows features of natural origin within a 215-km (134-mi) wide crater on Mars. Can you see this "face"? (a: NSSDC/NASA and Dr. Michael H. Carr; b, c: Malin Space Science Systems/NASA)

advanced *Mars Global Surveyor* spacecraft viewed the surface in 1998 using a superior camera (Figure 28-8*b*), it found no evidence for facial features.

Scientists are universally convinced that the "face" and other apparent patterns in the *Viking Orbiter* images were created by shadows on wind-blown hills. Microscopic life may once have existed on Mars, and may yet exist today, but there is no evidence that the red planet has ever been the home of intelligent beings.

28-3 Meteorites from Mars have been scrutinized for life-forms

While spacecraft can carry biological experiments to other worlds such as Mars, many astrobiologists look forward to the day when a spacecraft will return Martian samples to laboratories on Earth. Until that day arrives, we have the next best thing: A dozen meteorites that appear to have formed on Mars have been found at a variety of locations on the Earth.

These meteorites are called **SNC meteorites** after the names given to the first three examples found (Shergotty, Nakhla, and Chassigny). What identifies SNC meteorites as having come from Mars is the chemical composition of trace amounts of gas trapped within them. This composition is very different from that of the Earth's atmosphere, but is a nearly perfect match to the composition of the Martian atmosphere found by the *Viking Landers*.

How could a rock have traveled from Mars to Earth? When a large piece of space debris collides with a planet's surface and forms an impact crater, most of the material thrown upward by the impact falls back onto the planet's surface. But some extraordinarily powerful impacts produce large craters—on Mars, roughly 100 km in diameter or larger. These tremendous impacts eject some rocks with such speed that they escape the planet's gravitational attraction and fly off into space.

There are numerous large craters on Mars, so a good number of Martian rocks have probably been blasted into space over the planet's history. These ejected rocks then go into elliptical orbits around the Sun. A few such rocks will have orbits that put them on a collision course with the Earth, and these are the ones that scientists find as SNC meteorites.

Using the radioactive age-dating technique (see Section 8-3), scientists find that most SNC meteorites are between 200 million and 1.3 billion years old, much younger than the 4.56-billion-year age of the solar system. But one SNC meteorite, denoted by the serial number ALH 84001 and found in Antarctica in 1984, was discovered in 1993 to be 4.5 billion years old (**Figure 28-9***a*). Thus, ALH 84001 is a truly ancient piece of Mars. Analysis of ALH 84001 suggests that it was fractured by an impact between 3.8 and 4.0 billion years ago, was ejected from Mars by another impact 16 million years ago, and landed in Antarctica a mere 13,000 years ago.

ALH 84001 is the only known specimen of a rock that was on Mars during the era when liquid water existed on the planet's surface. Scientists have therefore investigated its chemical composition carefully, in the hope that this rock may contain clues to the amount of water that once flowed on the Martian surface. One such clue is the presence of rounded grains of minerals called carbonates, which can form only in the presence of water.

(a)

(b)

 Figure 28-9   R I V U X G

A Meteorite from Mars (a) This 1.9-kg meteorite, known as ALH 84001, formed on Mars some 4.5 billion years ago. About 16 million years ago a massive impact blasted it into space, where it drifted in orbit around the Sun until landing in Antarctica 13,000 years ago. The small cube at lower right is 1 cm (0.4 in.) across. **(b)** This electron microscope image, magnified some 100,000 times, shows tubular structures about 100 nanometers (10^{-7} m) in length found within the Martian meteorite ALH 84001. One controversial interpretation is that these are the fossils of microorganisms that lived on Mars billions of years ago. (a: NASA Johnson Space Center; b: *Science*, NASA)

In 1996, David McKay and Everett Gibson of the NASA Johnson Space Center, along with several collaborators, reported the results of a two-year study of the carbonate grains in ALH 84001. They made three remarkable findings. First, in and around the carbonate grains were large numbers of elongated, tubelike structures resembling fossilized microorganisms (Figure 28-9b). Second, the carbonate grains contain very pure crystals of iron sulfide and magnetite. These two compounds are rarely found together (especially in the presence of carbonates) but can be produced by certain types of bacteria. Indeed, about one-fourth of the magnetite crystals found in ALH 84001 are of a type that on Earth are formed only by bacteria. Third, the carbonates contain organic molecules—just the sort, in fact, that result from the decay of microorganisms.

McKay and Gibson concluded that the structures seen in Figure 28-9b are fossilized remains of microorganisms. If so, these organisms lived and died on Mars billions of years ago, during the era when liquid water was abundant.

Are McKay and Gibson's conclusions correct? Their claims of ancient life on Mars are extraordinary, and they require extraordinary proof. With only one rock like ALH 84001 known to science, however, such proof is hard to come by, and many scientists are skeptical. They argue that the structures found in ALH 84001 could have been formed in other ways that do not require the existence of Martian microorganisms. Future spacecraft may help resolve the controversy by examining rocks on the Martian surface. For now, the existence of microscopic life on Mars in the distant past remains an open question.

> **Claims that scientists have found Martian microorganisms are intriguing but very controversial**

28-4 The Drake equation helps scientists estimate how many civilizations may inhabit our galaxy

We have seen that only a few locations in our solar system may have been suitable for the origin of life. But what about other planetary systems? The development of life on the Earth seems to suggest that extraterrestrial life, including intelligent species, might evolve on terrestrial planets around other stars, given sufficient time and hospitable conditions. How can we learn whether such worlds exist, given the tremendous distances that separate us from them? This is the great challenge facing the **search for extraterrestrial intelligence, or SETI.**

> **Are we alone, or does the Galaxy teem with intelligent life? Or is the truth somewhere in between?**

Close Encounters Versus Remote Communication

A tenet of modern folklore is the belief that alien civilizations do exist, and that their spacecraft have visited Earth. Indeed, surveys show that between one-third and one-half of all Americans believe in unidentified flying objects (UFOs). A somewhat smaller percentage believes that aliens have landed on Earth. But, in fact, there is *no* scientifically verifiable evidence of alien visitations. As an example, many UFO proponents believe that the U.S. government is hiding evidence of an alien spacecraft that crashed near Roswell, New Mexico, in 1947. However, the bits of "spacecraft wreckage" found near Roswell turn out to be nothing more than remnants of an unmanned research balloon. To find real evidence of the presence or absence of intelligent civilizations on worlds orbiting other stars, we must look elsewhere.

With our present technology, sending even a small unmanned spacecraft to another star requires a flight time of tens of thousands of years. Speculative design studies have been made for unmanned probes that could reach other stars within a century or less, but these are prohibitively expensive. Instead, many astronomers hope to learn about extraterrestrial civilizations by detecting radio transmissions from them. Radio waves are a logical choice for interstellar communication because they can travel immense distances without being significantly degraded by the interstellar medium, the thin gas and dust found between the stars (see Section 8-1).

Over the past several decades, astronomers have proposed various ways to search for alien radio transmissions, and several searches have been undertaken. In 1960, Frank Drake first used a radio telescope at the National Radio Astronomy Observatory in West Virginia to listen to two Sunlike stars, Tau Ceti and Epsilon Eridani, without success. More than 60 more extensive SETI searches have taken place since then, using radio telescopes around the world. Occasionally, a search has detected an unusual or powerful signal. But none has ever repeated, as a signal of intelligent origin might be expected to do. To date, we have no confirmed evidence of radio transmissions from another world.

Are They Out There?

Should we be discouraged by this failure to make contact? What are the chances that a radio astronomer might someday detect radio signals from an extraterrestrial civilization? The first person to tackle this issue was Frank Drake, who proposed that the number of technologically advanced civilizations in the Galaxy could be estimated by a simple equation. This is now called the **Drake equation:**

Drake equation

$$N = R_* \, f_p \, n_e \, f_l \, f_i \, f_c \, L$$

N = number of technologically advanced civilizations in the Galaxy whose messages we might be able to detect

R_* = the rate at which solar-type stars form in the Galaxy

f_p = the fraction of stars that have planets

n_e = the number of planets per solar system that are Earthlike (that is, suitable for life)

f_l = the fraction of those Earthlike planets on which life actually arises

f_i = the fraction of those life-forms that evolve into intelligent species

f_c = the fraction of those species that develop adequate technology and then choose to send messages out into space

L = the lifetime of a technologically advanced civilization

The Drake equation is enlightening because it expresses the number of extraterrestrial civilizations in a simple series of terms. We can estimate some of these terms from what we know about stars, stellar evolution, and planetary orbits. The *Cosmic Connections* figure depicts some of these considerations.

For example, the first two factors, R_* and f_p, can be determined by observation. In estimating R_*, we should probably exclude stars with masses greater than about 1.5 times that of the Sun. These more massive stars use up the hydrogen in their cores in 3 billion (3×10^9) years or less. On Earth, by contrast, human intelligence developed only within the last million years or so, some 4.56 billion years after the formation of the solar system. If that is typical of the time needed to evolve higher life-forms, then a star of 1.5 solar masses or more probably fades away or explodes into a supernova before creatures as intelligent as we can evolve on any of that star's planets.

Although stars less massive than the Sun have much longer lifetimes, they, too, seem unsuited for life because they are so dim. Only planets very near a low-mass star would be sufficiently warm for life as we know it, and a planet that close is subject to strong tidal forces from its star. We saw in Section 4-8 how the Earth's tidal forces keep the Moon locked in synchronous rotation, with one face continually facing the Earth. In the same way, a planet that orbits too close to its star would have one hemisphere that always faced the star, while the other hemisphere would be in perpetual, frigid darkness.

This leaves us with stars not too different from the Sun. (Like Goldilocks sampling the three bears' porridge, we must have a star that is not too hot and not too cold, but just right.) Based on statistical studies of star formation in the Milky Way, some astronomers estimate that roughly one of these Sunlike stars forms in the Galaxy each year in the galactic **habitable zone** (see the *Cosmic Connections* figure). This sets R_* at 1 per year.

As we saw in Sections 8-4 and 8-5, the planets in our solar system formed as a natural consequence of the birth of the Sun. We have also seen evidence suggesting that planetary formation may be commonplace around single stars (see Figure 8-8). Many astronomers suspect that most Sunlike stars probably have planets, and so they give f_p a value of 1.

Unfortunately, the rest of the terms in the Drake equation are very uncertain. Let's play with some hypothetical values. The chances that a planetary system has an Earthlike world suitable for life are not known. Were we to consider our own solar system as representative, we could put n_e at 1. Let's be more conservative, however, and suppose that 1 in 10 solar-type stars is orbited by a habitable planet, making $n_e = 0.1$. From what we know about the evolution of life on the Earth, we might assume that, given appropriate conditions, the development of life is a certainty, which would make $f_l = 1$. This is an area of intense interest to astrobiologists.

For the sake of argument, we might also assume that evolution might naturally lead to the development of intelligence (a conjecture that is hotly debated) and also make $f_i = 1$. It's anyone's guess as to whether these intelligent extraterrestrial beings would attempt communication with other civilizations in the Galaxy, but were we to assume that they would, f_c would be put at 1 also.

The last variable, L, involving the longevity of a civilization, is the most uncertain of all. Looking at our own example, we see

COSMIC CONNECTIONS

Habitable Zones for Life

Intelligent civilizations in our Milky Way Galaxy can evolve only in a certain region called the galactic habitable zone. In that zone, a suitable planet must lie within the planetary habitable zone of its parent star. (After C. H. Lineweaver, Y. Fenner, and B. K. Gibson)

Galactic habitable zone

Too close to the center of the Milky Way Galaxy:
• The distances between stars are small, so there can be close encounters between stars that would disrupt a planetary system.
• There are also frequent outbursts of potentially lethal radiation from supernovae and from the supermassive black hole at the very center of the Galaxy.

Too far from the center of the Milky Way Galaxy:
• Stars are deficient in elements heavier than hydrogen and helium, so they lack both the materials needed to form Earthlike planets and the chemical substances required for life as we know it.

Planetary habitable zone

The Star:
• Must have a mass that is neither too large, or too small.
• If the star's mass is *too large*, it will use up its hydrogen fuel so rapidly that it will burn out before life can evolve on any of its planets.
• If the star's mass is *too small*, it will be too dim to provide the warmth needed for life.

The Planet:
• Must be a terrestrial planet with a solid surface.
• Must have enough mass to provide the gravity needed to retain an atmosphere and oceans.
• Must be at a comfortable distance from the star so that the temperatures are neither too high nor too low.
• Must be in a stable, nearly circular orbit. (A highly elliptical orbit would cause excessively large temperature swings as the planet moved toward and away from the star.)

The Neighborhood:
• There needs to be one or more large Jovian planets whose gravitational forces will clear away comets and meteors.

a planet whose atmosphere and oceans are increasingly polluted by creatures that possess nuclear weapons. If we are typical, perhaps L is as short as 100 years. Putting all these numbers together, we arrive at

$$N = 1/\text{year} \times 1 \times 0.1 \times 1 \times 1 \times 1 \times 100 \text{ years} = 10$$

In other words, out of the hundreds of billions of stars in the Galaxy, we would estimate that there are only 10 technologically advanced civilizations from which we might receive communications.

A wide range of values has been proposed for the terms in the Drake equation, and these various guesses produce vastly different estimates of N. Some scientists argue that there is exactly one advanced civilization in the Galaxy and that we are it. Others speculate that there may be hundreds or thousands of planets inhabited by intelligent creatures. If we wish to know whether our Galaxy is devoid of other intelligence, teeming with civilizations, or something in between, we must keep searching the skies.

28-5 Radio searches for alien civilizations are under way

Even if only a few alien civilizations are scattered across the Galaxy, we have the technology to detect radio transmissions from them. But if other civilizations are trying to communicate with us using radio waves, what frequency are they using? This is an important question, because if we fail to tune our radio telescopes to the right frequency, we might never know whether the aliens are out there.

A reasonable choice would be a frequency that is fairly free of interference from extraneous sources. SETI pioneer Bernard Oliver was the first to draw attention to a range of relatively noise-free frequencies in the neighborhood of the microwave emission lines of hydrogen (H) and hydroxide (OH) (Figure 28-10). This region of the microwave spectrum is called the **water hole**, because H and OH together make H_2O, or water.

In 1989, NASA began work on the High Resolution Microwave Survey (HRMS), an ambitious project to scan the entire sky at frequencies spanning the water hole from 10^3 to 10^4 MHz. HRMS would have observed more than 800 nearby solar-type stars over a narrower frequency range in the hope of detecting signals that were either pulsed (like Morse code) or continuous (like the carrier wave for a TV or radio broadcast). The sophisticated signal-processing technology of HRMS would have been able to sift through tens of millions of individual frequency channels simultaneously. It would even have been able to detect the minute Doppler shifts in a signal coming from an alien planet as that planet spun on its axis and moved around its star.

 Sadly, just one year after HRMS began operation in 1992, the U.S. Congress imposed a mandate requiring that NASA no longer support HRMS or any other radio searches for extraterrestrial intelligence. This decision, which was made on budgetary grounds, saved a few million dollars—an entirely negligible amount compared to the total NASA budget. Ironically, the senator who spearheaded this move was from the state of Nevada, where tax dollars have been spent to signpost a remote desert road as "The Extraterrestrial Highway."

Even though NASA funding is no longer available, several teams of scientists remain actively involved in SETI programs. Funding for these projects has come from nongovernmental organizations such as the Planetary Society and from private individuals. Since 1995 the SETI Institute in California has been carrying out Project Phoenix, the

Millions of personal computers have helped scan for alien radio transmissions

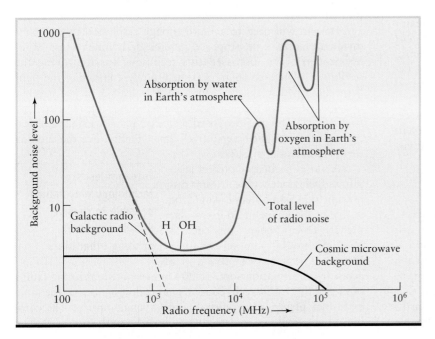

Figure 28-10

The Water Hole This graph shows the background noise level from the sky at various radio and microwave frequencies. The so-called water hole is a range of radio frequencies from about 10^3 to 10^4 megahertz (MHz) in which there is little noise and little absorption by the Earth's atmosphere. Some scientists suggest that this noise-free region would be well suited for interstellar communication. Within the water hole itself, the principal source of noise is the afterglow of the Big Bang, called the cosmic microwave background. To put this graph in perspective, a frequency of 100 MHz corresponds to "100" on an FM radio, and 10^3 MHz is a frequency used for various types of radar. (Adapted from C. Sagan and F. Drake) direct successor to HRMS. When complete, this project will have surveyed a thousand Sunlike stars within 200 light-years at millions of radio frequencies. At Harvard University, BETA (the *Billion-channel ExtraTerrestrial Assay*) is scanning the sky at even more individual frequencies within the water hole. Other multi-frequency searches are being carried out under the auspices of the University of Western Sydney in Australia and the University of California.

A major challenge facing SETI is the tremendous amount of computer time needed to analyze the mountains of data returned by radio searches. To this end, scientists at the University of California, Berkeley, have recruited nearly 5 million personal computer users to participate in a project called SETI@home. Each user receives actual data from a detector called SERENDIP IV (*Search for Extraterrestrial Radio Emissions from Nearby, Developed, Intelligent Populations*) and a data analysis program that also acts as a screensaver. When the computer's screensaver is on, the program runs, the data are analyzed, and the results are reported via the Internet to the researchers at Berkeley. The program then downloads new data to be analyzed. As of 2006, SETI@home users had provided as much computer time as a single computer working full-time for 2,000,000 years!

All current SETI projects make use of existing radio telescopes and must share telescope time with astronomical researchers. But by 2008, the SETI Institute plans to have completed and put into operation a radio telescope that will be dedicated solely to the search for intelligent signals. This telescope, called the Allen Telescope Array, will actually be hundreds of relatively small and inexpensive radio dishes working together. Perhaps this new array will be the first to detect a signal from a distant civilization.

28-6 Infrared telescopes in space will soon begin searching for Earthlike planets

Although no longer involved in SETI, NASA is planning a major effort to search for Earthlike planets suitable for the evolution of an advanced civilization. Such a search poses a major challenge. Astronomers have discovered many Jupiter-sized planets by detecting the "wobble" that these planets produce in their parent star (see Section 8-6). But a planet the size of the Earth would exert only a weak gravitational force on its parent star, and the resulting "wobble" is too small for us to detect with present technology. Earth-sized planets are also too dim to be seen in visible light against the glare of their parent star.

Searching for Planetary Transits

An alternative technique will be used by an orbiting telescope called Kepler, which is scheduled for launch in 2008. If a star is orbited by a planet whose orbital plane is oriented edge-on to our line of sight, once per orbit the planet will pass in front of the star in an event called a *transit*. This causes a temporary dimming of the light we see from that star. As we discussed in Section 8-6, astronomers have used Earth-based telescopes to observe dimming of this kind from Jupiter-sized planets transiting their parent stars. To detect the much slighter dimming caused by the transit of a small, Earth-sized planet, Kepler will use specialized detectors. Furthermore, by observing from orbit using a telescope with a wide field of view, Kepler will be able to continuously monitor thousands of stars at once.

If Kepler detects stars that dim slightly as expected, astronomers will have to make sure that the dimming really is due to a transiting planet and not some other cause. (One key test will be whether the dimming repeats with a definite period, as would be expected for a planet in a periodic orbit.) Once this is done, astronomers will be able to determine the transiting planets' sizes (which determines how much dimming takes place), as well as the sizes of their orbits (which can be calculated from Kepler's third law by using the orbital period of the planet, which is the same as the time interval between successive transits). Given the distance from the planet to its parent star and the star's luminosity, astronomers will even be able to estimate the planet's average temperature.

Searching Using Images and Spectra

The results from Kepler may help astronomers select stars to study in more detail using Darwin, a more advanced orbiting telescope under development by the European Space Agency. Planned for a 2015 launch, Darwin will search for Earthlike planets by detecting their infrared radiation. The rationale is that stars like the Sun emit much less infrared radiation than visible light, while planets are relatively strong emitters of infrared. Hence, observing in the infrared makes it less difficult (although still technically challenging) to detect planets orbiting a star.

Darwin will also analyze the infrared spectra of any planets that it finds, in the hope of seeing the characteristic absorption of atmospheric gases such as ozone, carbon dioxide, and water vapor (Figure 28-11). The relative amounts of these gases, as determined from a planet's spectrum, can reveal whether life is present on that planet.

Darwin will need to achieve enough resolution to detect individual planets. One proposed mission design makes use of interferometry. We discussed this technique for improving the resolution of telescopes in Section 6-6. By combining the light from three widely spaced dishes, each at least 3 m in diameter, Darwin will make the sharpest infrared images of any telescope in history. NASA has proposed a similar planet-finding interferometry mission called Terrestrial Planet Finder, but the funding for this mission is uncertain.

A more speculative project is an infrared telescope with sufficient resolution that some detail would be visible in the image of an extrasolar planet. One concept for such a mission would consist of five Darwin-type telescopes flying in a geometrical formation some 6000 km across (equal to the radius of the Earth). All five telescopes would collect light from the same extrasolar planet, then reflect it onto a single mirror. The combined light would go to detectors on board a sixth spacecraft. The

Future telescope technology may make it possible to resolve details on planets orbiting other stars

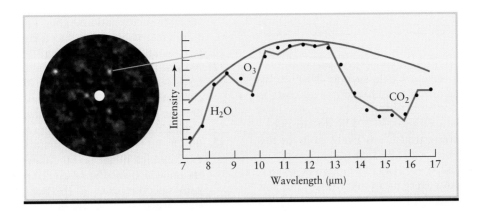

Figure 28-11

The Spectrum of a Simulated Planet The image on the left is a simulation of what the Terrestrial Planet Finder infrared telescope might see once it is launched. The white dot at the center is a nearby Sunlike star, and the smaller dots around it are planets orbiting the star. On the right is the simulated infrared spectrum of one of the planets, showing broad absorption lines of water vapor (H_2O), ozone (O_3), and carbon dioxide (CO_2). While all these molecules can be created by nonbiological processes, the presence of life will change the relative amounts of each molecule in the planet's atmosphere. Thus, the infrared spectrum of such planets will make it possible to identify worlds on which life may have evolved. (Jet Propulsion Laboratory)

technology needed for such an ambitious mission does not yet exist, but may become available within a few decades.

Sometime during the twenty-first century, missions such as Darwin may answer the question "Are there worlds like Earth orbiting other stars?" If the answer is yes, radio searches for intelligent signals will gain even more impetus.

The potential rewards from such searches are great. Detecting a message from an alien civilization could dramatically change the course of our own civilization, through the sharing of scientific information with another species or an awakening of social or humanistic enlightenment. In only a few years our technology, industry, and social structure might advance the equivalent of centuries into the future. Such changes would touch every person on the Earth. Mindful of these profound implications, scientists push ahead with the search for extraterrestrial intelligence.

Key Words

astrobiology, p. 28-1
carbonaceous chondrite, p. 28-3
Drake equation, p. 28-9
habitable zone, p. 28-9

organic molecules, p. 28-1
search for extraterrestrial intelligence (SETI), p. 28-8
SNC meteorite, p. 28-7
water hole, p. 28-11

Key Ideas

Organic Molecules in the Universe: All life on Earth, and presumably on other worlds, depends on organic (carbon-based) molecules. These molecules occur naturally throughout interstellar space.

• The organic molecules needed for life to originate were probably brought to the young Earth by comets or meteorites. Another likely source for organic molecules is chemical reactions in the Earth's primitive atmosphere. Similar processes may occur on other worlds.

Life in the Solar System: Besides Earth, only two worlds in our solar system—the planet Mars and Jupiter's satellite Europa—may have had the right conditions for the origin of life.

• Mars once had liquid water on its surface, though it has none today. Life may have originated on Mars during the liquid water era.

• The *Viking Lander* spacecraft searched for microorganisms on the Martian surface, but found no conclusive sign of their presence. The unsuccessful *Beagle 2* mission to Mars was to carry out a different set of biological experiments on samples taken from the interiors of rocks; these experiments may be attempted again on a future mission.

• An ancient Martian rock that came to Earth as a meteorite shows circumstantial evidence that microorganisms once existed on Mars. Additional rock samples are needed to provide corroboration.

• Europa appears to have extensive liquid water beneath its icy surface. Future missions may search for the presence of life there.

Radio Searches for Extraterrestrial Intelligence: Astronomers have carried out a number of searches for radio signals from other stars. No signs of intelligent life have yet been detected, but searches are continuing and using increasingly sophisticated techniques.

• The Drake equation is a tool for estimating the number of intelligent, communicative civilizations in our Galaxy.

Telescope Searches for Earthlike Planets: A new generation of orbiting telescopes may be able to detect terrestrial planets around nearby stars. If such planets are found, their infrared spectra may reveal the presence or absence of life.

Questions

Review Questions

1. Why are extreme life-forms on Earth, such as those shown in the photograph that opens this chapter, of interest to astrobiologists?

2. What is meant by "life as we know it"? Why do astrobiologists suspect that extraterrestrial life is likely to be of this form?

3. How have astronomers discovered organic molecules in interstellar space? Does this discovery mean that life of some sort exists in the space between the stars?

4. Mercury, Venus, and the Moon are all considered unlikely places to find life. Suggest why this should be.

5. Many science-fiction stories and movies—including *The War of the Worlds, Invaders from Mars, Mars Attacks!,* and *Martians, Go Home*—involve invasions of Earth by intelligent beings from Mars. Why Mars rather than any of the other planets?

6. Summarize the differences in philosophy between the biological experiments on board the *Viking Landers* and those on board *Beagle 2.*

7. What arguments can you give against the idea that the "face" on Mars (Figure 28-8) is of intelligent origin? What arguments can you give in favor of this idea?

8. Suppose someone brought you a rock that he claimed was a Martian meteorite. What scientific tests would you recommend be done to test this claim?

9. Why are most searches for extraterrestrial intelligence made using radio telescopes? Why are most of these carried out at frequencies between 10^3 MHz and 10^4 MHz?

10. Explain why infrared telescopes like those proposed for Darwin and Terrestrial Planet Finder need to be placed in space.

Advanced Questions

Problem-solving tips and tools

The small-angle formula, discussed in Box 1-1, will be useful. Section 5-2 gives the relationship between wavelength and frequency, while Section 5-9 and Box 5-6 discuss the Doppler effect. Section 6-3 gives the relationship between the angular resolution of a telescope, the telescope diameter, and the wavelength used. You will find useful data about the planets in Appendix 1.

11. In 1802, when it seemed likely to many scholars that there was life on Mars, the German mathematician Karl Friedrich Gauss proposed that we signal the Martian inhabitants by drawing huge geometric patterns in the snows of Siberia. His plan was never carried out. (a) Suppose patterns had been drawn that were 1000 km across. What minimum diameter would the objective of a Martian telescope need to have to be able to resolve these patterns? Assume that the observations are made at a wavelength of 550 nm, and assume that Earth and Mars are at their minimum separation. (b) Ideally, the patterns used would be ones that could not be mistaken

for natural formations. They should also indicate that they were created by an advanced civilization. What sort of patterns would you have chosen?

12. Assume that all the terms in the Drake equation have the values given in the text, except for *N* and *L*. (a) If there are 1000 civilizations in the Galaxy today, what must be the average lifetime of a technological civilization? (b) What if there are a million such civilizations?

13. (a) Of the visually brightest stars in the sky listed in Appendix 5, which might be candidates for having Earthlike planets on which intelligent civilizations have evolved? Explain your selection criteria. (b) Repeat part (a) for the nearest stars, listed in Appendix 4.

14. It has been suggested that extraterrestrial civilizations would choose to communicate at a wavelength of 21 cm. Hydrogen atoms in interstellar space naturally emit at this wavelength, so astronomers studying the distribution of hydrogen around the Galaxy would already have their radio telescopes tuned to receive extraterrestrial signals. (a) Calculate the frequency of this radiation in megahertz. Is this inside or outside the water hole? (b) Discuss the merits of this suggestion.

15. Imagine that a civilization in another planetary system is sending a radio signal toward Earth. As our planet moves in its orbit around the Sun, the wavelength of the signal we receive will change due to the Doppler effect. This gives SETI scientists a way to distinguish stray signals of terrestrial origin (which will not show this kind of wavelength change) from interstellar signals. (a) Use the data in Appendix 1 to calculate the speed of the Earth in its orbit. For simplicity, assume the orbit is circular. (b) If the alien civilization is transmitting at a frequency of 3000 MHz, what wavelength (in meters) would we receive if the Earth were moving neither toward nor away from their planet? (c) The maximum Doppler shift occurs if the Earth's orbital motion takes it directly toward or directly away from the alien planet. How large is that maximum wavelength shift? Express your answer both in meters and as a percentage of the unshifted wavelength you found in (b). (d) Discuss why it is important that SETI radio receivers be able to measure frequency and wavelength to very high precision.

16. Astronomers have proposed using interferometry to make an extremely high-resolution telescope. This proposal involves placing a number of infrared telescopes in space, separating them by thousands of kilometers, and combining the light from the individual telescopes. One design of this kind has an effective diameter of 6000 km and uses infrared radiation with a wavelength of 10 mm. If it is used to observe an Earthlike planet orbiting the star Epsilon Eridani, 3.22 parsecs (10.5 light-years) from Earth, what is the size of the smallest detail that this system will be able to resolve on the face of that planet? Give your answer in kilometers.

Discussion Questions

17. Suppose someone told you that the *Viking Landers* failed to detect life on Mars simply because the tests were designed to detect terrestrial life-forms, not Martian life-forms. How would you respond?

18. Science-fiction television shows and movies often depict aliens as looking very much like humans. Discuss the likelihood that intelligent creatures from another world would have (a) a biochemistry similar to our own, (b) two legs and two arms, and (c) about the same dimensions as a human.

19. The late, great science-fiction editor John W. Campbell exhorted his authors to write stories about organisms that think as well as humans, but not *like* humans. Discuss the possibility that an intelligent being from another world might be so alien in its thought processes that we could not communicate with it.

20. If a planet always kept the same face toward its star, just as the Moon always keeps the same face toward Earth, most of the planet's surface would be uninhabitable. Discuss why.

21. How do you think our society would respond to the discovery of intelligent messages coming from a civilization on a planet orbiting another star? Explain your reasoning.

22. What do you think will set the limit on the lifetime of our technological civilization? Explain your reasoning.

23. The first of all Earth spacecraft to venture into interstellar space were *Pioneer 10* and *Pioneer 11,* which were launched in 1972 and 1973, respectively. Their missions took them past Jupiter and Saturn and eventually beyond the solar system. Both spacecraft carry a metal plaque with artwork (reproduced below) that shows where the spacecraft is from and what sort of creatures designed it. If an alien civilization were someday to find one of these spacecraft, which of the features on the plaque do you think would be easily understandable to them? Explain.

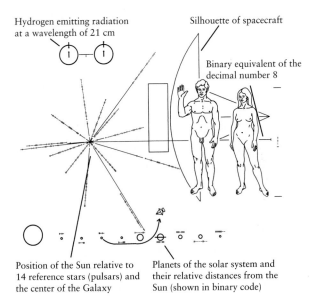

Hydrogen emitting radiation at a wavelength of 21 cm

Silhouette of spacecraft

Binary equivalent of the decimal number 8

Position of the Sun relative to 14 reference stars (pulsars) and the center of the Galaxy

Planets of the solar system and their relative distances from the Sun (shown in binary code)

Web/eBook Questions

24. Any living creatures in the subsurface ocean of Europa would have to survive without sunlight. Instead, they might obtain energy from Europa's inner heat. Search the World Wide Web for information about "black smokers," which are associated with high-temperature vents at the bottom of Earth's

oceans. What kind of life is found around black smokers? How do these life-forms differ from the more familiar organisms found in the upper levels of the ocean?

25. Search the World Wide Web for information about the *Mars Express* orbiter and the *Spirit* and *Opportunity* rovers. What discoveries have these missions made about water on Mars? Have they found any evidence that liquid water has existed on Mars in the recent past? Describe the evidence, if any.

26. Like other popular media, the World Wide Web is full of claims of the existence of "extraterrestrial intelligence"— namely, UFO sightings and alien abductions. (a) Choose a Web site of this kind and analyze its content using the idea of *Occam's razor,* the principle that if there is more than one viable explanation for a phenomenon, one should choose the simplest explanation that fits all the observed facts. (b) Read what a skeptical Web site has to say about UFO sightings. A good example is the Web site of the Committee for the Scientific Investigation of Claims of the Paranormal, or CSICOP. After considering what you have read on both sides of the UFO debate, discuss your opinions about whether aliens really have landed on Earth.

27. **The Drake Equation.** Access the Active Integrated Media Module "The Drake Equation" in Chapter 28 of the *Universe* Web site or eBook. (a) For each of the terms in the Drake equation, choose a value that seems reasonable to you. How did you choose these values? Using the module, what do you find for the number of civilizations in our Galaxy? From your calculation, are civilizations common or uncommon in our Galaxy? (b) Using the module, choose a set of values that give $N = 10^6$ (a million civilizations). What values did you use? Which of these seem reasonable to you, and why?

Activities

Observing Projects

28. Use the *Starry Night Enthusiast*™ program to view the Earth as it might be seen by a visiting spacecraft. First, select **Viewing Location . . .** in the **Options** menu and set the viewing location to your city or town from the list of cities provided or click on the **Map** tab in the **Viewing Location** pane and use the mouse to click on your approximate position on the world map. Then click the **Set Location** button. Set the local time to 12:00:00 P.M. (noon). To see the Earth from space, use the up and down elevation buttons on the toolbar to raise yourself above the surface until you can see the entire Earth. You can use the scrollbars (select **View > Show Scrollbars**) on the right side and bottom of the window to center the Earth in your view. The Earth can be rotated to allow you to see different locations by clicking and moving the mouse while its icon, a four-way arrow, is over the Earth's image. (a) Describe any features you see that suggest life could exist on Earth. Explain your reasoning. (b) Using the controls at the right-hand end of the toolbar, zoom in to show more detail around your city or town. The amount of detail is comparable to the view from a spacecraft a few million kilometers away. Can you see any

evidence that life does exist on Earth? (c) From a distance of a few million kilometers, are there any measurements that a spacecraft could carry out to prove that life exists on Earth? Explain your reasoning.

29. Use the *Starry Night Enthusiast*™ program to examine the planet Mars. Open the **Favourites** pane and double-click on **Solar System > Mars** to view this planet from about 6800 km above its surface. (Click on **View > Feet** to remove the astronaut's spacesuit from the view.) You can zoom in or out on Mars using the buttons in the Zoom section at the right of the toolbar. You can rotate Mars by placing the mouse cursor over the image and moving the mouse while holding down the mouse button. (On a two-button mouse, hold down the left mouse button.) Rotate Mars and zoom in and out to familiarize yourself with the different surface features. Based on what you observe, where on the Martian surface would you choose to land a spacecraft to search for the presence of life? Explain how you made your choice.

Collaborative Exercise

30. Imagine that astronomers have discovered intelligent life in a nearby star system. Your group is submitting a proposal for who on Earth should speak for the planet and what 50-word message should be conveyed. Prepare a maximum one-page proposal that states (**a**) who should speak for Earth and why; (**b**) what this person should say in 50 words; and (**c**) why this message is the most important compared to other things that could be said. Only serious responses receive full credit.

A Biologist's View of Astrobiology

by Kevin W. Plaxco

In the early 1960s, flush with the excitement of Sputnik and the first manned space missions, the Nobel Prize-winning biologist Joshua Lederberg coined the word "exobiology" to describe the scientific study of extraterrestrial life. But after a brief flurry of popularity in the 1960s and 1970s (due partly to the pioneering research and public outreach work of Carl Sagan), exobiology fell out of fashion among scientists; it was, after all, the only field of scientific study without any actual subject material to research.

In the last decade, however, the field of astrobiology has filled the void left by exobiology's demise by being simultaneously more encompassing and more practical. Astrobiology removes exobiology's distinction between life on earth and life "out there" and focuses on broader, but more tractable, questions regarding the relationship between life (any life, anywhere) and the universe. In a nutshell, astrobiology uses our significant (if incomplete) knowledge of life on Earth to address three broad questions about life in the universe:

- Which of the physical attributes of our universe were necessary in order for it to support the origins and further evolution of life?
- How did life arise and evolve on Earth, and how might it have arisen and evolved elsewhere?
- Where else might life have arisen in our universe, and how do we best search for it?

Much of the worth and appeal of astrobiology lie in the fact that these questions, which are among the most fundamental posed by science, address the most profound issues regarding our origins and our nature.

Perhaps not surprisingly, some aspects of astrobiology are far better understood than others. We understand, for example, the broad details of how the almost metal-free hydrogen and helium left over after the Big Bang was enriched in the metallic elements by fusion reactions occurring in the more massive stars and how these elements—absolutely critical for biology because only they support the complex chemistry required for life—were then returned to the cosmos in supernova explosions. Similarly, we understand in detail how rocky planets—which we see as potentially habitable—form out of the dust and gas of prestellar nebulae. In contrast, however, much of the *biology* in astrobiology remains speculative at best. There is, for example, nothing even remotely resembling consensus regarding how life first arose on Earth much less how—or if—it might have arisen elsewhere in the cosmos.

Perhaps equally unsurprisingly, some branches of the astrobiology community are more "optimistic" than others. Astronomers generally view astrobiology in terms of the processes by which potentially habitable planets form. As these processes are reasonably well understood, astronomers tend to be rather upbeat regarding the events surrounding life's origins. The radio astronomer Frank Drake, who pioneered the search for extraterrestrial intelligence (SETI) in the 1960s, is a relatively extreme example of this hopefulness: When trying to estimate the number of intelligent, "communicating" civilizations in the Galaxy, he assumed that rocky, water-soaked planets almost invariably give rise to life. Biochemists, in contrast, tend to view the key issues in astrobiology as relating to the first steps in the origins of life—that is, the detailed processes by which inanimate chemicals first combine to form a living, evolving, self-replicating system. To date, about a half dozen theories have been put forth to explain how this might have occurred on Earth, but all are fraught with enormous scientific difficulties. Specifically, every single theory of the origins of life postulated to date either utterly lacks experimental evidence to back it up or requires events that are so improbable they make even the "astronomical" number of planets in the universe look small by comparison. When faced with what is, in effect, a complete unknown, biochemists tend to be much more cautious about estimating the probability of life arising and the chances of our being alone in the cosmos. To quote Francis Crick, co-discoverer of the structure of DNA, "We cannot decide whether the origin of life on Earth was an extremely unlikely event or almost a certainty, or any possibility in between these two extremes."

Irrespective of our degree of personal optimism or pessimism, though, no one would be more thrilled than us biologists by the detection of life that had arisen independently of life on Earth. Such a find would revolutionize our field by providing the first, tangible evidence of whether there are viable routes to the evolution of life other than the DNA/proteins/cells path that it took on Earth. But even barring such a momentous discovery as the detection of extraterrestrial life, astrobiology still provides a significant service to biology: By placing life into a broader context, astrobiology provides the perspective that we need in order to truly understand who we are and where we come from.

Prior to becoming a professor of biochemistry at the University of California, Santa Barbara, Kevin Plaxco earned his B.Sc. in chemistry from the University of California at Riverside and his Ph.D. in molecular biology from the California Institute of Technology. While his research focuses predominantly on the study of proteins, the "macromolecular machines" that perform the lion's share of the action in our cells, he has also performed NASA-funded research into methods for the detection of extraterrestrial life. In addition, he teaches a course in astrobiology and has published a popular science book on the subject.

Appendices

Appendix 1   The Planets: Orbital Data

Planet	Semimajor axis (10⁶ km)	Semimajor axis (AU)	Sidereal period (years)	Sidereal period (days)	Synodic period (days)	Average orbital speed (km/s)	Orbital eccentricity	Inclination of orbit to ecliptic (°)
Mercury	57.9	0.387	0.241	87.969	115.88	47.9	0.206	7.00
Venus	108.2	0.723	0.615	224.70	583.92	35.0	0.007	3.39
Earth	149.6	1.000	1.000	365.256	—	29.79	0.017	0.00
Mars	227.9	1.524	1.88	686.98	779.94	24.1	0.093	1.85
Jupiter	778.3	5.203	11.86		398.9	13.1	0.048	1.30
Saturn	1429	9.554	29.46		378.1	9.64	0.053	2.48
Uranus	2871	19.194	84.10		369.7	6.83	0.043	0.77
Neptune	4498	30.066	164.86		367.5	5.5	0.010	1.77

Appendix 2   The Planets: Physical Data

Planet	Equatorial diameter (km)	Equatorial diameter (Earth = 1)	Mass (kg)	Mass (Earth = 1)	Average density (kg/m³)	Rotation period* (solar days)	Inclination of equator to orbit (°)	Surface gravity (Earth = 1)	Albedo	Escape speed (km/s)
Mercury	4880	0.383	3.302×10^{23}	0.0553	5430	58.646	0.5	0.38	0.12	4.3
Venus	12,104	0.949	4.868×10^{24}	0.815	5243	243.01^R	177.4	0.91	0.59	10.4
Earth	12,756	1.000	5.974×10^{24}	1.000	5515	0.997	23.45	1.000	0.39	11.2
Mars	6794	0.533	6.418×10^{23}	0.107	3934	1.026	25.19	0.38	0.15	5.0
Jupiter	142,984	11.209	1.899×10^{27}	317.8	1326	0.414	3.12	2.36	0.44	60.2
Saturn	120,536	9.449	5.685×10^{26}	95.16	687	0.444	26.73	1.1	0.47	35.5
Uranus	51,118	4.007	8.682×10^{25}	14.53	1318	0.718^R	97.86	0.92	0.56	21.3
Neptune	49,528	3.883	1.024×10^{26}	17.15	1638	0.671	29.56	1.1	0.51	23.5

*For Jupiter, Saturn, Uranus, and Neptune, the internal rotation period is given. A superscript R
means that the rotation is retrograde (opposite the planet's orbital motion).

Appendix 3 Satellites of the Planets

Planet	Satellite	Date of discovery	Average distance from center of planet (km)	Orbital (sidereal) period (days)*	Orbital eccentricity	Size of satellite (km)**	Mass (kg)
EARTH	Moon	—	384,400	27.322	0.0549	3476	7.349×10^{22}
MARS	Phobos	1877	9378	0.319	0.01	$28 \times 23 \times 20$	1.1×10^{16}
	Deimos	1877	23,460	1.263	0.00	$16 \times 12 \times 10$	1.8×10^{15}
JUPITER	Metis	1979	128,000	0.295	0.0012	44	1.2×10^{17}
	Adrastea	1979	129,000	0.298	0.0018	$24 \times 20 \times 16$	7.5×10^{15}
	Amalthea	1892	181,400	0.498	0.0031	$270 \times 200 \times 155$	2.1×10^{18}
	Thebe	1979	221,900	0.675	0.0177	98	1.5×10^{18}
	Io	1610	421,600	1.769	0.0041	3642	8.932×10^{22}
	Europa	1610	670,900	3.551	0.0094	3120	4.791×10^{22}
	Ganymede	1610	1,070,000	7.155	0.0011	5268	1.482×10^{23}
	Callisto	1610	1,883,000	16.689	0.0074	4800	1.077×10^{23}
	Themisto	2000	7,284,000	130.02	0.2426	9	6.9×10^{14}
	Leda	1974	11,165,000	240.92	0.1636	18	1.1×10^{16}
	Himalia	1904	11,461,000	250.56	0.1623	184	6.7×10^{18}
	Lysithea	1938	11,717,000	259.20	0.1124	38	6.3×10^{16}
	Elara	1905	11,741,000	259.64	0.2174	78	8.7×10^{17}
	Kallichore	2000	12,555,000	456.10	0.2480	4	1.5×10^{13}
	S/2003 J12	2003	15,912,000	489.52^R	0.6056	1	1.5×10^{12}
	Carpo	2003	16,989,000	456.10	0.4297	3	4.5×10^{13}
	Euporie	2001	19,304,000	550.74^R	0.1432	2	1.5×10^{13}
	S/2003 J3	2003	20,221,000	583.88^R	0.1970	2	1.5×10^{13}
	S/2003 J18	2003	20,514,000	596.59^R	0.0148	2	1.5×10^{13}
	Orthosie	2001	20,720,000	622.56^R	0.2808	2	1.5×10^{13}
	Euanthe	2001	20,797,000	620.49^R	0.2321	3	4.5×10^{13}
	Harpalyke	2000	20,858,000	623.31^R	0.2268	4	1.2×10^{14}
	Praxidike	2000	20,907,000	625.38^R	0.2308	7	4.3×10^{14}
	Thyone	2001	20,939,000	627.21^R	0.2286	4	9.0×10^{13}
	S/2003 J16	2003	20,963,000	616.36^R	0.2245	2	1.5×10^{13}
	Iocaste	2000	21,061,000	631.60^R	0.2160	5	1.9×10^{14}
	Mneme	2003	21,069,000	620.04^R	0.2273	2	1.5×10^{13}
	Hermippe	2001	21,131,000	633.90^R	0.2096	4	9.0×10^{13}
	Thelxinoe	2003	21,162,000	628.09^R	0.2206	2	1.5×10^{13}
	Helike	2003	21,263,000	634.77^R	0.1558	4	9.0×10^{13}
	Ananke	1951	21,276,000	629.77^R	0.2435	28	3.0×10^{16}
	S/2003 J15	2003	22,627,000	689.77^R	0.1916	2	1.5×10^{13}
	Eurydome	2001	22,865,000	717.33^R	0.2759	3	4.5×10^{13}
	Arche	2002	22,931,000	723.90^R	0.2588	3	4.5×10^{13}
	S/2003 J17	2003	23,001,000	714.47^R	0.2379	2	1.5×10^{13}
	Pasithee	2001	23,004,000	719.44^R	0.2675	2	1.5×10^{13}
	S/2003 J10	2003	23,042,000	716.25^R	0.4295	2	1.5×10^{13}
	Chaldene	2000	23,100,000	723.70^R	0.2519	4	7.5×10^{13}
	Isonoe	2000	23,155,000	726.25^R	0.2471	4	7.5×10^{13}

Appendix 3   Satellites of the Planets

Planet	Satellite	Date of discovery	Average distance from center of planet (km)	Orbital (sidereal) period (days)*	Orbital eccentricity	Size of satellite (km)**	Mass (kg)
JUPITER (continued)	Erinome	2000	23,196,000	728.51^R	0.2665	3	4.5×10^{13}
	Kale	2001	23,217,000	729.47^R	0.2599	2	1.5×10^{13}
	Aitne	2001	23,229,000	730.18^R	0.2643	3	4.5×10^{13}
	Taygete	2000	23,280,000	732.41^R	0.2525	5	1.6×10^{14}
	S/2003 J9	2003	23,384,000	733.29^R	0.2632	1	1.5×10^{12}
	Carme	1938	23,404,000	734.17^R	0.2533	46	1.3×10^{17}
	Sponde	2001	23,487,000	748.34^R	0.3121	2	1.5×10^{13}
	Megaclite	2000	23,493,000	752.88^R	0.4197	6	2.1×10^{14}
	S/2003 J5	2003	23,495,000	738.73^R	0.2478	4	9.0×10^{13}
	S/2003 J19	2003	23,533,000	740.42^R	0.2557	2	1.5×10^{13}
	S/2003 J23	2003	23,563,000	732.44^R	0.2714	2	1.5×10^{13}
	Kalyke	2000	23,566,000	742.03^R	0.2465	5	1.9×10^{14}
	S/2003 J14	2003	23,614,000	779.23^R	0.3439	2	1.5×10^{13}
	Pasiphaë	1908	23,624,000	743.63^R	0.4090	58	3.0×10^{17}
	Eukelade	2003	23,661,000	746.39^R	0.2721	4	9.0×10^{13}
	S/2003 J4	2003	23,930,000	755.24^R	0.3618	2	1.5×10^{13}
	Sinope	1914	23,939,000	758.90^R	0.2495	38	7.5×10^{16}
	Hegemone	2003	23,947,000	739.60^R	0.3276	3	4.5×10^{13}
	Aoede	2003	23,981,000	761.50^R	0.4322	4	9.0×10^{13}
	Kallichore	2003	24,043,000	764.74^R	0.2640	2	1.5×10^{13}
	Autonoe	2000	24,046,000	760.95^R	0.3168	4	9.0×10^{13}
	Callirrhoe	1999	24,103,000	758.77^R	0.2828	7	8.7×10^{14}
	Cyllene	2003	24,349,000	751.91^R	0.3189	2	1.5×10^{13}
	S/2003 J2	2003	29,541,000	979.99^R	0.2255	2	1.5×10^{13}
SATURN	Pan	1981	133,580	0.575	0	$35 \times 35 \times 23$	4.9×10^{15}
	Daphnis	2005	136,500	0.594	0	7	3×10^{14}
	Atlas	1980	137,670	0.602	0.0012	$46 \times 38 \times 19$	6.6×10^{15}
	Prometheus	1980	139,380	0.613	0.0022	$119 \times 87 \times 61$	1.6×10^{17}
	Pandora	1980	141,720	0.629	0.0042	$103 \times 80 \times 64$	1.4×10^{17}
	Epimetheus	1980	151,410	0.694	0.0098	$135 \times 108 \times 105$	5.3×10^{17}
	Janus	1980	151,460	0.695	0.0068	$193 \times 173 \times 137$	1.9×10^{18}
	Mimas	1789	185,540	0.942	0.0196	397	3.8×10^{19}
	Methone	2004	194,440	1.01	0.0001	3	2×10^{13}
	Pallene	2004	212,280	1.154	0.004	4	4×10^{13}
	Enceladus	1789	238,040	1.37	0.0047	504	1.1×10^{20}
	Tethys	1684	294,670	1.888	0.0001	1066	6.2×10^{20}
	Telesto	1980	294,710	1.888	0.0002	$29 \times 22 \times 20$	8×10^{15}
	Calypso	1980	294,710	1.888	0.0005	$30 \times 23 \times 14$	5×10^{15}
	Polydeuces	2004	377,200	2.737	0.0192	3.5	3×10^{13}
	Dione	1684	377,420	2.737	0.0022	1123	1.1×10^{21}
	Helene	1980	377,420	2.737	0.0071	$36 \times 32 \times 30$	2×10^{16}

Appendix 3   Satellites of the Planets

Planet	Satellite	Date of discovery	Average distance from center of planet (km)	Orbital (sidereal) period (days)*	Orbital eccentricity	Size of satellite (km)**	Mass (kg)
SATURN (continued)	Rhea	1672	527,070	4.518	0.001	1528	2.3×10^{21}
	Titan	1655	1,221,870	15.95	0.0288	5150	1.34×10^{23}
	Hyperion	1848	1,500,880	21.28	0.0274	$360 \times 280 \times 225$	5.7×10^{18}
	Iapetus	1671	3,560,840	79.33	0.0283	1472	2.0×10^{21}
	Kiviuq	2000	11,111,000	449.22	0.3288	16	3×10^{15}
	Ijiraq	2000	11,124,000	451.43	0.3163	12	1×10^{15}
	Phoebe	1898	12,947,780	550.31^{R}	0.1635	$230 \times 220 \times 210$	8.3×10^{18}
	Paaliaq	2000	15,200,000	686.93	0.3631	19	4×10^{15}
	Skathi	2000	15,541,000	728.21^{R}	0.2701	6	2×10^{14}
	Albiorix	2000	16,182,000	783.46	0.477	32	3×10^{16}
	S/2004 S11	2004	17,119,000	834.84	0.4691	6	2×10^{14}
	Erriapo	2000	17,343,000	871.18	0.4724	10	9×10^{14}
	Siarnaq	2000	17,531,000	895.55	0.2961	40	4×10^{16}
	S/2006 S8	2006	17,610,000	869^{R}	0.418	6	2×10^{14}
	Tarvos	2000	17,983,000	926.23	0.5305	15	3×10^{15}
	S/2006 S4	2006	18,105,000	905^{R}	0.374	6	2×10^{14}
	S/2004 S19	2004	18,217,125	912^{R}	0.36	8	8×10^{14}
	S/2004 S13	2004	18,403,000	933.45^{R}	0.2586	6	2×10^{14}
	S/2006 S6	2006	18,600,000	942^{R}	0.192	6	2×10^{14}
	Mundilfari	2000	18,685,000	952.67^{R}	0.21	7	3×10^{14}
	S/2006 S1	2006	18,981,135	970^{R}	0.13	6	2×10^{14}
	Narvi	2003	19,007,000	1003.93^{R}	0.4309	7	3×10^{14}
	S/2004 S15	2004	19,338,000	1005.93^{R}	0.1428	6	2×10^{14}
	S/2004 S17	2004	19,447,000	1014.7^{R}	0.1793	4	4×10^{13}
	Suttungr	2000	19,459,000	1016.67^{R}	0.114	7	3×10^{14}
	S/2004 S14	2004	19,856,000	1038.67^{R}	0.3715	6	2×10^{14}
	S/2004 S12	2004	19,878,000	1046.16^{R}	0.3261	5	2×10^{14}
	S/2004 S18	2004	20,129,000	1083.57^{R}	0.5214	7	3×10^{14}
	S/2004 S9	2004	20,390,000	1086.1^{R}	0.2397	5	2×10^{14}
	Thrymr	2000	20,474,000	1094.23^{R}	0.4652	7	3×10^{14}
	S/2004 S10	2004	20,735,000	1116.47^{R}	0.252	6	2×10^{14}
	S/2004 S7	2004	20,999,000	1140.28^{R}	0.5299	6	2×10^{14}
	S/2006 S3	2006	21,132,000	1142^{R}	0.471	6	2×10^{14}
	S/2006 S7	2006	22,290,000	1237^{R}	0.368	6	2×10^{14}
	S/2006 S2	2006	22,350,000	1245^{R}	0.341	7	3×10^{14}
	S/2004 S16	2004	22,453,000	1260.28^{R}	0.1364	4	4×10^{13}
	Ymir	2000	23,040,000	1315.21^{R}	0.335	18	3×10^{15}
	S/2006 S5	2006	23,190,000	1314^{R}	0.139	6	2×10^{14}
	S/2004 S8	2004	25,108,000	1490.87^{R}	0.2064	6	2×10^{14}
URANUS	Cordelia	1986	49,800	0.335	0.0003	50×36	4.4×10^{16}
	Ophelia	1986	53,800	0.376	0.0099	54×38	5.3×10^{16}

Appendix 3 Satellites of the Planets

Planet	Satellite	Date of discovery	Average distance from center of planet (km)	Orbital (sidereal) period (days)*	Orbital eccentricity	Size of satellite (km)**	Mass (kg)
URANUS (continued)	Bianca	1986	59,200	0.435	0.0009	64 × 46	9.2×10^{16}
	Cressida	1986	61,800	0.464	0.0004	92 × 74	3.4×10^{17}
	Desdemona	1986	62,700	0.474	0.0001	90 × 54	1.8×10^{17}
	Juliet	1986	64,400	0.493	0.0007	150 × 74	5.6×10^{17}
	Portia	1986	66,100	0.513	0.0001	156 × 126	1.7×10^{18}
	Rosalind	1986	69,900	0.558	0.0001	72	2.5×10^{17}
	Cupid	2003	74,800	0.618	0.0013	18	3.8×10^{15}
	Belinda	1986	75,300	0.624	0.0001	128 × 64	3.6×10^{17}
	Perdita	1986	76,420	0.638	0.003	26	1.3×10^{16}
	Puck	1985	86,000	0.762	0.0001	162	2.9×10^{18}
	Mab	2003	97,734	0.923	0.0025	24	1.0×10^{16}
	Miranda	1948	129,900	1.413	0.0013	471	6.59×10^{19}
	Ariel	1851	190,900	2.52	0.0012	1158	1.35×10^{21}
	Umbriel	1851	266,000	4.144	0.0039	1169	1.2×10^{21}
	Titania	1787	436,300	8.706	0.0011	1578	3.53×10^{21}
	Oberon	1787	583,500	13.46	0.0014	1522	3.01×10^{21}
	Francisco	2001	4,276,000	266.56	0.1459	22	1.4×10^{15}
	Caliban	1997	7,231,000	579.73^R	0.1587	72	7.4×10^{17}
	Stephano	1999	8,004,000	677.36^R	0.2292	32	6.0×10^{15}
	Trinculo	2001	8,504,000	749.24^R	0.22	18	7.5×10^{14}
	Sycorax	1997	12,179,000	1288.3^R	0.5224	150	5.4×10^{18}
	Margaret	2003	14,345,000	1687.01	0.6608	20	1.0×10^{15}
	Prospero	1999	16,256,000	1978.29^R	0.4448	25	2.1×10^{16}
	Setebos	1999	17,418,000	2225.21^R	0.5914	24	2.1×10^{16}
	Ferdinand	2003	20,901,000	2887.21^R	0.3682	21	4.4×10^{16}
NEPTUNE	Naiad	1989	48,227	0.294	0.0004	96 × 60 × 52	1.9×10^{17}
	Thalassa	1989	50,075	0.311	0.0002	108 × 100 × 52	3.5×10^{17}
	Despina	1989	52,526	0.335	0.0002	180 × 150 × 130	2.1×10^{18}
	Galatea	1989	61,953	0.429	0	204 × 184 × 144	2.1×10^{18}
	Larissa	1989	73,548	0.555	0.0014	216 × 204 × 164	4.2×10^{18}
	Proteus	1989	117,647	1.122	0.0005	440 × 416 × 404	4.4×10^{19}
	Triton	1846	354,800	5.877^R	0	2706	2.15×10^{22}
	Nereid	1949	5,513,400	360.14	0.7512	340	3.1×10^{19}
	S/2002 N1	2002	15,728,000	1879.71^R	0.5711	62	1.8×10^{17}
	S/2002 N2	2002	22,422,000	2914.07	0.2931	44	6.3×10^{16}
	S/2002 N3	2002	23,571,000	3167.85	0.4237	42	5.5×10^{16}
	Psamathe	2003	46,695,000	9115.91^R	0.4499	24	1.0×10^{16}
	S/2002 N4	2002	48,387,000	9373.99^R	0.4945	60	1.6×10^{17}

This table was compiled from data provided by the Jet Propulsion Laboratory.

*A superscript R means that the satellite orbits in a retrograde direction (opposite to the planet's rotation).

**The size of a spherical satellite is equal to its diameter.

Appendix 4 The Nearest Stars

Name	Parallax (arcsec)	Distance (parsecs)	Distance (light-years)	Spectral Type	Proper motion (arcsec/yr)	Apparent visual magnitude	Absolute visual magnitude	Mass (Sun = 1)
Proxima Centauri	0.772	1.30	4.22	M5.5 V	3.853	+11.09	+15.53	0.107
Alpha Centauri A	0.747	1.34	4.36	G2 V	3.710	−0.01	+4.36	1.144
Alpha Centauri B	0.747	1.34	4.36	K0 V	3.724	+1.34	+5.71	0.916
Barnard's Star	0.547	1.83	5.96	M4.0 V	10.358	+9.53	+13.22	0.166
Wolf 359	0.419	2.39	7.78	M6.0 V	4.696	+13.44	+16.55	0.092
Lalande 21185	0.393	2.54	8.29	M2.0 V	4.802	+7.47	+10.44	0.464
Sirius A	0.380	2.63	8.58	A1 V	1.339	−1.43	+1.47	1.991
Sirius B	0.380	2.63	8.58	white dwarf	1.339	+8.44	+11.34	0.500
UV Ceti	0.374	2.68	8.73	M5.5 V	3.368	+12.54	+15.40	0.109
BL Ceti	0.374	2.68	8.73	M6.0 V	3.368	+12.99	+15.85	0.102
Ross 154	0.337	2.97	9.68	M3.5 V	0.666	+10.43	+13.07	0.171
Ross 248	0.316	3.16	10.32	M5.5 V	1.617	+12.29	+14.79	0.121
Epsilon Eridani	0.310	3.23	10.52	K2 V	0.977	+3.73	+6.19	0.850
Lacaille 9352	0.304	3.29	10.74	M1.5 V	6.896	+7.34	+9.75	0.529
Ross 128	0.299	3.35	10.92	M4.0 V	1.361	+11.13	+13.51	0.156
EZ Aquarii A	0.290	3.45	11.27	M5.0 V	3.254	+13.33	+15.64	0.105
EZ Aquarii B	0.290	3.45	11.27	—	3.254	+13.27	+15.58	0.106
EZ Aquarii C	0.290	3.45	11.27	—	3.254	+14.03	+16.34	0.095
Procyon A	0.286	3.50	11.40	F5 IV-V	1.259	+0.38	+2.66	1.569
Procyon B	0.286	3.50	11.40	white dwarf	1.259	+10.70	+12.98	0.500
61 Cygni A	0.286	3.50	11.40	K5.0 V	5.281	+5.21	+7.49	0.703
61 Cygni B	0.286	3.50	11.40	K7.0 V	5.172	+6.03	+8.31	0.630
GJ725 A	0.283	3.53	11.53	M3.0 V	2.238	+8.90	+11.16	0.351
GJ725 B	0.283	3.53	11.53	M3.5 V	2.313	+9.69	+11.95	0.259
GX Andromedae	0.281	3.56	11.62	M1.5 V	2.918	+8.08	+10.32	0.486
GQ Andromedae	0.281	3.56	11.62	M3.5 V	2.918	+11.06	+13.30	0.163
Epsilon Indi A	0.276	3.63	11.82	K5 V	4.704	+4.69	+6.89	0.766
Epsilon Indi B	0.276	3.63	11.82	T1.0	4.823			0.044
Epsilon Indi C	0.276	3.63	11.82	T6.0	4.823			0.028
DX Cancri	0.276	3.63	11.83	M6.5 V	1.290	+14.78	+16.98	0.087
Tau Ceti	0.274	3.64	11.89	G8 V	1.922	+3.49	+5.68	0.921
RECONS 1	0.272	3.68	11.99	M5.5 V	0.814	+13.03	+15.21	0.113
YZ Ceti	0.269	3.72	12.13	M4.5 V	1.372	+12.02	+14.17	0.136
Luyten's Star	0.264	3.79	12.37	M3.5 V	3.738	+9.86	+11.97	0.257
Kapteyn's Star	0.255	3.92	12.78	M1.5 V	8.670	+8.84	+10.87	0.393
AX Microscopium	0.253	3.95	12.87	M0.0 V	3.455	+6.67	+8.69	0.600

This table, compiled from data reported by the Research Consortium on Nearby Stars, lists all known stars within 4.00 parsecs (13.05 light-years).

*Stars that are components of multiple star systems are labeled A, B, and C.

Appendix 5   The Visually Brightest Stars

Name	Designation	Distance (parsecs)	Distance (light-years)	Spectral type	Radial velocity (km/s)*	Proper motion (arcsec/year)	Apparent visual magnitude	Apparent visual brightness (Sirius = 1)**	Absolute visual magnitude
Sirius A	α CMa A	2.63	8.58	A1 V	−7.6	1.34	−1.43	1.000	+1.46
Canopus	α Car	95.9	313	F0 II	+20.5	0.03	−0.72	0.520	−5.63
Arcturus	α Boo	11.3	36.7	K1.5 III	−5.2	2.28	−0.04	0.278	−0.30
Alpha Centauri A	α Cen A	1.34	4.36	G2 V	−25	3.71	−0.01	0.270	+4.36
Vega	α Lyr	7.76	25.3	A0 V	−13.9	0.35	+0.03	0.261	+0.58
Capella	α Aur	12.9	42.2	G5 III	+30.2	0.43	+0.08	0.249	−0.48
Rigel	α Ori A	237	773	B8 Ia	+20.7	0.002	+0.12	0.240	−6.75
Procyon	α CMi A	3.50	11.4	F5 IV-V	−3.2	1.26	+0.34	0.196	+2.62
Achernar	α Eri	44.1	144	B3 V	+16	0.10	+0.50	0.169	−2.72
Betelgeuse	α Ori	131	427	M1 Iab	+21	0.03	+0.58	0.157	−5.01
Hadar	β Cen	161	525	B1 III	+5.9	0.04	+0.60	0.154	−5.43
Altair	α Aql	51.4	168	A7 V	−26.1	0.66	+0.77	0.132	−2.79
Aldebaran	α Tau A	20.0	65.1	K5 III	+54.3	0.20	+0.85	0.122	−0.65
Spica	α Vir	80.4	262	B1 III-IV	+1	0.05	+1.04	0.103	−3.49
Antares	α Sco A	185	604	M1.5 Iab	−3.4	0.03	+1.09	0.098	−5.25
Pollux	β Gem	10.3	33.7	K0 IIIb	+3.3	0.63	+1.15	0.093	+1.08
Fomalhaut	α PsA	7.69	25.1	A3 V	+6.5	0.37	+1.16	0.092	+1.73
Deneb	α Cyg	990	3230	A2 Ia	−4.5	0.002	+1.25	0.085	−8.73
Mimosa	β Cru	108	353	B0.5 IV	+15.6	0.05	+1.297	0.081	−3.87
Regulus	α Leo A	23.8	77.5	B7 V	+5.9	0.25	+1.35	0.077	−0.53

Data in this table were compiled from SIMBAD database operated at the Centre de Données Astronomiques de Strasbourg, France.

*A positive radial velocity means that the star is receding; a negative radial velocity means that the star is approaching.

**This is a ratio of the star's apparent brightness to that of Sirius, the brightest star in the night sky.

Note: Acrux, or α Cru (the brightest star in Crux, the Southern Cross) appears to the naked eye as a star of apparent magnitude +0.87, the same as Aldebaran. However, it does not appear in this table because Acrux is actually a binary star system. The blue-white component stars of this binary system have apparent magnitudes of +1.4 and +1.9, and so they are dimmer than any of the stars listed here.

Appendix 6   Some Important Astronomical Quantities

Astronomical Unit	$1 \text{ AU} = 1.4960 \times 10^{11} \text{ m}$
	$= 1.4960 \times 10^8 \text{ km}$
Light-year	$1 \text{ ly} = 9.4605 \times 10^{15} \text{ m}$
	$= 63,240 \text{ AU}$
Parsec	$1 \text{ pc} = 3.2616 \text{ ly}$
	$= 3.0857 \times 10^{16} \text{ m}$
	$= 206,265 \text{ AU}$
Year	$1 \text{ y} = 365.2564 \text{ days}$
	$= 3.156 \times 10^7 \text{ s}$
Solar mass	$1 \text{ M}_\odot = 1.989 \times 10^{30} \text{ kg}$
Solar Radius	$1 \text{ R}_\odot = 6.9599 \times 10^8 \text{ m}$
Solar Luminosity	$1 \text{ L}_\odot = 3.90 \times 10^{26} \text{ W}$

Appendix 7   Some Important Physical Constants

Speed of light	$c = 2.9979 \times 10^8 \text{ m/s}$
Gravitational constant	$G = 6.673 \times 10^{-11} \text{ N m}^2/\text{kg}^2$
Planck's constant	$h = 6.6261 \times 10^{-34} \text{ J s}$
	$= 4.1357 \times 10^{-15} \text{ eV s}$
Boltzmann constant	$k = 1.3807 \times 10^{-23} \text{ J/K}$
	$= 8.6173 \times 10^{-5} \text{ eV/K}$
Stefan-Boltzmann constant	$\sigma = 5.6704 \times 10^{-8} \text{ W m}^{-2} \text{ K}^{-4}$
Mass of electron	$m_e = 9.1094 \times 10^{-31} \text{ kg}$
Mass of proton	$m_p = 1.6726 \times 10^{-27} \text{ kg}$
Mass of neutron	$m_n = 1.6749 \times 10^{-27} \text{ kg}$
Mass of hydrogen atom	$m_H = 1.6735 \times 10^{-27} \text{ kg}$
Rydberg constant	$R = 1.0973 \times 10^7 \text{ m}^{-1}$
Electron volt	$1 \text{ eV} = 1.6022 \times 10^{-19} \text{ J}$

Appendix 8   Some Useful Mathematics

Area of a rectangle of sides a and b	$A = ab$
Volume of a rectangular solid of sides a, b, and c	$V = abc$
Hypotenuse of a right triangle whose other sides are a and b	$c = \sqrt{a^2 + b^2}$
Circumference of a circle of radius r	$C = 2\pi r$
Area of a circle of radius r	$A = \pi r^2$
Surface ares of a sphere of radius r	$A = 4\pi r^2$
Volume of a sphere of radius r	$V = 4\pi r^3/3$
Value of π	$\pi = 3.1415926536$

Glossary

You can find more information about each term in the indicated chapter or chapters. See the Key Words section in each chapter for the specific page on which the meaning of a given term is described.

A ring One of three prominent rings encircling Saturn. (Chapter 12)

absolute zero A temperature of $-273°C$ (or 0 K), at which all molecular motion stops; the lowest possible temperature. (Chapter 5)

absorption line spectrum Dark lines superimposed on a continuous spectrum. (Chapter 5)

acceleration The rate at which an object's velocity changes due to a change in speed, a change in direction, or both. (Chapter 4)

accretion The gradual accumulation of matter in one location, typically due to the action of gravity. (Chapter 8)

active optics A technique for improving a telescopic image by altering the telescope's optics to compensate for variations in air temperature or flexing of the telescope mount. (Chapter 6)

adaptive optics A technique for improving a telescopic image by altering the telescope's optics in a way that compensates for distortion caused by the Earth's atmosphere. (Chapter 6)

aerosol Tiny droplets of liquid dispersed in a gas. (Chapter 13)

albedo The fraction of sunlight that a planet, asteroid, or satellite reflects. (Chapter 9)

amino acids The chemical building blocks of proteins. (Chapter 15)

angle The opening between two lines that meet at a point. (Chapter 1)

angular diameter The angle subtended by the diameter of an object. (Chapter 1)

angular distance The angle between two points in the sky. (Chapter 1)

angular measure The size of an angle, usually expressed in degrees, arcminutes, and arcseconds. (Chapter 1)

angular momentum See *conservation of angular momentum*.

angular resolution The angular size of the smallest feature that can be distinguished with a telescope. (Chapter 6)

angular size See *angular diameter*.

annular eclipse An eclipse of the Sun in which the Moon is too distant to cover the Sun completely, so that a ring of sunlight is seen around the Moon at mid-eclipse. (Chapter 3)

anorthosite Rock commonly found in ancient, cratered highlands on the Moon. (Chapter 10)

Antarctic Circle A circle of latitude $23\frac{1}{2}°$ north of the Earth's south pole. (Chapter 2)

aphelion The point in its orbit where a planet is farthest from the Sun. (Chapter 4)

apogee The point in its orbit where a satellite or the Moon is farthest from the Earth. (Chapter 3)

apparent solar day The interval between two successive transits of the Sun's center across the local meridian. (Chapter 2)

apparent solar time Time reckoned by the position of the Sun in the sky. (Chapter 2)

arcminute One-sixtieth (1/60) of a degree, designated by the symbol ′. (Chapter 1)

arcsecond One-sixtieth (1/60) of an arcminute or 1/3600 of a degree, designated by the symbol ″. (Chapter 1)

Arctic Circle A circle of latitude $23\frac{1}{2}°$ south of the Earth's north pole. (Chapter 2)

asteroid One of tens of thousands of small, rocky, planetlike objects in orbit about the Sun. Also called minor planets. (Chapter 7, Chapter 15)

asteroid belt A region between the orbits of Mars and Jupiter that encompasses the orbits of many asteroids. (Chapter 7, Chapter 15)

asthenosphere A warm, plastic layer of the mantle beneath the lithosphere of the Earth. (Chapter 9)

astrobiology The study of life in the universe. (Chapter 28)

astrometric method A technique for detecting extrasolar planets by looking for stars that "wobble" periodically. (Chapter 8)

astronomical unit (AU) The semimajor axis of the Earth's orbit; the average distance between the Earth and the Sun. (Chapter 1)

atmosphere (atm) A unit of atmospheric pressure. (Chapter 9)

atmospheric pressure The force per unit area exerted by a planet's atmosphere. (Chapter 9)

atom The smallest particle of an element that has the properties characterizing that element. (Chapter 5)

atomic number The number of protons in the nucleus of an atom of a particular element. (Chapter 5, Chapter 8)

AU See *astronomical unit*.

aurora (plural aurorae) Light radiated by atoms and ions in the Earth's upper atmosphere, mostly in the polar regions. (Chapter 9)

aurora australis Aurorae seen from southern latitudes; the southern lights. (Chapter 9)

aurora borealis Aurorae seen from northern latitudes; the northern lights. (Chapter 9)

autumnal equinox The intersection of the ecliptic and the celestial equator where the Sun crosses the equator from north to south. Also used to refer to the date on which the Sun passes through this intersection. (Chapter 2)

average density The mass of an object divided by its volume. (Chapter 7)

B ring One of three prominent rings encircling Saturn. (Chapter 12)

Balmer line An emission or absorption line in the spectrum of hydrogen caused by an electron transition between the second and higher energy levels. (Chapter 5)

Balmer series The entire pattern of Balmer lines. (Chapter 5)

baseline In interferometry, the distance between two telescopes whose signals are combined to give a higher-resolution image. (Chapter 6)

belt A dark band in Jupiter's atmosphere. (Chapter 12)

Big Bang An explosion of all space that took place roughly 13.7 billion years ago and that marks the beginning of the universe. (Chapter 1)

biosphere The layer of soil, water, and air surrounding the Earth in which living organisms thrive. (Chapter 9)

black hole An object whose gravity is so strong that the escape speed exceeds the speed of light. (Chapter 1)

blackbody A hypothetical perfect radiator that absorbs and re-emits all radiation falling upon it. (Chapter 5)

blackbody curve The intensity of radiation emitted by a blackbody plotted as a function of wavelength or frequency. (Chapter 5)

blackbody radiation The radiation emitted by a perfect blackbody. (Chapter 5)

blueshift A decrease in the wavelength of photons emitted by an approaching source of light. (Chapter 5)

Bohr orbits In the model of the atom described by Niels Bohr, the only orbits in which electrons are allowed to move about the nucleus. (Chapter 5)

Bok globule A small, roundish, dark nebula. (Chapter 20)

bright terrain (Ganymede) Young, reflective, relatively crater-free terrain on the surface of Ganymede. (Chapter 13)

brown ovals Elongated, brownish features usually seen in Jupiter's northern hemisphere. (Chapter 12)

C ring One of three prominent rings encircling Saturn. (Chapter 12)

capture theory The hypothesis that the Moon was gravitationally captured by the Earth. (Chapter 10)

carbonaceous chondrite A type of meteorite that has a high abundance of carbon and volatile compounds. (Chapter 15, Chapter 28)

Cassegrain focus An optical arrangement in a reflecting telescope in which light rays are reflected by a secondary mirror to a focus behind the primary mirror. (Chapter 6)

Cassini division An apparent gap between Saturn's A and B rings. (Chapter 12)

CCD See *charge-coupled device*.

celestial equator A great circle on the celestial sphere 90° from the celestial poles. (Chapter 2)

celestial poles The points about which the celestial sphere appears to rotate. See also *north celestial pole* and *south celestial pole*. (Chapter 2)

celestial sphere An imaginary sphere of very large radius centered on an observer; the apparent sphere of the sky. (Chapter 2)

center of mass The point between a star and a planet, or between two stars, around which both objects orbit. (Chapter 8)

charge-coupled device (CCD) A type of solid-state device designed to detect photons. (Chapter 6)

chemical composition A description of which chemical substances make up a given object. (Chapter 7)

chemical differentiation The process by which the heavier elements in a planet sink toward its center while lighter elements rise toward its surface. (Chapter 8)

chemical element See *element*.

chondrule A glassy, roughly spherical blob found within meteorites. (Chapter 8)

chromatic aberration An optical defect whereby different colors of light passing through a lens are focused at different locations. (Chapter 6)

chromosphere A layer in the atmosphere of the Sun between the photosphere and the corona. (Chapter 16)

circumpolar A term describing a star that neither rises nor sets but appears to rotate around one of the celestial poles. (Chapter 2)

CNO cycle A series of nuclear reactions in which carbon is used as a catalyst to transform hydrogen into helium. (Chapter 16)

co-creation theory The hypothesis that the Earth and the Moon formed at the same time from the same material. (Chapter 10)

collisional ejection theory The hypothesis that the Moon formed from material ejected from the Earth by the impact of a large asteroid. (Chapter 10)

coma (of a comet) The diffuse gaseous component of the head of a comet. (Chapter 15)

coma (optical) The distortion of off-axis images formed by a parabolic mirror. (Chapter 6)

comet A small body of ice and dust in orbit about the Sun. While passing near the Sun, a comet's vaporized ices give rise to a coma and tail. (Chapter 7, Chapter 15)

compound A substance consisting of two or more chemical elements in a definite proportion. (Chapter 5)

condensation temperature The temperature at which a particular substance in a low-pressure gas condenses into a solid. (Chapter 8)

conduction The transfer of heat by directly passing energy from atom to atom. (Chapter 16)

conic section The curve of intersection between a circular cone and a plane; this curve can be a circle, ellipse, parabola, or hyperbola. (Chapter 4)

conjunction The geometric arrangement of a planet in the same part of the sky as the Sun, so that the planet is at an elongation of 0°. (Chapter 4)

conservation of angular momentum A law of physics stating that in an isolated system, the total amount of angular momentum—a measure of the amount of rotation—remains constant. (Chapter 8)

constellation A configuration of stars in the same region of the sky. (Chapter 2)

continuous spectrum A spectrum of light over a range of wavelengths without any spectral lines. (Chapter 5)

convection The transfer of energy by moving currents of fluid or gas containing that energy. (Chapter 9, Chapter 16)

convection cell A circulating loop of gas or liquid that transports heat from a warm region to a cool region. (Chapter 9)

convection current The pattern of motion in a gas or liquid in which convection is taking place. (Chapter 9)

convective zone The region in a star where convection is the dominant means of energy transport. (Chapter 16)

core (of the Earth) The iron-rich inner region of the Earth's interior. (Chapter 9)

core accretion model The hypothesis that each of the Jovian planets formed by accretion of gas onto a rocky core. (Chapter 8)

corona (of the Sun) The Sun's outer atmosphere, which has a high temperature and a low density. (Chapter 16)

coronal hole A region in the Sun's corona that is deficient in hot gases. (Chapter 16)

coronal mass ejection An event in which billions of tons of gas from the Sun's corona is suddenly blasted into space at high speed. (Chapter 16)

cosmic rays Fast-moving subatomic particles that enter our solar system from interstellar space. (Chapter 11)

coudé focus An optical arrangement with a reflecting telescope. A series of mirrors is used to direct light to a remote focus away from the moving parts of the telescope. (Chapter 6)

crater See *impact crater*.

crust (of a planet) The surface layer of a terrestrial planet. (Chapter 9)

crustal dichotomy (Mars) The contrast between the young northern lowlands and older southern highlands on Mars. (Chapter 11)

crystal A material in which atoms are arranged in orderly rows. (Chapter 9)

current sheet A broad, flat region in Jupiter's magnetosphere that contains an abundance of charged particles. (Chapter 12)

D ring One of several faint rings encircling Saturn. (Chapter 12)

dark terrain (Ganymede) Older, heavily cratered, dark-colored terrain on the surface of Ganymede. (Chapter 13)

decametric radiation Radiation from Jupiter whose wavelength is about 10 meters. (Chapter 12)

decimetric radiation Radiation from Jupiter whose wavelength is about a tenth of a meter. (Chapter 12)

declination Angular distance of a celestial object north or south of the celestial equator. (Chapter 2)

deferent A stationary circle in the Ptolemaic system along which another circle (an epicycle) moves, carrying a planet, the Sun, or the Moon. (Chapter 4)

degree A basic unit of angular measure, designated by the symbol °. (Chapter 1)

degree Celsius A basic unit of temperature, designated by the symbol °C and used on a scale where water freezes at 0° and boils at 100°. (Chapter 5)

degree Fahrenheit A basic unit of temperature, designated by the symbol °F and used on a scale where water freezes at 32° and boils at 212°. (Chapter 5)

density See *average density*.

differential rotation The rotation of a nonrigid object in which parts adjacent to each other at a given time do not always stay close together. (Chapter 12, Chapter 16)

differentiated asteroid An asteroid in which chemical differentiation has taken place, so that denser material is toward the asteroid's center. (Chapter 15)

differentiation See *chemical differentiation*.

diffraction The spreading out of light passing through an aperture or opening in an opaque object. (Chapter 6)

diffraction grating An optical device, consisting of thousands of closely spaced lines etched in glass or metal, that disperses light into a spectrum. (Chapter 6)

direct motion The apparent eastward movement of a planet seen against the background stars. (Chapter 4)

disk instability model The hypothesis that gases in the solar nebula coalesced rapidly to form the Jovian planets. (Chapter 8)

diurnal motion Any apparent motion in the sky that repeats on a daily basis, such as the rising and setting of stars. (Chapter 2)

Doppler effect The apparent change in wavelength of radiation due to relative motion between the source and the observer along the line of sight. (Chapter 5)

Drake equation An equation used to estimate the number of intelligent civilizations in the Galaxy with whom we might communicate. (Chapter 28)

dust devil Whirlwinds found in dry or desert areas on both Earth and Mars. (Chapter 11)

dust tail The tail of a comet that is composed primarily of dust grains. (Chapter 15)

dwarf planet A solar system body that is large enough to be spherical in shape and have a circular orbit around the Sun, but not large enough to clear its own path of other bodies. The term is used for Ceres, Pluto, and Eris. (Chapter 15)

dynamo The mechanism whereby electric currents within an astronomical body generate a magnetic field. (Chapter 7)

E ring A very broad, faint ring encircling Saturn. (Chapter 12)

earthquake A sudden vibratory motion of the Earth's surface. (Chapter 9)

eccentricity A number between 0 and 1 that describes the shape of an ellipse. (Chapter 4)

eclipse The cutting off of part or all the light from one celestial object by another. (Chapter 3)

eclipse path The track of the tip of the Moon's shadow along the Earth's surface during a total or annular solar eclipse. (Chapter 3)

eclipse year The interval between successive passages of the Sun through the same node of the Moon's orbit. (Chapter 3)

ecliptic The apparent annual path of the Sun on the celestial sphere. (Chapter 2)

ecliptic plane The plane of Earth's orbit around the Sun. (Chapter 2)

electromagnetic radiation Radiation consisting of oscillating electric and magnetic fields. Examples include gamma rays, X rays, visible light, ultraviolet and infrared radiation, radio waves, and microwaves. (Chapter 5)

electromagnetic spectrum The entire array of electromagnetic radiation. (Chapter 5)

electromagnetism Electric and magnetic phenomena, including electromagnetic radiation. (Chapter 5)

electron A subatomic particle with a negative charge and a small mass, usually found in orbits about the nuclei of atoms. (Chapter 5)

electron volt (eV) The energy acquired by an electron accelerated through an electric potential of one volt. (Chapter 5)

element A chemical that cannot be broken down into more basic chemicals. (Chapter 5)

ellipse A conic section obtained by cutting completely through a circular cone with a plane. (Chapter 4)

elongation The angular distance between a planet and the Sun as viewed from Earth. (Chapter 4)

emission line spectrum A spectrum that contains bright emission lines. (Chapter 5)

Encke gap A narrow gap in Saturn's A ring. (Chapter 12)

energy flux The rate of energy flow, usually measured in joules per square meter per second. (Chapter 5)

energy level In an atom, a particular amount of energy possessed by an atom above the atom's least energetic state. (Chapter 5)

energy-level diagram A diagram showing the arrangement of an atom's energy levels. (Chapter 5)

epicenter The location on the Earth's surface directly over the focus of an earthquake. (Chapter 9)

epicycle A moving circle in the Ptolemaic system about which a planet revolves. (Chapter 4)

epoch The date used to define the coordinate system for objects on the sky. (Chapter 2)

equal areas, law of See *Kepler's second law.*

equinox One of the intersections of the ecliptic and the celestial equator. Also used to refer to the date on which the Sun passes through such an intersection. (Chapter 2) See also *autumnal equinox* and *vernal equinox.*

escape speed The speed needed by an object (such as a spaceship) to leave a second object (such as a planet or star) permanently and to escape into interplanetary space. (Chapter 7)

eV See *electron volt.*

excited state A state of an atom, ion, or molecule with a higher energy than the ground state. (Chapter 5)

exponent A number placed above and after another number to denote the power to which the latter is to be raised, as n in 10^n. (Chapter 1)

extinction (interstellar) See *interstellar extinction.*

extrasolar planet A planet orbiting a star other the Sun. (Chapter 8)

eyepiece lens A magnifying lens used to view the image produced at the focus of a telescope. (Chapter 6)

F ring A thin, faint ring encircling Saturn just beyond the A ring. (Chapter 12)

false color In astronomical images, color used to denote different values of intensity, temperature, or other quantities. (Chapter 6)

far side (of the Moon) The side of the Moon that faces perpetually away from the Earth. (Chapter 10)

favorable opposition An opposition of Mars that affords good Earth-based views of the planet. (Chapter 11)

filament A portion of the Sun's chromosphere that arches to high altitudes. (Chapter 16)

first quarter moon The phase of the Moon that occurs when the Moon is 90° east of the Sun. (Chapter 3)

fission theory The hypothesis that the Moon was pulled out of a rapidly rotating proto-Earth. (Chapter 10)

flake tectonics A model of a planetary interior, particularly Venus, in which a thin crust remains stationary but wrinkles and flakes in response to interior convection currents. (Chapter 11)

flare See *solar flare.*

focal length The distance from a lens or mirror to the point where converging light rays meet. (Chapter 6)

focal plane The plane in which a lens or mirror forms an image of a distant object. (Chapter 6)

focal point The point at which a lens or mirror forms an image of a distant point of light. (Chapter 6)

focus (of an ellipse) (plural **foci**) One of two points inside an ellipse such that the combined distance from the two foci to any point on the ellipse is a constant. (Chapter 4)

focus (of a lens or mirror) The point to which light rays converge after passing through a lens or being reflected from a mirror. (Chapter 6)

force A push or pull that acts on an object. (Chapter 4)

frequency The number of crests or troughs of a wave that cross a given point per unit time. Also, the number of vibrations per unit time. (Chapter 5)

full moon A phase of the Moon during which its full daylight hemisphere can be seen from Earth. (Chapter 3)

fusion crust The coating on a stony meteorite caused by the heating of the meteorite as it descended through the Earth's atmosphere. (Chapter 15)

G ring A thin, faint ring encircling Saturn. (Chapter 12)

galaxy A large assemblage of stars, nebulae, and interstellar gas and dust. (Chapter 1)

Galilean satellites The four large moons of Jupiter. (Chapter 13)

gamma rays The most energetic form of electromagnetic radiation. (Chapter 5)

geocentric model An Earth-centered theory of the universe. (Chapter 4)

global warming The upward trend of the Earth's average temperature caused by increased amounts of greenhouse gases in the atmosphere. (Chapter 9)

granulation The rice grain-like structure found in the solar photosphere. (Chapter 16)

granule A convective cell in the solar photosphere. (Chapter 16)

grating See *diffraction grating.*

gravitational force See *gravity.*

gravity The force with which all matter attracts all other matter. (Chapter 4)

Great Red Spot A prominent high-pressure system in Jupiter's southern hemisphere. (Chapter 12)

greatest eastern elongation The configuration of an inferior planet at its greatest angular distance east of the Sun. (Chapter 4, Chapter 11)

greatest western elongation The configuration of an inferior planet at its greatest angular distance west of the Sun. (Chapter 4, Chapter 11)

greenhouse effect The trapping of infrared radiation near a planet's surface by the planet's atmosphere. (Chapter 9)

greenhouse gas A substance whose presence in a planet's atmosphere enhances the greenhouse effect. (Chapter 9)

ground state The state of an atom, ion, or molecule with the least possible energy. (Chapter 5)

habitable zone Regions of a galaxy or around a star in which conditions may be suitable for life to have developed. (Chapter 28)

half-life The time required for one-half of a quantity of a radioactive substance to decay. (Chapter 8)

heliocentric model A Sun-centered theory of the universe. (Chapter 4)

helioseismology The study of the vibrations of the Sun as a whole. (Chapter 16)

highlands (on Mars) See *southern highlands.*

highlands (on the Moon) See *lunar highlands.*

Hirayama family A group of asteroids that have nearly identical orbits about the Sun. (Chapter 15)

hot spot (on Jupiter) An unusually warm and cloud-free part of Jupiter's atmosphere. (Chapter 12)

hot-spot volcanism Volcanic activity that occurs over a hot region buried deep within a planet. (Chapter 11)

hydrocarbon Any one of a variety of chemical compounds composed of hydrogen and carbon. (Chapter 13)

hydrogen envelope A huge, tenuous sphere of gas surrounding the head of a comet. (Chapter 15)

hydrogen fusion The thermonuclear conversion of hydrogen into helium. (Chapter 16)

hydrostatic equilibrium A balance between the weight of a layer in a star and the pressure that supports it. (Chapter 16)

hyperbola A conic section formed by cutting a circular cone with a plane at an angle steeper than the sides of the cone. (Chapter 4)

hypothesis An idea or collection of ideas that seems to explain a specified phenomenon; a conjecture. (Chapter 1)

ice rafts (Europa) Segments of Europa's icy crust that have been moved by tectonic disturbances. (Chapter 13)

ices Solid materials with low condensation temperatures, including ices of water, methane, and ammonia. (Chapter 7)

igneous rock A rock that formed from the solidification of molten lava or magma. (Chapter 9)

imaging The process of recording the image made by a telescope of a distant object. (Chapter 6)

impact breccia A type of rock formed from other rocks that were broken apart, mixed, and fused together by a series of meteoritic impacts. (Chapter 10)

impact crater A circular depression on a planet or satellite caused by the impact of a meteoroid. (Chapter 7, Chapter 10)

inertia, law of See *Newton's first law.*

inferior conjunction The configuration when an inferior planet is between the Sun and Earth. (Chapter 4)

inferior planet A planet that is closer to the Sun than the Earth is. (Chapter 4)

infrared radiation Electromagnetic radiation of wavelength longer than visible light but shorter than radio waves. (Chapter 5)

inner core (of the Earth) The solid innermost portion of the Earth's iron-rich core. (Chapter 9)

interferometry A technique of combining the observations of two or more telescopes to produce images better than one telescope alone could make. (Chapter 6)

intermediate-period comet A comet with an orbital period between 20 and 200 years. (Chapter 15)

internal rotation period The period with which the core of a Jovian planet rotates. (Chapter 12)

interstellar medium Gas and dust in interstellar space. (Chapter 8)

Io torus A doughnut-shaped ring of gas circling Jupiter at the distance of Io's orbit. (Chapter 13)

ion tail The relatively straight tail of a comet produced by the solar wind acting on ions. (Chapter 15)

ionization The process by which a neutral atom becomes an electrically charged ion through the loss or gain of electrons. (Chapter 5)

iron meteorite A meteorite composed primarily of iron. (Chapter 15)

isotope Any of several forms for the same chemical element whose nuclei all have the same number of protons but different numbers of neutrons. (Chapter 5)

jet An extended line of fast-moving gas ejected from the vicinity of a star or a black hole. (Chapter 8)

joule (J) A unit of energy. (Chapter 5)

Jovian planet Low-density planets composed primarily of hydrogen and helium, including Jupiter, Saturn, Uranus, and Neptune. (Chapter 7)

Jupiter-family comet A comet with an orbital period of less than 20 years. (Chapter 15)

kelvin (K) A unit of temperature on the Kelvin temperature scale, equivalent to a degree Celsius. (Chapter 5)

Kelvin-Helmholtz contraction The contraction of a gaseous body, such as a star or nebula, during which gravitational energy is transformed into thermal energy. (Chapter 8)

Kepler's first law The statement that each planet moves around the Sun in an elliptical orbit with the Sun at one focus of the ellipse. (Chapter 4)

Kepler's second law The statement that a planet sweeps out equal areas in equal times as it orbits the Sun; also called the law of equal areas. (Chapter 4)

Kepler's third law A relationship between the period of an orbiting object and the semimajor axis of its elliptical orbit. (Chapter 4)

kiloparsec (kpc) One thousand parsecs; about 3260 light-years. (Chapter 1)

kinetic energy The energy possessed by an object because of its motion. (Chapter 7)

Kirchhoff's laws Three statements about circumstances that produce absorption lines, emission lines, and continuous spectra. (Chapter 5)

Kirkwood gaps Gaps in the spacing of asteroid orbits, discovered by Daniel Kirkwood. (Chapter 15)

Kuiper belt A region that extends from around the orbit of Pluto to about 500 AU from the Sun where many icy objects orbit the Sun. (Chapter 7, Chapter 14, Chapter 15)

lava Molten rock flowing on the surface of a planet. (Chapter 9)

law of equal areas See *Kepler's second law.*

law of inertia See *Newton's first law.*

law of universal gravitation A formula deduced by Isaac Newton that expresses the strength of the force of gravity that two masses exert on each other. (Chapter 4)

laws of physics A set of physical principles with which we can understand natural phenomena and the nature of the universe. (Chapter 1)

lens A piece of transparent material (usually glass) that can bend light and bring it to a focus. (Chapter 6)

libration An apparent rocking of the Moon whereby an Earth-based observer can, over time, see slightly more than one-half the Moon's surface. (Chapter 10)

light-gathering power A measure of the amount of radiation brought to a focus by a telescope. (Chapter 6)

light pollution Light from cities and towns that degrades telescope images. (Chapter 6)

light scattering The process by which light bounces off particles in its path. (Chapter 5, Chapter 12)

light-year (ly) The distance light travels in a vacuum in one year. (Chapter 1)

limb darkening The phenomenon whereby the Sun looks darker near its apparent edge, or limb, than near the center of its disk. (Chapter 16)

line of nodes The line where the plane of the Earth's orbit intersects the plane of the Moon's orbit. (Chapter 3)

liquid metallic hydrogen Hydrogen compressed to such a density that it behaves like a liquid metal. (Chapter 7, Chapter 12)

lithosphere The solid, upper layer of the Earth; essentially the Earth's crust. (Chapter 9)

local meridian See *meridian*.

long-period comet A comet that takes hundreds of thousands of years or more to complete one orbit of the Sun. (Chapter 15)

lower meridian The half of the meridian that lies below the horizon. (Chapter 2)

lowlands (on Mars) See *northern lowlands*.

luminosity The rate at which electromagnetic radiation is emitted from a star or other object. (Chapter 5, Chapter 16)

lunar eclipse An eclipse of the Moon by the Earth; a passage of the Moon through the Earth's shadow. (Chapter 3)

lunar highlands Ancient, high-elevation, heavily cratered terrain on the Moon. (Chapter 10)

lunar month See *synodic month*.

lunar phase The appearance of the illuminated area of the Moon as seen from Earth. (Chapter 3)

Lyman series A series of spectral lines of hydrogen produced by electron transitions to and from the lowest energy state of the hydrogen atom. (Chapter 5)

magma Molten rock beneath a planet's surface. (Chapter 9)

magnetic axis A line connecting the north and south magnetic poles of a planet or star possessing a magnetic field. (Chapter 14)

magnetic-dynamo model A theory that explains the solar cycle as a result of the Sun's differential rotation acting on the Sun's magnetic field. (Chapter 16)

magnetic reconnection An event where two oppositely directed magnetic fields approach and cancel, thus releasing energy. (Chapter 16)

magnetogram An image of the Sun that shows regions of different magnetic polarity. (Chapter 16)

magnetometer A device for measuring magnetic fields. (Chapter 7)

magnetopause That region of a planet's magnetosphere where the magnetic field counterbalances the pressure from the solar wind. (Chapter 9)

magnetosphere The region around a planet occupied by its magnetic field. (Chapter 9)

magnification The factor by which the apparent angular size of an object is increased when viewed through a telescope. (Chapter 6)

magnifying power See *magnification*.

major axis (of an ellipse) The longest diameter of an ellipse. (Chapter 4)

mantle (of a planet) That portion of a terrestrial planet located between its crust and core. (Chapter 9)

mare (plural maria) Latin for "sea"; a large, relatively crater-free plain on the Moon. (Chapter 10)

mare basalt A type of lunar rock commonly found in the mare basins. (Chapter 10)

mass A measure of the total amount of material in an object. (Chapter 4)

mean solar day The interval between successive meridian passages of the mean Sun; the average length of a solar day. (Chapter 2)

mean sun A fictitious object that moves eastward at a constant speed along the celestial equator, completing one circuit of the sky with respect to the vernal equinox in one tropical year. (Chapter 2)

medium (plural media) A material through which light travels. (Chapter 6)

medium, interstellar See *interstellar medium*.

megaparsec (Mpc) One million parsecs. (Chapter 1)

melting point The temperature at which a substance changes from solid to liquid. (Chapter 9)

meridian (or local meridian) The great circle on the celestial sphere that passes through an observer's zenith and the north and south celestial poles. (Chapter 2)

meridian transit The crossing of the meridian by any astronomical object. (Chapter 2)

mesosphere A layer in the Earth's atmosphere above the stratosphere. (Chapter 9)

metamorphic rock A rock whose properties and appearance have been transformed by the action of pressure and heat beneath the Earth's surface. (Chapter 9)

meteor The luminous phenomenon seen when a meteoroid enters the Earth's atmosphere; a "shooting star." (Chapter 15)

meteor shower Many meteors that seem to radiate from a common point in the sky. (Chapter 15)

meteorite A fragment of a meteoroid that has survived passage through the Earth's atmosphere. (Chapter 1, Chapter 8, Chapter 15)

meteoritic swarm A collection of meteoroids moving together along an orbit about the Sun. (Chapter 15)

meteoroid A small rock in interplanetary space. (Chapter 7, Chapter 15)

microwaves Short-wavelength radio waves. (Chapter 5)

mineral A naturally occurring solid composed of a single element or chemical combination of elements, often in the form of crystals. (Chapter 9)

minor planet See *asteroid*.

minute of arc See *arcminute*.

model A hypothesis that has withstood experimental or observational tests; or, the results of a theoretical calculation that gives the values of temperature, pressure, density, and so forth throughout the interior of an object such as a planet or star. (Chapter 1)

molecule A combination of two or more atoms. (Chapter 5)

moonquake Sudden, vibratory motion of the Moon's surface. (Chapter 10)

nanometer (nm) One billionth of a meter: 1 nm = 10^{-9} meter = 10^{-6} millimeter = 10^{-3} μm. (Chapter 5)

neap tide An ocean tide that occurs when the Moon is near first-quarter or third-quarter phase. (Chapter 4)

near-Earth object (NEO) An asteroid whose orbit lies wholly or partly within the orbit of Mars. (Chapter 15)

nebula A cloud of interstellar gas and dust. (Chapter 1)

nebular hypothesis The idea that the Sun and the rest of the solar system formed from a cloud of interstellar material. (Chapter 8)

nebulosity See *nebula*.

negative hydrogen ion A hydrogen atom that has acquired a second electron. (Chapter 16)

NEO See *near-Earth object*.

neutrino A subatomic particle with no electric charge and very little mass, yet one that is important in many nuclear reactions. (Chapter 16)

neutron A subatomic particle with no electric charge and with a mass nearly equal to that of the proton. (Chapter 5)

new moon The phase of the Moon when the dark hemisphere of the Moon faces the Earth. (Chapter 3)

Newtonian mechanics The branch of physics based on Newton's laws of motion. (Chapter 1, Chapter 4)

Newtonian reflector A reflecting telescope that uses a small mirror to deflect the image to one side of the telescope tube. (Chapter 6)

Newton's first law of motion The statement that a body remains at rest, or moves in a straight line at a constant speed, unless acted upon by a net outside force; the law of inertia. (Chapter 4)

Newton's form of Kepler's third law A relationship between the period of two objects orbiting each other, the semimajor axis of their orbit, and the masses of the objects. (Chapter 4)

Newton's second law of motion A relationship between the acceleration of an object, the object's mass, and the net outside force acting on the mass. (Chapter 4)

Newton's third law of motion The statement that whenever one body exerts a force on a second body, the second body exerts an equal and opposite force on the first body. (Chapter 4)

noble gas An element whose atoms do not combine into molecules. (Chapter 12)

node See *line of nodes*.

nonthermal radiation Radiation other than that emitted by a heated body. (Chapter 12)

north celestial pole The point directly above the Earth's north pole where the Earth's axis of rotation, if extended, would intersect the celestial sphere. (Chapter 2)

northern lights See *aurora borealis*.

northern lowlands (on Mars) Relatively young and crater-free terrain in the Martian northern hemisphere. (Chapter 11)

nucleus (of an atom) The massive part of an atom, composed of protons and neutrons, about which electrons revolve. (Chapter 5)

nucleus (of a comet) A collection of ices and dust that constitute the solid part of a comet. (Chapter 15)

objective lens The principal lens of a refracting telescope. (Chapter 6)

objective mirror The principal mirror of a reflecting telescope. (Chapter 6)

oblate Flattened at the poles. (Chapter 12)

oblateness A measure of how much a flattened sphere (or spheroid) differs from a perfect sphere. (Chapter 12)

Occam's razor The notion that a straightforward explanation of a phenomenon is more likely to be correct than a convoluted one. (Chapter 4)

occultation The eclipsing of an astronomical object by the Moon or a planet. (Chapter 13, Chapter 14)

oceanic rift A crack in the ocean floor that exudes lava. (Chapter 9)

1-to-1 spin-orbit coupling See *synchronous rotation*.

Oort cloud A presumed accumulation of comets and cometary material surrounding the Sun at distances of roughly 50,000 AU. (Chapter 7, Chapter 15)

opposition The configuration of a planet when it is at an elongation of 180° and thus appears opposite the Sun in the sky. (Chapter 4)

optical telescope A telescope designed to detect visible light. (Chapter 6)

optical window The range of visible wavelengths to which the Earth's atmosphere is transparent. (Chapter 6)

organic molecules Molecules containing carbon, some of which are the molecules of which living organisms are made. (Chapter 28)

outer core (of the Earth) The outer, molten portion of the Earth's iron-rich core. (Chapter 9)

outgassing The release of gases into a planet's atmosphere by volcanic activity. (Chapter 9)

ozone A type of oxygen whose molecules contain three oxygen atoms. (Chapter 9)

ozone hole A region of the Earth's atmosphere over Antarctica where the concentration of ozone is abnormally low. (Chapter 9)

ozone layer A layer in the Earth's upper atmosphere where the concentration of ozone is high enough to prevent much ultraviolet light from reaching the surface. (Chapter 9)

P wave One of three kinds of seismic waves produced by an earthquake; a primary wave. (Chapter 9)

parabola A conic section formed by cutting a circular cone at an angle parallel to one of the sides of the cone. (Chapter 4)

parallax The apparent displacement of an object due to the motion of the observer. (Chapter 4)

parsec (pc) A unit of distance; 3.26 light-years. (Chapter 1)

partial lunar eclipse A lunar eclipse in which the Moon does not appear completely covered. (Chapter 3)

partial solar eclipse A solar eclipse in which the Sun does not appear completely covered. (Chapter 3)

Paschen series A series of spectral lines of hydrogen produced by electron transitions between the third and higher energy levels. (Chapter 5)

penumbra (of a shadow) (plural **penumbrae**) The portion of a shadow in which only part of the light source is covered by an opaque body. (Chapter 3)

penumbral eclipse A lunar eclipse in which the Moon passes only through the Earth's penumbra. (Chapter 3)

perigee The point in its orbit where a satellite or the Moon is nearest the Earth. (Chapter 3)

perihelion The point in its orbit where a planet or comet is nearest the Sun. (Chapter 4)

period (of a planet) The interval of time between successive geometric arrangements of a planet and an astronomical object, such as the Sun. (Chapter 4)

periodic table A listing of the chemical elements according to their properties, invented by Dmitri Mendeleev. (Chapter 5)

photoelectric effect The phenomenon whereby certain metals emit electrons when exposed to short-wavelength light. (Chapter 5)

photometry The measurement of light intensities. (Chapter 6)

photon A discrete unit of electromagnetic energy. (Chapter 5)

photosphere The region in the solar atmosphere from which most of the visible light escapes into space. (Chapter 16)

photosynthesis A biochemical process in which solar energy is converted into chemical energy, carbon dioxide and water are absorbed, and oxygen is released. (Chapter 9)

physics, laws of See *laws of physics*.

pixel A picture element. (Chapter 6)

plage A bright region in the solar atmosphere as observed in the monochromatic light of a spectral line. (Chapter 16)

Planck's law The relationship between the energy of a photon and its wavelength or frequency; $E = hc/\lambda = h\nu$. (Chapter 5)

plane of the ecliptic The plane in which the Earth orbits the Sun. (Chapter 3)

planetesimal One of many small bodies of primordial dust and ice that combined to form the planets. (Chapter 8)

plasma A hot ionized gas. (Chapter 12, Chapter 16)

plastic The attribute of being nearly solid yet able to flow. (Chapter 9)

plate A large section of the Earth's lithosphere that moves as a single unit. (Chapter 9)

plate tectonics The motions of large segments (plates) of the Earth's surface over the underlying mantle. (Chapter 9)

plutino One of about 100 objects in the Kuiper belt that orbit the Sun with nearly the same semimajor axis as Pluto. (Chapter 14)

polymer A long molecule consisting of many smaller molecules joined together. (Chapter 13)

positional astronomy The study of the apparent positions of the planets and stars and how those positions change. (Chapter 2)

positron An electron with a positive rather than negative electric charge; the antiparticle of the electron. (Chapter 16)

power of ten The exponent n in 10^n. (Chapter 1)

powers-of-ten notation A shorthand method of writing numbers, involving 10 followed by an exponent. (Chapter 1)

precession (of the Earth) A slow, conical motion of the Earth's axis of rotation caused by the gravitational pull of the Moon and Sun on the Earth's equatorial bulge. (Chapter 2)

precession of the equinoxes The slow westward motion of the equinoxes along the ecliptic due to precession of the Earth. (Chapter 2)

primary mirror See *objective mirror*.

prime focus The point in a telescope where the objective focuses light. (Chapter 6)

primitive asteroid See *undifferentiated asteroid*.

prograde orbit An orbit of a satellite around a planet that is in the same direction as the rotation of the planet. (Chapter 13)

prograde rotation A situation in which an object (such as a planet) rotates in the same direction that it orbits around another object (such as the Sun). (Chapter 11)

prominence Flamelike protrusions seen near the limb of the Sun and extending into the solar corona. (Chapter 16)

proplyd See *protoplanetary disk*.

proton A heavy, positively charged subatomic particle that is one of two principal constituents of atomic nuclei. (Chapter 5)

proton-proton chain A sequence of thermonuclear reactions by which hydrogen nuclei are built up into helium nuclei. (Chapter 16)

protoplanet A Moon-sized object formed by the coalescence of planetesimals. (Chapter 8)

protoplanetary disk (proplyd) A disk of material encircling a protostar or a newborn star. (Chapter 8)

protosun The part of the solar nebula that eventually developed into the Sun. (Chapter 8)

Ptolemaic system The definitive version of the geocentric cosmogony of ancient Greece. (Chapter 4)

pulsar A pulsating radio source thought to be associated with a rapidly rotating neutron star. (Chapter 1)

quantum mechanics The branch of physics dealing with the structure and behavior of atoms and their constituents as well as their interaction with light. (Chapter 5)

quasar A very luminous object with a very large redshift and a starlike appearance. (Chapter 1)

radial velocity That portion of an object's velocity parallel to the line of sight. (Chapter 5)

radial velocity method A technique used to detect extrasolar planets by observing Doppler shifts in the spectrum of the planet's star. (Chapter 8)

radiant (of a meteor shower) The point in the sky from which meteors of a particular shower seem to originate. (Chapter 15)

radiation darkening The darkening of methane ice by electron impacts. (Chapter 14)

radiation pressure Pressure exerted on an object by radiation falling on the object. (Chapter 15)

radiative diffusion The random migration of photons from a star's center toward its surface. (Chapter 16)

radiative zone A region within a star where radiative diffusion is the dominant mode of energy transport. (Chapter 16)

radio telescope A telescope designed to detect radio waves. (Chapter 6)

radio waves The longest-wavelength electromagnetic radiation. (Chapter 5)

radio window The range of radio wavelengths to which the Earth's atmosphere is transparent. (Chapter 6)

radioactive dating A technique for determining the age of a rock sample by measuring the radioactive elements and their decay products in the sample. (Chapter 8)

radioactive decay The process whereby certain atomic nuclei spontaneously transform into other nuclei. (Chapter 8)

redshift The shifting to longer wavelengths of the light from remote galaxies and quasars; the Doppler shift of light from a receding source. (Chapter 5)

reflecting telescope A telescope in which the principal optical component is a concave mirror. (Chapter 6)

reflection The return of light rays by a surface. (Chapter 6)

reflector A reflecting telescope. (Chapter 6)

refracting telescope A telescope in which the principal optical component is a lens. (Chapter 6)

refraction The bending of light rays when they pass from one transparent medium to another. (Chapter 6)

refractor A refracting telescope. (Chapter 6)

refractory element An element with high melting and boiling points. (Chapter 10)

regolith The layer of rock fragments covering the surface of the Moon. (Chapter 10)

residual polar cap An ice-covered polar region on Mars that does not completely evaporate during the Martian summer. (Chapter 11)

respiration A biological process that produces energy by consuming oxygen and releasing carbon dioxide. (Chapter 9)

retrograde motion The apparent westward motion of a planet with respect to background stars. (Chapter 4)

retrograde orbit An orbit of a satellite around a planet that is in the direction opposite to which the planet rotates. (Chapter 13)

retrograde rotation A situation in which an object (such as a planet) rotates in the direction opposite to which it orbits around another object (such as the Sun). (Chapter 11)

rift valley A feature created when a planet's crust breaks apart along a line. (Chapter 11)

right ascension A coordinate for measuring the east-west positions of objects on the celestial sphere. (Chapter 2)

ring particles Small particles that constitute a planetary ring. (Chapter 12)

ringlet One of many narrow bands of particles of which Saturn's ring system is composed. (Chapter 12)

Roche limit The smallest distance from a planet or other object at which a second object can be held together by purely gravitational forces. (Chapter 12)

rock A mineral or combination of minerals. (Chapter 9)

runaway greenhouse effect A greenhouse effect in which the temperature continues to increase. (Chapter 11)

runaway icehouse effect A situation in which a decrease in atmospheric temperature causes a further decrease in temperature. (Chapter 11)

S wave One of three kinds of seismic waves produced by an earthquake; a secondary wave. (Chapter 9)

saros A particular cycle of similar eclipses that recur about every 18 years. (Chapter 3)

scarp A line of cliffs formed by the faulting or fracturing of a planet's surface. (Chapter 11)

scattering of light See *light scattering*.

scientific method The basic procedure used by scientists to investigate phenomena. (Chapter 1)

seafloor spreading The separation of plates under the ocean due to lava emerging in an oceanic rift. (Chapter 9)

search for extraterrestrial intelligence (SETI) The scientific search for evidence of intelligent life on other planets. (Chapter 28)

second of arc See *arcsecond*.

sedimentary rock A rock that is formed from material deposited on land by rain or winds, or on the ocean floor. (Chapter 9)

seeing disk The angular diameter of a star's image. (Chapter 6)

seismic wave A vibration traveling through a terrestrial planet, usually associated with earthquake-like phenomena. (Chapter 9)

seismograph A device used to record and measure seismic waves, such as those produced by earthquakes. (Chapter 9)

semimajor axis One-half of the major axis of an ellipse. (Chapter 4)

SETI See *search for extraterrestrial intelligence*.

shepherd satellite A satellite whose gravity restricts the motions of particles in a planetary ring, preventing them from dispersing. (Chapter 12)

shield volcano A volcano with long, gently sloping sides. (Chapter 11)

shock wave An abrupt, localized region of compressed gas caused by an object traveling through the gas at a speed greater than the speed of sound. (Chapter 9)

SI units The International System of Units, based on the meter (m), the second (s), and the kilogram (kg). (Chapter 1)

sidereal clock A clock that measures sidereal time. (Chapter 2)

sidereal day The interval between successive meridian passages of the vernal equinox. (Chapter 2)

sidereal month The period of the Moon's revolution about the Earth with respect to the stars. (Chapter 3)

sidereal period The orbital period of one object about another as measured with respect to the stars. (Chapter 4)

sidereal time Time reckoned by the location of the vernal equinox. (Chapter 2)

sidereal year The orbital period of the Earth about the Sun with respect to the stars. (Chapter 2)

small-angle formula A relationship between the angular and linear sizes of a distant object. (Chapter 1)

SNC meteorite A meteorite that came to Earth from Mars. (Chapter 28)

solar constant The average amount of energy received from the Sun per square meter per second, measured just above the Earth's atmosphere. (Chapter 5)

solar corona Hot, faintly glowing gases seen around the Sun during a total solar eclipse; the uppermost regions of the solar atmosphere. (Chapter 3)

solar cycle See *22-year solar cycle*.

solar eclipse An eclipse of the Sun by the Moon; a passage of the Earth through the Moon's shadow. (Chapter 3)

solar flare A sudden, temporary outburst of light from an extended region of the solar surface. (Chapter 16)

solar nebula The cloud of gas and dust from which the Sun and solar system formed. (Chapter 8)

solar neutrino A neutrino emitted from the core of the Sun. (Chapter 16)

solar neutrino problem The discrepancy between the predicted and observed numbers of solar neutrinos. (Chapter 16)

solar system The Sun, planets and their satellites, asteroids, comets, and related objects that orbit the Sun. (Chapter 1)

solar wind An outward flow of particles (mostly electrons and protons) from the Sun. (Chapter 8, Chapter 16)

south celestial pole The point directly above the Earth's south pole where the Earth's axis of rotation, if extended, would intersect the celestial sphere. (Chapter 2)

southern highlands (on Mars) Older, cratered terrain in the Martian southern hemisphere. (Chapter 11)

southern lights See *aurora australis*.

spectral analysis The identification of chemical substances from the patterns of lines in their spectra. (Chapter 5)

spectral line In a spectrum, an absorption or emission feature that is at a particular wavelength. (Chapter 5)

spectrograph An instrument for photographing a spectrum. (Chapter 6)

spectroscopy The study of spectra and spectral lines. (Chapter 5, Chapter 6, Chapter 7)

spectrum (plural spectra) The result of dispersing a beam of electromagnetic radiation so that components with different wavelengths are separated in space. (Chapter 5)

speed Distance traveled divided by the time elapsed to cover that distance. (Chapter 4)

spherical aberration The distortion of an image formed by a telescope due to differing focal lengths of the optical system. (Chapter 6)

spicule A narrow jet of rising gas in the solar chromosphere. (Chapter 16)

spin-orbit coupling See *1-to-1 spin-orbit coupling* and *3-to-2 spin-orbit coupling*.

spring tide An ocean tide that occurs at new moon and full moon phases. (Chapter 4)

stable Lagrange points Locations along Jupiter's orbit where the combined gravitational effects of the Sun and Jupiter cause asteroids to collect. (Chapter 15)

Stefan-Boltzmann law A relationship between the temperature of a blackbody and the rate at which it radiates energy. (Chapter 5)

stony iron meteorite A meteorite composed of both stone and iron. (Chapter 15)

stony meteorite A meteorite composed of stone. (Chapter 15)

stratosphere A layer in the Earth's atmosphere directly above the troposphere. (Chapter 9)

subduction zone A location where colliding tectonic plates cause the Earth's crust to be pulled down into the mantle. (Chapter 9)

subtend To extend over an angle. (Chapter 1)

summer solstice The point on the ecliptic where the Sun is farthest north of the celestial equator. Also used to refer to the date on which the Sun passes through this point. (Chapter 2)

sunspot A temporary cool region in the solar photosphere. (Chapter 16)

sunspot cycle The semiregular 11-year period with which the number of sunspots fluctuates. (Chapter 16)

sunspot maximum/minimum That time during the sunspot cycle when the number of sunspots is highest/lowest. (Chapter 16)

supergranule A large convective feature in the solar atmosphere, usually outlined by spicules. (Chapter 16)

superior conjunction The configuration of a planet being behind the Sun as viewed from the Earth. (Chapter 4)

superior planet A planet that is more distant from the Sun than the Earth is. (Chapter 4)

supernova (plural supernovae) A stellar outburst during which a star suddenly increases its brightness roughly a millionfold. (Chapter 1, Chapter 15)

surface wave A type of seismic wave that travels only over the Earth's surface. (Chapter 9)

synchronous rotation The rotation of a body with a period equal to its orbital period; also called 1-to-1 spin-orbit coupling. (Chapter 3, Chapter 10)

synchrotron radiation A type of nonthermal radiation emitted by charged particles moving through a magnetic field. (Chapter 12)

synodic month The period of revolution of the Moon with respect to the Sun; the length of one cycle of lunar phases. Also called the lunar month. (Chapter 3)

synodic period The interval between successive occurrences of the same configuration of a planet. (Chapter 4)

T Tauri wind A flow of particles away from a T Tauri star. (Chapter 8)

tail (of a comet) Gas and dust particles from a comet's nucleus that have been swept away from the comet's head by the radiation pressure of sunlight and the solar wind. (Chapter 15)

tangential velocity That portion of an object's velocity perpendicular to the line of sight. (Chapter 19)

temperature See *degree Celsius, degree Fahrenheit,* and *kelvin.*

terminator The line dividing day and night on the surface of the Moon or a planet; the line of sunset or sunrise. (Chapter 10)

terrae Cratered lunar highlands. (Chapter 10)

terrestrial planet High-density worlds with solid surfaces, including Mercury, Venus, Earth, and Mars. (Chapter 7)

theory A hypothesis that has withstood experimental or observational tests. (Chapter 1)

thermal equilibrium A balance between the input and outflow of heat in a system. (Chapter 16)

thermal radiation The radiation naturally emitted by any object that is not at absolute zero. Blackbody radiation is an idealized case of thermal radiation. (Chapter 12)

thermonuclear fusion The combining of nuclei under conditions of high temperature in a process that releases substantial energy. (Chapter 16)

thermosphere A region in the Earth's atmosphere between the mesosphere and the exosphere. (Chapter 9)

third quarter moon The phase of the Moon that occurs when the Moon is 90° west of the Sun. (Chapter 3)

3-to-2 spin-orbit coupling The rotation of Mercury, which makes three complete rotations on its axis for every two complete orbits around the Sun. (Chapter 11)

tidal force A gravitational force whose strength and/or direction varies over a body and thus tends to deform the body. (Chapter 4, Chapter 12)

tidal heating The heating of the interior of a satellite by continually varying tidal stresses. (Chapter 13)

time zone A region on the Earth where, by agreement, all clocks have the same time. (Chapter 2)

total lunar eclipse A lunar eclipse during which the Moon is completely immersed in the Earth's umbra. (Chapter 3)

total solar eclipse A solar eclipse during which the Sun is completely hidden by the Moon. (Chapter 3)

totality (lunar eclipse) The period during a total lunar eclipse when the Moon is entirely within the Earth's umbra. (Chapter 3)

totality (solar eclipse) The period during a total solar eclipse when the disk of the Sun is completely hidden. (Chapter 3)

transit An event in which an astronomical body moves in front of another. See also *meridian transit* and *solar transit.* (Chapter 8)

transit method A method for detecting extrasolar planets that come between us and their parent star, dimming the star's light. (Chapter 8)

Trans-Neptunian object Any small body of rock and ice that orbits the Sun within the solar system, but beyond the orbit of Neptune. (Chapter 7, Chapter 14)

Trojan asteroid One of several asteroids that share Jupiter's orbit about the Sun. (Chapter 15)

Tropic of Cancer A circle of latitude $23\frac{1}{2}°$ north of the Earth's equator. (Chapter 2)

Tropic of Capricorn A circle of latitude $23\frac{1}{2}°$ south of the Earth's equator. (Chapter 2)

tropical year The period of revolution of the Earth about the Sun with respect to the vernal equinox. (Chapter 2)

troposphere The lowest level in the Earth's atmosphere. (Chapter 9)

22-year solar cycle The semiregular 22-year interval between successive appearances of sunspots at the same latitude and with the same magnetic polarity. (Chapter 16)

ultramafic lava A type of lava enriched in magnesium and iron. These give the lava a higher melting temperature. (Chapter 13)

ultraviolet radiation Electromagnetic radiation of wavelengths shorter than those of visible light but longer than those of X rays. (Chapter 5)

umbra (of a shadow) (plural umbrae) The central, completely dark portion of a shadow. (Chapter 3)

undifferentiated asteroid An asteroid within which chemical differentiation did not occur. (Chapter 15)

universal constant of gravitation (*G*) The constant of proportionality in Newton's law of gravitation. (Chapter 4)

upper meridian The half of the meridian that lies above the horizon. (Chapter 2)

Van Allen belts Two doughnut-shaped regions around the Earth where many charged particles (protons and electrons) are trapped by the Earth's magnetic field. (Chapter 9)

velocity The speed and direction of an object's motion. (Chapter 4)

vernal equinox The point on the ecliptic where the Sun crosses the celestial equator from south to north. Also used to refer to the date on which the Sun passes through this intersection. (Chapter 2)

very-long-baseline interferometry (**VLBI**) A method of connecting widely separated radio telescopes to make very high-resolution observations. (Chapter 6)

visible light Electromagnetic radiation detectable by the human eye. (Chapter 5)

VLBI See *very-long-baseline interferometry*.

volatile element An element with low melting and boiling points. (Chapter 10)

waning crescent moon The phase of the Moon that occurs between third quarter and new moon. (Chapter 3)

waning gibbous moon The phase of the Moon that occurs between full moon and third quarter. (Chapter 3)

water hole A range of frequencies in the microwave spectrum suitable for interstellar radio communication. (Chapter 28)

watt A unit of power, equal to one joule of energy per second. (Chapter 5)

wavelength The distance between two successive wave crests. (Chapter 5)

wavelength of maximum emission The wavelength at which a heated object emits the greatest intensity of radiation. (Chapter 5)

waxing crescent moon The phase of the Moon that occurs between new moon and first quarter. (Chapter 3)

waxing gibbous moon The phase of the Moon that occurs between first quarter and full moon. (Chapter 3)

weight The force with which gravity acts on a body. (Chapter 4)

white ovals Round, whitish feature usually seen in Jupiter's southern hemisphere. (Chapter 12)

Widmanstätten patterns Crystalline structure seen in certain types of meteorites. (Chapter 15)

Wien's law A relationship between the temperature of a blackbody and the wavelength at which it emits the greatest intensity of radiation. (Chapter 5)

winter solstice The point on the ecliptic where the Sun reaches its greatest distance south of the celestial equator. Also used to refer to the date on which the Sun passes through this point. (Chapter 2)

X rays Electromagnetic radiation whose wavelength is between that of ultraviolet light and gamma rays. (Chapter 5)

Zeeman effect A splitting or broadening of spectral lines due to a magnetic field. (Chapter 16)

zenith The point on the celestial sphere directly overhead an observer. (Chapter 2)

zodiac A band of 12 constellations around the sky centered on the ecliptic. (Chapter 12)

zonal winds The pattern of alternating eastward and westward winds found in the atmospheres of Jupiter and Saturn. (Chapter 12)

zone A light-colored band in Jupiter's atmosphere. (Chapter 12)

Answers to Selected Questions

How to Get the Most from UNIVERSE

1. Paragraphs labeled **Caution! 2.** Paragraphs labeled **Analogy 3.** Tools of the Astronomer's Trade Boxes and **The Heavens on the Earth** Boxes **4.** On the *Universe* Web site **5.** It shows whether the image was made with Radio waves, Infrared radiation, Visible light, Ultraviolet light, **X** rays, or **G**amma rays **6.** On the *Universe* Web site **7.** (a) page A-8 (b) page A-1 (c) page A-4 (d) page A-6 **8.** (a) page 417 (b) page 130 (c) page 89 (d) page 74 **9.** 0.5 year **10.** At the back of the book, following the Index

CAUTION! Only mathematical answers are given in the following, not answers that require interpretation or discussion. Your instructor will expect you to show the steps required to reach each mathematical answer.

Chapter 1
24. 8.5×10^3 km **25.** 2.8×10^7 Suns **26.** About 3×10^{36} times larger **27.** 8.94×10^{56} hydrogen atoms **28.** (a) 1.581×10^{-5} ly (b) 4.848×10^{-6} pc **29.** 4.99×10^2 s **30.** 4.3×10^9 km **31.** (a) 1.59×10^{14} km (b) 16.8 years **32.** 4.32×10^{17} s **34.** (a) 1.5 m (b) 89 m (c) 5.4×10^3 m **35.** 6.9 m **36.** 3.4×10^3 km **37.** 3.7 km **38.** 0.320 arcmin

Chapter 2
28. Around 8:02 p.m. **33.** (a) About 9 hours **46.** October 25, 1917 **49.** 50° **50.** 3:50 a.m. local time **52.** (a) 6:00 p.m. (b) September 21

Chapter 3
31. (a) 0.91 hour (b) 11° **32.** 49 arcsec

Chapter 4
18. Semimajor axis = 0.25 AU, period = 0.125 year **19.** Average = 25 AU, farthest = 50 AU **23.** 6 newtons, 4 m/s² **26.** 1/9 as strong **33.** 87.97 days **38.** (a) 4 AU (b) 8 years **39.** (a) 16.0 AU (b) 0.5 AU **40.** (a) 1.26 AU (b) 258 days **41.** Earth exerts a force of 1.98×10^{20} newtons on the Moon **42.** Forces are approximately the same, Earth's acceleration is 100 times greater **43.** ¼ as much as on Earth **44.** 62 newtons, 0.13 **45.** 0.5 year **46.** (a) 24 hours (b) 43,200 km **47.** 119 minutes **49.** (a) 5 years (b) 2.92 AU **50.** (a) 3.43×10^{-5} newton (b) 3.21×10^{-5} newton (c) 2.2×10^{-6} newton

Chapter 5
2. 7.5 times **3.** 500 s **7.** 0.340 m **8.** 2.61×10^{14} Hz **14.** 135 nm **15.** 3400 K **24.** 9400 nm **25.** 9980°F **26.** About 10 μm **27.** 2.9 nm **28.** 3.9×10^{26} W **29.** 190 times more **30.** (a) 4890 nm = 4.89 μm (b) 540 times more **31.** (a) 5.75×10^8 W/m (b) 10,000 K **33.** 2.43×10^{-3} nm **34.** (a) 1005 nm **38.** Possible transitions: energy = 1 eV, λ = 1240 nm; energy = 2 eV, λ = 620 nm; energy = 3 eV, λ = 414 nm **39.** 656.6 nm **40.** Coming toward us at 13.0 km/s **41.** 8.6×10^4 km/s

Chapter 6
31. $^1/_{25} = 0.04$ **33.** (a) 222× (b) 100× (c) 36× (d) 0.75 arcsec **36.** 300 km (Hubble Space Telescope, Jupiter's moons); 110 km (human eye, our Moon) **37.** (a) 34 ly (b) 37 km **40.** (a) 5.39×10^{-4} m (c) 0.56 m

Chapter 7
22. Mass = 6.4×10^{23} kg, average density = 3900 kg/m³ **24.** (a) 1.0×10^{13} kg (b) 1.2 m/s **25.** (a) 3.3×10^{21} J (b) Equivalent to 3.9×10^7 Hiroshima-type weapons **26.** (a) 2.02 km/s (b) 19.4 km/s **27.** 1.2 km/s **28.** (a) 618 km/s **30.** 1.76 years **32.** (a) 1000 years (b) 17 days **34.** In

100 years, probability is 8.3×10^{-8} (one chance in 12 million); in 10^6 years, probability is 8.3×10^{-4} (one chance in 1200)

Chapter 8
30. 0.40 kg after 1.3 billion years; 0.20 kg after 2.6 billion years; 0.10 kg after 3.9 billion years **31.** 2.6 billion years **33.** (a) About 180 AU **34.** 2.9×10^7 AU = 140 parsecs = 460 light-years **35.** (a) About 600 AU (b) About 5×10^{40} cubic meters (c) About 10^{55} atoms (d) About 3×10^{14} atoms per cubic meter **36.** 860 years **37.** 2.2×10^{30} kg = 1.1 times the mass of the Sun **38.** (a) 12 m/s (b) 1.3×10^{-3} arcsec (c) 9.0×10^{-5} arcsec

Chapter 9
30. (a) 6.8×10^{16} W (b) 1.07×10^{17} W (c) 209 watts per square meter (d) 246 K = −27°C **31.** 32 K = −248°C **32.** 4 km **33.** Core: 17%; mantle: 82%; crust: 1% **34.** 0.020 (2% of the total mass) **35.** (b) About 15,000 kg/m³

Chapter 10
26. (a) 4671 km (b) 1707 km below the surface (c) 449 km **30.** 130 newtons on the Moon, 780 newtons on Earth **32.** (a) 5.78×10^{-5} newton (b) 4.15×10^{-5} newton (c) 1.39 **38.** 2.56 seconds **39.** 5.5×10^5 Km **41.** (b) 10^6 times greater

Chapter 11
46. 13.0 arcsec **48.** (a) 0.16 AU **51.** (a) 3.03 m/s (b) 1.26 nm **52.** 0.615 AU **55.** For T = 460°C, $\lambda_{max} = 4.0$ μm **63.** (a) 3770 km (b) 370 km **64.** Difference between the round-trip times is 2.0×10^{-4} s **66.** 920 m **73.** Phobos: about 16 arcminutes; Deimos: about 2.7 arcminutes **74.** (a) Radius = 20,400 km, altitude = 17,000 km

Chapter 12
37. 12.7 km/s **40.** 8.5×10^{53} hydrogen atoms, 7.1×10^{52} helium atoms **41.** Roughly 600 km/h **43.** 127 K **45.** 59.5 km/s **46.** 8300 newtons **47.** 7.1×10^4 kg/m³ **51.** (a) 14.4 hours for inner edge of A ring, 7.9 hours for inner edge of B ring

Chapter 13
40. 760 km **41.** 0.046 arcsec **41.** 2.8×10^{11} (280 billion) years **46.** About 4×10^{-11} (one part in 3×10^{10}) **47.** About 8.1 minutes **50.** Escape speed = 2.6 km/s; mass of molecule = 2.0×10^{-26} kg, corresponding to a molecular weight of 12 **53.** 0.09 arcsec **55.** (a) 65.7 hours (b) 27.6 arcsec

Chapter 14
29. Sun-Uranus force = 1.39×10^{11} newtons, Neptune-Uranus force = 2.24×10^{17} newtons; Neptune reduces the sunward gravitational pull on Uranus by 1.61×10^4, or 0.0161% **31.** (a) 2900 kg/m³ **32.** (a) 8.4 hours **35.** (a) 1.13×10^{-3} as bright as on Earth (b) 4.10×10^{-4} as bright as on Earth (c) 2.77 times as bright **36.** About 7200 km **38.** 0.95 arcsec **43.** Without the boost, a one-way trip would take 30.5 years

Chapter 15
33. 2.82 AU **35.** 30 km **37.** (a) 3.6×10^{12} kg (b) 0.83 m/s **39.** About 4% **40.** 3.6×10^7 km = 0.24 AU **41.** (a) Period = 350 years, lifetime = 3.5×10^4 years (b) Period = 1.1×10^4 years, lifetime = 1.1×10^6 years (c) Period = 3.5×10^5 years, lifetime = 3.5×10^7 years (d) Period = 1.1×10^7 years, lifetime = 1.1×10^9 years **44.** (a) 10^{15} kg (b) 10^{-16} kg/m³

Chapter 16

29. (a) 1.8×10^{-9} J (b) 9.0×10^{16} J (c) 5.4×10^{41} J **30.** (a) 4.6×10^{-36} s (b) 2.3×10^{10} s (c) 1.4×10^{15} s $= 4.4 \times 10^{7}$ years **31.** 0.048 (4.8%) of the Sun's mass will be converted from hydrogen to helium; chemical composition of the Sun (by mass) will be 69% hydrogen, 30% helium **32.** (a) 8.8×10^{29} kg of hydrogen consumed, 6.2×10^{26} kg lost **34.** (a) 1.64×10^{-13} J (b) 2.43×10^{-3} nm **35.** 1.4×10^{13} kg/s **39.** 98,600 kg/m^3 **40.** 1.9×10^{-7} nm **42.** (a) 1700 nm **44.** For the photosphere, 500 nm; for the chromosphere, 58 nm; for the corona, 1.9 nm **47.** For the umbra, 670 nm; for the penumbra, 580 nm **48.** (a) (Flux from patch of penumbra)/(flux from patch of photosphere) = 0.55 (b) (Flux from patch of penumbra)/(flux from patch of umbra) = 1.8

Chapter 28

11. (a) 3.7 cm **12.** (a) 10,000 years (b) 10 million years **14.** (a) 1430 MHz **15.** (a) 29.8 km/s (b) 0.10 m (c) 9.9×10^{-6} m, or 9.9×10^{-3}% of the unshifted wavelength **16.** 200 km

Index

Page numbers in *italics* indicate illustrations.

Star Charts

The following set of star charts, one for each month of the year, are from *Griffith Observer* magazine. They are useful in the northern hemisphere only. For a set of star charts suitable for use in the southern hemisphere, see the *Universe* Web site (**www.whfreeman.com/universe8e**).

To use these charts, first select the chart that best corresponds to the date and time of your observations. Hold the chart vertically as shown in the above illustration and turn it so that the direction you are facing is shown at the bottom.

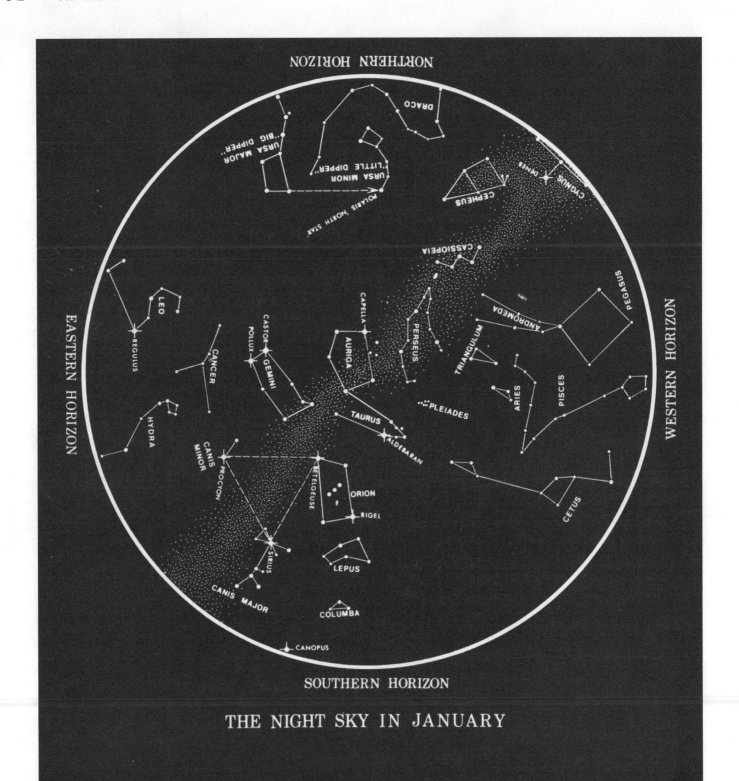

THE NIGHT SKY IN JANUARY

Chart time (Local Standard Time):

10 pm...First of January
9 pm...Middle of January
8 pm...Last of January

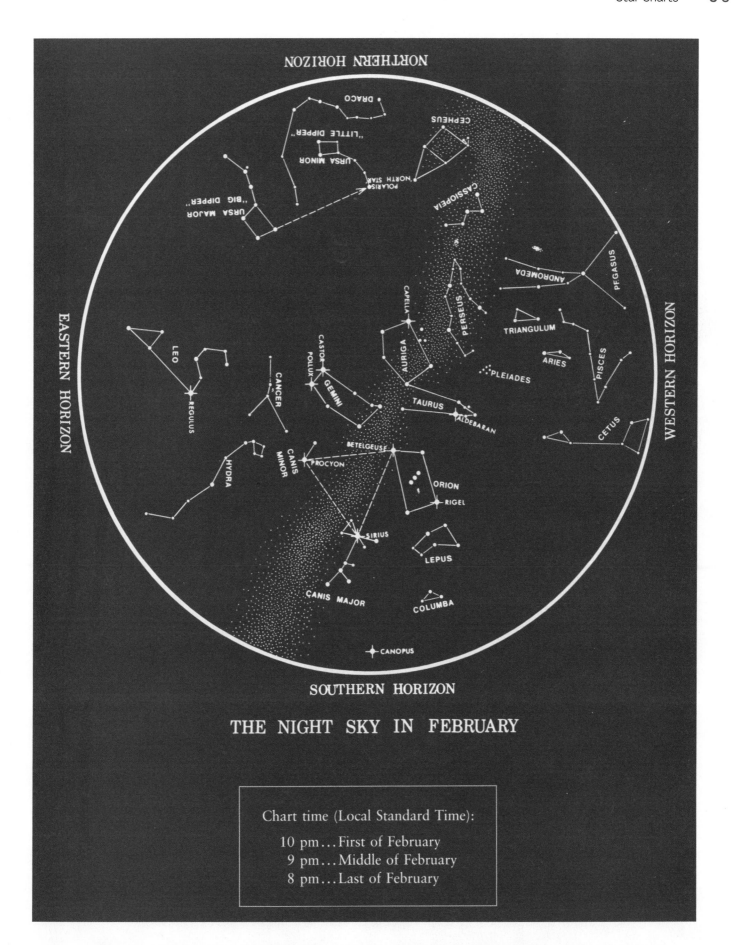

THE NIGHT SKY IN FEBRUARY

Chart time (Local Standard Time):

10 pm...First of February
9 pm...Middle of February
8 pm...Last of February

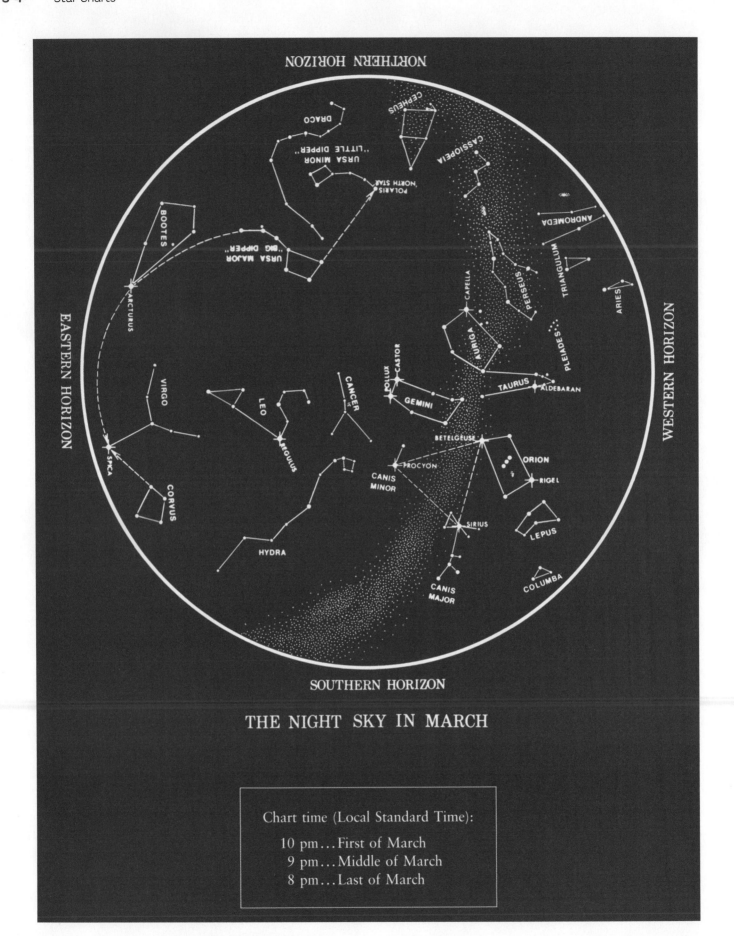

THE NIGHT SKY IN MARCH

Chart time (Local Standard Time):

10 pm...First of March
9 pm...Middle of March
8 pm...Last of March

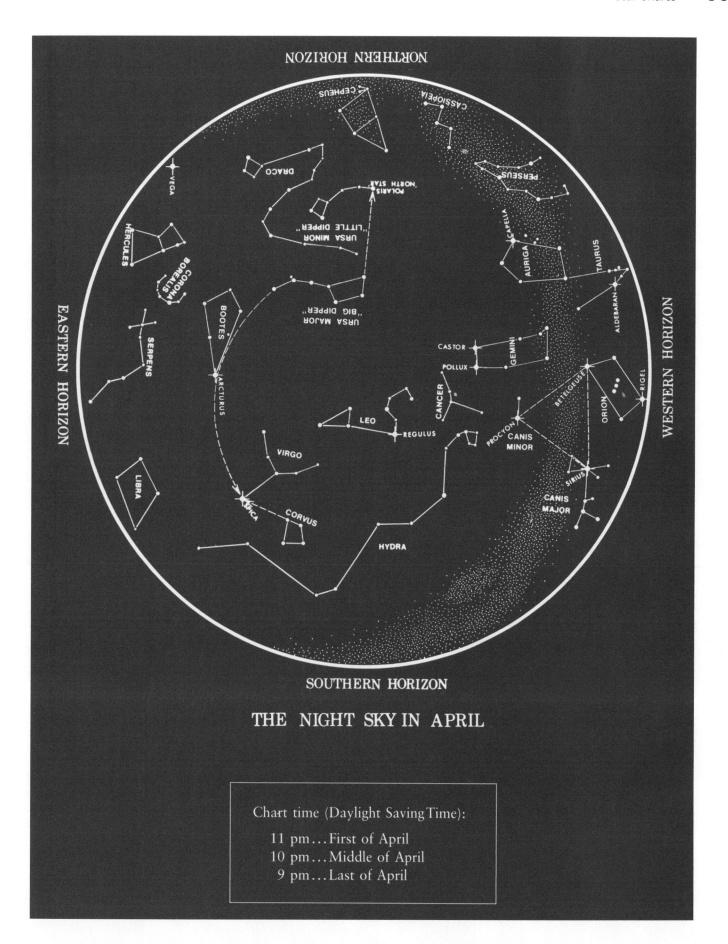

THE NIGHT SKY IN APRIL

Chart time (Daylight Saving Time):

11 pm...First of April
10 pm...Middle of April
9 pm...Last of April

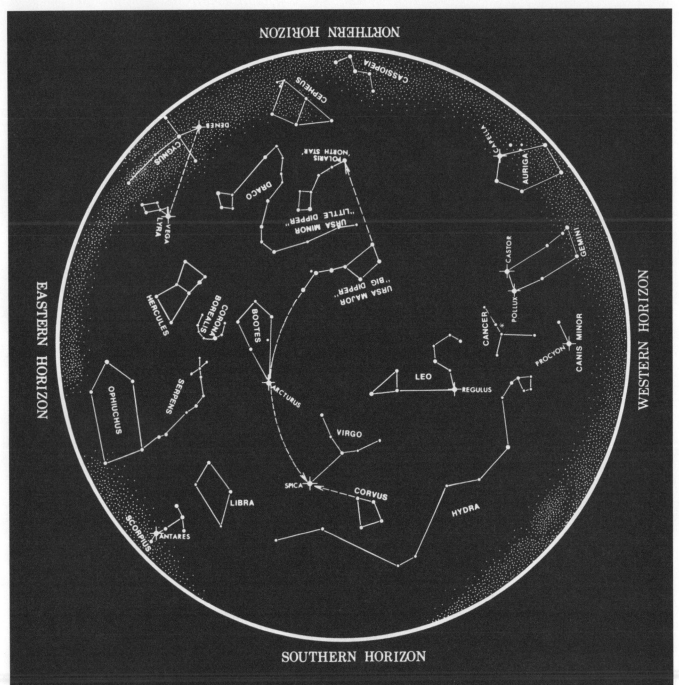

THE NIGHT SKY IN MAY

Chart time (Daylight Saving Time):

11 pm...First of May
10 pm...Middle of May
9 pm...Last of May

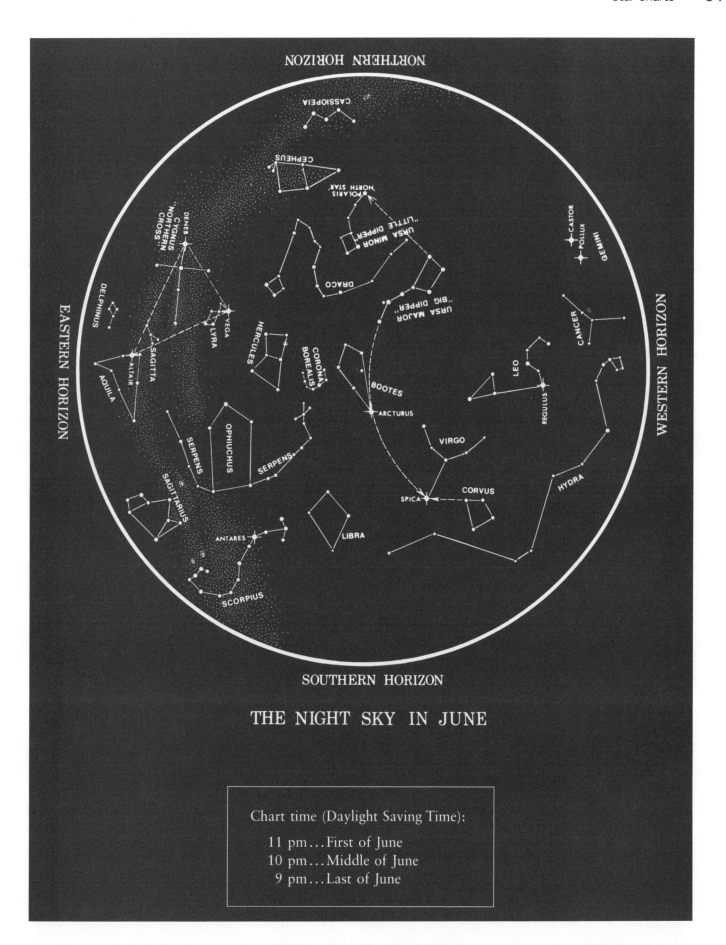

THE NIGHT SKY IN JUNE

Chart time (Daylight Saving Time):

11 pm...First of June
10 pm...Middle of June
9 pm...Last of June

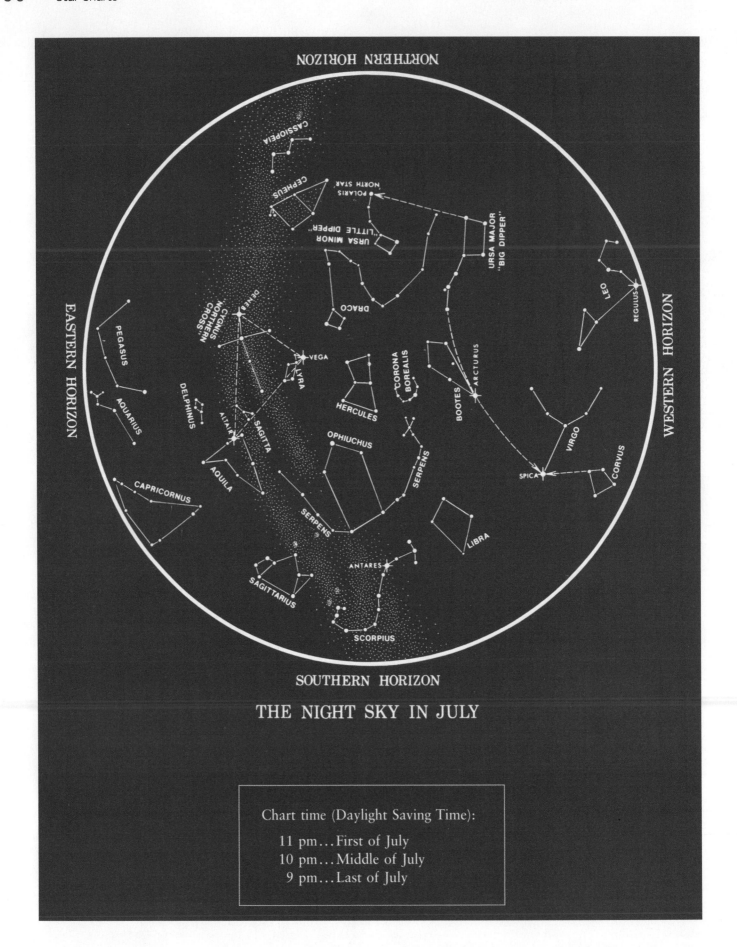

THE NIGHT SKY IN JULY

Chart time (Daylight Saving Time):

11 pm...First of July
10 pm...Middle of July
9 pm...Last of July

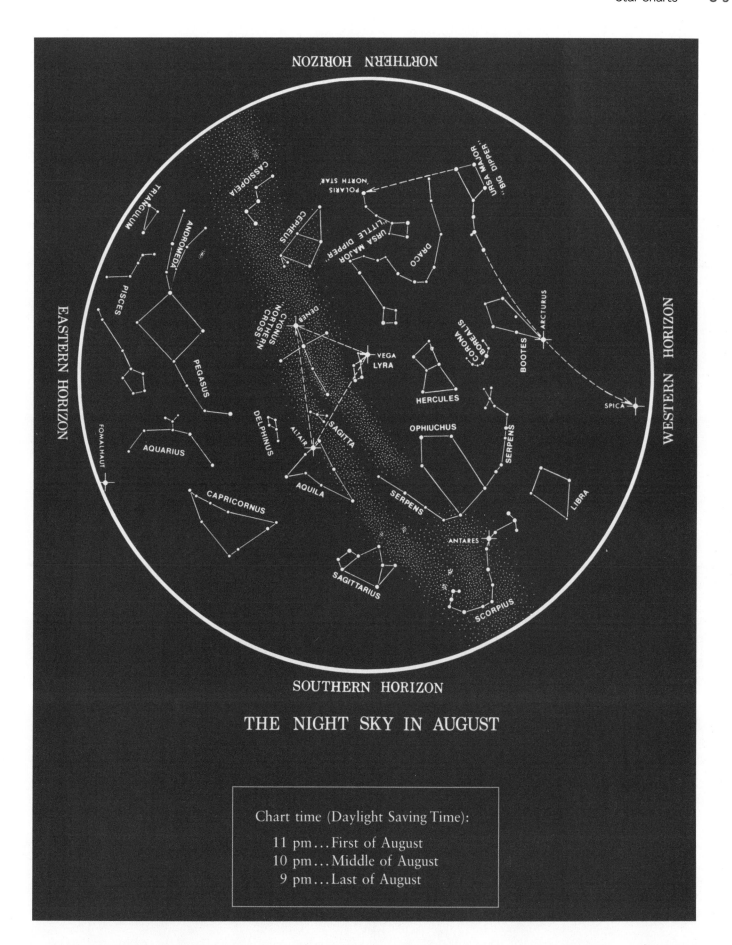

THE NIGHT SKY IN AUGUST

Chart time (Daylight Saving Time):

11 pm...First of August
10 pm...Middle of August
9 pm...Last of August

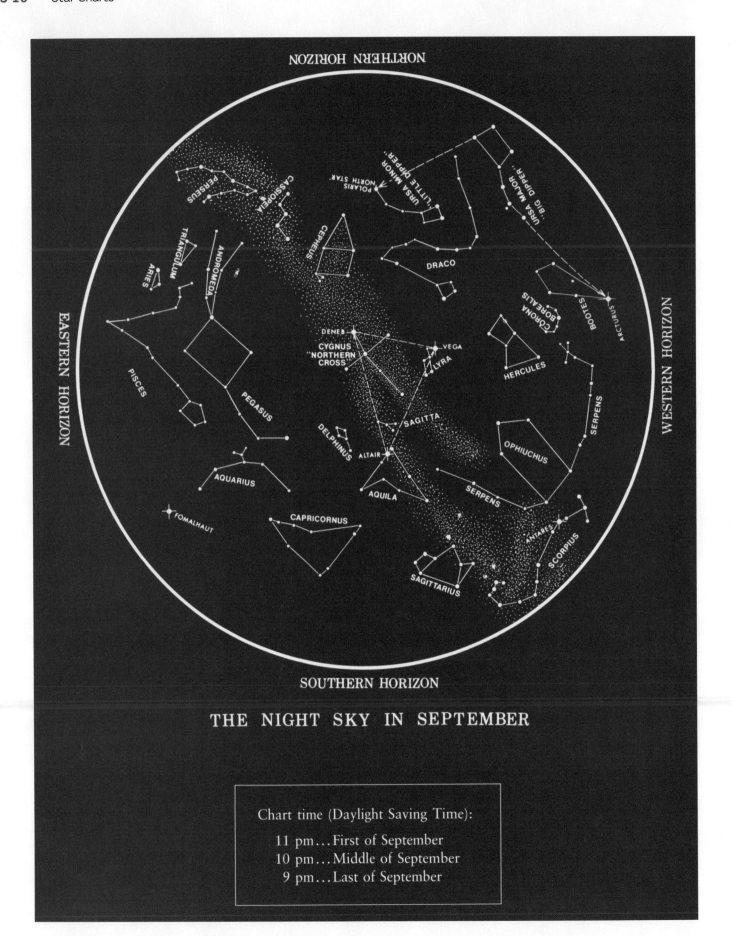

THE NIGHT SKY IN SEPTEMBER

Chart time (Daylight Saving Time):

11 pm...First of September
10 pm...Middle of September
9 pm...Last of September

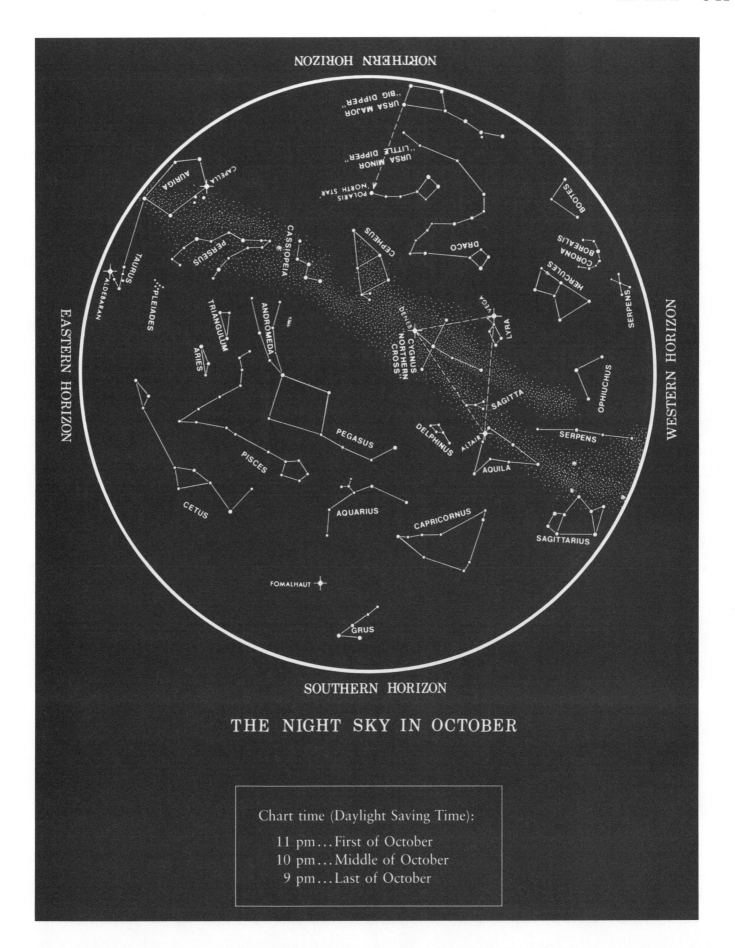

THE NIGHT SKY IN OCTOBER

Chart time (Daylight Saving Time):

11 pm...First of October
10 pm...Middle of October
9 pm...Last of October

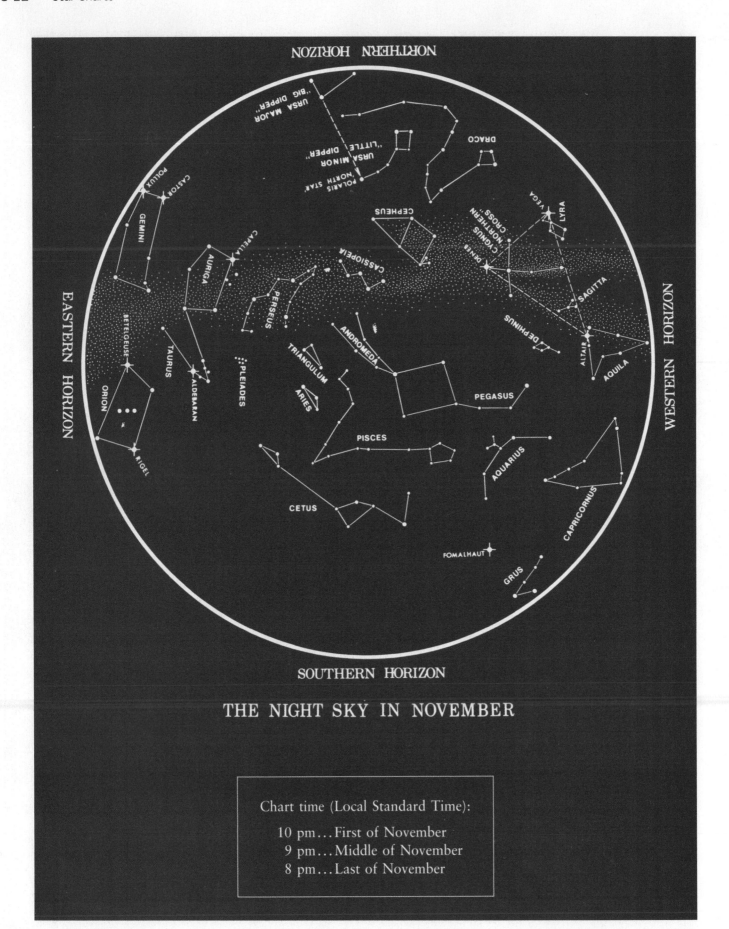

THE NIGHT SKY IN NOVEMBER

Chart time (Local Standard Time):

10 pm...First of November
9 pm...Middle of November
8 pm...Last of November

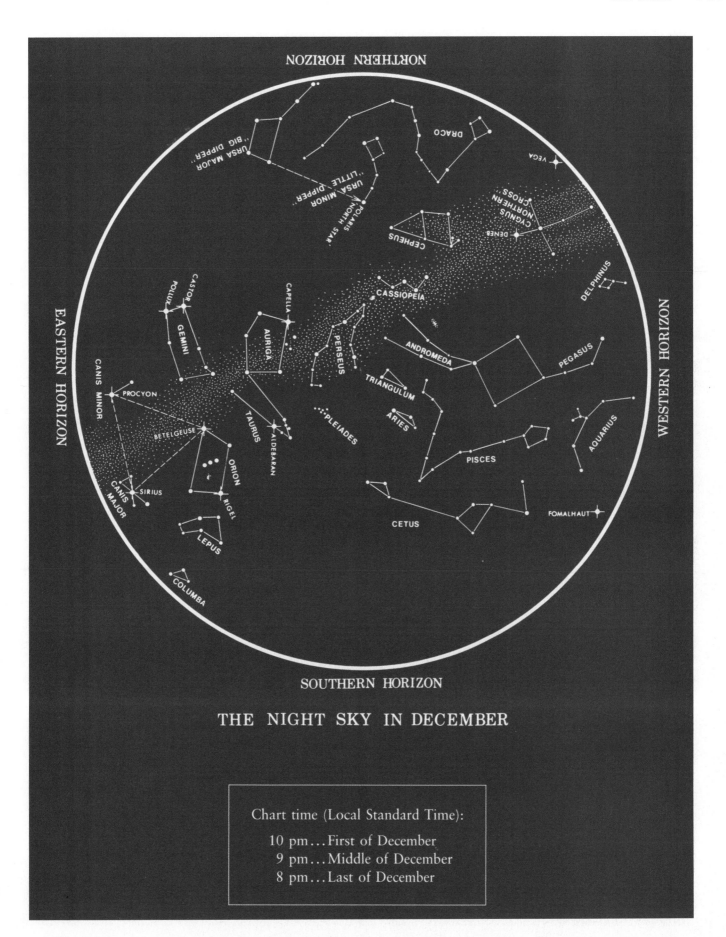

THE NIGHT SKY IN DECEMBER

Chart time (Local Standard Time):

10 pm...First of December
9 pm...Middle of December
8 pm...Last of December